# Springer Complexity

Springer Complexity is an interdisciplinary program publishing the best research and academic-level teaching on both fundamental and applied aspects of complex systems – cutting across all traditional disciplines of the natural and life sciences, engineering, economics, medicine, neuroscience, social and computer science.

Complex Systems are systems that comprise many interacting parts with the ability to generate a new quality of macroscopic collective behavior the manifestations of which are the spontaneous formation of distinctive temporal, spatial or functional structures. Models of such systems can be successfully mapped onto quite diverse "real-life" situations like the climate, the coherent emission of light from lasers, chemical reaction-diffusion systems, biological cellular networks, the dynamics of stock markets and of the internet, earthquake statistics and prediction, freeway traffic, the human brain, or the formation of opinions in social systems, to name just some of the popular applications.

Although their scope and methodologies overlap somewhat, one can distinguish the following main concepts and tools: self-organization, nonlinear dynamics, synergetics, turbulence, dynamical systems, catastrophes, instabilities, stochastic processes, chaos, graphs and networks, cellular automata, adaptive systems, genetic algorithms and computational intelligence.

The three major book publication platforms of the Springer Complexity program are the monograph series "Understanding Complex Systems" focusing on the various applications of complexity, the "Springer Series in Synergetics", which is devoted to the quantitative theoretical and methodological foundations, and the "SpringerBriefs in Complexity" which are concise and topical working reports, case-studies, surveys, essays and lecture notes of relevance to the field. In addition to the books in these two core series, the program also incorporates individual titles ranging from textbooks to major reference works.

## Editorial and Programme Advisory Board

# Springer Series in Synergetics

**Founding Editor: H. Haken**

The Springer Series in Synergetics was founded by Herman Haken in 1977. Since then, the series has evolved into a substantial reference library for the quantitative, theoretical and methodological foundations of the science of complex systems.

Through many enduring classic texts, such as Haken's *Synergetics and Information and Self-Organization*, Gardiner's *Handbook of Stochastic Methods*, Risken's *The Fokker Planck-Equation* or *Haake's Quantum Signatures of Chaos*, the series has made, and continues to make, important contributions to shaping the foundations of the field.

The series publishes monographs and graduate-level textbooks of broad and general interest, with a pronounced emphasis on the physico-mathematical approach.

More information about this series at http://www.springer.com/series/712

Jian-Jun Xu

# Interfacial Wave Theory of Pattern Formation in Solidification

## Dendrites, Fingers, Cells and Free Boundaries

Second Edition

Springer

Jian-Jun Xu
Department of Mathematics and Statistics
McGill University
Montreal
Québec, Canada

ISSN 0172-7389 ISSN 2198-333X (electronic)
Springer Series in Synergetics
ISBN 978-3-319-52662-1 ISBN 978-3-319-52663-8 (eBook)
DOI 10.1007/978-3-319-52663-8

Library of Congress Control Number: 2017932915



Printed on acid-free paper

This Springer imprint is published by Springer Nature
The registered company is Springer International Publishing AG
The registered company address is: Gewerbestrasse 11, 6330 Cham, Switzerland

*I dedicate this book to my family, my wife Xu Chen, my three sons: Wen-Bo, Wen-Tao, Wen-Lei, as well as to all persons who have helped me in my life*

# Preface of the Second Edition

This is a new extended version of the book that follows after its first edition published in 1998. After then, my studies on interfacial pattern formation in solidification have been extended into much broader areas in material science. Most results of these studies were scattered, published in some popular journals in physics, material science and applied mathematics. It seems that now is a proper time to present a new, extended version of book in the framework of synergetic theory, that summarizes all these advances in terms of a unified analytical approach. The present book covers three types of prominent pattern formation phenomena that are commonly observed in various systems of growth and solidification in material science: the single free dendritic growth from undercooled melt or binary mixture; the spatially-periodic cellular growth and the spatially-periodic eutectic growth from binary mixtures in directional solidification. It is found that the interfacial wave (IFW) theory that was previously established for mainly describing free dendritic growth is applicable for describing other two types of pattern formation phenomena as well.

The important characteristics of this book in comparison with other books on similar subjects available in the literature are the followings. First of all the book aims at the outstanding real-world problem, rather than some artificial mathematical model problem. It starts from the description of the phenomenon observed in experiments, then by making all possible simplifications, formulates the justifiable mathematical model for the phenomenon based on the first physical principles without artificial hypotheses. This process leads to the mathematical problems under studies.

Secondly the book uses the unified asymptotic approaches to explore all these pattern formation systems, rather then using numerical simulations. The analytical formulas obtained can be used for better and deeper elucidating the essence of the phenomena, as well as the mathematical similarities between various systems of growth and solidification with interfacial energy.

Thirdly, the book emphasizes the profound understanding of underlying physical mechanisms of the phenomena and the roles played by each of physical parameters in these mechanisms. The book reveals the intrinsic similarity in the essence of the dynamic phenomena observed in a variety of growth systems. In this way, it exhibits the unification and harmonics of the physical world, as well as the power and beauty of the mathematics.

Finally, this book emphasizes the idea that the reliable theoretical results and conclusions must undergo rigour verifications of the experimental data and observations. Hence, we made every effort to conduct the quantitative comparisons between the mathematical results and available experimental data without involving free parameter, and examine the theoretical predictions via experimental evidence. In fact, we attempt to establish a predictive mathematical theory for the physical phenomena under studies.

In the book, the readers will be guided to deal with a complicated dynamic system by the following procedure: determining the basic steady states for the systems; performing the linear stability analysis; deriving the local dispersion relationship for local normal modes in the outer region; finding out the singularity of the normal mode solutions; deriving the inner solutions near the singular point; matching the inner solution with the outer solution and, eventually, solving the linear eigenvalue problem (EVP). This thinking path runs through the whole book.

This procedure leads to the analytical forms of the global modes and the corresponding quantization conditions for the spectra of eigenvalues. It is shown that the interfacial perturbed states in all the systems under the study can be considered as the same type of interfacial waves subjecting a cubic complex dispersion relationship. The linear perturbed systems generally allow the real eigenvalue spectrum and the complex eigenvalue spectrum. It is therefore derived that the limiting state of these systems is either steady or oscillating; all oscillatory patterns observed are a wave phenomenon controlled by essentially the same mechanism—the so-called global trapped wave mechanism; the pattern selected at the later stage of experiments under steady growth conditions is eventually either described by the most dangerous global neutral mode of the system, or described by the steady state in the global stable region bounded by the most dangerous global neutral curve in the state space.

By the chance of the present edition, I refined the mathematical derivations performed in the first edition and included more details of derivations. Some algebraic errors in the derivations found in in previous edition are corrected and related numerical results are re-calculated. The changes and corrections do not alter the qualitative behaviours of the solutions and conclusions, but induce some quantitative, numerical changes.

The book presents the works that lie on the bounder between mathematics, physics and material science. The interfacial pattern formation itself has been an important subject in the field of interdisciplinary non-linear science. The book describes how the applied mathematics is used to give insight into the problems of growth and material processing, and how the physical problems can lead to new mathematical problems. As a consequence, for the readers from the area of material science and engineering, a complete understanding of all of the mathematics in the book will require at least a solid undergraduate training in mathematics, whereas for the readers from area of mathematics, a solid background in general physics, thermodynamics and statistical physics is necessary. Moreover, to follow the book, the readers will be also required the basic understanding of ordinary and partial differential equations, stability, bifurcation, and basic dynamic system theory. In particular, the readers are required to be familiar with the asymptotic analysis, singular perturbation theory, multiple variable expansion methods, since the book makes extensively use of these methods. For the readers whose mathematical background is lack of these topics, we recommend them to consult the excellent classical books by Kevorkian and Cole (1996) and Holmes (2013).

The researchers, post-doctoral and graduate students in the fields of condensed matter physics, material science, applied mathematics, mechanical engineering, and chemical engineering will find this book useful and beneficial. The book can be also used as a reference book for teaching a graduate course of mathematical theory of crystal growth, solidification and material processing, or a graduate course of advanced topics in material science or in applied mathematics.

Finally, for the completion of this project, credit should be given to a number of people, such as Prof. Martin Glicksman, Prof. Yoshinori Furukawa, Prof. Alain Pocheau and their co-workers, from whom I received and used their significant experimental results and photos. I thank many colleagues and friends for their invaluable advice, encouragement and discussions during the years of carrying on this project. On this regard, Prof. Stephen H. Davis and Robert E. O'Malley should be especially mentioned. I very much thank my colleague, former graduate student, Prof. Yong-Qiang Chen for his valuable contributions on this project and his extensive numerical computation work in preparing this edition of the book. I am grateful to all members of my family for their long term of tolerance, understanding and support; the book can never be completed without their help.

Montreal, QC, Canada Jian-Jun Xu
November 2016

# Preface of the First Edition

For the last several years, the study of interfacial instability and pattern formation phenomena has preoccupied many researchers in the broad area of nonlinear science. These phenomena occur in a variety of dynamical systems far from equilibrium. In many practically very important physical systems some fascinating patterns are always displayed at the interface between solid and liquid or between two liquids. Two prototypes of these phenomena are dendritic growth in solidification and viscous fingering in a Hele–Shaw cell. These two phenomena occur in completely different scientific fields, but both are described by similar nonlinear free boundary problems of partial-differential-equation systems; the boundary conditions on the interface for both cases contain a curvature operator involving the surface tension, which is nonlinear. Moreover, both cases raise the same challenging theoretical issues, interfacial instability mechanisms and pattern selection, and it is now found that these issues can be solved by the same analytical approach. Thus, these two phenomena are regarded as special examples of a class of nonlinear pattern formation phenomena in nature, and they are the prominent topics of the new interdisciplinary field of nonlinear science.

This research monograph is based on a series of lectures I have given at McGill University, Canada (1993–1994), Northwestern Polytechnical Institute, China (1994), Aachen University, Germany (1994), and the CRM summer school at Banff, Alberta, Canada (1995). I shall illustrate these phenomena, present the fundamental issues involved, and describe the recent theoretical developments. In particular, I shall discuss some long-standing problems in this field, such as the pattern selection principle and the interfacial instability mechanism, and systematically present a newly established, coherent, predictive theory of this subject: the interfacial wave theory.

This book will be useful for researchers, post-doctoral and graduate students in the fields of condensed matter physics, materials science, applied mathematics, mechanical engineering, and chemical engineering.

I am particularly indebted to Prof. J.D. Cole for teaching me the asymptotic analysis and various perturbation methods. I also thank him and Prof. M.E. Glicksman for introducing me to this challenging and exciting field and for their constant discussions and advice. During the past decade, I have also received invaluable help and advice from, and had discussions with, many other people in my investigation of this subject, as well as in the preparation of this book. This book could not have been written without their help; among them, Prof. C.C. Lin at MIT, Prof. S.H. Davis at Northwestern University, Prof. R.E. O'Malley at the University of Washington, and Dr. John R. Ockendon at Oxford University should be especially mentioned.

I thank my graduate student, Mr. Dong-Sheng Yu very much for carrying out the large amount of numerical computations and for preparing the figures. I also thank him and my graduate student, Mr. Mikhail Kharenko, for the careful verification of all the formulae. I also thank Miss Katina Michelis, a graduate student in my course, 'Topics in Applied Mathematics', for carefully proof-reading and checking the formulae in the manuscript. I would like to thank the production team at Springer-Verlag for their excellent editing work.

Montreal, QC, Canada
May 1997

Jian-Jun Xu

# Contents

# Chapter 1
# Introduction

## 1.1 Interfacial Pattern Formation in Dendritic Growth and Hele–Shaw Flow

Pattern formation in dynamical systems is a common phenomenon in nature. Generally speaking, when a system is driven away from its equilibrium state, a dynamic process will spontaneously take place. During this process, a certain type of pattern is often observed in the system. This pattern is characterized by the distribution of a certain physical quantity, such as streamlines, isothermal lines, isoconcentration lines for species, or isopotential lines. There are two major types of patterns that should be distinguished: (1) stationary patterns and (2) fractal, or chaotic, patterns. A stationary pattern may be either steady or time-periodic, while a fractal or chaotic pattern, such as turbulence, is characterized by many different length scales. Stationary and fractal patterns are often observed at the later stages of a process. The features of these patterns are determined by the nature of the corresponding dynamical system. These patterns are often irrelevant to the initial conditions of the system and can persist for a long time. Before reaching the final stage, some transient patterns will be observed at the early stages of a process. These transient patterns are related to the initial conditions of the system and are continually changing.

The essence and mechanism of fractal chaotic pattern formation is a wide-open subject in all scientific areas. Our understanding of this complex phenomenon is obviously far from complete, and in this book we will not deal with this topic. From a theoretical point of view, the formation of stationary patterns is more fundamental, as well as simpler. The essence and mechanism of such pattern formation is, of course, of great interest.

The most familiar examples of stationary patterns observed in fluid dynamics are the Rayleigh–Bénard convection cell and Taylor vortices in Taylor–Couette flow. But the most popular, prominent, and beautiful examples of pattern formation seen in mankind's experience are snowflakes and ice crystal growth. Figure 1.1 shows



J.-J. Xu, *Interfacial Wave Theory of Pattern Formation in Solidification*,
Springer Series in Synergetics, DOI 10.1007/978-3-319-52663-8_1

**Fig. 1.1** Photograph of a typical snowflake pattern taken by Yushinori Furukawa

**Fig. 1.2** Photograph of an ice crystal taken by Furukuwa et al., which is growing at the undercooled temperature $\Delta T = 0.37\,\text{K}$

a photo of a typical snowflake in nature, and Fig. 1.2 shows an ice crystal growing in undercooled water. Both photos were taken by Yushinori Furukawa at Hokkaido University, in Japan [14, 31].

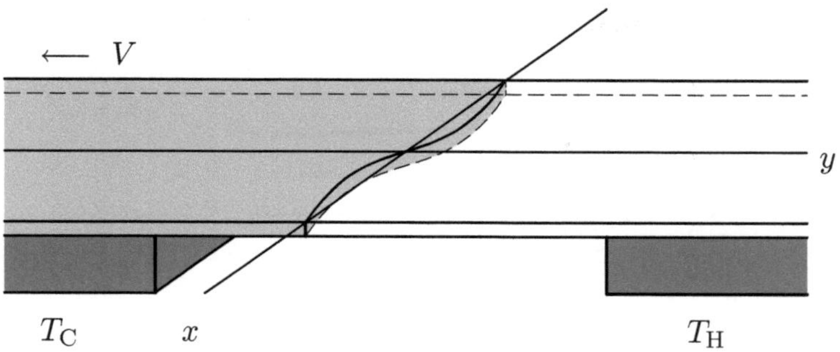


**Fig. 1.3** Sketch of a unidirectional solidification device—Hele–Shaw cell

The patterns formed in these cases are somewhat different from the above-mentioned Rayleigh–Bénard convection cell and Taylor vortices. The patterns of snowflakes and ice crystals are interfacial, formed at the interface between the solid and vapor phases or the solid and liquid phases, respectively. They are characterized by treelike structures along the interface, which are called dendritic patterns.

Dendritic patterns can appear in various growth systems for crystalline materials. Figures 1.1 and 1.2 describe a typical case of dendritic growth. The system is originally set up in a metastable state, such as a liquid state with a given uniform undercooled temperature. The dynamical process is initiated by introducing a tiny seed or in some other related way. Once it begins, crystal growth will proceed spontaneously according to physical laws. There is no way to artificially control its tip growth velocity or the formation of treelike microstructure at its interface. It is for this reason that in materials science such a system is often called free dendritic growth, while in condensed matter physics, it is frequently referred to as spontaneous pattern formation. In a realistic free dendritic growth system, there will be a set of dendrites with different orientations growing simultaneously and interacting with each other. Eventually, the system will form a so-called mushy zone.

Another typical example of dendritic growth occurs in unidirectional solidification. In Fig. 1.3, we show a sketch of a unidirectional solidification device, which consists of a thin sample in a Hele–Shaw cell and two uniform temperature zones at temperatures $T_1$ and $T_2$, which are, respectively, larger and smaller than the melting temperature of the material [28].

When the sample, which is made of a binary mixture, is at rest, the interface between solid and liquid is planar. The sample is pulled at a constant imposed velocity $V$. When the pulling velocity $V$ increases over a certain critical value, the interface will gradually develop the deep cellular structure shown in Fig. 1.4. When the pulling velocity $V$ increases further and exceeds another critical value, the deep cellular structure will transform into arrays of dendrites, as shown in Fig. 1.5.

Dendritic growth also frequently appears in many other forms of material processing, such as alloy casting, metal ingot formation, and welding. Dendritic

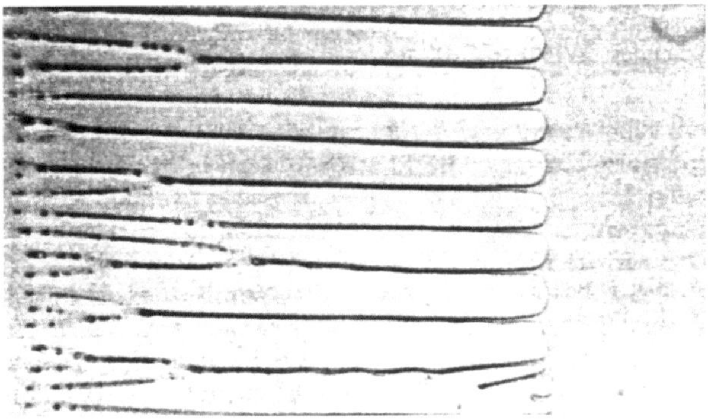

**Fig. 1.4** Deep cellular structure formed during unidirectional solidification

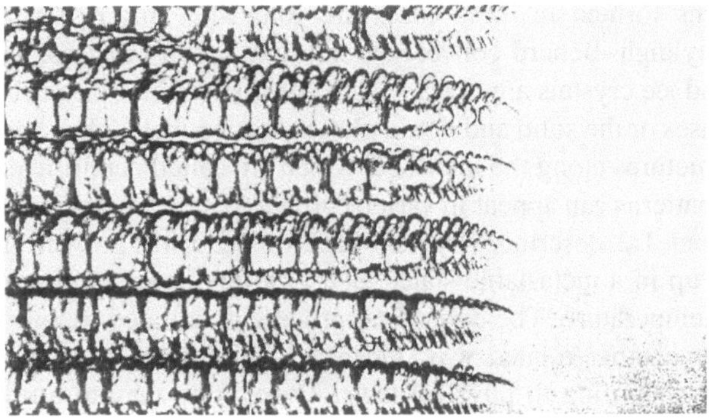

**Fig. 1.5** Dendritic array structure formed during unidirectional solidification

growth is one of the most profound subjects in the area of interfacial pattern formation. This is due not only to its underlying vital technical importance in the material processing industries, but also to the fact that dendritic growth represents a fascinating class of nonlinear phenomena occurring in inhomogeneous dynamical systems.

The theoretical investigation of dendritic growth at its preliminary stage has mainly focused on single free dendritic growth, which allows one to disregard the interactions between neighboring dendrites in more complicated systems of growth, such as the mushy zone and cellular dendrite formation.

A typical case of single free dendritic growth is shown in Fig. 1.6a, which is a photograph of single dendritic growth from a pure organic material melted succinonitrile (SCN) made by Huang and Glicksman [9], while Fig. 1.6b shows the experimental recording curves for two-dimensional dendritic growth from a supersaturated $NH_4Br$ solution, studied by Dougherty and Gollub [4].

Experimental observations show that at the later stages of growth, a dendrite appears to have a smooth tip moving with constant velocity. It emits a stationary wave train, propagating along the interface toward the root. The essence and origin

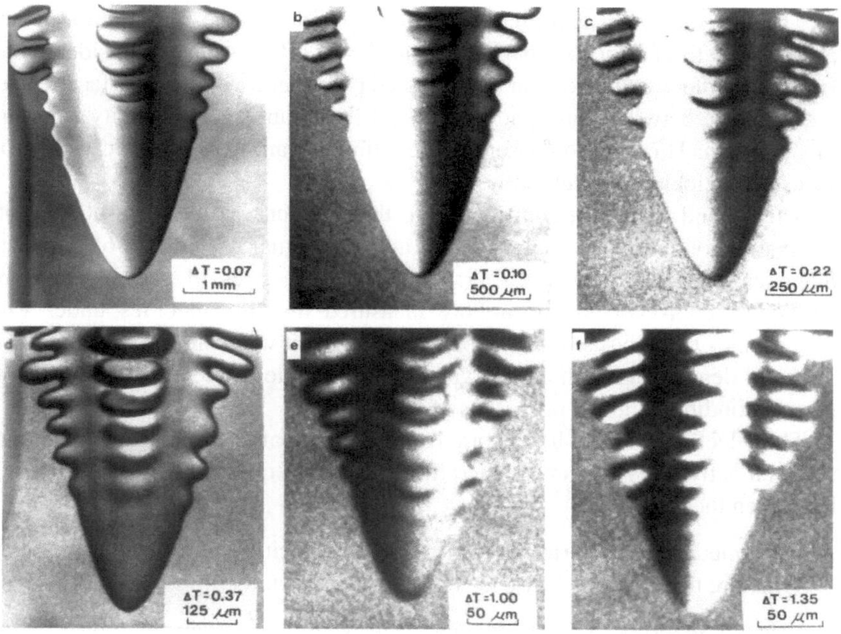


(a)

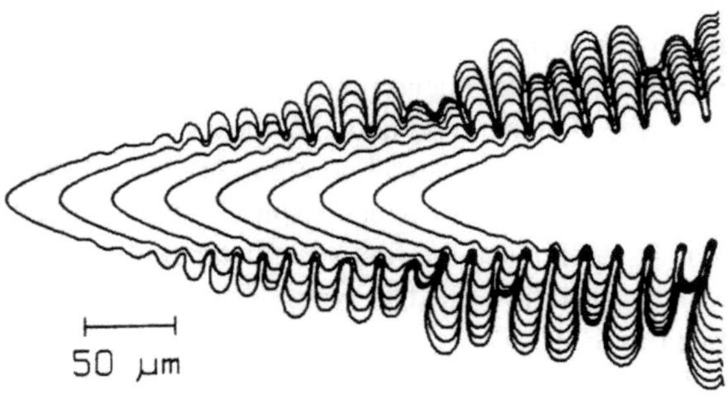


(b)

**Fig. 1.6** (**a**) A photograph of a single free dendritic growth from pure organic material melt succinonitrile (SCN). (**b**) Experimental curve of two-dimensional dendritic growth from a supersaturated ($NH_4Br$) solution

of this nonlinear interfacial phenomenon has been a fundamental subject in the field of condensed matter physics and materials science for a long time [1, 46, 49, 50]. It is believed that understanding this problem will have great significance for a much wider field, including fluid dynamics, chemical engineering, and the biological sciences, where similar pattern formation phenomena occur.

The first important contribution to this subject was the formulation of needle crystal growth and the Ivantsov solution (1947) [10]. Ivantsov considered steady smooth needle growth from an undercooled pure melt, with zero surface tension at the interface between solid and liquid, and found an exact similarity solution for this problem. This solution describes a continuous family of needle crystal growth, whose tip velocity is undetermined.

The second important contribution to this subject was the identification of the selection problem by Schaefer (1975) and Glicksman (1976) [5, 6]. Glicksman, Schaefer, and their coworkers first performed a series of careful experiments on dendritic growth and accurately measured the tip velocities under various conditions. On the basis of this experimental data, it was confirmed that at the later stages of dendritic growth, the tip velocity is a uniquely determined function of the growth condition and the properties of the material.

The selection problem has inspired a broad range of theoretical and experimental research activities for over a generation. The basic questions of dendritic growth have been these:

1. What mechanism determines the tip growth velocity?
2. What are the origin and essence of the microstructure?

Viscous fingering in the Hele–Shaw cell is another famous and classical interfacial phenomenon that occurs in the field of fluid dynamics [3, 5–9, 11, 12, 15, 17, 18, 21–23, 25, 27, 29, 47, 48, 50]. The Hele–Shaw cell is a thin, essentially "two-dimensional," rectangular container in which the intrusion of one fluid into another leads to the formation of what are commonly called "viscous fingers." This simple device is an excellent environment for observing some fundamental features of interfacial dynamics, and the theory developed for viscous fingers provides a foundation for understanding many fluid–fluid systems. Observation of viscous fingers in Hele–Shaw cells has inspired over a generation of experimental and theoretical research since the first systematic study by Saffman and Taylor in 1958 [29]. Experiments show that two types of stationary fingers can persist for long periods of time (see Fig. 1.7).

The first type, the subject of Saffman and Taylor's original 1958 research, consists of smooth, steady fingers, which occur if the surface tension parameter is large. When the surface tension parameter is small, another type of oscillatory fingers may form, which are narrow and dendrite-like. Oscillatory fingers were first discovered by Couder et al. [3], and later by Kopf-Sill and Homsy [15]. The oscillatory fingers found by Couder et al. had a tiny bubble at the tip, while the oscillatory fingers found by Kopf-Sill and Homsy were formed naturally without a nose bubble. The presence of even a small bubble has a noticeable effect on the finger. However, once the surface tension parameter becomes sufficiently small, stationary fingers of any kind can no longer exist, causing splitting, shielding, and spreading into a chaotic pattern throughout the Hele–Shaw cell.

The study of finger formation in a Hele–Shaw cell has followed a similar historical path to that of dendritic growth. The first solution to this problem was the solution for a steady smooth finger with zero surface tension. This solution

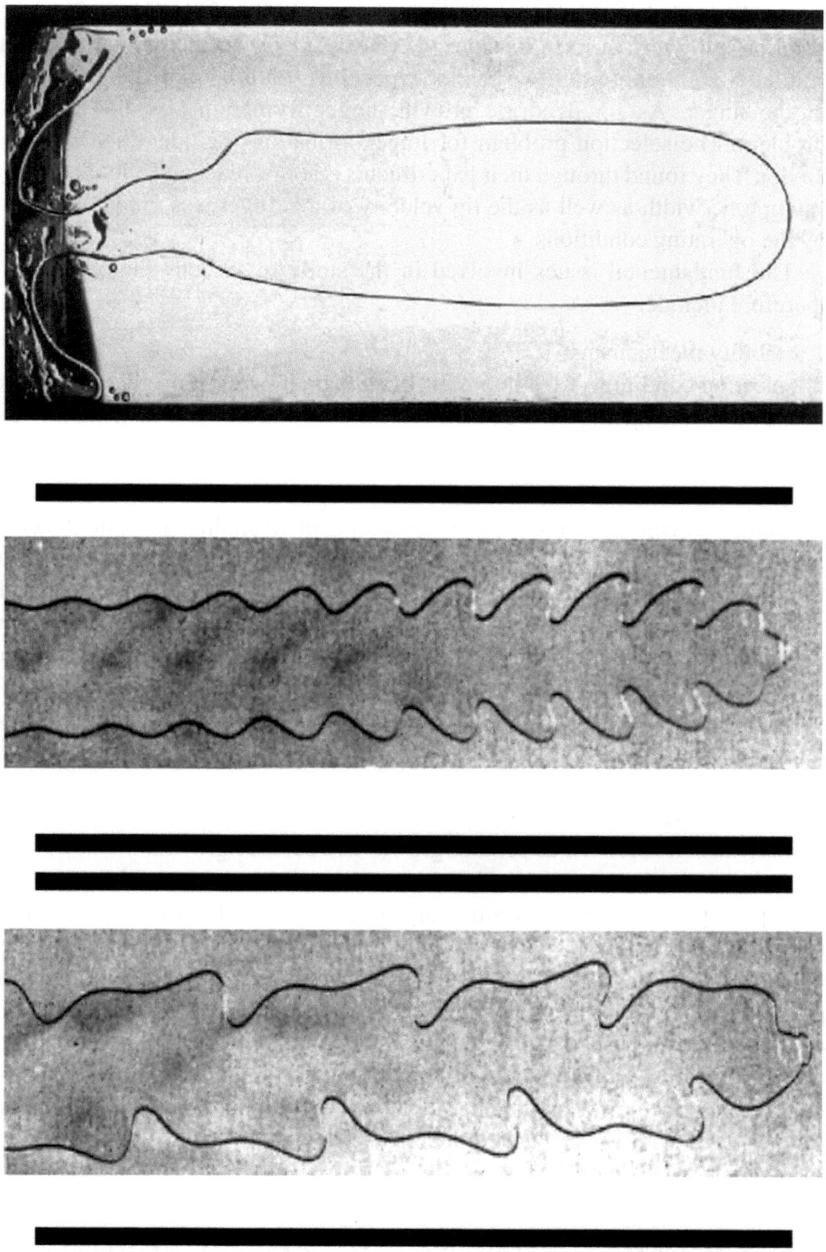

(a)

(b)

**Fig. 1.7** Finger formation in a Hele–Shaw cell: (**a**) A smooth Saffman–Taylor finger. (**b**) An oscillatory finger is usually called the Saffman–Taylor solution (1958) [29]. The Saffman–Taylor solution with zero surface tension, like the Ivantsov needle crystal solution, is a similarity solution. Hence, we cannot expect this solution to predict the tip velocity of the finger. As for dendritic growth, finger formation also has the selection problem. The selection problem for finger formation was identified by Saffman–Taylor. They found through their experiments that in a realistic Hele–Shaw flow, the asymptotic width as well as the tip velocity of the finger was uniquely determined by the operating conditions.

The fundamental issues involved in the study of viscous fingering formation therefore include:

1. stability mechanisms;
2. selection conditions for smooth fingers and both types of oscillatory fingers;
3. the transition point at which the oscillatory fingers with a nose bubble form chaotic or transient patterns.

The issues listed here for dendritic growth and Hele–Shaw flow have been long-standing challenging problems in the fields of applied mathematics, fluid mechanics, condensed matter physics, and chemical engineering, among others, for about half a century. These problems have preoccupied many investigators from various fields.

It has been recognized that these two distinct interfacial phenomena, dendritic growth in solidification and finger formation in a Hele–Shaw flow, are governed by some mathematically similar instability mechanisms. The understanding of these mechanisms evidently has vital theoretical significance. It is of great importance for practical applications.

The present research monograph is devoted to the study of these interfacial phenomena, as well as other prominent interfacial pattern formation phenomena in material science, such as cellular growth and eutectic growth.

A tremendous amount of work has been done in this field, and this book does not attempt to give an exhaustive summary of all significant work. Some well-written books and review articles with such a goal are available in the literature. Rather, this book intends to give a focused overview of the important theoretical results that have proved to be self-consistent and have been supported by experimental data. We will address the major issues involved, summarize the main theories on the subject, and demonstrate the current understanding of the underlying mechanisms. In particular, the present book will describe *interfacial wave theory* and compare the predictions of this theory with various experimental data.

## 1.2 A Brief Review of the Theories of Free Dendritic Growth

A first glance at dendritic growth phenomena reveals that the growth of the dendrite-tip region is steady, having a monotonic interface shape, so that it can be well described by a steady-state solution of the system; in contrast, the growth of the dendrite-stem region is oscillatory, forming a so-called side-branching structure, so that it is described by unsteady perturbed state solutions of the system. Therefore, in studying the phenomenon of dendritic growth, there are two major issues involved: (1) How does one specify the steady state of the system? (2) What are the stability properties of the system?

These issues have been investigated by many researchers for about a half a century, many of whom have made significant contributions to the subject. A number of theories have been proposed, attempting especially to resolve the above-mentioned selection problem. In the following, I shall review some major theories while ignoring others, although they have all played a role at some stage of the research.

### 1.2.1 Maximum Velocity Principle (1976)

In 1974, Nash and Glicksman investigated steady axially symmetric needle crystal growth with nonzero surface tension [25]. They attempted to find a selection principle for steady classical needle crystal growth by extending Ivantsov's problem to the case of nonzero isotropic surface tension ($\gamma \neq 0$). The classical needle crystal has a smooth tip and an infinitely long smooth root.

With the inclusion of surface tension, the needle crystal is nonisothermal due to the Gibbs–Thomson effect. The needle crystal growth problem becomes a highly nonlinear one, which is much more difficult to deal with than the isothermal case. Nash and Glicksman first attempted a complete mathematical formulation of the problem. They derived a nonlinear integrodifferential equation, and as the boundary conditions for the interface shape, they proposed that the tip of the needle must have no cusp, known as the smooth tip condition, while the root of the needle must approach Ivantsov's paraboloidal needle, which is the *Nash–Glicksman far-field condition*. This mathematical problem is called the *Nash–Glicksman problem*.

The Nash–Glicksman problem has subsequently been extensively examined by many investigators. In particular, the Nash–Glicksman far-field condition has been applied by many researchers as a basic assumption, without proper justification. Unfortunately, it has now been proven that this Nash–Glicksman far-field condition is inadequate.

Nash and Glicksman did not justify the existence of a solution to their problem. They solved this problem near the tip using an inaccurate numerical method. They found that under a given growth condition, the tip velocity has a unique maximum value. On the basis of this finding, they proposed that the selected tip velocity is just the maximum growth velocity. This selection mechanism was called the *maximum velocity principle* (MVP). The MVP was quickly ruled out by the authors themselves. In 1975–1976, Glicksman and his coworkers made precise measurements for dendritic growth of succinonitrile [5, 6]. They found that the calculated maximum velocity is larger than the real tip velocity by a factor of seven. They thought that the discrepancy was too large to be acceptable. However, Glicksman and his coworkers did not realize that the Nash–Glicksman problem actually had no mathematical solution. Their numerical solution actually gave the wrong information. The numerical work conducted by Nash and Glicksman on the Nash–Glicksman problem provided an excellent example in the literature of research on dendritic growth to demonstrate that numerical approaches alone might result in leading theoretical investigations astray. Conversely, it also shows the importance of the role that analytical solutions actually play in theoretical investigations.

### *1.2.2 Marginal Stability Hypothesis (1978)*

Not long after Nash and Glicksman abandoned the maximum velocity principle as a solution of the selection problem, Müller-Krumbhaar and Langer proposed the *marginal stability hypothesis* (MSH) in 1978 [21]. The idea of Müller-Krumbhaar and Langer, in contrast to that of Nash and Glicksman, is that the selection must have something to do with the stability mechanism of the system.

Based on this idea, the authors introduced a new dimensionless stability parameter $\sigma = 2\ell_T\ell_c/\ell_t^2$, where $\ell_T$ is the thermal diffusion length determined by the thermal diffusivity; $\ell_c$ is the capillary length determined by the surface tension at the interface; and $\ell_t$ is the tip radius of the dendrite. They further assumed that the system of dendritic growth will be linearly unstable for $\sigma < \sigma_*$ and claimed that the selected dendritic growth would be in a marginal stable state corresponding to such a critical number $\sigma_*$.

Müller-Krumbhaar and Langer also proposed that the parameter $\sigma_*$ could be written in the form

$$\sigma_* = \left(\frac{\lambda_s}{2\pi\ell_t}\right)^2, \quad \lambda_s = 2\pi\sqrt{2\ell_T\ell_c}, \tag{1.1}$$

where $\lambda_s$ is the shortest wavelength of a disturbance that could cause a plane interface moving at velocity $U$ to suffer a Mullins–Sekerka instability. They then made the important assumption that in the marginal state, the tip radius of the dendrite is equal to the marginal stability wavelength $\lambda_s$ of a plane interface. This assumption leads to the following stability criterion:

$$\ell_t \equiv \lambda_s\,; \tag{1.2}$$

or

$$\sigma_* \approx \frac{1}{4\pi^2} \approx 0.0253\,. \tag{1.3}$$

Müller-Krumbhaar and Langer conducted numerical computations, attempting to determine the critical number $\sigma_*$. In the authors' words, "Unfortunately, we have not been able to obtain much information from the general equation of motion of $F$ by analytic method, and have to resort to computational techniques. . . ." [22].

The basic idea of the MSH represents progress compared with the maximum velocity principle, and its results were in good agreement with the experimental data in the small undercooling regime available at that time. However, the MSH is not a self-consistent analytical theory. At that time, Müller-Krumbhaar and Langer were unable to perform a global linear stability analysis for an inhomogeneous dynamical system to obtain the stability criterion for growth systems with a curved front.

It is because of this theoretical weakness that the MSH was eventually abandoned by many researchers, including Langer himself, when a series of new challenges arose in the 1980s.

### *1.2.3 Microscopic Solvability Condition (MSC) Theory (1986–1990s)*

In the 1980s, several groups of researchers, such as Segur and Kruskal, Langer and coworkers, Kessler and Levine, Pelce and Pomeau, Ben Amer and Pomeau, Caroli and Caroli, and many others, reconsidered the Nash–Glicksman problem for steady needle crystal growth [12, 18], [26]. Some simplified versions of the Nash–Glicksman problem, such as the so-called boundary layer model and geometry model, were proposed and carefully examined using more powerful analytical approaches. It was very surprising when it emerged that for the case of isotropic surface tension, the Nash–Glicksman problem for a system of steady needle growth had no solution! It was shown that a solution satisfying the Nash–Glicksman condition at the far field must have a cusp at the tip. When the surface tension is very small, the slope of this cusp is transcendentally small; a solution satisfying the smooth tip condition can therefore never approach the Ivantsov paraboloid satisfying the Nash–Glicksman far-field condition. In order for the system to have a steady needle growth solution, one may include a certain type of anisotropy in the surface tension. These unexpected results led to what is known as *microscopic solvability condition* (MSC) theory. Hence, MSC theory began with the study of the existence of a steady needle solution, regarding side-branching structure formation as another subject, and was based on the acceptance of the Nash–Glicksman far-field condition. The MSC theory concluded the following conditions:

(i) There will be no needle solution to the Nash–Glicksman problem.
(ii) With the inclusion of a small amount of surface tension anisotropy, the system permits a discrete set of needle solutions to the Nash–Glicksman problem.
(iii) The solution with the largest tip velocity among these steady classical needle solutions is linearly stable, and so is selected.

Hence, MSC theory declares that the anisotropy of surface tension is a necessary condition for dendritic growth, and in essence, it considers dendritic growth to be smooth needle growth without side-branching. The side-branching structure of dendrites is induced by the external noise that is continuously acting on the system.

Although MSC theory raised some interesting mathematical issues on the role of exponentially small terms beyond all orders, in applying it to the original dendritic growth phenomenon, it cannot be considered a successful physical theory. Disregarding the problem of not receiving much experimental support (cf. [24]), the theory itself suffers some intrinsic difficulties. The key is that the artificial Nash–Glicksman far-field condition that MSC theory adopts is overprescribed for the dendritic growth system. Indeed, a dendritic growth system with isotropic surface tension does not allow the classical steady needle solution satisfying the Nash–Glicksman far-field condition. But the system does allow some "nearly" steady needle solutions without the constraint of the Nash–Glicksman far-field condition that are physically acceptable.

Moreover, the above conclusion (iii) made in MSC theory, that "the solution with the largest tip velocity among these steady classical needle solutions is linearly stable," is based solely on the numerical investigation by Kessler and Levine [11]. It was afterward accepted and followed by many researchers without reservation. Kessler and Levine claimed that the system did not allow any unstable growing oscillatory modes for the two-dimensional steady needle solution, since they did not find the complex spectrum of eigenvalues with positive real part in their numerical investigations. However, the range of the anisotropy parameter used in Kessler–Levine's numerical investigations was not wide enough, and in consequence, their numerical solutions actually missed the unstable growing oscillatory modes in the range of smaller values of the anisotropy parameter.

It has been proven by analytical solutions that in fact, *even in the two-dimensional case, the steady needle solution predicted by MSC theory is not always stable.* The steady needle solution becomes linearly unstable when the anisotropy of the surface tension is smaller than some critical value; cf. Chap. 7, Sect. 7.4. In the history of research on dendritic growth, the numerical investigation conducted by Kessler and Levine [11] provided another excellent example showing how numerical solutions alone, without guidance from analytical solutions, might lead to a misunderstanding and undesirable consequences in theoretical investigations.

### 1.2.4 Interfacial Wave (IFW) Theory (1990)

A more recent theory is *interfacial wave* (IFW) theory, which was developed using a unified matched asymptotic method and the multiple variable expansion approach and carrying forward the stability idea in the theory of dynamical systems. This IFW theory modifies the Nash–Glicksman classical needle crystal problem for the steady state and redefines the basic steady state of dendritic growth systems by introducing the new idea of a generalized steady needle solution. Moreover, it explores the instability mechanisms of the basic steady state, and on the basis of *global stability analysis*, establishes the mechanisms of selection and pattern formation for dendritic growth, as well as viscous fingering in a Hele–Shaw cell [38, 39, 46–48].

Interfacial wave theory concludes that:

(i) Dendritic growth is intrinsically a time-dependent wave phenomenon; a dendritic growth system is subject to entirely new instability mechanisms compared to a unidirectional solidification system. These instabilities are called *global instability mechanisms*, specifically global trapped wave instability and low-frequency instability.
(ii) As a consequence, the anisotropy of surface tension is not a necessary condition for dendritic growth. The selection condition can be found even in the case of isotropic surface tension.
(iii) At a later stage of growth, the system selects the unique most dangerous *global neutrally stable mode solution*. When the anisotropy of surface tension is small, the selected solution is not a steady stable solution, but rather a time-periodic, oscillatory, neutrally stable solution. Only when the value of anisotropy is larger than a critical number is the selected solution the steady needle solution.

Interfacial wave theory is a coherent and predictive analytical theory. It naturally follows from the ideas present in the general theory of dynamical systems. Interfacial wave theory was first established in studying systems of dendritic growth and viscous fingering formation, and it was found to be in good quantitative agreement with all the available experimental data. The ideas and approach of interfacial wave theory were then successfully applied to studying much broader phenomena of pattern formation in solidification, including spatially periodic pattern formation in systems of arrayed-cellular and eutectic growth.

## 1.3 Macroscopic Continuum Model

In this book, we use the macroscopic continuum model. This implies that the liquid and solid bulk phases will be treated as continuous media, while the interface is considered a geometric surface. This continuous medium model is applicable to the pattern formation phenomena under investigation. The size of the microstructure that we study is on the micrometer scale, which is far larger than the mean-free path of the molecules. Furthermore, the time scale of the interface movement is far longer than the mean collision period of particles. Let us now imagine that at first, we have a metastable undercooled liquid system. Then, due to nucleation or the presence of a seed, the solidification process begins. A front, which is described by the sharp interface between the solid phase and liquid phase, moves into the undercooled liquid phase and transforms more and more liquid into the solid phase. Eventually, the whole liquid phase is transformed into the more thermodynamically stable solid phase. During solidification, there may coexist several interactive macroscopic transport processes in the system. First of all, due to the phase transition, the latent heat is released at the interface. Hence, a solidification system intrinsically has an inhomogeneous temperature field, and heat transfer through conduction is unavoidable. Furthermore, if the system involves two species, if, for instance, it contains an impurity, then this binary mixture should be described not only by the temperature field, but also by the impurity concentration field. It is known that for a binary system with coexisting liquid and solid phase, the impurity concentration will have a jump at the interface because of the segregation effect. Therefore, the concentration field in a solidifying binary mixture system is always inhomogeneous. A mass transfer will be present through the mass diffusion in the concentration field. Finally, the liquid phase in a solidification system may be in motion induced by various driving forces. For instance, the density change during phase transition will cause convection in the bulk liquid, because the interface acts as a mass sink when the liquid density is smaller than the solid density, while it acts as a mass source when the liquid density is larger than the solid density. The buoyancy effect due to gravity may also cause convection, because the inhomogeneous temperature induces inhomogeneous density. Moreover, other forces on the body and external flow can all cause convective motion in the liquid phase. When convection exists in the liquid phase, there will be a momentum transfer in the system governed by fluid dynamics.

If the above three kinds of macroscopic transport processes coexist, they will obviously interact with one another.

In this section, we shall give a general mathematical description of solidification on the basis of the macroscopic continuum model.

### *1.3.1 Macroscopic Transport Equations*

Consider a front moving in an undercooled melt, with a characteristic velocity $V$ in the reference frame in which the solid phase is at rest. For convenience, it may be called the rest frame. The melt may be a pure substance or a binary mixture. If it is a mixture, one of the species is dilute and may be regarded as an impurity. We assume that the liquid phase of the binary mixture has mass density $\rho$, thermal diffusivity $\kappa_{\mathrm{T}}$, specific heat $c_{\mathrm{p}}$, mass diffusivity of the impurity $\kappa_{\mathrm{D}}$, while the corresponding thermal characteristics in the solid phase are denoted by $\rho_{\mathrm{S}}$, $\kappa_{\mathrm{TS}}$ and $c_{\mathrm{pS}}$, $\kappa_{\mathrm{DS}}$, respectively. Very often, as a simplification, it is assumed that the mass diffusivity in the solid phase, $\kappa_{\mathrm{DS}}$, is zero.

Convective motion may occur in the liquid phase for a number of reasons, as previously mentioned. We assume that the liquid can be considered an incompressible Newtonian fluid. The density inhomogeneity caused by the inhomogeneity of temperature is not important except in terms of a buoyancy effect. Thus, the Boussinesq approximation is applicable. The state of the system is described by the following macroscopic fields:

1. the temperature field in the liquid phase, $T(\mathbf{r}, t)$,
2. the temperature field in the solid phase, $T_S(\mathbf{r}, t)$,
3. the concentration field of the impurity in the liquid phase, $C(\mathbf{r}, t)$,
4. the concentration field of the impurity in the solid phase, $C_S(\mathbf{r}, t)$,
5. the absolute velocity field in the liquid phase, $\mathbf{U}(\mathbf{r}, \mathbf{t})$.

Here, we use $\mathbf{r}$ and $t$ to denote the space vector and time, respectively. The governing equations consist of the heat conduction equation, mass diffusion equation, and the Navier–Stokes equations in the Boussinesq model. These equations can be written as follows:

1. The heat conduction equation for the liquid phase. In the rest frame, the heat conduction equation can be written in the form

$$c_p\rho\left(\frac{\partial T}{\partial t} + \mathbf{U}\cdot\nabla T\right) = k_T\nabla^2 T, \tag{1.4}$$

or

$$\frac{\partial T}{\partial t} + \mathbf{U}\cdot\nabla T = \kappa_T\nabla^2 T, \tag{1.5}$$

where $k_T$ is the heat conduction coefficient, while $\kappa_T = \frac{k_T}{c_p\rho}$ is the thermal diffusivity in the liquid phase.
2. The heat conduction equation for the solid phase. In the rest frame, we similarly have

$$\frac{\partial T_S}{\partial t} = \kappa_{TS}\nabla^2 T_S, \tag{1.6}$$

where $\kappa_{TS}$ is the thermal diffusivity in the solid phase.
3. The mass diffusion equation in the liquid phase:

$$\frac{\partial C}{\partial t} + \mathbf{U}\cdot\nabla C = \kappa_D\nabla^2 C, \tag{1.7}$$

where $\kappa_D$ is mass diffusivity in the liquid phase.
4. The mass diffusion equation in the solid phase:

$$\frac{\partial C_S}{\partial t} = \kappa_{DS}\nabla^2 C, \tag{1.8}$$

where $\kappa_{DS}$ is mass diffusivity in the solid phase.
5. The continuity equation. For an incompressible liquid mixture, we have

$$\nabla\cdot\mathbf{U} = 0. \tag{1.9}$$

6. The momentum equation:

$$\frac{\partial \mathbf{U}}{\partial t} + (\mathbf{U} \cdot \nabla)\mathbf{U} = -\frac{1}{\rho}\nabla P + \nu \nabla^2 \mathbf{U} + \beta(T - T_*)\mathbf{g}, \tag{1.10}$$

where $P$ is the reduced pressure. If we introduce the vorticity $\Omega$ as

$$\mathbf{\Omega} = \nabla \times \mathbf{U}, \tag{1.11}$$

the momentum equation can be replaced by the vorticity equation:

$$\frac{\partial \mathbf{\Omega}}{\partial t} + \nabla \times (\mathbf{\Omega} \times \mathbf{U}) = \nu \nabla^2 \mathbf{\Omega} + \nabla \times \left\{\beta(T - T_*)\mathbf{g}\right\}. \tag{1.12}$$

Here, we assume that there is no body force in the system, except for gravity; $\mathbf{g}$ is the acceleration of gravity; $\nu$ is the kinematic viscosity; $\beta$ is the thermal expansion coefficient; and $T_*$ is a reference temperature.

### *1.3.2 The Interface Conditions*

The full set of boundary conditions must be specified for each specific problem. For instance, the far-field conditions will be different for different growth conditions. However, for the problems under investigation here, the same type of interface conditions are, in general, always applied. We denote the interface shape in the rest frame by $S(\mathbf{r}, t) = 0$. Since the interface location during solidification is unknown, determining this interface shape function is a part of the solution. These problems, in mathematics, are all called *free boundary problems* or *moving boundary problems.*

To determine the interface condition, we assume that the system is in local thermodynamic equilibrium at the interface. This implies the following. Let us take two control surfaces parallel and very close to the interface, one on the liquid side and the other on the solid side. The volume contained within these two control surfaces can be considered the "realistic interface domain," which we uniformly divide into many small volume elements by planes normal to the interface (cf. Fig. 1.8). At the same time, the interface is also divided into many small surface elements. As the division becomes finer and finer, the two control surfaces move closer and closer to each other, and the volume elements become smaller and smaller. As an approximation, one may assume that each subsystem, $(I)+(II)+(II)'$, consisting of the interface element $(I)$, the volume element in liquid $(II)$, and the volume element in solid $(II)'$, is in thermodynamic equilibrium. Certainly, different subsystems are in different thermodynamic equilibrium states. This *local thermodynamic equilibrium state* assumption is a basic assumption that we shall employ in most of this monograph.

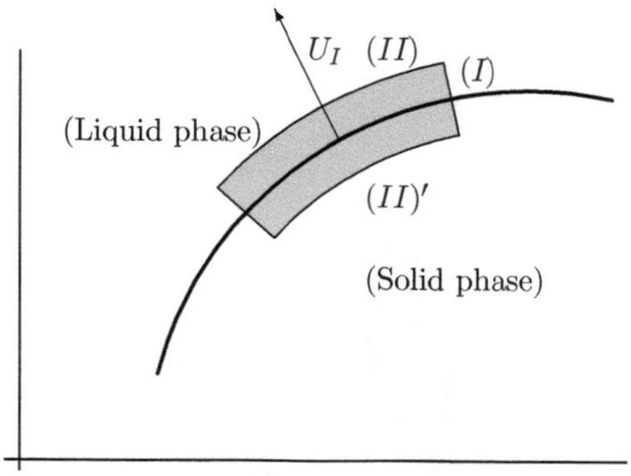


**Fig. 1.8** Sketch of a moving front in an undercooled melt

Based on this assumption, we can derive the following interface conditions in the rest frame used:

1. *The thermodynamic equilibrium condition for temperature:*

$$T = T_S\,. \tag{1.13}$$

2. Another thermodynamic equilibrium condition at the interface is the phase equilibrium condition. This implies that the chemical potential of the solid phase must equal the chemical potential of the liquid phase. This condition will lead to the *Gibbs–Thomson condition*, which determines the temperature of the interface. The Gibbs–Thomson condition states that the solidification temperature at a general curved interface is different from the solidification temperature of the pure melt at a flat interface, say $T_{M0}$. The correction comes from two sources. The first is from the effect of the curvature of the interface on the surface energy. To demonstrate this effect, first consider a pure substance with a curved interface between the liquid and solid phases. Assume that this system is in thermodynamic equilibrium. From the minimum free energy principle at the local equilibrium state, one finds that

$$T_S = T_{M0}\left(1 - \frac{\gamma}{\Delta H}\mathcal{K}\right), \tag{1.14}$$

where $\gamma$ is the surface tension constant, $\Delta H$ is the latent heat per unit volume of the solid, $T_{M0}$ is the melting temperature of a flat interface, and $\mathcal{K}$ is twice the local mean curvature of the interface.

The second source of the correction is the impurity at the interface. To study this effect, we consider a binary mixture with a flat interface between the liquid and solid phases, which is in thermodynamic equilibrium. From the thermodynamics, the equilibrium state of the system can be described by its phase diagram. A typical diagram is shown in Fig. 1.9. One can see that the melting temperature is determined by the liquidus curve and the concentration of the impurity in the liquid phase. When the impurity is dilute, the liquidus curve can be approximated by its tangent line at the impurity concentration

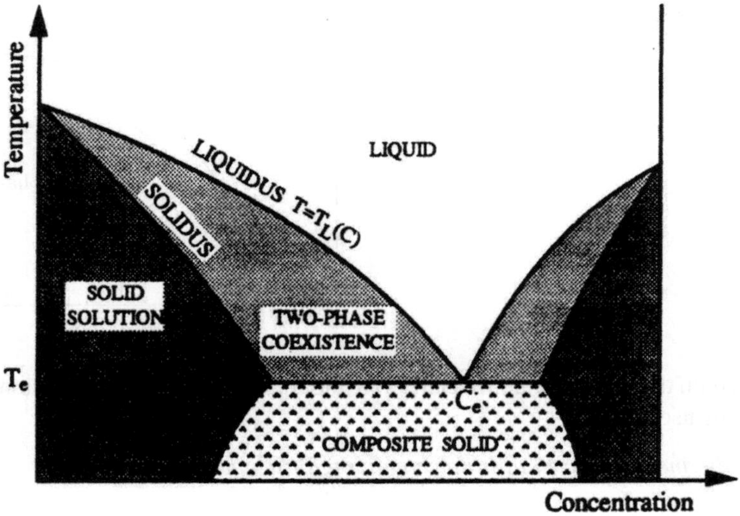


**Fig. 1.9** Typical phase diagram of a binary mixture system. The *shaded regions* indicate what phases are in thermodynamic equilibrium in a sample of given bulk composition and uniform temperature. When the temperature $T$ and bulk composition are between the liquidus and solidus curves, then liquid of the liquidus composition $C_{\mathrm{L}}(T)$ is in equilibrium with solid of the solidus composition given by $\kappa C_{\mathrm{L}}(T)$, which defines the local segregation coefficient $\kappa(C_{\mathrm{L}})$

$C = 0$. Suppose that the slope of the curve at $C = 0$ is $m < 0$. Then, since the impurity concentration on the liquid side of the interface is $C = C_{\mathrm{I}}$, the melting temperature will be different from the melting temperature of pure melt by an amount $mC_{\mathrm{I}}$.

Putting these two corrections together, we derive the interface temperature of a binary mixture as

$$T_{\mathrm{S}} = T_{\mathrm{M0}}\left(1 - \tfrac{\gamma}{\Delta H}\mathcal{K}\right) + mC_{\mathrm{I}}\,. \tag{1.15}$$

This is the Gibbs–Thomson condition for a binary mixture, which relates the local melting temperature to the surface tension, the local mean curvature of the interface, and the local impurity concentration. This physical condition is the most important cause of the rich variety of pattern formation phenomena.

3. *The energy conservation condition*: Enthalpy is conserved during the phase transition, so

$$\Delta H(U_{\mathrm{Sn}} - U_{\mathrm{I}}) = -\Delta H U_{\mathrm{I}} = \mathbf{n}\cdot\left[(k_{\mathrm{T}}\nabla T)_{\mathrm{liquid}} - (k_{\mathrm{T}}\nabla T)_{\mathrm{solid}}\right]$$
$$-\left[(c_{\mathrm{p}}T)_{\mathrm{liquid}} - (c_{\mathrm{p}}T)_{\mathrm{solid}}\right]\rho_{\mathrm{S}}U_{\mathrm{I}}\,, \tag{1.16}$$

where we have the following quantities:

$U_{\mathrm{Sn}}$ : the normal component of the solid phase velocity at the interface, observed in the rest frame. We have assumed that it is zero.
$U_{\mathrm{I}}$ : the local growth velocity of the interface in the rest frame.
$\mathbf{n}$ : the normal vector of the interface.

The left-hand side of the formula (1.16) is the latent heat release per unit volume of solid per unit time. The first part of the right-hand side of (1.16) is the total enthalpy flux away from the interface to both the liquid and solid sides due to heat conduction, while the second part is the total enthalpy flux due to the convective motion.

4. *The segregation condition for the impurity*: According to the thermodynamics, the concentration of impurity will have a jump at the interface. The relationship of concentrations of the impurity on the two sides of the interface is derived from the phase diagram for the given binary mixture, namely,

$$C_{\mathrm{S}} = \kappa C\,, \tag{1.17}$$

where $\kappa$ is a material constant, the so-called segregation coefficient. Normally, $0 < \kappa < 1$. Hence, the impurity is normally rejected from the solid phase. As a result, it must build up in front of the advancing interface.

5. *The mass conservation condition for the impurity*: As we have seen, the interface acts as a mass source. The quantity of the impurity rejected from the interface into the liquid must be balanced by the mass flux in the liquid and solid due to the concentration gradient near the interface. Thus, one has

$$\Delta C U_{\mathrm{I}} = C U_{\mathrm{Ln}} - \mathbf{n} \cdot \Big[ (\kappa_{\mathrm{D}} \nabla C)_{\mathrm{liquid}} - (\kappa_{\mathrm{D}} \nabla C)_{\mathrm{solid}} \Big]\,, \tag{1.18}$$

where $\Delta C = C - C_{\mathrm{S}} = (1-\kappa)C$ is the jump in the concentration of impurity at the interface and $U_{\mathrm{Ln}}$ is the normal component of the absolute velocity of a fluid element at the interface. In general, the mass diffusivity $\kappa_{\mathrm{DS}}$ in the solid is negligible.

6. *The total mass conservation condition*:

$$\rho(U_{\mathrm{Ln}} - U_{\mathrm{I}}) = \rho_{\mathrm{S}}(U_{\mathrm{Sn}} - U_{\mathrm{I}}) = -\rho_{\mathrm{S}} U_{\mathrm{I}}\,. \tag{1.19}$$

7. *The continuity condition for the tangential component of velocity*:

$$U_{\mathrm{L}\tau} = U_{\mathrm{S}\tau}\,. \tag{1.20}$$

Since $U_{\mathrm{S}\tau} = 0$, one has

$$U_{\mathrm{L}\tau} = \mathbf{U} \cdot \mathbf{e}_\tau = 0\,. \tag{1.21}$$

In the above, the symbols have the following meanings:

$\mathbf{e}_\tau$ : the local unit tangent vector at the interface;
$U_{L\tau}$ : the tangential component of the absolute velocity of a liquid element at the interface;
$U_{S\tau}$ : the tangential component of the absolute velocity of a solid element at the interface.

Besides the above local thermodynamic equilibrium conditions and kinematic conditions, we also have some mechanical equilibrium conditions at the interface. These are the dynamic interface conditions derived from the momentum balance along the normal and tangential directions at the interface. Since the pattern formation problem is mostly not concerned with the stresses and strains in the solid phase, these dynamic conditions will not be considered here.

One item in the above formulation that needs clarification is how to define and calculate the local growth velocity of the interface, $U_I$. Let the Cartesian coordinates $(X_1, X_2, X_3)$ be those of the rest frame. We consider the motion of an arbitrary reference point $P$:$\{X_1(t), X_2(t), X_3(t)\}$ that always remains on the interface $S(X_1, X_2, X_3) = 0$. As the time $t$ increases from $t \to (t + dt)$, the position of the point $P$ will change from $\{X_1(t), X_2(t), X_3(t)\} \to \{X_1(t+dt), X_2(t+dt), X_3(t+dt)\}$. These increments must be subject to the following relationship:

$$dS = \frac{\partial S}{\partial t}dt + \frac{\partial S}{\partial X_1}dX_1 + \frac{\partial S}{\partial X_2}dX_2 + \frac{\partial S}{\partial X_3}dX_3 = 0\,. \tag{1.22}$$

The velocity of the motion of the point $P$ is

$$\mathbf{U}_I(t) = \{X_1'(t), X_2'(t), X_3'(t)\} \tag{1.23}$$

and

$$\frac{\partial S}{\partial t} = -\nabla S \cdot \mathbf{U}_I\,. \tag{1.24}$$

We assume that the growth velocity of the interface remains along the direction normal to the interface, i.e., $\mathbf{U}_I \parallel \mathbf{n}$. Thus, from (1.24), we have

$$\frac{\partial S}{\partial t} = -|\nabla S| \cdot |\mathbf{U}_I| = -|\nabla S| U_I \tag{1.25}$$

with

$$U_I = -\frac{\frac{\partial S}{\partial t}}{|\nabla S|}\,. \tag{1.26}$$

For the sake of convenience, we often use a moving frame $(x_1, x_2, x_3)$ with velocity $\mathbf{V}$ as the reference frame and set one of the coordinate axes, for instance the $x_3$-axis, whose unit vector is $\hat{\mathbf{e}}_3$, parallel to $\mathbf{V}$. So we have that

$$\mathbf{U} = \mathbf{u} + V\hat{\mathbf{e}}_3 \,. \tag{1.27}$$

In the moving frame, we denote the relative velocity field of a fluid by $\mathbf{u}$ and the relative growth velocity of the interface by $\mathbf{u}_{\mathrm{I}}$. From the moving frame to the rest frame, the following Galilean transformation holds:

$$\begin{cases} X_1 = x_1, \\ X_2 = x_2, \\ X_3 = x_3 + Vt, \\ t = t\,. \end{cases} \tag{1.28}$$

The interface shape equation in the moving frame, $s(x_1, x_2, x_3, t) = 0$, can be obtained from that in the rest frame through the following relationships:

$$s(x_1, x_2, x_3, t) = S(x_1, x_2, x_3 + Vt, t) = 0 \tag{1.29}$$

and

$$S(X_1, X_2, X_3, t) = s(X_1, X_2, X_3 - Vt, t) = 0\,. \tag{1.30}$$

Thus, we can calculate the growth velocity of the interface in terms of the interface shape equation in the moving frame, namely,

$$U_{\mathrm{I}} = -\frac{\frac{\partial S}{\partial t}}{|\nabla S|} = -\frac{\frac{\partial s}{\partial t} - V\frac{\partial s}{\partial x_3}}{|\nabla s|} = \mathbf{V}\cdot\mathbf{n} - \frac{\frac{\partial s}{\partial t}}{|\nabla s|} = u_{\mathrm{I}} + V\hat{\mathbf{e}}_3 \cdot \mathbf{n}\,, \tag{1.31}$$

where

$$u_{\mathrm{I}} = -\frac{\frac{\partial s}{\partial t}}{|\nabla s|} \tag{1.32}$$

is the local growth velocity of the interface observed in the moving frame, and

$$\mathbf{n} = \frac{\nabla s}{|\nabla s|} \tag{1.33}$$

is a vector normal to the interface.

### 1.3.3 The Scaling and the Dimensionless System

As the first step of a mathematical treatment, one needs to make the system dimensionless by choosing a set of proper scales, which reflect the physical nature of the dynamical system. In many cases, using improper scales will cause difficulties in making either mathematical simplifications or physical interpretations.

For the problems formulated in the last section, we shall choose the characteristic growth velocity of the interface $V$ as the scale for the velocity. The length scale can be chosen from the intrinsic length scales of the system. There are two intrinsic length scales in the system determined by the macroscopic transport processes: the thermal diffusion length $\ell_T = \kappa_T/V$ and the mass diffusion length $\ell_D = \kappa_D/V$. In general, the mass diffusion length is always much smaller than the thermal diffusion length, since

$$\lambda = \frac{\kappa_D}{\kappa_T} \ll 1 \quad \text{and} \quad \lambda_s = \frac{\kappa_D}{\kappa_{TS}} \ll 1 \,. \tag{1.34}$$

Hence, when one deals with a binary mixture, the mass diffusion is the controlling factor for the macroscopic transport processes. Consequently, $\ell_D$ is chosen as the length scale. Of course, when one deals with a pure substance, where only thermal conduction is involved, the thermal diffusion length $\ell_T$ will be chosen as the length scale.

In the present system, the physical quantity $\Delta H/(c_p\rho)$ has the scale of temperature, so it is naturally chosen as the unit of temperature. Using these scales, we define the following dimensionless quantities and variables:

$$(\bar{X}_1, \bar{X}_2, \bar{X}_3, \bar{t}) = (X_1/\ell_D, X_2/\ell_D, X_3/\ell_D, tV/\ell_D)$$

$$(\bar{x}_1, \bar{x}_2, \bar{x}_3) = (x_1/\ell_D, x_2/\ell_D, x_3/\ell_D)$$

$$\bar{T} = \frac{T - T_{M0}}{\Delta H/(c_p\rho)}$$

$$\bar{\mathbf{u}} = \mathbf{u}/V$$

$$\bar{\omega} = \omega\kappa_T/V^2 \quad (\omega = \nabla \times \mathbf{u}) \,, \tag{1.35}$$

where $T_{M0}$ is the melting temperature at a flat interface. From (1.27), we have

$$\mathbf{U}/V = \bar{\mathbf{U}} = \bar{\mathbf{u}} + \hat{\mathbf{e}}_3 \,. \tag{1.36}$$

Then the governing equations and related boundary conditions in the moving frame can be derived from those in the rest frame by applying the transformation

$$\frac{\partial}{\partial t} \to \frac{\partial}{\partial t} - \mathbf{V} \cdot \nabla \tag{1.37}$$

or

$$\frac{\partial}{\partial \bar{t}} \to \frac{\partial}{\partial \bar{t}} - \hat{\mathbf{e}}_3 \cdot \nabla \,. \tag{1.38}$$

The dimensionless governing equations are obtained as follows.

1. Heat conduction equation in the liquid phase:

$$\lambda \left( \frac{\partial \bar{T}}{\partial \bar{t}} + \bar{\mathbf{u}} \cdot \nabla \bar{T} \right) = \nabla^2 \bar{T} \,. \tag{1.39}$$

2. Heat conduction equation in the solid phase:

$$\lambda \hat{\alpha}_{\mathrm{T}} \left( \frac{\partial \bar{T}_{\mathrm{S}}}{\partial \bar{t}} + \hat{\mathbf{e}}_3 \cdot \nabla \bar{T}_{\mathrm{S}} \right) = \nabla^2 \bar{T}_{\mathrm{S}} \,. \tag{1.40}$$

3. Mass diffusion equation in the liquid phase:

$$\frac{\partial \bar{C}}{\partial \bar{t}} + \bar{\mathbf{u}} \cdot \nabla \bar{C} = \nabla^2 \bar{C} \,. \tag{1.41}$$

4. Continuity equation for the liquid state:

$$\nabla \cdot \bar{\mathbf{u}} = 0 \,. \tag{1.42}$$

5. Momentum equation for the liquid state:

$$\frac{\partial \bar{\omega}}{\partial \bar{t}} + (\bar{\mathbf{u}} \cdot \nabla)\bar{\omega} - (\bar{\omega} \cdot \nabla)\bar{\mathbf{u}} = \mathrm{Pr}\nabla^2 \bar{\omega} - \frac{\mathrm{Gr}}{T_\infty} \nabla \times (\bar{T}\mathbf{e}_{\mathrm{g}}) \,. \tag{1.43}$$

In the above, $\mathbf{e}_{\mathrm{g}}$ is the unit vector along the direction of gravity, and we introduce several parameters: $\mathrm{Pr} = \nu/\kappa_{\mathrm{T}}$, the Prandtl number; $T_\infty = -\big[T_{\mathrm{M0}} - (T_\infty)_{\mathrm{D}}\big]c_{\mathrm{P}}\rho/\Delta H = -\mathrm{St}$, the undercooling parameter, where $(T_\infty)_{\mathrm{D}}$ denotes the dimensional undercooling temperature; the parameter St is sometimes called the Stefan number; $\mathrm{Gr} = g\beta(T_{\mathrm{M0}} - (T_\infty)_{\mathrm{D}})\kappa_{\mathrm{T}}/V^3$ is the Grashof number.

The dimensionless conditions at the interface $\bar{S}(\mathbf{r}, t) = 0$ are as follows:

(1) *Thermodynamic equilibrium for temperature:*

$$\bar{T} = \bar{T}_{\mathrm{S}} \,; \tag{1.44}$$

(2) *Gibbs–Thomson condition:*

$$\bar{T}_S = -\Gamma\{\bar{\mathcal{K}}\} - M\bar{C}\,, \tag{1.45}$$

where $\Gamma$ is the surface tension parameter

$$\Gamma = \frac{\ell_c}{\ell_D} \tag{1.46}$$

and

$$\ell_c = \frac{\gamma c_p T_{M0} \rho}{(\Delta H)^2} \tag{1.47}$$

is a length scale, usually called the capillary length, while $M$ is the morphological parameter

$$M = -\frac{mC_\infty}{\Delta H/(c_p \rho)}\,. \tag{1.48}$$

Since $\ell_c$ is determined by the interfacial energy $\gamma$, it is sometimes considered a microscopic length scale. It will be seen later that the parameter $\Gamma$, the ratio of the macroscopic length $\ell_D$ and the microscopic length $\ell_c$, is the most important parameter for the stability of the system.

(3) *Enthalpy conservation:*

$$\begin{aligned}&\Big[1 - \alpha_s(1-\beta_s)(\bar{T} + \bar{T}_{M0})\Big](\bar{u}_I + \hat{\mathbf{e}}_3 \cdot \mathbf{n}) \\ &\quad = \left[\frac{\beta_s \alpha_s}{\lambda_s}\nabla \bar{T}_S - \frac{1}{\lambda}\nabla \bar{T}\right] \cdot \mathbf{n}\,.\end{aligned} \tag{1.49}$$

Here we have defined $\bar{T}_{M0} = \dfrac{T_{M0}}{\Delta H/(c_p \rho)}$, and

$$\beta_s = \frac{c_{ps}}{c_p}, \quad \lambda_s = \frac{\kappa_D}{\kappa_{TS}}, \quad \lambda = \frac{\kappa_D}{\kappa_T}, \tag{1.50}$$

which measure the changes in the thermodynamic characteristics in the phase transition. We define $\hat{\alpha}_T = \kappa_T/\kappa_{TS}$ as the ratio of thermal diffusivities in the solid and liquid. Conventionally, the simplification with $\beta_s = \alpha_s = 1$ and $\hat{\alpha}_T = 1$ is called the *symmetric model*, while the simplification with $\beta_s = \alpha_s = 1$ but $\hat{\alpha}_T = 0$ is called the *one-sided model*.

(4) *Mass conservation for the impurity:*

$$(1-\kappa)\bar{C}(\bar{u}_I + \hat{\mathbf{e}}_3 \cdot \mathbf{n}) = \bar{C}(\bar{\mathbf{u}} + \hat{\mathbf{e}}_3) \cdot \mathbf{n} - \mathbf{n} \cdot \nabla \bar{C}\,; \tag{1.51}$$

(5) *Conservation of total mass:*

$$\bar{\mathbf{u}} \cdot \mathbf{n} + (\alpha_s - 1)\bar{u}_I + \alpha_s \hat{\mathbf{e}}_3 \cdot \mathbf{n} = 0\,; \tag{1.52}$$

(6) *Continuity of the tangential component of velocity:*

$$\bar{\mathbf{u}} \cdot \mathbf{e}_\tau = -\hat{\mathbf{e}}_3 \cdot \mathbf{e}_\tau\,. \tag{1.53}$$

Later, for the sake of simplicity, we shall omit the bar over all dimensionless quantities without confusion.

## References


1. D. Bensimon, P. Pelce, B.I. Shraiman, Dynamics of curved fronts and pattern selection. J. Phys. **48**, 2081–2087 (1987)
2. E.A. Brener, V.I. Melnikov, Pattern selection in two dimensional dendritic growth. Adv. Phys. **40**, 53–97 (1991)
3. Y. Couder, N. Gerard, M. Rabaud, Narrow fingers in the Saffman–Taylor instability. Phys. Rev. A **34**, 5175–5178 (1986)
4. A. Dougherty, J.P. Gollub, Steady-state dendritic growth of $NH_4Br$ from solution. Phys. Rev. A **38**, 3043–3053 (1988)
5. M.E. Glicksman, R.J. Schaefer, J.D. Ayers, High-confidence measurement of solid/liquid surface energy in a pure material. Philos. Mag. **32**, 725–743 (1975)
6. M.E. Glicksman, R.J. Schaefer, J.D. Ayers, Dendrite growth — a test of theory. Metall. Trans. **7A**, 1747–1757 (1976)
7. G.M. Homsy, Viscous fingering in porous media. Ann. Rev. Fluid Mech. **19**, 271–311 (1987)
8. G. Horvay, J.W. Cahn, Dendritic and spheroidal growth. Acta Metall. **9**, 695–705 (1961)
9. S.C. Huang, M.E. Glicksman, Fundamentals of dendritic solidification — I. Steady-state tip growth; II. Development of sidebranch structure. Acta Metall. **29**, 701–734 (1981)
10. G.P. Ivantsov, Temperature field around a spheroidal, cylindrical and acicular crystal growing in a supercooled melt. Dokl. Akad. Nauk. SSSR **58**(4), 567–569 (1947)
11. D.A. Kessler, H. Levine, Stability of dendritic crystals. Phys. Rev. Lett. **57**, 3069–3072 (1986)
12. D.A. Kessler, J. Koplik, H. Levine, Pattern formation far from equilibrium: the free space dendritic crystal, in *Proc. NATO A.R.W. on Patterns, Defects and Microstructures in Non-equilibrium Systems*, Austin, TX, March, 1986
13. J.S. Kirkaldy, Spontaneous evolution of spatiotemporal patterns in materials. Rep. Prog. Phys. **55**, 723–795 (1992)
14. T. Kobayashi, Y. Furukawa, *Snow Crystals* (Snow Crystal Museum, Asahikawa, Hokkaido, 1991)
15. A.R. Kopf-Sill, G.M. Homsy, Narrow fingers in a Hele–Shaw cell. Phys. Fluids **30**(9), 2607–2609 (1987)
16. M. Kruskal, H. Segur, Asymptotics beyond all orders in a model of crystal growth. Stud. Appl. Math. **85**, 129–181 (1991)
17. J.S. Langer, Instability and pattern formation in crystal growth. Rev. Mod. Phys. **52**, 1–28 (1980)
18. J.S. Langer, *Lectures in the Theory of Pattern Formation*. USMG NATO AS Les Houches Session XLVI 1986 — Le hasard et la matière/ chance and matter, ed. by J. Souletie, J. Vannimenus, R. Stora (Elsevier Science, Amsterdam, 1986)

19. J.S. Langer, Dendritic sidebranching in the three-dimensional symmetric model in the presence of noise. Phys. Rev. A **36**(7), 3350–3358 (1987)
20. J.S. Langer, Issues and opportunities in materials research. Phys. Today, 24–31 (1992)
21. J.S. Langer, H. Müller-Krumbhaar, Theory of dendritic growth — I. Elements of a stability analysis; II. Instabilities in the limit of vanishing surface tension; III. Effects of surface tension. Acta Metall. **26**, 1681–1708 (1978)
22. W.W. Mullins, R.F. Sekerka, Morphological stability of a particle growing by diffusion or heat flow. J. Appl. Phys. **34**, 323–329 (1963)
23. W.W. Mullins, R.F. Sekerka, Stability of a planar interface during solidification of a dilute binary alloy. J. Appl. Phys. **35**, 444–451 (1964)
24. M. Muschol, D. Liu, H.Z. Cummins, Surface-tension-anisotropy measurement of succinonitrile and pivalic acid: comparison with microscopic solvability theory. Phys. Rev. A **46**, 1038–1050 (1992)
25. G.E. Nash, M.E. Glicksman, Capillarity-limited steady-state dendritic growth I. Theoretical development. Acta Metall. **22**, 1283–1299 (1974)
26. P. Pelce, *Dynamics of Curved Front* (Academic, New York, 1988)
27. P. Pelce, Y. Pomeau, Dendrites in the small undercooling limit. Stud. Appl. Math. **74**, 245–258 (1986)
28. J.W. Rutter, B. Chalmers, A prismatic substructure formed during solidification of metals. Can. J. Phys. **31**, 15–39 (1953)
29. P.G. Saffman, G.I. Taylor, The penetration of a fluid into a porous medium or Hele–Shaw cell containing a more viscous liquid. Proc. R. Soc. London Ser. A **245**, 312–329 (1958)
30. H. Segur, S. Tanveer, H. Levine (eds.), *Asymptotics Beyond All Orders*. NATO ASI Series, Series B: Physics, vol. 284 (Plenum, New York, 1991)
31. K. Takahashi, Y. Furukawa, Y. Takahashi, *Story of Snow Crystals* (Koudansya, Tokyo, 1995)
32. S. Tanveer, Analytic theory for the selection of a symmetric Saffman–Taylor finger in a Hele–Shaw Cell. Phys. Fluid **30**(8), 1589–1605 (1987)
33. J.J. Xu, Global asymptotic solution for axi-symmetric dendrite growth with small undercooling, in *Structure and Dynamics of Partially Solidified System*, ed. by D.E. Loper, NATO ASI Series E. No. 125 (1987), pp. 97–109
34. J.J. Xu, PhD. Thesis, A global asymptotic theory of dendrite growth at small undercooling. Department of Mathematical Sciences, Rensselaer Polytechnic Institute, NY (1987)
35. J.J. Xu, Global wave mode theory for formation of dendritic structure on a growing needle crystal. Phys. Status Sol. (B) **157**, 577–591 (1990)
36. J.J. Xu, Global neutral stable state and selection condition of tip growth velocity. J. Crystal Growth **100**, 481–490 (1990)
37. J.J. Xu, Asymptotic theory of steady axisymmetric needle-like crystal growth. Stud. Appl. Math. **82**, 71–91 (1990)
38. J.J. Xu, Interfacial wave theory of solidification — dendritic pattern formation and selection of tip velocity. Phys. Rev. A15 **43**(2), 930–947 (1991)
39. J.J. Xu, Two-dimensional dendritic growth with anisotropy of surface tension. Physica (D) **51**, 579–595 (1991)
40. J.J. Xu, Interfacial wave theory of two-dimensional dendritic growth with anisotropy of surface tension. Can. J. Phys. **69**(7), 789–800 (1991)
41. J.J. Xu, Global instability of viscous fingering in Hele–Shaw cell (I) — formation of oscillatory fingers. Eur. J. Appl. Math. **2**, 105–132 (1991)
42. J.J. Xu, The effect of convection motion on dendritic growth, in *Interactive Dynamics of Convection and Solidification*, ed. by Davis et al. (Kluwer, Dordrecht, 1992), pp. 101–103
43. J.J. Xu, Global instability and pattern formation in dendritic solidification of dilute binary alloy system. Can. Appl. Math. Quart. **1**(2), 255–292 (1993)
44. J.J. Xu, Dendritic growth from melt with external flow: Uniformly valid asymptotic solution for the steady state. J. Fluid Mech. **263**, 227–243 (1994)
45. J.J. Xu, Effect of convection motion in melt induced by density-change on dendritic solidification. Can. J. Phys. **72**(3 & 4), 120–125 (1994)

46. J.J. Xu, Generalized needle solutions, interfacial instabilities and pattern formations. Phys. Rev. E **53**(5), 5051–5062 (1996)
47. J.J. Xu, Interfacial wave theory for oscillatory finger's formation in a Hele–Shaw cell: a comparison with experiments. Eur. J. Appl. Math. **7**, 169–199 (1996)
48. J.J. Xu, Interfacial instabilities and fingering formation in Hele–Shaw flow. IMA J. Appl. Math. **57**, 101–135 (1996)
49. J.J. Xu, Z.X. Pan, Interfacial wave theory of dendritic growth from a binary mixture: a comparison with experiments. J. Crystal Growth **129**, 666–682 (1993)
50. P.A. Zhuravlev, On the motion of a fluid in channels. Zap. Leningr. Gorn. In-ta. **33**(3), 54–61 (1956)

# Chapter 2
# Unidirectional Solidification and Mullins–Sekerka Instability

Before we begin the study of dendritic growth, it is appropriate to examine a simple case first: the instability of a planar interface in unidirectional solidification. Mullins and Sekerka were the first, in 1963, to perform a systematic analysis of this system. Their linear stability analysis is now called *Mullins–Sekerka instability* [6, 7].

In what follows, we attempt to review briefly the results of Mullins and Sekerka on linear stability of unidirectional solidification, both from a pure melt and from a binary mixture. Our main purpose is to explain the physical essence of the phenomenon and demonstrate the mathematical methods of solving the problem. Thus only the simplest model problem and very basic results will be discussed. The approach used below is somehow different from that originally used by Mullins and Sekerka; but its spirit can be extended to solve more difficult problems, such as the stability of a curved front.

## 2.1 Solidification with Planar Interface from a Pure Melt

Imagine that an originally flat solidifying front advances into a pure undercooled melt with velocity $V$. Suppose the system is two-dimensional and free of convection. This is the case of solidification in a thin Hele–Shaw cell. We employ moving two-dimensional Cartesian coordinates $(x, y)$ with the $x$-axis fixed on the interface, as shown in Fig. 2.1. The $y$-axis is along the growth direction of the interface. Furthermore, we assume that the mass density $\rho$, the thermal diffusivity constant $\kappa_T$, and the specific heat per unit volume $c_p$ are the same for both liquid and solid; namely, we adopt the *symmetric model* with $\beta_s = \alpha_s = \hat{\alpha}_T = 1$.

To make all physical properties dimensionless, we adopt the scale defined in Sect. 1.3.3 with the thermal length $\ell_T$ as the length scale. Thus, the dimensionless system can be derived from (1.39) in Sect. 1.3.1 by setting the relative velocity field



J.-J. Xu, *Interfacial Wave Theory of Pattern Formation in Solidification*, Springer Series in Synergetics, DOI 10.1007/978-3-319-52663-8_2

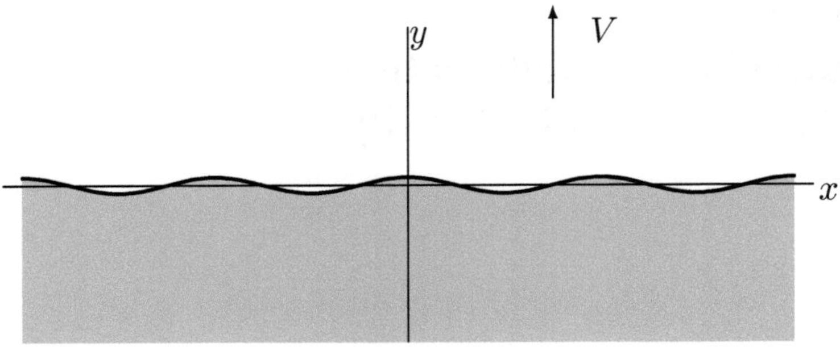


**Fig. 2.1** A simple model for unidirectional solidification

to $\mathbf{u} = -\hat{\mathbf{e}}_3 = -\mathbf{e}_y$ and $\lambda = \lambda_s = 1$, $\hat{\alpha}_T = 1$ as follows:

$$\nabla^2 \bar{T} = \frac{\partial \bar{T}}{\partial \bar{t}} - \frac{\partial \bar{T}}{\partial \bar{y}}, \tag{2.1}$$

where $\bar{T} = \bar{T}(x, y, t)$ is the temperature field in the liquid phase. For the solid phase, we replace $\bar{T}$ by $\bar{T}_S$.

The dimensionless boundary conditions are as follows:

1. The upstream far-field condition: as $\bar{y} \to \infty$,

$$\bar{T} \to T_\infty = \frac{(T_\infty)_D - T_{M0}}{\Delta H/(c_p \rho)}. \tag{2.2}$$

2. At the interface, $\bar{y} = \bar{h}(\bar{x}, \bar{t})$,

   (i) the thermodynamic equilibrium condition:

$$\bar{T} = \bar{T}_S, \tag{2.3}$$

   (ii) the Gibbs–Thomson condition:

$$\bar{T}_S = \Gamma \frac{\bar{h}_{\bar{x}\bar{x}}}{(1 + \bar{h}_{\bar{x}}^2)^{\frac{3}{2}}}, \tag{2.4}$$

   (iii) the enthalpy balance:

$$\frac{\partial}{\partial \bar{y}}(\bar{T} - \bar{T}_S) - \bar{h}_{\bar{x}} \frac{\partial}{\partial \bar{x}}(\bar{T} - \bar{T}_S) + \bar{h}_{\bar{t}} + 1 = 0. \tag{2.5}$$

For the sake of convenience, we shall hereinafter omit the bar over the dimensionless quantities.

### 2.1.1 Basic Steady-State Solution

The above system allows a one-dimensional steady solution with a flat interface. The interface equation is $y = h_B = 0$. The exact one-dimensional steady-state solution can be obtained by setting $\frac{\partial}{\partial x} = 0$ and $\frac{\partial}{\partial t=0}$ in the above system. One thus obtains

$$T_B = \begin{cases} e^{-y} - 1, & y \geq 0 \ \text{(for the liquid)}; \\ 0, & y \leq 0 \ \text{(for the solid)}, \end{cases} \tag{2.6}$$

where $T_B$ is the basic state solution for temperature. From (2.6), it is seen that as $y \to \infty$, $T_B(\infty) = T_\infty = -1$. Hence, this solution can only be applied to the special case $T_\infty = -1$. For $T_\infty \neq -1$, the system cannot have a steady solution with a flat interface. More precisely, for the case $0 > T_\infty > -1$, the interface of the steady-state solution will be curved, e.g., a parabola, while for the case $T_\infty < -1$, the system will have no solution, since the interface can no longer be considered to be in local thermodynamic equilibrium.

### 2.1.2 Unsteady Perturbed Solutions and Mullins–Sekerka Instability

The unsteady solutions can be expressed in the form

$$\begin{aligned} T(x, y, t) &= T_B(y) + \tilde{T}(x, y, t), \\ T_S(x, y, t) &= T_{SB}(y) + \tilde{T}_S(x, y, t), \\ h(x, t) &= h_B + \tilde{h}(x, t), \end{aligned} \tag{2.7}$$

where $\tilde{T}$, $\tilde{T}_S$, and $\tilde{h}$ are small perturbations around the steady state.

The governing equation for the perturbation is

$$\nabla^2 \tilde{T} = \frac{\partial \tilde{T}}{\partial t} - \frac{\partial \tilde{T}}{\partial y}. \tag{2.8}$$

The boundary conditions are as follows:

1. The upstream far-field condition: as $y \to \infty$,

$$\tilde{T} \to 0. \tag{2.9}$$

2. The interface conditions: Assume that the deformation $\tilde{h}(x, t)$ of the interface, measured using the characteristic amplitude $\delta \ll 1$, is very small. One can linearize the original boundary conditions on the interface $y = \tilde{h}(x, t)$ by

expanding them in a Taylor series in $\delta$ around $y = h_B = 0$. As the leading-order approximation, we obtain the following linear interface conditions at $y = 0$:

$$\tilde{T} = \tilde{T}_S - (\Delta G_1)\tilde{h}\,, \tag{2.10}$$

$$\tilde{T}_S = \Gamma \tilde{h}_{xx} - G_{1S}\tilde{h}\,, \tag{2.11}$$

$$\frac{\partial\left(\tilde{T} - \tilde{T}_S\right)}{\partial y} + (\Delta G_2)\,\tilde{h} + \frac{\partial \tilde{h}}{\partial t} = 0\,, \tag{2.12}$$

where

$$G_{1L} = \frac{\partial T_B}{\partial y}(0) = -1\,;\;\; G_{1S} = \frac{\partial T_{BS}}{\partial y}(0) = 0\,, \tag{2.13}$$

$$\Delta G_1 = (G_{1L} - G_{1S}) = -1\,, \tag{2.14}$$

$$\Delta G_2 = \left(\frac{\partial^2 T_B}{\partial y^2}\right)_{y=0} - \left(\frac{\partial^2 T_{SB}}{\partial y^2}\right)_{y=0} = 1\,. \tag{2.15}$$

This linear system contains one parameter $\Gamma$. In practice, $\Gamma$ is a very small parameter, and it appears in front of the second derivatives $\tilde{h}_{xx}$ in (2.11). Thus, the above system gives rise to a singular perturbation problem. One can let $\Gamma \to 0$ and look for an asymptotic expansion solution. In the limiting process $\Gamma \to 0$, all perturbed quantities $\{\tilde{T}, \tilde{T}_S, \tilde{h}\}$ should have the same order of magnitude. From the interface condition (2.11), it is seen that in order for $\tilde{T}_S$ to have the same order magnitude as $\tilde{h}$, one must have $\tilde{h}_{xx} = \mathcal{O}(\frac{\tilde{h}}{\Gamma})$. This is possible only when $\tilde{h}$ is also a function of the variable $\frac{x}{\varepsilon}$, where

$$\varepsilon = \sqrt{\Gamma}\,. \tag{2.16}$$

It will be seen that the parameter $\varepsilon$ plays a vital role in interfacial stability. We call this parameter the *interfacial stability parameter*. The above argument provides us with an important hint that the solution must have the structure of multiple length scales; so one may derive an asymptotic expansion form of the solution using the so-called *multiple variables expansion (MVE) method* (see [5]). The idea of the MVE method is that one may define a set of fast variables such that

$$\begin{aligned} x_+ &= \frac{k(\varepsilon)x}{\varepsilon} \\ y_+ &= \frac{g(\varepsilon)y}{\varepsilon} \\ t_+ &= \frac{\sigma(\varepsilon)t}{\varepsilon}\,, \end{aligned} \tag{2.17}$$

and consider the exact solution as a function of $(x_+, y_+, t_+, x, y, \varepsilon)$, where the fast variables $(x_+, y_+, t_+)$ and the slow variables $(x, y)$ are formally treated as the independent variables. Thus, as $\varepsilon \to 0$, the solution $\tilde{q} \equiv \{\tilde{T}, \tilde{T}_S, \tilde{h}\}$ is expanded in the following MVE form:

$$\begin{aligned}
&\tilde{q}(x, y, x_+, y_+, t_+, \varepsilon) \sim e^{t_+}\{\tilde{q}_0(x, y, x_+, y_+) + \varepsilon\tilde{q}_1(x, y, x_+, y_+) + \cdots\} \\
&k(\varepsilon) \sim k_0 + \varepsilon k_1 + \cdots \\
&g(\varepsilon) \sim k_0 + \varepsilon g_1 + \cdots \\
&g_s(\varepsilon) \sim k_0 + \varepsilon g_{s1} + \cdots \\
&\sigma(\varepsilon) \sim \sigma_0 + \varepsilon\sigma_1 + \cdots .
\end{aligned} \tag{2.18}$$

Note that in the above, we have used $g_s$ for the solutions in the solid state, which may be different from $g$ for the solutions in the liquid state. This implies that for the solutions in the solid state, we shall use the fast variable $y_+ = g_s y/\varepsilon$. Moreover, to obtain the asymptotic solution, one needs to make expansions in $\varepsilon$ for the parameter $\sigma$ and the wave numbers $k$ and $g$. One can verify that the leading terms of the expansions for $k(\varepsilon)$, $g(\varepsilon)$, and $g_s(\varepsilon)$ can be set to be the same, but their higher-order terms may be different. Now we need to transform the perturbed system (2.8)–(2.12) into a form with the above multiple variables. This can be done by replacing all the derivatives in (2.8)–(2.12) using the following transformation:

$$\begin{aligned}
\frac{\partial}{\partial x} &\Rightarrow \frac{\partial}{\partial x} + \frac{k}{\varepsilon}\frac{\partial}{\partial x_+}, \\
\frac{\partial}{\partial y} &\Rightarrow \frac{\partial}{\partial y} + \frac{g}{\varepsilon}\frac{\partial}{\partial y_+}, \\
\frac{\partial}{\partial t} &\Rightarrow \frac{\sigma}{\varepsilon}\frac{\partial}{\partial t_+}, \\
\frac{\partial^2}{\partial x^2} &\Rightarrow \frac{\partial^2}{\partial x^2} + \frac{2k}{\varepsilon}\frac{\partial^2}{\partial x \partial x_+} + \frac{k^2}{\varepsilon^2}\frac{\partial^2}{\partial x_+^2}, \\
\frac{\partial^2}{\partial y^2} &\Rightarrow \frac{\partial^2}{\partial y^2} + \frac{2g}{\varepsilon}\frac{\partial^2}{\partial y \partial y_+} + \frac{g^2}{\varepsilon^2}\frac{\partial^2}{\partial y_+^2}.
\end{aligned} \tag{2.19}$$

The multiple-variables form of the system is then obtained as follows:

$$\begin{aligned}
k^2\frac{\partial^2\tilde{T}}{\partial x_+^2} + g^2\frac{\partial^2\tilde{T}}{\partial y_+^2} &= \varepsilon\left(\sigma\tilde{T} - g\frac{\partial\tilde{T}}{\partial y_+} - 2k\frac{\partial^2\tilde{T}}{\partial x\partial x_+} - 2g\frac{\partial^2\tilde{T}}{\partial y\partial y_+}\right) \\
&\quad -\varepsilon^2\left(\frac{\partial^2}{\partial x^2} + \frac{\partial^2}{\partial y^2} + \frac{\partial}{\partial y}\right)\tilde{T};
\end{aligned} \tag{2.20}$$

$$k^2\frac{\partial^2 \tilde{T}_S}{\partial x_+^2} + g_s^2\frac{\partial^2 \tilde{T}_S}{\partial y_+^2} = \varepsilon\left(\sigma \tilde{T}_S - g_s\frac{\partial \tilde{T}_S}{\partial y_+} - 2k\frac{\partial^2 \tilde{T}_S}{\partial x \partial x_+} - 2g_s\frac{\partial^2 \tilde{T}_S}{\partial y \partial y_+}\right)$$

$$-\varepsilon^2\left(\frac{\partial^2}{\partial x^2} + \frac{\partial^2}{\partial y^2} + \frac{\partial}{\partial y}\right)\tilde{T}_S\,, \tag{2.21}$$

with the following boundary conditions:

1. As $y_+ \to \infty$,

$$\tilde{T} \to 0. \tag{2.22}$$

2. As $y_+ \to -\infty$,

$$\tilde{T}_S \to 0. \tag{2.23}$$

3. At the interface $y = y_+ = 0$,

(i)

$$\tilde{T} = \tilde{T}_S - (\Delta G_1)\tilde{h}\,, \tag{2.24}$$

(ii)

$$\tilde{T}_S = k^2\frac{\partial^2 \tilde{h}}{\partial x_+^2} + 2\varepsilon k\frac{\partial^2 \tilde{h}}{\partial x \partial x_+} + \varepsilon^2\frac{\partial^2 \tilde{h}}{\partial x^2} - G_{1S}\tilde{h}\,, \tag{2.25}$$

(iii)

$$g\frac{\partial \tilde{T}}{\partial y_+} - g_s\frac{\partial \tilde{T}_S}{\partial y_+} + \varepsilon\frac{\partial}{\partial y}(\tilde{T} - \tilde{T}_S) + \sigma\tilde{h} + \varepsilon(\Delta G_2)\tilde{h} = 0\,. \tag{2.26}$$

Here from (2.20) and (2.21), it is seen that as $\varepsilon \to 0$, in the leading-order approximation, both the convective and unsteady heat transfer terms in the bulk are negligible, whereas from (2.25) it is seen that the Gibbs–Thomson effect at the interface is important.

By substituting (2.19) into the above system (2.20)–(2.26), one can successively derive each order of approximation.

#### 2.1.2.1 Zeroth-Order Approximation Solutions

Assuming $\varepsilon \ll |k_0| = \mathcal{O}(1)$, in the leading-order approximation we obtain the governing equation

$$k_0^2\left(\frac{\partial^2}{\partial x_+^2} + \frac{\partial^2}{\partial y_+^2}\right)\tilde{T}_0 = 0\,, \tag{2.27}$$

and the following boundary conditions:

1. As $y_+ \to \infty$,

$$\tilde{T}_0 \to 0\,. \tag{2.28}$$

2. As $y_+ \to -\infty$,

$$\tilde{T}_{S0} \to 0\,. \tag{2.29}$$

3 At the interface $y = y_+ = 0$,

$$\tilde{T}_0 = \tilde{T}_{S0} - (\Delta G_1)\tilde{h}_0\,, \tag{2.30}$$

$$\tilde{T}_{S0} = k_0^2\frac{\partial^2\tilde{h}_0}{\partial x_+^2} - G_{1S}\tilde{h}_0\,, \tag{2.31}$$

$$k_0\frac{\partial}{\partial y_+}\left(\tilde{T}_0 - \tilde{T}_{S0}\right) + \sigma_0\tilde{h}_0 = 0\,. \tag{2.32}$$

The system (2.27)–(2.32) allows the following mode solutions:

$$\begin{aligned}\tilde{T}_0 &= A_0(x,y)\mathrm{e}^{\mathrm{i}x_+ - y_+}\,,\\ \tilde{T}_{S0} &= A_{S0}(x,y)\mathrm{e}^{\mathrm{i}x_+ + y_+}\,,\\ \tilde{h}_0 &= \hat{D}_0\mathrm{e}^{\mathrm{i}x_+}\,.\end{aligned} \tag{2.33}$$

One may define

$$\hat{A}_0 = A_0(x,0), \qquad \hat{A}_{S0} = A_{S0}(x,0). \tag{2.34}$$

By applying the boundary conditions (2.30)–(2.32), we derive

$$\begin{aligned}\hat{A}_0 &= \hat{A}_{S0} - \Delta G_1 D_0\,,\\ \hat{A}_{S0} &= -k_0^2\hat{D}_0 - G_{1S}\hat{D}_0\,,\\ -k_0(\hat{A}_0 &+ \hat{A}_{S0}) + \sigma_0\hat{D}_0 = 0\,.\end{aligned} \tag{2.35}$$

Note that since the coefficients of the system (2.27)–(2.32) are constants, one deduces that the amplitude functions $A_0, A_{S0}$ must be independent of the variables $x$ and $y$, so $\hat{A}_0, \hat{A}_{S0}, \hat{D}_0$ must be constants. Moreover, for a nontrivial solution, one must have

$$\Delta = \det \begin{pmatrix} 1 & -1 & \Delta G_l \\ 0 & 1 & k_0^2 + G_{S1} \\ -k_0 & -k_0 & \sigma_0 \end{pmatrix} = 0\,. \tag{2.36}$$

Thus the wave number $k_0$ and the eigenvalue $\sigma_0$ in the mode solutions cannot be arbitrarily chosen. These two quantities must be subject to the following dispersion relationship:

$$\sigma_0 = k_0\Big[-(G_{1L} + G_{1S}) - 2k_0^2\Big]\,. \tag{2.37}$$

For the system under consideration, we have

$$\sigma_0 = k_0\big(1 - 2k_0^2\big)\,. \tag{2.38}$$

This formula was first obtained by Mullins and Sekerka in 1963. It is now called the Mullins–Sekerka dispersion relationship. From this dispersion relationship, with a given eigenvalue $\sigma_0$, the corresponding wave numbers $k_0$ are determined (see Fig. 2.2). Note that if the wave number $k_0$ has an imaginary part, the perturbation will tend to infinity either as $x \to \infty$ or as $x \to -\infty$. Furthermore, if the wave number $k_0$ is a negative real number, the perturbed temperature $\tilde{T}_0$ will tend to infinity as $y \to \infty$. Evidently, these results are physically unacceptable.

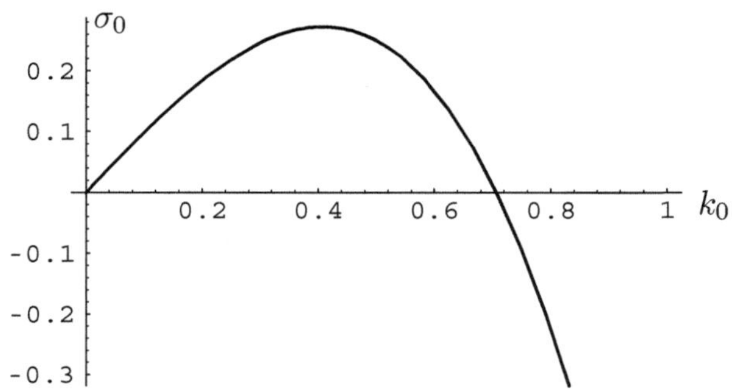


**Fig. 2.2** Mullins–Sekerka dispersion relationship for a pure melt system

Thus, the eigenvalue $\sigma_0$ of the system must be real. There are three cases:

(i) $0 < \sigma_0 < \frac{1}{3}\sqrt{\frac{2}{3}}$. From the local dispersion relationship (2.38), one solves for the three real roots: $k_0^{(1)} > 0$, $k_0^{(2)} < 0$, $k_0^{(3)} > 0$. Only the positive roots $k_0^{(1)}, k_0^{(3)}$ are physically meaningful. The temperature field corresponding to $k_0^{(2)}$ will increase exponentially as $y \to \infty$, violating the boundary condition (2.28).

The general solution of the perturbed states is then

$$\tilde{h} \approx \mathrm{Re}\left\{\left(D_0^{(1)} e^{\frac{i}{\varepsilon}k_0^{(1)}x} + D_0^{(3)} e^{\frac{i}{\varepsilon}k_0^{(3)}x}\right) e^{\frac{\sigma_0}{\varepsilon}t}\right\}. \tag{2.39}$$

These solutions are growing unstable modes. A special case is $\sigma_0 = 0$, for which we have $k_0^{(3)} = 0$ and $k_0^{(1)} > 0$. The corresponding solutions are neutrally stable modes.

(ii) If $\sigma_0 < 0$, the local dispersion relationship (2.38) allows only one real root $k_0^{(1)}$, and the system has only one corresponding decaying mode.

(iii) If $\sigma > \frac{1}{3}\sqrt{\frac{2}{3}}$, the local dispersion relationship (2.38) has no real root, so the system has no corresponding mode.

Returning to the original slow variables $(x, y, t)$, we see that the normal mode solution is written in the form

$$\begin{aligned} \tilde{T}_0 e^{\sigma_0 t +} &= A_0 e^{\tilde{k}_0(ix-y)+\tilde{\sigma}_0 t} \\ \tilde{T}_{S0} e^{\sigma_0 t +} &= A_{S0} e^{\tilde{k}_0(ix+y)+\tilde{\sigma}_0 t} \\ \tilde{h}_0 e^{\sigma_0 t +} &= \hat{D}_0 e^{i\tilde{k}_0 x+\tilde{\sigma}_0 t}, \end{aligned} \tag{2.40}$$

where

$$\tilde{k}_0 = \frac{k_0}{\varepsilon}; \quad \tilde{\sigma}_0 = \frac{\sigma_0}{\varepsilon}. \tag{2.41}$$

Thus, the dispersion relationship (2.37) is written in the form

$$\tilde{\sigma}_0 = \tilde{k}_0\Big[-(G_{1L} + G_{1S}) - 2\varepsilon^2\tilde{k}_0^2\Big]. \tag{2.42}$$

It is clear from this formula that when the mean temperature gradient satisfies $\bar{G}_1 = \frac{G_{1L}+G_{1S}}{2} < 0$, the heat conduction represented by the term $-2\bar{G}_1\tilde{k}_0$ is an unstable factor, while the surface tension, represented by the term $\varepsilon^2\tilde{k}_0^3$, is a stable factor. If the surface tension parameter is zero, the dispersion relationship reduces to

$$\tilde{\sigma}_0 = -2\bar{G}_1\tilde{k}_0. \tag{2.43}$$

Hence, for all $\tilde{k}_0 > 0$, one always has $\tilde{\sigma}_0 > 0$. This implies that the system will always be unstable. When the surface tension parameter is nonzero, the surface tension suppresses perturbations with a short wavelength. But perturbations with long wavelengths all grow with time. More precisely, for $\tilde{k}_0 > \sqrt{-\bar{G}_1}/\varepsilon$, $\tilde{\sigma}_0 < 0$, and the mode solutions are decaying and stable. For $0 < \tilde{k}_0 < \sqrt{-\bar{G}_1}/\varepsilon$, the mode solutions are growing and unstable. The critical number $\tilde{k}_c = \sqrt{-\bar{G}_1}/\varepsilon$ corresponds to the so-called neutral modes.

From the above results, one can conclude that in unidirectional solidification from a pure melt, due to the Mullins–Sekerka instability, every perturbed state is either purely growing or purely decaying. The system allows a unique neutrally stable state, which represents a steady state. It does not allow an oscillatory state. When $(G_{1L} + G_{1S}) > 0$, the system will be stable for all $k_0 > 0$. This implies that when the solidification interface advances to a high-temperature liquid region, the interface will always be smooth. It should be noted that so far, these conclusions are drawn from the leading-order approximation of the solution in the limit $\varepsilon \to 0$. The dispersion relationship (2.37) is not exact. For more accurate information, one needs to examine the higher-order approximation solutions.

#### 2.1.2.2 First-Order Approximation Solutions

In the first-order approximation, we obtain the solution

$$k_0^2 \left\{ \frac{\partial^2}{\partial x_+^2} + \frac{\partial^2}{\partial y_+^2} \right\} \tilde{T}_1 = \sigma_0 \tilde{T}_0 - k_0 \frac{\partial \tilde{T}_0}{\partial y_+} - 2k_0 \frac{\partial^2 \tilde{T}_0}{\partial x \partial x_+} - 2k_0 \frac{\partial^2 \tilde{T}_0}{\partial y \partial y_+}$$
$$-2k_0 k_1 \frac{\partial^2 \tilde{T}_0}{\partial x_+^2} - 2k_0 g_1 \frac{\partial^2 \tilde{T}_0}{\partial y_+^2}, \tag{2.44}$$

$$k_0^2 \left\{ \frac{\partial^2}{\partial x_+^2} + \frac{\partial^2}{\partial y_+^2} \right\} \tilde{T}_{S1} = \sigma_0 \tilde{T}_0 - k_0 \frac{\partial \tilde{T}_{S0}}{\partial y_+} - 2k_0 \frac{\partial^2 \tilde{T}_{S0}}{\partial x \partial x_+} - 2k_0 \frac{\partial^2 \tilde{T}_{S0}}{\partial y \partial y_+}$$
$$-2k_0 k_1 \frac{\partial^2 \tilde{T}_{S0}}{\partial x_+^2} - 2k_0 g_{s1} \frac{\partial^2 \tilde{T}_{S0}}{\partial y_+^2}. \tag{2.45}$$

Consequently,

$$\begin{cases} k_0^2 \left\{ \frac{\partial^2}{\partial x_+^2} + \frac{\partial^2}{\partial y_+^2} \right\} \tilde{T}_1 & = a_0 \mathrm{e}^{(\mathrm{i}x_+ - y_+)} \\ k_0^2 \left\{ \frac{\partial^2}{\partial x_+^2} + \frac{\partial^2}{\partial y_+^2} \right\} \tilde{T}_{S1} & = b_0 \mathrm{e}^{(\mathrm{i}x_+ + y_+)}, \end{cases} \tag{2.46}$$

where

$$\begin{cases} a_0 = 2k_0\Big[\frac{\partial A_0}{\partial y} + \big(\frac{1}{2} + \frac{\sigma_0}{2k_0} + k_1 - g_1\big)A_0\Big] \\ b_0 = -2k_0\Big[\frac{\partial A_{S0}}{\partial y} + \big(\frac{1}{2} - \frac{\sigma_0}{2k_0} - k_1 + g_{s1}\big)A_{S0}\Big] . \end{cases} \tag{2.47}$$

In order to obtain a uniformly valid asymptotic solution, one must eliminate the secular terms by setting $a_0 = b_0 = 0$. This leads to

$$\begin{cases} g_1 - k_1 = \frac{1}{2}\big(1 + \frac{\sigma_0}{k_0}\big) \\ g_{s1} = g_1 - 1 . \end{cases} \tag{2.48}$$

From $\nabla^2 \tilde{T}_1 = \nabla^2 \tilde{T}_{S1} = 0$, one may consider the normal mode solutions:

$$\begin{aligned} \tilde{T}_1 &= A_1(x, y)\mathrm{e}^{\mathrm{i}x_+ - y_+} , \\ \tilde{T}_{S1} &= A_{S1}(x, y)\mathrm{e}^{\mathrm{i}x_+ + y_+} , \\ \tilde{h}_1 &= \hat{D}_1\mathrm{e}^{\mathrm{i}x_+} . \end{aligned} \tag{2.49}$$

Likewise, one can deduce that the amplitude functions $A_1$ and $A_{S1}$ must be independent of the variables $x$ and $y$. Hence, one may write

$$A_1(x, y) = \hat{A}_1 ; \quad A_{S1}(x, y) = \hat{A}_{S1} . \tag{2.50}$$

For the first-order approximation, we derive the following boundary conditions at the interface, $y = y_+ = 0$:

$$\tilde{T}_1 = \tilde{T}_{S1} - (\Delta G_1)\tilde{h}_1 , \tag{2.51}$$

$$\tilde{T}_{S1} = k_0^2 \frac{\partial^2 \tilde{h}_1}{\partial x_+^2} + 2k_0 \frac{\partial^2 \tilde{h}_0}{\partial x \partial x_+} + 2k_0 k_1 \frac{\partial^2 \tilde{h}_0}{\partial x_+^2} - G_{1S}\tilde{h}_1 , \tag{2.52}$$

$$\begin{aligned} &k_0 \frac{\partial}{\partial y_+}\big(\tilde{T}_1 - \tilde{T}_{S1}\big) + g_1 \frac{\partial \tilde{T}_0}{\partial y_+} - g_{s1}\frac{\partial \tilde{T}_{S0}}{\partial y_+} + \frac{\partial}{\partial y}\big(\tilde{T}_0 - \tilde{T}_{S0}\big) \\ &+\sigma_0 \tilde{h}_1 + \sigma_1 \tilde{h}_0 + (\Delta G_2)\tilde{h}_0 = 0 . \end{aligned} \tag{2.53}$$

In order for the mode solution (2.50) to satisfy the boundary conditions (2.51)–(2.53), we must have

$$\hat{A}_1 = \hat{A}_{S1} - (\Delta G_1)\hat{D}_1 \,, \tag{2.54}$$

$$\hat{A}_{S1} = -(k_0^2 + G_{S1})\hat{D}_1 + I_2\hat{D}_0 \,, \tag{2.55}$$

$$\begin{aligned} &-k_0\big(\hat{A}_1 + \hat{A}_{S1}\big) - g_1\big(\hat{A}_0 + \hat{A}_{S0}\big) + \hat{A}_{S0} \\ &\quad +(\Delta G_2)\hat{D}_0 + \sigma_0\hat{D}_1 + \sigma_1\hat{D}_0 = 0 \,, \end{aligned} \tag{2.56}$$

or

$$-k_0\big(\hat{A}_1 + \hat{A}_{S1}\big) + \sigma_0\hat{D}_1 = I_3\hat{D}_0 \,. \tag{2.57}$$

In the above, we have used the notation

$$\begin{cases} I_2 = -2k_0k_1 \\ I_3 = k_0^2 + G_{1S} + \frac{\sigma_0 g_1}{k_0} - \sigma_1 - \Delta G_2 \,. \end{cases} \tag{2.58}$$

The determinant of the above inhomogeneous system is $\Delta = 0$. Hence, for a nontrivial solution $\{\hat{A}_1, \hat{A}_{S1}, \hat{D}_1\}$, the following solvability condition must hold:

$$\det\begin{pmatrix} 1 & -1 & 0 \\ 0 & 1 & I_2 \\ -k_0 & -k_0 & I_3 \end{pmatrix} = 0 \,, \tag{2.59}$$

or

$$I_3 + 2k_0 I_2 = 0 \,. \tag{2.60}$$

From (2.60), one obtains

$$\sigma_1 = g_1(1 - 2k_0^2) + k_0^2 - 1 - 4k_0^2k_1 \,. \tag{2.61}$$

We set $k_1 = 0$. Thus, with $\Delta G_1 = -1$ and $\Delta G_2 = 1$, we get

$$\sigma_1 = g_1(1 - 2k_0^2) + k_0^2 - 1 \,. \tag{2.62}$$

It follows from (2.48) that

$$g_1 = \frac{1}{2}\left(1 + \frac{\sigma_0}{k_0}\right) \,. \tag{2.63}$$

From the above, we obtain the modified dispersion relationship

$$\begin{aligned}\sigma &= \sigma_0 + \varepsilon\sigma_1 + \mathcal{O}(\varepsilon^2)\\ &= k_0(1-2k_0^2) - \varepsilon + \varepsilon k_0^2 + \varepsilon g_1(1-2k_0^2) + \mathcal{O}(\varepsilon^2)\\ &= -\varepsilon + \varepsilon k^2 + \Big(k + \frac{\varepsilon}{2} + \frac{\varepsilon\sigma_0}{2k}\Big)(1-2k^2) + \mathcal{O}(\varepsilon^2)\,. \end{aligned} \tag{2.64}$$

Note that we have replaced $k_0$ by $k$ in the final expression. The above procedure can be continued to even higher-order approximations if necessary.

However, as indicated before, the above MVE solution (2.19) is not applicable in the long-wavelength range $k_0 = \mathcal{O}(\varepsilon)$. In fact, for $k_0 \ll 1$, the effect of higher-order terms on the right-hand side of (2.20), as well as in the boundary condition (2.26), becomes important. These terms must be taken into account in the leading-order approximation. Moreover, the MVE solution (2.19) is not valid in the extremely short-wavelength range $k_0 = \mathcal{O}(\frac{1}{\varepsilon})$ either. For a given small $\varepsilon > 0$, for $k_0 \gg 1$, we have $|\sigma_0| \ll |\varepsilon\sigma_1|$ and $|g_0| \ll |\varepsilon g_1|$.

Therefore, in order to study the behavior of solutions in the ranges $k \ll 1$ and $k \gg 1$, one needs to take different asymptotic expansions. In what follows, we shall look into these cases separately.

### 2.1.3 Asymptotic Solutions in the Long-Wave Regime, $k = \mathcal{O}(\varepsilon)$

In the long-wavelength regime, $\{k = \mathcal{O}(\varepsilon); g = \mathcal{O}(\varepsilon); \sigma = \mathcal{O}(\varepsilon)\}$, the solution will not have the structure of multiple scales, since both the fast variables $x_+, y_+, t_+$ and the slow variables $x, y, t$ are of the same order. We set

$$\hat{k} = \frac{k}{\varepsilon}, \quad \hat{g} = \frac{g}{\varepsilon}, \quad \hat{\sigma} = \frac{\sigma}{\varepsilon}, \tag{2.65}$$

and introduce the new slow variables

$$\begin{cases} \hat{x} = \hat{k}x, \\ \hat{y} = \hat{g}y, \\ \hat{t} = \hat{\sigma}t\,. \end{cases} \tag{2.66}$$

We expand the solution $\tilde{q} \equiv \{\tilde{T}, \tilde{T}_S, \tilde{h}\}$ in the following asymptotic form:

$$\begin{aligned}
\tilde{q}(\hat{x}, \hat{y}, \hat{t}, \varepsilon) &\sim e^{\hat{t}}\left\{\tilde{q}_0(\hat{x}, \hat{y}) + \varepsilon^2 \tilde{q}_1(\hat{x}, \hat{y}) + \cdots\right\} \\
\hat{k}(\varepsilon) &\sim \hat{k}_0 + \varepsilon^2 \hat{k}_1 + \cdots \\
\hat{g}(\varepsilon) &\sim \hat{g}_0 + \varepsilon^2 \hat{g}_1 + \cdots \\
\hat{g}_s(\varepsilon) &\sim \hat{g}_{s0} + \varepsilon^2 \hat{g}_{s1} + \cdots \\
\hat{\sigma}(\varepsilon) &\sim \hat{\sigma}_0 + \varepsilon^2 \hat{\sigma}_1 + \cdots .
\end{aligned} \tag{2.67}$$

Note that for solutions of the temperature distribution in the solid phase, a different slow variable

$$\hat{y} = \hat{g}_s y \tag{2.68}$$

is used. In terms of the above variables, the linear system (2.8)–(2.12) is transformed to

$$\left\{\hat{k}^2 \frac{\partial^2}{\partial \hat{x}^2} + \hat{g}^2 \frac{\partial^2}{\partial \hat{y}^2}\right\} \tilde{T} = \hat{\sigma} \tilde{T} - \hat{g} \frac{\partial \tilde{T}}{\partial \hat{y}}, \tag{2.69}$$

with the following boundary conditions:

1. As $\hat{y} \to \infty$,

$$\tilde{T} \to 0. \tag{2.70}$$

2. As $\hat{y} \to -\infty$,

$$\tilde{T}_S \to 0. \tag{2.71}$$

3. At the interface, $\hat{y} = 0$,

(i)

$$\tilde{T} = \tilde{T}_S - (\Delta G_1)\tilde{h}, \tag{2.72}$$

(ii)

$$\tilde{T}_S = \varepsilon^2 \hat{k}^2 \frac{\partial^2 \tilde{h}}{\partial \hat{x}} - G_{1S} \tilde{h}, \tag{2.73}$$

(iii)

$$\hat{g} \frac{\partial \tilde{T}}{\partial \hat{y}} - \hat{g}_s \frac{\partial \tilde{T}_S}{\partial \hat{y}} + \hat{\sigma} \tilde{h} + \Delta G_2 \tilde{h} = 0. \tag{2.74}$$

From the above, it is seen that as $\varepsilon \to 0$, in the leading-order approximation for the solution in the long-wavelength regime, both heat convective and unsteady transfer terms in the bulk are important, but the Gibbs–Thomson effect on the interface is negligible.

By substituting (2.67) into the above system (2.69)–(2.74), one can successively derive each order of approximation.

#### 2.1.3.1 $\mathcal{O}(\varepsilon^0)$

In the leading-order approximation, we derive the governing equation

$$\hat{k}_0^2 \frac{\partial^2 \tilde{T}_0}{\partial \hat{x}^2} + \hat{g}_0^2 \frac{\partial^2 \tilde{T}_0}{\partial \hat{y}^2} = \hat{\sigma}_0 \tilde{T}_0 - \hat{g}_0 \frac{\partial \tilde{T}_0}{\partial \hat{y}} . \tag{2.75}$$

The above equation yields the solutions

$$\begin{aligned} \tilde{T}_0 &= \hat{A}_0 \mathrm{e}^{\mathrm{i}\hat{x} - \hat{y}}, \\ \tilde{T}_{S0} &= \hat{A}_{S0} \mathrm{e}^{\mathrm{i}\hat{x} + \hat{y}}, \\ \tilde{h}_0 &= \hat{D}_0 \mathrm{e}^{\mathrm{i}\hat{x}}, \end{aligned} \tag{2.76}$$

and we derive

$$\hat{g}_0^2 - \hat{g}_0 - (\hat{k}_0^2 + \hat{\sigma}_0) = 0 . \tag{2.77}$$

For the liquid phase,

$$\hat{g}_0 = \frac{1}{2} + \sqrt{\hat{k}_0^2 + \hat{\sigma}_0 + \frac{1}{4}} , \tag{2.78}$$

while for the solid phase,

$$\hat{g}_{s0} = \hat{g}_0 - 1 . \tag{2.79}$$

With $\Delta G_1 = -1$, $\Delta G_2 = 1$, and $G_{1S} = 0$, we derive from the boundary conditions (2.70)–(2.74),

$$\begin{cases} \hat{A}_0 = \hat{A}_{S0} + \hat{D}_0, \\ \hat{A}_{S0} = 0, \\ -\hat{g}_0 \hat{A}_0 - \hat{g}_{s0} \hat{A}_{S0} + (\hat{\sigma}_0 + 1)\hat{D}_0 = 0 . \end{cases} \tag{2.80}$$

Thus, it is found that

$$\hat{\sigma}_0 = -1 + \hat{g}_0 = -\frac{1}{2} + \sqrt{\hat{k}_0^2 + \hat{\sigma}_0 + \frac{1}{4}}, \tag{2.81}$$

or

$$\hat{\sigma}_0 = \hat{k}_0 \,. \tag{2.82}$$

#### 2.1.3.2 $\mathcal{O}(\varepsilon^2)$

In the first-order approximation, we derive the governing equations

$$\hat{k}_0^2 \frac{\partial^2 \tilde{T}_1}{\partial \hat{x}^2} + \hat{g}_0^2 \frac{\partial^2 \tilde{T}_1}{\partial \hat{y}^2} - \hat{\sigma}_0 \tilde{T}_1 + \hat{g}_0 \frac{\partial \tilde{T}_1}{\partial \hat{y}}$$
$$= \big(2\hat{k}_0\hat{k}_1 - 2\hat{g}_0\hat{g}_1 + \hat{\sigma}_1 + \hat{g}_1\big)\tilde{T}_0 \tag{2.83}$$

and

$$\hat{k}_0^2 \frac{\partial^2 \tilde{T}_{S1}}{\partial \hat{x}^2} + \hat{g}_{s0}^2 \frac{\partial^2 \tilde{T}_{S1}}{\partial \hat{y}^2} - \hat{\sigma}_0 \tilde{T}_{S1} + \hat{g}_{s0} \frac{\partial \tilde{T}_{S1}}{\partial \hat{y}}$$
$$= \big(2\hat{k}_0\hat{k}_1 - 2\hat{g}_{s0}\hat{g}_{s1} + \hat{\sigma}_1 - \hat{g}_{s1}\big)\tilde{T}_{S0} \,. \tag{2.84}$$

To eliminate the secular terms, we must set

$$\begin{cases} 2\hat{k}_0\hat{k}_1 - 2\hat{g}_0\hat{g}_1 + \hat{\sigma}_1 + \hat{g}_1 = 0, \\ 2\hat{k}_0\hat{k}_1 - 2\hat{g}_{s0}\hat{g}_{s1} + \hat{\sigma}_1 - \hat{g}_{s1} = 0 \,. \end{cases} \tag{2.85}$$

We put $k_1 = 0$. It follows that

$$\hat{g}_1 = \hat{g}_{s1} = \frac{\hat{\sigma}_1}{2\hat{g}_0 - 1} \,. \tag{2.86}$$

The first-order approximation yields the solutions

$$\begin{aligned} \tilde{T}_1 &= \hat{A}_1 e^{i\hat{x}-\hat{y}}, \\ \tilde{T}_{S1} &= \hat{A}_{S1} e^{i\hat{x}+\hat{y}}, \\ \tilde{h}_1 &= \hat{D}_1 e^{i\hat{x}} \,. \end{aligned} \tag{2.87}$$

From the boundary conditions (2.70)–(2.74), we derive

$$\begin{cases} \hat{A}_1 = \hat{A}_{S1} + \hat{D}_1 \\ \hat{A}_{S1} = -\hat{k}_0^2 \hat{D}_0, \\ -\hat{g}_0 \hat{A}_1 - \hat{g}_{s0} \hat{A}_{S1} + (\hat{\sigma}_0 + 1)\hat{D}_1 = \hat{A}_0 \hat{g}_1 + \hat{A}_{S0} \hat{g}_{s1} - \hat{\sigma}_1 \hat{D}_0 \, . \end{cases} \tag{2.88}$$

From this system, we obtain

$$\hat{\sigma}_1 = \hat{g}_1 + (1 - 2\hat{g}_0)\hat{k}_0^2, \tag{2.89}$$

or

$$\hat{\sigma}_1 = -\frac{\hat{k}_0}{2}(1 + 2\hat{k}_0)^2. \tag{2.90}$$

Combining (2.86) with (2.90), we derive

$$\hat{g}_1 = -\hat{k}_0^2 \left[ 1 + \frac{1}{2(\hat{g}_0 - 1)} \right], \tag{2.91}$$

or

$$\hat{g}_1 = -\frac{\hat{k}_0}{2}(1 + 2\hat{k}_0). \tag{2.92}$$

Thus, we have

$$\begin{aligned} \hat{\sigma} &= \hat{\sigma}_0 + \varepsilon^2 \hat{\sigma}_1 \\ &= (-1 + \hat{g}_0) + \varepsilon^2 \hat{g}_1 + \varepsilon^2 (1 - 2\hat{g}_0)\hat{k}_0^2 \\ &= -1 + \varepsilon^2 \hat{k}^2 + (\hat{g}_0 + \varepsilon^2 \hat{g}_1)(1 - 2\varepsilon^2 \hat{k}^2) + 2\varepsilon^4 \hat{g}_1 \hat{k}^2. \end{aligned} \tag{2.93}$$

Therefore, expressing the equations in terms of the original parameters $(k, \sigma)$, in the long-wavelength regime $(k \to 0)$, we have the dispersion relationship

$$\sigma = -\varepsilon + \varepsilon k^2 + (g_0 + \varepsilon^2 g_1)(1 - 2k^2) + 2\varepsilon^2 g_1 k^2, \tag{2.94}$$

where

$$\begin{cases} \sigma_0 = -\frac{\varepsilon}{2} + \sqrt{k^2 + \varepsilon\sigma_0 + \frac{\varepsilon^2}{4}}, \\ g_0 = \frac{\varepsilon}{2} + \sqrt{k^2 + \varepsilon\sigma_0 + \frac{\varepsilon^2}{4}}, \\ \varepsilon^2 g_1 = -\varepsilon k^2 \left\{ 1 + \frac{\varepsilon}{2(g_0 - \varepsilon)} \right\}. \end{cases} \tag{2.95}$$

In this case, one can write

$$g_0 + \varepsilon^2 g_1 = \frac{\varepsilon}{2} + \sqrt{k^2 + \varepsilon\sigma + \frac{\varepsilon^2}{4}} + \mathcal{O}(\varepsilon^3). \tag{2.96}$$

Comparing (2.64) with (2.94), it is seen that if one applies the approximate formula $\left(k + \frac{\varepsilon\sigma_0}{2k}\right) \approx \sqrt{k^2 + \varepsilon\sigma_0}$ to the dispersion relationship (2.64), then the resultant formula

$$\sigma = -\varepsilon + \varepsilon k^2 + \left(\frac{\varepsilon}{2} + \sqrt{k^2 + \varepsilon\sigma_0}\right)(1 - 2k^2) + \mathcal{O}(\varepsilon^2) \tag{2.97}$$

will be applicable to the whole range $0 \leq k \ll \frac{1}{\varepsilon}$.

### 2.1.4 Asymptotic Solutions in the Extremely Short-Wave Regime, $k = \mathcal{O}(\frac{1}{\varepsilon})$

In the extremely short-wave regime, $\{k = \mathcal{O}(\frac{1}{\varepsilon}); g = \mathcal{O}(\frac{1}{\varepsilon})\}$, one can still find an MVE solution with the fast variables $x_+, y_+, t_+$ and the slow variables $x, y, t$. We write

$$\begin{cases} \tilde{k} = \varepsilon k = \mathcal{O}(1), \\ \tilde{g} = \varepsilon g = \mathcal{O}(1), \\ \tilde{\sigma} = \varepsilon^{\alpha}\sigma = \mathcal{O}(1), \end{cases} \tag{2.98}$$

where the exponent $\alpha$ is to be determined. The MVE system (2.20)–(2.26) is still applicable to the present case if we change $k$, $g$, and $\sigma$ to $\tilde{k}, \tilde{g}$, and $\tilde{\sigma}$. Thus, we have

$$\left(\tilde{k}^2 \frac{\partial^2}{\partial x_+^2} + \tilde{g}^2 \frac{\partial^2}{\partial y_+^2}\right)\tilde{T} = \varepsilon^{3-\alpha}\tilde{\sigma}\tilde{T} - \varepsilon^2\tilde{g}\frac{\partial \tilde{T}}{\partial y_+} - 2\varepsilon^2\tilde{k}\frac{\partial^2 \tilde{T}}{\partial x \partial x_+}$$
$$-2\varepsilon^2\tilde{g}\frac{\partial^2 \tilde{T}}{\partial y \partial y_+} - \varepsilon^4\left(\frac{\partial^2}{\partial x^2} + \frac{\partial^2}{\partial y^2} - \frac{\partial}{\partial y}\right)\tilde{T}, \tag{2.99}$$

with the following boundary conditions:

1. As $y_+ \to \infty$,

$$\tilde{T} \to 0. \tag{2.100}$$

2. As $y_+ \to -\infty$,

$$\tilde{T}_S \to 0. \tag{2.101}$$

3. At the interface, $y = y_+ = 0$,

(i)

$$\tilde{T} = \tilde{T}_S - (\Delta G_1)\tilde{h}, \tag{2.102}$$

(ii)

$$\tilde{T}_S = \frac{\tilde{k}^2}{\varepsilon^2}\frac{\partial^2 \tilde{h}}{\partial x_+^2} + 2\tilde{k}\frac{\partial^2 \tilde{h}}{\partial x \partial x_+} + \varepsilon^2 \frac{\partial^2 \tilde{h}}{\partial x^2} - G_{1S}\tilde{h}, \tag{2.103}$$

(iii)

$$\tilde{g}\frac{\partial}{\partial y_+}\left(\tilde{T} - \tilde{T}_S\right) + \varepsilon^2 \frac{\partial}{\partial y}\left(\tilde{T} - \tilde{T}_S\right) + \varepsilon^{1-\alpha}\tilde{\sigma}\tilde{h}$$
$$+\varepsilon^2(\Delta G_2)\tilde{h} = 0\,. \tag{2.104}$$

One can see from (2.103) that $\tilde{h} = \mathcal{O}(\varepsilon^2 \tilde{T})$. Furthermore, from (2.104), one can prove that the exponent $\alpha$ must be chosen as $\alpha = 3$. The system allows the following asymptotic expansions:

$$\begin{aligned}
\tilde{T} &\sim e^{t_+}\left\{\tilde{T}_0(x, y, x_+, y_+) + \varepsilon^2 \tilde{T}_1(x, y, x_+, y_+) + \cdots\right\},\\
\tilde{T}_S &\sim e^{t_+}\left\{\tilde{T}_{S0}(x, y, x_+, y_+) + \varepsilon^2 \tilde{T}_{S1}(x, y, x_+, y_+) + \cdots\right\},\\
\tilde{h} &\sim \varepsilon^2 e^{t_+}\left\{\tilde{h}_0(x, y, x_+, y_+) + \varepsilon^2 \tilde{h}_1(x, y, x_+, y_+) + \cdots\right\},\\
\tilde{k}(\varepsilon) &\sim \tilde{k}_0 + \varepsilon^2 \tilde{k}_1 + \cdots,\\
\tilde{g}(\varepsilon) &\sim \tilde{g}_0 + \varepsilon^2 \tilde{g}_1 + \cdots,\\
\tilde{g}_s(\varepsilon) &\sim \tilde{g}_{s0} + \varepsilon^2 \tilde{g}_{s1} + \cdots,\\
\tilde{\sigma}(\varepsilon) &\sim \tilde{\sigma}_0 + \varepsilon^2 \tilde{\sigma}_1 + \cdots.
\end{aligned} \tag{2.105}$$

By substituting (2.105) into the system (2.99)–(2.104), one can successively derive each order of approximation.

#### 2.1.4.1 $\mathcal{O}(\varepsilon^0)$

In the zeroth-order approximation, we have

$$\tilde{k}_0^2 \frac{\partial^2 \tilde{T}_0}{\partial x_+^2} + \tilde{g}_0^2 \frac{\partial^2 \tilde{T}_0}{\partial y_+^2} = \tilde{\sigma}_0 \tilde{T}_0\,, \tag{2.106}$$

which allows the mode solutions

$$\begin{aligned}\tilde{T}_0 &= \hat{A}_0 e^{ix_+ - y_+},\\ \tilde{T}_{S0} &= \hat{A}_{S0} e^{ix_+ + y_+},\\ \tilde{h}_0 &= \hat{D}_0 e^{ix_+},\end{aligned} \tag{2.107}$$

with

$$\tilde{g}_0 = \sqrt{\tilde{\sigma}_0 + \tilde{k}_0^2}. \tag{2.108}$$

From the boundary conditions (2.102)–(2.104), we derive

$$\begin{cases} \hat{A}_0 = \hat{A}_{S0}, \\ \hat{A}_{S0} = -\tilde{k}_0^2 \hat{D}_0, \\ -\tilde{g}_0(\hat{A}_0 + \hat{A}_{S0}) + \tilde{\sigma}_0 \hat{D}_0 = 0. \end{cases} \tag{2.109}$$

It is seen that in the leading-order approximation for the solution in the extremely short-wavelength regime, the unsteady heat transfer term in the bulk is important, but the effect of interface displacement is negligible. We derive the following dispersion relationship:

$$\tilde{\sigma}_0 = -2\tilde{g}_0 \tilde{k}_0^2. \tag{2.110}$$

This may be rewritten as

$$\sigma_0 = -2k_0^2 \sqrt{k_0^2 + \varepsilon \sigma_0}. \tag{2.111}$$

### 2.1.4.2 $\mathcal{O}(\varepsilon^2)$

In the first-order approximation, we have

$$\begin{aligned}\tilde{k}_0^2 \frac{\partial^2 \tilde{T}_1}{\partial x_+^2} + \tilde{g}_0^2 \frac{\partial^2 \tilde{T}_1}{\partial y_+^2} - \tilde{\sigma}_0 \tilde{T}_1 &= \tilde{a}_0 e^{ix_+ - y_+},\\ \tilde{k}_0^2 \frac{\partial^2 \tilde{T}_{S1}}{\partial x_+^2} + \tilde{g}_0^2 \frac{\partial^2 \tilde{T}_{S1}}{\partial y_+^2} - \tilde{\sigma}_0 \tilde{T}_{S1} &= \tilde{b}_0 e^{ix_+ + y_+},\end{aligned} \tag{2.112}$$

where

$$\begin{cases} \tilde{a}_0 = (\tilde{\sigma}_1 + \tilde{g}_0 + 2\tilde{k}_0\tilde{k}_1 - 2\tilde{g}_0\tilde{g}_1)\hat{A}_0, \\ \tilde{b}_0 = (\tilde{\sigma}_1 - \tilde{g}_0 + 2\tilde{k}_0\tilde{k}_1 - 2\tilde{g}_0\tilde{g}_{s1})A_{S0}\,. \end{cases} \tag{2.113}$$

To eliminate the secular terms, we set $\tilde{a}_0 = \tilde{b}_0 = 0$. Thus, it follows that

$$\begin{cases} \tilde{g}_1 = \frac{1}{2} + \frac{\tilde{\sigma}_1}{2\tilde{g}_0}, \\ \tilde{g}_{s1} = \tilde{g}_1 - 1, \end{cases} \tag{2.114}$$

and

$$\tilde{\sigma}_1 = -\tilde{g}_0 - 2\tilde{k}_0\tilde{k}_1 + 2\tilde{g}_0\tilde{g}_1. \tag{2.115}$$

As before, we can let $\tilde{k}_1 = 0$. Hence, we have

$$\tilde{\sigma}_1 = \tilde{g}_0(2\tilde{g}_1 - 1). \tag{2.116}$$

The first-order system allows the mode solutions

$$\tilde{T}_1 = \hat{A}_1 e^{ix_+ - y_+}, \tag{2.117}$$

$$\tilde{T}_{S1} = \hat{A}_{S1} e^{ix_+ + y_+}, \tag{2.118}$$

$$\tilde{h}_1 = \hat{D}_1 e^{ix_+}\,. \tag{2.119}$$

In the above, the amplitude functions of the mode solutions have been set as constants. This is allowable, due to the fact that all the coefficients of the perturbed system are constants. From the boundary conditions (2.102)–(2.104), we derive

$$\begin{cases} \hat{A}_1 = \hat{A}_{S1} + \hat{D}_0, \\ \hat{A}_{S1} = -\tilde{k}_0^2\hat{D}_1 + I_2\hat{D}_0 \\ -\tilde{g}_0(\hat{A}_1 + \hat{A}_{S1}) + \tilde{\sigma}_0\hat{D}_1 = I_3\hat{D}_0\,, \end{cases} \tag{2.120}$$

where

$$\begin{cases} I_2 = -2\tilde{k}_0\tilde{k}_1, \\ I_3 = -2\tilde{k}_0^2\tilde{g}_1 + \tilde{k}_0^2 - \tilde{\sigma}_1. \end{cases} \tag{2.121}$$

For a nontrivial solution, the condition

$$I_3 + 2\tilde{g}_0 I_2 + \tilde{g}_0 = 0 \tag{2.122}$$

is necessary. Thus, it follows that

$$\tilde{k}_0^2 - \tilde{\sigma}_1 - 2\tilde{k}_0^2\tilde{g}_1 - 4\tilde{k}_0\tilde{g}_0\tilde{k}_1 + \tilde{g}_0 = 0\,. \tag{2.123}$$

For $\tilde{k}_1 = 0$, from (2.115)–(2.123) we derive

$$\tilde{g}_1 = 1 - \frac{\tilde{k}_0^2}{2(\tilde{g}_0 + \tilde{k}_0^2)} \tag{2.124}$$

and

$$\begin{aligned}\tilde{\sigma} &= \tilde{\sigma}_0 + \varepsilon^2\tilde{\sigma}_1 = -2\tilde{g}_0\tilde{k}_0^2 + \varepsilon^2(\tilde{k}_0^2 + \tilde{g}_0 - 2\tilde{k}_0^2\tilde{g}_1)\\ &= \varepsilon^2\tilde{k}_0^2 + (\tilde{g}_0 + \varepsilon^2\tilde{g}_1)(\varepsilon^2 - 2\tilde{k}_0^2) - \varepsilon^4\tilde{g}_1\,.\end{aligned} \tag{2.125}$$

Returning to the original parameters $\{k, \sigma\}$, we derive the dispersion relationship in the extremely short-wavelength regime as

$$\sigma = \varepsilon k^2 + (g_0 + \varepsilon^2 g_1)(1 - 2k^2) - \varepsilon^2 g_1, \tag{2.126}$$

where

$$\begin{cases} \sigma_0 = -2\left(\sqrt{k^2 + \varepsilon\sigma_0}\right)k^2, \\ g_0 = \sqrt{k^2 + \varepsilon\sigma_0}, \\ \varepsilon^2 g_1 = \varepsilon - \frac{1}{2}\frac{\varepsilon^2 k^2}{g_0 + \varepsilon k^2}. \end{cases} \tag{2.127}$$

In this case, one can show that

$$g_0 + \varepsilon^2 g_1 = \frac{\varepsilon}{2} + \sqrt{k^2 + \varepsilon\sigma} + \mathcal{O}(\varepsilon^3)\,. \tag{2.128}$$

It should be noted that the system (2.8)–(2.12) is a linear system with constant coefficients, so that it can be solved exactly. The exact solutions are

$$\tilde{T} = \hat{A}e^{(\frac{\sigma t}{\varepsilon} + \frac{ikx}{\varepsilon} - \frac{qy}{\varepsilon})}, \tag{2.129}$$

$$\tilde{T}_S = \hat{A}_S e^{(\frac{\sigma t}{\varepsilon} + \frac{ikx}{\varepsilon} + \frac{gy}{\varepsilon} - y)} \tag{2.130}$$

$$\tilde{h} = \hat{D}e^{(\frac{\sigma t}{\varepsilon} + \frac{ikx}{\varepsilon})}, \tag{2.131}$$

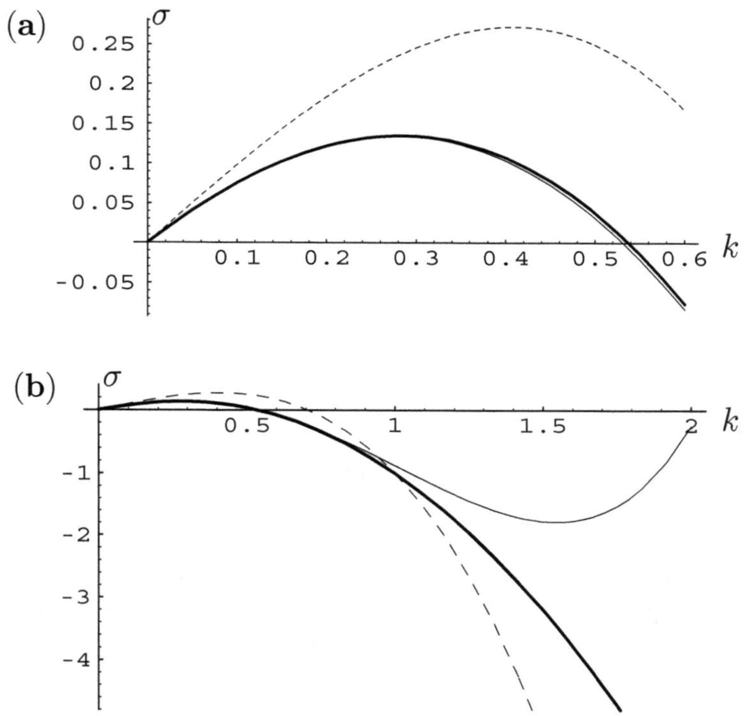


**Fig. 2.3** (**a**), (**b**) The dispersion curves for $\varepsilon = 0.5$. The *dashed line* is the zeroth-order approximation, the *thin solid line* is the modified first-order approximation, whereas the *bold solid line* is the exact solution: (**a**) for the range $0 \le k \le 0.6$; (**b**) for the range $0 \le k \le 2$

where

$$g = \frac{\varepsilon}{2} + \sqrt{k^2 + \varepsilon\sigma + \varepsilon^2/4}. \tag{2.132}$$

The exact dispersion relationship is

$$\sigma = -\varepsilon + \varepsilon k^2 + \left(\frac{\varepsilon}{2} + \sqrt{k^2 + \varepsilon\sigma + \varepsilon^2/4}\right)\left(1 - 2k^2\right). \tag{2.133}$$

By comparing (2.64) with (2.133), one sees that the dispersion formula of the MVE solution obtained above, in the regime ($\varepsilon \ll k \ll \frac{1}{\varepsilon}$), is the regular perturbation expansion of the exact solution as $\varepsilon \to 0$. In Fig. 2.3, we have shown all these dispersion curves in the $(\sigma, k)$-plane. It is seen that up to $\varepsilon = 0.5$, the agreement between the exact solution and the first-order MVE solution (2.97) is very good in the regime $0 \le k < 1$. However, for $k > 1.5$, the first-order MVE solution (2.64), as expected, is no longer close to the exact solution, even qualitatively. Nevertheless, the asymptotic solution in the extremely short-wavelength regime agrees with the exact solution (2.126) very well (see Fig. 2.4).

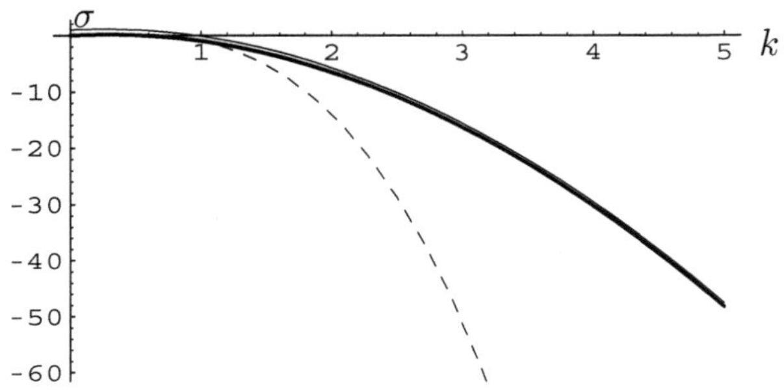


**Fig. 2.4** Dispersion curves: The *dashed line* is for the zeroth-order approximation, the *thin solid line* is the extremely short-wavelength approximation, whereas the *bold solid line* is the exact solution

## 2.2 Unidirectional Solidification from a Binary Mixture

Although the simple model problem discussed in the last section demonstrated the onset of the Mullins–Sekerka instability mechanism very well, it is difficult to implement precisely in experiment. This is because such a system is globally unstable, which always allows some unstable mode solutions.

A more practical model of unidirectional solidification, which can be easily examined in experiment, is solidification of a binary mixture in a Hele–Shaw cell. The Mullins–Sekerka instability mechanism that we have explored for the system with a pure melt is still valid for this binary system and can be derived with the same mathematical approach. Consider a unidirectional solidification device with two uniform temperature zones and a thin sample material, as sketched in Fig. 1.3. The distance between the two zones is set as $(L)_{\mathrm{D}}$. The temperature of the hot zone is $T_{\mathrm{H}}$, while the temperature of the cold zone is $T_{\mathrm{C}}$. Furthermore, we have $T_{\mathrm{C}} < T_{\mathrm{M0}} < T_{\mathrm{H}}$.

The sample is being pulled at a constant imposed velocity $V$ along the direction from the hot zone to the cold zone. The minor species in this binary mixture system, considered an impurity, is dilute. Hence, the mathematical formulation given in Sect. 2.1.3 is applicable.

Again, for the sake of simplicity, we neglect the effect of convection. Thus the whole system is governed by thermodynamics.

### 2.2.1 *Mathematical Formulation of the Problem*

When the pulling velocity $V$ is sufficiently small, the interface will be flat and located somewhere between the two zones. Assume that the distances from the interface to the hot zone and the cold zone are $L_1$ and $L_2$, respectively, which are to be determined. Adopt the same coordinate system $(x, y)$ as that defined in Sect. 2.2.1, whose origin is set at the interface. The rest frame is fixed to the solid phase, whereas the coordinate frame $(x, y)$ is considered to be a moving frame, moving together with the two zones with velocity $\mathbf{V}$ along the $y$-axis.

As specified in Sect. 1.3.3, we use the mass diffusion length $\ell_D$ as the length scale and the pulling velocity $V$ as the velocity scale. The scales of the temperature $T$ and concentration $C$ are set as $\Delta H/(c_p\rho)$ and $C_\infty$, respectively.

Since the cell is very thin, the whole process can be treated as two-dimensional. Thus, the dimensionless governing equations for the present system can be written in the following form:

$$\begin{aligned} \nabla^2\bar{T} &= \lambda\left(\frac{\partial\bar{T}}{\partial\bar{t}} - \frac{\partial\bar{T}}{\partial\bar{y}}\right), \\ \nabla^2\bar{C} &= \left(\frac{\partial\bar{C}}{\partial\bar{t}} - \frac{\partial\bar{C}}{\partial\bar{y}}\right). \end{aligned} \tag{2.134}$$

The dimensionless boundary conditions are as follows:

1. At the edge of the hot zone, $\bar{y} = \bar{L}_1$,

$$\bar{T} = \bar{T}_H = \frac{T_H - T_{M0}}{\Delta H/(c_p\rho)} > 0\,. \tag{2.135}$$

2. At the edge of the cold zone, $\bar{y} = -\bar{L}_2$,

$$\bar{T} = \bar{T}_C = \frac{T_H - T_{M0}}{\Delta H/(c_p\rho)} < 0\,. \tag{2.136}$$

3. In the far field, as $\bar{y} \to \infty$,

$$\bar{C} \to 1\,. \tag{2.137}$$

4. At the interface, $\bar{y} = \bar{h}(\bar{x}, \bar{t})$,

(i)

$$\bar{T} = \bar{T}_S, \tag{2.138}$$

(ii)

$$\bar{T}_S = \varepsilon^2 \frac{\bar{h}_{\bar{x}\bar{x}}}{(1 + \bar{h}_{\bar{x}}^2)^{\frac{3}{2}}} - M\bar{C}, \tag{2.139}$$

(iii)

$$\frac{\partial}{\partial \bar{y}}(\bar{T} - \bar{T}_S) - \bar{h}_{\bar{x}}\frac{\partial}{\partial \bar{x}}(\bar{T} - \bar{T}_S) + \lambda\left(\bar{h}_{\bar{t}} + 1\right) = 0, \tag{2.140}$$

(iv)

$$\frac{\partial \bar{C}}{\partial \bar{y}} - \bar{h}_{\bar{x}}\frac{\partial \bar{C}}{\partial \bar{x}} + \bar{C}(1 - \kappa)\left(\bar{h}_{\bar{t}} + 1\right) = 0\,. \tag{2.141}$$

In the above, $M$ is the morphological parameter defined by (1.48). In most cases, one has

$$\lambda = \frac{\ell_D}{\ell_T} \ll 1, \qquad \text{and} \;\; \bar{L}_1;\, \bar{L}_2 \gg 1\,. \tag{2.142}$$

For the sake of convenience, we shall hereinafter omit the bar over the dimensionless quantities.

## 2.2.2 Basic Steady State

The above system allows a one-dimensional steady-state solution with a flat interface for arbitrary temperature gradient and surface tension parameter $\varepsilon \geq 0$. The interface equation is taken to be $y = h_B = 0$. The exact solution of this one-dimensional steady state can be easily found as the following:

$$\begin{aligned} T_B &= A_1 + A_2 e^{-\lambda y}, \\ T_{BS} &= A_{S1} + A_{S2} e^{\lambda y}, \\ C_B &= B_1 + B_2 e^{-y}, \end{aligned} \tag{2.143}$$

where $T_B$ is the temperature distribution in the liquid phase, $T_{BS}$ is the temperature distribution in the solid phase, and $C_B$ is the concentration distribution in the liquid phase. From the boundary conditions (2.135), (2.136), we have

$$A_1 + A_2 e^{-\lambda L_1} = T_H, \tag{2.144}$$

$$A_{S1} + A_{S2} e^{-\lambda L_2} = T_C\,. \tag{2.145}$$

On the other hand, from (2.137), we have

$$B_1 = 1\,. \tag{2.146}$$

Moreover, from the interface conditions (2.139)–(2.141), we have

$$\begin{cases} A_1 + A_2 = A_{S1} + A_{S2}, \\ A_1 + A_2 = -M(B_1 + B_2), \\ A_2 + A_{S2} = 1, \\ B_2 - (1-\kappa)(1+B_2) = 0\,. \end{cases} \tag{2.147}$$

Finally, we have the relation

$$L_1 + L_2 = L\,. \tag{2.148}$$

From the above eight conditions one can determine the eight unknowns $A_1, A_2, A_{S1}, A_{S2}$, $B_1, B_2$, $L_1, L_2$, and thus completely determine the basic steady-state solution. We obtain

$$C_B(y) = 1 + \frac{1-\kappa}{\kappa}e^{-y}, \tag{2.149}$$

$$C_B(0) = \frac{1}{\kappa}, \tag{2.150}$$

$$T_B(0) = A_1 + A_2 = A_{S1} + A_{S2} = -\frac{M}{\kappa}, \tag{2.151}$$

$$\Delta G_1 = \frac{\partial T_B}{\partial y}(0) - \frac{\partial T_{BS}}{\partial y}(0) \quad = -\lambda\,, \tag{2.152}$$

$$\Delta G_2 = \frac{\partial^2 T_B}{\partial y^2}(0) - \frac{\partial^2 T_{BS}}{\partial y^2}(0) = -\lambda^2\,, \tag{2.153}$$

and

$$\begin{aligned} A_1 &= \frac{T_H + \frac{M}{\kappa}e^{-\lambda L_1}}{1 - e^{-\lambda L_1}}, \\ A_{S1} &= \frac{T_C + \frac{M}{\kappa}e^{\lambda L_2}}{1 - e^{-\lambda L_2}}, \\ A_2 &= -\frac{T_H + \frac{M}{\kappa}}{1 - e^{-\lambda L_1}}, \\ A_{S2} &= -\frac{T_C + \frac{M}{\kappa}}{1 - e^{-\lambda L_2}}\,. \end{aligned} \tag{2.154}$$

Thus, it follows that

$$\begin{aligned} G_{1\mathrm{L}} &= \frac{\partial T_{\mathrm{B}}}{\partial y}(0) = \frac{\lambda}{1-\mathrm{e}^{-\lambda L_1}}\Big(\frac{M}{\kappa}+T_{\mathrm{H}}\Big), \\ G_{1\mathrm{S}} &= \frac{\partial T_{\mathrm{BS}}}{\partial y}(0) = \frac{-\lambda}{1-\mathrm{e}^{-\lambda L_2}}\Big(\frac{M}{\kappa}+T_{\mathrm{C}}\Big). \end{aligned} \tag{2.155}$$

Furthermore, from (2.152), one obtains

$$-\frac{T_{\mathrm{H}}+\frac{M}{\kappa}}{1-\mathrm{e}^{-\lambda L_1}} - \frac{T_{\mathrm{C}}+\frac{M}{\kappa}}{1-\mathrm{e}^{-\lambda L_2}} = 1\,. \tag{2.156}$$

From (2.156) and (2.148), one can find $L_1$ and $L_2$ for any given $L$, and hence determine the location of the interface.

As mentioned before, for most binary systems, the parameter $\lambda = 10^{-2} - 10^{-4}$ is very small. On letting $\lambda \to 0$ and keeping the $\mathcal{O}(\lambda)$ terms, (2.156) becomes

$$\frac{T_{\mathrm{H}}+\frac{M}{\kappa}}{L_1} + \frac{T_{\mathrm{C}}+\frac{M}{\kappa}}{L_2} = 0\,. \tag{2.157}$$

On eliminating $L_1$ from (2.157) and (2.148), it follows that

$$\begin{cases} L_1 = \frac{L}{R}, \\ L_2 = \big(1-\frac{1}{R}\big)L\,, \end{cases} \tag{2.158}$$

where

$$R = \frac{T_{\mathrm{H}}-T_{\mathrm{C}}}{T_{\mathrm{H}}+\frac{M}{\kappa}}\,. \tag{2.159}$$

Furthermore, as $\lambda \to 0$, from (2.155) one obtains

$$\begin{aligned} G_{1\mathrm{L}} &\approx \frac{1}{L_1}\left(\frac{M}{\kappa}+T_{\mathrm{H}}\right) = \frac{T_{\mathrm{H}}-T_{\mathrm{C}}}{L}, \\ G_{1\mathrm{S}} &\approx -\frac{1}{L_2}\left(\frac{M}{\kappa}+T_{\mathrm{C}}\right) = \frac{T_{\mathrm{H}}-T_{\mathrm{C}}}{L}. \end{aligned} \tag{2.160}$$

As a result, we derive that

$$G = \frac{G_{1\mathrm{L}}+G_{1\mathrm{S}}}{2} \approx \frac{T_{\mathrm{H}}-T_{\mathrm{C}}}{L} \quad (\text{as } \lambda \to 0)\,. \tag{2.161}$$

### 2.2.3 Unsteady Perturbed Solutions

The unsteady solutions can be expressed in the forms

$$\begin{aligned} T(x,y,t) &= T_{\mathrm{B}}(y) + \tilde{T}(x,y,t), \\ T_{\mathrm{S}}(x,y,t) &= T_{\mathrm{SB}}(y) + \tilde{T}_{\mathrm{S}}(x,y,t), \\ C(x,y,t) &= C_{\mathrm{B}}(y) + \tilde{C}(x,y,t), \\ h(x,t) &= h_{\mathrm{B}} + \tilde{h}(x,t)\,, \end{aligned} \tag{2.162}$$

where $\tilde{T}$, $\tilde{T}_{\mathrm{S}}$, $\tilde{C}$, and $\tilde{h}$ are small perturbations around the basic steady state.

The governing equation for the perturbation part is

$$\begin{aligned} \nabla^2\tilde{T} &= \lambda\Big(\frac{\partial\tilde{T}}{\partial t} - \frac{\partial\tilde{T}}{\partial y}\Big), \\ \nabla^2\tilde{C} &= \Big(\frac{\partial\tilde{C}}{\partial t} - \frac{\partial\tilde{C}}{\partial y}\Big). \end{aligned} \tag{2.163}$$

The boundary conditions are as follows.

1. The external boundary condition:

$$\begin{cases} \tilde{C} \to 0, & \text{as } y \to \infty, \\ \tilde{T} = 0, & \text{as } y = L_1, \\ \tilde{T}_{\mathrm{S}} = 0, & \text{as } y = L_2\,. \end{cases} \tag{2.164}$$

2. The interface conditions: since the deformation of the interface, $\tilde{h}(x,t)$, is very small, the original boundary conditions on the interface $y = \tilde{h}(x,t)$ can be expanded in a Taylor series in $\tilde{h}$ around $y = h_{\mathrm{B}} = 0$. Retaining only the linear terms and omitting all higher-order small terms, we find that at $y = 0$,

$$\begin{aligned} &\tilde{T} = \tilde{T}_{\mathrm{S}} - (\Delta G_1)\tilde{h}, \\ &\tilde{T}_{\mathrm{S}} = \varepsilon^2\tilde{h}_{xx} - M\tilde{C} + \Big\{\Big(\frac{1-\kappa}{\kappa}\Big)M - G_{1\mathrm{S}}\Big\}\,\tilde{h}, \\ &\frac{\partial\,(\tilde{T} - \tilde{T}_{\mathrm{S}})}{\partial y} + (\Delta G_2)\,\tilde{h} + \lambda\frac{\partial\tilde{h}}{\partial t} = 0, \\ &\frac{\partial\tilde{C}}{\partial y} + \Big(\frac{1-\kappa}{\kappa}\Big)\frac{\partial\tilde{h}}{\partial t} + (1-\kappa)(\tilde{C} + \tilde{h}) = 0\,. \end{aligned} \tag{2.165}$$

We solve this system using the same MVE method described in the last section. The fast variables are still defined by (2.18), but the factor $g(\varepsilon)$ is used only for the solution of the concentration field. For the temperature field $T$ in the liquid phase and the temperature field $T_S$ in the solid phase, we use $q(\varepsilon)$ and $q_s(\varepsilon)$, respectively. The multiple variables form of this system is written as follows:

$$\left\{k^2\frac{\partial^2}{\partial x_+^2}+q^2\frac{\partial^2}{\partial y_+^2}\right\}\tilde{T}=\lambda\varepsilon\left(\sigma\tilde{T}-q\frac{\partial\tilde{T}}{\partial y_+}\right)-2\varepsilon\left(k\frac{\partial^2\tilde{T}}{\partial x\partial x_+}+q\frac{\partial^2\tilde{T}}{\partial y\partial y_+}\right)$$
$$-\varepsilon^2\left(\frac{\partial^2\tilde{T}}{\partial x^2}+\frac{\partial^2\tilde{T}}{\partial y^2}+\lambda\frac{\partial\tilde{T}}{\partial y}\right), \tag{2.166}$$

$$\left\{k^2\frac{\partial^2}{\partial x_+^2}+q_s^2\frac{\partial^2}{\partial y_+^2}\right\}\tilde{T}_S=\lambda\varepsilon\left(\sigma\tilde{T}_S-q_s\frac{\partial\tilde{T}_S}{\partial y_+}\right)-2\varepsilon\left(k\frac{\partial^2\tilde{T}_S}{\partial x\partial x_+}+q_s\frac{\partial^2\tilde{T}_S}{\partial y\partial y_+}\right)$$
$$-\varepsilon^2\left(\frac{\partial^2\tilde{T}_S}{\partial x^2}+\frac{\partial^2\tilde{T}_S}{\partial y^2}+\lambda\frac{\partial\tilde{T}_S}{\partial y}\right), \tag{2.167}$$

$$\left\{k^2\frac{\partial^2}{\partial x_+^2}+g^2\frac{\partial^2}{\partial y_+^2}\right\}\tilde{C}=\varepsilon\left(\sigma\tilde{C}-g\frac{\partial\tilde{C}}{\partial y_+}-2k\frac{\partial^2\tilde{C}}{\partial x\partial x_+}-2g\frac{\partial^2\tilde{C}}{\partial y\partial y_+}\right)$$
$$-\varepsilon^2\left(\frac{\partial^2\tilde{C}}{\partial x^2}+\frac{\partial^2\tilde{C}}{\partial y^2}+\frac{\partial\tilde{C}}{\partial y}\right), \tag{2.168}$$

with the following boundary conditions:

1. As $y_+=\frac{L_1}{\varepsilon}$,

$$\tilde{T}\to 0. \tag{2.169}$$

2. As $y_+=-\frac{L_2}{\varepsilon}$,

$$\tilde{T}_S\to 0. \tag{2.170}$$

3. As $y_+\to\infty$,

$$\tilde{C}\to 0. \tag{2.171}$$

4. At the interface $y=y_+=0$,

$$\tilde{T}=\tilde{T}_S-(\Delta G_1)\tilde{h}\,, \tag{2.172}$$

$$\tilde{T}_S=k^2\frac{\partial^2\tilde{h}}{\partial x_+^2}+2\varepsilon k\frac{\partial^2\tilde{h}}{\partial x\partial x_+}+\varepsilon^2\frac{\partial^2\tilde{h}}{\partial x^2}-M\tilde{C}$$

$$+\left(M\tfrac{1-\kappa}{\kappa}-G_{1S}\right)\tilde{h}=0\,, \tag{2.173}$$

$$\left\{g\tfrac{\partial}{\partial y_+}+\varepsilon\tfrac{\partial}{\partial y}\right\}\left(\tilde{T}-\tilde{T}_S\right)+\lambda\sigma\tilde{h}+\varepsilon(\Delta G_2)\tilde{h}=0\,, \tag{2.174}$$

$$g\tfrac{\partial\tilde{C}}{\partial y_+}+\varepsilon\tfrac{\partial\tilde{C}}{\partial y}+\tfrac{1-\kappa}{\kappa}\sigma\tilde{h}+\varepsilon(1-\kappa)(\tilde{C}+\tilde{h})=0\,. \tag{2.175}$$

We assume that as $\varepsilon\to 0$, the solution $\{\tilde{T},\tilde{T}_S,\tilde{C},\tilde{h}\}$ can be expanded in the following MVE form:

$$\begin{aligned}
\tilde{T}&\sim\left\{\tilde{T}_0(x,y,x_+,y_+)+\varepsilon\tilde{T}_1(x,y,x_+,y_+)+\cdots\right\}e^{\sigma t_+}\\
\tilde{T}_S&\sim\left\{\tilde{T}_{S0}(x,y,x_+,y_+)+\varepsilon\tilde{T}_{S1}(x,y,x_+,y_+)+\cdots\right\}e^{\sigma t_+}\\
\tilde{C}&\sim\left\{\tilde{C}_0(x,y,x_+,y_+)+\varepsilon\tilde{C}_1(x,y,x_+,y_+)+\cdots\right\}e^{\sigma t_+}\\
\tilde{h}&\sim\left\{\tilde{h}_0(x,x_+)+\varepsilon\tilde{h}_1(x,x_+)+\cdots\right\}e^{\sigma t_+}\\
k(\varepsilon)&\sim k_0+\varepsilon k_1+\cdots\\
g(\varepsilon)&\sim k_0+\varepsilon g_1+\cdots\\
q(\varepsilon)&\sim k_0+\varepsilon q_1+\cdots\\
q_s(\varepsilon)&\sim k_0+\varepsilon q_{s1}+\cdots\\
\sigma(\varepsilon)&\sim\sigma_0+\varepsilon\sigma_1+\cdots\,.
\end{aligned} \tag{2.176}$$

By substituting the above expansion into the system (2.166)–(2.175), one can derive the solutions to different orders of $\varepsilon$.

#### 2.2.3.1 Zeroth-Order Approximation Solutions

For the zeroth-order approximation, we have the governing equations

$$\begin{aligned}
\frac{\partial^2\tilde{T}_0}{\partial x_+^2}+\frac{\partial^2\tilde{T}_0}{\partial y_+^2}&=0\\
\frac{\partial^2\tilde{C}_0}{\partial x_+^2}+\frac{\partial^2\tilde{C}_0}{\partial y_+^2}&=0
\end{aligned} \tag{2.177}$$

and the following boundary conditions:

1. As $y_+\to\infty$,

$$\tilde{C}_0\to 0\,. \tag{2.178}$$

2. As $y_+ \to \infty$,

$$\tilde{T}_0 = 0 . \tag{2.179}$$

3. As $y_+ \to -\infty$,

$$\tilde{T}_{S0} = 0 . \tag{2.180}$$

Note that in deriving the above conditions (2.179) and (2.180), we have replaced the conditions (2.169) and (2.170) by

$$\text{as} \quad y_+ \to \infty, \quad \tilde{T} \to 0 \tag{2.181}$$

and

$$\text{as} \quad y_+ \to -\infty, \quad \tilde{T}_S \to 0 , \tag{2.182}$$

respectively. Such replacements will cause only a transcendentally small change in the asymptotic expansions.

4. At the interface, $y_+ = 0$,

$$\tilde{T}_0 = \tilde{T}_{S0} - (\Delta G_1)\tilde{h}_0 , \tag{2.183}$$

$$\tilde{T}_{S0} = k_0^2 \frac{\partial^2 \tilde{h}_0}{\partial x_+^2} - M\tilde{C}_0 + \left\{ M\frac{1-\kappa}{\kappa} - G_{1S} \right\} \tilde{h}_0 , \tag{2.184}$$

$$k_0 \frac{\partial}{\partial y_+}\left(\tilde{T}_0 - \tilde{T}_{S0}\right) + \lambda\sigma_0\tilde{h}_0 = 0 , \tag{2.185}$$

$$k_0 \frac{\partial \tilde{C}_0}{\partial y_+} + \left(\frac{1-\kappa}{\kappa}\right)\sigma_0\tilde{h}_0 = 0 . \tag{2.186}$$

The above system allows the following mode solutions:

$$\begin{aligned}
\tilde{T}_0 &= A_0(x,y)e^{ix_+ - y_+} , \\
\tilde{T}_{S0} &= A_{S0}(x,y)e^{ix_+ + y_+} , \\
\tilde{C}_0 &= B_0(x,y)e^{ix_+ - y_+} , \\
\tilde{h}_0 &= \hat{D}_0 e^{ix_+} .
\end{aligned} \tag{2.187}$$

Here, as we have seen in the last section, the functions $A_0(x,y)$, $A_{S0}(x,y)$, and $B_0(x,y)$ can be set as constants, namely

$$A_0(x,y) = \hat{A}_0, \quad A_{S0}(x,y) = \hat{A}_{S0}, \quad B_0(x,y) = \hat{B}_0, \tag{2.188}$$

since all the coefficients of the perturbed system are constants.

In order to satisfy the boundary conditions (2.183)–(2.186), one finds that the wave number $k_0$ and the eigenvalue $\sigma_0$ in the mode solutions must satisfy the following dispersion relationship:

$$\sigma_0 = \frac{k_0}{\frac{\lambda}{2} + M\frac{1-\kappa}{\kappa}} \left\{ \frac{M(1-\kappa)}{\kappa} - \frac{G_{1L} + G_{1S}}{2} - k_0^2 \right\}. \tag{2.189}$$

It is seen that

$$\sigma_0 \leq 0 \qquad \text{for all } k_0 \geq 0, \tag{2.190}$$

provided

$$G = \frac{G_{1L} + G_{1S}}{2} \geq M\frac{1-\kappa}{\kappa}. \tag{2.191}$$

Consequently, the system will be stable. If the above relation does not hold, the system will be unstable. This criterion was first found by Rutter and Chalmers in 1953 [8], and is referred to as the *constitutional supercooling* criterion.

The above dispersion relationship contains dimensionless parameters $(G, M, \lambda, \kappa)$. Note that both the length scale, $\ell_D$, and the time scale, $\ell_D/V$, are related to the pulling velocity $V$, and that the parameter $G$ depends on two operating conditions: the dimensional pulling velocity $V$ and the design of the zones, $(T_H, T_C, L)$. This is inconvenient, since one needs to consider how each of the operating conditions separately affects the stability of the system. In order to examine these effects separately, it is necessary to introduce some new dimensionless parameters, separately representing these two operating conditions, and reexpress the above asymptotic solution and dispersion relationship. Returning to the original dimensional quantities, it is easy to verify that

$$\begin{aligned} \frac{\varepsilon^2}{M\frac{1-\kappa}{\kappa}} &= \frac{d_c}{\ell_D}, \\ \frac{G}{M\frac{1-\kappa}{\kappa}} &= \frac{\ell_D}{\ell_G}, \end{aligned} \tag{2.192}$$

where $d_c$ and $\ell_G$ both have length scales and are defined as

$$d_c = -\frac{\gamma\kappa T_{M0}}{m\Delta C \Delta H}, \tag{2.193}$$

$$\ell_G = \frac{(G)_D}{m\Delta C}. \tag{2.194}$$

Since $d_c$ is proportional to the surface tension, it is also called the capillary length, like $\ell_c$, whereas $\ell_G$ is also called the thermal length, since it is proportional to the dimensional temperature gradient $(G)_\mathrm{D}$. Thus, it is very natural to define two new parameters

$$v = \frac{d_\mathrm{c}}{\ell_\mathrm{D}}, \tag{2.195}$$

$$\beta = \frac{d_\mathrm{c}}{\ell_\mathrm{G}}, \tag{2.196}$$

where $v$ depends only on the pulling velocity $V$, while $\beta$ depends only on the dimensional quantity $(G)_\mathrm{D}$, corresponding to the average temperature gradient $G = \frac{G_\mathrm{lL}+G_\mathrm{lS}}{2}$. Moreover, we replace the length scale $\ell_\mathrm{D}$ and the time scale $\ell_\mathrm{D}/V$ respectively by $d_\mathrm{c}$ and $d_\mathrm{c}/V$. Accordingly, the normal mode solutions are expressed in the form

$$\begin{aligned}
\tilde{T}_0 \mathrm{e}^{\tilde{\sigma}_0 \tilde{t}} &= \hat{A}_0 \mathrm{e}^{\tilde{k}(\mathrm{i}\tilde{x}-\tilde{y})+\tilde{\sigma}_0 \tilde{t}}, \\
\tilde{T}_{\mathrm{S}0} \mathrm{e}^{\tilde{\sigma}_0 \tilde{t}} &= \hat{A}_{\mathrm{S}0} \mathrm{e}^{\tilde{k}(\mathrm{i}\tilde{x}+\tilde{y})+\tilde{\sigma}_0 \tilde{t}}, \\
\tilde{C}_0 \mathrm{e}^{\tilde{\sigma}_0 \tilde{t}} &= \hat{B}_0 \mathrm{e}^{\tilde{k}(\mathrm{i}\tilde{x}-\tilde{y})+\tilde{\sigma}_0 \tilde{t}}, \\
\tilde{h}_0 &= \hat{D}_0 \mathrm{e}^{\tilde{\sigma}_0 \tilde{t}} \mathrm{e}^{\mathrm{i}\tilde{k}\tilde{x}+\tilde{\sigma}_0 \tilde{t}},
\end{aligned} \tag{2.197}$$

where

$$k = \varepsilon \tilde{k}/v; \quad \sigma_0 = \varepsilon \tilde{\sigma}_0/v. \tag{2.198}$$

Thus, the dispersion relationship (2.189) is written in the form

$$\tilde{\sigma}_0 = \frac{M\frac{1-\kappa}{\kappa}}{\frac{\lambda}{2} + M\frac{1-\kappa}{\kappa}} \tilde{k} \left(1 - \frac{\beta}{v} - \frac{\tilde{k}^2}{v}\right). \tag{2.199}$$

In the plane of operating parameters $(v, \beta)$, one has the stability criterion

$$\begin{cases} v > \beta : \text{unstable region;} \\ v < \beta : \text{globally, or absolutely stable region.} \end{cases} \tag{2.200}$$

These two regions in the $(\beta, v)$-plane are separated by the straight line $v = \beta$.

#### 2.2.3.2 First-Order Approximation Solutions

In the first-order approximation, we derive the equations

$$
\begin{aligned}
k_0^2\left\{\frac{\partial^2}{\partial x_+^2}+\frac{\partial^2}{\partial y_+^2}\right\}\tilde{T}_1 &= a_0 \mathrm{e}^{(\mathrm{i}x_+ - y_+)},\\
k_0^2\left\{\frac{\partial^2}{\partial x_+^2}+\frac{\partial^2}{\partial y_+^2}\right\}\tilde{T}_{S1} &= b_0 \mathrm{e}^{(\mathrm{i}x_+ + y_+)},\\
k_0^2\left\{\frac{\partial^2}{\partial x_+^2}+\frac{\partial^2}{\partial y_+^2}\right\}\tilde{C}_1 &= c_0 \mathrm{e}^{(\mathrm{i}x_+ - y_+)},
\end{aligned}
\tag{2.201}
$$

where

$$
\begin{cases}
a_0 = 2k_0\Big(\frac{\lambda}{2}+\frac{\lambda\sigma_0}{2k_0}+k_1-q_1\Big)\hat{A}_0,\\
b_0 = -2k_0\Big(\frac{\lambda}{2}-\frac{\lambda\sigma_0}{2k_0}-k_1+q_{s1}\Big)\hat{A}_{S0},\\
c_0 = 2k_0\Big(\frac{1}{2}+\frac{\sigma_0}{2k_0}+k_1-g_1\Big)\hat{B}_0\,.
\end{cases}
\tag{2.202}
$$

For the first-order approximation solution, we derive the following boundary conditions at the interface, $y = y_+ = 0$:

$$
\tilde{T}_1 = \tilde{T}_{S1} - (\Delta G_1)\tilde{h}_1\,, \tag{2.203}
$$

$$
\begin{aligned}
\tilde{T}_{S1} = {} & k_0^2\frac{\partial^2 \tilde{h}_1}{\partial x_+^2} + 2k_0\frac{\partial^2 \tilde{h}_0}{\partial x \partial x_+} + 2k_0 k_1 \frac{\partial^2 \tilde{h}_0}{\partial x_+^2} - M\tilde{C}_1\\
& + \Big(M\frac{1-\kappa}{\kappa} - G_{1S}\Big)\tilde{h}_1\,,
\end{aligned}
\tag{2.204}
$$

$$
\begin{aligned}
& k_0\frac{\partial}{\partial y_+}\big(\tilde{T}_1-\tilde{T}_{S1}\big) + q_1\frac{\partial \tilde{T}_0}{\partial y_+} - q_{s1}\frac{\partial \tilde{T}_{S0}}{\partial y_+} + \frac{\partial}{\partial y}\big(\tilde{T}_0-\tilde{T}_{S0}\big)\\
& + \lambda\sigma_0\tilde{h}_1 + \lambda\sigma_1\tilde{h}_0 + (\Delta G_2)\tilde{h}_0 = 0\,,
\end{aligned}
\tag{2.205}
$$

$$
\begin{aligned}
& k_0\frac{\partial \tilde{C}_1}{\partial y_+} + g_1\frac{\partial \tilde{C}_0}{\partial y_+} + \frac{\partial \tilde{C}_0}{\partial y} + \frac{1-\kappa}{\kappa}(\sigma_0\tilde{h}_1+\sigma_1\tilde{h}_0)\\
& + (1-\kappa)(\tilde{C}_0+\tilde{h}_0) = 0\,.
\end{aligned}
\tag{2.206}
$$

To obtain a uniformly valid asymptotic solution, one must eliminate the secular terms by setting $a_0 = b_0 = c_0 = 0$. Moreover, we assume $k_1 = 0$. Thus, we derive

$$\begin{cases} g_1 = \frac{1}{2}\left(1 + \frac{\sigma_0}{k_0}\right), \\ q_1 = \frac{\lambda}{2}\left(1 + \frac{\sigma_0}{k_0}\right), \\ q_{s1} = g_1 - \lambda, \end{cases} \tag{2.207}$$

and the mode solutions

$$\begin{aligned} \tilde{T}_1 &= \hat{A}_1 e^{ix_+ - y_+}, \\ \tilde{T}_{S1} &= \hat{A}_{S1} e^{ix_+ + y_+}, \\ \tilde{C}_1 &= \hat{B}_1 e^{ix_+ - y_+}, \\ \tilde{h}_0 &= \hat{D}_1 e^{ix_+}. \end{aligned} \tag{2.208}$$

Just as for the zeroth-order approximation, we have here set the amplitude functions of the mode solutions as constants. By substituting the mode solution (2.208) into the boundary conditions (2.203)–(2.206), one derives

$$\hat{A}_1 = \hat{A}_{S1} - (\Delta G_1)\hat{D}_1 , \tag{2.209}$$

$$\hat{A}_{S1} = -\left(k_0^2 - M\frac{1-\kappa}{\kappa} + G_{S1}\right)\hat{D}_1 - M\hat{B}_1 , \tag{2.210}$$

$$\begin{aligned} &-k_0(\hat{A}_1 + \hat{A}_{S1}) - q_1(\hat{A}_0 + \hat{A}_{S0}) + \lambda\sigma_0\hat{D}_1 + \lambda\sigma_1\hat{D}_0 \\ &+ (\Delta G_2)\hat{D}_0 + \lambda\hat{A}_{S0} = 0 , \end{aligned} \tag{2.211}$$

$$\begin{aligned} &-k_0\hat{B}_1 - g_1\hat{B}_0 + \frac{1-\kappa}{\kappa}(\sigma_0\hat{D}_1 + \sigma_1\hat{D}_0) \\ &+ (1-\kappa)(\hat{B}_0 + \hat{D}_0) = 0 . \end{aligned} \tag{2.212}$$

Noting that

$$\begin{aligned} \hat{B}_0 &= \frac{\sigma_0}{k_0}\frac{1-\kappa}{\kappa}\hat{D}_0, \\ (\hat{A}_0 + \hat{A}_{S0}) &= \frac{\lambda\sigma_0}{k_0}\hat{D}_0, \\ \lambda\hat{A}_{S0} &= \frac{\lambda^2}{2}\left(\frac{\sigma_0}{k_0} - 1\right)\hat{D}_0 , \end{aligned} \tag{2.213}$$

we obtain

$$\sigma_1\left(\lambda + 2M\frac{1-\kappa}{\kappa}\right) = -\Delta G_2 + \frac{1+\lambda}{4}\Delta G_1 - 2M(1-\kappa)$$
$$-2M\frac{1-\kappa}{\kappa}(1-\kappa-g_1)\frac{\sigma_0}{k_0}$$
$$+\frac{\lambda\sigma_0}{k_0}\left\{\frac{1+\lambda}{4} - \frac{\lambda}{2}\left(1+\frac{\sigma_0}{k_0}\right)\right\}. \tag{2.214}$$

The special case with $\lambda \to 0$ is particularly interesting. In this case, the results are greatly simplified, since $\Delta G_1 \to 0, \Delta G_2 \to 0$, and $(G)_D = (T_H - T_C)/L$. We obtain the following:

$$\sigma_1 = -\kappa - (1-\kappa-g_1)\frac{\sigma_0}{k_0}. \tag{2.215}$$

Thus, the modified dispersion relationship becomes

$$\sigma = \sigma_0 + \varepsilon\sigma_1 + \mathcal{O}(\varepsilon^2)$$
$$= -\varepsilon\kappa + \left\{k + \varepsilon\left(\kappa + \tfrac{\sigma}{2k} - \tfrac{1}{2}\right)\right\}\left(1 - \tfrac{G}{M\frac{1-\kappa}{\kappa}} - \tfrac{k^2}{M\frac{1-\kappa}{\kappa}}\right) \tag{2.216}$$
$$+\mathcal{O}(\varepsilon^2).$$

As we have indicated in the case of a pure melt, the MVE solution obtained above is not valid in the long-wavelength regime $k = \mathcal{O}(\varepsilon)$ and the extremely short-wavelength regime $k = \mathcal{O}(\frac{1}{\varepsilon})$. Therefore, for more accurate information about the behavior of the solution in the full range $0 \leq k < \infty$, one needs to seek different asymptotic expansions of solutions over different regimes, as we have done for the case of a pure melt. In what follows, we shall give the leading-order asymptotic solutions in these two regimes.

### 2.2.4 Asymptotic Solutions in the Long-Wave Regime, $k = \mathcal{O}(\varepsilon)$

As before, in the long-wavelength regime $\{k = \mathcal{O}(\varepsilon); g = \mathcal{O}(\varepsilon); \sigma = \mathcal{O}(\varepsilon)\}$, we write

$$\hat{k} = \frac{k}{\varepsilon}; \quad \hat{g} = \frac{g}{\varepsilon}; \quad \hat{\sigma} = \frac{\sigma}{\varepsilon} \tag{2.217}$$

and introduce the new variables

$$\begin{cases} \hat{x} = \hat{k}x, \\ \hat{y} = \hat{g}y, \\ \hat{t} = \hat{\sigma}t, \end{cases} \tag{2.218}$$

where $\hat{g}$ is used for the solution of the concentration field $\tilde{C}$. For the solution $\tilde{T}$ in the liquid phase and $\tilde{T}_S$ in the solid phase, we replace $\hat{g}$ by $\hat{q}$ and $\hat{q}_s$, respectively.

We expand the solution $\tilde{q} \equiv \{\tilde{T}, \tilde{T}_S, \tilde{C}, \tilde{h}\}$ into the following asymptotic form:

$$\begin{aligned} \tilde{q}(\hat{x}, \hat{y}, \hat{t}, \varepsilon) &\sim e^{\hat{t}}\Big\{\tilde{q}_0(\hat{x}, \hat{y}) + \varepsilon^2\tilde{q}_1(\hat{x}, \hat{y}) + \cdots\Big\}, \\ \hat{k}(\varepsilon) &\sim \hat{k}_0 + \varepsilon^2\hat{k}_1 + \cdots, \\ \hat{g}(\varepsilon) &\sim \hat{g}_0 + \varepsilon^2\hat{g}_1 + \cdots, \\ \hat{q}(\varepsilon) &\sim \hat{q}_0 + \varepsilon^2\hat{q}_1 + \cdots, \\ \hat{q}_s(\varepsilon) &\sim \hat{q}_{s0} + \varepsilon^2\hat{q}_{s1} + \cdots, \\ \hat{\sigma}(\varepsilon) &\sim \hat{\sigma}_0 + \varepsilon^2\hat{\sigma}_1 + \cdots . \end{aligned} \tag{2.219}$$

In terms of the above variables, the linear system (2.163)–(2.165) is transformed into

$$\left\{\hat{k}^2\frac{\partial^2}{\partial\hat{x}^2} + \hat{q}^2\frac{\partial^2}{\partial\hat{y}^2}\right\}\tilde{T} = \lambda\left(\hat{\sigma}\tilde{T} - \hat{q}\frac{\partial\tilde{T}}{\partial\hat{y}}\right), \tag{2.220}$$

$$\left\{\hat{k}^2\frac{\partial^2}{\partial\hat{x}^2} + \hat{q}_s^2\frac{\partial^2}{\partial\hat{y}^2}\right\}\tilde{T}_S = \lambda\left(\hat{\sigma}\tilde{T} - \hat{q}_s\frac{\partial\tilde{T}_S}{\partial\hat{y}}\right), \tag{2.221}$$

$$\left\{\hat{k}^2\frac{\partial^2}{\partial\hat{x}^2} + \hat{g}^2\frac{\partial^2}{\partial\hat{y}^2}\right\}\tilde{C} = \left(\hat{\sigma}\tilde{C} - \hat{g}\frac{\partial\tilde{C}}{\partial\hat{y}}\right), \tag{2.222}$$

with the same form of external boundary conditions as (2.164), and the interface conditions at $\hat{y} = 0$:

$$\begin{aligned} &\tilde{T} = \tilde{T}_S - (\Delta G_1)\tilde{h}, \\ &\tilde{T}_S = \varepsilon^2\hat{k}^2\tilde{h}_{\hat{x}\hat{x}} - M\tilde{C} + \left\{\left(\frac{1-\kappa}{\kappa}\right)M - G_{1S}\right\}\tilde{h}, \\ &\hat{q}\frac{\partial\tilde{T}}{\partial\hat{y}} - \hat{q}_s\frac{\partial\tilde{T}_S}{\partial\hat{y}} + (\Delta G_2 + \lambda\tilde{\sigma}\tilde{h}) = 0, \\ &\hat{g}\frac{\partial\tilde{C}}{\partial\hat{y}} + \left(\frac{1-\kappa}{\kappa}\right)\hat{\sigma}\tilde{h} + (1-\kappa)(\tilde{C} + \tilde{h}) = 0 . \end{aligned} \tag{2.223}$$

Substituting (2.219) into (2.222) and (2.223), one can successively derive each order of approximation. In the leading-order approximation, we obtain

$$\left\{\hat{k}_0^2\frac{\partial^2}{\partial\hat{x}^2}+\hat{q}_0^2\frac{\partial^2}{\partial\hat{y}^2}\right\}\tilde{T}_0=\lambda\left(\hat{\sigma}_0\tilde{T}_0-\hat{q}_0\frac{\partial\tilde{T}_0}{\partial\hat{y}}\right), \tag{2.224}$$

$$\left\{\hat{k}_0^2\frac{\partial^2}{\partial\hat{x}^2}+\hat{g}_0^2\frac{\partial^2}{\partial\hat{y}^2}\right\}\tilde{C}_0=\left(\hat{\sigma}_0\tilde{C}_0-\hat{g}_0\frac{\partial\tilde{C}_0}{\partial\hat{y}}\right), \tag{2.225}$$

and the interface conditions

$$\begin{aligned}
&\tilde{T}_0=\tilde{T}_{S0}-(\Delta G_1)\tilde{h}_0,\\
&\tilde{T}_{S0}=-M\tilde{C}_0+\left\{M\frac{1-\kappa}{\kappa}-G_{1S}\right\}\tilde{h}_0,\\
&\hat{q}_0\frac{\partial\tilde{T}_0}{\partial\hat{y}}-\hat{q}_{s0}\frac{\partial\tilde{T}_{S0}}{\partial\hat{y}}+(\Delta G_2)\tilde{h}_0+\lambda\sigma_0\tilde{h}_0=0,\\
&\hat{g}_0\frac{\partial\tilde{C}_0}{\partial\hat{y}}+\left(\frac{1-\kappa}{\kappa}\right)\hat{\sigma}_0\tilde{h}_0+(1-\kappa)(\tilde{C}_0+\tilde{h}_0)=0.
\end{aligned} \tag{2.226}$$

The zeroth-order approximation solutions are

$$\begin{aligned}
\tilde{T}_0&=\hat{A}_0\mathrm{e}^{\mathrm{i}\hat{x}-\hat{y}},\\
\tilde{T}_{S0}&=\hat{A}_{S0}\mathrm{e}^{\mathrm{i}\hat{x}+\hat{y}},\\
\tilde{C}_0&=\hat{B}_0\mathrm{e}^{\mathrm{i}\hat{x}-\hat{y}},\\
\tilde{h}_0&=\hat{D}_0\mathrm{e}^{\mathrm{i}\hat{x}},
\end{aligned} \tag{2.227}$$

with

$$\begin{cases}
\hat{g}_0=\frac{1}{2}+\sqrt{\hat{k}_0^2+\hat{\sigma}_0+\frac{1}{4}},\\
\hat{q}_0=\frac{\lambda}{2}+\bar{q}_0,\\
\hat{q}_{s0}=-\frac{\lambda}{2}+\bar{q}_0,\\
\bar{q}_0=\sqrt{\hat{k}_0^2+\lambda\hat{\sigma}_0+\frac{\lambda^2}{4}}\,.
\end{cases} \tag{2.228}$$

Substituting (2.227) into (2.226), we derive the dispersion relationship in the long-wavelength regime:

$$\hat{\sigma}_0 = -\frac{2\bar{q}_0 M(1-\kappa)}{\lambda(-1+\kappa+\hat{g}_0)+2\bar{q}_0 M\frac{1-\kappa}{\kappa}} + \frac{2\bar{q}_0(-1+\kappa+\hat{g}_0)}{\lambda(-1+\kappa+\hat{g}_0)+2\bar{q}_0 M\frac{1-\kappa}{\kappa}} \times \left\{\frac{M(1-\kappa)}{\kappa} - \left(1+\frac{\lambda}{2\bar{q}_0}\right)\frac{G_{1L}+G_{1S}}{2} + \frac{\lambda G_{1S}-\Delta G_2}{2\bar{q}_0}\right\}. \quad (2.229)$$

Returning to the parameters $(\sigma_0, k_0)$, we obtain

$$\sigma_0 = -\frac{2\varepsilon q_0 M(1-\kappa)}{\lambda g_0 - \varepsilon\lambda(1-\kappa)+2q_0 M\frac{1-\kappa}{\kappa}} + \frac{2q_0 g_0 - 2\varepsilon q_0(1-\kappa)}{\lambda g_0 - \varepsilon\lambda(1-\kappa)+2q_0 M\frac{1-\kappa}{\kappa}} \times \left\{M\frac{1-\kappa}{\kappa} - \left(1+\frac{\varepsilon\lambda}{2q_0}\right)\frac{G_{1L}+G_{1S}}{2} + \varepsilon\frac{\lambda G_{1S}-\Delta G_2}{2q_0}\right\}, \quad (2.230)$$

where

$$\begin{cases} g_0 = \varepsilon\hat{g}_0 = \frac{\varepsilon}{2} + \sqrt{k^2+\varepsilon\sigma_0+\frac{\varepsilon^2}{4}}, \\ q_0 = \varepsilon\bar{q}_0 = \sqrt{k^2+\varepsilon\lambda\sigma_0+\frac{\lambda^2\varepsilon^2}{4}}. \end{cases} \quad (2.231)$$

In particular, for the special case $\lambda \to 0$, we have

$$\sigma_0 = -\varepsilon\kappa + (-\varepsilon+\varepsilon\kappa+g_0)\left\{1 - \frac{G}{M(1-\kappa)/\kappa}\right\}. \quad (2.232)$$

### 2.2.5 Asymptotic Solutions in the Extremely Short-Wave Regime, $\{k = \mathcal{O}(\frac{1}{\varepsilon});\ g = \mathcal{O}(\frac{1}{\varepsilon})\}$

Similarly to the case of a pure melt, we can derive the MVE solution in the extremely short-wavelength regime. In doing so, we define

$$\begin{cases} \tilde{k} = \varepsilon k \\ \tilde{g} = \varepsilon g \\ \tilde{q} = \varepsilon q \\ \tilde{\sigma} = \varepsilon^3\sigma. \end{cases} \quad (2.233)$$

Assume that as $\varepsilon \to 0$, the solution $\{\tilde{T}, \tilde{T}_S, \tilde{C}, \tilde{h}\}$ is expanded in MVE form:

$$\begin{aligned}
\tilde{T} &\sim \Big\{\tilde{T}_0(x,y,x_+,y_+) + \varepsilon\tilde{T}_1(x,y,x_+,y_+) + \cdots\Big\}e^{\sigma t_+},\\
\tilde{T}_S &\sim \Big\{\tilde{T}_{S0}(x,y,x_+,y_+) + \varepsilon\tilde{T}_{S1}(x,y,x_+,y_+) + \cdots\Big\}e^{\sigma t_+},\\
\tilde{C} &\sim \Big\{\tilde{C}_0(x,y,x_+,y_+) + \varepsilon\tilde{C}_1(x,y,x_+,y_+) + \cdots\Big\}e^{\sigma t_+},\\
\tilde{h} &\sim \varepsilon^2\Big\{\tilde{h}_0(x,x_+) + \varepsilon\tilde{h}_1(x,x_+) + \cdots\Big\}e^{\sigma t_+},
\end{aligned} \tag{2.234}$$

and

$$\begin{aligned}
\tilde{k}(\varepsilon) &= \tilde{k}_0 + \varepsilon^2\tilde{k}_1 + \cdots,\\
\tilde{g}(\varepsilon) &= \tilde{g}_0 + \varepsilon^2\tilde{g}_1 + \cdots,\\
\tilde{q}(\varepsilon) &= \tilde{q}_0 + \varepsilon^2\tilde{q}_1 + \cdots,\\
\tilde{q}_s(\varepsilon) &= \tilde{q}_{s0} + \varepsilon^2\tilde{q}_{s1} + \cdots,\\
\tilde{\sigma}(\varepsilon) &= \tilde{\sigma}_0 + \varepsilon^2\tilde{\sigma}_1 + \cdots.
\end{aligned} \tag{2.235}$$

In leading order, we have the governing equations

$$\begin{aligned}
\tilde{k}_0^2\frac{\partial^2\tilde{T}_0}{\partial x_+^2} + \tilde{q}_0^2\frac{\partial^2\tilde{T}_0}{\partial y_+^2} &= \lambda\tilde{\sigma}_0\tilde{T}_0,\\
\tilde{k}_0^2\frac{\partial^2\tilde{T}_0}{\partial x_+^2} + \tilde{q}_{s0}^2\frac{\partial^2\tilde{T}_{S0}}{\partial y_+^2} &= \lambda\tilde{\sigma}_0\tilde{T}_{S0},\\
\tilde{k}_0^2\frac{\partial^2\tilde{C}_0}{\partial x_+^2} + \tilde{g}_0^2\frac{\partial^2\tilde{C}_0}{\partial y_+^2} &= \tilde{\sigma}_0\tilde{C}_0,
\end{aligned} \tag{2.236}$$

and the mode solutions

$$\begin{aligned}
\tilde{T}_0 &= \hat{A}_0 e^{ix_+ - y_+},\\
\tilde{T}_{S0} &= \hat{A}_{S0} e^{ix_+ + y_+},\\
\tilde{C}_0 &= \hat{B}_0 e^{ix_+ - y_+},\\
\tilde{h}_0 &= \hat{D}_0 e^{ix_+},
\end{aligned} \tag{2.237}$$

with $\tilde{q}_0 = \tilde{q}_{s0} = \sqrt{\lambda\tilde{\sigma}_0 + \tilde{k}_0^2}$ and $\tilde{g}_0 = \sqrt{\tilde{\sigma}_0 + \tilde{k}_0^2}$.

Thus, the boundary conditions of the leading-order approximation result in the following homogeneous system:

$$\begin{cases} \hat{A}_0 - \hat{A}_{S0} = 0, \\ \hat{A}_{S0} + \tilde{k}_0^2 \hat{D}_0 + M\hat{B}_0 = 0, \\ -\tilde{q}_0(\hat{A}_0 + \hat{A}_{S0}) + \lambda\tilde{\sigma}_0\hat{D}_0 = 0, \\ -\tilde{g}_0\hat{B}_0 + \frac{1-\kappa}{\kappa}\tilde{\sigma}_0\hat{D}_0 + (1-\kappa)\hat{B}_0 = 0, \end{cases} \tag{2.238}$$

which gives rise to the dispersion relationship

$$\tilde{\sigma}_0 = \frac{-2\tilde{q}_0\tilde{k}_0^2\left\{\tilde{g}_0 - (1-\kappa)\right\}}{2\tilde{q}_0 M\frac{1-\kappa}{\kappa} + \lambda(\tilde{g}_0 - 1 + \kappa)}. \tag{2.239}$$

Returning to the original notation, this may be rewritten as

$$\sigma_0 = \frac{-2k_0^2\sqrt{k_0^2 + \lambda\varepsilon\sigma_0}\left\{\varepsilon\sqrt{k_0^2 + \varepsilon\sigma_0} - (1-\kappa)\right\}}{-2\varepsilon M\frac{1-\kappa}{\kappa}\sqrt{k_0^2 + \lambda\varepsilon\sigma_0} + \lambda\left\{\varepsilon\sqrt{k_0^2 + \varepsilon\sigma_0} - (1-\kappa)\right\}}. \tag{2.240}$$

For the limiting case, $\lambda \to 0$, we have

$$\sigma_0 = \frac{-2k_0^2\left\{\varepsilon\sqrt{k_0^2 + \varepsilon\sigma_0} - (1-\kappa)\right\}}{2\varepsilon M\frac{1-\kappa}{\kappa}}. \tag{2.241}$$

The linear system with constant coefficients (2.163)–(2.165) can be solved exactly through various approaches. One can easily verify the following exact solutions for the special case $\lambda = 0$:

$$\begin{cases} \tilde{T} = \tilde{T}_S = 0, \\ \tilde{C} = \hat{B}\mathrm{e}^{(ikx - qy + \sigma t)/\varepsilon}, \\ \tilde{h} = \hat{D}\mathrm{e}^{(ikx + \sigma t)/\varepsilon}, \end{cases} \tag{2.242}$$

where

$$q = q_1 + \frac{1}{2}, \quad q_1 = \sqrt{k^2 + \varepsilon\sigma + \frac{\varepsilon^2}{2}}. \tag{2.243}$$

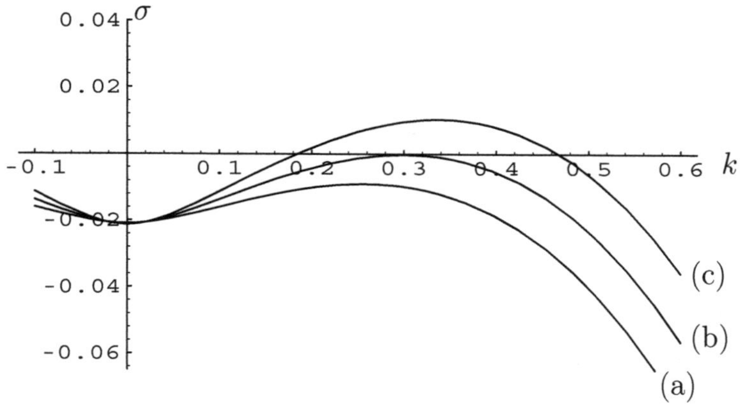


**Fig. 2.5** Mullins–Sekerka instability for a binary alloy system. Three typical dispersion relationships: (**a**) for the case $G = 1.572$, the system is absolutely stable; (**b**) for the case $G = 1.502$, the system is neutrally stable; (**c**) for the case $G = 1.432$, the system is unstable

The exact dispersion relationship is

$$\sigma = -\varepsilon\kappa + \left(q_1 - \frac{\varepsilon}{2} + \varepsilon\kappa\right)\left(1 - \frac{G}{M^{\frac{1-\kappa}{\kappa}}} - \frac{k^2}{M^{\frac{1-\kappa}{\kappa}}}\right). \tag{2.244}$$

By comparing the dispersion formula of the exact solution (2.244) with the dispersion formula of the MVE solution (2.216), one sees that (2.216) is the regular perturbation expansion of (2.244) as $\varepsilon \to 0$ with any fixed $k$. Such an agreement confirms the effectiveness of our asymptotic approach.

In Fig. 2.5, we show three typical curves of the exact dispersion relationship in the $(\sigma, k)$-plane, with fixed $\varepsilon = 0.2, \kappa = 0.103, M = 0.2$ and various values of $G$. Curve (a) corresponds to $G = 1.572$, and we see that the system is stable for the entire range of wave numbers $0 \leq k < \infty$. It is called *absolutely stable* or *globally stable*. Curve (b) corresponds to $G = 1.502$, and the system is neutrally stable. Curve (c) corresponds to $G = 1.432$, and the system is unstable.

In Fig. 2.6, we show the asymptotic solutions of the dispersion relationship in the $(\sigma, k)$-plane for the above three cases. In these figures, the dashed lines represent the zeroth-order MVE solution, the thin solid lines represent the first-order MVE solution, the dotted lines represent the zeroth-order long-wavelength approximate solution, and the bold solid lines represent the exact solutions. Evidently in the long-wavelength region $(0 \leq k < 0.1)$, as expected, the MVE solution is not a good approximation to the exact solution. It approximates the exact solution very well in the region $(0.2 < k < 1)$. The figure also shows that the long-wavelength asymptotic solution (2.232) agrees with the exact solution in the region $(0 \leq k < 0.1)$ very well.

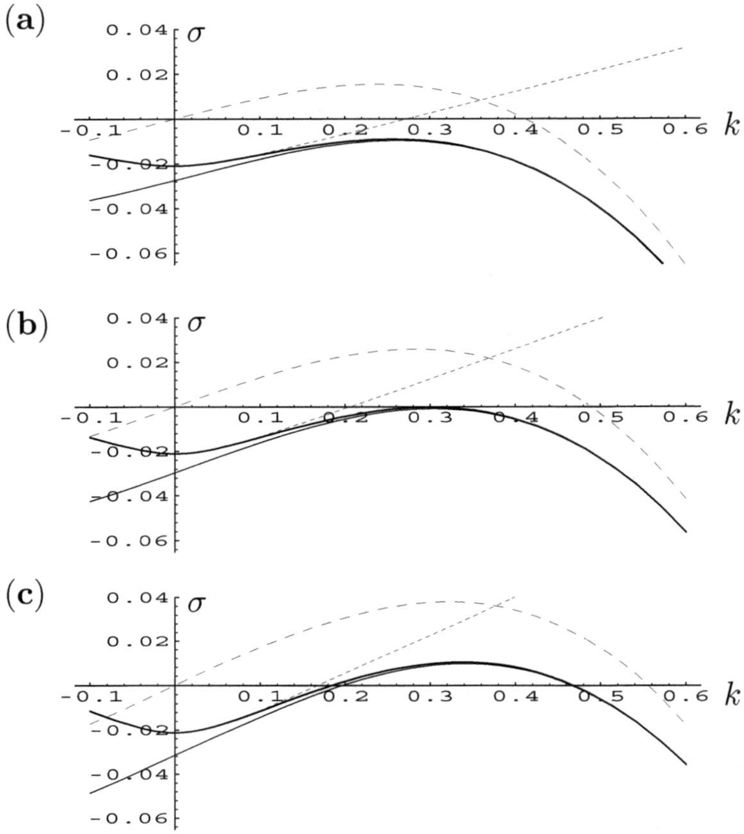


**Fig. 2.6** Dispersion curves for unidirectional solidification from a binary mixture: (**a**) for the absolutely stable case; (**b**) for the neutrally stable case; (**c**) for the unstable case. Here the *dashed lines* represent the zeroth-order MVE solution, the *thin solid lines* represent the first-order MVE solution, the *dotted lines* represent the zeroth-order long-wavelength approximate solution, and the *bold solid lines* represent the exact solutions

For the case of unidirectional solidification, which allows an exact analytical solution, the asymptotic solution obtained does not appear to be superior and does not provide any information beyond the exact solution. Nevertheless, it should be pointed out that the asymptotic approach demonstrated in this chapter has a profound significance. It can be applied to more general inhomogeneous dynamic systems with nonconstant coefficients. Growth systems with a nonuniform curved interface, such as dendritic growth and the evolution of viscous fingering, belong to this category. Evidently, for these more complicated systems, there is no way to find the exact solutions. Our asymptotic approach turns out to be the only viable analytical tool for tackling these problems.

### 2.2.6 Some Remarks on Unidirectional Solidification

At the start of applying the MVE method, we assumed the asymptotic expansion form with the multiple variables $(x, y)$ and $(x_+, y_+, t_+)$, as shown in (2.19) and (2.176). It finally turns out that the solutions to the system under investigation depend only on the fast variables $(x_+, y_+, t_+)$. However, this is not the case for more general inhomogeneous dynamical systems. Namely, for inhomogeneous dynamical systems, the MVE solutions may actually depend on the slow space variables $(x, y)$, as well as the fast variables $(x_+, y_+, t_+)$. In some cases, the MVE solutions may even depend on the slow time variable $t$.

Furthermore, our intention in this chapter was to demonstrate the asymptotic approach. Therefore, only some very basic features of unidirectional solidification at the early stage of interface evolution near the Mullins–Sekerka instability were discussed. Here, the unidirectional solidification implies that the basic steady state of the system has a planar interface. Many important aspects of this subject were left out. For instance, the effects of physical parameters, such as anisotropy and kinetic attachment on linear stability, the influence of convection flow in the liquid phase on solidification, the bifurcation theory, and pattern formation and selection in the weakly nonlinear regime are all issues of great interest that have been totally neglected. The scope of this book does not allow us to give a comprehensive review of this subject. Readers interested in studying these topics are referred to the excellent review articles [1–4]. It should be pointed out here that the results derived in this chapter have shown that for unidirectional solidification, the neutrally stable, or, say, marginally stable mode, is a steady mode, corresponding to the zero eigenvalue $\sigma = 0$. Generally speaking, the system of unidirectional solidification does not allow an oscillatory neutrally stable state with a pure imaginary eigenvalue unless some additional physical effect, for instance kinetic attachment, is included. In contrast, it will be seen later that a system with a curved front, such as dendritic growth, is quite different. The neutral mode of dendritic growth under the most interesting circumstances is not steady, but oscillatory with corresponding eigenvalue $\sigma = -i\omega \neq 0$.

## References

1. B. Billia, R. Trivedi, Pattern formation in crystal growth, Chap. 14, in *Handbook of Crystal Growth, Volume 1: Fundamentals, Part B: Transport and Stability*, ed. by D.T.J. Hurle (Elsevier Science, North–Holland, Amsterdam, 1993)
2. B. Caroli, C. Caroli, B. Roulet, Instability of planar solidification fronts, in *Solids Far from Equilibrium*, ed. by C. Godreche (Cambridge University Press, Cambridge, New York 1991)
3. S.H. Davis, Hydrodynamic interactions in directional solidification. J. Fluid Mech. **212**, 241–262 (1990)
4. S.H. Davis, Effect of flow on morphological stability, in *Handbook of Crystal Growth, Volume 1: Fundamentals, Part B: Transport and Stability*, ed. by D.T.J. Hurle (Elsevier Science, North–Holland, Amsterdam 1993)

5. J. Kevorkian, J.D. Cole, *Multiple Scale and Singular Perturbation Methods*. Applied Mathematical Sciences, vol. 114 (Springer, Berlin/Heidelberg 1996)
6. W.W. Mullins, R.F. Sekerka, Morphological stability of a particle growing by diffusion or heat flow. J. Appl. Phys. **34**, 323–329 (1963)
7. W.W. Mullins, R.F. Sekerka, Stability of a planar interface during solidification of a dilute binary alloy. J. Appl. Phys. **35**, 444–451 (1964)
8. J.W. Rutter, B. Chalmers, A prismatic substructure formed during solidification of metals. Can. J. Phys. **31**, 15–39 (1953)

# Chapter 3
# Mathematical Formulation of Free Dendritic Growth from a Pure Melt

We first study the problem of free dendritic growth from a pure substance. Dendritic growth from a binary mixture will be studied in Chap. 8. Typical free dendritic growth is shown in Fig. 1.6. In general, dendritic growth, starting from a tiny seed, proceeds via a very complex dynamic process until, at a later stage of evolution, a permanent pattern is displayed. The details of pattern formation during the entire evolution will obviously be affected by many factors, in particular by the details of the initial conditions. To describe and predict the whole history of growth is very difficult and not meaningful. From a theoretical point of view, the most interesting thing, which is also the objective of the present monograph, is to study the behavior of the system during the late stage of evolution. When the dendrite is fully developed with a sufficiently long stem, the effects of the initial growth conditions and the situation at the root will diminish to a minimum. Thus, one may consider that at the stage $t \gg 1$, a free dendrite grows into an undercooled pure melt with a constant velocity $U$ and stationary interfacial pattern. The undercooling temperature of the melt in the region away from the growing dendrite-tip is $T_\infty < T_{M0}$, where $T_{M0}$ is the melting temperature of a fat interface. For simplicity, we start by assuming the following:

1. The mass density $\rho$, the specific heat $c_p$, and other thermal characteristic constants of the solid state are the same as those of the liquid state, namely $\alpha_s = \beta_s = 1$. We may consider the effect of the ratio of thermal diffusivities, so we keep $\hat{\alpha}_T$ as a free parameter. Thus, for the symmetric model, we may set $\hat{\alpha}_T = 1$, whereas for the one-sided model, we may set $\hat{\alpha}_T = 0$.
2. Gravity is negligible, and no convection is involved.
3. The surface tension at the interface is isotropic.

The thermal diffusion length $\ell_T = \kappa_T/U$ is used as the length scale, and the quantity $\Delta H/(c_p\rho)$ is used as the temperature scale. Here, $\kappa_T$ is the thermal diffusivity, while $\Delta H$ is the latent heat released per unit volume of solid phase.



J.-J. Xu, *Interfacial Wave Theory of Pattern Formation in Solidification*, Springer Series in Synergetics, DOI 10.1007/978-3-319-52663-8_3

## 3.1 Three-Dimensional Free Dendritic Growth

Because the surface tension is assumed isotropic, one can consider three-dimensional axially symmetric dendritic growth. We adopt the paraboloidal coordinate system $(\xi, \eta, \theta)$, which is defined through the cylindrical coordinate system $(r, z, \theta)$ by (see Fig. 3.1)

$$\begin{cases} \dfrac{r}{\eta_0^2} = \xi\eta \\ \dfrac{z}{\eta_0^2} = \dfrac{1}{2}(\xi^2 - \eta^2) \end{cases} \tag{3.1}$$

or through the Cartesian coordinate system $(x, y, z)$ by

$$\begin{cases} \dfrac{x}{\eta_0^2} = \xi\eta\cos\theta \\ \dfrac{y}{\eta_0^2} = \xi\eta\sin\theta \\ \dfrac{z}{\eta_0^2} = \dfrac{1}{2}(\xi^2 - \eta^2), \end{cases} \tag{3.2}$$

where the constant $\eta_0^2$ is to be determined.

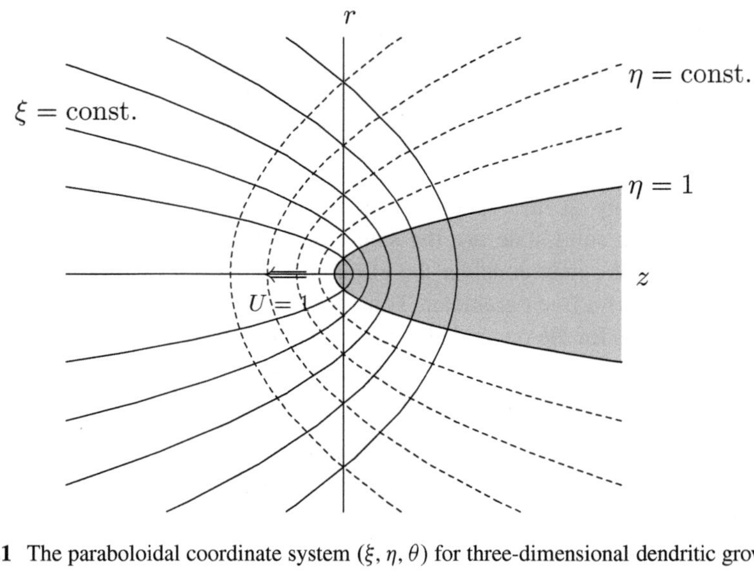


**Fig. 3.1** The paraboloidal coordinate system $(\xi, \eta, \theta)$ for three-dimensional dendritic growth

Note that the dynamical system under investigation is invariant under coordinate translation; we have the freedom to choose the origin of the paraboloidal system (3.1) at will. The constant $\eta_0^2$ in (3.1) can therefore be chosen so that the steady interface shape satisfies

$$\eta_s(0) = 1\,. \tag{3.3}$$

It will be seen later that this constant is just the Péclet number of the system with zero surface tension.

In order to describe the dendritic growth in the paraboloidal system, one needs to apply extensively some formulas of differential geometry. For the necessary background, readers are referred to [1].

The radius vector of any point with coordinates $(x, y, z)$ can be expressed in the form

$$\mathbf{r} = \eta_0^2 \left[ \xi\eta\cos\theta\mathbf{i} + \xi\eta\sin\theta\mathbf{j} + \frac{1}{2}(\xi^2 - \eta^2)\mathbf{k} \right], \tag{3.4}$$

where we have denoted the unit vectors along the $x, y, z$ directions by $\mathbf{i}, \mathbf{j}, \mathbf{k}$, respectively. In the moving frame, the relative velocity of the liquid phase is obviously $\mathbf{u} = \mathbf{k}$. We assume that the vectors $\mathbf{e}_1, \mathbf{e}_2, \mathbf{e}_3$ are unit vectors along the $\xi, \eta, \theta$ directions, respectively. Namely,

$$\mathbf{e}_1 \Big\| \frac{\partial \mathbf{r}}{\partial \xi}, \quad \mathbf{e}_2 \Big\| \frac{\partial \mathbf{r}}{\partial \eta}, \quad \mathbf{e}_3 \Big\| \frac{\partial \mathbf{r}}{\partial \theta}\,. \tag{3.5}$$

Thus, we derive that

$$\begin{aligned}
\mathbf{e}_1 &= \frac{1}{\sqrt{\xi^2 + \eta^2}}(\eta\cos\theta\mathbf{i} + \eta\sin\theta\mathbf{j} + \xi\mathbf{k}) \\
\mathbf{e}_2 &= \frac{1}{\sqrt{\xi^2 + \eta^2}}(\xi\cos\theta\mathbf{i} + \xi\sin\theta\mathbf{j} - \eta\mathbf{k}) \\
\mathbf{e}_3 &= (-\sin\theta\mathbf{i} + \cos\theta\mathbf{j}).
\end{aligned} \tag{3.6}$$

For an element of arc, $\mathrm{d}s$, we can write

$$(\mathrm{d}s)^2 = (\mathrm{d}x)^2 + (\mathrm{d}y)^2 + (\mathrm{d}z)^2 = (\alpha_1\mathrm{d}\xi)^2 + (\alpha_2\mathrm{d}\eta)^2 + (\alpha_3\mathrm{d}\theta)^2, \tag{3.7}$$

where $\alpha_i$ are Lamé constants. From (3.2), one has

$$\begin{cases}
\mathrm{d}x = \eta_0^2(\xi\mathrm{d}\eta\cos\theta + \eta\mathrm{d}\xi\cos\theta - \xi\eta\sin\theta\mathrm{d}\theta) \\
\mathrm{d}y = \eta_0^2(\xi\mathrm{d}\eta\sin\theta + \eta\mathrm{d}\xi\sin\theta + \xi\eta\cos\theta\mathrm{d}\theta) \\
\mathrm{d}z = \eta_0^2(\xi\mathrm{d}\xi - \eta\mathrm{d}\eta).
\end{cases} \tag{3.8}$$

Hence, it follows that the Lamé constants are

$$\alpha_1 = \alpha_2 = \eta_0^2\sqrt{\xi^2+\eta^2}, \quad \alpha_3 = \eta_0^2\xi\eta\,. \tag{3.9}$$

The unknown functions for the present problem are the temperature fields $T(\xi,\eta,t)$, $T_S(\xi,\eta,t)$, and the interface shape $\eta_s(\xi,t)$. The dimensionless governing equation is simply the heat conduction equation. In order to write this governing equation in the paraboloidal coordinate system, one needs the formulas

$$\begin{aligned}\nabla T &= \frac{\mathbf{e}_1}{\alpha_1}\frac{\partial T}{\partial \xi}+\frac{\mathbf{e}_2}{\alpha_2}\frac{\partial T}{\partial \eta}+\frac{\mathbf{e}_3}{\alpha_3}\frac{\partial T}{\partial \theta}\\ \nabla\cdot\mathbf{U} &= \frac{1}{\alpha_1\alpha_2\alpha_3}\left[\frac{\partial}{\partial\xi}\Big(\alpha_2\alpha_3U_1\Big)+\frac{\partial}{\partial\eta}\Big(\alpha_1\alpha_3U_2\Big)+\frac{\partial}{\partial\theta}\Big(\alpha_1\alpha_2U_3\Big)\right].\end{aligned} \tag{3.10}$$

We derive that

$$\begin{aligned}\nabla^2 T &= \nabla\cdot(\nabla T)\\ &= \frac{1}{\alpha_1\alpha_2\alpha_3}\left[\frac{\partial}{\partial\xi}\Big(\frac{\alpha_2\alpha_3}{\alpha_1}\frac{\partial}{\partial\xi}\Big)+\frac{\partial}{\partial\eta}\Big(\frac{\alpha_1\alpha_3}{\alpha_2}\frac{\partial}{\partial\eta}\Big)+\frac{\partial}{\partial\theta}\Big(\frac{\alpha_1\alpha_2}{\alpha_3}\frac{\partial}{\partial\theta}\Big)\right]T\\ &= \frac{1}{\eta_0^4(\xi^2+\eta^2)}\left[\frac{1}{\xi}\frac{\partial}{\partial\xi}\Big(\xi\frac{\partial T}{\partial\xi}\Big)+\frac{1}{\eta}\frac{\partial}{\partial\eta}\Big(\eta\frac{\partial T}{\partial\eta}\Big)+\frac{\xi^2+\eta^2}{\xi^2\eta^2}\frac{\partial^2 T}{\partial\theta^2}\right]\end{aligned} \tag{3.11}$$

and

$$\begin{aligned}\mathbf{k}\cdot\nabla T &= \frac{1}{\eta_0^2\sqrt{\xi^2+\eta^2}}\left[\frac{\partial T}{\partial\xi}(\mathbf{e}_1\cdot\mathbf{k})+\frac{\partial T}{\partial\eta}(\mathbf{e}_2\cdot\mathbf{k})\right]\\ &= \frac{1}{\eta_0^2(\xi^2+\eta^2)}\left[\xi\frac{\partial T}{\partial\xi}-\eta\frac{\partial T}{\partial\eta}\right].\end{aligned} \tag{3.12}$$

In terms of formulas (3.11) and (3.12), the governing equations (1.39) and (1.40) may be written in the form

$$\begin{aligned}&\frac{\partial^2 T}{\partial\xi^2}+\frac{\partial^2 T}{\partial\eta^2}+\frac{1}{\xi}\frac{\partial T}{\partial\xi}+\frac{1}{\eta}\frac{\partial T}{\partial\eta}+\Big(\frac{1}{\xi^2}+\frac{1}{\eta^2}\Big)\frac{\partial^2 T}{\partial\theta^2}\\ &= \eta_0^2\Big(\xi\frac{\partial T}{\partial\xi}-\eta\frac{\partial T}{\partial\eta}\Big)+\eta_0^4(\xi^2+\eta^2)\frac{\partial T}{\partial t}\,.\end{aligned} \tag{3.13}$$

and

$$\frac{\partial^2 T_S}{\partial \xi^2} + \frac{\partial^2 T_S}{\partial \eta^2} + \frac{1}{\xi}\frac{\partial T_S}{\partial \xi} + \frac{1}{\eta}\frac{\partial T_S}{\partial \eta} + \left(\frac{1}{\xi^2} + \frac{1}{\eta^2}\right)\frac{\partial^2 T_S}{\partial \theta^2}$$
$$= \hat{\alpha}_T \left[\eta_0^2\left(\xi\frac{\partial T_S}{\partial \xi} - \eta\frac{\partial T_S}{\partial \eta}\right) + \eta_0^4(\xi^2 + \eta^2)\frac{\partial T_S}{\partial t}\right] . \tag{3.14}$$

We now write the boundary conditions in the paraboloidal coordinate system. In doing so, one needs the expressions for the vector normal to the interface and the curvature. Assume that the general dimensionless asymmetric interface shape is denoted by $\eta_s(\xi, \theta, t)$, or $s(\xi, \eta, \theta, t) = \eta - \eta_s(\xi, \theta, t) = 0$. Then for every point on the interface, $(\xi, \eta_s, \theta)$, or say $(x_s, y_s, z_s)$, the radius vector is

$$\mathbf{r} = x_s\mathbf{i} + y_s\mathbf{j} + z_s\mathbf{k} = \eta_0^2\left[\xi\eta_s\cos\theta\mathbf{i} + \xi\eta_s\sin\theta\mathbf{j} + \frac{1}{2}(\xi^2 - \eta_s^2)\mathbf{k}\right] . \tag{3.15}$$

It follows that

$$\mathbf{r}_1 = \frac{\partial \mathbf{r}}{\partial \xi} = \eta_0^2\Big[(\xi\eta_s)'\cos\theta\mathbf{i} + (\xi\eta_s)'\sin\theta\mathbf{j} + (\xi - \eta_s\eta_s')\mathbf{k}\Big]$$
$$\mathbf{r}_2 = \frac{\partial \mathbf{r}}{\partial \theta} = \eta_0^2\Big[(-\xi\eta_s\sin\theta + \xi\eta_{s,\theta}\cos\theta)\mathbf{i}$$
$$+(\xi\eta_s\cos\theta + \xi\eta_{s,\theta}\sin\theta)\mathbf{j} - \eta_s\eta_{s,\theta}\mathbf{k}\Big]$$
$$\mathbf{r}_{11} = \frac{\partial^2 \mathbf{r}}{\partial \xi^2} = \eta_0^2(\xi\eta_s)''\cos\theta\mathbf{i} + (\xi\eta_s)''\sin\theta\mathbf{j} + (1 - \eta_s'^2 - \eta_s\eta_s'')\mathbf{k}$$
$$\mathbf{r}_{12} = \frac{\partial^2 \mathbf{r}}{\partial \xi\partial \theta} = \eta_0^2\Big\{\big[-(\xi\eta_s)'\sin\theta + (\xi\eta_{s,\theta})'\cos\theta\big]\mathbf{i}$$
$$+\big[(\xi\eta_s)'\cos\theta + (\xi\eta_{s,\theta})'\sin\theta\big]\mathbf{j} - (\eta_s\eta_{s,\theta})'\big]\mathbf{k}\Big\}$$
$$\mathbf{r}_{22} = \frac{\partial^2 \mathbf{r}}{\partial \theta^2} = \eta_0^2\Big\{\big[-\xi\eta_s\cos\theta - 2\xi\eta_{s,\theta}\sin\theta + \xi\eta_{s,\theta\theta}\cos\theta\big]\mathbf{i}$$
$$+\big[-\xi\eta_s\sin\theta + 2\xi\eta_{s,\theta}\cos\theta + \xi\eta_{s,\theta\theta}\sin\theta\big]\mathbf{j}$$
$$-\big[\eta_{s,\theta}^2 + \eta_s\eta_{s,\theta\theta}\big]\mathbf{k}\Big\} , \tag{3.16}$$

where the prime denotes the derivative $\frac{\partial}{\partial\xi}$. For an element of arc, d$s$, on the interface, we have

$$\mathrm{d}s^2 = \mathbf{r}_1^2\mathrm{d}\xi^2 + 2(\mathbf{r}_1\cdot\mathbf{r}_2)\mathrm{d}\xi\mathrm{d}\theta + \mathbf{r}_2^2\mathrm{d}\theta^2$$
$$= E\mathrm{d}\xi^2 + 2F\mathrm{d}\xi\mathrm{d}\theta + G\mathrm{d}\theta^2, \tag{3.17}$$

where

$$\begin{cases} E = \mathbf{r}_1 \cdot \mathbf{r}_1 = \eta_0^4 (1 + \eta_s'^2)(\xi^2 + \eta_s^2), \\ F = \mathbf{r}_1 \cdot \mathbf{r}_2 = \eta_0^4 \eta_{s,\theta} \eta_s' (\xi^2 + \eta_s^2), \\ G = \mathbf{r}_2 \cdot \mathbf{r}_2 = \eta_0^4 \big[(\xi\eta_s)^2 + \eta_{s,\theta}^2 (\xi^2 + \eta_s^2)\big]. \end{cases} \tag{3.18}$$

On the other hand, we can write the unit normal vector

$$\mathbf{n} = \frac{\mathbf{r}_1 \times \mathbf{r}_2}{|\mathbf{r}_1 \times \mathbf{r}_2|} = \frac{1}{\sqrt{(1 + \eta_s'^2 + \eta_{s,\theta}^2)(\xi^2 + \eta_s^2)}}$$

$$\times \Bigg\{ \Big[ (\eta_s \eta_s' - \xi) \cos\theta - \left( \frac{\xi}{\eta_s} + \frac{\eta_s}{\xi} \right) \eta_{s,\theta} \sin\theta \Big] \mathbf{i} \tag{3.19}$$

$$+ \Big[ (\eta_s \eta_s' - \xi) \sin\theta + \left( \frac{\xi}{\eta_s} + \frac{\eta_s}{\xi} \right) \eta_{s,\theta} \cos\theta \Big] \mathbf{j} + (\xi\eta_s)' \mathbf{k} \Bigg\} \tag{3.20}$$

and calculate

$$H = |\mathbf{r}_1 \times \mathbf{r}_2| = \eta_0^4 \xi \eta_s \sqrt{(1 + \eta_s'^2 + \eta_{s,\theta}^2)(\xi^2 + \eta_s^2)} \tag{3.21}$$

$$L = \mathbf{n} \cdot \mathbf{r}_{11} = \frac{\eta_0^2}{\sqrt{(\xi^2 + \eta_s^2)(1 + \eta_s'^2 + \eta_{s,\theta}^2)}}$$

$$\times \Big[ (\eta_s \eta_s' - \xi)(\xi\eta_s)'' + (1 - \eta_s'^2 - \eta_s \eta_s'')(\xi\eta_s)' \Big] \tag{3.22}$$

$$M = \mathbf{n} \cdot \mathbf{r}_{12} = \frac{\eta_0^2}{\sqrt{(\xi^2 + \eta_s^2)(1 + \eta_s'^2 + \eta_{s,\theta}^2)}}$$

$$\times \Big[ (\xi\eta_{s,\theta})' - (\eta_s \eta_s' - \xi) \Big] \tag{3.23}$$

$$N = \mathbf{n} \cdot \mathbf{r}_{22} = \frac{\eta_0^2}{\sqrt{(\xi^2 + \eta_s^2)(1 + \eta_s'^2 + \eta_{s,\theta}^2)}}$$

$$\times \Big[ \xi\eta_s (\xi - \eta_s \eta_s') + 2\xi \eta_{s,\theta} - (\xi^2 + \eta_s^2) \eta_{s,\theta\theta}$$

$$- (\xi\eta_s)' \eta_{s,\theta}^2 \Big]. \tag{3.24}$$

Thus, according to differential geometry, one derives twice the mean curvature of the interface as follows:

$$\mathcal{K}_c = \frac{1}{H^2}(EN - 2FM + GL) = \frac{1}{\eta_0^2}\mathcal{K}_3\left\{\frac{\mathrm{d}}{\mathrm{d}\xi}, \frac{\mathrm{d}^2}{\mathrm{d}\xi^2}; \frac{\mathrm{d}}{\mathrm{d}\theta}, \frac{\mathrm{d}^2}{\mathrm{d}\theta^2}\right\}\eta_s, \quad (3.25)$$

where

$$\mathcal{K}_3\Big\{\eta_s(\xi,\theta,\varepsilon)\Big\} = -\frac{1}{[\xi^2+\eta_s^2]^{\frac{3}{2}}[\xi^2\eta_s^2(1+\eta_s'^2)+\eta_{s,\theta}^2(\xi^2+\eta_s^2)]^{\frac{3}{2}}}$$

$$\times\Bigg\{\Big[(1+\eta_s'^2)(\xi^2+\eta_s^2)\Big]\Big[-(\xi\eta_s)^2(\eta_s\eta_s'-\xi)+2\xi\eta_{s,\theta}^2(\xi^2+\eta_s^2)$$

$$+\xi^2\eta_s\eta_{s,\theta\theta}(\eta_s\eta_s'-\xi)-\xi\eta_s(\xi\eta_s)'(\eta_{s,\theta}^2+\eta_s\eta_{s,\theta\theta})\Big]$$

$$-2\Big[\eta_{s,\theta}\eta_s'(\xi^2+\eta_s^2)\Big]\Big[(\xi\eta_s)'\eta_{s,\theta}(\xi^2+\eta_s^2)+\xi\eta_s(\eta_s\eta_s'-\xi)(\xi\eta_{s,\theta})'$$

$$-\xi\eta_s(\eta_s\eta_{s,\theta})'(\xi\eta_s)'\Big]+\xi\eta_s\Big[(\xi\eta_s)^2+\eta_{s,\theta}^2(\xi^2+\eta_s^2)\Big]$$

$$\times\Big[(\xi\eta_s)''(\eta_s\eta_s'-\xi)+(\xi\eta_s)'(1-\eta_s'^2-\eta_s\eta_s'')\Big]\Bigg\}. \quad (3.26)$$

The boundary condition are as follows:

1. The upstream condition: as $\eta \to \infty$,

$$T \to T_\infty. \quad (3.27)$$

2. The regularity condition in the solid state: as $\eta \to 0$,

$$T_S = \mathcal{O}(1). \quad (3.28)$$

3. The interface conditions: at $\eta = \eta_s(\xi, \theta, t)$:

   (a) the thermodynamic equilibrium condition

$$T = T_S, \quad (3.29)$$

   (b) the Gibbs–Thomson condition

$$T = -\frac{\Gamma}{\eta_0^2}\mathcal{K}_3\Big\{\eta_s(\xi,\theta,\varepsilon)\Big\}, \quad (3.30)$$

(c) the heat balance condition

$$\Big(\frac{\partial}{\partial\eta}-\eta_s'\frac{\partial}{\partial\xi}\Big)(T-\hat{\alpha}_T T_S)-\eta_{s\theta}\Big(\frac{1}{\xi^2}+\frac{1}{\eta_s^2}\Big)\frac{\partial}{\partial\theta}(T-\hat{\alpha}_T T_S)$$
$$+\eta_0^4(\xi^2+\eta_s^2)\frac{\partial\eta_s}{\partial t}+\eta_0^2(\xi\eta_s)'=0\,. \tag{3.31}$$

## 3.2 Axisymmetric Free Dendrite Growth

From now on, we shall mainly consider the symmetric case ($\hat{\alpha}_T = 1$), unless we specify otherwise. For the special case of axisymmetric dendritic growth, since $T_\theta = 0$, Eq. (3.13) reduces to

$$\frac{\partial^2 T}{\partial\xi^2}+\frac{\partial^2 T}{\partial\eta^2}+\frac{1}{\xi}\frac{\partial T}{\partial\xi}+\frac{1}{\eta}\frac{\partial T}{\partial\eta}=\eta_0^2\Big(\xi\frac{\partial T}{\partial\xi}-\eta\frac{\partial T}{\partial\eta}\Big)$$
$$+\eta_0^4(\xi^2+\eta^2)\frac{\partial T}{\partial t}\,. \tag{3.32}$$

On the other hand, due to $\eta_{s,\theta} = 0$, one has

$$\mathcal{K}_c=\frac{1}{H^2}(EN-2FM+GL)=\frac{1}{\eta_0^2}\mathcal{K}\Big\{\frac{\mathrm{d}}{\mathrm{d}\xi},\frac{\mathrm{d}^2}{\mathrm{d}\xi^2}\Big\}\eta_s, \tag{3.33}$$

where

$$\mathcal{K}=-\frac{1}{\sqrt{\xi^2+\eta_s^2}}\Bigg[\frac{\eta_s''}{(1+\eta_s'^2)^{\frac{3}{2}}}-\frac{1}{\eta_s(1+\eta_s'^2)^{\frac{1}{2}}}$$
$$+\frac{\eta_s'(\eta_s^2+2\xi^2)-\xi\eta_s}{\xi(\xi^2+\eta_s^2)(1+\eta_s'^2)^{\frac{1}{2}}}\Bigg]. \tag{3.34}$$

In terms of the above formulas, we can write the boundary conditions as follows:

1. The upstream far-field condition:

$$T\to T_\infty=\frac{(T_\infty)_D-T_{M0}}{\Delta H/(c_p\rho)}<0 \quad \text{as } \eta\to\infty\,. \tag{3.35}$$

2. The regularity condition:

$$T_S=\mathcal{O}(1) \quad \text{as } \eta\to 0\,. \tag{3.36}$$

3. The interface conditions at $\eta = \eta_s(\xi, t)$:

   (a) the thermodynamic equilibrium condition:

$$T = T_S \, ; \tag{3.37}$$

   (b) the Gibbs–Thomson condition:

$$T_S = -\frac{\Gamma}{\eta_0^2}\mathcal{K}\left\{\frac{d}{d\xi}, \frac{d^2}{d\xi^2}\right\}\eta_s \, ; \tag{3.38}$$

   (c) the heat balance condition:

$$\left(\frac{\partial}{\partial \eta} - \eta_s' \frac{\partial}{\partial \xi}\right)(T - T_S) + \eta_0^4(\xi^2 + \eta^2)\frac{\partial \eta_s}{\partial t} + \eta_0^2(\xi \eta_s)' = 0 \, , \tag{3.39}$$

   where, as defined in Chap. 1,

$$\Gamma = \frac{\ell_c}{\ell_T} \tag{3.40}$$

   is the surface tension parameter,

$$\ell_c = \frac{\gamma c_p T_{M0}}{(\Delta H)^2} \tag{3.41}$$

   is the capillary length, and the constant $\gamma$ is the coefficient of isotropic surface tension.

For reasons that will become clear later, in performing the stability analysis, the interfacial stability parameter of isotropic surface tension $\varepsilon$ will often be used. The relationship between the two parameters $\varepsilon$ and $\Gamma$ is

$$\varepsilon = \frac{\sqrt{\Gamma}}{\eta_0^2} \, . \tag{3.42}$$

At this point, the mathematical formulation remains incomplete. A complete mathematical formulation should also include the boundary conditions that describe the behavior of the solution at the tip of the dendrite $\xi = 0$, as well as at its root, $\xi = L \gg 1$. In addition to the above boundary conditions, one may also need to impose the initial conditions.

A complete mathematical formulation of the pattern formation problem at the late stage of evolution will be given later.

## 3.3 Two-Dimensional Free Dendritic Growth

Two-dimensional free dendritic growth can be obtained in a Hele–Shaw cell. The mathematical formulations and the treatments for the two-dimensional case are very similar to those for the two-dimensional case. We adopt the parabolic cylindrical coordinate system $(\xi, \eta)$ moving with the characteristic velocity $U$ defined as

$$\begin{cases} \dfrac{x}{\eta_0^2} = \dfrac{1}{2}(\xi^2 - \eta^2) \\ \dfrac{y}{\eta_0^2} = \xi\eta \\ \dfrac{z}{\eta_0^2} = \zeta \end{cases} \tag{3.43}$$

and illustrated in Fig. 3.2.

The radius vector of every point $(x, y, z)$ can be expressed in the form

$$\mathbf{r} = \eta_0^2 \left[ \frac{1}{2}(\xi^2 - \eta^2)\mathbf{i} + \xi\eta\mathbf{j} + \zeta\mathbf{k} \right]. \tag{3.44}$$

In the moving frame, the relative velocity of the liquid phase for this case is now $\mathbf{u} = \mathbf{i}$.

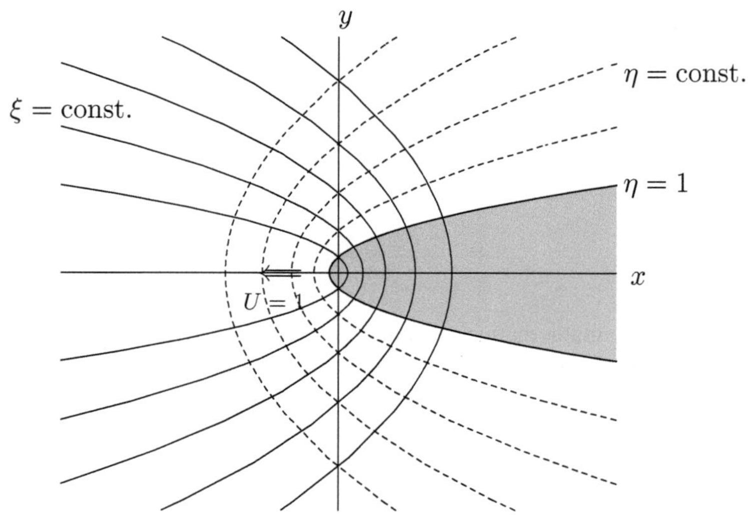


**Fig. 3.2** The parabolic cylindrical coordinate system $(\xi, \eta)$ for two-dimensional dendritic growth

Assume that the unit vectors are

$$\mathbf{e}_1 \parallel \frac{\partial \mathbf{r}}{\partial \xi}, \quad \mathbf{e}_2 \parallel \frac{\partial \mathbf{r}}{\partial \eta}, \quad \text{and} \quad \mathbf{e}_3 \parallel \frac{\partial \mathbf{r}}{\partial \zeta}. \tag{3.45}$$

Then it follows that

$$\begin{aligned} \mathbf{e}_1 &= \frac{1}{\sqrt{\xi^2+\eta^2}}(\xi \mathbf{i} + \eta \mathbf{j}), \\ \mathbf{e}_2 &= \frac{1}{\sqrt{\xi^2+\eta^2}}(-\eta \mathbf{i} + \xi \mathbf{j}), \\ \mathbf{e}_3 &= \mathbf{k}. \end{aligned} \tag{3.46}$$

For the element of arc, d$s$, we can write

$$(\mathrm{d}s)^2 = (\mathrm{d}x)^2 + (\mathrm{d}y)^2 + (\mathrm{d}z)^2 = (\alpha_1 \mathrm{d}\xi)^2 + (\alpha_2 \mathrm{d}\eta)^2 + (\alpha_3 \mathrm{d}\zeta)^2, \tag{3.47}$$

where the $\alpha_i$ are Lamé constants. From (3.43), one has

$$\begin{cases} \mathrm{d}x = \eta_0^2(\xi \mathrm{d}\xi - \eta \mathrm{d}\eta), \\ \mathrm{d}y = \eta_0^2(\eta \mathrm{d}\xi + \xi \mathrm{d}\eta), \\ \mathrm{d}z = \eta_0^2 \mathrm{d}\zeta. \end{cases} \tag{3.48}$$

We then derive that the Lamé constants are

$$\alpha_1 = \alpha_2 = \eta_0^2\sqrt{\xi^2+\eta^2}, \quad \alpha_3 = \eta_0^2. \tag{3.49}$$

Moreover, in the parabolic cylindrical system, we have

$$\begin{aligned} \nabla T &= \frac{\mathbf{e}_1}{\alpha_1}\frac{\partial T}{\partial \xi} + \frac{\mathbf{e}_2}{\alpha_2}\frac{\partial T}{\partial \eta} + \frac{\mathbf{e}_3}{\alpha_3}\frac{\partial T}{\partial \zeta} \\ \nabla \cdot \mathbf{U} &= \frac{1}{\alpha_1\alpha_2\alpha_3}\left[\frac{\partial}{\partial \xi}\left(\alpha_2\alpha_3 U_1\right) + \frac{\partial}{\partial \eta}\left(\alpha_1\alpha_3 U_2\right) + \frac{\partial}{\partial \zeta}\left(\alpha_1\alpha_2 U_3\right)\right], \end{aligned} \tag{3.50}$$

so that

$$\begin{aligned} \nabla^2 T &= \nabla \cdot (\nabla T) \\ &= \frac{1}{\alpha_1\alpha_2\alpha_3}\left[\frac{\partial}{\partial \xi}\left(\frac{\alpha_2\alpha_3}{\alpha_1}\right)\frac{\partial}{\partial \xi} + \frac{\partial}{\partial \eta}\left(\frac{\alpha_1\alpha_3}{\alpha_2}\right)\frac{\partial}{\partial \eta} + \frac{\partial}{\partial \zeta}\left(\frac{\alpha_1\alpha_2}{\alpha_3}\right)\frac{\partial}{\partial \zeta}\right] T \\ &= \frac{1}{\eta_0^4(\xi^2+\eta^2)}\left(\frac{\partial^2 T}{\partial \xi^2} + \frac{\partial^2 T}{\partial \eta^2}\right) + \frac{1}{\eta_0^2}\frac{\partial^2 T}{\partial \zeta^2} \end{aligned} \tag{3.51}$$

and

$$\mathbf{i}\cdot\nabla T=\frac{1}{\eta_0^2(\xi^2+\eta^2)}\left(\xi\frac{\partial T}{\partial\xi}-\eta\frac{\partial T}{\partial\eta}\right). \tag{3.52}$$

With (3.51) and (3.52), the governing equation is written in the form

$$\left(\frac{\partial^2 T}{\partial\xi^2}+\frac{\partial^2 T}{\partial\eta^2}\right)=\eta_0^2\left(\xi\frac{\partial T}{\partial\xi}-\eta\frac{\partial T}{\partial\eta}\right)+\eta_0^4(\xi^2+\eta^2)\frac{\partial T}{\partial t}. \tag{3.53}$$

To write the boundary conditions in the parabolic cylindrical coordinate system, one needs to obtain expressions for the vector normal to the interface and the mean curvature operator. Let the two-dimensional interface shape be denoted by $\eta_s(\xi,t)$, or, $s(\xi,\eta,t)=\eta-\eta_s(\xi,\eta,t)=0$. Then

$$\mathbf{n}=\frac{\nabla s}{|\nabla s|}=\frac{1}{\sqrt{1+\eta_s'^2}}(-\eta_s'\mathbf{e}_1+\mathbf{e}_2). \tag{3.54}$$

On the other hand, for every point on the interface, we can write

$$\mathbf{r}=\eta_0^2\left[\frac{1}{2}(\xi^2-\eta_s^2)\mathbf{i}+\xi\eta_s\mathbf{j}\right]+\eta_0^2\zeta\mathbf{k}. \tag{3.55}$$

Thus, it follows that

$$\begin{aligned}
\mathbf{r}_1&=\frac{\partial\mathbf{r}}{\partial\xi}=\eta_0^2\Big[(\xi-\eta_s\eta_s')\mathbf{i}+(\eta_s+\xi\eta_s')\mathbf{j}\Big]\\
\mathbf{r}_2&=\frac{\partial\mathbf{r}}{\partial\zeta}=\eta_0^2\mathbf{k},\\
\mathbf{r}_{11}&=\frac{\partial^2\mathbf{r}}{\partial\xi^2}=\eta_0^2\Big[(1-\eta_s'^2-\eta_s\eta_s'')\mathbf{i}+(\xi\eta_s)''\mathbf{j}\Big],\\
\mathbf{r}_{12}&=\frac{\partial^2\mathbf{r}}{\partial\xi\partial\zeta}=0,\\
\mathbf{r}_{22}&=\frac{\partial^2\mathbf{r}}{\partial\zeta^2}=0.
\end{aligned} \tag{3.56}$$

Therefore, we obtain

$$\begin{cases}
E=\mathbf{r}_1\cdot\mathbf{r}_1=\eta_0^4(\xi^2+\eta_s^2)(1+\eta_s'^2),\\
F=\mathbf{r}_1\cdot\mathbf{r}_2=0,\\
G=\mathbf{r}_2\cdot\mathbf{r}_2=\eta_0^4,
\end{cases} \tag{3.57}$$

and

$$
\begin{aligned}
L = -\mathbf{n}\cdot\mathbf{r}_{11} = {} & \frac{\eta_0^2}{\sqrt{(\xi^2+\eta_s^2)(1+\eta_s'^2)}} \\
& \times\Big[-(\xi^2+\eta_s^2)\eta_s'' - \xi\eta_s' + \eta_s\eta_s'^2 - \xi\eta_s'^3 + \eta_s\Big], \\
M = \mathbf{n}\cdot\mathbf{r}_{12} = {} & 0, \\
N = \mathbf{n}\cdot\mathbf{r}_{22} = {} & 0\,.
\end{aligned}
\tag{3.58}
$$

We now apply some differential geometry to derive the mean curvature operator as

$$
\begin{aligned}
\mathcal{K}_c &= \frac{L}{E} = \frac{1}{\eta_0^2}\mathcal{K}\left\{\frac{d}{d\xi},\frac{d^2}{d\xi^2}\right\}\eta_s, \\
\mathcal{K}_2 &= -\frac{1}{\sqrt{\xi^2+\eta_s^2}}\left(\frac{\eta_s''}{(1+\eta_s'^2)^{\frac{3}{2}}} + \frac{\xi\eta_s'-\eta_s}{(\xi^2+\eta_s^2)(1+\eta_s'^2)^{\frac{1}{2}}}\right).
\end{aligned}
\tag{3.59}
$$

In terms of the above formulas, the boundary conditions for the problem are expressed as follows:

1. The upstream far-field condition:

$$
T \to T_\infty = \frac{(T_\infty)_D - T_{M0}}{\Delta H/(c_p\rho)} < 0 \quad \text{as } \eta \to \infty\,. \tag{3.60}
$$

2. The regularity condition:

$$
T_S = \mathcal{O}(1) \quad \text{as } \eta \to 0\,. \tag{3.61}
$$

3. The interface conditions, at $\eta = \eta_s(\xi, t)$:
   (a) the thermodynamic equilibrium condition:

$$
T = T_S\,, \tag{3.62}
$$

   (b) the Gibbs–Thomson condition:

$$
T = -\frac{\Gamma}{\eta_0^2}\mathcal{K}_2\left\{\frac{d}{d\xi},\frac{d^2}{d\xi^2}\right\}\eta_s\,, \tag{3.63}
$$

(c) the heat balance condition:

$$\left(\frac{\partial}{\partial \eta} - \eta_s' \frac{\partial}{\partial \xi}\right)(T - T_S) + \eta_0^4(\xi^2 + \eta^2)\frac{\partial \eta_s}{\partial t} + \eta_0^2(\xi \eta_s)' = 0\,. \tag{3.64}$$

The tip smoothness conditions at $\xi = 0$ and the root conditions at $\xi = L \gg 1$ will be specified later.

## Reference

1. M.M. Lipshutz, *Schaum's Outline Series: Theory and Problems of Differential Geometry* (McGraw-Hill, New York, 1969)

# Chapter 4
# Basic Steady State of Axisymmetric Dendritic Growth and Its Regular Perturbation Expansion

## 4.1 The Ivantsov Solution and Unsolved Fundamental Problems

For the case of zero surface tension ($\varepsilon = 0$) and arbitrary undercooling $|T_\infty| < 1$, the three-dimensional system (3.13), (3.27)–(3.31) allows the following steady similarity solution for needle crystal growth:

$$
\begin{aligned}
T &= T_*(\eta) = T_\infty + \frac{\eta_0^2}{2} e^{\frac{\eta_0^2}{2}} E_1\left(\frac{\eta_0^2 \eta^2}{2}\right), \\
T_S &= T_{S*} = T_*(1) = 0, \\
\eta_* &= 1, \\
T_\infty &= -\frac{\eta_0^2}{2} e^{\frac{\eta_0^2}{2}} E_1\left(\frac{\eta_0^2}{2}\right) \\
& \qquad (0 \leq \xi < \infty),
\end{aligned}
\tag{4.1}
$$

where $E_1(x)$ is the exponential function defined as

$$
E_1(x) = \int_x^\infty \frac{e^{-t}}{t} dt \tag{4.2}
$$

(see [1]). This solution was first found by Ivantsov in 1946 (cf. [9] and [8]) and is now called the *Ivantsov needle solution*.

In the above, the constant $\eta_0^2$ is uniquely determined as a function of the undercooling $T_\infty$. The radius of curvature of the parabolic interface $\eta_* = 1$ at the tip $\xi = 0$ is calculated as

$$
\ell_t = \eta_0^2 \ell_T. \tag{4.3}
$$



J.-J. Xu, *Interfacial Wave Theory of Pattern Formation in Solidification*, Springer Series in Synergetics, DOI 10.1007/978-3-319-52663-8_4

Some investigators define the Péclet number as the ratio of the tip radius to the thermal diffusion length, i.e., $\mathrm{Pe} = \ell_t/\ell_T$. Obviously, the Péclet number is generally a function of $\varepsilon$ as well as $T_\infty$, i.e., $\mathrm{Pe} = \mathrm{Pe}(\varepsilon)$. Equation (4.3) shows that $\eta_0^2$ for the Ivantsov problem is actually the Péclet number with zero surface tension, i.e., $\eta_0^2 = \mathrm{Pe}_0 = \mathrm{Pe}(0)$. Returning to the dimensional tip radius $\ell_t$, we see that formula (4.3) shows that

$$\ell_t U = (\mathrm{Pe}_0)\kappa_T. \tag{4.4}$$

The Ivantsov solution has the following properties:

1. Being a smooth needle crystal solution, its interface has no microstructure.
2. Being a similarity solution, it represents a continuous family of dimensional physical solutions for needle growth, with arbitrary tip velocity for given growth conditions and material properties.

In comparing the Ivantsov solution with the original dendritic growth phenomenon, it is immediately seen that the Ivantsov solution cannot be applied to the dendrite stem region, since it gives no information about microstructure formation at the interface. The Ivantsov solution also cannot, as was hoped, be locally applied to the dendrite tip region to obtain sufficient information on the growth of the dendrite tip, since it cannot fully determine the tip velocity.

However, when the tip velocity is determined correctly through experiments, the Ivantsov solution does describe the shape of the dendrite tip with high accuracy. Therefore, the Ivantsov solution has turned out to be one of the most significant results on this subject. It has provided an important backdrop to all further related research.

Following Ivantsov, many researchers studied dendritic growth phenomena. The most important issue they faced was whether in practice, the tip velocity of a dendrite at the later stages of growth was actually uniquely determined by the growth condition and material properties. If the answer is yes, the following selection problem arises: how is the tip velocity actually selected? The answer to this question is not trivial. Many earlier researchers in this field presumed, without theoretical justification or accurate experimental evidence, that at the later stages of dendritic growth, the tip velocity would be uniquely determined. For instance, in the 1960s and 1970s, Temkin [10], Bolling and Tiller [2], Trivedi [11], Sekerka, and others all recognized that in realistic dendritic growth, the surface tension would give a sharp upper bound to the dendrite tip velocity. They made great efforts to determine this upper bound, using one or another ad hoc assumption. Their conjecture cannot be considered as having identified the selection problem of the tip velocity—one of the most significant scientific issues in pattern formation.

The selection problem was logically proposed based on two independent experimental results by Schaefer and coworkers in 1975 [3] and by Glicksman and coworkers in 1976 [4]. Using the transparent organic material SCN, these authors first performed a series of careful experiments on dendritic growth and accurately measured the tip velocities under various conditions. On the basis of their experimental data, it was confirmed that at the later stages of dendritic growth, the dendrite's tip velocity is a uniquely determined function of the growth conditions and the properties of the material.

This selection problem has been at the center of broad theoretical and experimental research activities in the field during the past decades. Since then, research on dendritic growth has focused on the following basic problems:

1. What is the mechanism that determines the tip growth velocity?
2. What are the origin and essence of the microstructure, specifically the highly visible lateral foliation?

These problems are of long standing in the fields of condensed matter physics and materials science. It was recognized early on that the key to resolving these problems is understanding the role of surface tension at the interface. The surface tension at the interface between the solid and liquid phases is usually a very small quantity. However, it is precisely this extremely small quantity that plays a vital role in interfacial pattern formation phenomena. It has been recognized that the above two problems are related to each other. To resolve them, one needs to study the steady state of the system with nonzero surface tension and its linear stability ([5–7, 12–18]).

## 4.2 Axially Symmetric Steady Needle Growth with Nonzero Surface Tension

This chapter will discuss the effect of isotropic surface tension on steady needle crystal growth. We shall first study the three-dimensional axially symmetric case.

### *4.2.1 Mathematical Formulation*

For steady three-dimensional axially symmetric needle crystal growth, we have the equation

$$\frac{\partial^2 T}{\partial \xi^2} + \frac{\partial^2 T}{\partial \eta^2} + \frac{1}{\xi}\frac{\partial T}{\partial \xi} + \frac{1}{\eta}\frac{\partial T}{\partial \eta} = \eta_0^2 \left( \xi \frac{\partial T}{\partial \xi} - \eta \frac{\partial T}{\partial \eta} \right) \tag{4.5}$$

with the following boundary conditions:

1. The upstream condition:

$$T \to T_\infty \quad \text{(exponentially)}, \qquad \text{as } \eta \to \infty . \tag{4.6}$$

2. The regularity condition:

$$T_S \text{ is regular}, \quad \text{as } \eta \to 0. \tag{4.7}$$

3. At the interface, $\eta = \eta_s(\xi)$, we must satisfy
   (i) the thermodynamic equilibrium condition

$$T = T_S\,, \tag{4.8}$$

   (ii) the Gibbs–Thomson condition

$$T_S = -\varepsilon^2\eta_0^2\mathcal{K}\{\eta_s(\xi)\}\,, \tag{4.9}$$

   (iii) and the heat-balance condition

$$\frac{\partial}{\partial\eta}(T - T_S) - \eta_s'\frac{\partial}{\partial\xi}(T - T_S) + \eta_0^2(\xi\eta_s)' = 0\,. \tag{4.10}$$

   This problem is not yet well posed. In order to specify a certain type of solution, one also needs to introduce additional conditions at the tip as well as in the root field. Different tip conditions and root conditions will give rise to different types of solutions. Based on physical reasoning, we can first specify the following tip condition:
4. The smooth tip conditions: at $\xi = 0$,

$$\eta_s'(0) = 0\,, \tag{4.11}$$

$$\eta_s(0) = 1\,. \tag{4.12}$$

   The condition (4.11) describes the smoothness of the tip, while the condition (4.12) specifies the location of the tip.
5. The root conditions: how to specify the far-field conditions is a subtle issue that must be handled carefully. A definite form of the root condition will be specified in the next chapter.

### 4.2.2 Regular Perturbation Expansion Solutions (RPE) as $\varepsilon \to 0$

The interfacial stability parameter $\varepsilon$ is in practice very small. Its numerical magnitude is generally between 0.1 and 0.2. Thus, in order to examine the steady perturbation induced by the surface tension, it is very natural to consider a regular perturbation expansion (RPE) around the Ivantsov solution in the limit $\varepsilon \to 0$.

Assume that

$$
\begin{aligned}
T &= T(\xi,\eta) = T_0(\eta) + \varepsilon^2\eta_0^2 T_1(\xi,\eta) + \ldots, \\
T_S &= T_S(\xi,\eta) = T_{S0} + \varepsilon^2\eta_0^2 T_{S1}(\xi,\eta) + \ldots, \\
\eta_B(\xi) &= 1 + \varepsilon^2\eta_1(\xi) + \ldots .
\end{aligned}
\tag{4.13}
$$

By substituting (4.13) into the system (4.5)–(4.12) and equating the coefficients of like powers of $\varepsilon$, one can derive the approximations at each order of $\varepsilon$.

#### 4.2.2.1 The Zeroth-Order Approximation $\mathcal{O}(\varepsilon^0)$

The zeroth-order approximation solution is the Ivantsov solution.

#### 4.2.2.2 The First-Order Approximation $\mathcal{O}(\varepsilon^2)$

In the first-order approximation, we derive

$$
L\{T_1\} = \left\{ \frac{\partial^2}{\partial\xi^2} + \frac{\partial^2}{\partial\eta^2} + \left(\frac{1}{\xi} - \eta_0^2\xi\right)\frac{\partial}{\partial\xi} + \left(\frac{1}{\eta} + \eta_0^2\eta\right)\frac{\partial}{\partial\eta} \right\} T_1 = 0, \tag{4.14}
$$

with the following boundary conditions:

1. As $\eta \to \infty$,

$$
T_1 \to 0 \ \text{(exponentially)}. \tag{4.15}
$$

2. Far-field condition in the upstream: we do not impose any specific condition, but we assume that the solution cannot grow faster than algebraically. Thus, as $\xi \to \infty$, we require that

$$
T_1 \quad \text{(increase at most algebraically)}. \tag{4.16}
$$

3. As $\eta \to 0$,

$$
T_{S1} \quad \text{regular}. \tag{4.17}
$$

4. At $\eta = 1$,

$$
T_1 = T_{S1} + \eta_1\,, \tag{4.18}
$$

$$
T_{S1} = -\mathcal{K}_0(\xi) = -\frac{\xi^2+2}{(1+\xi^2)^{\frac{3}{2}}}\,, \tag{4.19}
$$

$$\frac{\partial}{\partial \eta}\big(T_1 - T_{S1}\big) + \big(2 + \eta_0^2\big)\eta_1 + \xi \frac{d\eta_1}{d\xi} = 0\,. \tag{4.20}$$

5. The tip-regularity condition at $\xi = 0$:

$$\eta_1'(0) = 0 \tag{4.21}$$

and

$$\eta_1(0) = 0. \tag{4.22}$$

The solution for (4.14)–(4.22) can be derived using separation of variables. Let

$$T_1(\xi, \eta) = X(\xi)Y(\eta). \tag{4.23}$$

From (4.14), it follows that

$$X'' + \Big(\frac{1}{\xi} - \eta_0^2\xi\Big)X' + \eta_0^2{\lambda_1}^2 X = 0\,, \tag{4.24}$$

$$Y'' + \Big(\frac{1}{\eta} + \eta_0^2\eta\Big)Y' - \eta_0^2{\lambda_1}^2 Y = 0\,. \tag{4.25}$$

Letting

$$\sigma = \frac{\eta_0^2\xi^2}{2}, \quad X = X(\sigma), \tag{4.26}$$

Eq. (4.24) is transformed into the Kummer equation [1]

$$\sigma X''(\sigma) + (1 - \sigma)X'(\sigma) + \frac{\lambda_1^2}{2}X = 0\,, \tag{4.27}$$

whose fundamental solutions are

$$X(\sigma) = \begin{cases} M\Big(-\frac{\lambda_1^2}{2}, 1, \sigma\Big) & \text{(regular at } \quad \xi = \sigma = 0)\,; \\ U\Big(-\frac{\lambda_1^2}{2}, 1, \sigma\Big) & \text{(with a logarithmic singularity} \\ & \quad \text{at } \xi = \sigma = 0)\,. \end{cases} \tag{4.28}$$

Here $M$ and $U$ are the confluent hypergeometric functions (see [1]). We choose

$$X(\sigma) = M\Big(-\frac{\lambda_1^2}{2}, 1, \sigma\Big). \tag{4.29}$$

In the far field, as $\xi \to \infty$ it is assumed that the solution cannot be allowed to increase faster than algebraically. However, as $\sigma \to \infty$, $M\left(-\frac{\lambda_1^2}{2}, 1, \sigma\right)$ grows algebraically only for

$$\frac{\lambda_1^2}{2} = n = 0, 1, 2, \ldots . \tag{4.30}$$

For other values of $\frac{\lambda_1^2}{2}$, $M\left(-\frac{\lambda_1^2}{2}, 1, \sigma\right)$ grows exponentially as $\sigma \to \infty$. From the boundary condition (4.16), one has to set $\lambda_1^2/2 = n$, so that

$$X(\sigma) = M(-n, 1, \sigma) = L_n\left(\frac{\eta_0^2\xi^2}{2}\right), \tag{4.31}$$

where $L_n$ is the Laguerre polynomial. We also need to solve (4.25). In the solid phase region, we let

$$\hat{\tau} = -\frac{\eta_0^2\eta^2}{2}, \tag{4.32}$$

so that (4.25) is transformed into

$$\hat{\tau}Y''(\hat{\tau}) + (1-\hat{\tau})Y'(\hat{\tau}) + nY = 0. \tag{4.33}$$

The solution that is regular at $\hat{\tau} = \eta = 0$ is

$$Y(\hat{\tau}) = L_n\left(-\frac{\eta_0^2\eta^2}{2}\right). \tag{4.34}$$

Therefore, the general solution for the temperature in the solid phase is

$$T_{S1}(\xi, \eta) = \sum_{n=0}^{\infty} \frac{\alpha_n}{L_n\left(-\frac{\eta_0^2}{2}\right)} L_n\left(\frac{\eta_0^2\xi^2}{2}\right) L_n\left(-\frac{\eta_0^2\eta^2}{2}\right). \tag{4.35}$$

In the liquid phase region, we introduce

$$\tau = \frac{\eta_0^2\eta^2}{2}, \quad Y(\eta) = \frac{e^{-\frac{\tau}{2}}}{\tau^{\frac{1}{2}}} Z(\tau). \tag{4.36}$$

Equation (4.25) can be transformed into the Whittaker equation [1]

$$Z''(\tau) + \left(-\frac{1}{4} + \frac{\lambda}{\tau} + \frac{\frac{1}{4} - \mu^2}{\tau^2}\right) Z(\tau) = 0, \tag{4.37}$$

where

$$\lambda = -\left(n + \frac{1}{2}\right), \quad \mu = 0. \tag{4.38}$$

The fundamental solutions of (4.37) are known:

$$W_{\lambda,\mu}(\tau) = \begin{cases} e^{-\frac{\tau}{2}}\tau^{\mu+\frac{1}{2}}M\left(\mu - \lambda + \frac{1}{2}, \quad 2\mu + 1, \quad \tau\right), \\ e^{-\frac{\tau}{2}}\tau^{\mu+\frac{1}{2}}U\left(\mu - \lambda + \frac{1}{2}, \quad 2\mu + 1, \quad \tau\right). \end{cases} \tag{4.39}$$

Accordingly, we obtain the following two basic solutions for $Y(\eta)$:

$$Y(\eta) = \begin{cases} e^{-\frac{\eta_0^2\eta^2}{2}} M\left(n + 1, \quad 1, \quad \frac{\eta_0^2\eta^2}{2}\right) \text{ (vanishes algebraically} \\ \qquad\qquad \text{as } \quad \eta \to \infty), \\ e^{-\frac{\eta_0^2\eta^2}{2}} U\left(n + 1, \quad 1, \quad \frac{\eta_0^2\eta^2}{2}\right) \text{ (vanishes exponentially} \\ \qquad\qquad \text{as } \quad \eta \to \infty). \end{cases} \tag{4.40}$$

The boundary condition (4.15) determines

$$Y(\eta) = e^{-\frac{\eta_0^2\eta^2}{2}} U\left(n + 1, \quad 1, \quad \frac{\eta_0^2\eta^2}{2}\right). \tag{4.41}$$

Therefore, the general solution for the temperature in the liquid is

$$T_1(\xi, \eta) = \sum_{n=0}^{\infty} \beta_n L_n\left(\frac{\eta_0^2\xi^2}{2}\right) \frac{e^{-\frac{\eta_0^2\eta^2}{2}} U\left(n + 1, 1, \frac{\eta_0^2\eta^2}{2}\right)}{e^{-\frac{\eta_0^2}{2}} U\left(n + 1, 1, \frac{\eta_0^2}{2}\right)}. \tag{4.42}$$

We can expand the function $\eta_1(\xi)$ as the following Laguerre series:

$$\eta_1(\xi) = \sum_{n=0}^{\infty} \gamma_n L_n\left(\frac{\eta_0^2\xi^2}{2}\right). \tag{4.43}$$

Thus we must determine the coefficients $\{\alpha_n, \beta_n, \gamma_n \ (n = 0, 1, 2 \ldots)\}$ such that the boundary conditions (4.18)–(4.20) are satisfied. In what follows, we shall give the analytical forms for these coefficients.

(1) It follows from the boundary condition (4.18) that

$$\beta_n = \alpha_n + \gamma_n. \tag{4.44}$$

(2) The boundary condition (4.19) gives

$$\sum_{n=0}^{\infty} \alpha_n L_n\left(\frac{\eta_0^2 \xi^2}{2}\right) = -\mathcal{K}_0(\xi) = -\left[\frac{1}{(1+\xi^2)^{\frac{1}{2}}} + \frac{1}{(1+\xi^2)^{\frac{3}{2}}}\right]. \tag{4.45}$$

Due to the orthogonality of the functions $L_n(x)$ $(n = 0, 1, 3, \ldots)$, the coefficients $\alpha_n$ in the above expansion can be determined from the following integral:

$$\alpha_n = -\int_0^{\infty} \mathrm{e}^{-x} L_n(x) \mathcal{K}_0\left(\sqrt{2x}/\eta_0\right) dx. \tag{4.46}$$

By applying the formulas [1]

$$\mathrm{e}^{-x} L_n(x) = \frac{2}{n!} \int_0^{\infty} \mathrm{e}^{-z^2} z^{2n+1} J_0(2z\sqrt{x}) dz,$$

$$\tag{4.47}$$

$$\int_0^{\infty} \frac{J_0(2z\sqrt{x})}{(x+\frac{\eta_0^2}{2})^{\frac{1}{2}}} dx = 2\int_0^{\infty} \frac{J_0(2zy)}{(y^2+\frac{\eta_0^2}{2})^{\frac{1}{2}}} y dy = \frac{\mathrm{e}^{-\sqrt{2}\eta_0 z}}{z},$$

$$\int_0^{\infty} \frac{J_0(2z\sqrt{x})}{(x+\frac{\eta_0^2}{2})^{\frac{3}{2}}} dx = 2\int_0^{\infty} \frac{J_0(2zy)}{(y^2+\frac{\eta_0^2}{2})^{\frac{3}{2}}} y dx = \frac{2\sqrt{2}}{\eta_0} \mathrm{e}^{-\sqrt{2}\eta_0 z},$$

we find that

$$\alpha_n = -\frac{\mathrm{e}^{\frac{\eta_0^2}{2}}}{n!}\left[\frac{\sqrt{2}}{\eta_0}\int_{\frac{\eta_0}{\sqrt{2}}}^{\infty} \mathrm{e}^{-t^2}\left(t-\frac{\eta_0}{\sqrt{2}}\right)^{2n} dt + 2\int_{\frac{\eta_0}{\sqrt{2}}}^{\infty} \mathrm{e}^{-t^2}\left(t-\frac{\eta_0}{\sqrt{2}}\right)^{2n+1} dt\right]$$

$$= -\frac{\sqrt{\pi}}{2}\frac{(2n)!}{n!}\eta_0^2 \mathrm{e}^{\frac{\eta_0^2}{2}}$$

$$\times\left[\frac{\sqrt{2}}{\eta_0} i^{2n} \operatorname{erfc}\left(\frac{\eta_0}{\sqrt{2}}\right) + 2(2n+1)\, i^{2n+1} \operatorname{erfc}\left(\frac{\eta_0}{\sqrt{2}}\right)\right], \tag{4.48}$$

where

$$i^n \operatorname{erfc}(a) = \frac{2}{\sqrt{\pi}} \int_a^\infty \frac{(t-a)^n}{n!} e^{-t^2} dt . \tag{4.49}$$

Therefore, given $\eta_0^2$, the numerical values of $\alpha_n$ can then be calculated in terms of the recurrence formulas [1]

$$\begin{aligned} i^{-1} \operatorname{erfc}(a) &= \frac{2}{\sqrt{2}} e^{-a^2}, \\ i^0 \operatorname{erfc}(a) &= \operatorname{erfc}(a) \\ i^n \operatorname{erfc}(a) &= -\frac{a}{n} i^{n-1} \operatorname{erfc}(a) + \frac{1}{2n} i^{n-2} \operatorname{erfc}(a) \\ & (n = 1, 2, 3, \ldots) . \end{aligned} \tag{4.50}$$

(3) We now apply the boundary condition (4.20). From (4.35) and (4.42), one can write

$$\frac{\partial T_{S1}}{\partial \eta}(\xi, 1) = \sum_{n=0}^{\infty} b_n L_n\left(\frac{\eta_0^2 \xi^2}{2}\right) \tag{4.51}$$

and

$$\frac{\partial T_1}{\partial \eta}(\xi, 1) = \sum_{n=0}^{\infty} a_n L_n\left(\frac{\eta_0^2 \xi^2}{2}\right), \tag{4.52}$$

where

$$a_n = 2\beta_n A_n , \qquad b_n = 2n\alpha_n B_n , \tag{4.53}$$

$$\begin{cases} A_n = (n+1)^2 \dfrac{U\left(n+2, 1, \frac{\eta_0^2}{2}\right)}{U\left(n+1, 1, \frac{\eta_0^2}{2}\right)} - (n+1) - \dfrac{\eta_0^2}{2} \\ B_n = 1 - \dfrac{L_{n-1}\left(\frac{-\eta_0^2}{2}\right)}{L_n\left(\frac{-\eta_0^2}{2}\right)} . \end{cases} \tag{4.54}$$

In the above, we have used the following formulas:

$$\begin{aligned} xU'(n+1, 1, x) &= (n+1)^2 U(n+2, 1, x) - (n+1) U(n+1, 1, x), \\ xL_n'(x) &= n\left\{L_n(x) - L_{n-1}(x)\right\} . \end{aligned} \tag{4.55}$$

Thus, from the boundary condition (4.20), we have

$$\sum_{n=0}^{\infty}(a_n-b_n)L_n\Big(\frac{\eta_0^2\xi^2}{2}\Big)+(2+\eta_0^2)\sum_{n=0}^{\infty}\gamma_nL_n\Big(\frac{\eta_0^2\xi^2}{2}\Big)$$

$$+2\sum_{n=0}^{\infty}n\gamma_n\Big\{L_n\Big(\frac{\eta_0^2\xi^2}{2}\Big)-L_{n-1}\Big(\frac{\eta_0^2\xi^2}{2}\Big)\Big\}=0 \tag{4.56}$$

and

$$(a_n-b_n)+(2+\eta_0^2+2n)\gamma_n-2(n+1)\gamma_{n+1}=0, \tag{4.57}$$

$$(n=0,1,2,3,\ldots).$$

From (4.53) and (4.58), we find that

$$\gamma_{n+1}=g_n\gamma_n+f_n\alpha_n \quad (n=0,1,2\ldots), \tag{4.58}$$

where

$$g_n=(n+1)\frac{U\big(n+2,1,\frac{\eta_0^2}{2}\big)}{U\big(n+1,1,\frac{\eta_0^2}{2}\big)}, \tag{4.59}$$

$$f_n=\left\{(n+1)\frac{U\big(n+2,1,\frac{\eta_0^2}{2}\big)}{U\big(n+1,1,\frac{\eta_0^2}{2}\big)}-1-\frac{\eta_0^2}{2(n+1)}\right\}-\frac{nB_n}{(n+1)\eta_0^2}$$

$$=\left\{\frac{1-\frac{\eta_0^2}{2}}{n+1}-2+\frac{n}{n+1}\frac{L_{n-1}\big(-\frac{\eta_0^2}{2}\big)}{L_n\big(-\frac{\eta_0^2}{2}\big)}+g_n\right\}. \tag{4.60}$$

For given $\gamma_0$, the formula (4.58) allows us to generate the series

$$\{\gamma_0,\gamma_1,\gamma_2,\ldots\gamma_n\ldots\}.$$

Thus, the function $\eta_1(\gamma_0,\xi)$ can be evaluated as

$$\eta_1(\gamma_0,\xi)=\sum_{n=0}^{\infty}\gamma_nL_n\Big(\frac{\xi^2}{2}\Big). \tag{4.61}$$

The tip-regularity condition (4.21) is automatically satisfied. The value of $\gamma_0$ is then uniquely determined by the tip condition (4.22):

$$\eta_1(\gamma_0, 0) = \sum_{n=0}^{\infty} \gamma_n = 0. \tag{4.62}$$

In the above, through (4.44), (4.45), and (4.58)–(4.62), we have obtained an analytical solution to the problem.

### 4.2.3 The Asymptotic Behavior of the Regular Perturbation Expansion Solution as $\xi \to \infty$

The Laguerre polynomial expansion for $\eta_1(\xi)$ obtained in the last section, (4.61) is valid for $0 \leq \xi < \infty$. However, numerically, it can be used only in the region near the tip ($\xi \leq \xi_{\max}$). Its partial summation with any large number of terms, $N$, begins to oscillate rapidly when $\xi \gg 1$, since the sum becomes dominated by the last term $\gamma_N L_N(\xi^2/2)$ (see Fig. 4.1). To describe the asymptotic behavior of the solution in the limit $\xi \to \infty$, one needs to find a different asymptotic form.

As we have deduced, the solution is an analytic function at $\xi = \infty$. Hence, it can be expanded as a Taylor series

$$\begin{aligned}
T_1(\xi, \eta) &= \frac{A_1(\eta)}{\xi} + \frac{A_2(\eta)}{\xi^2} + \cdots, \\
T_{S1}(\xi, \eta) &= \frac{A_{S1}(\eta)}{\xi} + \frac{A_{S2}(\eta)}{\xi^2} + \cdots \\
\eta_1(\xi) &= \frac{C_1}{\xi} + \frac{C_2}{\xi^2} + \cdots .
\end{aligned} \tag{4.63}$$

The first terms in the Taylor series are subject to the system

$$\left\{ \frac{d^2}{d\eta^2} + \left( \frac{1}{\eta} + \eta_0^2 \eta \right) \frac{d}{d\eta} + \eta_0^2 \right\} \begin{pmatrix} A_1 \\ A_{S1} \end{pmatrix} = 0 \tag{4.64}$$

with the following boundary conditions at $\eta = 1$:

$$A_1(1) = A_{S1}(1) + C_1 , \tag{4.65}$$

$$A_{S1}(1) = -1 , \tag{4.66}$$

$$\frac{d}{d\eta}\big(A_1 - A_{S1}\big) + \big(1 + \eta_0^2\big) C_1 = 0 . \tag{4.67}$$

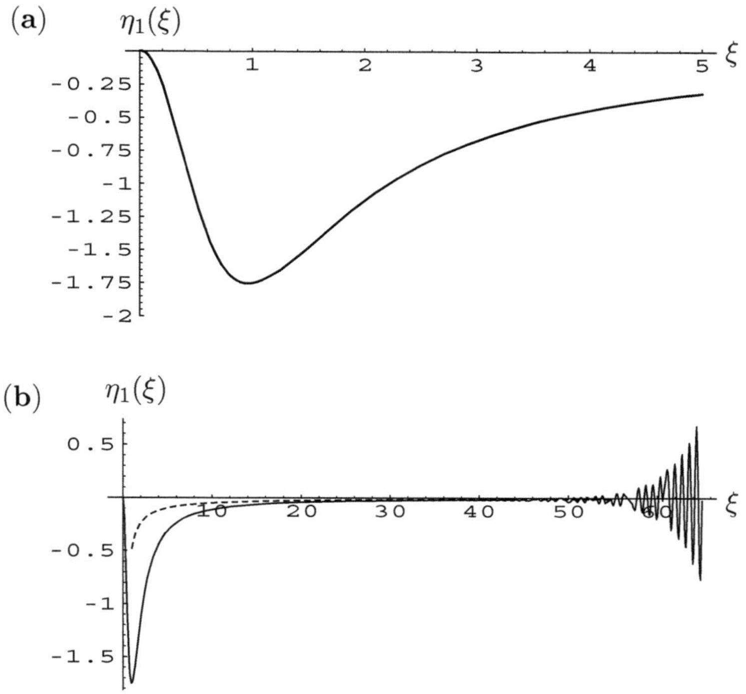


**Fig. 4.1** The solution $\eta_1(\xi)$ for $\frac{\eta_0^2}{2} = 0.005$: (**a**) in the region of $0 \leq \xi < 5$; (**b**) in the region of $0 \leq \xi < 65$. The *dashed line* represents the asymptotic solution in the far field, as $\xi \to \infty$. The *solid line* represents the Laguerre series solution. It begins to oscillate increasingly for $\xi > 50$, due to the presence of numerical errors

Equation (4.64) is of the same type as (4.25). On letting

$$\tau = \frac{\eta_0^2 \eta^2}{2} \qquad \text{and} \qquad A_1(\eta) = \frac{e^{-\frac{\tau}{2}}}{\tau^{\frac{1}{2}}} Z(\tau), \tag{4.68}$$

Eq. (4.64) is transformed to (4.37) with

$$\lambda = \mu = 0. \tag{4.69}$$

Thus, we obtain the solutions

$$A_1(\eta) = \sqrt{\pi} \hat{A}_1 e^{-\frac{\eta_0^2 \eta^2}{2}} M\Big(\frac{1}{2}, 1, \frac{\eta_0^2 \eta^2}{2}\Big) = \hat{A}_1 e^{-\frac{\eta_0^2 \eta^2}{4}} K_0\Big(\frac{\eta_0^2 \eta^2}{4}\Big), \tag{4.70}$$

$$A_{S1}(\eta) = \sqrt{\pi} \hat{A}_{S1} e^{-\frac{\eta_0^2 \eta^2}{2}} U\Big(\frac{1}{2}, 1, \frac{\eta_0^2 \eta^2}{2}\Big) = \hat{A}_{S1} e^{-\frac{\eta_0^2 \eta^2}{4}} I_0\Big(\frac{\eta_0^2 \eta^2}{4}\Big). \tag{4.71}$$

The boundary conditions (4.65)–(4.67) uniquely determine the three unknown constants $\{\hat{A}_1, \hat{A}_{S1}, C_1\}$. The results are as follows:

$$\hat{A}_1 = \frac{I_1 + (1-a_0)I_0}{K_1 + (1-a_0)K_0}\hat{A}_{S1}, \tag{4.72}$$

$$\hat{A}_{S1} = -\frac{1}{I_0}e^{\frac{\eta_0^2}{4}}, \tag{4.73}$$

$$C_1 = \frac{I_0K_1 - K_0I_1}{I_0\Big\{(1-a_0)K_0 + K_1\Big\}}, \tag{4.74}$$

where

$$K_{0,1} = K_{0,1}\left(\frac{\eta_0^2}{4}\right),$$

$$I_{0,1} = I_{0,1}\left(\frac{\eta_0^2}{4}\right), \tag{4.75}$$

$$a_0 = 2 + \frac{2}{\eta_0^2},$$

and the functions $K_\nu(x)$ and $I_\nu(x)$ are $\nu$th-order modified Bessel functions.

The above procedure can be continued to higher-order approximations with no difficulty. The asymptotic solution for $\xi \to \infty$ is shown by the dashed line in Fig. 4.1.

### 4.2.4 *Some Numerical Results of the Interface Shape*

With the above results, to carry out the numerical calculations for the solution $\eta_1(\xi, \gamma_0)$, one may divide the interval $0 \leq \xi < \infty$ into two subintervals: (*I*) : $(0 \leq \xi \leq \xi_{max}$ and (*II*) : $(\xi_{max} \leq \xi < \infty)$. In the subinterval (*I*), one may calculate the solution with the Laguerre expansion form (4.61), while in the subinterval (*II*) it can be done with the asymptotic form (4.63). By joining both solutions numerically at $\xi = \xi_{max}$, one obtains the full solution. Moreover, using the formula (4.62) and truncating the series at $n = N$, one can determine an approximate value of the unknown constant $\gamma_0$.

The solutions depend on the number $\eta_0^2$, which is related to $T_\infty$. For $\frac{\eta_0^2}{2} = 0.005$, we obtain $\gamma_0 = -0.1681$. The solution $\eta_1(\xi)$ is displayed in Fig. 4.1. Thus, for $\varepsilon > 0$, we have

$$\eta_s(\xi, \varepsilon) \approx 1 + \varepsilon^2\eta_1(\xi, \eta_0^2). \tag{4.76}$$

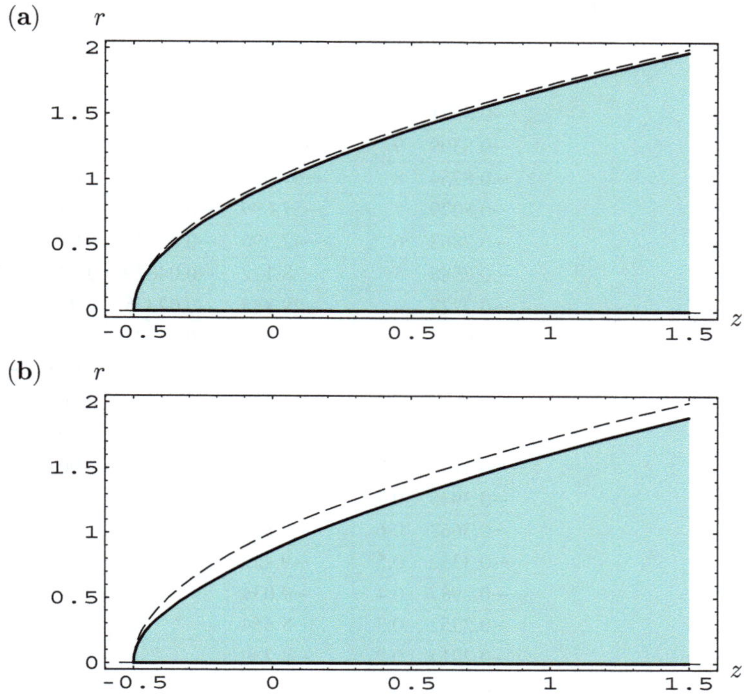


**Fig. 4.2** Dendrite shape with $\frac{\eta_0^2}{2} = 0.005$: (**a**) for $\varepsilon = 0.1$; (**b**) for $\varepsilon = 0.2$

The dendrite shapes with $\varepsilon = 0.1$ and 0.2 are shown in Fig. 4.2a, b, respectively, in which the dashed lines represent the Ivantsov paraboloid with $\varepsilon = 0$.

The mean curvature of the interface at the tip is

$$\mathcal{K}(0) \approx 2 - \varepsilon^2 \eta_1''(0). \tag{4.77}$$

On the other hand, the dimensionless tip radius $\mathrm{Pe} = \ell_t/\ell_T$ can be calculated with the formula of curvature, namely

$$\frac{\mathcal{K}(0)}{\eta_0^2} = \frac{2}{\mathrm{Pe}}. \tag{4.78}$$

Therefore, we obtain

$$\mathrm{Pe} = \frac{\ell_t}{\ell_T} = \frac{\eta_0^2}{1 - \varepsilon^2 \eta_1''(0)/2}. \tag{4.79}$$

**Table 4.1** The values of $\eta_1''(0)$

| $T_\infty$ | $\eta_0^2$ | $\eta_1''(0)/2$ | $T_\infty$ | $\eta_0^2$ | $\eta_1''(0)/2$ |
|---|---|---|---|---|---|
| −0.9806 | 100 | −5402.5 | −0.1297 | 0.09 | −7.883 |
| −0.9629 | 50 | −1452.9 | −0.1116 | 0.08 | −7.846 |
| −0.8521 | 10 | −94.732 | −0.1018 | 0.07 | −7.809 |
| −0.8399 | 9 | −81.371 | −0.0915 | 0.06 | −7.773 |
| −0.8254 | 8 | −69.025 | −0.0803 | 0.05 | −7.737 |
| −0.8079 | 7 | −57.699 | −0.06845 | 0.04 | −7.699 |
| −0.7863 | 6 | −47.396 | −0.05539 | 0.03 | −7.662 |
| −0.7588 | 5 | −38.122 | −0.04079 | 0.02 | −7.624 |
| −0.7227 | 4 | −29.885 | −0.02375 | 0.01 | −7.585 |
| −0.6724 | 3 | −22.695 | −0.02184 | 0.009 | −7.581 |
| −0.5963 | 2 | −16.569 | −0.01987 | 0.008 | −7.577 |
| −0.4614 | 1 | −11.529 | −0.01785 | 0.007 | −7.573 |
| −0.4413 | 0.9 | −11.086 | −0.01575 | 0.006 | −7.568 |
| −0.4191 | 0.8 | −10.654 | −0.01358 | 0.005 | −7.562 |
| −0.3945 | 0.7 | −10.233 | | | |
| −0.3668 | 0.6 | −9.822 | | | |
| −0.3352 | 0.5 | −9.423 | | | |
| −0.2987 | 0.4 | −9.034 | | | |
| −0.2552 | 0.3 | −8.654 | | | |
| −0.2014 | 0.2 | −8.280 | | | |
| −0.1297 | 0.1 | −7.919 | | | |

The derivatives $\eta_1''(0, \eta_0^2)$ can be numerically computed as

$$\eta_1''(0) = \lim_{\xi \to 0} \frac{2\eta_1(\xi)}{\xi^2}. \tag{4.80}$$

This solution has been evaluated over a large range of undercooling temperatures $T_\infty$, say $\frac{\eta_0^2}{2}$. It should be pointed out that as $\frac{\eta_0^2}{2}$ becomes smaller and smaller, the numerical calculations become increasingly difficult, because the convergence of the Laguerre series expansions, such as in (4.45) and (4.61), becomes increasingly slow. As a consequence, to guarantee numerical accuracy, more and more terms in the Laguerre series must be included.

For the case $\frac{\eta_0^2}{2} = 0.005$, we obtain $\eta_1''(0) = -15.838$. The variations of $\eta_1''(0)$ with the parameters $T_\infty$ are shown in Table 4.1, as well as in Fig. 4.3.

The variations of the Péclet number $\mathrm{Pe} = \ell_t/\ell_T$ with the parameters $T_\infty$ and $\varepsilon$ are shown in Fig. 4.4. This steady correction for the Ivantsov needle solution due to the isotropic surface tension is of practical interest. Such a correction has been observed experimentally.

One can continue the above procedure further to determine higher-order approximations. For the higher-order approximate system, the governing equations are the same as (4.14), so that the general solutions remain of the form (4.35), (4.42),

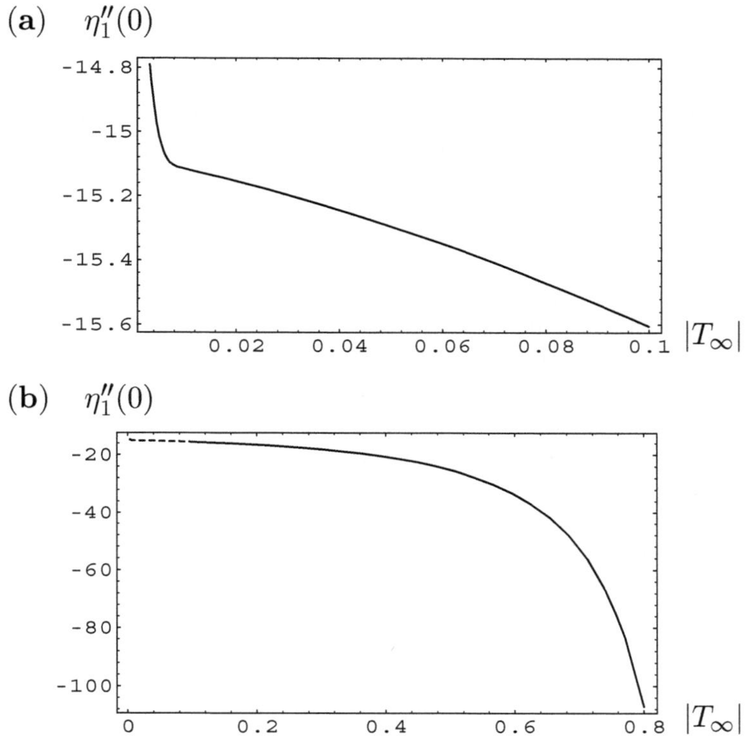


**Fig. 4.3** Variations of $\eta_1''(0)$ with the parameters $|T_\infty|$: (**a**) for the small undercooling temperature regime $0 < |T_\infty| < 0.1$; (**b**) for the large undercooling temperature regime $0.1 < |T_\infty| < 0.8$

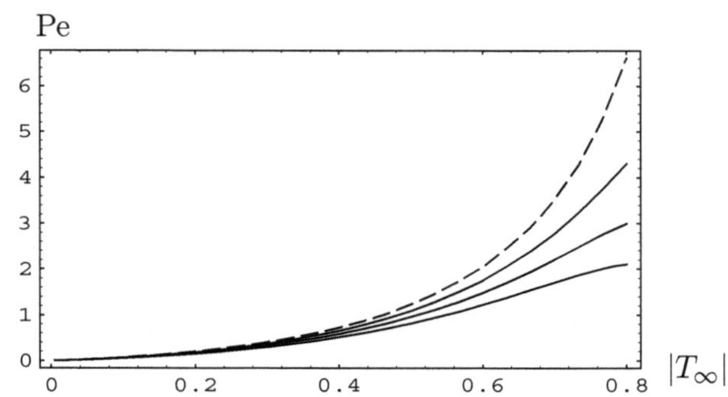


**Fig. 4.4** Variations of the Péclet number $\text{Pe} = \ell_t/\ell_T$ with the parameters $|T_\infty|$ and $\varepsilon$ for the cases $\varepsilon = 0.0, 0.1, 0.15$, and 0.2 from top to bottom. The *dashed line* is the Ivantsov solution with $\varepsilon = 0$

and (4.43). The boundary conditions on the interface yield similar formulas to (4.44), (4.45), and (4.58). Consequently, the coefficients in the general solutions can be determined in the same manner as was just carried out for the first-order approximation. The conclusions drawn from the first-order approximation, namely that the solution satisfying the smooth tip conditions can be analytically extended to the complex $\xi$-plane and that it decays algebraically as $\xi \to \infty$, will also be valid for the higher-order approximate solutions $\{\eta_2(\xi); \eta_3(\xi)\ldots\}$.

Several remarks should be made here.

1. The solutions $T_1(\xi,\eta)$, $T_{S1}(\xi,\eta)$, and $\eta_1(\xi)$ were fully determined without the root conditions. This implies that the (RPE) solution is actually representing all the steady states that may have different behaviors at the root.
2. In the downstream far field, the behavior of the RPE solution $T_1(\xi) \to 0$ as $\xi \to \infty$ is derived directly from the form of solution obtained. Moreover, one can also derive that in the higher order of approximations, the RPE solutions $T_n(\xi)$ have similar behavior. Namely, $T_n(\xi) \to 0$ $(n = 1, 2, \ldots)$ as $\xi \to \infty$, which is irrelevant to the behavior of the steady state itself in the root region.
3. The function $\mathcal{K}_0(\xi)$ is an analytic function of $\xi$ in the whole complex $\xi$-plane, except for $\xi = \pm \mathrm{i}$. Therefore, it is expected that the solutions $\{\eta_1(\xi), T_1(\xi,\eta)\}$ are also analytic functions of $\xi$ in the complex $\xi$-plane, with isolated singular points at $\xi = \pm \mathrm{i}$.

## 4.3 Summary and Discussion

In the previous sections, we obtained the regular perturbation expansion (RPE) for axisymmetric steady needle growth. Using

$$\mathcal{R}_N = 1 + \sum_{n=1}^{N} \varepsilon^{2n}\eta_n \tag{4.81}$$

to denote the summation of the first $N$ terms of the RPE, it can be concluded that the $2N$th-order asymptotic approximation solution, $\mathcal{R}_N$ $(N = 1, 2, 3, \ldots)$, has the following properties:

1. It is an analytic function of $\xi$ in the complex $\xi$-plane, with isolated singular points at $\xi = \pm \mathrm{i}$.
2. It has a regular tip satisfying the smooth tip condition.
3. It can be extended to the region $0 \le \xi < \infty$. Furthermore, as $\xi \to \infty$, it approaches the Ivantsov solution, $\eta_s = 1$.

However, the RPE obtained above is not a convergent Taylor series in $\varepsilon$. For fixed $0 < \varepsilon \ll 1$, the RPE diverges at every point $(\xi, \eta)$. The behavior of $\mathcal{R}_N$ as $\xi \to \infty$ does not give any information about the behavior of the true steady-state solution $\eta_s(\xi)$ for $\xi \gg 1$. In other words, the fact that $\eta_n(\infty) = 0$ $(n = 1, 2, 3, \ldots)$, or

$\mathcal{R}_N(\infty) = 1$ $(N = 1, 2, 3, \ldots)$ does not imply that for given $\varepsilon \neq 0$, the exact steady needle crystal growth solution itself also satisfies the limit

$$\lim_{\xi \to \infty} \eta_s(\xi, \varepsilon) = 1.$$

This situation can be easily illustrated by the following example. Suppose the exact solution is

$$\eta_s(\xi, \varepsilon) = 1 + e^{\xi - \frac{1}{\varepsilon}} + \varepsilon \frac{\xi^2 e^{-\xi}}{1 - \varepsilon e^{-\xi}}. \tag{4.82}$$

As $\varepsilon \to 0$, $\eta_s$ has the following RPE:

$$\begin{aligned} \eta_s &\sim \{1 + \varepsilon \eta_1 + \varepsilon^2 \eta_2 + \cdots\} \\ &= 1 + \varepsilon \xi^2 e^{-\xi} \{1 + \varepsilon e^{-\xi} + \varepsilon^2 e^{-2\xi} + \cdots\}. \end{aligned} \tag{4.83}$$

Thus,

$$\begin{aligned} \eta_1'(0) &= \eta_2'(0) &&= \cdots = 0, \\ \eta_1(\infty) &= \eta_2(\infty) &&= \cdots = 0. \end{aligned} \tag{4.84}$$

However, for fixed $\varepsilon > 0$,

$$\eta_s(\infty) = \infty; \quad \eta_s'(0) = e^{-\frac{1}{\varepsilon}}. \tag{4.85}$$

This example shows that as an asymptotic expansion solution, the RPE may miss some exponentially small component of the true solution as $\varepsilon \to 0$. Such an exponentially small component may be significant for describing the solution behavior with a fixed $\varepsilon > 0$ at the far field $\xi \to \infty$, as well as at the tip. This possibility relates to some very subtle mathematical issues and has caused some controversy concerning the problems of dendritic growth and other related subjects.

## References


1. M. Abramovitz, I.A. Stegun (eds.), *Handbook of Mathematical Functions* (Dover, New York, 1964)
2. G.F. Bolling, W.A.J. Tiller, Growth from the melt. III. Dendrite growth. J. Appl. Phys. **32**(12), 2587–2605 (1961)
3. M.E. Glicksman, R.J. Schaefer, J.D. Ayers, High-confidence measurement of solid/liquid surface energy in a pure material. Philos. Mag. **32**, 725–743 (1975)
4. M.E. Glicksman, R.J. Schaefer, J.D. Ayers, Dendrite growth — a test of theory. Metall. Trans. **7A**, 1747–1759 (1976)

5. M.E. Glicksman, S.P. Marsh, The dendrite, in *Handbook of Crystal Growth, Volume 1: Fundamentals, Part B: Transport and Stability*, ed. by D.T.J. Hurle, (Elsevier Science, North–Holland, Amsterdam 1993)
6. K.F. Gurski, J.B. McFadden, The effect of anisotropic surface energy on Rayleigh instability. Proc. R. Soc. Lond. A **459**, 1–24 (2003)
7. C. Herring, *Structure and Properties of Solid Surface*, ed. by R. Gomer, C.S. Smith (University Chicago, Chicago, 1952)
8. G. Horvay, J.W. Cahn, Dendritic and spheroidal growth. Acta Metall. **9**, 695–705 (1961)
9. G.P. Ivantsov, Temperature field around a spheroidal, cylindrical and acicular crystal growing in a supercooled melt. Dokl. Akad. Nauk SSSR **58**(4), 567–569 (1947)
10. D.E. Temkin, Growth rate of the needle-crystal formed in a supercooled melt. Dokl. Akad. Nauk. SSSR **132**, 1307–1310 (1960)
11. R. Trivedi, Growth of dendritic needles from a supercooled melt. Acta Metall. **18**, 287–296 (1970)
12. J.J. Xu, Global asymptotic solution for axisymmetric dendrite growth with small undercooling, in *Structure and Dynamics of Partially Solidified System*, ed. by D.E. Loper, Proc. of ARW, Fallen Leaf, Tahoe (U.S.A.), ISBN 90-247-3500-9, published by: NIJHOFF, NATO ASI Series E. vol. 125 (1987), pp. 97–109
13. J.J. Xu, Asymptotic theory of steady axisymmetric needle-like crystal growth. Stud. Appl. Math. **82**, 71–91 (1990)
14. J.J. Xu, Interfacial wave theory of two-dimensional dendritic growth with anisotropy of surface tension. Can. J. Phys. **69**(7), 789–800 (1991)
15. J.J. Xu, Generalized needle solutions, interfacial instabilities, and pattern formation. Phys. Rev. E **53**(5), 5031–5062 (1996)
16. J.J. Xu, *Interfacial Wave Theory of Pattern Formation: Selection of Dendrite Growth and Viscous Fingering in a Hele–Shaw Flow* (Springer, Berlin, 1997)
17. J.J. Xu, Stability and selection of dendritic growth with anisotropic kinetic attachment. J. Cryst. Growth **245**, 134–148 (2002)
18. J.J. Xu, D.S. Yu, Regular perturbation expansion solution for generalized needle crystal growth. J. Cryst. Growth **187**, 314–326 (1998)

# Chapter 5
# The Steady State for Dendritic Growth with Nonzero Surface Tension

In the last chapter we derived the asymptotic form of a steady solution that we called the RPE solution. However, so far we have not specified the steady-state solution itself. How to specify the steady state solution in dendritic growth is an important issue that one must clarify. In the literature, many researchers have looked for the steady classical needle solution for dendritic growth. Such efforts have not been successful for the case of isotropic surface tension.

The so-called steady classical needle solution, like Ivantsov's solution, has both a smooth tip and an infinitely long smooth, nonoscillating tail. Given the fact that the system with $\varepsilon = 0$ allows a steady classical needle solution, what about nonzero surface tension? The question whether the system with $\varepsilon \neq 0$ still allows a steady classical needle solution is not trivial. It involves the subtle mathematical issue of how to catch the exponentially small terms missed by the regular asymptotic expansion. This issue is sometimes called *asymptotics beyond all orders*. At the early stage of research on dendritic growth, most researchers thought that when a small isotropic surface tension is included, the steady classical needle solution would still persist with a small perturbation from the Ivantsov needle solution in the whole infinite region. It was with this idea that Nash and Glicksman formulated the needle crystal growth problem. It is now recognized that the above idea is incorrect.

What is wrong with this idea? What theoretical difficulty has occurred? How do we overcome this difficulty? In this chapter, we attempt to discuss these problems.

## 5.1 The Nash–Glicksman Problem and the Classical Needle Crystal Solution

The steady classical needle crystal growth problem was first formulated mathematically by Nash and Glicksman in 1974 [1]. Nash and Glicksman took the isotropic surface tension into account, neglecting the anisotropy. Furthermore, as a boundary



J.-J. Xu, *Interfacial Wave Theory of Pattern Formation in Solidification*, Springer Series in Synergetics, DOI 10.1007/978-3-319-52663-8_5

condition in the far field for fixed $\varepsilon > 0$, they assumed that the needle crystal solution must approach the Ivantsov solution as $\xi \to \infty$.

In the mathematical formulation of three-dimensional axially symmetric steady needle crystal growth proposed by Nash and Glicksman, the temperature field is described by the heat equation in infinite space,

$$\left(\frac{\partial^2 T}{\partial \xi^2} + \frac{\partial^2 T}{\partial \eta^2} + \frac{1}{\xi}\frac{\partial T}{\partial \xi} + \frac{1}{\eta}\frac{\partial T}{\partial \eta}\right) = \eta_0^2\left(\xi\frac{\partial T}{\partial \xi} - \eta\frac{\partial T}{\partial \eta}\right) \tag{5.1}$$

$$(0 \le \xi < \infty;\ 0 \le \eta < \infty).$$

The boundary conditions are as follows:

1. The upstream far-field condition

$$T \to T_\infty = \frac{(T_\infty)_D - T_{M0}}{\Delta H/(c_p \rho)} < 0 \quad \text{as } \eta \to \infty\,. \tag{5.2}$$

2. The regularity condition:

$$T_S = \mathcal{O}(1) \quad \text{as } \eta \to 0\,. \tag{5.3}$$

3. The interface conditions at $\eta = \eta_s(\xi)$:

   (a) the thermodynamic equilibrium condition

$$T = T_S\,, \tag{5.4}$$

   (b) the Gibbs–Thomson condition

$$T_S = -\varepsilon^2\eta_0^2\mathcal{K}\left\{\frac{d}{d\xi}, \frac{d^2}{d\xi^2}\right\}\eta_s\,, \tag{5.5}$$

   (c) the heat balance condition

$$\frac{\partial}{\partial \eta}(T - T_S) - \eta_s'\frac{\partial}{\partial \xi}(T - T_S) + \eta_0^2(\xi\eta_s)' = 0\,. \tag{5.6}$$

4. The tip smoothness condition: at the tip of a dendrite,

$$\frac{\partial}{\partial \xi}\{T, T_S, \eta_s\}(0) = 0\,, \quad \eta_s(0) = 1\,. \tag{5.7}$$

5. The downstream far-field conditions:

$$\{T, T_S, \eta_s\} = \{T_*, T_{S*}, \eta_*\} \quad \text{as} \quad \xi \to \infty\,. \tag{5.8}$$

Obviously, the downstream far-field condition (5.8) imposed by Nash and Glicksman is not easy to justify. In practice, the situation at the root region is rather complicated. A realistically growing dendrite is always connecting with neighboring dendrites at its root. Although an isolated dendrite at the later stages of evolution may be considered sufficiently long compared with the thermal length scale, it is never infinitely long. Therefore, the Nash–Glicksman far-field condition certainly is a great simplification of reality. Nash and Glicksman defined the solution of the above problem as the "steady state" of dendritic growth. This definition was also accepted by many other researchers. However, it was later suggested by a number of investigators, and rigorously proved by Segur and Kruskal in terms of a simplified local model equation, that such a Nash–Glicksman problem did not have a solution [2]!

The result of Segur and Kruskal is quite understandable. It implies that for the problem of steady dendritic growth with nonzero surface tension, the solution cannot always be monotonically extended to infinity, since the dynamical system may have an intrinsic mathematical singularity at infinity. It is possible that for fixed $(\xi, \eta)$, steady dendritic growth indeed approaches the Ivantsov solution as $\varepsilon \to 0$; however, for a fixed $\varepsilon > 0$, as $\xi \to \infty$ the solution for steady growth may be far from the Ivantsov solution. The solution may have a nonsmooth tail, oscillatory with a rapidly increasing amplitude.

The subsequent theoretical difficulty caused by the above definition of the steady state is evident. Because the system does not allow such a steady state, the regular perturbation expansion (RPE) solutions for the steady state of needle crystal growth derived in the last chapter are meaningless. Moreover, the stability analysis of the steady state is also not applicable.

Further investigations for decades after Segur–Kruskal have been progressing in two different directions. Some investigators continue to adopt the Nash–Glicksman formulation, looking for the steady classical needle crystal solution, but include an additional physical effect in the problem, the anisotropy of surface tension. This approach leads to the theory of the microscopic solvability condition (MSC) [2–9]. It is found that in order for the system of two-dimensional dendritic growth to have a steady classical needle solution, a small amount of anisotropy of surface tension must be taken into account.

The other direction modifies the Nash–Glicksman formulation and redefines the steady state for the system. This alternative approach appeared to be very promising and led to the interfacial wave theory [10]. It is certain that the nonexistence of a mathematical solution for steady classical needle growth does not imply that a dendritic growth system with isotropic surface tension does not permit a physically acceptable steady nonclassical needle solution.

The key is how to formulate a physical problem as a proper mathematical problem. From the physical point of view, a mathematically infinitely large system is a simplification of a sufficiently large realistic system, whereas a mathematical steady state is also a simplification of a sufficiently slowly time-evolving state. Hence, in order to guarantee the existence of a mathematical solution, one may modify the classical Nash–Glicksman problem in these two respects: consider the system to be finite and/or take a small time dependence into account. More specifically, we can formulate the following two problems:

1. Assume that the needle grows steadily with a constant, long but finite, length, $L_0 = \frac{C}{\varepsilon^\nu}$ $(C = \mathcal{O}(1), \nu > 0)$, and at the root $\xi = L_0$, as a boundary condition, the solution is assumed to be close to the Ivantsov solution, say with an error of $\mathcal{O}(\varepsilon)$. We call this problem the *steady nonclassical needle growth problem.*
2. Assume that the needle is at the later stages of growth with $t \geq t_0 \gg 1$, so that it is "nearly" steady and evolves with a slow time variable defined as $\tau = \varepsilon(t - t_0)$. In this case, the total length of the needle may be expressed in the form

$$L(\tau, \varepsilon) = \frac{1}{\varepsilon}\Big(C + \int_0^\tau \bar{U}(\tau')d\tau'\Big), \tag{5.9}$$

where $\bar{U}(\tau, \varepsilon)$ is the growth velocity of the needle's tip. At the root, which is now moving, the solution is also assumed to be close to the Ivantsov solution. For the sake of convenience, we shall call this problem the *needle crystal formation problem.*

We call the solutions for both types of problems the "generalized steady needle solution." Obviously, the Ivantsov solution is the special solution to both of the above problems as $\varepsilon \to 0$.

In the next section, we shall use a simple local model to further demonstrate these ideas.

## 5.2 The Geometric Model and Solutions of Needle Crystal Growth

### 5.2.1 Geometric Model of Dendritic Growth

In the literature, the geometric model equation

$$\varepsilon^2\theta'''(s) + \theta'(s) = \cos\theta \quad (0 \leq s < \infty), \tag{5.10}$$

with the boundary conditions

$$\theta(0) = 0, \quad \theta(\infty) = \frac{\pi}{2}, \tag{5.11}$$

has often been used for the investigation of free dendritic growth. In (5.10), the variable $s$ represents the dimensionless arc length of the interface starting from the tip, while $\theta$ is the angle between the local tangent to the interface and the vertical line. The parameter $\varepsilon$ represents the effect of surface tension.

It is evident that for the case of zero surface tension ($\varepsilon = 0$), Eq. (5.10) reduces to the first-order equation

$$\theta'(s) = \cos\theta \, . \tag{5.12}$$

This allows an analytical solution,

$$\theta = \theta_*(s) = -\pi/2 + 2\tan^{-1}(\mathrm{e}^s) \, , \tag{5.13}$$

which is symmetric, i.e.,

$$\theta(-s) = -\theta(s) \, , \tag{5.14}$$

and satisfies the boundary condition (5.11).

This solution is a needle-like solution, which, like the Ivantsov solution, has the following properties:

1. It has a smooth tip.
2. It is symmetric and has a smooth and infinitely long tail.

### 5.2.2 *The Segur–Kruskal Problem*

For the case of nonzero surface tension ($\varepsilon \neq 0$), whether the system still allows a symmetrical needle solution satisfying the boundary conditions (5.11) is not a trivial problem. As $\varepsilon \to 0$, for fixed $s$, $0 \leq s < \infty$, one can derive the following formal regular perturbation expansion (RPE) solution for (5.10):

$$\theta(s, \varepsilon) \sim \theta_*(s) + \varepsilon^2\theta_1(s) + \varepsilon^4\theta_2(s) + \cdots \, . \tag{5.15}$$

By substituting (5.15) into (5.10), one can derive

$$\begin{aligned}
\frac{\mathrm{d}\theta_*}{\mathrm{d}s} &= \cos\theta_*, \\
\frac{\mathrm{d}\theta_1}{\mathrm{d}s} &= -(\sin\theta_*)\theta_1 - \frac{\mathrm{d}^3\theta_*}{\mathrm{d}s^3}, \\
\frac{\mathrm{d}\theta_2}{\mathrm{d}s} &= -(\sin\theta_*)\theta_2 - \frac{\mathrm{d}^3\theta_1}{\mathrm{d}s^3} - \frac{\cos\theta_*}{2}\theta_1^2, \\
&\vdots
\end{aligned} \tag{5.16}$$

Each order of solution $\theta_n(s)$ can be easily solved analytically, and it is uniquely determined by the boundary condition at the tip:

$$\theta_n(0) = 0 \quad (n = 0, 1, 2, \cdots). \tag{5.17}$$

Moreover, it can be proved that all the solutions $\theta_n(s)$ $(n = 0, 1, 2, \cdots)$ vanish at the far field, as $s \to \infty$, i.e.,

$$\theta_n(\infty) = 0 \quad (n = 0, 1, 2, \cdots). \tag{5.18}$$

Thus, if one defines

$$\mathcal{R}_N = \sum_{n=1}^{n=N} \theta_n \tag{5.19}$$

as the partial summation of the formal RPE solution, the function $\theta(s, \varepsilon) = \theta_*(s) + \mathcal{R}_N(s, \varepsilon)$ must satisfy (5.10) and the boundary conditions (5.11) and (5.14) up to any order of approximation $\mathcal{O}(\varepsilon^{2N})$. However, up to this point, it is still uncertain whether this formal solution is mathematically meaningful. Evidently, if the true solution for the above problem does not exist, the above formal RPE solution will have no mathematical meaning. Around 1996, the problem of the existence of a solution to (5.10) for $(-\infty < s < \infty)$ with the conditions (5.11) and (5.14) was extensively studied by Segur and Kruskal in terms of asymptotic methods, which we call the "Segur–Kruskal problem."

Segur and Kruskal first proved that this problem has no solution for any $0 \leq \varepsilon \ll 1$ (refer to [2]). The system may allow the unique solution satisfying the boundary conditions (5.11). This solution, however, cannot satisfy the symmetrical condition (5.14), which is equivalent to the condition

$$\theta''(0) = 0. \tag{5.20}$$

In fact, Segur and Kruskal proved that

$$\theta''(0^+) = \frac{\Gamma_0}{4\varepsilon^{\frac{5}{2}}} e^{-\frac{\pi}{2\varepsilon}}, \tag{5.21}$$

where the constant $\Gamma_0$ is $\mathcal{O}(1)$. The results obtained by Kruskal and Segur have had an important impact on the applied mathematics community. They raise issues such as exponential asymptotics, or asymptotics "beyond all orders," of solutions to a nonlinear equation. Their work motivated a series of research efforts to develop new approaches, either numerical or analytical, exploring exponentially small factors for similar nonlinear problems arising in different applied areas that may be missed by the formal RPE solutions [8].

Moreover, Segur and Kruskal found that in order for the system to allow a needle solution, one may include an additional parameter in the system, so that (5.10) is modified to

$$\varepsilon^2 \theta'''(s) + \theta'(s) = \frac{\cos\theta}{1 + \alpha \cos(4\theta)} \quad (0 \leq s < \infty). \tag{5.22}$$

Here $\alpha$ can be interpreted as the surface tension anisotropy.

Segur and Kruskal's results were soon applied by some investigators to the original nonlocal dendritic growth system, and interpreted as indicating the nonexistence of a "steady state" for a system of dendritic growth with isotropic surface tension and that such a steady state is allowable only when a small amount of anisotropy of surface tension is included in the system. This is the basis of the so-called microscopic solvability condition (MSC) theory, which influenced the physics community for a long time. As pointed out in the last section, in interpreting the physical implications of Segur and Kruskal's results, one needs to recognize the connection and differences between the given physical phenomenon and the resulting simplified mathematical formulation. The nonexistence of the true solution for the Segur–Kruskal problem does not imply that the system has no physically acceptable steady-state solution. To ensure the existence of a mathematical solution, one may modify the Segur–Kruskal problem in two ways, either by assuming that the needle growth is steady and the length of needle is long but finite, of order $\mathcal{O}(\frac{1}{\varepsilon})$, or by assuming that the needle growth is "nearly" steady with an increasingly long length as time $t$ goes to infinity.

In the next sections we shall formulate these modified problems.

### 5.2.3 Steady Nonclassical Needle Growth Problem

Consider the following modified Segur–Kruskal problem:

$$\varepsilon^2\theta'''(s) + \theta'(s) = \cos\theta \quad \big(0 \le s < L_0(\varepsilon)\big)\,, \tag{5.23}$$

with the following boundary conditions:

(1) The tip conditions

$$\theta(0) = \theta''(0) = 0\,. \tag{5.24}$$

(2) The root condition

$$\theta(L_0) = \theta_*(L_0) + f(\varepsilon)\,. \tag{5.25}$$

In the above, we assume that $L_0(\varepsilon)$ represents the total length of the dendrite's stem, satisfying $\lim_{\varepsilon\to 0} L(\varepsilon) \to \infty$, $f(\varepsilon)$ is an analytic function of $\varepsilon$, and $f(0) = 0$, while $\theta_*(s)$ is the "Ivantsov solution" for the reduced problem:

$$\theta'(s) = \cos\theta \quad (0 \le s < \infty), \quad \theta(0) = 0\,. \tag{5.26}$$

For $0 < \varepsilon \ll 1$, we attempt to find a solution that is a monotonic function in the region $\big(0 \le s \le L_0(\varepsilon)\big)$. It can be proven that a solution to the problem (5.23)–(5.25) exists for some properly chosen $L_0(\varepsilon)$ and can be extended by reflection to the region $\big(0 \ge s \ge -L_0(\varepsilon)\big)$, with no singularity at $s = 0$ up to second-order derivatives. It implies that the solution has continuous curvature at $s = 0$, but *it may be not analytic at* $s = 0$. We call such a solution a "steady nonclassical needle solution," and define it as the steady state of the system. In order to prove the existence of the solution to the above two-point boundary value problem, one may use the standard shooting method. In doing so, let

$$\theta'(0, \varepsilon) = \beta(\varepsilon) . \tag{5.27}$$

Since the system has no singularity in the entire interval $(0 < s < \infty)$, a solution to the initial value problem (IVP) of the ordinary differential equation (5.23) with the initial conditions (5.24) and (5.27) is guaranteed. Thus, in order to have a solution to the modified Segur–Kruskal problem, one needs to show further that the function $\beta(\varepsilon)$ can be determined in terms of the root condition (5.25). This monograph, however, is not concerned with a rigorous proof of the existence of a solution. Rather, in what follows, we shall explore the behavior of the solutions and derive their asymptotic expansion form.

#### 5.2.3.1 The Outer Expansion

As $\varepsilon \to 0$, for fixed $s$ $(0 < s < L_0)$, one can derive the following outer expansion solution:

$$\theta(s, \varepsilon) \sim \theta_*(s) + \varepsilon^2\theta_1(s) + \varepsilon^4\theta_2(s) + \cdots . \tag{5.28}$$

This outer expansion solution is just the regular perturbation expansion (RPE) solution for (5.15). This solution satisfies the boundary conditions (5.24) at the tip $s = 0$. However, it does not satisfy the root condition (5.25) or the general initial condition (5.27). In order to find a uniformly valid asymptotic solution, normally one should use the matched asymptotic expansion method. In other words, look for the inner solution near the root region, match the inner solution and the outer solution in the intermediate region, and construct the composite solution.

#### 5.2.3.2 The Composite Solution

Let us define

$$\tilde{\theta}(s, \varepsilon) = \theta(s, \varepsilon) - \theta_*(s) . \tag{5.29}$$

The $N$th-order composite solution $\theta^{(N)}(s, \varepsilon)$ should be a *general asymptotic expansion of the solution* $\tilde{\theta}(s, \varepsilon)$ in the region $\big(0 \le s \le L_0(\varepsilon)\big)$ (refer to [11]). Namely, as $\varepsilon \to 0$,

$$\frac{|\tilde{\theta}(s, \varepsilon) - \theta^{(N)}(s, \varepsilon)|}{\varepsilon^N} \to 0 , \tag{5.30}$$

for $0 \le s = s(\varepsilon) \le L_0(\varepsilon)$. Note that $s$ is not fixed in the limit $\varepsilon \to 0$. The variable $s$, during the limit process, can vary with $\varepsilon$ arbitrarily in the region of definition of the solution: $\big(0, L_0(\varepsilon)\big)$. Evidently, the summation of the first $N$ terms of the RPE,

$$\mathcal{R}_N = \varepsilon^2\theta_1(s) + \varepsilon^4\theta_2(s) + \cdots + \varepsilon^{2N}\theta_N(s), \tag{5.31}$$

is the $2N$th-order asymptotic approximation of the solution $\tilde{\theta}(s,\varepsilon)$ in the ordinary asymptotic sense. But it is not the $N$th-order asymptotic approximation in the general asymptotic sense, since it does not satisfy the root condition. Of course, it should be noted that given a solution $\tilde{\theta}(s,\varepsilon)$, its general asymptotic expansion solution is not uniquely determined. Adding some exponentially small terms to a general asymptotic expansion solution will still result in a general asymptotic expansion solution.

To find the aforementioned composite solution for the problem under consideration, a better method is to apply the *multiple variables expansion* (MVE) technique.

In doing so, we define a stretched fast variable $s_+$ in the form

$$s_+ = \frac{1}{\varepsilon}\int_0^s k(s',\varepsilon)ds' \tag{5.32}$$

and assume that the perturbation function $\tilde{\theta}(s,\varepsilon)$ can be considered as a function of the two variables $s$ and $s_+$, and as $\varepsilon \to 0$ it can be expanded in a multiple variable expansion form. According to MVE method, the variables $(s; s_+)$ are formally treated as independent variables. Hence, the derivatives in (5.23) must be replaced by

$$\begin{aligned}
\frac{\mathrm{d}}{\mathrm{d}s} &= \frac{k}{\varepsilon}\frac{\partial}{\partial s_+} + \frac{\partial}{\partial s} \\
\frac{\mathrm{d}^2}{\mathrm{d}s^2} &= \frac{2k}{\varepsilon}\frac{\partial^2}{\partial s\partial s_+} + \frac{k'}{\varepsilon}\frac{\partial}{\partial s_+} + \frac{k^2}{\varepsilon^2}\frac{\partial^2}{\partial s_+^2} + \frac{\partial^2}{\partial s^2} \\
\frac{\mathrm{d}^3}{\mathrm{d}s^3} &= \frac{1}{\varepsilon}\left(3k'\frac{\partial^2}{\partial s\partial s_+} + 3k\frac{\partial^3}{\partial s^2\partial s_+} + k''\frac{\partial}{\partial s_+}\right) \\
&\quad + \frac{1}{\varepsilon^2}\left(3kk'\frac{\partial^2}{\partial s_+^2} + 3k^2\frac{\partial^3}{\partial s\partial s_+^2}\right) + \frac{k^3}{\varepsilon^3}\frac{\partial^3}{s_+^3} + \frac{\partial^3}{\partial s^3}.
\end{aligned} \tag{5.33}$$

Equation (5.23) is then converted into a multivariable form:

$$\begin{aligned}
&\frac{k^3}{\varepsilon}\frac{\partial^3\tilde{\theta}}{\partial s_+^3} + \frac{k}{\varepsilon}\frac{\partial\tilde{\theta}}{\partial s_+} + 3kk'\frac{\partial^2\tilde{\theta}}{\partial s_+^2} + 3k^2\frac{\partial^3\tilde{\theta}}{\partial s\partial s_+^2} + \frac{\partial\tilde{\theta}}{\partial s} \\
&+\varepsilon\left(3k'\frac{\partial^2\tilde{\theta}}{\partial s\partial s_+} + 3k\frac{\partial^3\tilde{\theta}}{\partial s^2\partial s_+} + k''\frac{\partial\tilde{\theta}}{\partial s_+}\right) + \varepsilon^2\frac{\partial^3\tilde{\theta}}{\partial s^3}
\end{aligned}$$

$$= \cos\theta_*\Big(\cos\tilde{\theta} - 1\Big) - \sin\theta_* \sin\tilde{\theta} - \varepsilon^2\theta_*'''(s)$$

$$= F(\tilde{\theta}, s)\,. \tag{5.34}$$

The boundary conditions are as follows:

(1)

$$\begin{cases} \tilde{\theta} = 0, \\ \dfrac{\mathrm{d}^2\tilde{\theta}}{\mathrm{d}s^2} = 0, \end{cases} \quad \text{as } s = 0\,. \tag{5.35}$$

(2) The additional initial condition

$$\frac{\mathrm{d}\tilde{\theta}}{\mathrm{d}s} = \beta(\varepsilon) - 1, \quad \text{as } s = 0\,, \tag{5.36}$$

or (2)′ the boundary condition

$$\tilde{\theta} = f(\varepsilon), \quad \text{at } s = L_0. \tag{5.37}$$

To find this general asymptotic approximate solution $\tilde{\theta}(s, \varepsilon)$, we assume that

$$\tilde{\theta}(s, \varepsilon) = \mathcal{R} + \mathcal{S}, \tag{5.38}$$

in which $\mathcal{R}$ is the regular perturbation expansion (RPE), and the additional part $\mathcal{S}$ is called the singular perturbation expansion (SPE). Let $\mathcal{S}$ be expanded in the following form:

$$\mathcal{S} = \alpha_0(\varepsilon)\tilde{h}_0(s, s_+) + \alpha_1(\varepsilon)\tilde{h}_1(s, s_+) + \cdots \alpha_N(\varepsilon)\varepsilon^{2N}\tilde{h}_{2N}(s, s_+)$$
$$+ \cdots \quad (\text{as } \varepsilon \to 0)\,, \tag{5.39}$$

where the asymptotic sequences $\alpha_0(\varepsilon) \ll \alpha_1(\varepsilon) \ll \cdots$ are to be determined. By substituting (5.38) and (5.39) into (5.34), one can successively derive each order of approximation. Here, we are interested in only the leading-order approximation. The results are summarized as follows:

(1) $\mathcal{O}(\varepsilon^0)$ : The equation is

$$k_0^3\frac{\partial^3\tilde{h}_0}{\partial s_+^3} + k_0\frac{\partial\tilde{h}_0}{\partial s_+} = 0\,. \tag{5.40}$$

Letting

$$\tilde{h}_0(s, s_+) = A(s)H(s_+) = A(s)e^{is_+}, \tag{5.41}$$

it follows that

$$-k_0^3 + k_0 = 0. \tag{5.42}$$

Here, one gets three roots:

$$k_0^{(1)} = 0; \ k_0^{(2)} = 1; \ k_0^{(3)} = -1. \tag{5.43}$$

Hence, the general solution of (5.40) is

$$\tilde{h}_0(s, s_+) = a_0(s)H_1(s_+) + b_0(s)H_2(s_+) + c_0(s)H_3(s_+), \tag{5.44}$$

where

$$H_i(s_+) = e^{is_+} = e^{\frac{i}{\varepsilon}\int_0^s k^{(i)} ds} \quad (i = 1, 2, 3). \tag{5.45}$$

(2) $\mathcal{O}(\varepsilon)$ : We set $\tilde{h}_0(s, s_+) = A_0(s)H(s_+)$, where $A_0(s)H(s_+)$ represents any of the three solutions $\{a_0(s)H_1(s_+); b_0(s)H_2(s_+); c_0(s)H_3(s_+)\}$. The equation for $\tilde{h}_1$ is then

$$k_0^3\frac{\partial^3 \tilde{h}_1}{\partial s_+^3} + k_0\frac{\partial \tilde{h}_1}{\partial s_+} = \Big[(3k_0^2 - 1)A_0'(s) + A_0(s)\Big(-\sin\theta_* +$$

$$+i3k_0^2 k_1 - ik_1 + 3k_0 k_0'\Big)\Big]H(s_+). \tag{5.46}$$

To eliminate the secular terms, one must set

$$(3k_0^2 - 1)A_0'(s) - A_0(s)\Big(-\sin\theta_* + i3k_0^2 k_1 - ik_1 + 3k_0 k_0'\Big) = 0. \tag{5.47}$$

Letting

$$ik_1(1 - 3k_0^2) + \sin\theta_* - 3k_0 k_0' = 0 \tag{5.48}$$

or

$$k_1 = \frac{i\sin\theta_*(s)}{(1 - 3k_0^2)}, \tag{5.49}$$

we have

$$k_1^{(1)} = \mathrm{i}\sin\theta_*(s),\ \ k_1^{(2)} = -\frac{\mathrm{i}}{2}\sin\theta_*(s),\ \ k_1^{(3)} = -\frac{\mathrm{i}}{2}\sin\theta_*(s), \tag{5.50}$$

and

$$A_0(s) = \hat{A} = \text{const.} \tag{5.51}$$

Thus, we derive three fundamental solutions:

$$\begin{aligned} H_1(s) &= \mathrm{e}^{-\int_0^s \sin\theta_*(s)ds}, \\ H_2(s) &= \mathrm{e}^{\frac{\mathrm{i}s}{\varepsilon}+\frac{1}{2}\int_0^s \sin\theta_*(s)ds}, \\ H_3(s) &= \mathrm{e}^{-\frac{\mathrm{i}s}{\varepsilon}+\frac{1}{2}\int_0^s \sin\theta_*(s)ds}. \end{aligned} \tag{5.52}$$

The general solution (5.44) becomes

$$\begin{aligned} \tilde{h}_0(s,s_+) = \hat{a}_0\mathrm{e}^{-\int_0^s \sin\theta_*(s)ds} &+ \hat{b}_0\mathrm{e}^{\frac{\mathrm{i}s}{\varepsilon}+\frac{1}{2}\int_0^s \sin\theta_*(s)ds} \\ &+\hat{c}_0\mathrm{e}^{-\frac{\mathrm{i}s}{\varepsilon}+\frac{1}{2}\int_0^s \sin\theta_*(s)ds}. \end{aligned} \tag{5.53}$$

The above procedure can be continued to higher-order approximations. The coefficients $(\hat{a}_0, \hat{b}_0, \hat{c}_0)$ and the prefactor $\alpha_0(\varepsilon)$ can be determined by the initial condition (5.35) with (5.36) or the boundary condition (5.37).

To leading order, we have

$$\begin{cases} \tilde{h}_0 = 0 \\ \dfrac{\mathrm{d}^2\tilde{h}_0}{\mathrm{d}s^2} = 0 \end{cases} \quad \text{as } s = 0\,, \tag{5.54}$$

and

$$\begin{aligned} &\alpha_0(\varepsilon)\Big\{\tilde{h}_0 + \varepsilon\tilde{h}_1\cdots\Big\} + \cdots \\ &\quad\sim f(\varepsilon) - \Big\{\varepsilon^2\theta_1(L) + \cdots\Big\} = \tilde{f}(\varepsilon) \quad \text{at } s = L_0\,. \end{aligned} \tag{5.55}$$

From (5.54), we derive that

$$\hat{a}_0 + \hat{b}_0 + \hat{c}_0 = 0 \tag{5.56}$$

and

$$-\hat{a}_0 + \frac{1}{4}\big(\hat{b}_0 + \hat{c}_0\big) = 0\,. \tag{5.57}$$

Hence, we have

$$\hat{a}_0 = 0 \quad \text{and} \quad \hat{b}_0 = -\hat{c}_0\,. \tag{5.58}$$

Thus, one can determine $\alpha_0(\varepsilon)$ through (5.55):

$$\alpha_0(\varepsilon)\tilde{h}_0\big(L(\varepsilon)\big) = \tilde{f}(\varepsilon)\,. \tag{5.59}$$

This results in

$$\alpha_0(\varepsilon)\hat{b}_0 \sin\left(\frac{L_0(\varepsilon)}{\varepsilon}\right) e^{\frac{1}{2}\int_0^{L_0(\varepsilon)} \sin\tilde{\theta}_*(s)ds} = \tilde{f}(\varepsilon). \tag{5.60}$$

Noting that

$$\int_0^s \sin\theta_*(s_1)ds_1 = \ln(e^s + e^{-s}) - \ln 2\,, \tag{5.61}$$

we obtain

$$\hat{b}_0 = \frac{1}{\sqrt{2}} \tag{5.62}$$

and

$$\alpha_0(\varepsilon) = \frac{\tilde{f}(\varepsilon)}{\cosh^{\frac{1}{2}} L_0(\varepsilon)\sin\left(\frac{L_0}{\varepsilon}\right)}\,. \tag{5.63}$$

It is seen that in the case $\sin\big(L_0/\varepsilon\big) = 0$ or $\varepsilon = L_0/(n\pi)$, $\alpha_0(\varepsilon)$ does not exist. Consequently, the system does not admit a solution to the two-point boundary value problem under consideration. However, if the root $L_0$ is assumed to be a proper function of $\varepsilon$, the existence of the solution is guaranteed. We choose $L_0(\varepsilon)$ such that $\sin\big(L_0/\varepsilon\big) = 1$. In this case,

$$L_0(\varepsilon) = \big(2M + \frac{1}{2}\big)\pi\varepsilon = \frac{C}{\varepsilon^{\nu}} \quad (\nu > 0), \tag{5.64}$$

where the integer $M$ is defined as the integer part of $\frac{1}{4\pi\varepsilon^{(1+\nu)}}$, namely

$$M = \left[\frac{1}{4\pi\varepsilon^{(1+\nu)}}\right], \tag{5.65}$$

and the constant $C$ satisfies $0 < C < 1$. We derive

$$\alpha_0(\varepsilon) = \frac{\tilde{f}(\varepsilon)}{\cosh^{\frac{1}{2}} L_0(\varepsilon)} \tag{5.66}$$

and finally obtain

$$\theta(s,\varepsilon) = \theta_*(s) + \varepsilon^2\theta_1(s) + \varepsilon^4\theta_2(s) + \cdots + \varepsilon^{2N}\theta_N(s)$$
$$+\left[\tilde{f}(\varepsilon)\sin\left(\frac{s}{\varepsilon}\right)\sqrt{\frac{\cosh(s)}{2\cosh(L_0)}}\right] + \cdots. \tag{5.67}$$

By substituting the solution (5.67) into the condition (5.36), one can easily determine the parameter $\beta(\varepsilon)$.

In (5.67), the natural number $N$ can be set arbitrarily large. Hence, for convenience, one may express (5.67) in the form

$$\theta(s,\varepsilon) = \theta_*(s) + \{\text{RPE}\} + \{\text{SPE}\}. \tag{5.68}$$

We can say that the nonclassical needle crystal solution consists of three parts: (1) the Ivantsov solution; (2) the regular perturbation expansion; and (3) the singular perturbation expansion.

#### 5.2.3.3 The Properties of the Composite Solutions

The above solutions are a family of solutions containing some undefined function $f(\varepsilon)$ and free constants, such as $C$. These functions and constants depend on how one sets the root condition. However, it can be seen that this root condition has very little effect on the behavior of the solution in the tip region. More precisely, one can make the following statements:

(1) All the members of this family of solutions with different root conditions have the same regular perturbation expansion (RPE) as $\varepsilon \to 0$, with the classical needle crystal solution $\theta_*(s)$ as its leading term. Thus, the RPE part has a more profound physical significance than the exact solution itself. It remains invariant for a variety of possible root conditions.

(2) At every fixed point $s$ in the region $0 \le s < L_0$, as $\varepsilon \to 0$, correction terms, such as $|\alpha_0(\varepsilon)\tilde{h}_0(s,s_+)| = O\left(\varepsilon e^{-\frac{L(\varepsilon)}{\varepsilon}}\right)$, are transcendentally small. Therefore, in the physical region ($0 \le s < L_0$) in which we are most interested, our solution will be very close to the needle solution $\theta_*(s)$ and the accompanying RPE.

(3) Though in the tip region the SPE part is exponentially small beyond all orders as $\epsilon \to 0$, it is the dominant part of the solution in the root region.

The root conditions (5.25) cannot be satisfied without it. In fact, we have $|\alpha_0(\varepsilon)\tilde{h}_0(s, s_+)| = O\big(\sin(s/\varepsilon)e^{s/2}\big)$. If one extends the solution analytically to the nonphysical region beyond the root, i.e., $s > L_0(\varepsilon)$, then the SPE part of the solution will be oscillatory with an exponentially growing amplitude as $s \to \infty$. This behavior is consistent with the conclusion of nonexistence of the classical needle solution for the Segur–Kruskal problem.

In realistic dendritic growth experiments, the physical phenomenon observed in the tip region at the later stages of growth will be determined by two classes of conditions:

(1) The macroscopic growth conditions, which include the material properties and undercooling temperature.
(2) The changeable detailed conditions of operation, the history of growth, and the initial settings, such as the selection of seed and device. All these factors will eventually affect the root conditions.

For different runs of experiments, one can set the first class of macroscopic growth conditions to be the same by employing the same materials and establishing the same undercooling temperature. However, the second class of conditions for different runs, which are reflected in the root conditions at the late stages, will never be exactly the same. Nevertheless, according to our findings, the differences between the members in this family of solutions are transcendentally small in the outer regions far from the root. Such small differences are indistinguishable in experiments, so that all these solutions may be practically considered to be identical. It is sensible, therefore, to call the entire family of nonclassical needle crystal solutions for the modified Segur–Kruskal problem "*the steady state solution*." One should note that when the system allows the classical needle solution, this classical needle solution will be a special member of the nonclassical needle solutions defined above.

### 5.2.4 *Needle Crystal Formation Problem*

Consider a more general unsteady system obtained by modifying the geometric model to include a time derivative term as follows:

$$\frac{\partial\theta}{\partial t} + \varepsilon^2\frac{\partial^3\theta}{\partial s^3} + \frac{\partial\theta}{\partial s} = \cos\theta \quad \big(0 \le s < L(t,\varepsilon)\big), \tag{5.69}$$

with the following boundary conditions:

(1) The tip smoothness condition

$$\theta(0) = \theta''(0) = 0\,. \tag{5.70}$$

(2) The root condition

$$\theta(L) = \theta_*(L) + \mathcal{O}(\varepsilon)\,. \tag{5.71}$$

Assume that the total length of the needle, $L(t, \varepsilon)$, increases as its tip grows with a speed $U(t)$. so that

$$L(t, \varepsilon) = L_0 + \int_{t_0}^{t} U(t')dt'. \tag{5.72}$$

In addition to the above boundary conditions, one may impose various initial conditions. We restrict ourselves to the investigation of a special class of unsteady solutions describing the needle evolution at the later stage $t \geq t_0 \gg 1$. To describe such a solution, we introduce the slow time variable

$$\tau = \varepsilon(t - t_0) \tag{5.73}$$

and assume that the speed function $U$ can be written in the form

$$U = \bar{U}(\tau, \varepsilon)\,. \tag{5.74}$$

Thus, the governing equation is transformed to

$$\varepsilon\frac{\partial\theta}{\partial\tau} + \varepsilon^2\frac{\partial^3\theta}{\partial s^3} + \frac{\partial\theta}{\partial s} = \cos\theta, \tag{5.75}$$

and the root condition is transformed to that at $s = \bar{L}(\tau, \varepsilon) = \frac{1}{\varepsilon}\big(C + \int_0^{\tau} \bar{U}(\tau')d\tau'\big)$,

$$\theta(\bar{L}, \tau) = \theta_*(\bar{L}) + \mathcal{O}(\varepsilon). \tag{5.76}$$

In addition, we impose the following initial condition: As $\tau = 0$,

$$\theta(s, 0, \varepsilon) = \Theta(s, \varepsilon), \tag{5.77}$$

where the initial function $\Theta(s, \varepsilon)$ is the steady solution to the above system with $\tau = 0$ in the root condition. Evidently, this function is just the steady nonclassical needle solution defined in the last section.

Note that the Ivantsov solution is the exact solution of this system in the special case $\varepsilon = 0$. As $\varepsilon \to 0$, for fixed $(s, \tau)$ we can derive the asymptotic expansion of the solution:

$$\theta(s, \tau, \varepsilon) \sim \theta_*(s) + \varepsilon^2\theta_1(s) + \varepsilon^4\theta_2(s) + \cdots\,. \tag{5.78}$$

Therefore, one can write

$$\theta(s,\tau,\varepsilon) = \mathcal{R}_N(s,\varepsilon) + \tilde{\theta}(s,\tau,\varepsilon), \tag{5.79}$$

where $\tilde{\theta}(s,\tau,\varepsilon) \ll \mathcal{O}(\varepsilon^{2N})$ for every large natural number $N$. This implies that the solution under investigation is a "nearly" steady needle solution with a transcendentally small time-dependent tail $\tilde{\theta}(s,\tau,\varepsilon)$.

From the above discussion, we see that although in some cases the classical steady needle solution for the Segur–Kruskal problem does not exist, the above-defined two kinds of solutions do exist, and they both have the same regular perturbation expansion solution (RPE). We may define all these solutions as the *generalized steady state* of the system and take it as the basic state for the stability analysis. Evidently, the conventional steady state defined by the classical needle solution, if it exists, is also the generalized steady state. It would therefore be governed by the same instability mechanisms to be derived in the following chapters.

From the point of view of pattern formation, we are interested in the behavior of the "nearly" steady solution or the solutions for the needle formation problem in the limit $\tau \to \infty$. The exact form of a uniformly valid asymptotic expansion as $\varepsilon \to 0$ for the "nearly" steady needle solutions, or the needle formation solution $\tilde{\theta}(s,\tau,\varepsilon)$, in the range of $0 \leq \tau < \infty$, is generally very difficult to obtain, if not impossible. However, for a dissipative dynamic system, as $t \to \infty$ all limiting solutions will settle down in a low-dimensional subspace of the phase space, and the following three possibilities may be anticipated:

1. The solution may have a steady limit solution, and hence the system tends to a fixed-point state in a somehow defined state space.
2. The solution may have a time-periodic limit solution, so that the system approaches a limit circle in the state space.
3. The solution may have a chaotic limit state, and the system therefore exhibits chaotic behavior.

Only in case (1) does a solution to the Segur–Kruskal problem exist. But the result obtained by Segur and Kruskal has ruled out this case. Achieving a good understanding of the long-term behavior of the solutions of a nonlinear dynamical system is one of the basic and most important tasks in the broad field of nonlinear science. For a full exploration of this subject, readers who are interested in studying this general topic are referred to books such as [5, 12]. One way to tackle this problem for systems of dendritic growth is to study the linear stability of the above-defined solutions. The following chapters will proceed in this direction.

## 5.3 The Nonclassical Steady State of Dendritic Growth with Nonzero Surface Tension

We now turn to the original dendrite growth system. The ideas of the steady nonclassical needle solution demonstrated in the last section can be successfully applied to this nonlocal model.

### 5.3.1 *The Complete Mathematical Formulation for Free Dendrite Growth*

Consider the nonclassical needle growth problem. It describes free dendritic growth as $t \gg 1$. The solution is assumed have a long but finite stem $L$.

The coordinate system $(\xi, \eta, \theta)$ fixed at the dendrite-tip is assumed to move with the constant velocity $U_0$, which is used as the scale of velocity. In the moving frame, the location root corresponds to $\xi = \xi_{\max}(\varepsilon)$. Here, one may assume that $\xi_{\max}$ is a monotonically increasing function of the slow time variable $\tau = \varepsilon(t - t_0)$, and $\xi_{\max} \to \infty$ as $\varepsilon \to \infty$.

The steady three-dimensional axially symmetric dendritic growth is then subject to the following equation:

$$\left(\frac{\partial^2 T}{\partial \xi^2} + \frac{\partial^2 T}{\partial \eta^2} + \frac{1}{\xi}\frac{\partial T}{\partial \xi} + \frac{1}{\eta}\frac{\partial T}{\partial \eta}\right) = \eta_0^2\left(\xi\frac{\partial T}{\partial \xi} - \eta\frac{\partial T}{\partial \eta}\right),$$

$$\left(0 \le \xi \le \xi_{\max},\ 0 \le \eta < \infty\right). \tag{5.80}$$

The boundary conditions are as follows:

1. The upstream far-field condition: as $\eta \to \infty$,

$$T \to T_\infty = \frac{(T_\infty)_D - T_{M0}}{\Delta H/(c_p\rho)} < 0\,. \tag{5.81}$$

2. The regularity condition: as $\eta \to 0$,

$$\frac{\partial T_S}{\partial \eta} \to 0 \ \ T_S = \mathcal{O}(1)\,. \tag{5.82}$$

3. The interface conditions: at $\eta = \eta_s(\xi)$:

   (a) the thermodynamic equilibrium condition

$$T = T_S\,, \tag{5.83}$$

(b) the Gibbs–Thomson condition

$$T = -\eta_0^2\varepsilon^2\mathcal{K}\left\{\frac{d}{d\xi}, \frac{d^2}{d\xi^2}\right\}\eta_s\,, \tag{5.84}$$

(c) the heat balance condition

$$\left(\frac{\partial}{\partial\eta} - \eta_s'\frac{\partial}{\partial\xi}\right)(T - T_S) + \varepsilon\eta_0^4(\xi^2+\eta^2)\frac{\partial\eta_s}{\partial\tau}$$
$$+\eta_0^2(\xi\eta_s)' = 0\,. \tag{5.85}$$

4. The tip smoothness condition:

$$\eta_s(0) = 1, \qquad \frac{\partial}{\partial\xi}\{T, T_S, \eta_s\} = 0\,. \tag{5.86}$$

5. The root condition: at $\xi = \xi_{\max}(\varepsilon)$,

$$\begin{aligned} q \equiv \{T, T_S, \eta_s\} &= \{T_*, T_{S*}, \eta_*\} + \{F(\eta,\varepsilon), G(\eta,\varepsilon), R(\varepsilon)\} \\ &= q_* + f(\varepsilon)\,, \end{aligned} \tag{5.87}$$

where $f(\varepsilon) \to 0$ as $\varepsilon \to 0$.

In the root condition, we assume that the functions $F$, $G$, and $R$ are sufficiently smooth functions so that the existence of a solution can be guaranteed for all $\varepsilon \geq 0$.

It is evident that the Ivantsov solution (4.1) is the solution to the above system for the special case $\varepsilon = 0$. For the general case $\varepsilon > 0$, for fixed $(\xi, \eta)$, the solution to the above nonclassical needle growth problem has the steady regular perturbation expansion discussed in Chap. 4:

$$q(\xi,\eta,\tau) \sim q_* + \varepsilon^2\bar{q}_1(\xi,\eta) + \varepsilon^4\bar{q}_2(\xi,\eta) + \cdots \quad (\varepsilon \to 0). \tag{5.88}$$

This steady RPE solution satisfies all boundary conditions except for the root condition (5.87). Let $\mathcal{R}$ denote the regular perturbation expansion (RPE), and let $\mathcal{R}_N$ denote the summation of the first $N$ terms of the RPE. To further satisfy the root condition, one may write a more accurate general asymptotic expansion form of the solution $q$ as

$$q \sim \mathcal{R} + \mathcal{S}, \tag{5.89}$$

where the second term on the right-hand side of the expression is the singular perturbation expansion (SPE) part, which correct the errors in the root condition introduced by the RPE. One can deduce that at fixed $0 < \xi < \xi_{\max}$, these modifications of the interface shape made by the terms in the singular perturbation expansion (SPE) part are exponentially small beyond all orders of $\varepsilon$, as $\varepsilon \to 0$. Therefore, the solution of the nonclassical needle growth problem $q$ may be called a "nearly" steady needle solution. As mentioned before, we shall define such a nonclassical "nearly" steady needle solution $q$ as the basic steady state of dendritic growth. This definition overcomes the difficulty encountered due to the nonexistence of a solution for the Nash–Glicksman problem. Evidently, with this definition, the basic steady state of dendritic growth with the inclusion of surface tension does not provide a mechanism for the selection of the tip velocity. The selection problem of dendritic growth is essentially related to the behavior of the unsteady solution in the limit $t \to \infty$. For the dynamical system under investigation, the exact form of the unsteady solution is very difficult to obtain. Nevertheless, it will be seen in the following chapters that insofar as the long-term behavior of system has a close connection with its stability properties of the basic steady state, the selection problem may be resolved on the basis of linear stability analysis in terms of a unified asymptotic approach. Moreover, in most cases, knowing the leading term of the RPE part of the basic steady state is sufficient for us to determine the stability properties of the system.

## References


1. G.E. Nash, M.E. Glicksman, Capillarity–limited steady-state dendritic growth I. Theoretical development. Acta Metall. **22**, 1283–1299 (1974)
2. M. Kruskal, H. Segur, Asymptotics beyond all orders in a model of crystal growth. Stud. Appl. Math. **85**, 129–181 (1991)
3. J.S. Langer, Lectures in the theory of pattern formation, in *USMG NATO AS Les Houches Session XLVI 1986 — Le hasard et la matière/Chance and Matter*, ed. by J. Souletie, J. Vannimenus, R. Stora (Elsevier Science, Amsterdam, 1986)
4. D.A. Kessler, J. Koplik, H. Levine, Pattern formation far from equilibrium: the free space dendritic crystal, in *Proc. NATO A.R.W. on Patterns, Defects and Microstructures in Non-equilibrium Systems*, Austin (1986)
5. S. Wiggins, *Global Bifurcations and Chaos: Analytical Methods*, ed. by F. John, J.E. Marsden, L. Sirovich. Series of Applied Mathematical Sciences, vol. 73 (Springer, New York, 1988)
6. P. Pelce, *Dynamics of Curved Front* (Academic, New York, 1988)
7. E.A. Brener, V.I. Melnikov, Pattern selection in two dimensional dendritic growth. Adv. Phys. **40**, 53–97 (1991)
8. H. Segur, S. Tanveer, H. Levine (eds.), *Asymptotics Beyond All Orders*. NATO ASI Series, Series B: Physics, vol. 284 (Plenum, New York, 1991)
9. Y. Pomeau, M. Ben Amer, Dendrite growth and related topics, in *Solids Far from Equilibrium*, ed. by C. Godreche (Cambridge University Press, Cambridge, 1991)
10. J.J. Xu, Interfacial wave theory of solidification — dendritic pattern formation and selection of tip velocity. Phys. Rev. A15 **43**(2), 930–947 (1991)
11. J. Kevorkian, J.D. Cole, *Multiple Scale and Singular Perturbation Methods*. Applied Mathematical Sciences, vol. 114 (Springer, Berlin/Heidelberg, 1996)
12. J.K. Hale, L.T. Magalhaes, W.M. Oliva, *An Introduction to Infinite Dimensional Dynamical Systems — Geometric Theory*, ed. by F. John, J.E. Marsden, L. Sirovich. Series of Applied Mathematical Sciences, vol. 47 (Springer, New York, 1984)

# Chapter 6
# Global Interfacial Wave Instability of Dendritic Growth from a Pure Melt

We now turn to study global linear stability of the steady or "nearly" steady needle crystal solutions that were demonstrated in the last chapter and hereinafter defined as the basic states of the system. We shall first deal with the 3D axially symmetric dendrite with nonzero isotropic surface tension. The effect of anisotropy will be discussed in the next chapter. For global linear stability, one needs to investigate the evolution of infinitesimal perturbations around the basic state solutions.

There are two approaches that have been used in the study of the evolution of perturbations. The first approach is to solve an initial value problem: Assume that a given initial disturbance is introduced into the system and then consider the evolution of the initial disturbance by solving the initial value problem. The second approach is a normal mode analysis, whereby we assume that the perturbations are in the form of quasistationary waves and then investigate their evolution by solving an eigenvalue problem under a certain set of boundary conditions. The normal mode approach has been used to solve stability problems for various inhomogeneous dynamical systems from a broad area of physics and engineering. Famous examples include the critical layer instability theory in fluid dynamics and the density wave theory for the spiral structure of galaxies by C.C. Lin in the 1970s. See, for example, [10, 15, 17].

In this book, we adopt the normal mode approach. The major aim here is to derive the so-called global mode solutions. The eigenvalue corresponding to a global mode gives both the growth rate of the amplitude of the perturbation and the frequency of the oscillation.

To find the global mode solutions, we shall apply the matched asymptotic expansion method. Specifically, we shall divide the entire complex plane into three regions: the outer region, the turning point region, and the tip inner region. The outer solutions in the outer region will be derived by the *multiple variables expansion* (MVE) method. These outer solutions can be interpreted as some special interfacial traveling waves, propagating along the interface of the basic state. The local dispersion relationship for these interfacial waves is obtained in the zeroth-



J.-J. Xu, *Interfacial Wave Theory of Pattern Formation in Solidification*, Springer Series in Synergetics, DOI 10.1007/978-3-319-52663-8_6

order approximation, while the amplitude functions of the waves are determined in the first-order approximation. It is in the first-order approximation that some singularities of normal mode form of outer solutions are unveiled. It was first found in 1989 (e.g., [21, 24]) that in addition to the singularity at the tip of a dendrite, there existed a special simple turning point in the complex plane that played a crucial role in understanding the dynamics of dendritic growth. Since the outer solutions are found to be invalid near the turning point and the leading edge of the dendrite tip, it is necessary to find a different form of asymptotic expansion for the exact solution in the vicinity of each singular point. For this purpose, one must choose proper new length scales and derive the inner solutions in the inner regions of the turning point and dendrite tip, respectively. The inner equation in the vicinity of the turning point can be reduced to the Airy equation with complex coefficients, whereas the inner solutions in the vicinity of the tip can be expressed by Hankel functions. Finally, all asymptotic expansion solutions must be matched in the intermediate regions. The global mode solutions and a quantization condition for the eigenvalues are then obtained. Given $\varepsilon > 0$, the system permits a discrete sets of complex eigenvalues $\sigma_n = (\sigma_{\rm R} - {\rm i}\omega)_n$ $(n = 0, 1, 2, \ldots)$ and corresponding global wave modes [24]. The global instability mechanism discovered here is called the *global trapped-wave* (GTW) instability, and its presence explains the origin and persistence of the pattern formation in the solidification process. In a certain sense, the global instability mechanism discovered in dendritic growth is similar to the so-called *overreflection mechanism* explored in the critical layer theory of shear flow of fluid dynamics and in the density wave theory of galactic dynamics.

It will be shown that when $\varepsilon$ equals a critical number $\varepsilon_*$, the system permits a uniquely determined *global neutrally stable mode*; its corresponding eigenvalue $\sigma$ has zero real part. This global neutrally stable mode will be selected at the later stages of growth. Therefore, the stability criterion $\varepsilon = \varepsilon_*$ is also the selection criterion for the tip speed.

## 6.1 Linear Perturbed System Around the Basic State of Axisymmetric Dendritic Growth

We write the basic state in the form

$$\begin{aligned}
T_{\rm B}(\xi, \eta, \tau, \varepsilon) &= T_*(\eta) + ({\rm h.o.t.}),\\
T_{\rm SB}(\xi, \eta, \tau, \varepsilon) &= ({\rm h.o.t.}),\\
\eta_{\rm B}(\xi, \tau, \varepsilon) &= 1 + ({\rm h.o.t.}),
\end{aligned} \tag{6.1}$$

and separate the general unsteady solutions into two parts:

$$\begin{aligned} T &= T_{\mathrm{B}} + \tilde{T}(\xi, \eta, t, \varepsilon), \\ T_{\mathrm{S}} &= T_{\mathrm{SB}} + \tilde{T}_{\mathrm{S}}(\xi, \eta, t, \varepsilon) \\ \eta_{\mathrm{s}} &= \eta_{\mathrm{B}} + \tilde{h}(\xi, t, \varepsilon)/\eta_0^2 . \end{aligned} \tag{6.2}$$

Assume that the above perturbations around the basic state are caused by initially infinitesimal perturbations with a characteristic amplitude $\delta \ll 1$. Thus, within a sufficiently short time period, we will have $|\tilde{q}| =: \{|\tilde{T}|, |\tilde{h}|\} \ll \{|T_{\mathrm{B}}|, |\eta_{\mathrm{B}}|\}$. Hence, the linearization around the basic state solution in the small-amplitude parameter $\delta \ll 1$ is applicable. The linearized perturbed system must be a homogeneous system. It is shown below that

$$\begin{aligned} &\left(\frac{\partial^2 \tilde{T}}{\partial \xi^2} + \frac{\partial^2 \tilde{T}}{\partial \eta^2} + \frac{1}{\xi}\frac{\partial \tilde{T}}{\partial \xi} + \frac{1}{\eta}\frac{\partial \tilde{T}}{\partial \eta}\right) \\ &-\eta_0^2\left(\xi\frac{\partial \tilde{T}}{\partial \xi} - \eta\frac{\partial \tilde{T}}{\partial \eta}\right) - \eta_0^4(\xi^2 + \eta^2)\frac{\partial \tilde{T}}{\partial t} = 0, \\ &\big(0 \le \xi \le \xi_{\max} = \mathcal{O}(1/\varepsilon);\ 0 \le \eta < \infty\big) \end{aligned} \tag{6.3}$$

with the following boundary conditions:

1. As $\eta \to \infty$,

$$\tilde{T} \to 0 . \tag{6.4}$$

2. As $\eta \to 0$,

$$\tilde{T}_{\mathrm{S}} = \mathcal{O}(1) . \tag{6.5}$$

3. The interface conditions: from the Taylor expansions around the interface of the basic state, it follows that for the case of isotropic surface tension, at $\eta = \eta_{\mathrm{B}}(\xi, \varepsilon)$, the following hold:

   (i)

$$\tilde{T} - \tilde{T}_{\mathrm{S}} = -\left[\frac{\partial T_{\mathrm{B}}}{\partial \eta} - \frac{\partial T_{\mathrm{SB}}}{\partial \eta}\right]\frac{\tilde{h}}{\eta_0^2} , \tag{6.6}$$

   (ii)

$$\tilde{T}_{\mathrm{S}} + \frac{\partial T_{\mathrm{SB}}}{\partial \eta}\frac{\tilde{h}}{\eta_0^2} = \frac{\varepsilon^2}{\hat{S}(\xi)}\left[\frac{\partial^2 \tilde{h}}{\partial \xi^2} + \frac{(1+2\xi^2)}{\xi \hat{S}^2(\xi)}\frac{\partial \tilde{h}}{\partial \xi} - \frac{\tilde{h}}{\hat{S}^2}\right] , \tag{6.7}$$

(iii)

$$\frac{\partial}{\partial\eta}\left(\tilde{T}-\tilde{T}_{\mathrm{S}}\right)+\eta_0^2\hat{S}^2(\xi)\frac{\partial\tilde{h}}{\partial t}+\xi\frac{\partial\tilde{h}}{\partial\xi}+\tilde{h}+\left[\frac{\partial^2}{\partial\eta^2}\left(T_{\mathrm{B}}-T_{\mathrm{SB}}\right)\right]\frac{\tilde{h}}{\eta_0^2}$$

$$-\frac{1}{\eta_0^2}\frac{\partial\tilde{h}}{\partial\xi}\frac{\partial}{\partial\xi}\left(T_{\mathrm{B}}-T_{\mathrm{SB}}\right)-\frac{\eta_{\mathrm{B}}'\tilde{h}}{\eta_0^2}\frac{\partial^2}{\partial\xi\partial\eta}\left(T_{\mathrm{B}}-T_{\mathrm{SB}}\right)$$

$$-\eta_{\mathrm{B}}'\frac{\partial}{\partial\xi}\left(\tilde{T}-\tilde{T}_{\mathrm{S}}\right)=0\,, \tag{6.8}$$

where

$$\hat{S}(\xi)=\sqrt{\xi^2+\eta_{\mathrm{B}}^2}\,. \tag{6.9}$$

4. The root condition: at $\xi=\xi_{\max}(\varepsilon)$,

$$\{\tilde{T},\tilde{T}_{\mathrm{S}},\tilde{h}\}=0\,. \tag{6.10}$$

5. The tip smoothness condition: at $\xi=0,\ \eta=\eta_{\mathrm{B}}(0,\varepsilon)$,

$$\frac{\partial}{\partial\xi}\{\tilde{T},\tilde{T}_{\mathrm{S}},\tilde{h}\}=0\,. \tag{6.11}$$

The above system (6.3)–(6.11) contains two parameters $\varepsilon$ and $\eta_0^2$. The slow time variable $\tau$ in the basic steady state may be formally treated as a constant. The system leads to a linear eigenvalue problem when one looks for solutions of the type $\tilde{q}=\hat{q}\mathrm{e}^{\sigma t}$. The eigenvalue $\sigma$ must be a function of $(\eta_0^2,\varepsilon)$. We shall solve this eigenvalue problem in two steps. We first solve the system (6.3)–(6.10) for given $(\sigma,\eta_0^2,\varepsilon)$. In doing so, we shall apply the multiple variables expansion (MVE) method to look for a uniformly valid asymptotic solution in the limit $\varepsilon\to 0$. The solution should satisfy all boundary conditions except the tip condition (6.11). Then we apply the tip condition (6.11) to the asymptotic solution obtained. Thus, the parameter $\sigma$ must be chosen as a function of $\eta_0^2$ and $\varepsilon$.

## 6.2 Outer Solution in the Outer Region Away from the Tip

In the outer region, $|\xi|\gg\varepsilon$, we shall look for the asymptotic expansion of the solutions $\tilde{q}(\xi,\eta,t,\varepsilon)$ in terms of the multiple variables expansion method, similar to our approach in Chap. 2. However, the dynamical system under investigation is now inhomogeneous with variable coefficients, which is much more difficult to deal with than the one encountered in Chap. 2. Thus, there are significant differences between the approach applied here and that used in Chap. 2. The first difference is that in this case, the solutions will depend on all the fast variables $\xi_+, \eta_+$ and the slow variables $\xi, \eta$. The second difference is that in this case, the fast variable along the $\xi$-direction, $\xi_+$, cannot be homogeneous. It must be stretched in the $\xi$-direction, and it may slowly vary in the $\eta$-direction. We define the fast variable $\xi_+$ along the $\xi$-direction as

$$\xi_+ = \frac{\phi(\xi, \eta, \varepsilon)}{\varepsilon} = \frac{1}{\varepsilon}\int_{\xi_0}^{\xi} k(\xi_1, \eta, \varepsilon)\mathrm{d}\xi_1, \quad k(\xi, \eta, \varepsilon) = \frac{\partial \phi}{\partial \xi}. \tag{6.12}$$

Here the lower limit $\xi_0$ is an arbitrary constant, which can be determined later. Furthermore, we introduce the fast variable $\eta_+$ along the $\eta$-direction as

$$\eta_+ = \frac{\psi(\xi, \eta, \varepsilon)}{\varepsilon} = \frac{1}{\varepsilon}\int_{1}^{\eta} k(\xi, \eta_1, \varepsilon)\mathrm{d}\eta_1. \tag{6.13}$$

The fast time variable $t_+$ is defined as

$$t_+ = \frac{t}{\eta_0^2 \varepsilon}. \tag{6.14}$$

The fast variable $\eta_+$ determines a boundary layer with the thickness of $\mathcal{O}(\varepsilon)$ near the interface $\eta = 1$, which we call the *interfacial wave* (IW) layer. It will be seen that the perturbations of the temperature field are restricted to the IW layer. It vanishes exponentially away from this layer.

At the interface $\eta = 1$, the fast variable $\xi_+$ is given as

$$\bar{\xi}_+(\xi, \varepsilon) = \xi_+(\xi, 1, \varepsilon) = \frac{\bar{\phi}(\xi, \varepsilon)}{\varepsilon} = \frac{1}{\varepsilon}\int_{\xi_0}^{\xi} \bar{k}(\xi_1, \varepsilon)\mathrm{d}\xi_1, \tag{6.15}$$

where

$$\bar{\phi}(\xi, \varepsilon) = \phi(\xi, 1, \varepsilon), \quad \bar{k}(\xi, \varepsilon) = \frac{\partial \bar{\phi}(\xi, \varepsilon)}{\partial \xi} = k(\xi, 1, \varepsilon). \tag{6.16}$$

We assume that the pair of functions $(\phi, \psi)$ are subject to the following Cauchy–Riemann conditions:

$$\begin{array}{ll} \phi_\xi = \psi_\eta, & \psi_\xi = -\phi_\eta \\ (\mathrm{Re}\{\phi_\xi > 0\}, & \mathrm{Re}\{\psi_\eta\} > 0). \end{array} \tag{6.17}$$

It follows that

$$\phi_\xi \psi_\xi + \phi_\eta \psi_\eta = 0, \tag{6.18}$$

and

$$\nabla^2\phi(\xi,\eta,\varepsilon) = \nabla^2\psi(\xi,\eta,\varepsilon) = 0. \tag{6.19}$$

Since we have $\psi_\eta = k(\xi,\eta,\varepsilon)$, let us further define

$$\phi_\eta(\xi,\eta,\varepsilon) = g(\xi,\eta,\varepsilon). \tag{6.20}$$

From the boundary conditions at $\eta = 1$, one deduces $\psi(\xi,1,\varepsilon) = 0$ and

$$\begin{aligned}\phi(\xi,1,\varepsilon) &= \bar{\phi}(\xi,\varepsilon) = \textstyle\int_{\xi_0}^{\xi} \bar{k}(\xi_1,\varepsilon)\mathrm{d}\xi_1,\\ \phi_\eta(\xi,1,\varepsilon) &= -\psi_\xi(\xi,1,\varepsilon) = g(\xi,1,\varepsilon) = 0,\end{aligned} \tag{6.21}$$

where $\bar{k}(\xi,\varepsilon)$ is the wave number function on the interface. We can solve the solution $\phi(\xi,\eta)$ from (6.19) in the whole $(\xi,\eta)$-plane. For this purpose, we first extend the function $\bar{k}(\xi)$ analytically to the function $\bar{k}(\zeta)$ defined in the complex plane $\zeta = \xi + \mathrm{i}\xi_{\mathrm{I}}$. Furthermore, we may rewrite (6.19) in the form

$$\frac{\partial^2\phi(\xi,\eta)}{\partial\xi^2} + \frac{\partial^2\phi(\xi,\eta)}{\partial\eta^2} = \left(\frac{\partial}{\partial\xi} - \mathrm{i}\frac{\partial}{\partial\eta}\right)\left(\frac{\partial}{\partial\xi} + \mathrm{i}\frac{\partial}{\partial\eta}\right)\phi(\xi,\eta) = 0, \tag{6.22}$$

and derive the solution

$$\begin{aligned}\phi(\xi,\eta) &= \frac{1}{2}\Big\{\bar{\phi}[\xi+\mathrm{i}(\eta-1)] + \bar{\phi}[\xi+\mathrm{i}(\eta-1)]\Big\}\\ &= \frac{1}{2}\left\{\int_0^{\xi+\mathrm{i}(\eta-1)} \bar{k}(t)\mathrm{d}t + \int_0^{\xi-\mathrm{i}(\eta-1)} \bar{k}(t)\mathrm{d}t\right\}.\end{aligned} \tag{6.23}$$

It follows that

$$\begin{aligned}\phi_\xi(\xi,\eta) &= \psi_\eta(\xi,\eta) = k(\xi,\eta)\\ &= \frac{1}{2}\Big\{\bar{k}[\xi+\mathrm{i}(\eta-1)] + \bar{k}[\xi-\mathrm{i}(\eta-1)]\Big\}.\end{aligned} \tag{6.24}$$

Furthermore, from the condition (6.17), we derive

$$\begin{aligned}\phi_\eta(\xi,\eta) &= -\psi_\xi(\xi,\eta) = g(\xi,\eta)\\ &= \frac{\mathrm{i}}{2}\Big\{\bar{k}[\xi+\mathrm{i}(\eta-1)] - \bar{k}[\xi-\mathrm{i}(\eta-1)]\Big\},\end{aligned} \tag{6.25}$$

and

$$\psi(\xi,\eta) = -\frac{\mathrm{i}}{2}\int_{\xi-\mathrm{i}(\eta-1)}^{\xi+\mathrm{i}(\eta-1)} \bar{k}(t)\mathrm{d}t. \tag{6.26}$$

It should be remarked that the functions $(\phi, \psi)$ and $\bar{k}(\xi, \varepsilon)$ under discussion are not necessarily real-valued. They may be complex functions of the real variables $(\xi, \eta)$. It is seen from the above that the phase functions $\psi(\xi, \eta)$ and $\phi(\xi, \eta)$ are fully determined as soon as the wave function $\bar{k}(\xi, \varepsilon)$ is solved.

We now apply the multiple variables expansion (MVE) method, by formally treating the fast and slow variables $(\xi, \eta, \xi_+, \eta_+, t_+)$ as independent variables. To covert the original system (6.3)–(6.8) to one with multiple variables, one must make the following replacements for the derivatives in (6.3)–(6.8):

$$
\begin{aligned}
\frac{\partial}{\partial \xi} &\Longrightarrow \frac{\phi_\xi}{\varepsilon}\frac{\partial}{\partial \xi_+} + \frac{\psi_\xi}{\varepsilon}\frac{\partial}{\partial \eta_+} + \frac{\partial}{\partial \xi}, \\
\frac{\partial}{\partial \eta} &\Longrightarrow \frac{\psi_\eta}{\varepsilon}\frac{\partial}{\partial \eta_+} + \frac{\phi_\eta}{\varepsilon}\frac{\partial}{\partial \xi_+} + \frac{\partial}{\partial \eta}, \\
\frac{\partial^2}{\partial \xi^2} &\Longrightarrow \frac{\phi_\xi^2}{\varepsilon^2}\frac{\partial^2}{\partial \xi_+^2} + \frac{2\phi_\xi}{\varepsilon}\frac{\partial^2}{\partial \xi_+ \partial \xi} + \frac{\phi_{\xi\xi}}{\varepsilon}\frac{\partial}{\partial \xi_+} + \frac{\partial^2}{\partial \xi^2} \\
&\quad + \frac{\psi_\xi^2}{\varepsilon^2}\frac{\partial^2}{\partial \eta_+^2} + \frac{2\phi_\xi\psi_\xi}{\varepsilon^2}\frac{\partial^2}{\partial \xi_+ \partial \eta_+} + \frac{\psi_{\xi\xi}}{\varepsilon}\frac{\partial}{\partial \eta_+} + \frac{2\psi_\xi}{\varepsilon}\frac{\partial^2}{\partial \eta_+ \partial \xi}, \\
\frac{\partial^2}{\partial \eta^2} &\Longrightarrow \frac{\psi_\eta^2}{\varepsilon^2}\frac{\partial^2}{\partial \eta_+^2} + \frac{2\psi_\eta}{\varepsilon}\frac{\partial^2}{\partial \eta_+ \partial \eta} + \frac{\psi_{\eta\eta}}{\varepsilon}\frac{\partial}{\partial \eta_+} + \frac{\partial^2}{\partial \eta^2} \\
&\quad + \frac{\phi_\eta^2}{\varepsilon^2}\frac{\partial^2}{\partial \xi_+^2} + \frac{2\phi_\eta\psi_\eta}{\varepsilon^2}\frac{\partial^2}{\partial \xi_+ \partial \eta_+} + \frac{\phi_{\eta\eta}}{\varepsilon}\frac{\partial}{\partial \xi_+} + \frac{2\phi_\eta}{\varepsilon}\frac{\partial^2}{\partial \xi_+ \partial \eta}.
\end{aligned}
\tag{6.27}
$$

The converted system with multiple variables is as follows: The equation in the liquid phase is

$$
\begin{aligned}
&\left[(\phi_\xi^2 + \phi_\eta^2)\frac{\partial^2}{\partial \xi_+^2} + (\psi_\xi^2 + \psi_\eta^2)\frac{\partial^2}{\partial \eta_+^2} + 2(\phi_\xi\psi_\xi + \phi_\eta\psi_\eta)\frac{\partial^2}{\partial \xi_+ \partial \eta_+}\right]\tilde{T} \\
&= -\varepsilon\left(2\phi_\xi \frac{\partial^2 \tilde{T}}{\partial \xi_+ \partial \xi} + \phi_{\xi\xi}\frac{\partial \tilde{T}}{\partial \xi_+} + \psi_{\xi\xi}\frac{\partial \tilde{T}}{\partial \eta_+} + 2\psi_\xi \frac{\partial^2 \tilde{T}}{\partial \eta_+ \partial \xi}\right) \\
&\quad -\varepsilon\left(2\psi_\eta \frac{\partial^2 \tilde{T}}{\partial \eta_+ \partial \eta} + \psi_{\eta\eta}\frac{\partial \tilde{T}}{\partial \eta_+} + \phi_{\eta\eta}\frac{\partial \tilde{T}}{\partial \xi_+} + 2\phi_\eta \frac{\partial^2 \tilde{T}}{\partial \xi_+ \partial \eta}\right) \\
&\quad -\varepsilon\left(\frac{1}{\xi} - \eta_0^2\xi\right)\left(\phi_\xi\frac{\partial \tilde{T}}{\partial \xi_+} + \psi_\xi\frac{\partial \tilde{T}}{\partial \eta_+}\right) - \varepsilon\left(\frac{1}{\eta} + \eta_0^2\eta\right)\left(\phi_\eta\frac{\partial \tilde{T}}{\partial \xi_+} + \psi_\eta\frac{\partial \tilde{T}}{\partial \eta_+}\right) \\
&\quad +\varepsilon\eta_0^2(\xi^2 + \eta^2)\frac{\partial \tilde{T}}{\partial t_+} + \mathcal{O}(\varepsilon^2).
\end{aligned}
\tag{6.28}
$$

For the symmetric model under discussion, the equation for the temperature field in the region of solid phase is the same as the above for the region of liquid phase, except that the function $\tilde{T}$ is replaced by $\tilde{T}_S$.

With the condition (6.17), the multivariable equation (6.28) can be simplified as

$$
\begin{aligned}
&(\phi_\xi^2 + \phi_\eta^2)\left(\frac{\partial^2}{\partial \xi_+^2} + \frac{\partial^2}{\partial \eta_+^2}\right)\tilde{T} \\
&= -\varepsilon\left(2\phi_\xi \frac{\partial^2 \tilde{T}}{\partial \xi_+ \partial \xi} + 2\psi_\xi \frac{\partial^2 \tilde{T}}{\partial \eta_+ \partial \xi}\right) - \varepsilon\left(2\psi_\eta \frac{\partial^2 \tilde{T}}{\partial \eta_+ \partial \eta} + 2\phi_\eta \frac{\partial^2 \tilde{T}}{\partial \xi_+ \partial \eta}\right) \\
&\quad -\varepsilon\left(\frac{1}{\xi} - \eta_0^2 \xi\right)\left(\phi_\xi \frac{\partial \tilde{T}}{\partial \xi_+} + \psi_\xi \frac{\partial \tilde{T}}{\partial \eta_+}\right) - \varepsilon\left(\frac{1}{\eta} + \eta_0^2 \eta\right)\left(\phi_\eta \frac{\partial \tilde{T}}{\partial \xi_+} + \psi_\eta \frac{\partial \tilde{T}}{\partial \eta_+}\right) \\
&\quad +\varepsilon \eta_0^2(\xi^2 + \eta^2)\frac{\partial \tilde{T}}{\partial t_+} + \mathcal{O}(\varepsilon^2). \qquad (6.29)
\end{aligned}
$$

The boundary conditions for the perturbed system with multiple variables are as follows:

1. As $\eta_+ \to \infty$,

$$\tilde{T} \to 0\,. \qquad (6.30)$$

2. As $\eta_+ \to -\infty$,

$$\tilde{T}_S \to 0\,. \qquad (6.31)$$

3. We now impose the interface conditions at the interface of the Ivantsov solution, $\eta_+ = 0, \eta = 1$. To do so, one needs to express the Taylor expansion with a single variable

$$\frac{\partial \tilde{T}}{\partial \eta}(1 + \tilde{\eta}) = \frac{\partial \tilde{T}}{\partial \eta}(1) + \frac{\partial^2 \tilde{T}}{\partial \eta^2}(1)\tilde{\eta}_s + \cdots$$

in a form with multiple variables. Using multiple variables, we may write

$$\frac{\partial \tilde{T}}{\partial \eta}(1) = \frac{\partial \tilde{T}}{\partial \eta}(1) + \frac{\bar{k}}{\varepsilon}\frac{\partial \tilde{T}}{\partial \eta_+}(1)$$

$$\frac{\partial^2 \tilde{T}}{\partial \eta^2}(1) = \frac{\partial^2 \tilde{T}}{\partial \eta^2}(1) + \frac{\bar{k}^2}{\varepsilon^2}\frac{\partial^2 \tilde{T}}{\partial \eta_+^2}(1) + 2\frac{\bar{k}}{\varepsilon}\frac{\partial^2 \tilde{T}}{\partial \eta_+ \partial \eta}(1),$$

so that

$$\frac{\partial \tilde{T}}{\partial \eta}(1+\tilde{\eta}) = \frac{\partial \tilde{T}}{\partial \eta}(1) + \frac{\bar{k}}{\varepsilon}\frac{\partial \tilde{T}}{\partial \eta_+}(1) + \left[\frac{\partial^2 \tilde{T}}{\partial \eta^2}(1) + \frac{\bar{k}^2}{\varepsilon^2}\frac{\partial^2 \tilde{T}}{\partial \eta_+^2}(1) + 2\frac{\bar{k}}{\varepsilon}\frac{\partial^2 \tilde{T}}{\partial \eta_+ \partial \eta}(1)\right]\tilde{\eta}_s + \cdots . \tag{6.32}$$

Thus, in terms of formula (6.32), the interface conditions can be written as follows. At $\eta_+ = 0, \eta = 1$:

(i) The thermodynamic equilibrium condition:

$$\tilde{T} = \tilde{T}_S + \tilde{h} - \varepsilon \eta_1 \bar{k} \frac{\partial}{\partial \eta_+}(\tilde{T} - \tilde{T}_S) + \mathcal{O}(\varepsilon^2) , \tag{6.33}$$

(ii) The Gibbs–Thomson condition:

$$\tilde{T}_S + \varepsilon \eta_1 \bar{k} \frac{\partial \tilde{T}_S}{\partial \eta_+} = \frac{1}{\hat{S}(\xi)} \left\{ \left( \bar{k}^2 \frac{\partial^2}{\partial \xi_+^2} + 2\varepsilon \bar{k} \frac{\partial^2}{\partial \xi_+ \partial \xi} + \varepsilon \bar{k}' \frac{\partial}{\partial \xi_+} \right) + \varepsilon \left( \frac{1}{\xi} + \frac{\xi}{\hat{S}^2(\xi)} \right) \left( \bar{k} \frac{\partial}{\partial \xi_+} + \varepsilon \frac{\partial}{\partial \xi} \right) \right\} \tilde{h} + \mathcal{O}(\varepsilon^2), \tag{6.34}$$

(iii) The heat balance condition:

$$\bar{k}\frac{\partial(\tilde{T} - \tilde{T}_S)}{\partial \eta_+} + \varepsilon \eta_1 \bar{k}^2 \frac{\partial^2(\tilde{T} - \tilde{T}_S)}{\partial \eta_+^2} + \varepsilon \frac{\partial(\tilde{T} - \tilde{T}_S)}{\partial \eta} + \hat{S}^2(\xi)\frac{\partial \tilde{h}}{\partial t_+} + \xi \left( k \frac{\partial \tilde{h}}{\partial \xi_+} + \varepsilon \frac{\partial \tilde{h}}{\partial \xi} \right) + \varepsilon \left( 2 + \eta_0^2 \right) \tilde{h} = \mathcal{O}(\varepsilon^2). \tag{6.35}$$

Note that in the above, we have applied the conditions (6.21).

4. The tip smoothness condition: at $\xi = 0,\ \eta = 1$,

$$\frac{\partial}{\partial \xi}\{\tilde{T}, \tilde{T}_S, \tilde{h}\} = 0 . \tag{6.36}$$

5. The root condition: we reasonably require that the perturbed state not grow exponentially as $\xi \to \infty$. It will be seen that this requirement is equivalent to the following condition: at $\xi \to \infty$

$$\{\tilde{T};\ \tilde{T}_S;\ \tilde{h}\} = 0 . \tag{6.37}$$

In terms of the multiple variables $(\xi, \eta, \xi_+, \eta_+, t_+)$, we now make the following multiple variables expansion (MVE) for perturbed states:

$$\begin{aligned}
\tilde{T} &= \tilde{\mu}_0(\varepsilon)\left\{\tilde{T}_0(\xi, \eta, \xi_+, \eta_+) + \varepsilon\tilde{T}_1(\xi, \eta, \xi_+, \eta_+) + \cdots\right\} e^{\sigma t_+}, \\
\tilde{T}_S &= \tilde{\mu}_0(\varepsilon)\left\{\tilde{T}_{S0}(\xi, \eta, \xi_+, \eta_+) + \varepsilon\tilde{T}_{S1}(\xi, \eta, \xi_+, \eta_+) + \cdots\right\} e^{\sigma t_+}, \\
\tilde{h} &= \tilde{\mu}_0(\varepsilon)\left\{\tilde{h}_0(\xi, \bar{\xi}_+) + \varepsilon\tilde{h}_1(\xi, \bar{\xi}_+) + \cdots\right\} e^{\sigma t_+},
\end{aligned} \tag{6.38}$$

where $\sigma = \sigma_R - i\omega \ (\omega \geq 0)$ is generally a complex number.

In addition to the expansion form (6.38), we make the following asymptotic expansions for the functions $(\phi, \psi, k)$ and the parameter $\sigma$:

$$\begin{aligned}
\phi(\xi, \eta, \varepsilon) &= \phi_0(\xi, \eta) + \varepsilon\phi_1(\xi, \eta) + \cdots, \\
\psi(\xi, \eta, \varepsilon) &= \psi_0(\xi, \eta) + \varepsilon\psi_1(\xi, \eta) + \cdots, \\
k(\xi, \eta, \varepsilon) &= k_0(\xi, \eta) + \varepsilon k_1(\xi, \eta) + \cdots, \\
\bar{k}(\xi, \varepsilon) &= \bar{k}_0(\xi) + \varepsilon\bar{k}_1(\xi) + \cdots, \\
g(\xi, \eta, \varepsilon) &= g_0(\xi, \eta) + \varepsilon g_1(\xi, \eta) + \cdots, \\
\bar{g}(\xi, \varepsilon) &= 0, \\
\sigma &= \sigma_0 + \varepsilon\sigma_1 + \varepsilon^2\sigma_2 + \cdots.
\end{aligned} \tag{6.39}$$

By substituting (6.38)–(6.39) into the system (6.29) and (6.30)–(6.37), one can successively derive each order of approximation.

### 6.2.1 *Zeroth-Order Approximation*

As the zeroth-order approximation, we derive the following quasisteady heat conduction equation:

$$\begin{aligned}
\left(\frac{\partial^2}{\partial\xi_+^2} + \frac{\partial^2}{\partial\eta_+^2}\right)\tilde{T}_0 &= 0, \\
\left(\frac{\partial^2}{\partial\xi_+^2} + \frac{\partial^2}{\partial\eta_+^2}\right)\tilde{T}_{S0} &= 0.
\end{aligned} \tag{6.40}$$

It has the following boundary conditions:

1. In the liquid region $\eta > 1$, as $\eta_+ \to \infty$,

$$\tilde{T}_0 \to 0\,. \tag{6.41}$$

2. In the solid region $\eta < 1$, as $\eta_+ \to -\infty$,

$$\tilde{T}_{S0} \to 0\,. \tag{6.42}$$

3. At the interface $\eta_+ = 0$ or $\eta = 1$,

$$\tilde{T}_0 = \tilde{T}_{S0} + \tilde{h}_0, \tag{6.43}$$

$$\tilde{T}_{S0} = \frac{\bar{k}_0^2}{\hat{S}(\xi)} \frac{\partial^2 \tilde{h}_0}{\partial \bar{\xi}_+^2}, \tag{6.44}$$

$$\frac{\partial}{\partial \eta_+}(\tilde{T}_0 - \tilde{T}_{S0}) + \sigma_0 \hat{S}^2(\xi)\tilde{h}_0 + \bar{k}_0 \xi \frac{\partial \tilde{h}_0}{\partial \bar{\xi}_+} = 0 . \tag{6.45}$$

4. The tip smoothness condition: at $\xi = 0,\ \eta = 1$,

$$\frac{\partial}{\partial \xi}\{\tilde{T}, \tilde{T}_S, \tilde{h}\} = 0 . \tag{6.46}$$

5. The root condition: as $\xi \to \infty$,

$$\tilde{h}_0 = 0 . \tag{6.47}$$

The above system allows the following normal mode solutions:

$$\begin{aligned} \tilde{T}_0 &= A_0(\xi, \eta) \exp\{i\xi_+ - \eta_+\} , \\ \tilde{T}_{S0} &= A_{S0}(\xi, \eta) \exp\{i\xi_+ + \eta_+\} , \\ \tilde{h}_0 &= \hat{D}_0 \exp\left\{i\bar{\xi}_+\right\} , \end{aligned} \tag{6.48}$$

where $\bar{\xi}_+ = \frac{1}{\varepsilon} \int_{\xi_0}^{\xi} \bar{k}_0 d\xi_1$ and the coefficient $\hat{D}_0$ is set as a constant. Note that here, by introducing the form of the solutions (6.48), the solution has been extended analytically to the complex space $(\xi_+, \eta_+)$. Setting

$$\begin{cases} \hat{A}_0(\xi) = A_0(\xi, 1), \\ \hat{A}_{S0}(\xi) = A_{S0}(\xi, 1) , \end{cases} \tag{6.49}$$

From (6.43) to (6.45) we derive the following system of homogeneous equations:

$$\begin{cases} \hat{A}_0 - \hat{A}_{S0} - \hat{D}_0 = 0, \\ \hat{A}_{S0} + \frac{\bar{k}_0^2}{\hat{S}} \hat{D}_0 = 0 \\ -\bar{k}_0(\hat{A}_0 + \hat{A}_{S0}) + (\sigma_0 \hat{S}^2 + i\xi \bar{k}_0)\hat{D}_0 = 0 . \end{cases} \tag{6.50}$$

Obviously, for a nontrivial solution, one must have

$$\Delta = \det \begin{pmatrix} 1 & -1 & -1 \\ 0 & 1 & \bar{k}_0^2/\hat{S} \\ -\bar{k}_0 & -\bar{k}_0 & \sigma_0 \hat{S}^2 + i\xi \bar{k}_0 \end{pmatrix} = 0 , \tag{6.51}$$

which gives the local dispersion relationship

$$\sigma_0 = \Sigma(\xi, \bar{k}_0) = \frac{\bar{k}_0}{\hat{S}^2}\left(1 - \frac{2\bar{k}_0^2}{\hat{S}}\right) - \frac{\mathrm{i}\xi}{\hat{S}^2}\bar{k}_0. \tag{6.52}$$

Then one solves

$$\begin{cases} \hat{A}_0 + \hat{A}_{S0} = \left(1 - \frac{2\bar{k}_0^2}{\hat{S}}\right)\hat{D}_0, \\ \hat{A}_0 - \hat{A}_{S0} = \hat{D}_0. \end{cases} \tag{6.53}$$

The amplitude functions $\hat{A}(\xi), \hat{A}_S(\xi)$ are fully determined up to an arbitrary constant $\hat{D}$.

The local dispersion formula (6.52) obtained above is the generalization of the well-known Mullins–Sekerka formula in unidirectional solidification. The dispersion formula (6.52) is a cubic polynomial of $\bar{k}_0$, from which for fixed parameter $\sigma_0$ and variable $\xi$ one can solve for three roots $\bar{k}_0$. In doing so, we express (6.52) in the form

$$a\bar{k}_0^3 - b\bar{k}_0 = c, \tag{6.54}$$

where

$$a = \frac{2}{\hat{S}}, \quad b = 1 - \mathrm{i}\xi, \quad c = -\hat{S}^2\sigma_0.$$

Assume that $\bar{k}_0 = M\cos\theta$. Hence, (6.54) is changed to

$$M^3 a\left(\cos\theta^3 - \frac{b}{M^2 a}\cos\theta\right) = c, \tag{6.55}$$

which can be further rewritten in the form

$$\frac{1}{4}M^3 a\left(4\cos\theta^3 - 3\cos\theta\right) = \frac{1}{4}M^3 a\cos 3\theta = c, \tag{6.56}$$

provided we set $\frac{b}{M^2 a} = \frac{3}{4}$, or $M = 2\sqrt{\frac{b}{3a}}$. From the above, one may obtain the three wave numbers $\bar{k}_0$ as functions of $\xi$ (Fig. 6.1). Namely

$$\begin{cases} \bar{k}_0^{(1)}(\xi) = M(\xi)\cos\left\{\frac{1}{3}\cos^{-1}\left(\frac{\sigma_0}{N(\xi)}\right)\right\} \\ \bar{k}_0^{(2)}(\xi) = M(\xi)\cos\left\{\frac{1}{3}\cos^{-1}\left(\frac{\sigma_0}{N(\xi)}\right) + \frac{2\pi}{3}\right\} \\ \bar{k}_0^{(3)}(\xi) = M(\xi)\cos\left\{\frac{1}{3}\cos^{-1}\left(\frac{\sigma_0}{N(\xi)}\right) + \frac{4\pi}{3}\right\} \end{cases}, \tag{6.57}$$

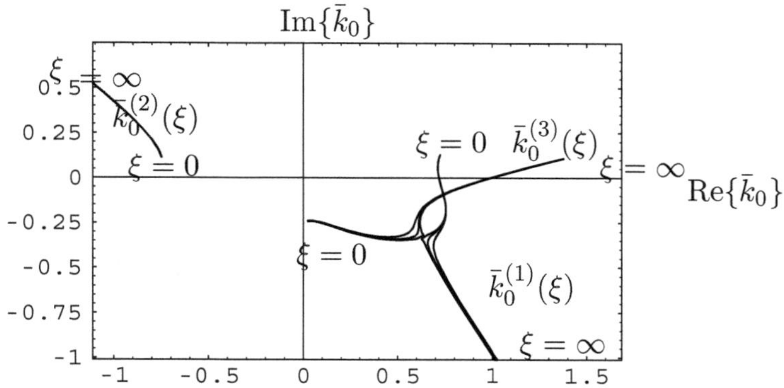


**Fig. 6.1** The variations of the wave number functions $\{\bar{k}_0^{(1)}(\xi), \bar{k}_0^{(2)}(\xi), \bar{k}_0^{(3)}(\xi)\}$ in the complex $k$-plane with $\xi$ for given $\sigma_0 = (0.035, -0.265)$; $(0.035, -0.269)$; $(0.035, -0.270)$

where

$$\begin{cases} M(\xi) = \sqrt{\dfrac{2\hat{S}(\xi)}{3}}(1 - \mathrm{i}\xi)^{\frac{1}{2}} \\ N(\xi) = -\dfrac{M(\xi)}{3\hat{S}^2(\xi)}(1 - \mathrm{i}\xi). \end{cases} \tag{6.58}$$

Given $\sigma_0$, the system allows three fundamental wave solutions corresponding to the wave number functions $\{\bar{k}_0^{(1)}, \bar{k}_0^{(2)}, \bar{k}_0^{(3)}\}$:

$$\begin{aligned} \tilde{h}_0 &= H_1(\xi) = \mathrm{e}^{\mathrm{i}\bar{\xi}_+^{(1)}}, \\ \tilde{h}_0 &= H_2(\xi) = \mathrm{e}^{\mathrm{i}\bar{\xi}_+^{(2)}}, \\ \tilde{h}_0 &= H_3(\xi) = \mathrm{e}^{\mathrm{i}\bar{\xi}_+^{(3)}}, \end{aligned} \tag{6.59}$$

where

$$\bar{\xi}_+^{(i)} = \frac{1}{\xi}\int_{\xi_0}^{\xi} \bar{k}^{(i)}(\xi_1)\mathrm{d}\xi_1 \quad (i = 1, 2, 3).$$

We shall call $\tilde{h}_0 = H_1(\xi)$ the $H_1$-wave solution, $\tilde{h}_0 = H_2(\xi)$ the $H_2$-wave solution, and $\tilde{h}_0 = H_3(\xi)$ the $H_3$-wave solution. Noting that $\mathrm{Re}\{\bar{k}_0^{(1)}(\xi)\} > \mathrm{Re}\{\bar{k}_0^{(3)}(\xi)\} > 0 > \mathrm{Re}\{\bar{k}_0^{(2)}(\xi)\}$, one can deduce that among these three fundamental solutions, the $H_2$-wave solution must be ruled out. Since with a negative real part of its wave number, the $H_2$-wave has $\mathrm{Re}\{\eta_+\} < 0$ in the liquid region ($\eta > 1$). Hence, as $\eta_+ \to \infty$, in the liquid region away from the interface its corresponding perturbed temperature field $\tilde{T}$ will grow exponentially, violating the boundary condition (6.41). In other words, only the $H_1$-wave with the larger positive number $\mathrm{Re}\{\bar{k}_0^{(1)}\}$, which may be called the short-wavelength branch, and the $H_3$-wave with the smaller positive number $\mathrm{Re}\{\bar{k}_0^{(3)}\}$, which may be called the long-wavelength branch, are physically meaningful. One may describe the above statements as follows:

$$\begin{aligned} &H_1\text{-wave (short wave branch)} && \text{with larger } \mathrm{Re}\{\bar{k}_0^{(1)}\} > 0, \\ &H_2\text{-wave (no physical meaning)} && \text{with } \mathrm{Re}\{\bar{k}_0^{(2)}\} < 0, \\ &H_3\text{-wave (long wave branch)} && \text{with smaller } \mathrm{Re}\{\bar{k}_0^{(3)}\} > 0\,. \end{aligned}$$

As a consequence, the general $H$-wave solution of the interface shape function has two components:

$$\tilde{h}_0(\xi) = H(\xi) = D_1 H_1(\xi) + D_3 H_3(\xi), \tag{6.60}$$

and the general outer solution may be written in the form

$$\begin{aligned} \tilde{h} = D_1 \exp\left[\frac{\sigma t}{\varepsilon \eta_0^2} + \frac{\mathrm{i}}{\varepsilon}\int_{\xi_c}^{\xi}\left(\bar{k}_0^{(1)} + \varepsilon \bar{k}_1^{(1)} + \cdots\right)\mathrm{d}\xi_1\right] \\ +D_3 \exp\left[\frac{\sigma t}{\varepsilon \eta_0^2} + \frac{\mathrm{i}}{\varepsilon}\int_{\xi_c}^{\xi}\left(\bar{k}_0^{(3)} + \varepsilon \bar{k}_1^{(3)} + \cdots\right)\mathrm{d}\xi_1\right] + \cdots, \end{aligned} \tag{6.61}$$

where the coefficients $\{D_1, D_3\}$ are arbitrary constants independent of $\varepsilon$.

Having obtained the solution for the function $\bar{k}_0(\xi)$ defined in the region $(0 \le \xi < \infty)$, we still need to determine the functions $k_0(\xi,\eta)$, $\phi_0(\xi,\eta)$ and $\psi_0(\xi,\eta)$ in the whole $(\xi,\eta)$-plane. For this purpose, we extend $\bar{k}_0(\xi)$ by analytical continuation to the function $\bar{k}_0(\zeta)$ defined in the complex plane $\zeta = \xi + \mathrm{i}\xi_{\mathrm{I}}$. It is evident that the extended function $\bar{k}_0(\zeta)$ is subject to the extended dispersion formula

$$\sigma_0 = \Sigma(\zeta, \bar{k}_0) = \frac{\bar{k}_0}{S(\zeta)^2}\left[1 - \frac{2\bar{k}_0^2}{S(\zeta)}\right] - \frac{\mathrm{i}\zeta}{S(\zeta)^2}\bar{k}_0, \tag{6.62}$$

and that it can be expressed in the same form as (6.57) with the replacement of $\xi$ by $\zeta$.

As demonstrated above, one may derive the functions $k_0(\xi,\eta)$, $\phi_0(\xi,\eta)$, and $\psi_0(\xi,\eta)$ in the whole $(\xi,\eta)$-plane as follows:

$$\begin{aligned} \phi_0(\xi,\eta) &= \frac{1}{2}\left\{\int_0^{\xi+\mathrm{i}(\eta-1)} \bar{k}_0(\zeta)\mathrm{d}\zeta + \int_0^{\xi-\mathrm{i}(\eta-1)} \bar{k}_0(\zeta)\mathrm{d}\zeta\right\}, \\ \psi_0(\xi,\eta) &= -\frac{\mathrm{i}}{2}\int_{\xi-\mathrm{i}(\eta-1)}^{\xi+\mathrm{i}(\eta-1)} \bar{k}_0(\zeta)\mathrm{d}\zeta, \end{aligned} \tag{6.63}$$

and

$$\begin{aligned} k_0(\xi,\eta) &= \phi_{0,\xi}(\xi,\eta) = \psi_{0,\eta}(\xi,\eta) \\ &= \frac{1}{2}\Big\{\bar{k}_0\big[\xi + \mathrm{i}(\eta-1)\big] + \bar{k}_0\big[\xi - \mathrm{i}(\eta-1)\big]\Big\} \\ g_0(\xi,\eta) &= \phi_{0,\eta}(\xi,\eta) = -\psi_{0,\xi}(\xi,\eta) \\ &= \frac{\mathrm{i}}{2}\Big\{\bar{k}_0\big[\xi + \mathrm{i}(\eta-1)\big] - \bar{k}_0\big[\xi - \mathrm{i}(\eta-1)\big]\Big\}. \end{aligned} \tag{6.64}$$

At this point, the two fast variables $(\xi_+,\eta_+)$ are fully determined.

Now we turn to the determination of the coefficients $\{D_1, D_3\}$. From Fig. 6.1, it is noted that for fixed $0 < \varepsilon \ll 1$, we have $\xi \gg 1$ near the root. Due to that fact that $\mathrm{Im}\{\bar{k}_0^{(1)}\} < 0$ and $\mathrm{Im}\{\bar{k}_0^{(3)}\} > 0$, one deduces that near the root, the solution $H_1$ increases exponentially, while $H_3$ decreases exponentially to zero. Thus in order to satisfy the root condition in the far field, the following asymptotic form must hold:

$$\tilde{h}_0 \sim D_3' H_3(\xi,\varepsilon) \quad \text{as } \xi \to \infty \ (\varepsilon > 0). \tag{6.65}$$

This asymptotic condition in the far field is called the *radiation condition*. Thus, to satisfy the radiation condition, the coefficients of the outer solution (6.60) in the far field must be set as follows:

$$D_1 = 0, D_3 = D_3'. \tag{6.66}$$

This implies that the outer solution must be described solely by a long wave $H_3$.

However, a single wave solution can never satisfy the tip smoothness condition (6.46) at $\xi = 0$. To satisfy the tip smoothness condition (6.46), the two terms in the outer solution (6.60) must be comparable, say the two coefficients $(D_1, D_3)$ must both be nonzero, namely

$$D_1 \neq 0, D_3 = D_3'. \tag{6.67}$$

In other words, near the tip region, the outer solution must be described by two waves interacting with each other, the $H_1$-wave and the $H_3$-wave. With the above analysis, one can deduce that along the $\xi$-axis from $\xi = 0$ to $\xi = \infty$, the pair of coefficients $(D_1, D_3)$ must may be piecewise constant. It may be change values from (6.67) to (6.66) at some point $\xi = \xi_c$, and as a consequence, the outer solution (6.60) must have a jump discontinuity at that point. Now the following questions arise: Where is the discontinuity point $\xi_c$ located? How can one determine the nonzero constant $D_1$? To answer these questions, one needs to study the RPE form of the outer solution in the extended complex plane $\zeta = \xi + \xi_I$ and to explore the singular points of the RPE solution in the $\zeta$-plane.

To do so, let us look into the first-order approximation of the RPE solution.

### 6.2.2 First-Order Approximation

The first-order approximation solution will fully determine the amplitude $\{A_0(\xi,\eta), A_{S0}(\xi,\eta)\}$, the functions $\bar{k}_1(\xi)$, and $\sigma_1$. The equation for the first-order approximation can be obtained from (6.29):

$$\begin{aligned}
&(k_0^2+g_0^2)\left(\frac{\partial^2}{\partial\xi_+^2}+\frac{\partial^2}{\partial\eta_+^2}\right)\tilde{T}_1\\
&=-\left(2k_0\frac{\partial^2\tilde{T}_0}{\partial\xi_+\partial\xi}-2g_0\frac{\partial^2\tilde{T}_0}{\partial\eta_+\partial\xi}\right)-\left(2k_0\frac{\partial^2\tilde{T}_0}{\partial\eta_+\partial\eta}+2g_0\frac{\partial^2\tilde{T}_0}{\partial\xi_+\partial\eta}\right)\\
&\quad-\left(\frac{1}{\xi}-\eta_0^2\xi\right)\left(k_0\frac{\partial\tilde{T}_0}{\partial\xi_+}-g_0\frac{\partial\tilde{T}_0}{\partial\eta_+}\right)-\left(\frac{1}{\eta}+\eta_0^2\eta\right)\left(g_0\frac{\partial\tilde{T}_0}{\partial\xi_+}+k_0\frac{\partial\tilde{T}_0}{\partial\eta_+}\right)\\
&\quad+\eta_0^2(\xi^2+\eta^2)\frac{\partial\tilde{T}_0}{\partial t_+}.
\end{aligned}\tag{6.68}$$

We may write

$$\begin{aligned}
(k_0^2+g_0^2)\left(\frac{\partial^2}{\partial\xi_+^2}+\frac{\partial^2}{\partial\eta_+^2}\right)\tilde{T}_1&=a_0(\xi,\eta)e^{i\xi_+-\eta_+},\\
(k_0^2+g_0^2)\left(\frac{\partial^2}{\partial\xi_+^2}+\frac{\partial^2}{\partial\eta_+^2}\right)\tilde{T}_{S1}&=a_{S0}(\xi,\eta)e^{i\xi_++\eta_+},
\end{aligned}\tag{6.69}$$

where

$$\begin{aligned}
a_0=2(k_0-ig_0)&\left(\frac{\partial A_0}{\partial\eta}-i\frac{\partial A_0}{\partial\xi}\right)+A_0\left\{\sigma_0\eta_0^2(\xi^2+\eta^2)\right.\\
&\left.-(ik_0+g_0)\left[\left(\frac{1}{\xi}-\eta_0^2\xi\right)+i\left(\frac{1}{\eta}+\eta_0^2\eta\right)\right]\right\}
\end{aligned}\tag{6.70}$$

and

$$\begin{aligned}
a_{S0}=-2(k_0+ig_0)&\left(\frac{\partial A_{S0}}{\partial\eta}+i\frac{\partial A_{S0}}{\partial\xi}\right)+A_{S0}\left\{\sigma_0\eta_0^2(\xi^2+\eta^2)\right.\\
&\left.-(ik_0-g_0)\left[\left(\frac{1}{\xi}-\eta_0^2\xi\right)-i\left(\frac{1}{\eta}+\eta_0^2\eta\right)\right]\right\}.
\end{aligned}\tag{6.71}$$

To eliminate the secular terms on the right-hand side of (6.69), one must set

$$a_0=a_{S0}=0\,.\tag{6.72}$$

Then from (6.72), it follows that

$$\begin{aligned}\left(\frac{\partial}{\partial\eta}-\mathrm{i}\frac{\partial}{\partial\xi}\right)\ln A_0(\xi,\eta) &= \Phi_0(\xi,\eta)\\ \left(\frac{\partial}{\partial\eta}+\mathrm{i}\frac{\partial}{\partial\xi}\right)\ln A_{S0}(\xi,\eta) &= \Phi_{S0}(\xi,\eta),\end{aligned} \tag{6.73}$$

where

$$\begin{aligned}\Phi_0(\xi,\eta) &= -\frac{\sigma_0\eta_0^2(\xi^2+\eta^2)}{2\bar{k}_0[\xi+\mathrm{i}(\eta-1)]}+\frac{\mathrm{i}}{2}\left[\left(\frac{1}{\xi}-\eta_0^2\xi\right)+\mathrm{i}\left(\frac{1}{\eta}+\eta_0^2\eta\right)\right]\\ \Phi_{S0}(\xi,\eta) &= \frac{\sigma_0\eta_0^2(\xi^2+\eta^2)}{2\bar{k}_0[\xi-\mathrm{i}(\eta-1)]}-\frac{\mathrm{i}}{2}\left[\left(\frac{1}{\xi}-\eta_0^2\xi\right)-\mathrm{i}\left(\frac{1}{\eta}+\eta_0^2\eta\right)\right].\end{aligned} \tag{6.74}$$

Equations (6.73) are first-order hyperbolic equations. We can solve $A_0(\xi,\eta)$ and $A_{S0}(\xi,\eta)$ in the $(\xi,\eta)$-plane as an initial value problem given the initial conditions on the line $\eta = 1$:

$$A_0(\xi,1)=\hat{A}_0(\xi);\quad A_{S0}(\xi,1)=\hat{A}_{S0}(\xi).$$

Moreover, with the above results, we may derive that at $\eta = 1$,

$$\begin{aligned}\hat{D}_0 Q_0 = \frac{\partial}{\partial\eta}(A_0-A_{S0})\Big|_{\eta=1} &= \mathrm{i}\frac{\partial}{\partial\xi}(\hat{A}_0+\hat{A}_{S0})+\hat{A}_0\Phi_0-\hat{A}_{S0}\Phi_{S0}\\ &= \mathrm{i}\frac{\partial}{\partial\xi}(\hat{A}_0+\hat{A}_{S0})-(\hat{A}_0+\hat{A}_{S0})\frac{\sigma_0\eta_0^2(1+\eta^2)}{2\bar{k}_0(\xi)}\\ &\quad+\frac{\mathrm{i}}{2}(\hat{A}_0+\hat{A}_{S0})\left(\frac{1}{\xi}-\eta_0^2\xi\right)-\frac{1}{2}(1+\eta_0^2)(\hat{A}_0-\hat{A}_{S0}).\end{aligned} \tag{6.75}$$

It follows that

$$\begin{aligned}Q_0(\xi) = -\mathrm{i}\frac{\partial}{\partial\xi}\left(\frac{2\bar{k}_0^2}{\hat{S}}\right)+\left(1-\frac{2\bar{k}_0^2}{\hat{S}}\right)\left[-\frac{\sigma_0\eta_0^2\hat{S}^2}{2\bar{k}_0}+\frac{\mathrm{i}}{2}\left(\frac{1}{\xi}-\eta_0^2\xi\right)\right]\\ -\frac{1}{2}(1+\eta_0^2).\end{aligned} \tag{6.76}$$

These formulas will be needed later to solve for $\bar{k}_1(\xi)$.

We now turn to solving the equation of the temperature field with the conditions (6.72):

$$\left(\frac{\partial^2}{\partial\xi_+^2}+\frac{\partial^2}{\partial\eta_+^2}\right)\tilde{T}_1=0,$$
$$\left(\frac{\partial^2}{\partial\xi_+^2}+\frac{\partial^2}{\partial\eta_+^2}\right)\tilde{T}_{S1}=0\,. \tag{6.77}$$

The interface conditions at the interface $\eta_+=0, \eta=1$ are as follows:

1. The thermodynamic equilibrium condition:

$$\tilde{T}_1=\tilde{T}_{S1}+\tilde{h}_1-\eta_1\bar{k}_0\frac{\partial}{\partial\eta_+}\left(\tilde{T}_0-\tilde{T}_{S0}\right). \tag{6.78}$$

2. The Gibbs–Thomson condition:

$$\tilde{T}_{S1}+\eta_1\bar{k}_0\frac{\partial\tilde{T}_{S0}}{\partial\eta_+}=\frac{1}{\hat{S}(\xi)}\left\{\bar{k}_0^2\frac{\partial^2\tilde{h}_1}{\partial\xi_+^2}+2\bar{k}_0\bar{k}_1\frac{\partial^2\tilde{h}_0}{\partial\xi_+^2}+2\bar{k}_0\frac{\partial^2\tilde{h}_0}{\partial\xi_+\partial\xi}\right.$$
$$\left.+\bar{k}_0'\frac{\partial\tilde{h}_0}{\partial\xi_+}+\bar{k}_0\left[\frac{1}{\xi}+\frac{\xi}{\hat{S}^2(\xi)}\right]\frac{\partial\tilde{h}_0}{\partial\xi_+}\right\}. \tag{6.79}$$

3. The heat balance condition:

$$\bar{k}_0\left(\frac{\partial\tilde{T}_1}{\partial\eta_+}-\frac{\partial\tilde{T}_{S1}}{\partial\eta_+}\right)+\bar{k}_1\left(\frac{\partial\tilde{T}_0}{\partial\eta_+}-\frac{\partial\tilde{T}_{S0}}{\partial\eta_+}\right)+\eta_1\bar{k}_0^2\frac{\partial^2}{\partial\eta_+^2}\left(\tilde{T}_0-\tilde{T}_{S0}\right)$$
$$+\left(\frac{\partial\tilde{T}_0}{\partial\eta}-\frac{\partial\tilde{T}_{S0}}{\partial\eta}\right)+\hat{S}^2(\xi)(\sigma_0\tilde{h}_1+\sigma_1\tilde{h}_0)$$
$$+\xi\left(\bar{k}_0\frac{\partial\tilde{h}_1}{\partial\bar{\xi}_+}+\bar{k}_1\frac{\partial\tilde{h}_0}{\partial\xi_+}\right)+\left(2+\eta_0^2\right)\tilde{h}_0=0. \tag{6.80}$$

The first-order approximate solutions can be obtained in the form of normal modes:

$$\tilde{T}_1=A_1(\xi,\eta)\exp\{\mathrm{i}\xi_+-\eta_+\}\,,$$
$$\tilde{T}_{S1}=A_{S1}(\xi,\eta)\exp\{\mathrm{i}\xi_++\eta_+\}\,, \tag{6.81}$$
$$\tilde{h}_1=\hat{D}_1\exp\left\{\mathrm{i}\bar{\xi}_+\right\}.$$

By setting

$$\begin{cases} \hat{A}_1(\xi) = A_1(\xi, 1), \\ \hat{A}_{S1}(\xi) = A_{S1}(\xi, 1), \end{cases} \tag{6.82}$$

from (6.78) to (6.80), it is derived that

$$\begin{aligned} &\hat{A}_1 - \hat{A}_{S1} - \hat{D}_1 = I_1\hat{D}_0 = \bar{k}_0\eta_1\Big(1 - \tfrac{2\bar{k}_0^2}{\hat{S}}\Big)\hat{D}_0, \\ &\hat{A}_{S1} + \tfrac{\bar{k}_0^2}{\hat{S}}\hat{D}_1 = I_2\hat{D}_0, \\ &-\bar{k}_0(\hat{A}_1 + \hat{A}_{S1}) + (\sigma_0\hat{S}^2 + \mathrm{i}\xi\bar{k}_0)\hat{D}_1 = I_3\hat{D}_0, \end{aligned} \tag{6.83}$$

where we have defined

$$I_1 = \bar{k}_0\eta_1\Big(1 - \frac{2\bar{k}_0^2}{\hat{S}}\Big), \tag{6.84}$$

$$I_2 = \frac{1}{\hat{S}}\left[\mathrm{i}\bar{k}_0' + \mathrm{i}\bar{k}_0\Big(\frac{1}{\xi} + \frac{\xi}{\hat{S}^2}\Big) - 2\bar{k}_0\bar{k}_1\right] + \frac{\bar{k}_0^3}{\hat{S}}\eta_1, \tag{6.85}$$

and

$$I_3 = -\sigma_1\hat{S}^2 + \bar{k}_1\left(1 - \mathrm{i}\xi - \frac{2\bar{k}_0^2}{\hat{S}}\right) - Q_0 - (2 + \eta_0^2) - \eta_1\bar{k}_0^2 . \tag{6.86}$$

Since the determinant $\Delta$ of the coefficient matrix of the above inhomogeneous system is zero, a necessary condition for the existence of a nontrivial solution for $\{\hat{A}_1, \hat{A}_{S1}, \hat{D}_1\}$ is

$$\det\begin{pmatrix} 1 & -1 & I_1 \\ 0 & 1 & I_2 \\ -\bar{k}_0 & -\bar{k}_0 & I_3 \end{pmatrix} = 0, \tag{6.87}$$

which results in the solvability condition

$$I_3 + 2\bar{k}_0 I_2 + I_1\bar{k}_0 = 0, \tag{6.88}$$

or

$$\bar{k}_1\left[1 - \mathrm{i}\xi - \frac{6\bar{k}_0^2}{\hat{S}}\right] = \hat{S}^2\sigma_1 + (2 + \eta_0^2) - \mathrm{i}\frac{2\bar{k}_0\bar{k}_0'}{\hat{S}} - \mathrm{i}\frac{2\bar{k}_0^2}{\hat{S}}\left(\frac{1}{\xi} + \frac{\xi}{\hat{S}^2}\right) + Q_0.$$

With (6.76), one may rewrite the above formula in the following form:

$$\bar{k}_1(\xi) = \frac{\mathrm{i}}{2\xi} + \frac{R_1(\xi)}{F(\xi)} - \mathrm{i}\frac{R_2(\xi)}{F(\xi)}\frac{\bar{k}_0'(\xi)}{\bar{k}_0(\xi)}, \tag{6.89}$$

where we have defined

$$\begin{aligned} F(\xi) &= \left[1 - \mathrm{i}\xi - \frac{6\bar{k}_0^2}{\hat{S}}\right], \\ R_1(\xi) &= \hat{S}^2\sigma_1 + \left(1 + \frac{\eta_0^2}{2}\right) - \frac{\eta_0^2}{2}\left(1 - \frac{2\bar{k}_0^2}{\hat{S}}\right)^2, \\ R_2(\xi) &= \frac{6\bar{k}_0^2}{\hat{S}}. \end{aligned} \tag{6.90}$$

We extend all the functions such as $\{\bar{k}_1(\xi), F(\xi), \ldots\}$ to the complex plane $\zeta = \xi + \mathrm{i}\xi_I$ by analytic continuation. From (6.89), it is seen that the solution $k_1(\zeta)$ has a singularity at the tip $\zeta = 0$. As $\zeta \to 0$,

$$k_1 \sim \frac{\mathrm{i}}{2\zeta}. \tag{6.91}$$

Moreover, from the dispersion formula (6.52), one finds that $F(\zeta) = \hat{S}^2\partial\Sigma/\partial\bar{k}_0$. Assume that $\zeta_c$ is a root of the functions $\Sigma(\zeta, \bar{k}_0)$ and $F(\zeta)$. Namely, at $\zeta = \zeta_c$,

$$\frac{\partial\Sigma(\zeta_c, \bar{k}_0)}{\partial\bar{k}_0} = 0, \quad F(\zeta_c) = \left[1 - \mathrm{i}\zeta_c - \frac{6\bar{k}_0^2(\zeta_c)}{\hat{S}(\zeta_c)}\right] = 0. \tag{6.92}$$

Then $\zeta_c$ is a singular point of the function $k_1(\zeta)$. To examine the behavior of $\bar{k}_1$ in the vicinity of $\zeta_c$, one may find from the local dispersion formula (6.52) with fixed $\sigma_0$ that

$$\begin{aligned} \frac{\bar{k}_0'(\zeta)}{\bar{k}_0(\zeta)} &= \frac{R_0(\zeta)}{F(\zeta)}, \\ F'(\zeta) &= -\mathrm{i} + \frac{6\zeta\bar{k}_0^2}{\hat{S}^3} - \frac{12\bar{k}_0^2}{\hat{S}}\frac{\bar{k}_0'}{\bar{k}_0}, \end{aligned} \tag{6.93}$$

where

$$R_0(\zeta) = \mathrm{i} + \frac{2\zeta(1 - \mathrm{i}\zeta)}{\hat{S}^2} - \frac{6\zeta\bar{k}_0^2}{\hat{S}^3}. \tag{6.94}$$

Noting that $R_0(\zeta_c) = \mathcal{O}(1)$, we deduce that as $\zeta \to \zeta_c$, $F(\zeta)F'(\zeta) \sim m_0$ holds, where

$$m_0 = -\frac{12\bar{k}_0^2}{\hat{S}} R_0(\zeta_c) \neq 0. \tag{6.95}$$

It follows that as $\zeta \to \zeta_c$,

$$F(\zeta) \sim 2^{\frac{1}{2}} m_0^{\frac{1}{2}} (\zeta - \zeta_c)^{\frac{1}{2}},$$

and

$$-\mathrm{i}\frac{R_2(\zeta)}{F(\zeta)} \frac{\bar{k}_0'(\zeta)}{\bar{k}_0(\zeta)} = -\mathrm{i}\frac{R_2(\zeta)R_0(\zeta)}{F^2(\zeta)} \sim \frac{\mathrm{i}}{4(\zeta - \zeta_c)}.$$

As a result, we find that as $\zeta \to \zeta_c$,

$$\bar{k}_1 \sim \frac{R_1(\zeta_c)}{2^{\frac{1}{2}} m_0^{\frac{1}{2}} (\zeta - \zeta_c)^{\frac{1}{2}}} + \frac{\mathrm{i}}{4(\zeta - \zeta_c)} + \mathcal{O}(1). \tag{6.96}$$

According to the theory of analytic functions, near an isolated singular point $\zeta_c$ in the complex plane, the meromorphic function $\bar{k}_1(\zeta)$ can be expanded into a Laurent series. This implies that

$$R_1(\zeta_c) = 0 \tag{6.97}$$

must hold. It is therefore deduced that the point $\zeta = \zeta_c$, as well as $\zeta = 0$, is actually a simple pole of the function $\bar{k}_1(\zeta)$ in the complex $\zeta$-plane.

From the condition (6.97), one obtains

$$\sigma_1 = -\frac{1}{(1+\zeta_c^2)} \left\{ \left(1 + \frac{\eta_0^2}{2}\right) - \frac{\eta_0^2}{2}\left[1 - \frac{2\bar{k}_0^2(\zeta_c)}{S(\zeta_c)}\right]^2 \right\}. \tag{6.98}$$

Furthermore, from $F(\zeta_c) = 0$, we obtain

$$\left[1 - \frac{2\bar{k}_0^2(\zeta_c)}{S(\zeta_c)}\right] = \frac{2 + \mathrm{i}\zeta_c}{3}. \tag{6.99}$$

Hence, (6.98) may be rewritten as

$$\sigma_1 = -\frac{1}{(1+\zeta_c^2)} \left[ \left(1 + \frac{\eta_0^2}{2}\right) - \frac{\eta_0^2}{18}(2 + \mathrm{i}\zeta_c)^2 \right]. \tag{6.100}$$

With the formula (6.100), $\sigma_1$ is found to be a function of $\sigma_0$. Once the parameter $\sigma_0$ is obtained, then $\sigma_1$ follows. The above analysis shows that in the first-order approximation, though the interface shape correction term $\eta_1(\xi, \theta)$, which may be induced by isotropic/anisotropic surface tension, affects the eigenfunctions of perturbed states, it does not affect the corresponding eigenvalue $\sigma$. Therefore, for the purpose of stability analysis, up to the accuracy of $\mathcal{O}(\varepsilon)$ one may use Ivantsov's solution with zero surface tension as the basic state of dendritic growth.

### 6.2.3 Singular Point $\zeta_c$ of the Outer Solution

From (6.96), we see that the normal modes of the solutions (6.60) fail to be valid at $\zeta_c$, so that the solution is not uniformly valid in the whole complex $\zeta$-plane. The singular point $\zeta_c$ did not appear directly in the zeroth-order approximation, but it is now clearly seen in the first-order approximation. The existence of the singular point $\zeta_c$ plays a crucial role in pattern formation and selection in dendritic growth. This singularity was unknown by all previous researchers in the field until its discovery in 1989 (see, e.g., [22, 24]).

The location of the singular point $\zeta_c$ in the complex $\xi$-plane depends on the value of $\sigma_0$.

Combining (6.92) and (6.52), one finds that for given $\sigma_0$, $\zeta_c$ is determined by the equation

$$\sigma_0 = \sqrt{\frac{2}{27}}\left[\frac{1-\mathrm{i}\zeta_c}{\hat{S}(\zeta_c)}\right]^{3/2} = \mathrm{e}^{-\mathrm{i}\frac{3\pi}{4}}\sqrt{\frac{2}{27}}\left(\frac{\zeta_c+\mathrm{i}}{\zeta_c-\mathrm{i}}\right)^{3/4}, \tag{6.101}$$

or

$$\sigma_0 = \sqrt{\frac{2}{27}}\left(\frac{r_2}{r_1}\right)^{3/4}\mathrm{e}^{\mathrm{i}\frac{3}{4}(\theta_2-\theta_1-\pi)}, \tag{6.102}$$

where

$$\zeta_c+\mathrm{i} = r_2\mathrm{e}^{\mathrm{i}\theta_2}; \quad \zeta_c-\mathrm{i} = r_1\mathrm{e}^{\mathrm{i}\theta_1}.$$

Evidently, if $\zeta_c$ is on the real axis, then $|\sigma_0| = \sqrt{\frac{2}{27}}$, so that the eigenvalue $\sigma_0$ will be on a circle in the complex $\sigma_0$-plane. This will be illustrated further in the next section.

Having identified the location of the singular point $\zeta_c$ of the outer solution (6.60) in the complex $\zeta$-plane, one can better understand the characteristics of the two fundamental solutions $H_1$-wave and $H_3$-wave on the right-hand side of (6.60) as $\varepsilon \to 0$. Note that as exponential functions of $1/\varepsilon$, these solutions have different orders of magnitude in the complex $\zeta$-plane, except at some isolated lines known as the Stokes lines. To be clearer, let us choose the singular point $\zeta_c$ as the lower limit of the integral in the definition of $\xi_+$, so that the fundamental solutions in the $\zeta$-plane are defined as

$$\begin{aligned} H_1(\zeta,\varepsilon) &= \exp\left(\frac{\mathrm{i}}{\varepsilon}\int_{\zeta_c}^{\zeta}\bar{k}_0^{(1)}\mathrm{d}\zeta'\right), \\ H_3(\zeta,\varepsilon) &= \exp\left(\frac{\mathrm{i}}{\varepsilon}\int_{\zeta_c}^{\zeta}\bar{k}_0^{(3)}\mathrm{d}\zeta'\right). \end{aligned} \tag{6.103}$$

The Stokes lines in the $\zeta$-plane are defined by the integral

$$\mathrm{Im}\left\{\int_{\zeta_c}^{\zeta}\left(\bar{k}_0^{(1)}-\bar{k}_0^{(3)}\right)\mathrm{d}\zeta'\right\}=0. \tag{6.104}$$

The anti-Stokes lines are defined by

$$\mathrm{Re}\left\{\int_{\zeta_c}^{\zeta}\left(\bar{k}_0^{(1)}-\bar{k}_0^{(3)}\right)\mathrm{d}\zeta'\right\}=0. \tag{6.105}$$

A sketch of the Stokes lines $(L_1)$, $(L_2)$, $(L_3)$ and anti-Stokes lines $(A_1)$, $(A_2)$, $(A_3)$ of our system are shown in Fig. 6.2. The anti-Stokes line $(A_2)$ emitted from $\zeta_c$ divides the entire complex $\zeta$-plane into the sectors $(S_1)$ and $(S_2)$. The root of the dendrite, $\xi=\infty$, belongs to the sector $(S_2)$, while the tip of the dendrite, $\zeta=0$, belongs to the sector $(S_1)$.

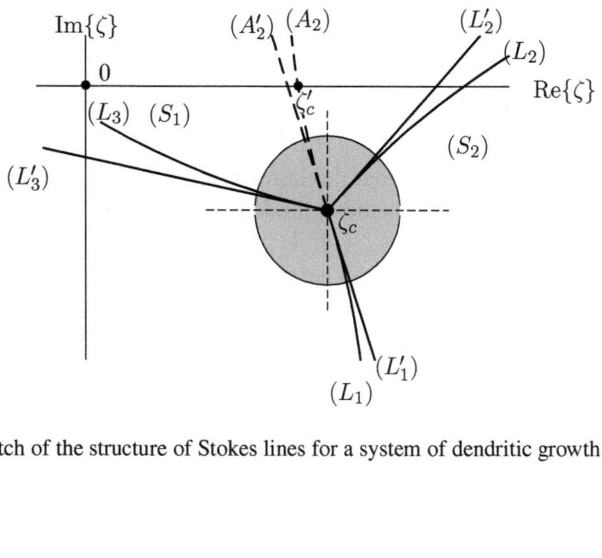


**Fig. 6.2** A sketch of the structure of Stokes lines for a system of dendritic growth

As indicated above, in order to satisfy the radiation condition, the asymptotic form $\tilde{h}_0 \sim D_3' H_3(\zeta, \varepsilon)$ must hold in the root region, whereas to satisfy the tip smoothness condition, $\tilde{h}_0 \sim D_1 H(\zeta, \varepsilon) + D_3' H_3(\zeta, \varepsilon)$ must hold in the tip region. It can be proved for a general ODE system that the coefficients $(D_1, D_3)$ remain constant in the respective sectors $(S_1)$ and $(S_2)$. As a consequence, the jump discontinuity of size $D_1 H(\xi, \varepsilon)$ in the solution $\tilde{h}_0$ occurs along the anti-Stokes line $(A_2)$, which intersects the $\xi$-axis at the point $\xi_c'$. This is the so-called *Stokes phenomenon*.

Why must the discontinuous line, but not any other line, be an anti-Stokes line? And how can one determine the constant $D_1$ in the sector $(S_1)$? Or rather, what is the condition connecting the coefficients $\{D_1, D_3 = D_3'\}$ in the sector $(S_1)$ and $\{D_1' = 0, D_3'\}$ in the sector $(S_2)$? This connection condition will be derived by matching the outer asymptotic solution (6.60) with the inner asymptotic solution in the intermediate region.

The simplest answer to the first question might be that **it is only along the anti-Stokes line that the ratio** $|H_3/H_1|$ has the maximum value with a fixed $\varepsilon \ll 1$, **and as a consequence, the relative portion of the discontinuity caused by adding** $D_1 H_1(\zeta, \varepsilon)$ **to the solution will have the minimum value**.

There are different ways to demonstrate such a Stokes phenomenon systematically and rigorously for a general ODE system. One way is through the WKB (Wentzel–Kramers–Brillouin) method and contour integral; another is through the resurgent function and Borel–Laplace transformation. For a detailed discussion, the reader is referred to [2–6, 9, 11, 12, 14, 16–27].

For the specific system of dendritic growth under study, in the next section we shall prove that the discontinuity line is indeed the anti-Stakes line $(A_2)$ by matching the inner solution in the vicinity of $\zeta_c$ and the outer solution in different directions in the complex $\zeta$-plane. At the same time, we shall derive the connection condition.

## 6.3 The Inner Solutions near the Singular Point $\zeta_c$

As previously indicated, the normal mode form of the MVE solution (6.60) has a singularity at $\zeta = \zeta_c; \eta = 1$ in the extended space $(\zeta, \eta)$. This implies that the solutions in the neighborhood $|\zeta - \zeta_c| \ll 1; |\eta - 1| \ll 1$, cannot be expressed in the form of normal modes. To derive a general form of solutions in the vicinity of the singular point, one must start with the perturbed system (6.3)–(6.8) and construct a different asymptotic expansion. For this purpose, we introduce the inner variables

$$\begin{aligned} \zeta_* &= \frac{\zeta - \zeta_c}{\varepsilon^\alpha}, \\ \eta_* &= \frac{\eta - 1}{\varepsilon^\alpha}, \end{aligned} \tag{6.106}$$

where $\alpha$ is to be determined. In terms of the inner variables, Eq. (6.3) for perturbed states can be expressed in the form

$$\left(\frac{\partial^2}{\partial\zeta_*^2}+\frac{\partial^2}{\partial\eta_*^2}\right)\tilde{T}=\varepsilon^{2\alpha-1}\sigma\eta_0^2(\zeta^2+\eta^2)\tilde{T}+\varepsilon^{\alpha}\eta_0^2\left(\zeta\frac{\partial}{\partial\zeta_*}-\eta\frac{\partial}{\partial\eta_*}\right)\tilde{T}$$
$$-\varepsilon^{\alpha}\left(\frac{1}{\zeta}\frac{\partial}{\partial\zeta_*}+\frac{1}{\eta}\frac{\partial}{\partial\eta_*}\right)\tilde{T}\,. \tag{6.107}$$

When $\alpha > \frac{1}{2}$ (which can be verified later), the above equation can be reduced, as $\varepsilon \to 0$, to

$$\left(\frac{\partial^2}{\partial\zeta_*^2}+\frac{\partial^2}{\partial\eta_*^2}\right)\tilde{T}\approx 0\,. \tag{6.108}$$

With the inner variables, we set

$$\begin{aligned}\tilde{T}(\zeta,\eta,\theta,t)&=\varepsilon^{\alpha}\hat{T}(\zeta_*,\eta_*,\theta,t);\\ \tilde{T}_S(\zeta,\eta,\theta,t)&=\varepsilon^{\alpha}\hat{T}_S(\zeta_*,\eta_*,\theta,t);\\ \tilde{h}(\zeta,\theta,t)&=\varepsilon^{\alpha}\hat{h}(\zeta_*,\theta,t).\end{aligned} \tag{6.109}$$

The interface shape function is

$$\eta_s(\zeta,\theta,t)=1+\tilde{h}(\zeta,\theta,t)/\eta_0^2=1+\varepsilon^{\alpha}\hat{h}(\zeta_*,\theta,t)/\eta_0^2. \tag{6.110}$$

The boundary conditions (6.4)–(6.8) are transformed to the following:

1. As $\eta_* \to \infty$,

$$\hat{T}\to 0\,. \tag{6.111}$$

2. As $\eta_* \to -\infty$,

$$\hat{T}_S\to 0\,. \tag{6.112}$$

3. At the interface $\eta_* = 0$,

 (i)

$$\hat{T}=\hat{T}_S+\hat{h}+\text{ (h.o.t)}\,, \tag{6.113}$$

 (ii)

$$\hat{T}_S=\frac{\varepsilon^{2-2\alpha}}{\hat{S}(\zeta)}\frac{\partial^2\hat{h}}{\partial\zeta_*^2}+\text{ (h.o.t)}\,, \tag{6.114}$$

(iii)

$$\varepsilon^{1-\alpha}\frac{\partial}{\partial\eta_*}\left(\hat{T}-\hat{T}_S\right)+\sigma\hat{S}^2(\zeta)\hat{h}+\varepsilon^{1-\alpha}\zeta\frac{\partial\hat{h}}{\partial\zeta_*}$$
$$= \text{(h.o.t)}. \tag{6.115}$$

We seek the mode solutions and make the following inner expansions:

$$\begin{aligned}
\hat{T}(\zeta_*,\eta_*,t) &= \left[\hat{\nu}_0(\varepsilon)\hat{T}_0(\zeta_*,\eta_*)+\nu_1(\varepsilon)\hat{T}_1(\zeta_*,\eta_*)+\cdots\right]e^{\frac{\sigma t}{\varepsilon\eta_0^2}},\\
\hat{T}_S(\zeta_*,\eta_*,t) &= \left[\hat{\nu}_0(\varepsilon)\hat{T}_{S0}(\zeta_*,\eta_*)+\nu_1(\varepsilon)\hat{T}_{S1}(\zeta_*,\eta_*)+\cdots\right]e^{\frac{\sigma t}{\varepsilon\eta_0^2}},\\
\hat{h}(\zeta_*,t) &= \left[\hat{\nu}_0(\varepsilon)\hat{h}_0+\nu_1(\varepsilon)\hat{h}_1+\cdots\right]e^{\frac{\sigma t}{\varepsilon\eta_0^2}}.
\end{aligned} \tag{6.116}$$

Letting $\varepsilon\to 0$, the above inner equations can be further simplified into the following third-order complex ODE for the unsteady interface perturbations. As a matter of fact, from the zeroth-order approximate equation,

$$\left(\frac{\partial^2}{\partial\zeta_*^2}+\frac{\partial^2}{\partial\eta_*^2}\right)\hat{T}_0=0, \tag{6.117}$$

it can be deduced that

1. in the liquid phase region, we have

$$\left(\mathrm{i}\frac{\partial}{\partial\zeta_*}-\frac{\partial}{\partial\eta_*}\right)\hat{T}_0=0; \tag{6.118}$$

2. in the solid phase region, we have

$$\left(\mathrm{i}\frac{\partial}{\partial\zeta_*}+\frac{\partial}{\partial\eta_*}\right)\hat{T}_{S0}=0. \tag{6.119}$$

Consequently, at the interface ($\eta=1,\eta_*=0$), we have

$$\frac{\partial}{\partial\eta_*}\left(\hat{T}_0-\hat{T}_{S0}\right)=\mathrm{i}\frac{\partial}{\partial\zeta_*}\left(\hat{T}_0+\hat{T}_{S0}\right). \tag{6.120}$$

On the other hand, from the boundary conditions (6.113) and (6.114), one can obtain

$$\left(\hat{T}_0+\hat{T}_{S0}\right)=\hat{h}_0+\frac{2\varepsilon^{2-2\alpha}}{\hat{S}}\frac{\mathrm{d}^2\hat{h}_0}{d\zeta_*^2}. \tag{6.121}$$

Combining (6.115) with (6.120) and (6.121), we derive the equation for the perturbed interface as follows:

$$\mathrm{i}\frac{2\varepsilon^{3-3\alpha}}{\hat{S}}\frac{\mathrm{d}^3\hat{h}_0}{\mathrm{d}\zeta_*^3} + \varepsilon^{1-\alpha}\big(\zeta + \mathrm{i}\big)\frac{\mathrm{d}\hat{h}_0}{\mathrm{d}\zeta_*} + \sigma\hat{S}^2\hat{h}_0 = 0\,. \tag{6.122}$$

Returning to the outer variable $\zeta$, this equation may be written in the form

$$\mathrm{i}\frac{2\varepsilon^{3}}{\hat{S}}\frac{\mathrm{d}^3\tilde{h}_0}{\mathrm{d}\zeta^3} + \varepsilon\big(\zeta + \mathrm{i}\big)\frac{\mathrm{d}\tilde{h}_0}{\mathrm{d}\zeta} + \sigma_0\hat{S}^2\tilde{h}_0 = 0\,. \tag{6.123}$$

It should be pointed out that one may apply (6.123) to the entire outer region and derive its asymptotic solutions by the WKB method. By doing so, one can regain the local dispersion relationship (6.52). Hence, in the leading-order approximation, the WKB solution obtained for the ODE system (6.123) is exactly the same as the outer solution (6.60) obtained for the original PDE system.

One may also formally derive the inner equation (6.123) in a simpler way starting with the dispersion relationship (6.52). To do so, let us extend the dispersion formula to the complex $\zeta$-plane as

$$\left[\hat{S}^2\sigma_0 - \bar{k}_0(1 - \mathrm{i}\zeta) + \frac{2\bar{k}_0^3}{\hat{S}}\right]\tilde{h}_0 = 0, \tag{6.124}$$

and then rewrite it in operator form by making the replacements:

$$\bar{k}_0 \Longleftrightarrow -\mathrm{i}\varepsilon\frac{\mathrm{d}}{\mathrm{d}\zeta}.$$

This leads directly to Eq. (6.123).

To solve (6.122), we introduce the transformation

$$H(\zeta) = W(\zeta)\exp\left[\frac{\mathrm{i}}{\varepsilon}\int_{\zeta_c}^{\zeta} k_c(\zeta_1)\,\mathrm{d}\zeta_1\right], \tag{6.125}$$

which transforms an $H$-wave

$$H(\zeta) = \exp\left[\frac{\mathrm{i}}{\varepsilon}\int_{\zeta_c}^{\zeta} k(\zeta_1)\,\mathrm{d}\zeta_1\right]$$

into a $W$-wave

$$W(\zeta) = D_0\exp\left[\frac{\mathrm{i}}{\varepsilon}\int_{\zeta_c}^{\zeta}\big[\bar{k}(\zeta_1) - k_c(\zeta_1)\big]\,\mathrm{d}\zeta_1\right]. \tag{6.126}$$

In (6.125), the function $k_c(\zeta)$ is called the reference wave number function, which is to be chosen later, but it will be seen that $k_c(\zeta)$ will satisfy the inequalities

$$\mathrm{Re}\{\bar{k}_0^{(3)}\} < \mathrm{Re}\{k_c\} < \mathrm{Re}\{\bar{k}_0^{(1)}\}. \tag{6.127}$$

With the transformation (6.125), in the outer region we may change the $H$-wave representation of the outer solution to the $W$-wave representation as follows:

$$\tilde{h}_0 = H_1 = W_0^{(+)}(\zeta)\exp\left[\frac{\mathrm{i}}{\varepsilon}\int_{\zeta_c}^{\zeta} k_c(\zeta_1)\,\mathrm{d}\zeta_1\right] \tag{6.128}$$

and

$$\tilde{h}_0 = H_3 = W_0^{(-)}(\zeta)\exp\left[\frac{\mathrm{i}}{\varepsilon}\int_{\zeta_c}^{\zeta} k_c(\zeta_1)\,\mathrm{d}\zeta_1\right]. \tag{6.129}$$

In the form of the $W$-wave representation, the general outer solutions may be written as

$$W_0 = D_1 W_0^{(+)} + D_3 W_0^{(-)}. \tag{6.130}$$

The wave number function for the $W_0^{(+)}$ wave is $\bar{k}_0^{(+)} = \bar{k}_0^{(1)} = \bar{k}_0^{(1)} - k_c$. Since $\mathrm{Re}\{\bar{k}_0^{(+)}\} > 0$, the phase velocity of the $W_0^{(+)}$ wave is positive, and as such, it is an outgoing wave. On the other hand, the wave number function for the $W^{(-)}$ wave is $\bar{k}_0^{(-)} = \bar{k}_0^{(3)} = \bar{k}_0^{(3)} - k_c$, and $\mathrm{Re}\{\bar{k}_0^{(-)}\} < 0$. Hence, the $W_0^{(-)}$ wave is an incoming wave with negative phase velocity. One may draw the following diagram showing the relationship between the $H$-wave representation and the $W$-wave representation:

$H_1$-wave $\Longleftrightarrow$ $W^{(+)}$
(short-wave branch) (outgoing wave);

$H_3$-wave $\Longleftrightarrow$ $W^{(-)}$
(long-wave branch) (incoming wave).

In accordance with the above transformation, in the inner region we can set the inner solution to

$$\hat{h}_0 = \hat{W}_0(\zeta_*)\exp\left\{\frac{\mathrm{i}}{\varepsilon}\int_{\zeta_c}^{\zeta} k_c(\zeta_1)\,\mathrm{d}\zeta_1\right\}. \tag{6.131}$$

Hence one derives

$$\begin{aligned}
\frac{d\hat{h}_0}{d\zeta} &= \left(\frac{d\hat{W}_0}{d\zeta} + \frac{ik_c}{\varepsilon}\hat{W}_0\right)\exp\left[\frac{i}{\varepsilon}\int_{\zeta_c}^{\zeta} k_c(\zeta_1)\,d\zeta_1\right] \\
\frac{d^2\hat{h}_0}{d\zeta^2} &= \left[\frac{d^2\hat{W}_0}{d\zeta^2} + \frac{2ik_c}{\varepsilon}\frac{d\hat{W}_0}{d\zeta} + \left(\frac{ik_c'}{\varepsilon} - \frac{k_c^2}{\varepsilon^2}\right)\hat{W}_0\right] \\
&\quad \times \exp\left\{\frac{i}{\varepsilon}\int_{\zeta_c}^{\zeta} k_c(\zeta_1)\,d\zeta_1\right\} \\
&\approx \left[\frac{d^2\hat{W}_0}{d\zeta^2} + \frac{2ik_c}{\varepsilon}\frac{d\hat{W}_0}{d\zeta} - \frac{k_c^2}{\varepsilon^2}\hat{W}_0\right]\exp\left\{\frac{i}{\varepsilon}\int_{\zeta_c}^{\zeta} k_c(\zeta_1)\,d\zeta_1\right\} \\
\frac{d^3\hat{h}_0}{d\zeta^3} &= \left[\frac{d^3\hat{W}_0}{d\zeta^3} + \frac{3ik_c}{\varepsilon}\frac{d^2\hat{W}_0}{d\zeta^2} + \left(\frac{3ik_c'}{\varepsilon} - \frac{3k_c^2}{\varepsilon^2}\right)\frac{d\hat{W}_0}{d\zeta}\right. \\
&\quad \left. - \left(\frac{ik_c^3}{\varepsilon^3} + \frac{3k_ck_c'}{\varepsilon^2} - \frac{ik_c''}{\varepsilon}\right)\hat{W}_0\right]\exp\left\{\frac{i}{\varepsilon}\int_{\zeta_c}^{\zeta} k_c(\zeta_1)\,d\zeta_1\right\} \\
&\approx \left[\frac{d^3\hat{W}_0}{d\zeta^3} + \frac{3ik_c}{\varepsilon}\frac{d^2\hat{W}_0}{d\zeta^2} - \frac{3k_c^2}{\varepsilon^2}\frac{d\hat{W}_0}{d\zeta} - \frac{ik_c^3}{\varepsilon^3}\hat{W}_0\right] \\
&\quad \times \exp\left\{\frac{i}{\varepsilon}\int_{\zeta_c}^{\zeta} k_c(\zeta_1)\,d\zeta_1\right\}.
\end{aligned} \tag{6.132}$$

Equation (6.123) is now changed to

$$\varepsilon^3\Omega_3\frac{d^3\hat{W}_0}{d\zeta^3} + i\varepsilon^2\Omega_2\frac{d^2\hat{W}_0}{d\zeta^2} - \varepsilon\Omega_1\frac{d\hat{W}_0}{d\zeta} + i\hat{S}^2\left\{\sigma - \Sigma_c(\zeta)\right\}\hat{W}_0 = 0\,, \tag{6.133}$$

where

$$\begin{aligned}
\Omega_3 &= \tfrac{1}{3!}\left[\tfrac{\partial^3(\hat{S}^2\Sigma)}{\partial\bar{k}_0^3}\right]_{\bar{k}_0=k_c} = -\tfrac{2}{\hat{S}}, \\
\Omega_2 &= \tfrac{1}{2!}\left[\tfrac{\partial^2(\hat{S}^2\Sigma)}{\partial\bar{k}_0^2}\right]_{\bar{k}_0=k_c} = -\tfrac{6k_c}{\hat{S}}, \\
\Omega_1 &= \left[\tfrac{\partial(\hat{S}^2\Sigma)}{\partial\bar{k}_0}\right]_{\bar{k}_0=k_c} = -\tfrac{6k_c^2}{\hat{S}} + (1 - i\zeta), \\
\Sigma_c(\zeta) &= \Sigma(k_c,\zeta) = e^{-i\frac{3\pi}{4}}\sqrt{\tfrac{2}{27}}\left(\tfrac{\zeta+i}{\zeta-i}\right)^{\frac{3}{4}}.
\end{aligned} \tag{6.134}$$

If we set

$$\Omega_1 = -\frac{6k_c^2}{\hat{S}} + (1 - i\zeta) = 0 \tag{6.135}$$

or

$$k_c(\zeta) = \mathrm{e}^{-\mathrm{i}\frac{\pi}{4}} \sqrt{\frac{\hat{S}}{6}(\zeta + \mathrm{i})} \qquad \left(\mathrm{Re}\{k_c\} > 0\right), \tag{6.136}$$

then (6.123) can be changed to

$$\varepsilon^3 \Omega_3 \frac{\mathrm{d}^3 \hat{W}_0}{\mathrm{d}\zeta^3} + \mathrm{i}\varepsilon^2 \Omega_2 \frac{\mathrm{d}^2 \hat{W}_0}{\mathrm{d}\zeta^2} + \mathrm{i}\hat{S}^2 \Big\{\sigma - \Sigma_c(\zeta)\Big\} \hat{W}_0 = 0\,. \tag{6.137}$$

Using the relationship $\zeta = \zeta_c + \varepsilon^\alpha \zeta_*$, Eq. (6.137) may be written in the form

$$\varepsilon^{3-3\alpha} \Omega_3 \frac{\mathrm{d}^3 \hat{W}_0}{\mathrm{d}\zeta_*^3} + \mathrm{i}\varepsilon^{2-2\alpha} \Omega_2 \frac{\mathrm{d}^2 \hat{W}_0}{\mathrm{d}\zeta_*^2} + \mathrm{i}\hat{S}^2 \Big\{\sigma - \Sigma_c(\zeta)\Big\} \hat{W}_0 = 0\,. \tag{6.138}$$

It is now clear that this equation has a turning point singularity in the complex $\zeta$-plane at the root of the equation

$$\sigma_0 - \Sigma_c(\zeta) = \sigma_0 - \mathrm{e}^{-\mathrm{i}\frac{3\pi}{4}} \sqrt{\frac{2}{27}} \left(\frac{\zeta + \mathrm{i}}{\zeta - \mathrm{i}}\right)^{\frac{3}{4}} = 0\,. \tag{6.139}$$

This simple turning point coincides with the singular point $\zeta_c$ that was previously determined from the equations

$$\frac{\partial \Sigma(\zeta, \bar{k}_0)}{\partial \bar{k}_0} = 0\,, \quad F(\zeta) = 0\,, \quad \text{or} \quad \bar{k}_0^{(3)} = \bar{k}_0^{(1)}. \tag{6.140}$$

To obtain a single-valued analytical solution in the complex $\zeta$-plane, we make a branch cut along the Stokes line ($L_1$) as shown in Fig. 6.2. For given $\sigma_0 = |\sigma_0|\mathrm{e}^{\mathrm{i}\theta}$, from (6.139) one can obtain the solution

$$\zeta_c = \mathrm{i} \left( \frac{|\sigma_0|^{\frac{4}{3}} \mathrm{e}^{\mathrm{i}\frac{4\theta}{3}} - \frac{2^{\frac{2}{3}}}{9}}{|\sigma_0|^{\frac{4}{3}} \mathrm{e}^{\mathrm{i}\frac{4\theta}{3}} + \frac{2^{\frac{2}{3}}}{9}} \right). \tag{6.141}$$

Note that $\zeta_c = \zeta_c(\sigma_0)$ is a multivalued analytic function with three branches. The conformal mapping from the complex $\sigma_0$-plane to the complex $\zeta_c$-plane by one of its branches is illustrated in Figs. 6.3 and 6.4. The semicircle ($\gamma$) in the complex $\sigma_0$-plane corresponding to $|\sigma_0| = \sigma_{\max} = \sqrt{\frac{2}{27}} = 0.2722$, where $\sigma_{\max}$ is the maximum growth rate of the Mullins–Sekerka instability for a flat interface, plays a special role. It is seen that as $\sigma_0 \to 0$, this turning point approaches the point $\zeta = -\mathrm{i}$, while as $\sigma_0 \to \infty$, $\zeta_c \to \mathrm{i}$. Furthermore, when $\sigma_0$ is on the semicircle ($\gamma$) in the complex $\sigma_0$-plane, the corresponding critical point is on the real axis ($-\infty < \zeta_c < \infty$); when the eigenvalue $\sigma_0$ belongs to the domain outside the semicircle ($\gamma$), the critical

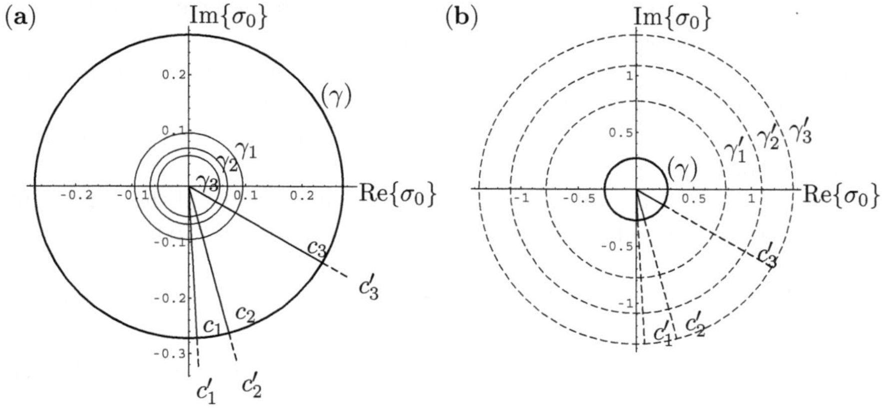


**Fig. 6.3** The *rays* and *circles* in the complex $\sigma_0$-plane, where the *circle* $(\gamma)$ corresponds to $|\sigma_0| = 0.2722$, all the *lines* outside $(\gamma)$ are shown as *dashed lines*, and all *lines* inside $(\gamma)$ are shown as *solid lines*

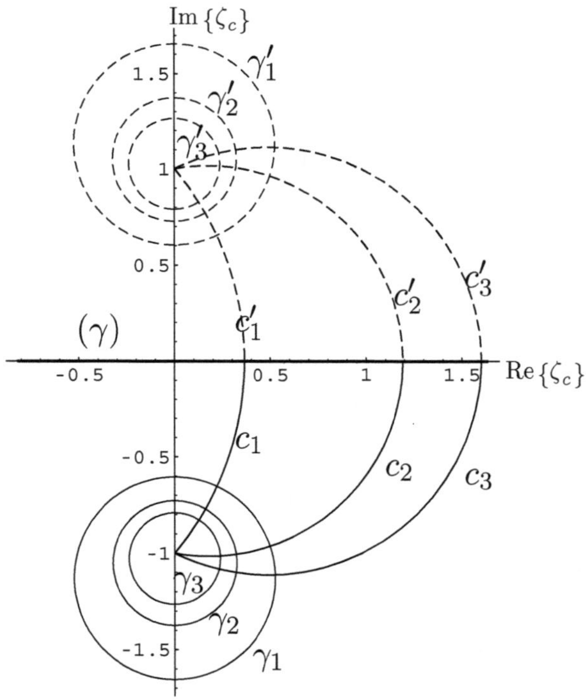


**Fig. 6.4** The images in the complex $\zeta_c$-plane under the conformal mapping $\zeta_c = \zeta_c(\sigma_0)$, transformed from the *lines* in the $\sigma_0$-plane, where *dashed lines* correspond to *dashed lines*, *solid lines to solid lines*

point $\zeta_c$ under discussion will be in the upper half of the complex $\zeta$-plane. Thus, the branch cut will intersect the real axis ($\zeta > 0$). In this case, the system has no physically acceptable continuous solution. Thus, as a necessary condition, the eigenvalue $\sigma$ must be in the domain inside the semicircle ($\gamma$) in order for the turning point $\zeta_c$ to be in the lower half of the complex $\zeta$-plane. This condition was called the *pattern formation condition* in [24].

In the vicinity of $\zeta_c$, one can take Taylor expansions for all the coefficients of (6.138) as functions of $\zeta$. By balancing the leading terms, it is found that

$$\alpha = \frac{2}{3} \quad \text{and} \quad \zeta_* = \frac{(\zeta - \zeta_c)}{\varepsilon^{\frac{2}{3}}}. \tag{6.142}$$

Letting $\varepsilon \to 0$, in the leading-order approximation, the inner equation reduces to the Airy equation:

$$\frac{d^2 \hat{W}_0}{d\zeta_*^2} + A^2 \zeta_* \hat{W}_0 = 0, \tag{6.143}$$

where

$$A = \left( \frac{S^3}{6k_c} \frac{\partial \Sigma_c}{\partial \zeta} \right)^{\frac{1}{2}}_{\zeta = \zeta_c} = e^{i\pi/4} \sqrt{\frac{1}{6}} \left( \frac{\zeta_c + i}{\zeta_c - i} \right)^{\frac{1}{4}} \tag{6.144}$$
$$\left( -\frac{\pi}{4} < \arg\{A\} < 0 \right).$$

This is written as the standard Airy equation

$$\frac{d^2 \hat{W}_0}{d\hat{\zeta}_*^2} + \hat{\zeta}_* \hat{W}_0 = 0, \tag{6.145}$$

by introducing the new inner variable $\hat{\zeta}_*$:

$$\hat{\zeta}_* = \frac{A^{\frac{2}{3}}}{\varepsilon^{\frac{2}{3}}} (\zeta - \zeta_c). \tag{6.146}$$

The general solution of the above Airy equation is

$$\hat{W}_0 = D_{*1} \hat{\zeta}_*^{\frac{1}{2}} H^{(1)}_{\frac{1}{3}}(z) + D_{*2} \hat{\zeta}_*^{\frac{1}{2}} H^{(2)}_{\frac{1}{3}}(z) \tag{6.147}$$
$$\left( z = \frac{2}{3} \hat{\zeta}_*^{\frac{3}{2}} \right),$$

where $H^{(1)}_\nu(z)$ is the $\nu$th-order Hankel function of the first kind, while $H^{(2)}_\nu(z)$ is the Hankel function of the second kind. In order to match this with the outer solution satisfying the downstream far-field condition (6.65), the inner solution must be

$$\hat{W}_0 = D_* \hat{\zeta}_*^{\frac{1}{2}} H_{\frac{1}{3}}^{(2)}(z)\,. \tag{6.148}$$

As $\hat{\zeta}_* \to \infty$ , the above Airy equation has two asymptotic solutions expressible in the form

$$\hat{W}_0 \sim \begin{cases} D_* \exp\left\{ \mathrm{i} \int_0^{\hat{\zeta}_*} k_* \mathrm{d}\zeta_* \right\}, \\ D_* \exp\left\{ -\mathrm{i} \int_0^{\hat{\zeta}_*} k_* \mathrm{d}\zeta_* \right\}, \end{cases} \tag{6.149}$$

where

$$k_* = \hat{\zeta}_*^{\frac{1}{2}}\,. \tag{6.150}$$

From this we find that

$$\int_0^{\hat{\zeta}_*} k_* \mathrm{d}\zeta_* = \frac{2}{3} \hat{\zeta}_*^{\frac{3}{2}} = z\,. \tag{6.151}$$

The Stokes lines of the asymptotic solutions (6.149) for the Airy equation are found to be the rays of $\arg(\zeta) = 0,\ \pi$, and $2\pi$ (or $\arg(\hat{\zeta}_*) = 0,\ \frac{2\pi}{3}$, and $\frac{4\pi}{3}$), which we label by $(L_1')$, $(L_2')$, $(L_3')$, respectively. The anti-Stokes lines of the asymptotic solutions (6.149) are the rays of $\arg(\zeta) = \frac{\pi}{2},\ \frac{3\pi}{2}$, and $\frac{5\pi}{2}$ (or $\arg(\hat{\zeta}_*) = \frac{\pi}{3},\ \pi$, and $\frac{5\pi}{3}$), which we label by $(A_1')$, $(A_2')$, $(A_3')$, respectively. These lines are sketched in Fig. 6.2. The Stokes and anti-Stokes lines of the inner equation (6.148) are respectively tangential lines of the Stokes and anti-Stokes lines of the outer equation determined by (6.104) and (6.105), which are also sketched in Fig. 6.2 and are labeled $(L_1), (L_2), (L_3)$ and $(A_1), (A_2), (A_3)$. As mentioned in the last section, along the anti-Stokes line $(A_2)$, which crosses the real $\zeta$-axis at the point $\zeta_c'$, the coefficient pair of the outer solution will have a minimum jump discontinuity.

We now turn to matching the inner solution with the outer solution. In doing so, we need the asymptotic expansions of the inner solution (6.148) in the far field, $\zeta_* \to \infty$. From Bessel function theory, one can find the asymptotic expansions of (6.148) along different directions in the different sectors, and the connections between these expansions (refer to [1]). Namely,

$$\begin{aligned} & H_\nu^{(1)}(z) \sim \sqrt{\frac{2}{\pi z}} \mathrm{e}^{\mathrm{i}z - \mathrm{i}(\nu/2+1/4)\pi} \left\{ 1 + \mathcal{O}(1/z) \right\} \\ & \quad (-\pi < \arg(z) \leq 2\pi), \\ & H_\nu^{(2)}(z) \sim \sqrt{\frac{2}{\pi z}} \mathrm{e}^{-\mathrm{i}z + \mathrm{i}(\nu/2+1/4)\pi} \left\{ 1 + \mathcal{O}(1/z) \right\} \\ & \quad (-2\pi < \arg(z) \leq \pi)\,. \end{aligned} \tag{6.152}$$

When $z_1 \in (S_1)$,

$$H_\nu^{(2)}(z_1) = H_\nu^{(2)}(z e^{i\pi}) = 2\cos(\nu\pi) H_\nu^{(2)}(z) + e^{i\nu\pi} H_\nu^{(1)}(z) \quad (-\pi \le \arg(z) \le \pi). \tag{6.153}$$

In terms of these formulas, it is seen that in order to match the inner solution with the outer solution in the sector $(S_2)$ as $|\zeta_*| \to \infty$, one needs to balance

$$\hat{\nu}_0(\varepsilon)\hat{W}_0 \Longleftrightarrow \tilde{\mu}_0(\varepsilon) D_3' W_0^{(-)}(\zeta, \varepsilon),$$

or

$$\hat{\nu}_0(\varepsilon)\varepsilon^{\frac{1}{6}}\sqrt{\frac{2}{\pi}} e^{i(\nu/2+1/4)\pi} D_* \frac{e^{-iz}}{(\zeta - \zeta_c)^{\frac{1}{4}}}$$
$$\Updownarrow$$
$$\tilde{\mu}_0(\varepsilon) D_3' H_3(\zeta, \varepsilon) \exp\left\{-\frac{i}{\varepsilon}\int_{\zeta_c}^{\zeta} \bar{k}_c(\zeta_1)\, d\zeta_1\right\}.$$

Noting that in the intermediate region between the outer region and the inner region near the singular point $\zeta_c$, we have $\bar{k}_0 \approx k_c$, the dispersion relationship of the outer solutions is then reduced to the following form:

$$0 = \Big[\sigma_0 - \Sigma(\bar{k}_0, \zeta)\Big] \approx \Big[\sigma_0 - \Sigma_c(\zeta)\Big] - \frac{S^3}{6k_c}(\bar{k}_0 - k_c)^2,$$

or

$$(\bar{k}_0 - k_c)^2 \approx \frac{S^3}{6k_c}\Big[\sigma_0 - \Sigma_c(\zeta)\Big].$$

Hence, it may be derived that, as $\zeta \to \zeta_c$ in $(S_2)$,

$$\bar{k}_0^{(1)} - k_c \propto k_*, \quad \text{and} \quad \bar{k}_0^{(3)} - k_c \propto -k_*, \tag{6.154}$$

subsequently,

$$\begin{cases} W_0^{(+)}(\zeta, \varepsilon) = e^{\frac{i}{\varepsilon}\int_{\zeta_c}^{\zeta}\left(k_0^{(1)} - k_c\right) d\zeta} \Leftrightarrow e^{iz} \\ W_0^{(-)}(\zeta, \varepsilon) e^{\frac{i}{\varepsilon}\int_{\zeta_c}^{\zeta}\left(k_0^{(3)} - k_c\right) d\zeta} \Leftrightarrow e^{-iz}. \end{cases} \tag{6.155}$$

In this case, the matching condition leads to

$$D_* = D_3'\sqrt{\frac{2}{\pi}} e^{-i(\nu/2+1/4)\pi}, \quad \text{and} \quad \tilde{\mu}_0(\varepsilon) = \varepsilon^{\frac{1}{6}}\hat{\nu}_0(\varepsilon). \tag{6.156}$$

On the other hand, in the sector $(S_1)$, we have

$$\hat{W}_0(\zeta_*) \sim D_3'\zeta_*^{-\frac{1}{4}}\Big\{e^{-iz}\big(1+\mathcal{O}(1/z)\big) - ie^{iz}\big(1+\mathcal{O}(1/z)\big)\Big\}, \tag{6.157}$$

as $\zeta_* \to \infty$. Thus in order to match the inner solution with the outer solution in the intermediate region in the sector $(S_1)$ as $\varepsilon \to 0$, one needs to balance

$$\hat{\nu}_0(\varepsilon)\hat{W}_0 \Longleftrightarrow \tilde{\mu}_0(\varepsilon)\left\{D_1 W_0^{(+)}(\zeta,\varepsilon) + D_3 W_0^{(-)}(\zeta,\varepsilon)\right\},$$

or

$$\begin{gathered}\frac{D_3'}{(\zeta-\zeta_c)^{\frac{1}{4}}}\left\{e^{-iz} - ie^{iz} + \mathcal{O}(1/z)\right\} \\ \Updownarrow \\ \tilde{\mu}_0(\varepsilon)\left\{D_1 W_0^{(+)}(\zeta,\varepsilon) + D_3 W_0^{(-)}(\zeta,\varepsilon)\right\}.\end{gathered} \tag{6.158}$$

It should be remarked that $\hat{\zeta}_* = 0$ is the branch point of the function $k_*(\hat{\zeta}_*)$. Hence, as $\zeta$ moves from sector $(S_2)$ to $(S_1)$, the function $z(\hat{\zeta}_*) = \int_0^{\hat{\zeta}_*} k_* d\zeta_*$ will be analytically extended to the function $-z(\hat{\zeta}_*)$. Accordingly, as $\zeta \to \zeta_c$ in $(S_1)$, we have

$$\begin{aligned} W_0^{(+)}(\zeta,\varepsilon) &= e^{\frac{i}{\varepsilon}\int_{\zeta_c}^{\zeta}\left(k_0^{(1)}-k_c\right)d\zeta} \Leftrightarrow e^{-iz}, \\ W_0^{(-)}(\zeta,\varepsilon) &= e^{\frac{i}{\varepsilon}\int_{\zeta_c}^{\zeta}\left(k_0^{(3)}-k_c\right)d\zeta} \Leftrightarrow e^{iz}. \end{aligned} \tag{6.159}$$

Therefore, for matching, the connection condition

$$\frac{D_1}{D_3} = e^{\frac{i\pi}{2}}, \quad D_3 = D_3' \tag{6.160}$$

must hold. Moreover, we point out that since $k_1 \sim \frac{i}{4(\zeta-\zeta_c)}$ as $\zeta \to \zeta_c$, the magnitude functions of the inner solution and outer solution are automatically matched with each other:

$$\frac{1}{(\zeta-\zeta_c)^4} \quad \Longleftrightarrow \quad e^{i\int_{\zeta_c}^{\zeta} k_1 d\zeta'}.$$

## 6.4 Tip Inner Solution in the Tip Region

As previously indicated, the MVE solution is also not valid in the tip inner region $(|\xi| = \mathcal{O}(\varepsilon); |1-\eta| = \mathcal{O}(\varepsilon))$, for in this region, the terms containing the factor $\varepsilon/\xi$ on the left-hand side of the linear system (6.3)–(6.11) are no longer higher-order terms. They must all be included in the leading-order approximation. Moreover, the first-order approximation solution also shows a singularity at $\xi = 0$. Thus, in the tip region, we need to rescale the variables and make different asymptotic expansions for the solution. We define the tip inner variables

$$\begin{aligned} \hat{\xi} &= \tfrac{\hat{k}\xi}{\varepsilon}, \\ \hat{\eta} &= \tfrac{\hat{k}^2(\eta-1)}{\varepsilon}, \\ \hat{t} &= \tfrac{t}{\eta_0^2\varepsilon}, \end{aligned} \tag{6.161}$$

where $|\xi| \ll \varepsilon$ and $|\eta - 1| \ll 1$. The tip solution can be expressed as a function of these inner variables and expanded in the following asymptotic form as $\varepsilon \to 0$:

$$\begin{aligned} \hat{T} &= \left\{\hat{\mu}_0(\varepsilon)\hat{T}_0 + \hat{\mu}_1(\varepsilon)\hat{T}_1 + \cdots\right\}e^{\sigma\hat{t}} \\ \hat{T}_S &= \left\{\hat{\mu}_0(\varepsilon)\hat{T}_{S0} + \hat{\mu}_1(\varepsilon)\hat{T}_{S1} + \cdots\right\}e^{\sigma\hat{t}} \\ \hat{h} &= \left\{\hat{\mu}_0(\varepsilon)\hat{h}_0 + \hat{\mu}_1(\varepsilon)\hat{h}_1 + \cdots\right\}e^{\sigma\hat{t}} \\ \hat{k} &= \hat{k}_0 + \varepsilon\hat{k}_1 + \cdots. \end{aligned} \tag{6.162}$$

In the zeroth-order approximation, in the tip region the system (6.3)–(6.8) can be reduced to

$$\begin{aligned} \hat{k}_0^2\frac{\partial^2\hat{T}_0}{\partial\hat{\xi}^2} + \hat{k}_0^2\frac{\partial^2\hat{T}_0}{\partial\hat{\eta}^2} + \frac{\hat{k}_0^2}{\hat{\xi}}\frac{\partial\hat{T}_0}{\partial\hat{\xi}} &= 0, \\ \hat{k}_0^2\frac{\partial^2\hat{T}_{S0}}{\partial\hat{\xi}^2} + \hat{k}_0^2\frac{\partial^2\hat{T}_{S0}}{\partial\hat{\eta}^2} + \frac{\hat{k}_0^2}{\hat{\xi}}\frac{\partial\hat{T}_{S0}}{\partial\hat{\xi}} &= 0, \end{aligned} \tag{6.163}$$

with the following boundary conditions at $\hat{\eta}$=0:

$$\hat{T}_0 = \hat{T}_{S0} + \hat{h}_0, \tag{6.164}$$

$$\hat{T}_{S0} = \hat{k}_0^2\left(\frac{\partial^2\hat{h}_0}{\partial\hat{\xi}^2} + \frac{1}{\hat{\xi}}\frac{\partial\hat{h}_0}{\partial\hat{\xi}}\right), \tag{6.165}$$

$$\frac{\partial}{\partial\hat{\eta}}\left(\hat{T}_0 - \hat{T}_{S0}\right) + \sigma_0\hat{h}_0 = 0. \tag{6.166}$$

This system admits the inner solutions

$$\begin{aligned} \hat{T}_0 &= \hat{a}_0 H_0^{(1)}(\hat{\xi})e^{-\hat{k}_0\hat{\eta}}, \\ \hat{T}_{S0} &= \hat{a}_{S0} H_0^{(1)}(\hat{\xi})e^{\hat{k}_0\hat{\eta}}, \\ \hat{h}_0 &= \hat{d}_0 H_0^{(1)}(\hat{\xi}), \end{aligned} \tag{6.167}$$

where $H_0^{(1)}(\hat{\xi})$ is the zeroth-order Hankel function of the first kind.

From the boundary conditions (6.164)–(6.166), one obtains

$$\hat{a}_0 = (1 - \hat{k}_0^2)\hat{d}_0 , \quad \hat{a}_{S0} = -\hat{k}_0^2\hat{d}_0 , \tag{6.168}$$

and the local dispersion relationship in the tip region

$$\sigma_0 = \hat{k}_0\big(1 - 2\hat{k}_0^2\big). \tag{6.169}$$

For fixed $\sigma_0$, (6.169) has three roots for $k_0$. Comparing the local dispersion relationship in the tip region (6.169) with the local dispersion relationship in the outer region (6.52), one can evidently write

$$\begin{cases} \hat{k}_0^{(1)} = k_0^{(1)}(0), \\ \hat{k}_0^{(2)} = k_0^{(2)}(0), \\ \hat{k}_0^{(3)} = k_0^{(3)}(0). \end{cases} \tag{6.170}$$

The root $\hat{k}_0^{(2)}$ must be ruled out due to the fact that $\mathrm{Re}\{\hat{k}_0^{(2)}\} < 0$. Therefore, the general solution of $\hat{h}_0$ in the tip region is

$$\hat{h}_0 = \hat{d}_0^{(1)} H_0^{(1)}\left(\hat{k}_0^{(1)}\hat{\xi}\right) + \hat{d}_0^{(3)} H_0^{(1)}\left(\hat{k}_0^{(3)}\hat{\xi}\right). \tag{6.171}$$

The above tip inner solution must match the outer solution. By setting

$$\hat{d}_0^{(1)} = d_0, \quad \hat{d}_0^{(3)} = -Q_0 d_0, \tag{6.172}$$

we can express the tip inner solution as

$$\hat{h}_0 = d_0 \left\{ H_0^{(1)}\left(\hat{k}_0^{(1)}\hat{\xi}\right) - Q_0 H_0^{(1)}\left(\hat{k}_0^{(3)}\hat{\xi}\right)\right\} . \tag{6.173}$$

As $\hat{\xi} \to \infty$, one has

$$\hat{h}_0 \approx d_0 \sqrt{\frac{2}{\pi}} \left\{ \sqrt{\frac{\varepsilon}{k_0^{(1)}(0)\xi}}\, \mathrm{e}^{\frac{\mathrm{i}k_0^{(1)}(0)\xi}{\varepsilon}} - Q_0 \sqrt{\frac{\varepsilon}{k_0^{(3)}(0)\xi}}\, \mathrm{e}^{\frac{\mathrm{i}k_0^{(3)}(0)\xi}{\varepsilon}} \right\} . \tag{6.174}$$

We now match the tip inner solution (6.173) with the outer solution derived in the last section. It implies that we must balance

$$\hat{\mu}_0(\varepsilon) \sqrt{\frac{2\varepsilon}{\pi}} d_0 \left[ \sqrt{\frac{1}{\hat{k}_0^{(1)}\hat{\xi}}}\mathrm{e}^{\frac{\mathrm{i}\hat{k}_0^{(1)}\xi}{\varepsilon}} - Q_0 \sqrt{\frac{1}{\hat{k}_0^{(3)}\hat{\xi}}}\mathrm{e}^{\frac{\mathrm{i}\hat{k}_0^{(3)}\xi}{\varepsilon}} \right]$$

$$\Updownarrow$$

$$\tilde{\mu}_0(\varepsilon) \left\{ D_1 \exp\left[\frac{i}{\varepsilon}\int_{\xi_c}^{\xi} k_0^{(1)} \mathrm{d}\xi'\right] + D_3 \exp\left[\frac{i}{\varepsilon}\int_{\xi_c}^{\xi} k_0^{(3)} \mathrm{d}\xi'\right] \right\} .$$

To satisfy this matching condition, the parameter $\sigma_0$ must be a proper function of $\varepsilon$ such that as $\varepsilon \to 0$, the two functions, $\left\{e^{i\chi_1(\varepsilon)}, e^{i\chi_3(\varepsilon)}\right\}$ are of the same order of magnitude. Here we have defined

$$\chi_1(\varepsilon) = \frac{1}{\varepsilon}\int_0^{\zeta_c} k_0^{(1)}(\xi')d\xi',$$
$$\chi_3(\varepsilon) = \frac{1}{\varepsilon}\int_0^{\zeta_c} k_0^{(3)}(\xi')d\xi'. \tag{6.175}$$

In other words, the parameter $\sigma_0$ must be properly chosen so that the turning point $\zeta_c$ and the structure of Stokes lines, as functions of $\sigma_0$, are arranged in such a way that the tip $\xi = 0$ is located on the Stokes line $(L_3)$.

Thus it follows from the matching condition that

$$\hat{\mu}_0(\varepsilon)\varepsilon^{\frac{1}{2}} = \tilde{\mu}_0(\varepsilon), \tag{6.176}$$

and

$$\frac{D_3 e^{-i\chi_3}}{D_1 e^{-i\chi_1}} = -ie^{i(\chi_1-\chi_3)} = -Q_0\left[\frac{k_0^{(1)}(0)}{k_0^{(3)}(0)}\right]^{\frac{1}{2}}, \tag{6.177}$$

or

$$Q_0 = Q_0(\sigma_0, \varepsilon) = ie^{i(\chi_1-\chi_3)}\left[\frac{k_0^{(1)}(0)}{k_0^{(3)}(0)}\right]^{\frac{1}{2}}. \tag{6.178}$$

## 6.5 Global Trapped-Wave Modes and the Quantization Condition

We turn to the second step: constructing the global eigenmodes and deriving the quantization condition for the eigenvalues by applying the tip smoothness condition (6.11) to the asymptotic solutions obtained above.

For axially symmetric modes, the tip smoothness condition is that

$$\text{as } \xi \to 0, \quad \hat{h}_0(0) < \infty, \quad \text{and} \quad \hat{h}_0'(0) = 0. \tag{6.179}$$

In terms of the formula (6.173), this tip condition leads to

$$d_0(1 - Q_0) = 0, \tag{6.180}$$

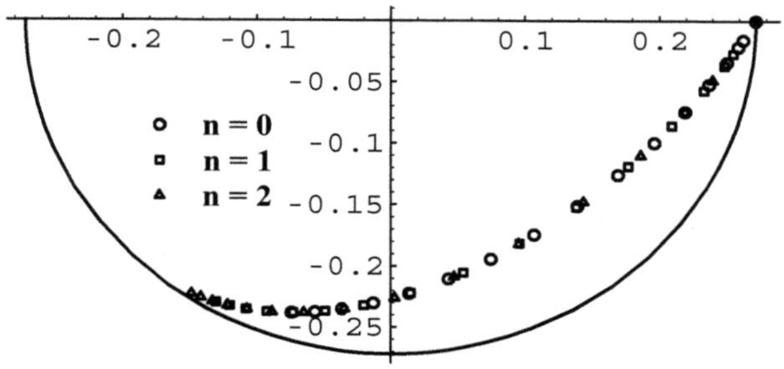


**Fig. 6.5** The variation of eigenvalues $\sigma_0$ with $\varepsilon$ in the complex $\sigma_0$-plane

or

$$Q_0(\sigma_0, \varepsilon) = 1. \tag{6.181}$$

Formula (6.181) is the quantization condition for the eigenvalues. Combining it with (6.178), it may be rewritten in the form

$$\frac{1}{\varepsilon}\int_0^{\xi_c}\left(k_0^{(1)} - k_0^{(3)}\right)\mathrm{d}\xi = \left(2n + 1 + \tfrac{1}{2} + \tfrac{\theta_0}{2}\right)\pi - \tfrac{\mathrm{i}}{2}\ln\alpha_0 \tag{6.182}$$

$$\left(n = 0, \pm1, \pm2, \pm3, \ldots\right),$$

where

$$\alpha_0\, \mathrm{e}^{\mathrm{i}\theta_0\pi} = \frac{k_0^{(1)}(0)}{k_0^{(3)}(0)}. \tag{6.183}$$

This quantization condition gives a discrete set of complex eigenvalues

$$\sigma_0^{(n)} \quad (n = 0, \pm1, \pm2, \ldots),$$

which are functions of $\varepsilon$ (Fig. 6.5). In Figs. 6.6 and 6.7, we show the variations of $\sigma_{0\mathrm{R}}$ and $\omega_0$ of global trapped-wave (GTW) modes $n = 0, 1, 2, 3$, for the case $\hat{\alpha}_\mathrm{T} = 1$. The system under consideration has no real spectrum. It is seen that the system allows a unique neutral $n$ mode ($\sigma_\mathrm{R} = 0$) with the eigenvalue $\sigma = -\mathrm{i}\omega_{*n}$ when $\varepsilon = \varepsilon_{*n}$. It is calculated that

$$\varepsilon_{*0} = \varepsilon_* > \varepsilon_{*1} > \varepsilon_{*2} > \cdots.$$

Here, the critical number $\varepsilon_*$ corresponds to the neutrally stable mode with the index $n = 0$. Obviously, when $\varepsilon > \varepsilon_*$, the system will be absolutely stable. When $\varepsilon_{*1} < \varepsilon < \varepsilon_{*0}$, the system has one growing mode and infinitely many decaying modes; when $\varepsilon_{*2} < \varepsilon < \varepsilon_{*1}$, the system has two growing modes, and in general, when

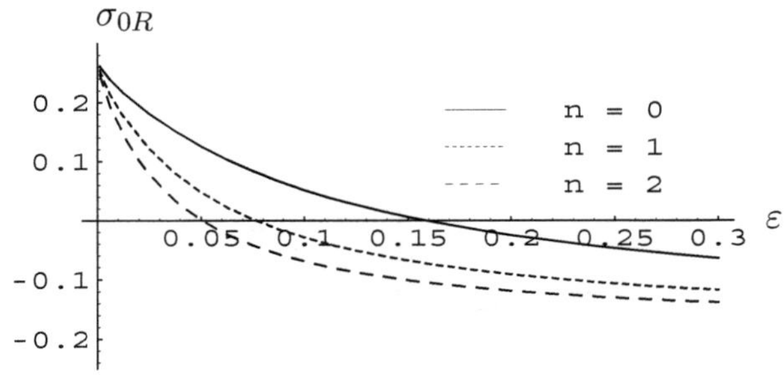


**Fig. 6.6** The variations of the real part of the zeroth-order approximation of eigenvalues, $\sigma_{0R}$, of 3D axially symmetric GTW modes ($n = 0, 1, 2$) with $\varepsilon$

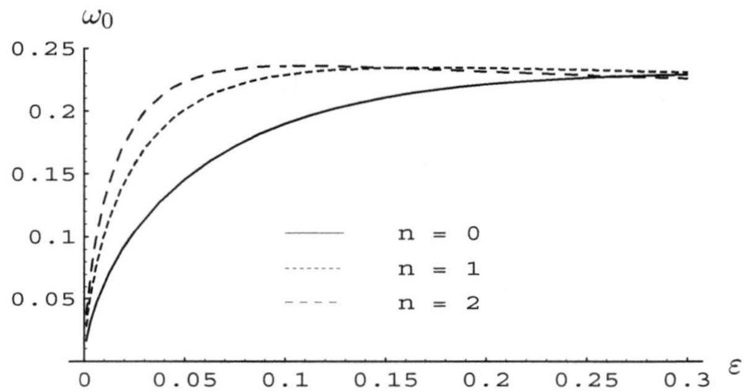


**Fig. 6.7** The variations of the imaginary part of the zeroth-order approximation of eigenvalues, $\omega_0$, of 3D axially symmetric GTW modes ($n = 0, 1, 2$) with $\varepsilon$

$\varepsilon_{*m} < \varepsilon < \varepsilon_{*(m-1)}$, the system has $m$ growing modes. As $\varepsilon \to 0$, the eigenvalues of these growing modes apparently tend to the limit $\sigma_0 = (0.2722, 0.0)$, which corresponds to the maximum growth rate of the Mullins–Sekerka instability. We also show the variation of the eigenvalues in the complex $\sigma_0$-plane with $\varepsilon$ in Fig. 6.5. It is very interesting to see that in the leading-order approximation, all eigenvalues of the modes $n = 0, 1, 2, \ldots$ lie on the same curve in the complex $\sigma_0$-plane.

In Table 6.1, we list both the zeroth- and first-order approximate eigenvalues of the first four modes for the case $\varepsilon = 0.1, \eta_0^2 = 0.01$.

In the leading-order approximation, the eigenvalues $\sigma \approx \sigma_0$ are independent of the Péclet number $\eta_0^2$. We have calculated that the global neutrally stable mode with the index $n = 0$ has the eigenvalue $\sigma = -\mathrm{i}\omega_*^{(0)} = -0.2129\mathrm{i}$. It corresponds to the critical number

$$\varepsilon = \varepsilon_*^{(0)} = 0.1590. \tag{6.184}$$

In the first-order approximation, the eigenvalues $\sigma \approx \sigma_0 + \varepsilon\sigma_1$ will be a function of the Péclet number $\eta_0^2$. Consequently, the eigenvalue $\sigma = -\mathrm{i}\omega_*^{(1)}$ of the neutral

**Table 6.1** The eigenvalues of axisymmetric GTW modes ($\varepsilon = 0.1$; $\eta_0^2 = 0.01$; $T_\infty = -0.2375$; $\hat{\alpha}_{\mathrm{T}} = 1$)

| $n$ | $\sigma_0$ | $\sigma = \sigma_0 + \varepsilon\sigma_1$ | $V_{\mathrm{p}}$ |
|---|---|---|---|
| 0 | (0.05205, −0.18957) | (0.01098, −0.21175) | 1.03327 |
| 1 | (−0.02853, −0.22840) | (−0.04602, −0.23714) | 0.97429 |
| 2 | (−0.06899, −0.23553) | (−0.07939, −0.23983) | 0.92588 |
| 3 | (−0.09294, −0.23568) | (−0.10006, −0.23818) | 0.89086 |

mode ($n = 0$), as well as the corresponding critical number $\varepsilon = \varepsilon_*^{(1)}$, is a function of the Péclet number $\mathrm{Pe}_0 = \eta_0^2$. Here $\varepsilon = \varepsilon_*^{(1)}$ is calculated from the condition $\mathrm{Re}\{\sigma_0 + \varepsilon\sigma_1\} = 0$.

Such a dependence is insensitive to a small undercooling ( $|T_\infty| \ll 1$ ).

Recall that the analysis carried out and the results obtained so far are for the case of a symmetric model with $\hat{\alpha}_{\mathrm{T}} = 1$. The results can be easily extended to the more general case with $0 \le \hat{\alpha}_{\mathrm{T}} \le 1$, in which case only the heat balance condition (6.35) will be changed. It is found that the form of the quantization condition (6.182) remains valid for all $\hat{\alpha}_{\mathrm{T}} \in [0, 1]$, but the local dispersion relationship is modified as

$$\sigma_0 = \Sigma(\xi, \bar{k}_0) = \frac{\bar{k}_0}{\hat{S}^2}\left[1 - \frac{(1+\hat{\alpha}_{\mathrm{T}})\hat{k}_0^2}{\hat{S}}\right] - \frac{\mathrm{i}\xi}{\hat{S}^2}\bar{k}_0 \tag{6.185}$$

and the formula for parameter $\sigma_1$ is changed to

$$\begin{aligned}\sigma_1 = -\frac{1}{(1+\zeta_c^2)}\bigg\{1 + \frac{1}{2}\eta_0^2 + \frac{\eta_0^2}{6}\Big[(1-\hat{\alpha}_{\mathrm{T}})(1-\mathrm{i}\xi_c) \\ -(2+\mathrm{i}\zeta_c)\Big(1 - \frac{1+\hat{\alpha}_{\mathrm{T}}^2}{1+\hat{\alpha}_{\mathrm{T}}}\frac{1-\mathrm{i}\zeta_c}{3}\Big)\Big]\bigg\}.\end{aligned} \tag{6.186}$$

It is calculated that as $|T_\infty| = 0.002$, for the symmetric model ($\hat{\alpha}_{\mathrm{T}} = \kappa_{\mathrm{TS}}/\kappa_T = 1$),

$$\varepsilon_*^{(1)} \approx 0.1108, \quad \omega_*^{(1)} \approx 0.2183, \tag{6.187}$$

whereas for the one-sided model ($\hat{\alpha}_{\mathrm{T}} = \kappa_{\mathrm{TS}}/\kappa_T = 0$), it is found that

$$\varepsilon_*^{(1)} \approx 0.1567, \quad \omega_*^{(1)} \approx 0.3087. \tag{6.188}$$

However, for large undercooling, say $\mathrm{Pe}_0 > 1.0$, the situation is changed. The critical number $\varepsilon_*^{(1)}$ decreases significantly as the undercooling temperature increases (see Table 6.2, and Fig. 6.8).

**Table 6.2** The zeroth- and first-order approximations of the critical numbers $\varepsilon_*^{(0)}$, $\varepsilon_*^{(1)}$ and the corresponding frequencies $\omega_*^{(0)}$, $\omega_*^{(1)}$ of the GTW neutral modes ($n = 0$, $\hat{\alpha}_T = 1$) for various Péclet number $Pe_0 = \eta_0^2$

| $Pe_0 = \eta_0^2$ | $T_\infty$ | $\varepsilon_*^{(0)}$ | $\omega_*^{(0)}$ | $\varepsilon_*^{(1)}$ | $\omega_*^{(1)}$ | $V_p$ |
|---|---|---|---|---|---|---|
| 0.0005 | −0.1930E−2 | 0.1590 | 0.2129 | 0.11081 | 0.21828 | 1.02414 |
| 0.0045 | −0.1245E−1 | 0.1590 | 0.2129 | 0.11073 | 0.21826 | 1.02413 |
| 0.01 | −0.2375E−1 | 0.1590 | 0.2129 | 0.11066 | 0.21830 | 1.02411 |
| 0.1 | −0.1297 | 0.1590 | 0.2129 | 0.10878 | 0.21826 | 1.02410 |
| 1.0 | −0.4615 | 0.1590 | 0.2129 | 0.09167 | 0.21610 | 1.02331 |
| 10 | −0.8521 | 0.1590 | 0.2129 | 0.03157 | 0.14557 | 1.00525 |

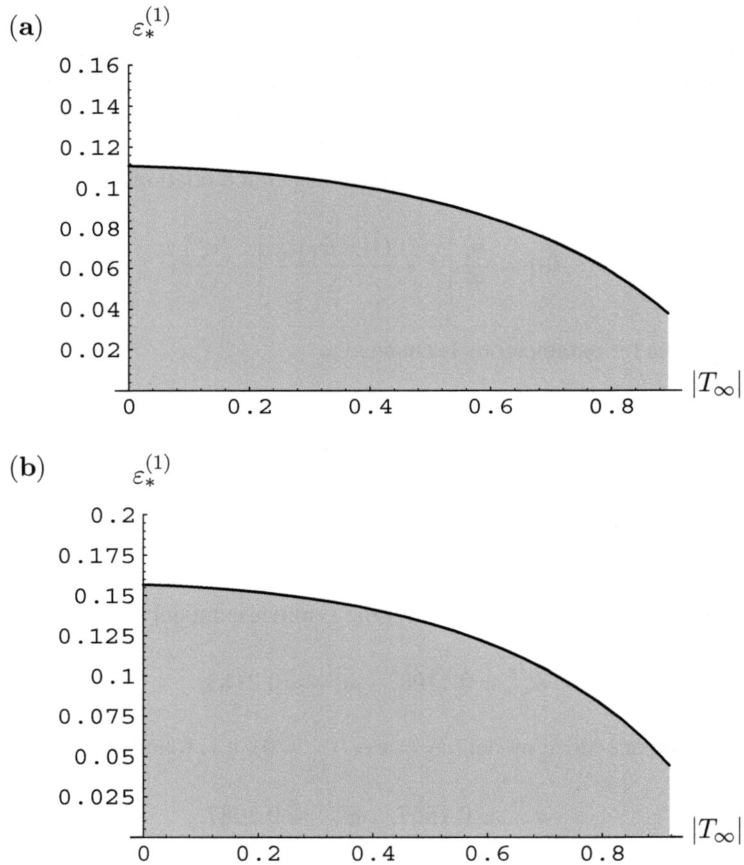


**Fig. 6.8** Variation of the critical number $\varepsilon_*^{(1)}$ for the cases (**a**) $\hat{\alpha} = 1$ (symmetric model); (**b**) $\hat{\alpha} = 0$ (one-sided model). The *shaded region* is the linearly unstable region of steady solutions

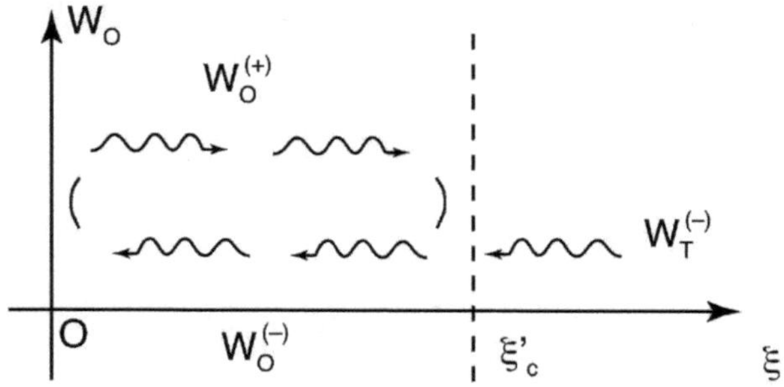


**Fig. 6.9** Wave diagram of the GTW mechanism

The global mode solutions obtained above have important physical implications. A wave diagram for these global modes is sketched in Fig. 6.9. It is seen that an incident outgoing wave $W_0^{(+)}$ from the tip collides with an incoming wave $W_0^{(T)}$ from the far field at the point $\xi_c'$ on the anti-Stokes line ($A_2$); the collision generates an incoming wave $W_0^{(-)}$ propagating toward the tip region. This incoming wave $W_0^{(-)}$ is then reflected at the tip region, and again becomes an outgoing wave $W_0^{(+)}$. The waves appear trapped in the sector ($S_2$) between the tip point and the point $\xi_c'$. No wave escapes beyond the anti-Stokes line ($A_2$). This is why we call these global modes the *global trapped-wave* (GTW) modes. In the far field, the solution $\tilde{h}(\xi, t)$ describes a long outgoing $H_3$ wave.

It is of interest to examine the phase velocity of the traveling wave in the far field $\xi \to \infty$. In doing so, we express the global mode solution in the far field, (6.65) in the form

$$\tilde{h} \sim A(\xi, t)\mathrm{e}^{\frac{\mathrm{i}}{\varepsilon}\Phi(\xi,t)}\,, \tag{6.189}$$

where

$$\begin{aligned} \Phi(\xi, t) &= \textstyle\int_{\xi_c}^{\xi} \mathrm{Re}\{k_0^{(3)}\}\mathrm{d}\xi - \frac{\omega_* t}{\eta_0^2}, \\ A(\xi, t) &= \exp\left\{ \tfrac{\sigma_{\mathrm{R}} t}{\varepsilon \eta_0^2} - \tfrac{1}{\varepsilon} \textstyle\int_{\xi_c}^{\xi} \mathrm{Im}\{k_0^{(3)}\}\mathrm{d}\xi \right\}. \end{aligned} \tag{6.190}$$

As $\xi \to \infty$, for a fixed eigenvalue $\sigma$, the wave number functions $k_0^{(i)}$ $(i = 1, 2, 3)$ have the asymptotic expansion

$$k_0^{(i)}(\xi) \sim a_0^{(i)}\xi + a_1^{(i)} + \frac{a_2^{(i)}}{\xi} + \cdots. \tag{6.191}$$

The phase velocity of the wave along the interface $\eta = 1$ is calculated as

$$V_{\mathrm{p}} = -\Big(\frac{\partial \Phi}{\partial t}\Big)\Big(\frac{\partial \Phi}{\partial \ell}\Big) = \frac{\omega_*(1+\xi^2)^{\frac{1}{2}}}{\mathrm{Re}\{k_0^{(3)}(\xi)\}}, \tag{6.192}$$

and in the far field, we have

$$V_{\mathrm{p}} \approx \frac{\omega_*}{Re\{a_0^{(3)}\}} \qquad (\xi \to \infty). \tag{6.193}$$

In the above, $\ell$ represents the arc length measured along the interface $\eta = 1$ starting from the tip. We have $d\ell/d\xi = 1/\sqrt{1+\xi^2}$. Numerical computations show that for the GTW neutral modes, the phase velocity $V_{\mathrm{p}}$ is approximately equal to 1.0 (see Table 6.2). This implies that the phase velocity of the GTW modes in the moving frame fixed at the tip is approximately equal to the tip velocity in the laboratory frame. This result is in agreement with experimental observations.

The existence of growing GTW modes explains the origin and essence of the dendritic structure in the solidifying system. Any initial perturbation in the growth process will stimulate a spectrum of the above global modes. As $t \to \infty$, all decaying modes will vanish, while the amplitudes of the growing modes increase exponentially. Eventually, the GTW mode with the largest growth rate dominates the features of the microstructure of the dendrite. From the above analysis, one sees that to form the GTW mechanism, properly imposing the boundary conditions at both the tip and the root is important. In linear stability theory, the amplitude of a growing mode increases exponentially with time. In the real system, however, one can anticipate that when the amplitude becomes large, further increase will be suppressed by the nonlinearity and other dissipative effects that may be involved. It appears that with the GTW mode, the head of the dendrite eventually persistently emits a long outgoing interfacial wave train propagating along the interface of the Ivantsov needle solution toward the far field with a phase velocity near unity. This forms the fantastic patterns that are observed in experiments. The selection problem for realistic dendritic growth addressed in Chap. 4 actually implies the question of what limit solution will be naturally approached by the basic state under investigation as $t \to \infty$. Since the spectrum of the system does not contain the zero eigenvalue ($\sigma = 0$), dendritic growth cannot approach a steady state. However, with a stationary pattern at the later stages of evolution, realistic dendritic growth may approach a nonlinear periodic solution as $t \to \infty$. Indeed, it is observed in experiments that in the frame moving with the tip and at the later stages of the process, the amplitude of oscillation at any point on the interface is apparently time independent. In the scope of linear theory, a nonlinear limit circle solution will correspond to the neutral point of linear stability. Thus we deduce that as time tends to infinity, dendritic growth will approach the neutrally stable GTW mode. As a consequence, the selection condition of dendritic growth can be expressed as

$$\sigma_{\mathrm{R}}^*(\varepsilon_*) = 0\,. \tag{6.194}$$

We call the selection condition (6.194) the *neutrally stable mode* (NSM) selection principle.

The critical number $\varepsilon_*$ is directly connected with the selected dendrite's tip velocity, tip radius, and oscillation frequency of the dendrite. In fact, if one uses the capillary length $\ell_c$ as the length scale, one can write the dimensionless tip velocity as

$$U_{\text{tip}} = \frac{U\ell_c}{\kappa_T} = \frac{\ell_c}{\ell_T} = \varepsilon_*^2 \text{Pe}_0^2 , \tag{6.195}$$

and the dimensionless tip radius and frequency of oscillation as

$$R_{\text{tip}} = \frac{\ell_t}{\ell_c} = \frac{\ell_t}{\ell_T}\frac{\ell_T}{\ell_c} = \frac{\text{Pe}}{\varepsilon_*^2 \eta_0^4} = \frac{\text{Pe}}{\varepsilon_*^2 \text{Pe}_0^2}, \tag{6.196}$$

$$\Omega_* = \frac{\omega_*}{\eta_0^2 \varepsilon_*} = \frac{\omega_*}{\text{Pe}_0 \varepsilon_*} . \tag{6.197}$$

We recall that $\text{Pe}_0 = \eta_0^2$ is the Péclet number for the case of zero surface tension.

The eigenfunction of a typical selected global neutrally stable mode and its interface shape in a time sequence are shown in Fig. 6.10. A 3D graphic of axisymmetric dendritic growth is shown in Fig. 6.11.

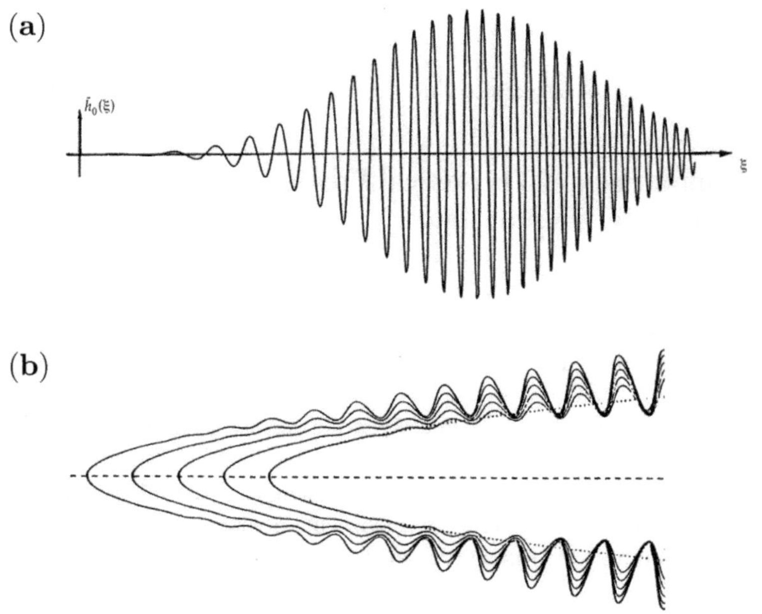


**Fig. 6.10** A typical GTW neutral mode: (**a**) the graphics of the eigenfunction; (**b**) the interface shape in a time sequence

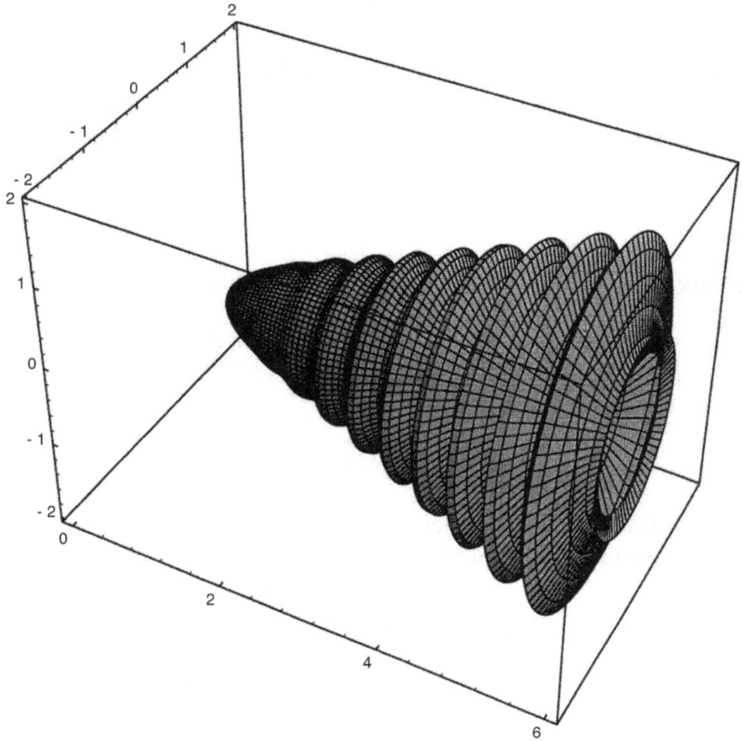


**Fig. 6.11** Three-dimensional graphics of the interface shape of a typical GTW neutral mode

## 6.6 Comparison of Theoretical Predictions with Experimental Data

The dendritic growth theory developed above shows that dendritic growth is essentially a wave phenomenon involving the propagation and interaction of interfacial waves along the interface. This theory is therefore called interfacial wave (IFW) theory.

interfacial wave theory states that when the surface tension is isotropic, in the later stages of growth, dendritic growth is not described by a steady state. Instead, it is described by a time-periodic oscillatory state, the so-called global neutrally stable (GNS) state. Such a GNS state consists mathematically of three parts: (1) the Ivantsov solution, (2) the steady regular perturbation expansion (RPE) part due to the surface tension, and (3) the unsteady singular perturbation expansion (SPE) part. The interface shape of the dendrite can be expressed approximately in the form

$$\eta_s(\xi, t) \approx 1 + \varepsilon^2 \eta_1(\xi) + \hat{h}_0(\xi) e^{\frac{\sigma t}{\eta_0^2 \varepsilon}} \quad (\text{as } \varepsilon \to 0). \tag{6.198}$$

In the above, $\eta_1(\xi)$ is the first-order correction term in the RPE part, while $\hat{h}_0(\xi)e^{\frac{\sigma t}{\eta_0^2 \varepsilon}}$ is the leading term in the SPE part. The parameter $\sigma = \sigma_R - i\omega$ is the eigenvalue with the expansion form

$$\sigma = \sigma_0 + \varepsilon\sigma_1 + \cdots \quad (\text{as } \varepsilon \to 0). \tag{6.199}$$

For the selected solution, $\sigma_R = 0$.

In 1994, a series of careful experiments on free dendritic growth in pure succinonitrile (SCN) were conducted for the first time in the space shuttle Columbia by a research team headed by Glicksman. Under the microgravity environment, convective motion in the melt is greatly reduced [11, 12]. As expected, the new data for tip velocity and tip radius during dendritic growth obtained by Glicksman et al. were more accurate than terrestrially collected data.

Since the material SCN has a very small surface tension anisotropy, one can expect that the theoretical results for 3D axially symmetric dendritic growth from a pure melt obtained in this chapter are comparable with the experimental results obtained by Glicksman et al.

As indicated above, according to the IFW theory, the selected dimensionless tip velocity $U_{\text{tip}}$ and dimensionless tip radius $R_{\text{tip}}$ are determined by the critical number $\varepsilon_*$. In the leading-order approximation, one has $\varepsilon_* = \varepsilon_*^{(0)} = 0.1590$, which is independent of the undercooling $T_\infty$. Furthermore, in the first-order approximation, one has $\varepsilon_* = \varepsilon_*^{(1)}(\eta_0^2)$, which rapidly decreases as the undercooling $T_\infty$ approaches $-1$. But in the range of small undercooling $|T_\infty| = 0.001 - 0.2$ under discussion, $\varepsilon_* = \varepsilon_*^{(1)} \approx 0.1108$.

The above theoretical prediction is free of adjustable parameters and can be directly tested in terms of the experimental data of Glicksman et al. It is therefore of great interest to compare quantitatively the predictions of this theory with these newest experimental data. In this section, we attempt to make such direct comparisons.

### 6.6.1 The Dendrite Tip Velocity and Tip Radius

The thermodynamic data for SCN provided by Glicksman and his coworkers are listed in Table 6.3. Based on these data, one can calculate that the capillary length for SCN is $\ell_c = 2.804\times10^{-7}$cm, the velocity unit is 4069.19 cm/s, and the temperature unit is 23.067 K.

**Table 6.3** The thermodynamic properties of SCN

| | |
|---|---|
| mol | 0.092 g/mol |
| $\Delta H$ | 11.051 cal/g |
| $c_p$L | 0.4791 cal $g^{-1}$ $K^{-1}$ |
| $c_p$S | 0.4468 cal $g^{-1}$ $K^{-1}$ |
| $c_p = (c_p\mathrm{L} + c_p\mathrm{S})/2$ | 0.4630 cal $g^{-1}$ $K^{-1}$ |
| $\kappa_{\mathrm{TL}}$ | $1.127 \times 10^{-3}$ $m^2s^{-1}$ |
| $\kappa_{\mathrm{TS}}$ | $1.155 \times 10^{-3}$ $m^2s^{-1}$ |
| $\kappa_{\mathrm{T}} = (\kappa_{\mathrm{TL}} + \kappa_{\mathrm{TS}})/2$ | $1.141 \times 10^{-3}$ $m^2s^{-1}$ |
| $T_{\mathrm{M0}}$ | 331.233 K |
| $\gamma'$ (Gibbs–Thomson coefficient) | $6.480 \times 10^{-6}$ cm K |
| $\gamma$ | $2.136 \times 10^{-7}$ cal $cm^{-2}$ |

In Fig. 6.12, we show the tip velocity $U_{\mathrm{tip}}$ versus the undercooling temperature $T_\infty$ and compare the experimental data with the theoretical curve calculated by (6.195). In Fig. 6.13 we show the tip radius $R_{\mathrm{tip}}$ calculated by (6.196) versus the undercooling $T_\infty$. It is seen that the overall agreement between the theoretical curve and the experimental data is very satisfactory, especially considering that the experimental data of the tip radius has an error of about 10%. Only in the regime of small undercooling, $|\Delta T| < 0.4$ K, or $|T_\infty| < 0.01$, can one see a growing deviation between the theoretical curve and the experimental data.

### 6.6.2 The Critical Number $\varepsilon_*$

To further compare the theoretical results with the experimental data, we calculate $U_{\mathrm{tip}}^{\frac{1}{2}}/\mathrm{Pe}_0 = (\varepsilon_*)_{\mathrm{exp}}$ in terms of the data obtained from both the flight experiments and ground experiments (shown in Fig. 6.14 by two sets of round dots, data I and data II).

It is seen that these two sets of data are well separated. The data $(\varepsilon_*)_{\mathrm{exp}}$ derived from the experiments in microgravity remain approximately constant at 0.094 within the regime of large undercooling. On the other hand, in the regime of small $T_\infty$, starting from $|T_\infty| = 0.01$, the data $(\varepsilon_*)_{\mathrm{exp}}$ increase as $|T_\infty|$ decreases. The data derived from the ground experiments show a similar tendency, but the increase starts at a higher undercooling temperature $|T_\infty| = 0.06$. This fact clearly shows that the increase of $(\varepsilon_*)_{\mathrm{exp}}$ in the low undercooling regime is due to the convection caused by the buoyancy effect, since gravity is greatly reduced in the flight experiment but not completely eliminated.

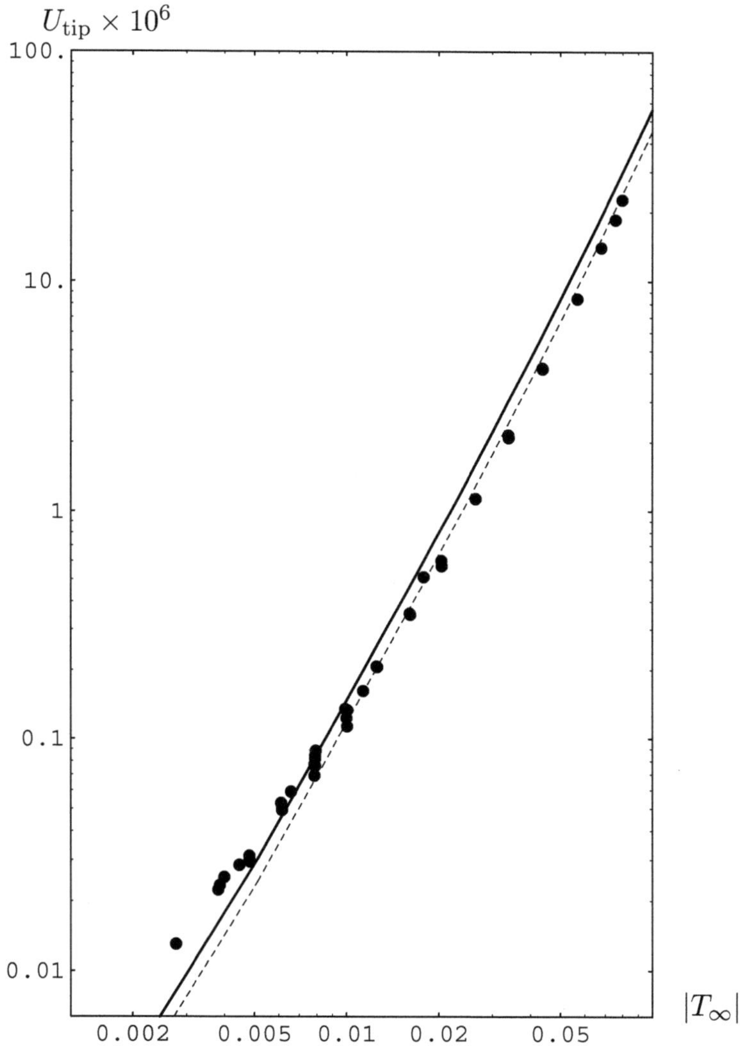


**Fig. 6.12** The variation of $U_{tip}$ with $|T_\infty|$. The *solid line* is given by the IFW theory with zero anisotropy, $\alpha_4 = 0$. The *dotted line* is the modified IFW theory's results with the inclusion of the axial Herring anisotropy of surface tension $\alpha_4 = 0.075$ in terms of the formula given in Chap. 7. The *dots* are the microgravity experimental data

As a comparison, in Fig. 6.14 we also show the theoretical value of $\varepsilon_*$ by the thin solid line. It is seen that in the entire undercooling regime under discussion, the theoretical curve remains flat with $\varepsilon_* \approx 0.1108$, which agrees with the experimental data in the high $T_\infty$ regime reasonably well. The quantitative discrepancy between the theoretical and experimental data is of order $\mathcal{O}(\varepsilon_*^2)$, which can be attributed to

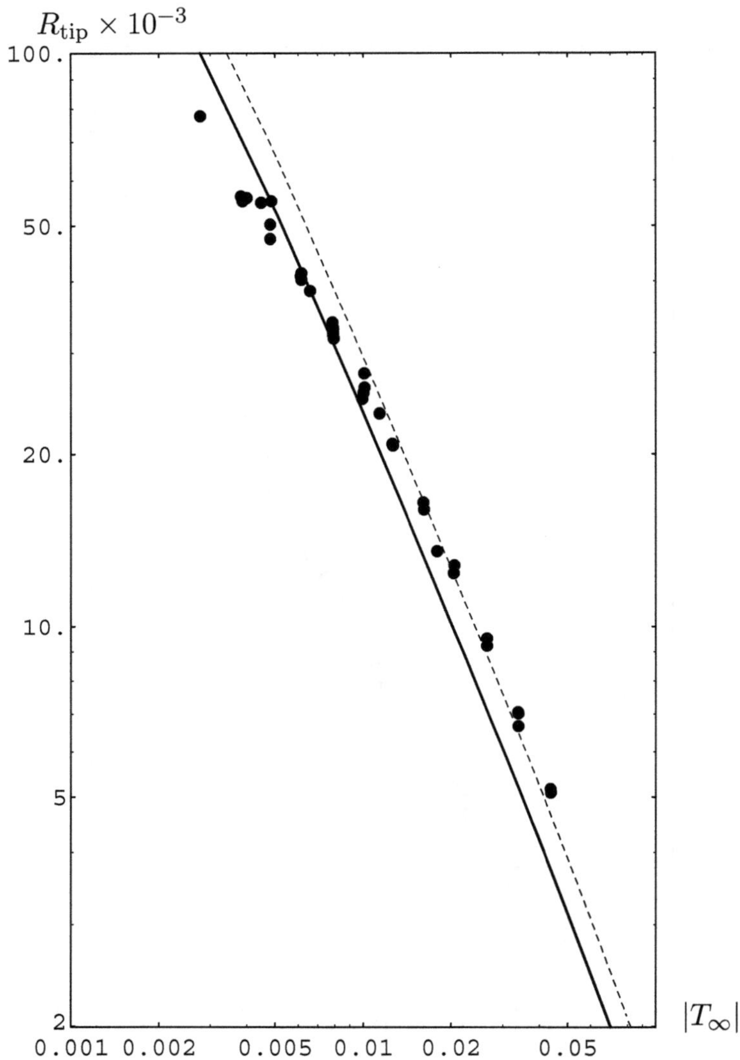


**Fig. 6.13** The variation of $R_{\text{tip}}$ with $T_\infty$. The *solid line* is given by the IFW theory with zero anisotropy, $\alpha_4 = 0$. The *dotted line* is the modified IFW theory's results with the inclusion of the axial anisotropy of surface tension $\alpha_4 = 0.075$ in terms of the formula given in Chap. 7. The *dots* are the microgravity experimental data

the following sources.

1. In the calculation of the asymptotic solution (6.38), all small terms of of $\mathcal{O}(\varepsilon^2)$ are neglected. As a consequence, the critical number $\varepsilon_*$ obtained is expected to have an error up to $\mathcal{O}(\varepsilon_*^2)$.

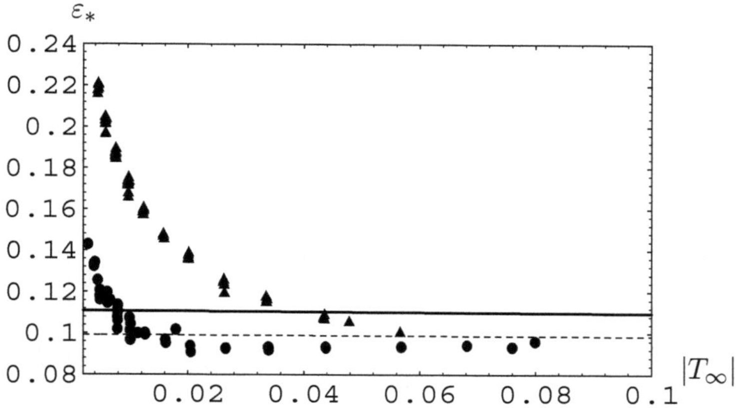


**Fig. 6.14** Comparison of $\varepsilon_*$ with $(\varepsilon_*)_{\rm exp}$ within the region of undercooling, $0.002 < |T_\infty| < 0.1$. The *dots* are the microgravity experimental data, while the *triangles* are the ground experimental data. The *solid line* is the stability criterion, $\varepsilon_* = 0.01108$, predicted by IFW theory, with zero anisotropy, $\alpha_4 = 0$. The *dashed line* is the stability criterion with the anisotropy correction, $\varepsilon_* = 0.0991$, assuming that SCN has an axial anisotropy $\alpha_4 = 0.075$

2. The theory given above neglected the effects of many physical parameters such as the anisotropy of surface tension $\alpha_4$, kinetic attachment, convection motion, and nonlinearity. Though these effects may be presumed to be of secondary importance for the selection criterion for the materials under investigation, they do have small quantitative effects on the numerical values of $\varepsilon_*$. For instance, the anisotropy for SCN is $\alpha_4 \approx 0.075$. Taking this effect into account, we calculate $\varepsilon_* \approx 0.0991$ in terms of the formula to be derived in Chap. 7. This number is shown by the dashed line in Fig. 6.14, which displays an excellent quantitative agreement with experimental data in the regime of undercooling temperature $|T_\infty| > 0.004$.

### 6.6.3 The Universal Scaling Parameter, $\varepsilon_*$ or $\sigma_*$?

It is now necessary and appropriate to make some remarks about the parameter $\sigma_*$. In the literature, for historical reasons, some researchers used to employ the parameter

$$\sigma_* = \frac{2}{U_{\rm tip} R_{\rm tip}^2} \tag{6.200}$$

as the selection criterion. As we mentioned in Chap. 1, the parameter $\sigma_*$ was first introduced by Langer and Müller-Krumbhaar in the so-called marginal stability hypothesis (MSH) in 1979 as the selection criterion. On the basis of scaling analysis,

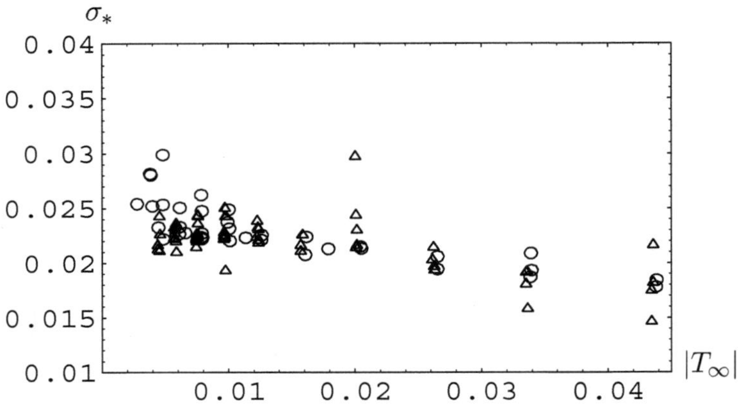


**Fig. 6.15** The variation of $\sigma_*$ calculated with the experimental data versus $T_\infty$. The *circles* are the microgravity data, while the *triangles* are the terrestrial experimental data. These data of $\sigma_*$ are totally mixed

Langer and Müller-Krumbhaar proposed the parameter $\sigma = 2\ell_T\ell_c/\ell_t^2$. They assumed that the system of dendritic growth is linearly unstable for $\sigma < \sigma_*$ and that the selected dendritic growth would be in the marginal stable state corresponding to this critical number $\sigma_*$. As a hypothesis, Langer and Müller-Krumbhaar further proposed that the critical number $\sigma_* \approx 0.0253$ is a universal constant, independent of the undercooling temperature.

Glicksman et al. calculated $\sigma_*$ using the experimental data of $U_{\text{tip}}$ and $R_{\text{tip}}$ obtained in their own flight experiments and the ground experiments, and plotted the result against undercooling, as shown in Fig. 6.15. It is seen that the two sets of data of $\sigma_*$ are well mixed up, in contrast to the data $\varepsilon_*$ versus $|T_\infty|$ shown in Fig. 6.14. This result clearly shows that at a higher level of precision, the parameter $\sigma_*$, unlike the parameter $\varepsilon_*$, is not suitable for use as a criterion for dendritic growth, since it remains insensitive to a great change in the dendritic system.

Another significant difference between the parameter $\sigma_*$ and the critical number $\varepsilon_*$ is that $\sigma_*$ was claimed by MSH as a universal constant independent of the undercooling, whereas $\varepsilon_*$ is derived as a function of the undercooling $T_\infty$. Especially in the large undercooling regime, the numerical value of $\varepsilon_*$ is quite sensitive to the variation of $T_\infty$. Only for small undercooling is the critical number $\varepsilon_*$ approximately a constant, as we have seen in previous sections. In conclusion, strictly speaking, the so-called universal constant $\sigma_*$ cannot be considered the selection criterion parameter for the system of dendritic growth, but $\varepsilon_*$ is. It is only in the regime of small undercooling that the hypothesis of constant $\sigma_*$ is approximately correct and $\sigma_* \approx 2\varepsilon_*^2$.

### 6.6.4 *The Nature of the Dendrite-Tip: Steady or Oscillatory?*

For a long time, researchers in materials science have phenomenologically considered that a dendrite consists of two different parts: the steady and smooth tip part, and the unsteady side-branching part. However, nobody seems to have the answer to such questions as where the two parts are connected, how to determine the location of connection, and what the physical implication of the connection point is.

From the IFW theory, one sees that the dendrite tip part is not really steady or at all smooth. Every point on the entire interface oscillates with the same frequency $\Omega_*$ but with a different amplitude. Hence, in this regard, there is no difference between the tip part of the dendrite and the remaining part of the interface. The tip part appears "steady" and "smooth" only because the amplitude of oscillation in the tip part is much smaller than that in the other part. If one wants, based on the wave diagram of the GTW mode, one may separate the whole interfacial pattern into two parts connected at the point $\xi_c'$: one is the superficial "smooth" tip part. It is oscillating with very small amplitude and involves two interacting interfacial waves on the Ivantsov paraboloid. The other is the side-branching part. It is oscillating with large amplitude and describes a single outgoing interfacial wave.

On the other hand, from the MSC theory, the dendritic growth is a steady needle-like crystal growth, whose interface including the tip region is essentially steady and smooth; the side-branching is generated by continuously acting noises. Once the noises are fully screened, the whole interface will be steady and smooth.

Thus, is the tip of the dendrite actually steady or oscillating? This is the question to be resolved.

The difference in nature between the steady smooth tip and oscillating tip with very small amplitude is hard to visualize. The weak signal describing the oscillation of the tip is submerged into the white noise; hence it is very hard to detect in the ordinary ways. Nevertheless, the signal of the tip oscillation can still be detected in a way that filters the noise components. Glicksman at his coworkers carefully analyzed the data of dendrite-tip movement that they collected in the space experiments of PVA dendritic growth and drew the power spectrum of the tip location. They found that after the noise components were filtered, the power spectrum showed a sequence of peaks. Surprisingly, the frequencies at these peaks appear to be very close to $\omega_*, 2\omega_*, \ldots$, and $\frac{1}{2}\omega_*, \frac{1}{3}\omega_*, \ldots$, as shown in Fig. 6.16, where $\omega_*$ is the eigenfrequency predicted by IFW theory [13].

The nature of tip oscillation may be also detected dynamically via specially designed experiments, by setting the system under a well-controlled weak external disturbance that oscillates with a constant frequency $\Omega$. For instance, one may introduce a weak external signal to the system by applying a small oscillation in

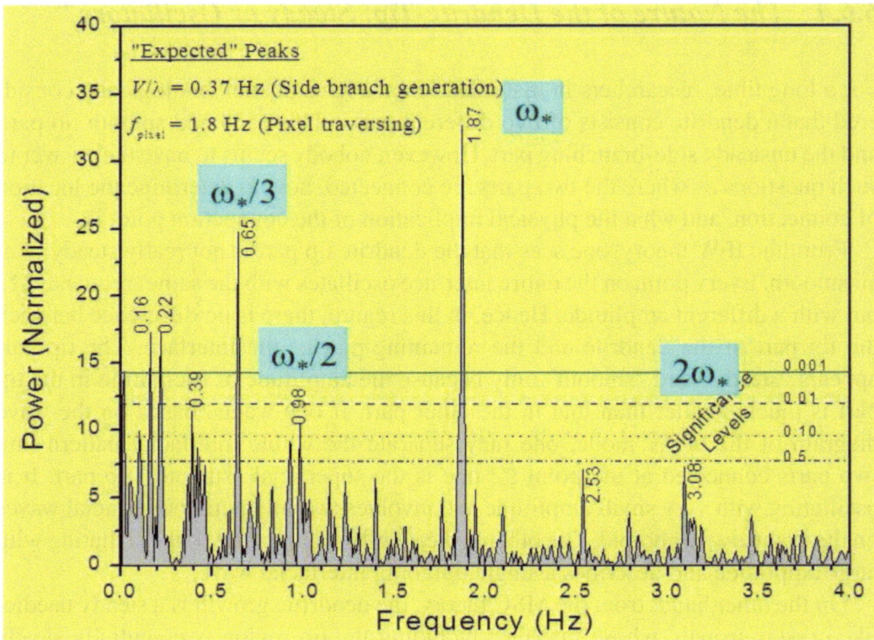


**Fig. 6.16** The power spectrum of the tip oscillation discovered by Glicksman's team from the data measured in their PVA space experiments for dendritic growth (refer to [12, 13]). As shown in the figure, the primary peak occurs at the frequency $f = 1.87$ Hz, which very closely corresponds to the eigenfrequency $\omega_*$; other secondary peaks occur at frequencies close to $2\omega_*$, $\frac{1}{2}\omega_*$, $\frac{1}{3}\omega_*$

the far field as sketched in Fig. 6.17, or by focusing a laser beam with different frequencies on a certain point of a growing dendrite and then examining the responses of the system to the frequencies of the signals and plotting the power spectrum curve. If the dendrite-tip is steady, the system will not resonate with these external signals. The response curve of the power spectrum versus the frequency will show no peak. However, if the dendrite-tip is oscillating with an eigenfrequency $\omega_*$, the system will resonate with the external signals with certain frequencies, which leads to some peaks in the response curve of the power spectrum. The observed resonance may be the primary resonance, the subharmonic/superharmonic resonance, or both. There are a number of experiments in the literature designed in such a way that have detected some resonance in the growth system. Moreover, the resonance frequencies observed in the experiments are in very good quantitative agreement with the eigenfrequency $\omega_*$ calculated by IFW theory. Hence these experiments have confirmed the oscillation nature of the dendrite-tip [5, 18, 20]. For the analyses, the reader is referred to [28].

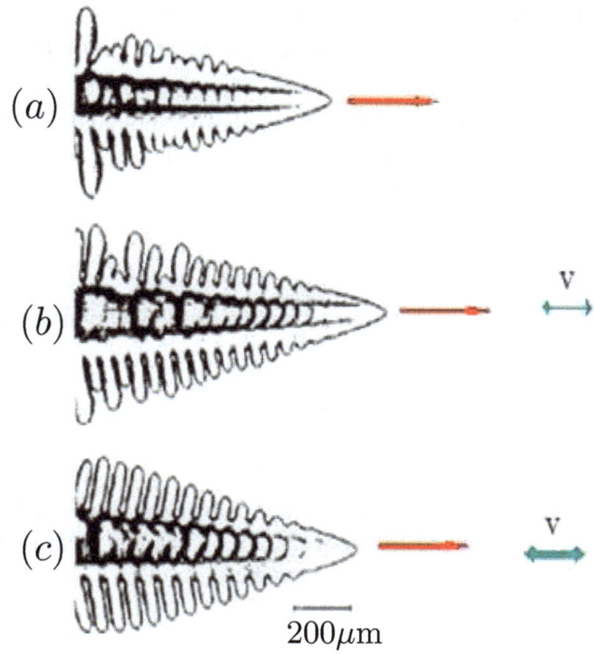


**Fig. 6.17** The resonance of dendritic growth with the oscillating external signals in the far field (see [18]): (**a**) Free dendritic growth without external oscillatory disturbance generated in the far field. (**b**) Free growing dendrite slightly reacts with external oscillatory disturbance imposed in the far field. (**c**) A subharmonic resonance appears between the free dendritic growth and the weak external oscillatory disturbance at some special frequency imposed in the far field

## 6.7 Three-Dimensional Nonaxisymmetric Spiral Dendritic Modes of Perturbed States

So far, we have investigated only the axisymmetric modes of perturbed states. In reality, although its basic steady states are axisymmetric, the system of dendritic growth may allow nonaxisymmetric modes of perturbed states. This is because the system always undergoes various asymmetric disturbances.

In this section, we attempt to explore the forms of 3D nonaxisymmetric modes of perturbed states and their stability properties. The ideas and approaches addressed in the previous sections are fully applicable to this case.

### 6.7.1 Mathematical Formulation of General Three-Dimensional Unsteady Dendritic Growth

We adopt the symmetric model and employ the paraboloidal coordinate system $(\xi, \eta, \theta)$ moving with velocity $U$ as before (see Fig. 3.1):

$$\frac{x}{\eta_0^2} = \xi\eta\cos\theta, \qquad \frac{y}{\eta_0^2} = \xi\eta\sin\theta, \qquad \frac{z}{\eta_0^2} = \frac{1}{2}(\xi^2 - \eta^2), \tag{6.201}$$

where the constant $\eta_0^2$ is to be determined by setting the location of the dendrite-tip so that the steady interface shape satisfies

$$\eta_s(0) = 1\,. \tag{6.202}$$

In the moving paraboloidal system, the 3D dimensionless governing equation for the temperature field in both solid and liquid phases and the boundary conditions were derived in Chap. 3 in the forms of (3.12), (3.14), and (3.26)–(3.31). For the convenience of the reader, we have rewritten these as follows:

$$\frac{\partial^2 T}{\partial \xi^2} + \frac{\partial^2 T}{\partial \eta^2} + \frac{1}{\xi}\frac{\partial T}{\partial \xi} + \frac{1}{\eta}\frac{\partial T}{\partial \eta} + \left(\frac{1}{\xi^2} + \frac{1}{\eta^2}\right)\frac{\partial^2 T}{\partial \theta^2}$$

$$= \eta_0^2\left(\xi\frac{\partial T}{\partial \xi} - \eta\frac{\partial T}{\partial \eta}\right) + \eta_0^4(\xi^2 + \eta^2)\frac{\partial T}{\partial t}\,. \tag{6.203}$$

The boundary conditions are as follows:

1. The far-field condition:

$$T \to T_\infty = \frac{(T_\infty)_D - T_{M0}}{\Delta H/(c_p\rho)} < 0 \quad \text{as } \eta \to \infty\,, \tag{6.204}$$

   where $(T_\infty)_D$ is the dimensional temperature in the far field ahead of the dendrite-tip, while $T_\infty$ is the dimensionless undercooling parameter.
2. The regularity condition:

$$T_S = \mathcal{O}(1) \quad \text{as } \eta \to 0\,. \tag{6.205}$$

3. The interface conditions, at $\eta = \eta_s(\xi, \theta, t)$:

   (i) the thermodynamic equilibrium condition:

$$T = T_S\,, \tag{6.206}$$

(ii) the Gibbs–Thomson condition: for simplicity, we neglect the effect of anisotropy of surface tension. Thus, we have

$$T = T_S = -\varepsilon^2 \eta_0^2 \mathcal{K}_3 \Big\{ \eta_s(\xi, \theta) \Big\}, \tag{6.207}$$

where $\mathcal{K}_3/\eta_0^2$ is twice of the local mean curvature of the interface $\eta_s(\xi, \theta)$, as shown by (3.25)–(3.26), while $\varepsilon$ is the interfacial stability parameter of isotropic surface tension defined as before.

(iii) the heat balance condition:

$$\left(\frac{\partial}{\partial \eta} - \eta_s' \frac{\partial}{\partial \xi}\right)(T - T_S) - \eta_{s\theta}\left(\frac{1}{\xi^2} + \frac{1}{\eta_s^2}\right)\frac{\partial}{\partial \theta}(T - T_S)$$

$$+\eta_0^4(\xi^2 + \eta_s^2)\frac{\partial \eta_s}{\partial t} + \eta_0^2(\xi \eta_s)' = 0\,, \tag{6.208}$$

4. The tip-smoothness condition: at $\xi = 0$,

$$\eta_s(0, \theta) = 1 \qquad \eta_s'(0, \theta) = 0. \tag{6.209}$$

5. The root condition is demonstrated as in the last section.

### 6.7.2 *The Basic State for Dendritic Growth with Nonzero Surface Tension*

The basic state of 3D dendritic growth at the late stage of growth is given by the so-called nonclassical needle solutions as before. For the special case of $\varepsilon = 0$, it is still the Ivantsov solution. Moreover, since we do not take into account anisotropy of surface tension and nonaxisymmetric boundary conditions, for given $(\xi, \eta)$ as $\varepsilon \to 0$, the basic state solution can be expanded in the regular perturbation expansion (RPE) as addressed in the previous chapter:

$$\begin{aligned} T &= T_B(\xi, \eta, \varepsilon) = T_*(\eta) + \varepsilon^2 \eta_0^2 T_1(\xi, \eta) + \cdots\,, \\ T_S &= T_{BS}(\xi, \eta, \varepsilon) = \varepsilon^2 \eta_0^2 T_{S1}(\xi, \eta) + \cdots\,, \\ \eta &= \eta_B(\xi, \varepsilon) = 1 + \varepsilon^2 \eta_1(\xi) + \cdots\,. \end{aligned} \tag{6.210}$$

The correction terms to the Ivantsov solution in the basic steady state are of order $\mathcal{O}(\varepsilon^2)$.

### 6.7.3 3D Linear Perturbed System

We write the 3D nonaxisymmetric perturbed state in two parts:

$$\begin{aligned} T &= T_{\mathrm{B}} + \tilde{T}(\xi, \eta, \theta, t, \varepsilon), \\ T_{\mathrm{S}} &= T_{\mathrm{SB}} + \tilde{T}_{\mathrm{S}}(\xi, \eta, \theta, t, \varepsilon), \\ \eta_{\mathrm{s}} &= \eta_{\mathrm{B}} + \tilde{h}(\xi, \theta, t, \varepsilon)/\eta_0^2 . \end{aligned} \tag{6.211}$$

The linearized perturbed system can be obtained from Eqs. (6.203)–(6.209). The results are as follows:

$$\frac{\partial^2 \tilde{T}}{\partial \xi^2} + \frac{\partial^2 \tilde{T}}{\partial \eta^2} + \left(\frac{1}{\xi} - \eta_0^2 \xi\right) \frac{\partial \tilde{T}}{\partial \xi} + \left(\frac{1}{\eta} + \eta_0^2 \eta\right) \frac{\partial \tilde{T}}{\partial \eta}$$
$$+ \left(\frac{1}{\xi^2} + \frac{1}{\eta^2}\right) \frac{\partial^2 \tilde{T}}{\partial \theta^2} = \eta_0^4 (\xi^2 + \eta^2) \frac{\partial \tilde{T}}{\partial t}, \tag{6.212}$$

with the following boundary conditions:

1. As $\eta \to \infty$,

$$\tilde{T} \to 0 . \tag{6.213}$$

2. As $\eta \to 0$,

$$\tilde{T}_{\mathrm{S}} = \mathcal{O}((1) . \tag{6.214}$$

3. The interface conditions: taking the Taylor expansions around the interface of the basic state yields the following at $\eta = \eta_{\mathrm{B}}(\xi, \theta, \varepsilon)$:

   (i)

$$\tilde{T} - \tilde{T}_{\mathrm{S}} = -\left\{ \frac{\partial T_{\mathrm{B}}}{\partial \eta} - \frac{\partial T_{\mathrm{SB}}}{\partial \eta} \right\} \frac{\tilde{h}}{\eta_0^2} , \tag{6.215}$$

   (ii) Denote $\eta_{\mathrm{s}} = 1 + \tilde{\eta}_{\mathrm{s}}$, where ($\tilde{\eta}_{\mathrm{s}} \ll 1$). Note that after linearization, (3.26) is reduced to

$$\mathcal{K}_3 \left\{ \tilde{\eta}_{\mathrm{s}}(\xi, \theta, \varepsilon) \right\} = -\frac{2 + \xi^2}{(1 + \xi^2)^{\frac{3}{2}}} + \frac{\varepsilon^2}{(1 + \xi^2)^{\frac{1}{2}}}$$
$$\times \left\{ \tilde{\eta}_{\mathrm{s}}'' + \frac{1 + 2\xi^2}{\xi(1 + \xi^2)} \tilde{\eta}_{\mathrm{s}}' + \frac{\xi^2(2 + \xi^2) + 4}{(1 + \xi^2)^2} \tilde{\eta}_{\mathrm{s}} + \frac{(1 + \xi^2)}{\xi^2} \tilde{\eta}_{\mathrm{s},\theta\theta} \right\}$$
$$+ \mathcal{O}(\varepsilon^4). \tag{6.216}$$

Thus, the Gibbs–Thomson condition for the perturbed states can be written as

$$\tilde{T}_{\rm S} = \frac{\varepsilon^2}{\hat{S}(\xi)}\left\{\frac{\partial^2\tilde{h}}{\partial\xi^2} + \left(1+\frac{1}{\xi^2}\right)\frac{\partial^2\tilde{h}}{\partial\theta^2} + \frac{1+2\xi^2}{\xi\hat{S}^2(\xi)}\frac{\partial\tilde{h}}{\partial\xi} + \frac{\xi^2(2+\xi^2)+4}{\hat{S}^2(\xi)}\tilde{h}\right\} + (\text{h.o.t.})\,, \tag{6.217}$$

(iii)

$$\frac{\partial}{\partial\eta}\left(\tilde{T}-\tilde{T}_{\rm S}\right) + \eta_0^2\hat{S}^2(\xi)\frac{\partial\tilde{h}}{\partial t} + \xi\frac{\partial\tilde{h}}{\partial\xi} + \tilde{h} + \left\{\frac{\partial^2}{\partial\eta^2}\left(T_{\rm B}-T_{\rm SB}\right)\right\}\frac{\tilde{h}}{\eta_0^2} - \frac{1}{\eta_0^2}\frac{\partial\tilde{h}}{\partial\xi}\frac{\partial}{\partial\xi}\left(T_{\rm B}-T_{\rm SB}\right) - \frac{\eta_{\rm B}'\tilde{h}}{\eta_0^2}\frac{\partial^2}{\partial\xi\partial\eta}\left(T_{\rm B}-T_{\rm SB}\right) - \eta_{\rm B}'\frac{\partial}{\partial\xi}\left(\tilde{T}-\tilde{T}_{\rm S}\right) + (\text{h.o.t.}) = 0\,, \tag{6.218}$$

where

$$\hat{S}(\xi) = \sqrt{\xi^2+\eta_{\rm B}^2}\,. \tag{6.219}$$

4. The root condition: at $\xi \to \infty$,

$$\{\tilde{T}, \tilde{T}_{\rm S}, \tilde{h}\} = 0\,. \tag{6.220}$$

5. The tip smoothness condition: at $\xi = 0,\ \eta = \eta_{\rm B}(0,\varepsilon)$,

$$\frac{\partial}{\partial\xi}\{\tilde{T}, \tilde{T}_{\rm S}, \tilde{h}\} = 0\,. \tag{6.221}$$

In the system (6.212)–(6.221), the variables $\theta, t$ appear only in the derivatives, not in the coefficients, so that one may look for solutions of the type $\tilde{q}(\xi,\eta,\theta,t) = \hat{q}(\xi,\eta)e^{\sigma t + i\theta}$, resulting in a linear eigenvalue problem with two parameters $\varepsilon$ and $\eta_0^2$. In following the approach developed in the previous section, one may solve the above eigenvalue problem. First, solve the system (6.212)–(6.220) for given $(\sigma, \eta_0^2, \varepsilon)$ using the multiple variables expansion (MVE) method. The solution is required to satisfy all boundary conditions except the tip smoothness condition (6.221). Then apply the tip condition (6.221) to the asymptotic solution obtained, which will lead to the quantization conditions and yield the parameter $\sigma$ as a function of $\eta_0^2$ and $\varepsilon$.

### 6.7.4 Multiple Variables Expansion Solution in the Outer Region

In the outer region, away from all possible singular points, we now look for the asymptotic expansion of the solutions $\tilde{q}(\xi, \eta, \theta, t, \varepsilon)$ in terms of the multiple variables expansion method. We shall introduce the same fast variables in both the $\xi$- and $\eta$-directions, as in the previous section. However, we assume that the perturbed state solutions do not have short length scales along the $\theta$-direction. Accordingly, there is no need for the fast variable along the $\theta$-direction.

As a consequence, we have the following fast variables:

$$\begin{aligned} \eta_+ &= \frac{\psi(\xi, \eta, \varepsilon)}{\varepsilon} = \frac{1}{\varepsilon}\int_1^{\eta} k(\xi, \eta_1, \varepsilon)\mathrm{d}\eta_1, \\ \xi_+ &= \frac{\phi(\xi, \eta, \varepsilon)}{\varepsilon} = \frac{1}{\varepsilon}\int_{\xi_0}^{\xi} k(\xi_\xi, \eta, \varepsilon)\mathrm{d}\xi_1, \\ t_+ &= \frac{t}{\eta_0^2 \varepsilon}. \end{aligned} \tag{6.222}$$

The formulas (6.15)–(6.26) given in the previous subsections for the axisymmetric case are fully applicable to the present case.

We apply the multiple variables expansion (MVE) method ([5, 8] in Chap. 8) by treating all the fast and slow variables $(\xi, \eta, \theta, \xi_+, \eta_+, t_+)$ as independent variables and making the replacements (6.27) for the derivatives in the system (6.212)–(6.218). The converted equation with multiple variables in the liquid phase is obtained as

$$\begin{aligned} &(\phi_\xi^2 + \phi_\eta^2)\left(\frac{\partial^2}{\partial \xi_+^2} + \frac{\partial^2}{\partial \eta_+^2}\right)\tilde{T} + \varepsilon^2\left(\frac{1}{\xi^2} + \frac{1}{\eta^2}\right)\frac{\partial^2 \tilde{T}}{\partial \theta^2} \\ &= -\varepsilon\left(2\phi_\xi \frac{\partial^2 \tilde{T}}{\partial \xi_+ \partial \xi} + 2\psi_\xi \frac{\partial^2 \tilde{T}}{\partial \eta_+ \partial \xi}\right) - \varepsilon\left(2\psi_\eta \frac{\partial^2 \tilde{T}}{\partial \eta_+ \partial \eta} + 2\phi_\eta \frac{\partial^2 \tilde{T}}{\partial \xi_+ \partial \eta}\right) \\ &\quad -\varepsilon\left(\frac{1}{\xi} - \eta_0^2 \xi\right)\left(\phi_\xi \frac{\partial \tilde{T}}{\partial \xi_+} + \psi_\xi \frac{\partial \tilde{T}}{\partial \eta_+}\right) - \varepsilon\left(\frac{1}{\eta} + \eta_0^2 \eta\right)\left(\phi_\eta \frac{\partial \tilde{T}}{\partial \xi_+} + \psi_\eta \frac{\partial \tilde{T}}{\partial \eta_+}\right) \\ &\quad +\varepsilon \eta_0^2(\xi^2 + \eta^2)\frac{\partial \tilde{T}}{\partial t_+} + \mathcal{O}((\varepsilon^2). \end{aligned} \tag{6.223}$$

The heat equation for the region of the solid phase is the same as the above, except that the function $\tilde{T}$ is replaced by $\tilde{T}_S$.

The boundary conditions for the system with multiple variables are as follows:

1. As $\eta_+ \to \infty$,

$$\tilde{T} \to 0\,. \tag{6.224}$$

2. As $\eta_+ \to -\infty$,

$$\tilde{T}_S \to 0\,. \tag{6.225}$$

3. At the interface, $\eta_+ = 0, \eta = 1$:

(i) the thermodynamic equilibrium condition

$$\tilde{T} = \tilde{T}_S + \tilde{h} - \varepsilon\eta_1\bar{k}\frac{\partial}{\partial\eta_+}(\tilde{T} - \tilde{T}_S) + \mathcal{O}((\varepsilon^2)\,, \tag{6.226}$$

(ii) the Gibbs–Thomson condition

$$\tilde{T}_S + \varepsilon\eta_1\bar{k}\frac{\partial\tilde{T}_S}{\partial\eta_+} = \frac{1}{S(\xi)}\Bigg\{\left(\bar{k}^2\frac{\partial^2}{\partial\xi_+^2} + 2\varepsilon\bar{k}\frac{\partial^2}{\partial\xi_+\partial\xi} + \varepsilon\bar{k}'\frac{\partial}{\partial\xi_+}\right)$$
$$+\varepsilon\left(\frac{1}{\xi} + \frac{\xi}{S^2(\xi)}\right)\left(\bar{k}\frac{\partial}{\partial\xi_+} + \varepsilon\frac{\partial}{\partial\xi}\right)\Bigg\}\tilde{h} + \frac{\varepsilon^2}{\xi^2}\frac{\partial^2\tilde{h}}{\partial\theta^2} + \mathcal{O}(\varepsilon^2), \tag{6.227}$$

where we have defined

$$S = \sqrt{1 + \xi^2}, \tag{6.228}$$

(iv) the heat balance condition

$$\bar{k}\left(\frac{\partial\tilde{T}}{\partial\eta_+} - \frac{\partial\tilde{T}_S}{\partial\eta_+}\right) + \varepsilon\eta_1\bar{k}^2\frac{\partial^2}{\partial\eta_+^2}(\tilde{T} - \tilde{T}_S) + \varepsilon\left(\frac{\partial\tilde{T}}{\partial\eta} - \frac{\partial\tilde{T}_S}{\partial\eta}\right)$$
$$+S^2(\xi)\frac{\partial\tilde{h}}{\partial t_+} + \xi\left(k\frac{\partial\tilde{h}}{\partial\bar{\xi}_+} + \varepsilon\frac{\partial\tilde{h}}{\partial\xi}\right)$$
$$+\varepsilon\left(2 + \eta_0^2\right)\tilde{h} + \mathcal{O}((\varepsilon^2) = 0\,, \tag{6.229}$$

(iii) The root condition: as $\xi \to \infty$, one has

$$\{\tilde{T};\ \tilde{T}_S;\ \tilde{h}\} = 0\,. \tag{6.230}$$

In terms of the multiple variables, $(\xi, \eta, \theta, \xi_+, \eta_+, t_+)$, we now make the following multiple variables expansion (MVE) for general 3D perturbed states:

$$\begin{aligned}
\tilde{T} &= \tilde{\mu}_0(\varepsilon)\left\{\tilde{T}_0(\xi,\eta,\xi_+,\eta_+) + \varepsilon\tilde{T}_1(\xi,\eta,\xi_+,\eta_+) + \cdots\right\}\mathrm{e}^{im\theta+\sigma t_+},\\
\tilde{T}_S &= \tilde{\mu}_0(\varepsilon)\left\{\tilde{T}_{S0}(\xi,\eta,\xi_+,\eta_+) + \varepsilon\tilde{T}_{S1}(\xi,\eta,\xi_+,\eta_+) + \cdots\right\}\mathrm{e}^{im\theta+\sigma t_+},\\
\tilde{h} &= \tilde{\mu}_0(\varepsilon)\left\{\tilde{h}_0(\xi,\bar{\xi}_+) + \varepsilon\tilde{h}_1(\xi,\bar{\xi}_+) + \cdots\right\}\mathrm{e}^{im\theta+\sigma t_+},
\end{aligned} \tag{6.231}$$

where $m = 0, 1, 2, \ldots$ and $\sigma = \sigma_R - i\omega$ $(\omega \geq 0)$ is generally a complex number. Furthermore, we make the following asymptotic expansions:

$$\begin{aligned}
\phi(\xi, \eta, \varepsilon) &= \phi_0(\xi, \eta) + \varepsilon\phi_1(\xi, \eta) + \cdots, \\
\psi(\xi, \eta, \varepsilon) &= \psi_0(\xi, \eta) + \varepsilon\psi_1(\xi, \eta) + \cdots, \\
k(\xi, \eta, \varepsilon) &= k_0(\xi, \eta) + \varepsilon k_1(\xi, \eta) + \cdots, \\
g(\xi, \eta, \varepsilon) &= g_0(\xi, \eta) + \varepsilon g_1(\xi, \eta) + \cdots, \\
\bar{k}(\xi, \varepsilon) &= \bar{k}_0(\xi) + \varepsilon\bar{k}_1(\xi) + \cdots, \\
\sigma &= \sigma_0 + \varepsilon\sigma_1 + \varepsilon^2\sigma_2 + \cdots.
\end{aligned} \tag{6.232}$$

Note from the above system that the variable $\theta$ is not contained in all the coefficients of the system; the 3D effect is represented only by the term $\left\{\frac{\varepsilon^2}{\xi^2}\frac{\partial^2\tilde{T}}{\partial\theta^2}\right\}$ contained in the heat equation and the term $\left\{\frac{\varepsilon^2}{\xi^2}\frac{\partial^2\tilde{h}}{\partial\theta^2}\right\}$ contained in the Gibbs–Thomson condition. These terms are important only in the region near the dendrite-tip, where $\xi \approx 0$. In the outer region away from the tip, these terms are negligible up to order $\mathcal{O}(\varepsilon)$. As a consequence, one can treat $\theta$ as a parameter and copy all the zeroth- and first-order outer solutions for the axisymmetric system obtained in the last section to the present system.

It what follows, we just summarize these results without derivations.

### 6.7.5 Zeroth-Order Approximation of Outer Solution

The system of the zeroth-order approximation is

$$\begin{aligned}
\left(\frac{\partial^2}{\partial\xi_+^2} + \frac{\partial^2}{\partial\eta_+^2}\right)\tilde{T}_0 &= 0, \\
\left(\frac{\partial^2}{\partial\xi_+^2} + \frac{\partial^2}{\partial\eta_+^2}\right)\tilde{T}_{S0} &= 0,
\end{aligned} \tag{6.233}$$

with the following boundary conditions:

1. In the region of liquid phase $\eta > 1$, as $\eta_+ \to \infty$,

$$\tilde{T}_0 \to 0\,. \tag{6.234}$$

2. In the region of solid phase $\eta < 1$, as $\eta_+ \to -\infty$,

$$\tilde{T}_{S0} \to 0\,. \tag{6.235}$$

3. At the interface $\eta_+ = 0$ and $\eta = 1$, we have

$$\tilde{T}_0 = \tilde{T}_{S0} + \tilde{h}_0, \tag{6.236}$$

$$\tilde{T}_{S0} = \frac{\bar{k}_0^2}{S(\xi)} \frac{\partial^2 \tilde{h}_0}{\partial \bar{\xi}_+^2}, \tag{6.237}$$

$$\bar{k}_0 \frac{\partial}{\partial \eta_+} \left( \tilde{T}_0 - \tilde{T}_{S0} \right) + \sigma_0 S^2(\xi) \tilde{h}_0 + \bar{k}_0 \xi \frac{\partial \tilde{h}_0}{\partial \bar{\xi}_+} = 0\,. \tag{6.238}$$

4. As $\xi \to \infty$,

$$\tilde{h}_0 = 0\,. \tag{6.239}$$

The system allows the following normal mode solutions:

$$\begin{aligned} \tilde{T}_0 &= A_0(\xi, \eta) \exp\{i\xi_+ - \eta_+\}, \\ \tilde{T}_{S0} &= A_{S0}(\xi, \eta) \exp\{i\xi_+ + \eta_+\}, \\ \tilde{h}_0 &= \hat{D}_0 \exp\left\{i\bar{\xi}_+\right\}, \end{aligned} \tag{6.240}$$

where the coefficient $\hat{D}_0$ is set as a constant. Setting

$$\begin{cases} \hat{A}_0(\xi) = A_0(\xi, 1), \\ \hat{A}_{S0}(\xi) = A_{S0}(\xi, 1)\,. \end{cases} \tag{6.241}$$

From (6.236) to (6.238) we derive the local dispersion relationship

$$\sigma_0 = \Sigma(\xi, \bar{k}_0) = \frac{\bar{k}_0}{S^2} \left( 1 - \frac{2\bar{k}_0^2}{S} \right) - \frac{i\xi}{S^2} \bar{k}_0 \quad (0 \le \xi < \infty), \tag{6.242}$$

and

$$\begin{aligned} \hat{A}_0 + \hat{A}_{S0} &= \left(1 - \tfrac{2\bar{k}_0^2}{S}\right) \hat{D}_0, \\ \hat{A}_0 - \hat{A}_{S0} &= \hat{D}_0. \end{aligned} \tag{6.243}$$

The amplitude functions $\hat{A}(\xi), \hat{A}_S(\xi)$ are fully determined up to an arbitrary constant $\hat{D}$.

From the local dispersion formula (6.242) for a fixed parameter $\sigma_0$, one can solve for three wave numbers as functions of $\xi$ (Fig. 6.1). Namely,

$$\begin{cases} \bar{k}_0^{(1)}(\xi) = M(\xi) \cos\left\{\frac{1}{3} \cos^{-1}\left(\frac{\sigma_0}{N(\xi)}\right)\right\} & \text{(short-wave branch)}, \\ \bar{k}_0^{(2)}(\xi) = M(\xi) \cos\left\{\frac{1}{3} \cos^{-1}\left(\frac{\sigma_0}{N(\xi)}\right) + \frac{2\pi}{3}\right\}, & \\ \bar{k}_0^{(3)}(\xi) = M(\xi) \cos\left\{\frac{1}{3} \cos^{-1}\left(\frac{\sigma_0}{N(\xi)}\right) + \frac{4\pi}{3}\right\} & \text{(long-wave branch)}, \end{cases} \tag{6.244}$$

where

$$\begin{cases} M(\xi) = \sqrt{\dfrac{2S(\xi)}{3}}(1 - \mathrm{i}\xi)^{\frac{1}{2}}, \\ N(\xi) = -\dfrac{M(\xi)}{3S^2(\xi)}(1 - \mathrm{i}\xi). \end{cases} \tag{6.245}$$

Having obtained the solution for the function $\bar{k}_0(\xi)$, one may extend it to the complex $\zeta = (\xi + \mathrm{i}\xi_{\mathrm{I}})$-plane and derive the functions $k_0(\xi, \eta)$, $\phi_0(\xi, \eta)$ and $\psi_0(\xi, \eta)$ on the whole $(\xi, \eta)$-plane as follows:

$$\begin{aligned} \phi_0(\xi, \eta) &= \frac{1}{2}\left\{\int_0^{\xi+\mathrm{i}(\eta-1)} \bar{k}_0(t)\mathrm{d}t + \int_0^{\xi-\mathrm{i}(\eta-1)} \bar{k}_0(t)\mathrm{d}t\right\}, \\ \psi_0(\xi, \eta) &= -\frac{\mathrm{i}}{2}\int_{\xi-\mathrm{i}(\eta-1)}^{\xi+\mathrm{i}(\eta-1)} \bar{k}_0(t)\mathrm{d}t, \end{aligned} \tag{6.246}$$

and

$$\begin{aligned} k_0(\xi, \eta) &= \phi_{0,\xi}(\xi, \eta) = \psi_{0,\eta}(\xi, \eta) \\ &= \frac{1}{2}\Big\{\bar{k}_0\big[\xi + \mathrm{i}(\eta - 1)\big] + \bar{k}_0\big[\xi - \mathrm{i}(\eta - 1)\big]\Big\}, \\ g_0(\xi, \eta) &= \phi_{0,\eta}(\xi, \eta) = -\psi_{0,\xi}(\xi, \eta) \\ &= \frac{\mathrm{i}}{2}\Big\{\bar{k}_0\big[\xi + \mathrm{i}(\eta - 1)\big] - \bar{k}_0\big[\xi - \mathrm{i}(\eta - 1)\big]\Big\}. \end{aligned} \tag{6.247}$$

Thus, the two fast variables $(\xi_+, \eta_+)$ are fully determined.

Given $\sigma_0$, the system has three fundamental wave solutions of the system, which respectively correspond to the wave number functions $\{\bar{k}_0^{(1)}, \bar{k}_0^{(2)}, \bar{k}_0^{(3)}\}$:

$$\begin{aligned} &H_1 \text{ wave (short wave branch)} && \text{with larger } \mathrm{Re}\{\bar{k}_0^{(1)}\} > 0, \\ &H_2 \text{ wave (no physical meaning)} && \text{with } \mathrm{Re}\{\bar{k}_0^{(2)}\} < 0, \\ &H_3 \text{ wave (long wave branch)} && \text{with smaller } \mathrm{Re}\{\bar{k}_0^{(3)}\} > 0. \end{aligned}$$

Among these solutions, the $H_2$ wave solution must be ruled out. Therefore, in the zeroth-order approximation, the general $H$-wave solution is

$$\tilde{h}_0 = D_1 \mathrm{e}^{\mathrm{i}(\bar{\xi}_+^{(1)} + m\theta)} + D_3 \mathrm{e}^{\mathrm{i}(\bar{\xi}_+^{(3)} + m\theta)}, \tag{6.248}$$

where the coefficients $\{D_1, D_3\}$ are arbitrary constants independent of $\varepsilon$.

### 6.7.6 First-Order Approximation of the Outer Solution

The system of first-order approximation is

$$
\begin{aligned}
(k_0^2+g_0^2)\left(\frac{\partial^2}{\partial \xi_+^2}+\frac{\partial^2}{\partial \eta_+^2}\right)\tilde{T}_1 &= a_0(\xi,\eta)e^{i\xi_+-\eta_+}\,,\\
(k_0^2+g_0^2)\left(\frac{\partial^2}{\partial \xi_+^2}+\frac{\partial^2}{\partial \eta_+^2}\right)\tilde{T}_{S1} &= a_{S0}(\xi,\eta)e^{i\xi_++\eta_+}\,,
\end{aligned}
\tag{6.249}
$$

where

$$
\begin{aligned}
a_0 = 2(k_0-ig_0)\left(\frac{\partial A_0}{\partial \eta}-i\frac{\partial A_0}{\partial \xi}\right)+A_0\Big\{\sigma_0\eta_0^2(\xi^2+\eta^2)\\
-(ik_0+g_0)\left[\left(\frac{1}{\xi}-\eta_0^2\xi\right)+i\left(\frac{1}{\eta}+\eta_0^2\eta\right)\right]\Big\}
\end{aligned}
\tag{6.250}
$$

and

$$
\begin{aligned}
a_{S0} = -2(k_0+ig_0)\left(\frac{\partial A_{S0}}{\partial \eta}+i\frac{\partial A_{S0}}{\partial \xi}\right)+A_{S0}\Big\{\sigma_0\eta_0^2(\xi^2+\eta^2)\\
-(ik_0-g_0)\left[\left(\frac{1}{\xi}-\eta_0^2\xi\right)-i\left(\frac{1}{\eta}+\eta_0^2\eta\right)\right]\Big\}.
\end{aligned}
\tag{6.251}
$$

To eliminate the secular terms on the right-hand side of (6.249), one must set

$$
a_0 = a_{S0} = 0\,. \tag{6.252}
$$

Then Eq. (6.249) reduces to

$$
\begin{aligned}
\left(\frac{\partial^2}{\partial \xi_+^2}+\frac{\partial^2}{\partial \eta_+^2}\right)\tilde{T}_1 &= 0,\\
\left(\frac{\partial^2}{\partial \xi_+^2}+\frac{\partial^2}{\partial \eta_+^2}\right)\tilde{T}_{S1} &= 0\,.
\end{aligned}
\tag{6.253}
$$

The interface conditions are as follows: at the interface, $\eta_+ = 0, \eta = 1$,

1. the thermodynamic equilibrium condition

$$
\tilde{T}_1 = \tilde{T}_{S1}+\tilde{h}_1-\eta_1\bar{k}_0\frac{\partial}{\partial \eta_+}\big(\tilde{T}_0-\tilde{T}_{S0}\big), \tag{6.254}
$$

2. the Gibbs–Thomson condition

$$\tilde{T}_{S1} + \eta_1 \bar{k}_0 \frac{\partial \tilde{T}_{S0}}{\partial \eta_+} = \frac{1}{S(\xi)} \left\{ \bar{k}_0^2 \frac{\partial^2 \tilde{h}_1}{\partial \xi_+^2} + 2\bar{k}_0 \bar{k}_1 \frac{\partial^2 \tilde{h}_0}{\partial \xi_+^2} + 2\bar{k}_0 \frac{\partial^2 \tilde{h}_0}{\partial \xi_+ \partial \xi} \right.$$

$$\left. + \bar{k}_0' \frac{\partial \tilde{h}_0}{\partial \xi_+} + \bar{k}_0 \left( \frac{1}{\xi} + \frac{\xi}{S^2(\xi)} \right) \frac{\partial \tilde{h}_0}{\partial \xi_+} \right\}, \tag{6.255}$$

3. the heat balance condition

$$\bar{k}_0 \left( \frac{\partial \tilde{T}_1}{\partial \eta_+} - \frac{\partial \tilde{T}_{S1}}{\partial \eta_+} \right) + \bar{k}_1 \left( \frac{\partial \tilde{T}_0}{\partial \eta_+} - \frac{\partial \tilde{T}_{S0}}{\partial \eta_+} \right) + \eta_1 \bar{k}_0^2 \frac{\partial^2}{\partial \eta_+^2} (\tilde{T}_0 - \tilde{T}_{S0})$$

$$+ \left( \frac{\partial \tilde{T}_0}{\partial \eta} - \frac{\partial \tilde{T}_{S0}}{\partial \eta} \right) + S^2(\xi)(\sigma_0 \tilde{h}_1 + \sigma_1 \tilde{h}_0)$$

$$+ \xi \left( \bar{k}_0 \frac{\partial \tilde{h}_1}{\partial \bar{\xi}_+} + \bar{k}_1 \frac{\partial \tilde{h}_0}{\partial \xi_+} \right) + \left( 2 + \eta_0^2 \right) \tilde{h}_0 = 0. \tag{6.256}$$

The first-order approximate solutions can be obtained in the form of normal modes:

$$\begin{aligned} \tilde{T}_1 &= A_1(\xi, \eta) \exp\{\mathrm{i}\xi_+ - \eta_+\}, \\ \tilde{T}_{S1} &= A_{S1}(\xi, \eta) \exp\{\mathrm{i}\xi_+ + \eta_+\}, \\ \tilde{h}_1 &= \hat{D}_1 \exp\{\mathrm{i}\bar{\xi}_+\}, \end{aligned} \tag{6.257}$$

with

$$\begin{cases} \hat{A}_1(\xi) = A_1(\xi, 1), \\ \hat{A}_{S1}(\xi) = A_{S1}(\xi, 1). \end{cases} \tag{6.258}$$

One then obtains

$$\bar{k}_1(\xi) = \frac{\mathrm{i}}{2\xi} + \frac{R_1(\xi)}{F(\xi)} - \mathrm{i} \frac{R_2(\xi)}{F(\xi)} \frac{\bar{k}_0'(\xi)}{\bar{k}_0(\xi)}, \tag{6.259}$$

where we have defined

$$\begin{aligned} F(\xi) &= \left[ 1 - \mathrm{i}\xi - \tfrac{6\bar{k}_0^2}{S} \right], \\ R_1(\xi) &= S^2 \sigma_1 + \left( 1 + \tfrac{\eta_0^2}{2} \right) - \tfrac{\eta_0^2}{2} \left( 1 - \tfrac{2\bar{k}_0^2}{S} \right)^2, \\ R_2(\xi) &= \tfrac{6\bar{k}_0^2}{S}. \end{aligned} \tag{6.260}$$

The function $\bar{k}_1(\xi)$ can be extended to $\bar{k}_1(\zeta)$ defined in the complex plane $\zeta = \xi + \mathrm{i}\xi_I$ by analytic continuation. Then one may write

$$k_0(\zeta,\eta) = \tfrac{1}{2}\Big\{[\bar{k}_0[\zeta + \mathrm{i}(\eta-1)] + \bar{k}_0[\zeta - \mathrm{i}(\eta-1)]\Big\},$$
$$k_1(\zeta,\eta) = \tfrac{1}{2}\Big\{\bar{k}_1[\zeta + \mathrm{i}(\eta-1)] + \bar{k}_1[\zeta - \mathrm{i}(\eta-1)]\Big\},$$

and

$$\phi(\zeta,\eta) = \int_0^{\zeta}(k_0 + \varepsilon k_1 + \cdots)\mathrm{d}\zeta_1,$$
$$\psi(\zeta,\eta) = \int_1^{\eta}(k_0 + \varepsilon k_1 + \cdots)\mathrm{d}\eta_1.$$

Furthermore, one may deduce that the wave function $\bar{k}_1(\zeta)$, like the MVE solution, has a singularity at the root $\zeta = \zeta_c$ of the function $F(\zeta)$, so that $\partial\Sigma(\zeta_c,\bar{k}_0)/\partial\bar{k}_0 = F(\zeta_c) = 0$. Since the meromorpic function $\bar{k}_1(\zeta)$ can be expanded into a Laurent series near the isolated singular point $\zeta = \zeta_c$, it follows that $R_1(\zeta_c) = 0$ and

$$\sigma_1 = -\frac{1}{(1+\zeta_c^2)}\left[\left(1+\frac{\eta_0^2}{2}\right) - \frac{\eta_0^2}{18}(2+\mathrm{i}\zeta_c)^2\right]. \tag{6.261}$$

### 6.7.7 *The Inner Solution near the Singular Point $\zeta_c$ of the Outer Solution*

The location of the singular point $\zeta = \zeta_c$ on the interface $\eta = 1$ is the same as that obtained in the previous section. It is uniquely determined by the parameter $\sigma_0$ as

$$\zeta_c = \zeta(\sigma_0) = \mathrm{i}\left(\frac{|\sigma_0|^{\frac{4}{3}}\mathrm{e}^{\mathrm{i}\frac{4\theta}{3}} - \frac{2^{\frac{2}{3}}}{9}}{|\sigma_0|^{\frac{4}{3}}\mathrm{e}^{\mathrm{i}\frac{4\theta}{3}} + \frac{2^{\frac{2}{3}}}{9}}\right). \tag{6.262}$$

The inner equation near the singular point is also unchanged as derived in the previous section. With the inner variables

$$\zeta_* = \frac{\zeta-\zeta_c}{\varepsilon^{\frac{3}{2}}}, \quad \eta_* = \frac{\eta-1}{\varepsilon^{\frac{3}{2}}}, \tag{6.263}$$

we set

$$\tilde{T}(\zeta,\eta,\theta,t) = \varepsilon^{\frac{3}{2}}\hat{T}(\zeta_*,\eta_*,\theta,t);$$

$$\tilde{T}_S(\zeta,\eta,\theta,t) = \varepsilon^{\frac{3}{2}}\hat{T}_S(\zeta_*,\eta_*,\theta,t), \tag{6.264}$$

$$\tilde{h}(\zeta,\theta,t) = \varepsilon^{\frac{3}{2}}\hat{h}(\zeta_*,\theta,t), \tag{6.265}$$

and the interface shape function is described as

$$\eta_s(\zeta,\theta,t) = 1 + \tilde{h}(\zeta,\theta,t)/\eta_0^2 = 1 + \varepsilon^{\frac{3}{2}}\hat{h}(\zeta_*,\theta,t)/\eta_0^2. \tag{6.266}$$

By applying the transformation

$$H(\zeta) = W(\zeta)\exp\left[\frac{\mathrm{i}}{\varepsilon}\int_{\zeta_c}^{\zeta} k_c(\zeta_1)\,\mathrm{d}\zeta_1\right], \tag{6.267}$$

where

$$k_c(\zeta) = \mathrm{e}^{-\mathrm{i}\frac{\pi}{4}}\sqrt{\frac{\hat{S}}{6}(\zeta+\mathrm{i})} \qquad \left(\mathrm{Re}\{k_c\} > 0\right), \tag{6.268}$$

one may write the inner solution in the form

$$\hat{h}_0(\zeta_*) = \hat{W}_0(\zeta_*)\exp\left\{\frac{\mathrm{i}}{\varepsilon}\int_{\zeta_c}^{\zeta} k_c(\xi_1)\,\mathrm{d}\xi_1\right\}. \tag{6.269}$$

In the leading-order approximation, the inner equation is simplified to the Airy equation:

$$\frac{\mathrm{d}^2\hat{W}_0}{\mathrm{d}\zeta_*^2} + A^2\zeta_*\hat{W}_0 = 0, \tag{6.270}$$

where

$$A = \left(\frac{S^3}{6k_c}\frac{\partial\Sigma_c}{\partial\zeta}\right)^{\frac{1}{2}}_{\zeta=\zeta_c} = \mathrm{e}^{\mathrm{i}}\pi/4\sqrt{\frac{1}{6}}\left(\frac{\zeta_c+\mathrm{i}}{\zeta_c-\mathrm{i}}\right)^{\frac{1}{4}} \tag{6.271}$$

$$\left(\frac{-\pi}{4} < \arg\{A\} < 0\right).$$

The Stokes lines $(L_1)$, $(L_2)$, $(L_3)$ defined by the integral

$$\mathrm{Im}\left\{\int_{\zeta_c}^{\zeta}\left(\bar{k}_0^{(1)} - k_0^{(3)}\right)\mathrm{d}\zeta'\right\} = 0 \tag{6.272}$$

and the anti-Stokes lines $(A_1)$, $(A_2)$, $(A_3)$ defined by

$$\mathrm{Re}\left\{\int_{\zeta_c}^{\zeta}\left(\bar{k}_0^{(1)} - \bar{k}_0^{(3)}\right)\mathrm{d}\zeta'\right\} = 0 \tag{6.273}$$

are the same as those shown in Fig. 6.2. The anti-Stokes line ($A_2$) divides the entire complex $\zeta$-plane into the sectors ($S_1$) and ($S_2$). Due to the Stokes phenomenon, the pair of coefficients ($D_1, D_3$) of the outer solution

$$\tilde{h} \approx D_1 \mathrm{e}^{\sigma_0 t_+ + \mathrm{i}(\bar{\xi}_+^{(1)} + m\theta)} + D_3 \mathrm{e}^{\sigma_0 t_+ + \mathrm{i}(\bar{\xi}_+^{(3)} + m\theta)} \tag{6.274}$$

have different values in different sectors. Hence one may denote them by $(D_1', D_3')$ in sector ($S_2$) and by ($D_1, D_3$) in sector ($S_1$).

By matching the inner solution with the outer solution in different sectors, we derive the connection condition

- In sector ($S_2$):

$$D_1' = 0, \quad D_3' \text{ (free)}.$$

- In sector ($S_1$):

$$\frac{D_1}{D_3} = \mathrm{e}^{\frac{\mathrm{i}\pi}{2}}, \quad D_3 = D_3'. \tag{6.275}$$

### 6.7.8 *Tip Inner Solution in the Tip Region*

As indicated in the previous section, the MVE solution is also invalid in the vicinity of the point ($\xi = 0, \eta = 1$), which may be called the tip region. Especially, as indicated before, the tip region is the region in which the 3D effect plays an important role.

As in the last section, we define the tip inner variables

$$\begin{aligned} \hat{\xi} &= \frac{\hat{k}\xi}{\varepsilon}, \quad \hat{\zeta} = \frac{\hat{k}\zeta}{\varepsilon}, \\ \hat{\eta} &= \frac{\hat{k}(\eta - 1)}{\varepsilon}, \\ \hat{t} &= \frac{t}{\eta_0^2 \varepsilon}, \end{aligned} \tag{6.276}$$

where $|\zeta| \ll \varepsilon$ and $|\eta - 1| \ll 1$. The tip solution can be expressed as a function of these inner variables and expanded in the following asymptotic form as $\varepsilon \to 0$:

$$\begin{aligned} \hat{T}(\hat{\xi}, \hat{\eta}, \theta, \hat{t}) &= \left\{ \hat{\mu}_0(\varepsilon)\hat{T}_0 + \hat{\mu}_1(\varepsilon)\hat{T}_1 + \cdots \right\} \mathrm{e}^{im\theta + \sigma\hat{t}}, \\ \hat{T}_S(\hat{\xi}, \hat{\eta}, \theta, \hat{t}) &= \left\{ \hat{\mu}_0(\varepsilon)\hat{T}_{S0} + \hat{\mu}_1(\varepsilon)\hat{T}_{S1} + \cdots \right\} \mathrm{e}^{im\theta + \sigma\hat{t}}, \\ \hat{h}(\hat{\xi}, \theta, \hat{t}) &= \left\{ \hat{\mu}_0(\varepsilon)\hat{h}_0 + \hat{\mu}_1(\varepsilon)\hat{h}_1 + \cdots \right\} \mathrm{e}^{im\theta + \sigma\hat{t}} \\ \hat{k} &= \hat{k}_0 + \varepsilon\hat{k}_1 + \cdots . \end{aligned} \tag{6.277}$$

At the zeroth-order approximation in the tip region, the system (6.212)–(6.221) can be reduced to

$$\begin{aligned} \hat{k}_0^2\frac{\partial^2\hat{T}_0}{\partial\hat{\xi}^2}+\hat{k}_0\frac{\partial^2\hat{T}_0}{\partial\hat{\eta}^2}+\frac{\hat{k}_0^2}{\hat{\xi}}\frac{\partial\hat{T}_0}{\partial\hat{\xi}}-\frac{m^2\hat{k}_0^2}{\hat{\xi}^2}\hat{T}_0=0 \\ \hat{k}_0^2\frac{\partial^2\hat{T}_{S0}}{\partial\hat{\xi}^2}+\hat{k}_0\frac{\partial^2\hat{T}_{S0}}{\partial\hat{\eta}^2}+\frac{\hat{k}_0^2}{\hat{\xi}}\frac{\partial\hat{T}_{S0}}{\partial\hat{\xi}}-\frac{m^2\hat{k}_0^2}{\hat{\xi}^2}\hat{T}_{S0}=0 \end{aligned} \tag{6.278}$$

with the boundary conditions at $\hat{\eta}$=0

$$\hat{T}_0=\hat{T}_{S0}+\hat{h}_0, \tag{6.279}$$

$$\hat{T}_{S0}=\hat{k}_0^2\left(\frac{\partial^2\hat{h}_0}{\partial\hat{\xi}^2}+\frac{1}{\hat{\xi}}\frac{\partial\hat{h}_0}{\partial\hat{\xi}}-\frac{m^2}{\hat{\xi}^2}\hat{h}_0\right), \tag{6.280}$$

$$\frac{\partial}{\partial\hat{\eta}}(\hat{T}_0-\hat{T}_{S0})+\sigma_0\hat{h}_0=0. \tag{6.281}$$

This system admits the inner solutions

$$\begin{aligned} \hat{T}_0&=\hat{a}_0H_m^{(1)}(\hat{\xi})\mathrm{e}^{-\hat{k}_0\hat{\eta}}, \\ \hat{T}_{S0}&=\hat{a}_{S0}H_m^{(1)}(\hat{\xi})\mathrm{e}^{\hat{k}_0\hat{\eta}}, \\ \hat{h}_0&=\hat{d}_0H_m^{(1)}(\hat{\xi}), \end{aligned} \tag{6.282}$$

where $H_m^{(1)}(\hat{\xi})$ is the $m$th-order Hankel function of the first kind, which satisfies the equation

$$\frac{\partial^2H_m}{\partial\hat{\xi}^2}+\frac{1}{\hat{\xi}}\frac{\partial H_m}{\partial\hat{\xi}}+\frac{1-m^2}{\hat{\xi}^2}H_m=0.$$

From the boundary conditions (6.279)–(6.281), one obtains

$$\hat{a}_0=(1-\hat{k}_0^2)\hat{d}_0, \quad \hat{a}_{S0}=-\hat{k}_0^2\hat{d}_0, \tag{6.283}$$

and the local dispersion relationship in the tip region

$$\sigma_0=\hat{k}_0\left(1-2\hat{k}_0^2\right). \tag{6.284}$$

For fixed $\sigma_0$, (6.284) has three roots for $\hat{k}_0$. Comparing the local dispersion relationship in the tip region (6.284) with the local dispersion relationship in the outer region (6.242), one can evidently write

$$\begin{cases} \hat{k}_0^{(1)} = \bar{k}_0^{(1)}(0), \\ \hat{k}_0^{(2)} = \bar{k}_0^{(2)}(0), \\ \hat{k}_0^{(3)} = \bar{k}_0^{(3)}(0). \end{cases} \tag{6.285}$$

The root $\hat{k}_0^{(2)}$ must be ruled out, due to the fact that $\mathrm{Re}\{\hat{k}_0^{(2)}\} < 0$. Therefore, the general solution of $\hat{h}_0$ in the tip region is

$$\hat{h}_0 = \hat{d}_0^{(1)} H_m^{(1)}\left(\frac{\hat{k}_0^{(1)}\xi}{\varepsilon}\right) + \hat{d}_0^{(3)} H_m^{(1)}\left(\frac{\hat{k}_0^{(3)}\xi}{\varepsilon}\right). \tag{6.286}$$

The above tip inner solution must match the outer solution. By setting

$$\hat{d}_0^{(1)} = d_0, \quad \hat{d}_0^{(3)} = -\hat{Q}_0 d_0, \tag{6.287}$$

we can express the tip inner solution as

$$\hat{h}_0 = d_0 \left\{ H_m^{(1)}\left(\frac{\hat{k}_0^{(1)}\xi}{\varepsilon}\right) - \hat{Q}_0 H_m^{(1)}\left(\frac{\hat{k}_0^{(3)}\xi}{\varepsilon}\right) \right\}. \tag{6.288}$$

As $\hat{\xi} \to \infty$, one has

$$\hat{h}_0 \approx d_0 \sqrt{\frac{2}{\pi}} \mathrm{e}^{-(m+\frac{1}{2})\frac{\pi}{2}} \times \left\{ \sqrt{\frac{\varepsilon}{k_0^{(1)}(0)\xi}}\, \mathrm{e}^{\frac{\mathrm{i}k_0^{(1)}(0)\xi}{\varepsilon}} - \hat{Q}_0 \sqrt{\frac{\varepsilon}{k_0^{(3)}(0)\xi}}\, \mathrm{e}^{\frac{\mathrm{i}k_0^{(3)}(0)\xi}{\varepsilon}} \right\}. \tag{6.289}$$

Now we need to match the tip inner solution (6.288) with the outer solution derived in the last section, namely,

$$\hat{\mu}_0(\varepsilon)\sqrt{\frac{2\varepsilon}{\pi}} d_0 \left\{ \sqrt{\frac{1}{\hat{k}_0^{(1)}\hat{\xi}}}\mathrm{e}^{\frac{\mathrm{i}\hat{k}_0^{(1)}\xi}{\varepsilon}} - \hat{Q}_0 \sqrt{\frac{1}{\hat{k}_0^{(3)}\hat{\xi}}}\mathrm{e}^{\frac{\mathrm{i}\hat{k}_0^{(3)}\xi}{\varepsilon}} \right\}$$

$$\Updownarrow$$

$$\tilde{\mu}_0(\varepsilon)\left( D_1 \exp\left\{\frac{\mathrm{i}}{\varepsilon}\int_{\zeta_c}^{\xi} \bar{k}_0^{(1)} \mathrm{d}\xi'\right\} + D_3 \exp\left\{\frac{\mathrm{i}}{\varepsilon}\int_{\zeta_c}^{\xi} \bar{k}_0^{(3)} \mathrm{d}\xi'\right\}\right)$$

$$= \tilde{\mu}_0(\varepsilon)\left[ D_1 \mathrm{e}^{-\mathrm{i}\chi_1} \exp\left\{\frac{\mathrm{i}}{\varepsilon}\int_0^{\xi} \bar{k}_0^{(1)} \mathrm{d}\xi'\right\} + D_3 \mathrm{e}^{-\mathrm{i}\chi_3} \exp\left\{\frac{\mathrm{i}}{\varepsilon}\int_0^{\xi} \bar{k}_0^{(3)} \mathrm{d}\xi'\right\}\right].$$

Here we have used the notation

$$\chi_1(\varepsilon) = \frac{1}{\varepsilon}\int_0^{\xi_c} \bar{k}_0^{(1)}(\xi')\mathrm{d}\xi', \quad \chi_3(\varepsilon) = \frac{1}{\varepsilon}\int_0^{\xi_c} \bar{k}_0^{(3)}(\xi')\mathrm{d}\xi'. \tag{6.290}$$

In order to satisfy the above matching condition, the parameter $\sigma_0$ must be a proper function of $\varepsilon$ such that as $\varepsilon \to 0$, the two functions $\exp\left\{-\mathrm{i}\chi_1 + \frac{\mathrm{i}}{\varepsilon}\int_0^{\xi}\bar{k}_0^{(1)}\mathrm{d}\xi'\right\}$ and $\exp\left\{-\mathrm{i}\chi_3 + \frac{\mathrm{i}}{\varepsilon}\int_0^{\xi}\bar{k}_0^{(3)}\mathrm{d}\xi'\right\}$ are of the same order of magnitude. Therefore, the origin $\zeta = 0$ will not lie on a Stokes line, unless $\left|\bar{k}_0^{(1)}(0)\right| = \left|\bar{k}_0^{(3)}(0)\right|$.

Thus it follows from the matching condition that

$$\hat{\mu}_0(\varepsilon)\varepsilon^{\frac{1}{2}} = \tilde{\mu}_0(\varepsilon) \tag{6.291}$$

and

$$\frac{D_3 \mathrm{e}^{-\mathrm{i}\chi_3}}{D_1 \mathrm{e}^{-\mathrm{i}\chi_1}} = -\mathrm{i}\mathrm{e}^{\mathrm{i}(\chi_1-\chi_3)} = -\hat{Q}_0\left(\frac{\bar{k}_0^{(1)}(0)}{\bar{k}_0^{(3)}(0)}\right)^{\frac{1}{2}}, \tag{6.292}$$

or

$$\hat{Q}_0 = \hat{Q}_0(\sigma_0, \varepsilon) = \mathrm{i}\mathrm{e}^{\mathrm{i}(\chi_1-\chi_3)}\left(\frac{\bar{k}_0^{(3)}(0)}{\bar{k}_0^{(1)}(0)}\right)^{\frac{1}{2}}. \tag{6.293}$$

Up to this point, the first-order asymptotic solution contains a free parameter, $\sigma_0$.

### 6.7.9 *Global Trapped-Wave (GTW) Modes and Quantization Condition*

We now turn to the second step: constructing the global eigenmodes and deriving the quantization condition for the eigenvalues by applying the tip smoothness condition

$$\frac{\partial}{\partial\hat{\xi}}\{\hat{T}, \hat{T}_S, \hat{h}\} = 0 \tag{6.294}$$

to the asymptotic solutions obtained above. Note that as $\hat{\xi} \to 0$,

$$H_m^{(1)}\left(\frac{\hat{k}_0\xi}{\varepsilon}\right) \sim \begin{cases} -\mathrm{i}\frac{(m-1)!}{\pi}\left(\frac{2\varepsilon}{\hat{k}_0\xi}\right)^m + O\left(\frac{\hat{k}_0\xi}{2\varepsilon}\right)^m, & (m>0), \\ \mathrm{i}\frac{2}{\pi}\ln\left(\frac{\hat{k}_0\xi}{\varepsilon}\right) + 1 + \mathrm{i}\frac{2}{\pi}(\gamma - \ln 2) + O\left(\frac{\hat{k}_0\xi}{\varepsilon}\right), & (m=0). \end{cases}$$

(see p. 360 in [1]).

To satisfy the tip smoothness condition $\hat{h}_0(0) = \mathcal{O}(1), \hat{h}_0'(0) = 0$, let us investigate the behavior of the function $\hat{h}_0(\hat{\xi})$ as $\hat{\xi} \to 0$ for each of the cases $m = 0, 1, 2, \ldots$.

- For the case $m = 0$, we have

$$\hat{h}_0(\xi) \approx d_0 \left\{ \mathrm{i}\frac{2}{\pi} \ln\left(\frac{\hat{k}_0^{(1)}\xi}{\varepsilon}\right) - \mathrm{i}\hat{Q}_0 \frac{2}{\pi} \ln\left(\frac{\hat{k}_0^{(3)}\xi}{\varepsilon}\right) \right\} + \mathcal{O}(1).$$

To ensure that $\hat{h}_0(0) = \mathcal{O}(1)$, one must set $\hat{Q}_0 = 1$. Meanwhile, we have

$$\hat{h}_0'(0) = 0.$$

Thus the tip smoothness conditions are fully satisfied.

- For the case $m = 1$, we have

$$\hat{h}_0(\hat{\xi}) \approx d_0 \left\{ \left( \frac{\hat{k}_0^{(1)}\xi}{2\varepsilon} - \mathrm{i}\frac{2\varepsilon}{\pi \hat{k}_0^{(1)}\xi} \right) - \hat{Q}_0 \left( \frac{\hat{k}_0^{(3)}\xi}{2\varepsilon} - \mathrm{i}\frac{2\varepsilon}{\pi \hat{k}_0^{(3)}\xi} \right) \right\}.$$

To ensure that $\hat{h}_0(0) = \mathcal{O}(1)$, one must set $\hat{Q}_0 = \hat{k}_0^{(3)}/\hat{k}_0^{(1)}$. However, with such a value of $\hat{Q}_0$, we have

$$\hat{h}_0'(0) = \frac{d_0 \hat{k}_0^{(3)}}{2} \left\{ \frac{\hat{k}_0^{(1)}}{\hat{k}_0^{(3)}} - \frac{\hat{k}_0^{(3)}}{\hat{k}_0^{(1)}} \right\} \neq 0.$$

Therefore, the tip smoothness conditions can never be fully satisfied, since $\hat{k}_0^{(1)} \neq \hat{k}_0^{(3)}$.

- For the case $m > 1$, we have

$$\hat{h}_0(\hat{\xi}) \approx d_0 \left\{ \left[ \frac{1}{m!} \left( \frac{\hat{k}_0^{(1)}\xi}{2\varepsilon} \right)^m - \mathrm{i}\frac{(m-1)!}{\pi} \left( \frac{2\varepsilon}{\hat{k}_0^{(1)}\xi} \right)^m \right] \right.$$
$$\left. -\hat{Q}_0 \left[ \frac{1}{m!} \left( \frac{\hat{k}_0^{(3)}\xi}{2\varepsilon} \right)^m - \mathrm{i}\frac{(m-1)!}{\pi} \left( \frac{2\varepsilon}{\hat{k}_0^{(3)}\xi} \right)^m \right] \right\}.$$

To ensure that $\hat{h}_0(0) = \mathcal{O}(1)$, one must set $\hat{Q}_0 = \left(\hat{k}_0^{(3)}/\hat{k}_0^{(1)}\right)^m$. Meanwhile, one has $\hat{h}_0'(0) = 0$. Hence the tip smoothness conditions are fully satisfied.

In summary, we derive that in order for the tip solution (6.288) to satisfy the tip condition, one must set

$$\hat{Q}_0 = \left(\frac{\hat{k}_0^{(3)}}{\hat{k}_0^{(1)}}\right)^m = \left(\frac{\bar{k}_0^{(3)}(0)}{\bar{k}_0^{(1)}(0)}\right)^m \tag{6.295}$$

for the values $m = 0$ and $m = 2, 3, \ldots$. The mode $m = 0$ gives an axisymmetric interfacial traveling wave, while the modes $m = 2, 3, \ldots$ give multiarm spiral interfacial waves.

Formula (6.295) is the quantization condition for the eigenvalues. Combining it with (6.293), one may rewrite (6.295) in the form

$$\begin{aligned}\frac{1}{\varepsilon}\int_0^{\zeta_c}\left(\bar{k}_0^{(1)} - \bar{k}_0^{(3)}\right)\mathrm{d}\zeta &= \left(2n + 1 + \frac{1}{2} - \frac{2m-1}{2}\theta_0\right)\pi \\ &\quad + \mathrm{i}\,\frac{2m-1}{2}\ln\alpha_0 \\ &\quad \left(m = 0, 2, 3, \ldots; \quad n = 0, 1, 2, \ldots\right),\end{aligned} \tag{6.296}$$

where

$$\alpha_0\, \mathrm{e}^{\mathrm{i}\theta_0\pi} = \frac{\bar{k}_0^{(1)}(0)}{\bar{k}_0^{(3)}(0)}\,. \tag{6.297}$$

For the special axisymmetric case (m=0), we regain

$$\frac{1}{\varepsilon}\int_0^{\zeta_c}\left(\bar{k}_0^{(1)} - \bar{k}_0^{(3)}\right)\mathrm{d}\zeta = \left(2n + 1 + \frac{1}{2} + \frac{1}{2}\theta_0\right)\pi - \mathrm{i}\,\frac{1}{2}\ln\alpha_0. \tag{6.298}$$

The quantization (6.296) condition gives a discrete set of complex eigenvalues

$$\sigma_0^{(m,n)} \quad \left(m = 0, 2, 3, \ldots; \quad n = 0, 1, 2, \ldots\right),$$

as functions of the parameters $\varepsilon$. Here it should be kept in mind that as a necessary condition in solving the quantization condition for the eigenvalue $\sigma_0$, the corresponding singular point $\zeta_c$ must be located in the fourth quadrant, satisfying $\mathrm{Re}\{\zeta_c\} > 0, \mathrm{Im}\{\zeta_c\} < 0$.

We derive that the system allows two types of global modes:

- The axisymmetric modes:

$$(0,0), (0,1), (0,2), (0,3), \ldots,$$

  which we found in the last section.

- The multiarm spiral modes:

$$(2,1),(2,2);(3,2),(3,3);(4,2),(4,3),(4,4),\ldots.$$

In the zeroth-order approximation, the variations of the real part and imaginary part of the eigenvalues $\sigma_{0R}$ and $\omega_0$ with $\varepsilon$ are shown in Figs. 6.18 and 6.19, respectively. The distribution of the eigenvalue $\sigma_0$ for different values of $\varepsilon$ is shown in Fig. 6.20.

Once the zeroth-order approximation of the eigenvalue $\sigma_0$ is obtained, the first approximation of the eigenvalue $\sigma_1$ can be determined using the formula (6.261). Then one can obtain the first-order approximation of the eigenvalue $\sigma \approx \sigma_0 + \varepsilon\sigma_1$.

The variations of the real part and imaginary part of $\sigma_R$ and $\omega$ with $\varepsilon$ in the first-order approximation of the eigenvalues are shown in Figs. 6.21 and 6.22, respectively. The distribution of the eigenvalue $\sigma$ in the first-order approximation with different values of $\varepsilon$ is shown in Fig. 6.23.

Our numerical investigation has shown that there is a unique neutral mode with $\sigma_R = 0$ for each of these acceptable $(m,n)$ modes corresponding to the critical numbers $\varepsilon_*^{(m,n)}$. In the first-order approximation, these critical numbers are denoted by $\varepsilon_{*1}^{(m,n)}$.

The shape of the dendritic interface corresponding to a global neutral mode can be written in the form

$$\eta_s(\xi,\theta,t) \approx 1 + \varepsilon^2\eta_1(\xi,\theta) + \hat{h}_0(\xi)\mathrm{e}^{-\mathrm{i}\frac{\omega_* t}{\eta_0^2\varepsilon_*}+\mathrm{i}m\theta}, \tag{6.299}$$

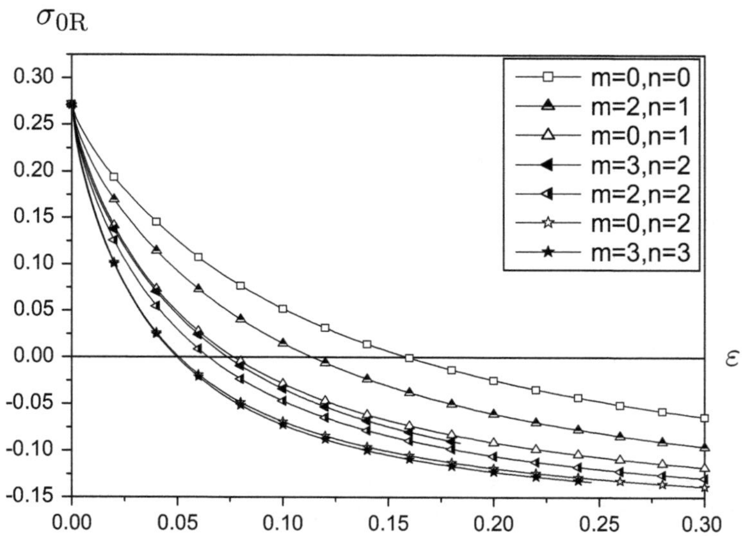


**Fig. 6.18** Variations of the real part of the zeroth-order approximation of eigenvalues $\sigma_{0R}$ with $\varepsilon$ of the modes (0, 0),(0, 1),(0, 2); (2, 1), (2, 2); (3, 2), (3, 3) for the case $\eta_0^2 = 0.01$

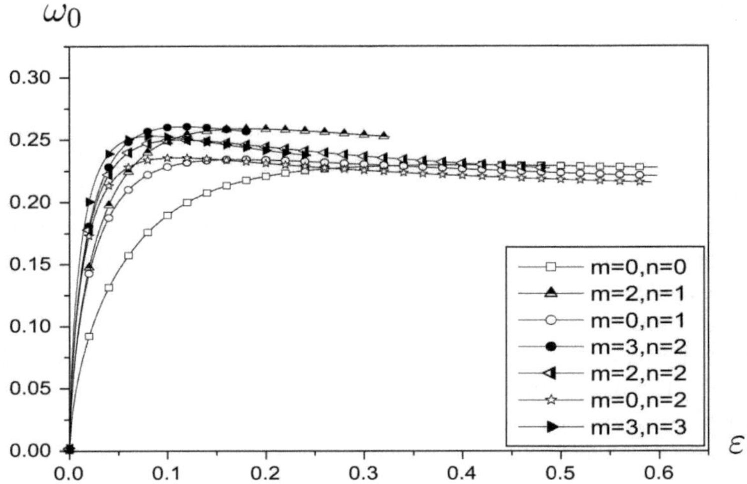


**Fig. 6.19** Variations of the imaginary part of the zeroth-order approximation of eigenvalues, $\omega_0$ with $\varepsilon$ of the modes $(0,0)$,$(0,1)$,$(0,2)$; $(2,1)$, $(2,2)$; $(3,2)$, $(3,3)$ for the case $\eta_0^2 = 0.01$

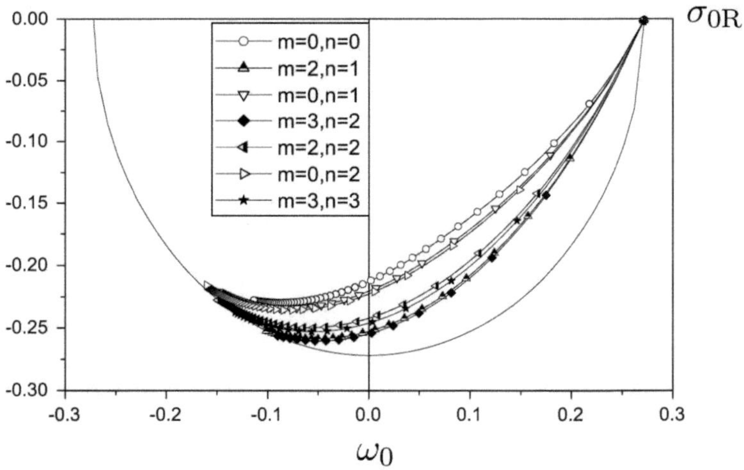


**Fig. 6.20** The distribution of the zeroth-order approximate eigenvalues, $\sigma_0 = \sigma_{0R} - \mathrm{i}\omega_0$, in the complex $\sigma_0$-plane for various values of $\varepsilon$

as $\varepsilon \to 0$. Here $\eta_1(\xi)$ is the leading term in the RPE part, while $\hat{h}_0(\xi)\mathrm{e}^{-\mathrm{i}\frac{\omega_* t}{\eta_0^2\varepsilon_*}+\mathrm{i}m\theta}$ is the global neutral mode $(m,n)$ in the SPE part. The parameter $\sigma = -\mathrm{i}\omega_*$ is the eigenvalue corresponding to the critical number $\varepsilon_* = \varepsilon_*^{(m,n)}$.

The shape of the eigenfunction of the selected global neutral mode $(0,0)$ has been shown in the previous section (see Fig. 6.10). The 3D-interface shape of the

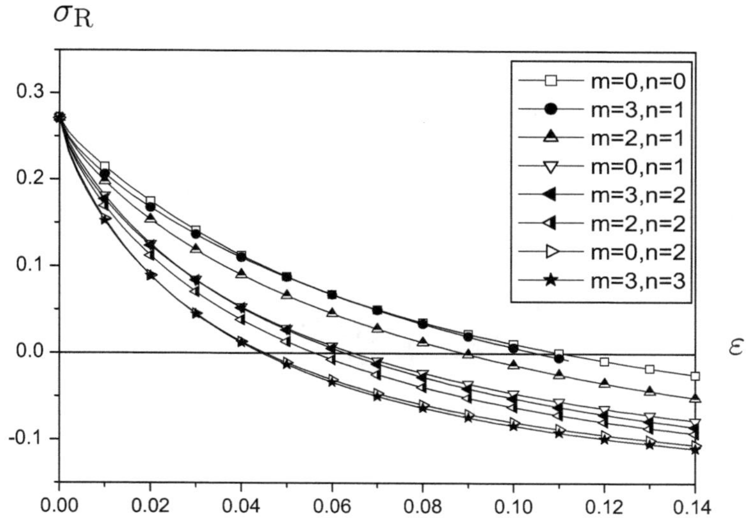


**Fig. 6.21** Variations of the real part of the first-order approximation of eigenvalues $\sigma_R$ with $\varepsilon$ of the modes (0, 0),(0, 1),(0, 2); (2, 1), (2, 2); (3, 2), (3, 3) for the case $\eta_0^2 = 0.01$

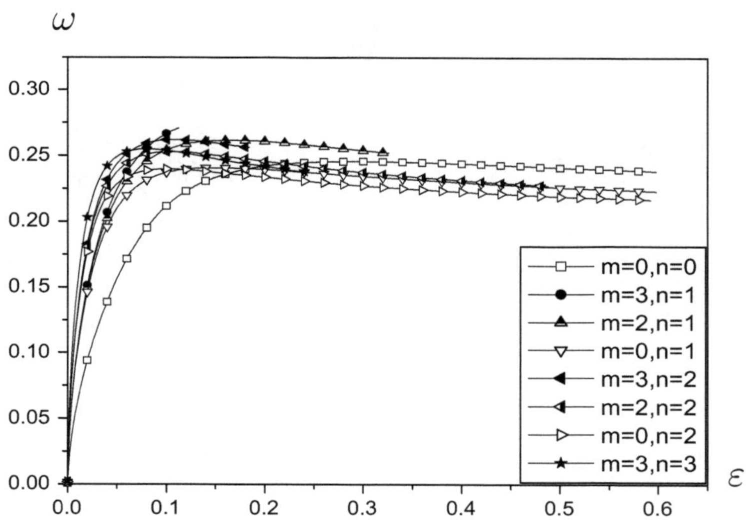


**Fig. 6.22** Variations of the imaginary part of the first-order approximation of eigenvalues $\omega$ with $\varepsilon$ of the modes (0, 0),(0, 1),(0, 2); (2, 1), (2, 2); (3, 2), (3, 3) for the case $\eta_0^2 = 0.01$

selected global mode $(0,0)$ and its projection onto the $(x,y)$-plane are also shown in Fig. 6.24.

The projections of the interface onto the $(x,y)$-plane and the 3D graph of the interface shapes for the nonaxisymmetric global neutral modes $(2,1)$ and $(3,2)$

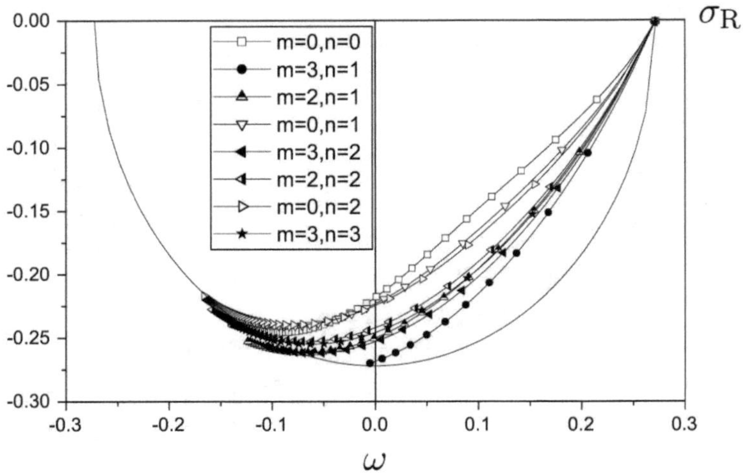


**Fig. 6.23** The distribution of the first-order approximate eigenvalues, $\sigma = \sigma_R - i\omega$, in the complex $\sigma$-plane for various values of $\varepsilon$

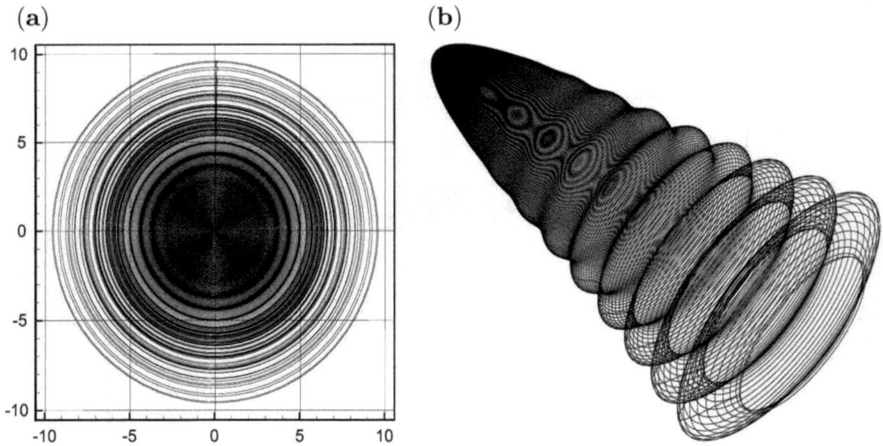


**Fig. 6.24** (**a**) The projection of the shape of the interface global neutral mode (0,0) onto the $(x, y)$-plane; (**b**) the shape of the interface of the selected global neutral mode $(0, 0)$

are respectively shown in Figs. 6.25 and 6.26, which describe the propagation of multiarm spiral interfacial waves along the Ivantsov paraboloid.

Numerical computations show that in the regime of small undercooling, the critical number $\varepsilon_{*1}^{(m,n)}$ is approximately a constant. Table 6.4 lists the numerical values of $\varepsilon_{*1}^{(m,n)}$ for some typical small undercooling cases. The numerical results show that

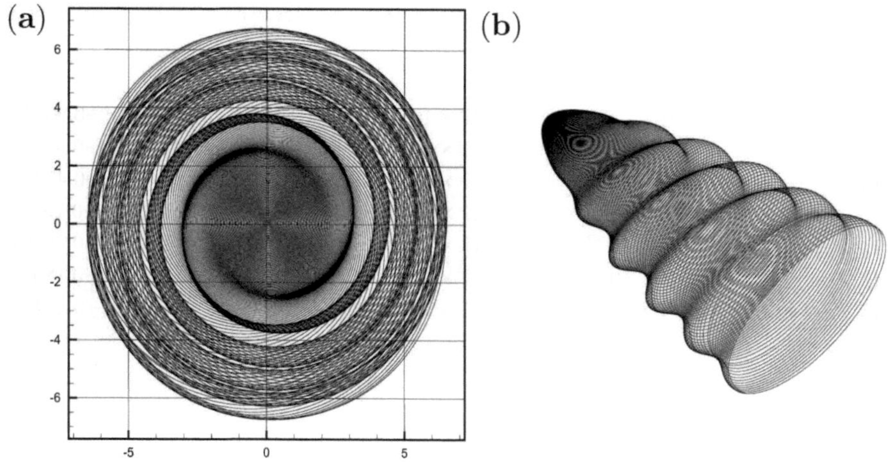

**Fig. 6.25** The shape of the interface of a dendrite for the global neutral mode $(2, 1)$: (**a**) the projection of the interface onto the $(x, y)$-plane; (**b**) the 3D graph of the interface

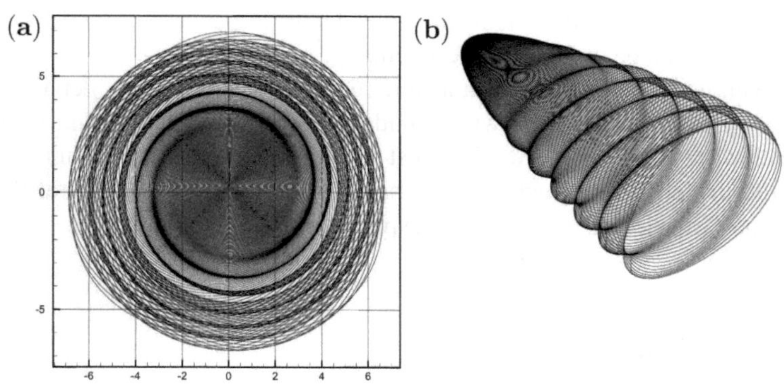

**Fig. 6.26** The shape of the interface of a dendrite for the global neutral mode $(3, 2)$: (**a**) the projection of the interface onto the $(x, y)$-plane; (**b**) the 3D graph of the interface

$$\varepsilon_{*1}^{(0,0)} > \varepsilon_{*1}^{(3,1)} > \varepsilon_{*1}^{(2,1)} > \varepsilon_{*1}^{(0,1)} > \varepsilon_{*1}^{(3,2)} > \varepsilon_{*1}^{(2,2)} > \varepsilon_{*1}^{(3,3)} > \varepsilon_{*1}^{(0,2)} > \cdots .$$

The most important conclusion that can be drawn from these results is that the most dangerous global neutral mode, which has the largest critical number $\varepsilon_*$, is the axisymmetric mode $(0, 0)$. In the regime of small undercooling, as shown in the last section, the critical number of this neutral mode is $\varepsilon_{*1}^{(0,0)} = 0.1107$.

It is therefore deduced that only the axisymmetric global neutral mode can be selected in the system under investigation; all the nonaxisymmetric unstable modes existing in the system do not affect the stability property of the system. The stability criterion of the system of 3D dendritic growth remains as given in the previous

**Table 6.4** The critical numbers $\varepsilon_{*1}^{(m,n)}$ and the corresponding eigenvalues of the global neutral modes for the case $\eta_0^2 = 0.01$

| $(m,n)$ | $\varepsilon_{*1}^{(m,n)}$ | $\sigma = (\sigma_R - i\omega_*)$ | $\zeta_c$ |
|---|---|---|---|
| (0, 0) | 0.1107 | $(2 \times 10^{-6}, -0.2181)$ | (1.1547, −0.5074) |
| (0, 1) | $6.4700 \times 10^{-2}$ | $(1 \times 10^{-6}, -0.2231)$ | (1.3990, −0.4924) |
| (0, 2) | $4.5153 \times 10^{-2}$ | $(1 \times 10^{-6}, -0.2245)$ | (1.4787, −0.4835) |
| (2, 1) | $8.8970 \times 10^{-2}$ | $(1 \times 10^{-5}, -0.2500)$ | (1.4394, −0.2068) |
| (2, 2) | $5.6130 \times 10^{-2}$ | $(1 \times 10^{-6}, -0.2420)$ | (1.5220, −0.3008) |
| (3, 2) | $6.2901 \times 10^{-2}$ | $(1 \times 10^{-6}, -0.2530)$ | (1.5346, −0.1863) |
| (3, 3) | $4.4457 \times 10^{-2}$ | $(1 \times 10^{-6}, -0.2457)$ | (1.5717, −0.2647) |

section, $\varepsilon_* \approx \varepsilon_{*1}^{(0,0)}$, so that

$$\begin{cases} \text{as} \quad \varepsilon > \varepsilon_*, & \text{the system is globally stable;} \\ \text{as } 0 < \varepsilon < \varepsilon_*, & \text{the system is unstable.} \end{cases}$$

The present analysis predicts that no spiral dendritic pattern of dendritic growth from pure melt can be observed experimentally.

However, it should be noted that a spiral dendritic pattern might persist in some special systems of dendritic growth. If under some circumstances, it turns out that the spiral wave mode becomes the most dangerous mode due to some physical effects that have been neglected in the present analysis, then the spiral dendritic patterns may be selected and observed experimentally.

## 6.8 A Brief Summary

In this chapter we have investigated a single-free dendritic growth from undercooled pure melt. The ideas and procedure of treatment are summarized as follows:

1. First of all, we specify the basic states of the system as the nonclassical steady needle solutions defined previously. Instead of the exact form of such basic state solutions, one may utilize the RPE form of the basic state solutions in the limit of $\varepsilon \to 0$ as an approximation of the basic state solutions in the finite region behind the dendrite-tip but not close to the root. The form of RPE of the basic state is independent of the root conditions.
2. For the case $\varepsilon = 0$, many dendritic growth systems allow an analytical form of solution $\{T = T_*(\eta), T_S = 0, \eta = 1\}$ called Ivantsov's needle solution. in the region not close to the dendrite's root, the basic state solutions are well approximated by the Ivantsov solution with the accuracy of $(\varepsilon^2)$.
3. Consider the stability of the basic states. Separate the general unsteady solutions into two parts—the basic steady state and the unsteady perturbed states—

and linearize around the basic state solution in terms of the small-amplitude parameter of the initially infinitesimal perturbations $\delta \ll 1$. This leads to a linear eigenvalue problem.
4. Solve the eigenvalue problem in two steps. First, find the uniformly valid asymptotic solutions for the linear perturbed system with fixed $\sigma$ and related parameters. This system consists of the governing equations and all boundary conditions except for the smooth tip condition. Then apply the smooth tip condition to obtain the quantization condition for the eigenvalues.
5. Apply the MVE method to derive the outer solutions. This gives the local dispersion relationship for the normal mode solutions in the zeroth-order approximation. Some singularities of the normal mode solutions may be identified in the first-order approximation.
6. In the inner region near the singular point, one needs to introduce new length scales and derive and solve the inner equations. Then one must match the outer solutions to these inner solutions.
7. Finally, we apply the tip conditions.

The above recipe gives uniformly valid global modes and the quantization conditions of the eigenvalues. we shall follow this recipe in the remainder of this book.

## References


1. M. Abramovitz, I.A. Stegun (eds.), *Handbook of Mathematical Functions* (Dover, New York, 1964)
2. R. Ananth, W.N. Gill, Dendritic growth of an elliptical paraboloid with forced convection in the melt. J. Fluid Mech. **208**, 575–593 (1989)
3. R. Ananth, W.N. Gill, Self-consistent theory of dendritic growth with convection. J. Cryst. Growth **108**, 173–189 (1991)
4. C.M. Bender, S.A. Orszag, *Advanced Mathematical Methods for Scientists and Engineers* (McGraw-Hill, New York, 1978)
5. P.H. Bouissou, A. Chiffaudel, B. Perrin, P. Tabling, Europhys. Lett. **13**, 89 (1990)
6. D. Canright, S.H. Davis, Buoyancy effect of a growing, isolated dendrite. J. Cryst. Growth **114**, 153–185 (1991)
7. Y.Q. Chen, X.X. Tang, J.J. Xu, Three-dimensional interfacial wave theory of dendritic growth: (I). Multiple variables expansion solutions. Chin. Phys. B **18**(2), 1674-1056/2009/18(02)/0671-15 (2009)
8. Y.Q. Chen, X.X. Tang, J.J. Xu, Three-dimensional interfacial wave theory of dendritic growth: (II). Non-axisymmetric global wave modes and selection criterion of pattern formation. Chin. Phys. B **18**(2), 1674-1056/2009/18(02)/0686-13 (2009)
9. R.B. Dingle, *Asymptotic Expansions: Their Derivation and Interpretation* (Academic, London, 1973)
10. P.G. Drazin, W.H. Reid, *Hydrodynamic Stability* (Cambridge University Press, Cambridge, 1971)
11. M.E. Glicksman, M.B. Koss, E.A. Winsa, Dendritic growth velocities in microgravity. Phys. Rev. Lett. **73**(4), 573–576 (1994)
12. M.E. Glicksman, M.B. Koss, L.T. Bushnell, J.C. Lacombe, E.A. Winsa, Dendritic growth of succinonitrile in terrestrial and microgravity conditions as a test of theory. ISIJ Int. **35**(6), 604–610 (1995)

13. J.C. LaCombe, M.B. Koss, J.E. Frei, C. Giummarra, A.O. Lupulecscu, M.E. Glicksman, Evidence for tip velocity oscillations in dendritic solidification. Phys. Rev. E **65**, 031604 (2012)
14. J.S. Langer, H. Müller-Krumbhaar, Theory of dendritic growth — I. Elements of a stability analysis; II. Instabilities in the limit of vanishing surface tension; III. Effects of surface tension. Acta Metall. **26**, 1681–1708 (1978)
15. C.C. Lin, *The Theory of Hydrodynamic Stability* (Cambridge University Press, Cambridge, 1955)
16. C.C. Lin, Y.Y. Lau, On spiral waves in galaxies—a gas dynamic approach. SIMA. J. Appl. Math. **29**(2), 352–370 (1975)
17. C.C. Lin, Y.Y. Lau, Density wave theory of spiral structure of galaxies, in *Studies in Applied Mathematics*, vol. 60 (MIT Press, Cambridge, MA, 1979), pp. 97–163
18. M. Rabaud, Y. Couder, N. Gerard, Phys. Rev. A **37**, 935 (1988)
19. B. Steinin, V. Shatalov, *Borel–Laplace Transform and Asymptotic Theory: Introduction to Resurgent Analysis* (CRC-Press, Boca Raton, 1995)
20. L.M. Williams, M. Muschol, X. Qian, W. Losert, H.Z. Cummins, Phys. Rev. E **48**, 489–498 (1993)
21. J.J. Xu, Interfacial wave theory for dendritic structure of a growing needle crystal (I): local instability mechanism. Phys. Rev. A **40**(3), 1599–1608 (1989)
22. J.J. Xu, Interfacial wave theory for dendritic structure of a growing needle crystal (II): wave-emission mechanism at the turning point. Phys. Rev. A **40**(3), 1609–1614 (1989)
23. J.J. Xu, Global neutral stable state and selection condition of tip growth velocity. J. Cryst. Growth **100**, 481–490 (1990)
24. J.J. Xu, Interfacial wave theory of solidification—dendritic pattern formation and selection of tip velocity. Phys. Rev. A15 **43**(2), 930–947 (1991)
25. J.J. Xu, Two-dimensional dendritic growth with anisotropy of surface tension. Phys. (D) **51**, 579–595 (1991)
26. J.J. Xu, Interfacial wave theory of two-dimensional dendritic growth with anisotropy of surface tension. Can. J. Phys. **69**(7), 789–800 (1991)
27. J.J. Xu, Dendritic growth from a melt in an external flow: uniformly valid asymptotic solution for the steady state. J. Fluid Mech. **263**, 227–243 (1994)
28. J.J. Xu, D.S. Yu, Selection and resonance of dendritic growth with interference of oscillatory external sources. J. Cryst. Growth **226**, 378–392 (2001)

# Chapter 7
# Free Dendritic Growth with Anisotropy

So far, we have studied only the system of dendritic growth including the isotropic surface tension. The anisotropy of surface tension may play an important role in free dendritic growth. To explore the effect of anisotropy, for the sake of simplicity, we study two-dimensional free dendritic growth from undercooled pure melt, which can be considered a simplified mathematical model for demonstration of the dendritic pattern formation observed in a thin Hele–Shaw cell. The typical experimental record is shown in Fig. 7.1. We are concerned with the behavior of the system at the late stage of growth and retain all the assumptions, scales, and notation that were introduced in the previous chapters for the 3D case, such as that the undercooling away from the growing dendrite-tip is uniform and that there is no convection or external forces involved in the system.

It may be presumed that the 2D system may allow global trapped-wave (GTW) instability, similar to the 3D system. Since the isotropic surface tension plays a principal role in GTW instability, with the inclusion of a small anisotropy, GTW instability can be altered only slightly. However, it is very interesting to see that introducing anisotropy will activate a new instability mechanism in the system that we call low-frequency (LF) instability, so that the system will be subject to two distinct coexisting instability mechanisms: global trapped-wave (GTW) instability and -frequency (LF) instability. When the anisotropy parameter is smaller than a critical number, the GTW instability dominates, but when the anisotropy is sufficiently large, the LF instability will dominate the system.

This chapter is devoted to an exploration of these issues.



J.-J. Xu, *Interfacial Wave Theory of Pattern Formation in Solidification*,
Springer Series in Synergetics, DOI 10.1007/978-3-319-52663-8_7

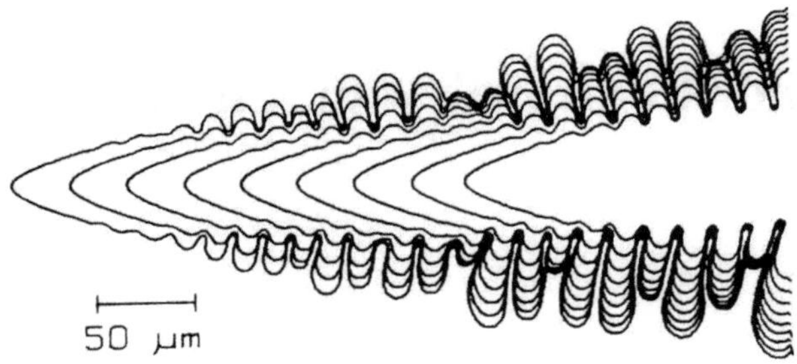


**Fig. 7.1** Experimental curve of two-dimensional dendritic growth from a supersaturated ($NH_4Br$) solution. From Dougherty and Gollub (1988)

## 7.1 Mathematical Formulation for 2D Dendritic Growth with Anisotropy of Surface Tension

For the 2D system, we adopt the parabolic coordinate system as shown Fig. 3.2. The tip of the dendrite is set at $\xi = 0, \eta = 1$.

By taking the anisotropy of surface tension into account, the Gibbs–Thomson condition at the interface will be changed to $\eta = \eta_s(\xi, t, \varepsilon)$,

$$T = -\frac{\Gamma}{\eta_0^2}\mathcal{K}_2\left\{\frac{\mathrm{d}}{\mathrm{d}\xi}, \frac{\mathrm{d}^2}{\mathrm{d}\xi^2}\right\}\eta_s.$$

Here the surface tension parameter $\Gamma$ and capillary length as previously defined,

$$\Gamma = \frac{\ell_c}{\ell_T}, \qquad \ell_c = \frac{\gamma c_p T_{M0}}{(\Delta H)^2}, \tag{7.1}$$

are still applicable, but with the inclusion of anisotropy, the coefficient of surface tension, $\gamma$, is no longer a constant. It is now a function of the local orientation of the interface. In a commonly used form, $\gamma$ is expressed as

$$\begin{cases} \gamma = \gamma_0 \gamma_s(\hat{\theta}) \\ \gamma_s(\hat{\theta}) = 1 + \gamma_m \cos(m\hat{\theta}), \end{cases} \tag{7.2}$$

where $\gamma_0$ is the coefficient of isotropic surface tension, $\gamma_m$ is the anisotropy of surface tension, and $\hat{\theta}$ is the local orientation angle between the tangent line of the interface and the vertical line. From the Herring formula [1], in the 2D case one may write

$$\Gamma = \hat{\Gamma} A_s(\hat{\theta}), \tag{7.3}$$

where

$$\begin{cases} A_s(\hat{\theta}) = \gamma_s(\hat{\theta}) + \gamma_s''(\hat{\theta}) = 1 - \alpha_m \cos(m\hat{\theta}), \\ \alpha_m = (m^2 - 1)\gamma_m, \quad \hat{\Gamma} = \dfrac{\hat{\ell}_c}{\ell_T}, \quad \hat{\ell}_c = \dfrac{\gamma_0 c_p T_{M0}}{(\Delta H)^2}. \end{cases} \tag{7.4}$$

We call the parameter $\alpha_m$ the Herring anisotropy of surface tension to distinguish it from the static anisotropy parameter $\gamma_m$. Sometimes, however, for simplicity we may call $\alpha_m$ simply the anisotropy parameter of surface tension without confusion.

In general, we have

$$\tan\hat{\theta} = \frac{\mathrm{d}x}{\mathrm{d}y} = \frac{\xi - \eta_s\eta_s'}{\eta_s + \xi\eta_s'}, \tag{7.5}$$

where $\eta = \eta_s(\xi, t)$ is the interface shape function. For the interface $\eta_s = 1 + \tilde{\eta}_s$ with $\tilde{\eta}_s, \tilde{\eta}_s' \ll 1$, the above formula can be linearized to obtain

$$\begin{cases} \tan\hat{\theta} \approx \xi - \xi\tilde{\eta}_s - \hat{S}^2\tilde{\eta}_s', \\ \cos^2\hat{\theta} \approx \dfrac{1}{\hat{S}^2}\left(1 + \dfrac{2\xi^2}{\hat{S}^2}\tilde{\eta}_s + 2\xi\tilde{\eta}_s'\right), \end{cases} \tag{7.6}$$

where $\hat{S} = \sqrt{1+\xi^2}$. Typically, for fourfold anisotropy, one can set $m = 4$. Using

$$\cos 4\hat{\theta} = 8\cos^4\hat{\theta} - 8\cos^2\hat{\theta} + 1, \tag{7.7}$$

we obtain

$$A_s(\hat{\theta}) = \bar{A}_s(\xi) + \tilde{A}_{s0}(\xi)\tilde{\eta}_s(\xi) + \tilde{A}_{s1}(\xi)\tilde{\eta}_s'(\xi), \tag{7.8}$$

where

$$\begin{aligned} \bar{A}_s(\xi) &= \frac{f(\xi)}{\hat{S}^4}, \\ \tilde{A}_{s0}(\xi) &= -16\alpha_4\frac{\xi^2(1-\xi^2)}{\hat{S}^6}, \\ \tilde{A}_{s1}(\xi) &= -16\alpha_4\tfrac{\xi(1-\xi^2)}{\hat{S}^4}, \end{aligned} \tag{7.9}$$

and

$$f(\xi) = (1-\alpha_4)(1+\xi^2)^2 + 8\alpha_4\xi^2. \tag{7.10}$$

Note that the function $f(\xi)$ has four imaginary zeros $(\pm \mathrm{i}a_1, \pm \mathrm{i}a_2)$, where

$$\begin{cases} a_1^2 = 1 + 3\alpha_4 - 2\sqrt{2}\alpha_4^{\frac{1}{2}}(1+\alpha_4)^{\frac{1}{2}}, \\ a_2^2 = 1 + 3\alpha_4 + 2\sqrt{2}\alpha_4^{\frac{1}{2}}(1+\alpha_4)^{\frac{1}{2}}, \end{cases} \tag{7.11}$$

one can rewrite (7.10) as

$$f(\xi) = (1 - \alpha_4)(\xi^2 + a_1^2)(\xi^2 + a_2^2)\,. \tag{7.12}$$

Especially, for $\alpha_4 \ll 1$,

$$a_1 \approx 1 - \sqrt{2\alpha_4}, \quad a_2 \approx 1 + \sqrt{2\alpha_4}. \tag{7.13}$$

For the case with anisotropic surface tension, the interfacial stability parameter of isotropic surface tension, $\varepsilon$, is expressed as

$$\varepsilon^2 = \frac{\hat{\Gamma}}{\eta_0^2}\,. \tag{7.14}$$

In this chapter, we consider the general case ($0 \leq \hat{\alpha}_{\mathrm{T}} \leq 1$). The equations for the temperature field of the liquid and solid phases are

$$\left(\frac{\partial^2 T}{\partial \xi^2} + \frac{\partial^2 T}{\partial \eta^2}\right) = \eta_0^2\left(\xi\frac{\partial T}{\partial \xi} - \eta\frac{\partial T}{\partial \eta}\right) + \eta_0^4(\xi^2 + \eta^2)\frac{\partial T}{\partial t} \tag{7.15}$$

and

$$\frac{\partial^2 T_{\mathrm{S}}}{\partial \xi^2} + \frac{\partial^2 T_{\mathrm{S}}}{\partial \eta^2} = \hat{\alpha}_{\mathrm{T}}\left[\eta_0^4(\xi^2 + \eta^2)\frac{\partial T_{\mathrm{S}}}{\partial t} + \eta_0^2\left(\xi\frac{\partial T_{\mathrm{S}}}{\partial \xi} - \eta\frac{\partial T_{\mathrm{S}}}{\partial \eta}\right)\right], \tag{7.16}$$

respectively. The boundary conditions are as follows:

1. The upstream far-field condition

$$T \to T_\infty\,, \qquad \text{as } \eta \to \infty\,. \tag{7.17}$$

2. The regularity condition

$$T_{\mathrm{S}} = \mathcal{O}(1)\,, \qquad \text{as } \eta \to 0\,. \tag{7.18}$$

3. The interface conditions at $\eta = \eta_{\mathrm{s}}(\xi, \varepsilon)$:

   (i) the thermodynamic equilibrium condition

$$T = T_{\mathrm{S}}\,, \tag{7.19}$$

   (ii) the Gibbs–Thomson condition

$$T_{\mathrm{S}} = -\eta_0^2\varepsilon^2\frac{f(\xi)}{\hat{S}^4}\mathcal{K}_2\left\{\frac{d}{d\xi}, \frac{d^2}{d\xi^2}\right\}\eta_{\mathrm{s}}\,. \tag{7.20}$$

(iii) the heat balance condition

$$\left(\frac{\partial}{\partial\eta}-\eta_s'\frac{\partial}{\partial\xi}\right)(T-\hat{\alpha}_T T_S)+\eta_0^4(\xi^2+\eta^2)\frac{\partial\eta_s}{\partial t}+\eta_0^2(\xi\eta_s)'=0\,. \tag{7.21}$$

4. The tip smoothness conditions

$$\eta_s'(0)=0\,,\quad \eta_s(0)=1\,. \tag{7.22}$$

5. The root conditions are not specified.

## 7.2 RPE for Basic Steady-State Solutions

As $\varepsilon \to 0$, we form the regular perturbation expansion (RPE) for the generalized steady needle solution as

$$\begin{aligned} T=T(\xi,\eta)&=T_*(\eta)+\varepsilon^2\eta_0^2T_1(\xi,\eta)+\cdots\,,\\ T_S=T_S(\xi,\eta)&=T_{S*}+\varepsilon^2\eta_0^2T_{S1}(\xi,\eta)+\cdots\,,\\ \eta_s(\xi)&=1+\varepsilon^2\eta_1(\xi)+\cdots\,. \end{aligned} \tag{7.23}$$

By substituting the above RPE into the system (7.15)–(7.22), one derives the approximate solutions of each order successively.

### 7.2.1 *The Zeroth-Order Approximation* ($\varepsilon = 0$)

The zeroth-order approximate solution is the 2D Ivantsov needle solution with zero surface tension:

For $\varepsilon = 0$, the system (7.15)–(7.22) allows the steady similarity solution

$$\begin{aligned} &T=T_*(\eta)=T_\infty+\eta_0^2 e^{\frac{\eta_0^2}{2}}\int_\eta^\infty e^{-\frac{\eta_0^2\eta_1^2}{2}}\,d\eta_1,\\ &T_S=T_{S*}=T_*(1)=0,\\ &\eta_*=1,\\ &T_\infty=-\eta_0^2 e^{\frac{\eta_0^2}{2}}\int_1^\infty e^{-\frac{\eta_0^2\eta_1^2}{2}}\,d\eta_1\\ &\qquad(0\le\xi<\infty)\,. \end{aligned} \tag{7.24}$$

### 7.2.2 The First-Order Approximation, $\mathcal{O}(\varepsilon^2)$

The first-order approximate solutions will depend on the parameters $\hat{\alpha}_{\mathrm{T}}$ and $\alpha_4$. Here, for the sake of mathematical simplicity, we give the solutions only for the case $\hat{\alpha}_{\mathrm{T}} = 1$ and $\alpha_4 = 0$. These solutions can be easily extended to the general case, $0 \leq \hat{\alpha}_{\mathrm{T}} \leq 1$ and $\alpha_4 \neq 0$, without significant difficulty. Note that for our main purposes in this book, the first-order approximation of the RPE is not important, since it does not affect the stability properties of the system up to $\mathcal{O}(\varepsilon)$. For the analytical solution of the first-order approximation of the RPE for the case of the 3D steady needle solution with anisotropy of surface tension, the reader is referred to [2].

The system of first-order approximation is

$$L\{T_1\} = \left[\frac{\partial^2}{\partial \xi^2} + \frac{\partial^2}{\partial \eta^2} - \eta_0^2 \xi \frac{\partial}{\partial \xi} + \eta_0^2 \eta \frac{\partial}{\partial \eta}\right] T_1 = 0 \tag{7.25}$$

with the following boundary conditions:

1. As $\eta \to \infty$,

$$T_1 \to 0 \text{ (exponentially).} \tag{7.26}$$

2. As $\xi \to \infty$,

$$T_1 \to 0 \text{ (algebraically).} \tag{7.27}$$

3. As $\eta \to 0$,

$$T_{\mathrm{S}1} \text{ regular.} \tag{7.28}$$

4. At $\eta = 1$,

$$T_1 = T_{\mathrm{S}1} + \eta_1 , \tag{7.29}$$

$$T_{\mathrm{S}1} = -\mathcal{K}_0(\xi) = -\frac{1}{\left(1 + \xi^2\right)^{\frac{3}{2}}} , \tag{7.30}$$

$$\frac{\partial}{\partial \eta}\left(T_1 - T_{\mathrm{S}1}\right) + \left(1 + \eta_0^2\right)\eta_1 + \xi \frac{d\eta_1}{d\xi} = 0 . \tag{7.31}$$

5. The tip smoothness condition at $\xi = 0$,

$$\eta_1'(0) = 0 , \tag{7.32}$$

$$\eta_1(0) = 0 . \tag{7.33}$$

Let

$$T_1(\xi,\eta) = X(\xi)Y(\eta), \tag{7.34}$$

where

$$X'' - \eta_0^2\xi X' + \eta_0^2{\lambda_1}^2 X = 0, \tag{7.35}$$

$$Y'' + \eta_0^2\eta Y' - \eta_0^2{\lambda_1}^2 Y = 0\,. \tag{7.36}$$

By letting

$$\sigma = \frac{\eta_0^2\xi^2}{2},\quad X = X(\sigma), \tag{7.37}$$

equation (7.35) is transformed into the Kummer equation

$$\sigma X''(\sigma) + \left(\frac{1}{2} - \sigma\right) X'(\sigma) + \frac{\lambda_1^2}{2} X = 0, \tag{7.38}$$

whose fundamental solutions are

$$X(\sigma) = \begin{cases} M\left(-\frac{\lambda_1^2}{2}, \frac{1}{2}, \sigma\right) & \text{(regular at } \xi = \sigma = 0), \\ U\left(-\frac{\lambda_1^2}{2}, \frac{1}{2}, \sigma\right) & \text{(with a logarithmic singularity} \\ & \quad \text{at } \xi = \sigma = 0). \end{cases} \tag{7.39}$$

Here $M$ and $U$ are the confluent hypergeometric functions. For the same reasons as given for the three-dimensional case, we must choose

$$X(\sigma) = M\left(-\frac{\lambda_1^2}{2}, 1, \sigma\right) \tag{7.40}$$

and

$$\frac{\lambda_1^2}{2} = n = 0, 1, 2, \ldots\,. \tag{7.41}$$

Thus,

$$X(\sigma) = M\Big(-n, \frac{1}{2}, \sigma\Big) = H_{2n}(\eta_0\xi)\,, \tag{7.42}$$

where $H_n$ is the Hermite polynomial. To solve (7.36) in the solid phase region, we let

$$\hat{\tau} = -\frac{\eta_0^2\eta^2}{2}. \tag{7.43}$$

Thus (7.36) is transformed into

$$\hat{\tau}Y''(\hat{\tau}) + (1-\hat{\tau})Y'(\hat{\tau}) + nY = 0. \tag{7.44}$$

The solution regular at $\hat{\tau} = \eta = 0$ is

$$Y(\hat{\tau}) = M_n\left(-n, \frac{1}{2}, -\frac{\eta_0^2\eta^2}{2}\right) = H_{2n}(\mathrm{i}\eta_0\eta). \tag{7.45}$$

Therefore, the temperature in the solid phase is

$$T_{S1}(\xi,\eta) = \sum_{n=0}^{\infty} \alpha_n H_{2n}(\eta_0\xi)H_{2n}(\mathrm{i}\eta_0\eta). \tag{7.46}$$

In the liquid phase region, we set

$$\tau = \frac{\eta_0^2\eta^2}{2}, \qquad Y(\eta) = \frac{1}{\tau^{\frac{1}{4}}}\mathrm{e}^{-\frac{\tau}{2}}Z(\tau). \tag{7.47}$$

Equation (7.36) can now be transformed into the Whittaker equation

$$Z''(\tau) + \left(-\frac{1}{4} + \frac{\lambda}{\tau} + \frac{\frac{1}{4}-\mu^2}{\tau^2}\right)Z(\tau) = 0 \tag{7.48}$$

with

$$\lambda = -\left(n + \frac{1}{4}\right), \qquad \mu = \frac{1}{4}. \tag{7.49}$$

Since it is known that $Y(\eta)$ vanishes exponentially as $\eta \to \infty$, it follows that

$$Y(\eta) = \mathrm{e}^{-\frac{\eta_0^2\eta^2}{2}}\eta\, U\left(n+1, \frac{3}{2}, \frac{\eta_0^2\eta^2}{2}\right). \tag{7.50}$$

Therefore, the temperature in the liquid is

$$T_1(\xi,\eta) = \sum_{n=0}^{\infty} \beta_n H_{2n}(\eta_0\xi)\frac{\mathrm{e}^{-\frac{\eta_0^2\eta^2}{2}}\eta\, U\left(n+1, \frac{3}{2}, \frac{\eta_0^2\eta^2}{2}\right)}{\mathrm{e}^{-\frac{\eta_0^2}{2}}U\left(n+1, \frac{3}{2}, \frac{\eta_0^2}{2}\right)}. \tag{7.51}$$

We use the following Hermite series for the function $\eta_1(\xi)$:

$$\eta_1(\xi) = \sum_{n=0}^{\infty} \gamma_n H_{2n}(\eta_0 \xi). \tag{7.52}$$

We now need to determine the coefficients $\{\alpha_n, \beta_n, \gamma_n\ (n = 0, 1, 2 \ldots)\}$ in terms of the boundary conditions (7.29)–(7.31).

It follows from the boundary condition (7.29) that

$$\beta_n = \alpha_n + \gamma_n. \tag{7.53}$$

Furthermore, the boundary condition (7.30) gives

$$\sum_{n=0}^{\infty} \alpha_n H_{2n}(\eta_0 \xi) = -\mathcal{K}_0(\xi) = -\frac{1}{(1+\xi^2)^{\frac{3}{2}}}. \tag{7.54}$$

Due to the orthogonality of the set of functions $H_n(x)$ $(n = 0, 1, 3, \ldots)$, the coefficients $\alpha_n$ can be determined by the integrals:

$$\begin{aligned} \alpha_n &= -\frac{\eta_0^3}{2^n \pi^{\frac{1}{2}} n!} \hat{\alpha}_n, \\ \hat{\alpha}_n &= \int_{\infty}^{\infty} \frac{\mathrm{e}^{-x^2} H_{2n}(x)}{\left(x^3 + \eta_0^2\right)^{\frac{3}{2}}} dx. \end{aligned} \tag{7.55}$$

Finally, from (7.31), we get

$$(a_n - b_n) + \left(1 + \eta_0^2 + 2n\right)\gamma_n + 8(n+1)\left(n + \frac{1}{2}\right)\gamma_{n+1} = 0 \tag{7.56}$$

$$(n = 0, 1, 2, 3, \ldots),$$

where

$$a_n = 2\beta_n A_n, \qquad b_n = 2\alpha_n B_n, \tag{7.57}$$

and

$$A_n = n - \frac{U\left(n, \frac{3}{2}, \frac{\eta_0^2}{2}\right)}{U\left(n+1, \frac{3}{2}, \frac{\eta_0^2}{2}\right)}, \tag{7.58}$$

$$B_n = 1 - 2(2n-1)\frac{H_{2n-2}(i\eta_0)}{H_{2n}(i\eta_0)}. \tag{7.59}$$

From the above, we find that

$$\gamma_{n+1} = -g_n\gamma_n + \frac{\eta_0^3 f_n}{2^n\pi^{\frac{1}{2}}n!}\hat{\alpha}_n \quad (n = 0, 1, 2 \ldots), \tag{7.60}$$

where

$$g_n = \frac{1 + 2n + \eta_0^2 + 2A_n}{8(n+1)(n+\frac{1}{2})}, \quad f_n = \frac{A_n - B_n}{4(n+1)(n+\frac{1}{2})}. \tag{7.61}$$

We may set

$$\gamma_n = \frac{\eta_0^3}{2^n\pi^{\frac{1}{2}}n!}\hat{\gamma}_n . \tag{7.62}$$

Thus, we rewrite (7.60) as

$$\hat{\gamma}_{n+1} = -2(n+1)g_n\hat{\gamma}_n + 2(n+1)f_n\hat{\alpha}_n \qquad (n = 0, 1, 2 \ldots). \tag{7.63}$$

For a given value of $\hat{\gamma}_0$, Eq. (7.63) generates the sequence

$$\{\hat{\gamma}_0, \hat{\gamma}_1, \hat{\gamma}_2, \ldots \hat{\gamma}_n \ldots\}.$$

Thus, the function $\eta_1(\hat{\gamma}_0, \xi)$ can be evaluated as

$$\eta_1(\hat{\gamma}_0, \xi) = \sum_{n=0}^{\infty} \gamma_n H_{2n}(\eta_0\xi) . \tag{7.64}$$

The value of $\hat{\gamma}_0$ is then uniquely determined by the tip condition (7.33), namely

$$\eta_1(\hat{\gamma}_0, 0) = \sum_{n=0}^{\infty} \frac{\eta_0^3}{2^n\pi^{\frac{1}{2}}n!}\hat{\gamma}_n H_{2n}(0) = 0 . \tag{7.65}$$

Note that

$$H_{2n}(0) = (-1)^n 2^n (2n-1)!! . \tag{7.66}$$

Thus,

$$\eta_1(\hat{\gamma}_0, 0) = \sum_{n=0}^{\infty}(-1)^n \frac{\eta_0^3}{\pi^{\frac{1}{2}}} \frac{(2n-1)!!}{n!}\hat{\gamma}_n = 0. \tag{7.67}$$

The dimensionless tip radius Pe is calculated by the formula

$$\frac{\mathcal{K}_2(0)}{\eta_0^2} = \frac{1}{\mathrm{Pe}}. \tag{7.68}$$

Finally, we obtain

$$\mathrm{Pe} = \frac{\ell_t}{\ell_T} = \frac{\eta_0^2}{1 - \varepsilon^2 \eta_1''(0)}. \tag{7.69}$$

### 7.2.3 Asymptotic Behavior of the Regular Perturbation Expansion Solution as $\xi \to \infty$

In the far field, the function $\mathcal{K}_0(\xi)$ is asymptotic to $\frac{1}{\xi^3}$ as $\xi \to \infty$. Thus, we make the expansions

$$\begin{aligned} T_1(\xi,\eta) &= \frac{A_1(\eta)}{\xi^3} + \frac{A_2(\eta)}{\xi^4} + \cdots, \\ T_{S1}(\xi,\eta) &= \frac{A_{S1}(\eta)}{\xi^3} + \frac{A_{S2}(\eta)}{\xi^4} + \cdots, \\ \eta_1(\xi) &= \frac{C_1}{\xi^3} + \frac{C_2}{\xi^4} + \cdots. \end{aligned} \tag{7.70}$$

The first-order approximate system is

$$\left\{ \frac{d^2}{d\eta^2} + \eta_0^2 \eta \frac{d}{d\eta} + 3\eta_0^2 \right\} \begin{pmatrix} A_1 \\ A_{S1} \end{pmatrix} = 0, \tag{7.71}$$

and at $\eta = 1$,

$$A_1(1) = A_{S1}(1) + C_1; \tag{7.72}$$

$$A_{S1}(1) = -1; \tag{7.73}$$

$$\frac{d}{d\eta}\big(A_1 - A_{S1}\big) + \big(\eta_0^2 - 2\big)C_1 = 0. \tag{7.74}$$

Similarly to the way we treated (7.36) in the liquid region, we let

$$\tau = \frac{\eta_0^2 \eta^2}{2}; \quad \begin{pmatrix} A_{S1}(\eta) \\ A_1(\eta) \end{pmatrix} = \mathrm{e}^{-\frac{\tau}{2}} \tau^{-\frac{1}{4}} Z(\tau). \tag{7.75}$$

Then (7.71) is transformed into the Whittaker equation with

$$\lambda = \frac{5}{4}, \qquad \mu = \frac{1}{4}. \tag{7.76}$$

Applying the boundary conditions

$$A_1(\eta) \to 0, \quad \text{as} \quad \eta \to \infty \quad \text{(exponentially)}, \tag{7.77}$$

$$A_{S1}(\eta) \quad \text{is regular at} \quad \eta = 0, \tag{7.78}$$

we obtain

$$A_1(\eta) = \hat{A}_1 \eta \mathrm{e}^{-\frac{\eta_0^2\eta^2}{2}} U\left(-\frac{1}{2}, \frac{3}{2}, \frac{\eta_0^2\eta^2}{2}\right), \tag{7.79}$$

$$A_{S1}(\eta) = \hat{A}_{S1} \eta \mathrm{e}^{-\frac{\eta_0^2\eta^2}{2}} M\left(-\frac{1}{2}, \frac{3}{2}, \frac{\eta_0^2\eta^2}{2}\right). \tag{7.80}$$

From (7.72)–(7.74), one can uniquely determine the three unknown constants $\{\hat{A}_1, \hat{A}_{S1}, C_1\}$ to be

$$\begin{cases} \hat{A}_{S1} = -\frac{1}{M}\mathrm{e}^{\frac{\eta_0^2}{2}}, \\ \hat{A}_1 = \frac{M - M'}{U - U'}\hat{A}_{S1}, \\ C_1 = \frac{U\hat{A}_1 - M\hat{A}_{S1}}{\mathrm{e}^{-\frac{\eta_0^2}{2}}}, \end{cases} \tag{7.81}$$

where

$$\begin{aligned} M &= M\left(-\tfrac{1}{2}, \tfrac{3}{2}, \tfrac{\eta_0^2}{2}\right), \\ M' &= \tfrac{d}{d\eta}M\left(-\tfrac{1}{2}, \tfrac{3}{2}, \tfrac{\eta_0^2\eta^2}{2}\right)\Big|_{\eta=1} = -\tfrac{1}{3}M\left(\tfrac{1}{2}, \tfrac{5}{2}, \tfrac{\eta_0^2}{2}\right), \\ U &= U\left(-\tfrac{1}{2}, \tfrac{3}{2}, \tfrac{\eta_0^2}{2}\right), \\ U' &= \tfrac{d}{d\eta}U\left(-\tfrac{1}{2}, \tfrac{3}{2}, \tfrac{\eta_0^2\eta^2}{2}\right)\Big|_{\eta=1} = -\tfrac{1}{2}U\left(\tfrac{1}{2}, \tfrac{5}{2}, \tfrac{\eta_0^2}{2}\right). \end{aligned} \tag{7.82}$$

It is seen that the first-order solution is solved without applying any root conditions. The solution, however, automatically has the following behavior at the far field:

$$\eta_1(\xi) \to 0, \quad \text{as} \quad \xi \to \infty.$$

We do not intend to proceed any further to find higher-order approximate solutions. It should be noted that all approximate solutions in the RPE can be solved without imposing any root conditions as $\xi \to \infty$. Moreover, without giving the proof, it can be deduced that not only the first approximation, but approximations of all orders of the RPE for the interface shape vanish as $\xi \to \infty$. Or one could say that the partial sums with any large number of terms of the RPE for the interface shape function all approach the Ivantsov parabola as $\xi \to \infty$. Therefore, the solutions of the RPE for the basic steady state do not resolve the selection problem.

## 7.3 Global Interfacial Wave Instabilities of Two-Dimensional Dendritic Growth

We now consider the evolution of infinitesimal perturbations around the basic steady state.

### 7.3.1 *Linear Perturbed System Around the Basic State*

As before, we separate the general unsteady solutions into two parts,

$$\begin{aligned} T &= T_{\mathrm{B}} + \tilde{T}(\xi, \eta, t, \varepsilon), \\ T_{\mathrm{S}} &= T_{\mathrm{SB}} + \tilde{T}_{\mathrm{S}}(\xi, \eta, t, \varepsilon) \\ \eta_{\mathrm{s}} &= \eta_{\mathrm{B}} + \tilde{h}(\xi, t, \varepsilon)/\eta_0^2, \end{aligned} \tag{7.83}$$

and linearize around the basic state solution. Here we consider the general case $(0 \leq \hat{\alpha}_{\mathrm{T}} \leq 1)$. The linearized perturbed equations are derived as

$$\frac{\partial^2 \tilde{T}}{\partial \xi^2} + \frac{\partial^2 \tilde{T}}{\partial \eta^2} = \eta_0^4(\xi^2 + \eta^2)\frac{\partial \tilde{T}}{\partial t} + \eta_0^2\left(\xi\frac{\partial \tilde{T}}{\partial \xi} - \eta\frac{\partial \tilde{T}}{\partial \eta}\right) \tag{7.84}$$

and

$$\frac{\partial^2 \tilde{T}_{\mathrm{S}}}{\partial \xi^2} + \frac{\partial^2 \tilde{T}_{\mathrm{S}}}{\partial \eta^2} = \hat{\alpha}_{\mathrm{T}}\left[\eta_0^4(\xi^2 + \eta^2)\frac{\partial \tilde{T}_{\mathrm{S}}}{\partial t} + \eta_0^2\left(\xi\frac{\partial \tilde{T}_{\mathrm{S}}}{\partial \xi} - \eta\frac{\partial \tilde{T}_{\mathrm{S}}}{\partial \eta}\right)\right]. \tag{7.85}$$

The interface conditions are as follows: at $\eta = 1$,

1.

$$(\tilde{T} - \tilde{T}_{\mathrm{S}}) + \varepsilon^2\eta_1\frac{\partial(\tilde{T} - \tilde{T}_{\mathrm{S}})}{\partial \eta} = \tilde{h} + (\text{h.o.t}), \tag{7.86}$$

2.

$$\tilde{T}_S + \varepsilon^2\eta_1 \frac{\partial \tilde{T}_S}{\partial \eta} = \frac{\varepsilon^2 \bar{A}_s(\xi)}{\hat{S}(\xi)} \left[ \frac{\partial^2 \tilde{h}}{\partial \xi^2} + \frac{\xi}{\hat{S}^2(\xi)} \frac{\partial \tilde{h}}{\partial \xi} - \frac{\tilde{h}}{\hat{S}^2(\xi)} \right]$$

$$-\varepsilon^2 \mathcal{K}_0(\xi) \Big( \tilde{A}_{s0} \tilde{h} + \tilde{A}_{s1} \frac{\partial \tilde{h}}{\partial \xi} \Big) + \text{(h.o.t)}\,, \tag{7.87}$$

where

$$\mathcal{K}_0(\xi) = \frac{1}{\hat{S}^3}, \tag{7.88}$$

3.

$$\frac{\partial}{\partial \eta}\Big(\tilde{T} - \hat{\alpha}_T \tilde{T}_S\Big) + \varepsilon^2 \eta_1 \frac{\partial^2}{\partial \eta^2}\Big(\tilde{T} - \hat{\alpha}_T \tilde{T}_S\Big) + \eta_0^2 \hat{S}^2(\xi) \frac{\partial \tilde{h}}{\partial t} + \xi \frac{\partial \tilde{h}}{\partial \xi}$$

$$+\varepsilon(1 + \eta_0^2)\tilde{h} + \text{(h.o.t)} = 0. \tag{7.89}$$

### *7.3.2 Multivariable Expansion Solution in the Outer Region*

The stretched fast variables are defined as for the 3D dendritic growth system:

$$\begin{aligned}
\xi_+ &= \frac{\phi(\xi, \eta, \varepsilon)}{\varepsilon} = \frac{1}{\varepsilon} \int_{\xi_0}^{\xi} k(\xi_1, \eta, \varepsilon) \mathrm{d}\xi_1, \\
\eta_+ &= \frac{\psi(\xi, \eta, \varepsilon)}{\varepsilon} = \frac{1}{\varepsilon} \int_{1}^{\eta} k(\xi, \eta_1, \varepsilon) \mathrm{d}\eta_1, \\
t_+ &= \frac{t}{\eta_0^2 \varepsilon},
\end{aligned} \tag{7.90}$$

where the pair of phase functions $\psi$ and $\psi$ satisfy the Cauchy–Riemann conditions:

$$\phi_\xi = \psi_\eta = k(\xi, \eta, \varepsilon);\ \phi_\eta = -\psi_\xi = g(\xi, \eta, \varepsilon).$$

As was done before, we write that at the interface $\eta = 1$, $\bar{k}(\xi, \varepsilon) = k(\xi, 1, \varepsilon)$, $\bar{\phi}(\xi, \varepsilon) = \phi(\xi, 1, \varepsilon)$, $\bar{\psi}(\xi, \varepsilon) = \psi(\xi, 1, \varepsilon)$, and

$$\bar{\xi}_+(\xi, \varepsilon) = \xi_+(\xi, 1, \varepsilon) = \frac{\bar{\phi}(\xi, \varepsilon)}{\varepsilon} = \frac{1}{\varepsilon} \int_{\xi_0}^{\xi} \bar{k}(\xi_1, \varepsilon) \mathrm{d}\xi_1.$$

The formulas (6.23)–(6.26) derived in the last chapter for the axisymmetric case are still applicable here. Namely, we may write

$$
\begin{aligned}
\phi(\xi,\eta) &= \frac{1}{2}\left\{\int_0^{\xi+\mathrm{i}(\eta-1)}\bar{k}(t)\mathrm{d}t+\int_0^{\xi-\mathrm{i}(\eta-1)}\bar{k}(t)\mathrm{d}t\right\},\\
\psi(\xi,\eta) &= -\frac{\mathrm{i}}{2}\int_{\xi-\mathrm{i}(\eta-1)}^{\xi+\mathrm{i}(\eta-1)}\bar{k}(t)\mathrm{d}t,\\
k(\xi,\eta) &= \frac{1}{2}\left\{\bar{k}[\xi+\mathrm{i}(\eta-1)]+\bar{k}[\xi-\mathrm{i}(\eta-1)]\right\},\\
g(\xi,\eta) &= \frac{\mathrm{i}}{2}\left\{\bar{k}[\xi+\mathrm{i}(\eta-1)]-\bar{k}[\xi-\mathrm{i}(\eta-1)]\right\}.
\end{aligned}
\tag{7.91}
$$

The converted system with multiple variables is as follows:

$$
\begin{aligned}
&(k^2+g^2)\left(\frac{\partial^2}{\partial\xi_+^2}+\frac{\partial^2}{\partial\eta_+^2}\right)\tilde{T}\\
&=-\varepsilon\left(2k\frac{\partial^2\tilde{T}}{\partial\xi_+\partial\xi}+2g\frac{\partial^2\tilde{T}}{\partial\eta_+\partial\xi}\right)-\varepsilon\left(2k\frac{\partial^2\tilde{T}}{\partial\eta_+\partial\eta}-2g\frac{\partial^2\tilde{T}}{\partial\xi_+\partial\eta}\right)\\
&+\varepsilon\eta_0^2\xi\left(k\frac{\partial\tilde{T}}{\partial\xi_+}+g\frac{\partial\tilde{T}}{\partial\eta_+}\right)+\varepsilon\eta_0^2\eta\left(g\frac{\partial\tilde{T}}{\partial\xi_+}-k\frac{\partial\tilde{T}}{\partial\eta_+}\right)\\
&+\varepsilon\eta_0^2(\xi^2+\eta^2)\frac{\partial\tilde{T}}{\partial t_+}+\mathcal{O}(\varepsilon^2).
\end{aligned}
\tag{7.92}
$$

$$
\begin{aligned}
&(k^2+g^2)\left(\frac{\partial^2}{\partial\xi_+^2}+\frac{\partial^2}{\partial\eta_+^2}\right)\tilde{T}_\mathrm{S}\\
&=-\varepsilon\left(2k\frac{\partial^2\tilde{T}}{\partial\xi_+\partial\xi}+2g\frac{\partial^2\tilde{T}_\mathrm{S}}{\partial\eta_+\partial\xi}\right)-\varepsilon\left(2k\frac{\partial^2\tilde{T}_\mathrm{S}}{\partial\eta_+\partial\eta}-2g\frac{\partial^2\tilde{T}_\mathrm{S}}{\partial\xi_+\partial\eta}\right)\\
&+\varepsilon\hat{\alpha}_\mathrm{T}\eta_0^2\xi\left(k\frac{\partial\tilde{T}_\mathrm{S}}{\partial\xi_+}+g\frac{\partial\tilde{T}_\mathrm{S}}{\partial\eta_+}\right)+\varepsilon\hat{\alpha}_\mathrm{T}\eta_0^2\eta\left(g\frac{\partial\tilde{T}_\mathrm{S}}{\partial\xi_+}-k\frac{\partial\tilde{T}_\mathrm{S}}{\partial\eta_+}\right)\\
&+\varepsilon\hat{\alpha}_\mathrm{T}\eta_0^2(\xi^2+\eta^2)\frac{\partial\tilde{T}_\mathrm{S}}{\partial t_+}+\mathcal{O}(\varepsilon^2).
\end{aligned}
\tag{7.93}
$$

The multivariable forms of the upstream condition, tip smoothness condition, and root condition are trivial and formally the same as those given in the previous section. Here we give only the interface conditions as follows: the interface conditions at $\eta_+ = 0$ or $\eta = 1$,

1. the thermodynamic equilibrium condition

$$\tilde{T} = \tilde{T}_S + \tilde{h} - \varepsilon\eta_1\bar{k}\frac{\partial}{\partial\eta_+}(\tilde{T} - \tilde{T}_S) + \mathcal{O}(\varepsilon^2)\,, \tag{7.94}$$

2. the Gibbs–Thomson condition

$$\begin{aligned}\tilde{T}_S + \varepsilon\eta_1\bar{k}\frac{\partial\tilde{T}_S}{\partial\eta_+} &= \frac{\bar{A}_s(\xi)}{\hat{S}(\xi)}\left[\left(k^2\frac{\partial^2\tilde{h}}{\partial\bar{\xi}_+^2} + 2\varepsilon k\frac{\partial^2\tilde{h}}{\partial\bar{\xi}_+\partial\xi} + \varepsilon k'\frac{\partial\tilde{h}}{\partial\bar{\xi}_+} + \varepsilon^2\frac{\partial^2\tilde{h}}{\partial\xi^2}\right)\right.\\ &\quad\left. + \frac{\varepsilon\xi}{\hat{S}^2(\xi)}\left(k\frac{\partial\tilde{h}}{\partial\bar{\xi}_+} + \varepsilon\frac{\partial\tilde{h}}{\partial\xi}\right) - \frac{\varepsilon^2}{\hat{S}^2(\xi)}\tilde{h}\right]\\ &\quad - \varepsilon\tilde{A}_{s1}\mathcal{K}_0(\xi)\left(k\frac{\partial\tilde{h}}{\partial\bar{\xi}_+} + \varepsilon\frac{\partial\tilde{h}}{\partial\xi}\right) + \mathcal{O}(\varepsilon^3)\,,\end{aligned} \tag{7.95}$$

3. the heat balance condition

$$\begin{aligned}&\bar{k}\left(\frac{\partial\tilde{T}}{\partial\eta_+} - \hat{\alpha}_T\frac{\partial\tilde{T}_S}{\partial\eta_+}\right) + \varepsilon\eta_1\bar{k}^2\frac{\partial^2}{\partial\eta_+^2}(\tilde{T} - \hat{\alpha}_T\tilde{T}_S) + \varepsilon\left(\frac{\partial\tilde{T}}{\partial\eta} - \hat{\alpha}_T\frac{\partial\tilde{T}_S}{\partial\eta}\right)\\ &+\hat{S}^2(\xi)\frac{\partial\tilde{h}}{\partial t_+} + \xi\left(k\frac{\partial\tilde{h}}{\partial\bar{\xi}_+} + \varepsilon\frac{\partial\tilde{h}}{\partial\xi}\right)\\ &+\varepsilon\left(1 + \eta_0^2\right)\tilde{h} + \mathcal{O}(\varepsilon^2) = 0\,.\end{aligned} \tag{7.96}$$

The outer solution can be expanded in the same MVE form given in the previous section:

$$\begin{aligned}\tilde{T} &= \tilde{\mu}_0(\varepsilon)\left\{\tilde{T}_0(\xi,\eta,\xi_+,\eta_+) + \varepsilon\tilde{T}_1(\xi,\eta,\xi_+,\eta_+) + \cdots\right\}e^{\sigma t_+},\\ \tilde{T}_S &= \tilde{\mu}_0(\varepsilon)\left\{\tilde{T}_{S0}(\xi,\eta,\xi_+,\eta_+) + \varepsilon\tilde{T}_{S1}(\xi,\eta,\xi_+,\eta_+) + \cdots\right\}e^{\sigma t_+},\\ \tilde{h} &= \tilde{\mu}_0(\varepsilon)\left\{\tilde{h}_0(\xi,\bar{\xi}_+) + \varepsilon\tilde{h}_1(\xi,\bar{\xi}_+) + \cdots\right\}e^{\sigma t_+},\end{aligned} \tag{7.97}$$

with

$$\begin{aligned}\phi(\xi,\eta,\varepsilon) &= \phi_0(\xi,\eta) + \varepsilon\phi_1(\xi,\eta) + \cdots\,,\\ \psi(\xi,\eta,\varepsilon) &= \psi_0(\xi,\eta) + \varepsilon\psi_1(\xi,\eta) + \cdots\,,\\ k(\xi,\eta,\varepsilon) &= k_0(\xi,\eta) + \varepsilon k_1(\xi,\eta) + \cdots\,,\\ \bar{k}(\xi,\varepsilon) &= \bar{k}_0(\xi) + \varepsilon\bar{k}_1(\xi) + \cdots\,,\end{aligned} \tag{7.98}$$

$$g(\xi,\eta,\varepsilon) = g_0(\xi,\eta) + \varepsilon g_1(\xi,\eta) + \cdots,$$
$$\bar{g}(\xi,\varepsilon) = 0,$$
$$\sigma = \sigma_0 + \varepsilon\sigma_1 + \varepsilon^2\sigma_2 + \cdots.$$

One can then successively derive the higher-order approximations of the outer solutions.

#### 7.3.2.1 Zeroth-Order Approximation

In the zeroth-order approximation, the system is

$$\begin{aligned} \left(\frac{\partial^2}{\partial\xi_+^2} + \frac{\partial^2}{\partial\eta_+^2}\right)\tilde{T}_0 = 0, \\ \left(\frac{\partial^2}{\partial\xi_+^2} + \frac{\partial^2}{\partial\eta_+^2}\right)\tilde{T}_{S0} = 0, \end{aligned} \tag{7.99}$$

with the following boundary conditions:

1. As $\eta_+ \to \infty$,

$$\tilde{T}_0 \to 0. \tag{7.100}$$

2. As $\eta_+ \to -\infty$,

$$\tilde{T}_{S0} \to 0. \tag{7.101}$$

3. The interface conditions at $\eta_+ = 0$ and $\eta = 1$,

$$\tilde{T}_0 = \tilde{T}_{S0} + \tilde{h}_0, \tag{7.102}$$

$$\tilde{T}_{S0} = \frac{\bar{A}_s(\xi)\bar{k}_0^2}{\hat{S}(\xi)}\frac{\partial^2\tilde{h}_0}{\partial\bar{\xi}_+^2}, \tag{7.103}$$

$$\frac{\partial}{\partial\eta_+}(\tilde{T}_0 - \hat{\alpha}_T\tilde{T}_{S0}) + \sigma_0\hat{S}^2(\xi)\tilde{h}_0 + \bar{k}_0\xi\frac{\partial\tilde{h}_0}{\partial\bar{\xi}_+} = 0. \tag{7.104}$$

4. The root conditions as $\xi$ and $\xi_+ \to \infty$,

$$\{\tilde{T}_0,\ \tilde{T}_{S0},\ \tilde{h}_0\} = 0. \tag{7.105}$$

5. The tip smoothness conditions at $\xi = \eta_+ = 0$,

$$\frac{\partial}{\partial\xi}\{\tilde{T}_0,\ \tilde{T}_{S0},\ \tilde{h}_0\} = 0 \qquad \text{for the S-modes;} \tag{7.106}$$

$$\{\tilde{T}_0,\ \tilde{T}_{S0},\ \tilde{h}_0\} = 0 \qquad \text{for the A-modes.} \tag{7.107}$$

The system allows the following normal mode solutions:

$$\begin{aligned} \tilde{T}_0 &= A_0(\xi,\eta)\exp\{\mathrm{i}\xi_+ - \eta_+\}, \\ \tilde{T}_{S0} &= A_{S0}(\xi,\eta_+)\exp\{\mathrm{i}\xi_+ + \eta_+\}, \\ \tilde{h}_0 &= \hat{D}_0 \exp\{\mathrm{i}\bar{\xi}_+\}\ , \end{aligned} \tag{7.108}$$

where $\hat{D}_0$ is an arbitrary constant. The local dispersion relationship is derived as

$$\sigma_0 = \frac{\bar{k}_0}{S^2}\left[1 - \frac{(1+\hat{\alpha}_{\mathrm{T}})\bar{A}_s(\xi)\bar{k}_0^2}{\hat{S}}\right] - \mathrm{i}\frac{\xi\bar{k}_0}{S^2}. \tag{7.109}$$

It is also derived that

$$\begin{aligned} &\hat{A}_0 - \hat{\alpha}_{\mathrm{T}}^2\hat{A}_{S0} = \left(1 - \tfrac{1-\hat{\alpha}_{\mathrm{T}}^2}{\hat{S}}\bar{A}_s\bar{k}_0^2\right)\hat{D}_0, \\ &\hat{A}_0 + \hat{\alpha}_{\mathrm{T}}^2\hat{A}_{S0} = \left(1 - \tfrac{1+\hat{\alpha}_{\mathrm{T}}^2}{\hat{S}}\bar{A}_s\bar{k}_0^2\right)\hat{D}_0 \\ &\hat{A}_0 + \hat{A}_{S0} = \left(1 - \tfrac{2\bar{A}_s\bar{k}_0^2}{\hat{S}}\right)\hat{D}_0, \\ &\hat{A}_0 = \left(1 - \tfrac{\bar{A}_s\bar{k}_0^2}{\hat{S}}\right)\hat{D}_0, \quad \hat{A}_{S0} = -\tfrac{\bar{A}_s\bar{k}_0^2}{\hat{S}}\hat{D}_0. \end{aligned} \tag{7.110}$$

When the surface tension anisotropy is zero ($\alpha_4 = 0$), we recover

$$\sigma_0 = \frac{\bar{k}_0}{\hat{S}^2}\left[1 - \frac{(1+\hat{\alpha}_{\mathrm{T}})\bar{k}_0^2}{\hat{S}}\right] - \mathrm{i}\frac{\xi}{\hat{S}^2}\bar{k}_0. \tag{7.111}$$

On the other hand, for the symmetric model with $\hat{\alpha}_{\mathrm{T}} = 0$, we have

$$\sigma_0 = \Sigma(\xi,\bar{k}_0) = \frac{\bar{k}_0}{\hat{S}^2}\left[1 - \frac{2\bar{A}_s(\xi)\bar{k}_0^2}{\hat{S}}\right] - \mathrm{i}\frac{\xi\bar{k}_0}{\hat{S}^2}. \tag{7.112}$$

Many pattern formation systems studied in the present book have a cubic dispersion formula similar to (7.109). For convenience, we express the above local dispersion relationship in a more general form:

$$\sigma_0 = \Sigma(\xi,k_0) = \frac{k_0}{\hat{S}^2}\left(1 - \frac{k_0^2}{\Lambda_1\hat{S}}\right) - \mathrm{i}\frac{\xi k_0}{S^2}. \tag{7.113}$$

Obviously, for the system with anisotropy, we have

$$\Lambda_1 = \frac{1}{(1+\hat{\alpha}_{\mathrm{T}})\bar{A}_s(\xi)}. \tag{7.114}$$

For a fixed eigenvalue $\sigma_0$, one can solve for three wave numbers as functions of $\xi$, yielding results similar to those presented in Fig. 6.1, namely

$$\begin{cases} \bar{k}_0^{(1)}(\xi) = \hat{M}(\xi)\cos\left[\frac{1}{3}\cos^{-1}\left(\frac{\sigma_0}{\hat{N}(\xi)}\right)\right] & \text{(short-wavelength branch)} \\ \bar{k}_0^{(2)}(\xi) = \hat{M}(\xi)\cos\left[\frac{1}{3}\cos^{-1}\left(\frac{\sigma_0}{\hat{N}(\xi)}\right) + \frac{2\pi}{3}\right] & \\ \bar{k}_0^{(3)}(\xi) = \hat{M}(\xi)\cos\left[\frac{1}{3}\cos^{-1}\left(\frac{\sigma_0}{\hat{N}(\xi)}\right) + \frac{4\pi}{3}\right] & \text{(long-wavelength branch),} \end{cases} \tag{7.115}$$

where we define

$$\begin{cases} \hat{M}(\xi) = \sqrt{\dfrac{4\Lambda_1\hat{S}(\xi)}{3}}(1 - \mathrm{i}\xi)^{\frac{1}{2}} \\ \hat{N}(\xi) = -\dfrac{\hat{M}(\xi)}{3\hat{S}^2(\xi)}(1 - \mathrm{i}\xi)\,. \end{cases} \tag{7.116}$$

As we have discussed before, only the short-wavelength branch, the $H_1$ wave, with the larger $\mathrm{Re}\{k_0^{(1)}\} > 0$, and the long-wavelength branch, the $H_3$ wave, with the smaller $\mathrm{Re}\{k_0^{(3)}\} > 0$, are physically meaningful. The $H_2$ wave, with $\mathrm{Re}\{k_0^{(2)}\} < 0$, must be ruled out. The general solution of the $H$ wave is

$$\begin{cases} \tilde{h}_0 = D_1H_1(\xi) + D_3H_3(\xi) \\ \tilde{h} = D_1\exp\left\{\dfrac{\sigma t}{\varepsilon\eta_0^2} + \dfrac{\mathrm{i}}{\varepsilon}\displaystyle\int_{\xi_c}^{\xi}\left[\bar{k}_0^{(1)} + \varepsilon\bar{k}_1^{(1)} + \cdots\right]\mathrm{d}\xi_1\right\} \\ \quad + D_3\exp\left\{\dfrac{\sigma t}{\varepsilon\eta_0^2} + \dfrac{\mathrm{i}}{\varepsilon}\displaystyle\int_{\xi_c}^{\xi}\left[\bar{k}_0^{(3)} + \varepsilon\bar{k}_1^{(3)} + \cdots\right]\mathrm{d}\xi_1\right\} + \cdots. \end{cases} \tag{7.117}$$

The coefficients $\{D_1, D_3\}$ are arbitrary constants to be determined.

#### 7.3.2.2 First-Order Approximation

Similar to the 2D case studied in the last section, one can derive

$$\begin{aligned} (k_0^2 + g_0^2)\left(\frac{\partial^2}{\partial\xi_+^2} + \frac{\partial^2}{\partial\eta_+^2}\right)\tilde{T}_1 &= a_0(\xi,\eta)\mathrm{e}^{\mathrm{i}\xi_+ - \eta_+}\,, \\ (k_0^2 + g_0^2)\left(\frac{\partial^2}{\partial\xi_+^2} + \frac{\partial^2}{\partial\eta_+^2}\right)\tilde{T}_{S1} &= a_{S0}(\xi,\eta)\mathrm{e}^{\mathrm{i}\xi_+ + \eta_+}\,, \end{aligned} \tag{7.118}$$

where

$$a_0 = 2(k_0 - \mathrm{i}g_0)\left(\frac{\partial A_0}{\partial\eta} - \mathrm{i}\frac{\partial A_0}{\partial\xi}\right) + A_0\Big[\sigma_0\eta_0^2(\xi^2+\eta^2) \\ +\eta_0^2(\mathrm{i}k_0 + g_0)(\xi - \mathrm{i}\eta)\Big] \tag{7.119}$$

and

$$a_{S0} = -2(k_0 + \mathrm{i}g_0)\left(\frac{\partial A_{S0}}{\partial\eta} + \mathrm{i}\frac{\partial A_{S0}}{\partial\xi}\right) + A_{S0}\Big[\sigma_0\hat{\alpha}_T\eta_0^2(\xi^2+\eta^2) \\ +\eta_0^2\hat{\alpha}_T(\mathrm{i}k_0 - g_0)(\xi + \mathrm{i}\eta)\Big]. \tag{7.120}$$

To eliminate the secular terms on the right-hand side of (7.118), one must set

$$a_0 = a_{S0} = 0\,. \tag{7.121}$$

Then from (7.121), it follows that

$$\begin{aligned}\left(\frac{\partial}{\partial\eta} - \mathrm{i}\frac{\partial}{\partial\xi}\right)\ln A_0(\xi,\eta) &= \Phi_0(\xi,\eta)\\ \left(\frac{\partial}{\partial\eta} + \mathrm{i}\frac{\partial}{\partial\xi}\right)\ln A_{S0}(\xi,\eta) &= \Phi_{S0}(\xi,\eta),\end{aligned} \tag{7.122}$$

where

$$\begin{aligned}\Phi_0(\xi,\eta) &= -\frac{\sigma_0\eta_0^2(\xi^2+\eta^2)}{2\bar{k}_0[\xi + \mathrm{i}(\eta-1)]} - \frac{\mathrm{i}\eta_0^2}{2}\big(\xi - \mathrm{i}\eta\big)\\ \Phi_{S0}(\xi,\eta) &= \frac{\sigma_0\eta_0^2(\xi^2+\eta^2)}{2\bar{k}_0[\xi - \mathrm{i}(\eta-1)]} + \frac{\mathrm{i}\eta_0^2}{2}\big(\xi + \mathrm{i}\eta\big).\end{aligned} \tag{7.123}$$

Equations (7.122) are the first-order hyperbolic equations, which can used to solve the coefficient functions $A_0(\xi,\eta)$ and $A_{S0}(\xi,\eta)$ in the $(\xi,\eta)$-plane with the initial conditions imposed on the line $\eta = 1$:

$$A_0(\xi,1) = \hat{A}_0(\xi); \quad A_{S0}(\xi,1) = \hat{A}_{S0}(\xi).$$

For the general case $(0 \le \hat{\alpha}_T \le 1$, letting

$$\hat{D}_0 Q_0 = \frac{\partial}{\partial\eta}(A_0 - \hat{\alpha}_T A_{S0})\bigg|_{\eta=1},$$

we derive that at $\eta = 1$,

$$\begin{aligned}
\hat{D}_0 Q_0 &= \frac{\partial}{\partial \eta}(A_0 - \hat{\alpha}_{\mathrm{T}} A_{\mathrm{S}0})\bigg|_{\eta=1} \\
&= \mathrm{i}\frac{\partial}{\partial \xi}(\hat{A}_0 + \hat{\alpha}_{\mathrm{T}} \hat{A}_{\mathrm{S}0}) - (\hat{A}_0 + \hat{\alpha}_{\mathrm{T}}^2 \hat{A}_{\mathrm{S}0})\frac{\sigma_0 \eta_0^2 (1+\eta^2)}{2\bar{k}_0(\xi)} \\
&\quad -\frac{\mathrm{i}\eta_0^2 \xi}{2}(\hat{A}_0 + \hat{\alpha}_{\mathrm{T}}^2 \hat{A}_{\mathrm{S}0}) - \frac{1}{2}\eta_0^2(\hat{A}_0 - \hat{\alpha}_{\mathrm{T}}^2 \hat{A}_{\mathrm{S}0}),
\end{aligned} \tag{7.124}$$

and

$$\begin{aligned}
Q_0 &= -\mathrm{i}(1+\hat{\alpha}_{\mathrm{T}})\bar{A}_s \frac{\partial}{\partial \xi}\left(\frac{\bar{k}_0^2}{\hat{S}}\right) - \left(1 - \frac{1+\hat{\alpha}_{\mathrm{T}}^2}{\hat{S}}\bar{A}_s \bar{k}_0^2\right)\left[\frac{\eta_0^2}{2\bar{k}_0}\sigma_0 \hat{S}^2 + \frac{\mathrm{i}}{2}\xi \eta_0^2\right] \\
&\quad -\frac{\eta_0^2}{2}\left(1 - \frac{1-\hat{\alpha}_{\mathrm{T}}^2}{\hat{S}}\bar{A}_s \bar{k}_0^2\right).
\end{aligned} \tag{7.125}$$

With the solvability conditions (7.121), the equations for the first-order approximation are then reduced to

$$\begin{aligned}
\left(\frac{\partial^2}{\partial \xi_+^2} + \frac{\partial^2}{\partial \eta_+^2}\right)\tilde{T}_1 &= 0, \\
\left(\frac{\partial^2}{\partial \xi_+^2} + \frac{\partial^2}{\partial \eta_+^2}\right)\tilde{T}_{\mathrm{S}1} &= 0.
\end{aligned} \tag{7.126}$$

The interface conditions are

(a)

$$\tilde{T}_1 = \tilde{T}_{\mathrm{S}1} + \tilde{h}_1 - \eta_1 \bar{k}_0 \frac{\partial}{\partial \eta_+}(\tilde{T}_0 - \tilde{T}_{\mathrm{S}0}), \tag{7.127}$$

(b)

$$\begin{aligned}
\tilde{T}_{\mathrm{S}1} + \eta_1 \bar{k}_0 \frac{\partial \tilde{T}_{\mathrm{S}0}}{\partial \eta_+} &= \frac{\bar{A}_s(\xi)}{\hat{S}(\xi)}\left[\left(\bar{k}_0^2 \frac{\partial^2 \tilde{h}_1}{\partial \bar{\xi}_+^2} + 2\bar{k}_0 \bar{k}_1 \frac{\partial^2 \tilde{h}_0}{\partial \bar{\xi}_+^2} + \bar{k}_0' \frac{\partial \tilde{h}_0}{\partial \bar{\xi}_+}\right)\right. \\
&\quad \left. + \frac{\xi \bar{k}_0}{\hat{S}^2(\xi)}\frac{\partial \tilde{h}_0}{\partial \bar{\xi}_+}\right] - \tilde{A}_{s1}(\xi)\mathcal{K}_0(\xi)\bar{k}_0 \frac{\partial \tilde{h}_0}{\partial \bar{\xi}_+},
\end{aligned} \tag{7.128}$$

(c)

$$
\begin{aligned}
&\bar{k}_0\left(\frac{\partial \tilde{T}_1}{\partial \eta_+}-\hat{\alpha}_{\mathrm{T}}\frac{\partial \tilde{T}_{\mathrm{S}1}}{\partial \eta_+}\right)+\bar{k}_1\left(\frac{\partial \tilde{T}_0}{\partial \eta_+}-\hat{\alpha}_{\mathrm{T}}\frac{\partial \tilde{T}_{\mathrm{S}0}}{\partial \eta_+}\right)+\eta_1\bar{k}_0^2\frac{\partial^2}{\partial \eta_+^2}\left(\tilde{T}_0-\hat{\alpha}_{\mathrm{T}}\tilde{T}_{\mathrm{S}0}\right)\\
&+\left(\frac{\partial \tilde{T}_0}{\partial \eta}-\hat{\alpha}_{\mathrm{T}}\frac{\partial \tilde{T}_{\mathrm{S}0}}{\partial \eta}\right)+\hat{S}^2(\xi)(\sigma_0\tilde{h}_1+\sigma_1\tilde{h}_0)\\
&+\xi\left(\bar{k}_0\frac{\partial \tilde{h}_1}{\partial \bar{\xi}_+}+\bar{k}_1\frac{\partial \tilde{h}_0}{\partial \xi_+}\right)+\left(1+\eta_0^2\right)\tilde{h}_0=0. \qquad (7.129)
\end{aligned}
$$

Consider the normal mode of the first-order approximate solutions:

$$
\begin{aligned}
\tilde{T}_1 &= A_1(\xi,\eta)\exp\{\mathrm{i}\xi_+-\eta_+\},\\
\tilde{T}_{S1} &= A_{S1}(\xi,\eta)\exp\{\mathrm{i}\xi_++\eta_+\},\\
\tilde{h}_1 &= \hat{D}_1\exp\{\mathrm{i}\bar{\xi}_+\}.
\end{aligned} \qquad (7.130)
$$

By setting

$$
\begin{cases}
\hat{A}_1(\xi)=A_1(\xi,1),\\
\hat{A}_{S1}(\xi)=A_{S1}(\xi,1),
\end{cases} \qquad (7.131)
$$

one may derive the following system from the corresponding boundary conditions at the interface:

$$
\begin{aligned}
&\hat{A}_1-\hat{A}_{S1}-\hat{D}_1=I_1\hat{D}_0,\\
&\hat{A}_{S1}+\tfrac{\bar{A}_s\bar{k}_0^2}{S}\hat{D}_1=I_2\hat{D}_0,\\
&-\bar{k}_0(\hat{A}_1+\hat{\alpha}_{\mathrm{T}}\hat{A}_{S1})+(\sigma_0\hat{S}^2+\mathrm{i}\xi\bar{k}_0)\hat{D}_1=I_3\hat{D}_0,
\end{aligned} \qquad (7.132)
$$

where we define

$$
\begin{aligned}
I_1&=\eta_1\bar{k}_0\left(1-\tfrac{2\bar{A}_s\bar{k}_0^2}{\hat{S}}\right),\\
I_2&=\tfrac{\bar{A}_s}{\hat{S}}\left[\mathrm{i}\bar{k}_0'+\mathrm{i}\tfrac{\xi\bar{k}_0}{\hat{S}^2}-2\bar{k}_0\bar{k}_1-\mathrm{i}\bar{A}_{s1}(\xi)\mathcal{K}_0(\xi)\bar{k}_0\right]+\tfrac{\bar{A}_s\bar{k}_0^3}{\hat{S}}\eta_1,\\
I_3&=-(\sigma_1\hat{S}^2+\mathrm{i}\xi\bar{k}_1)-Q_0-(1+\eta_0^2)+\bar{k}_1\left(1-\tfrac{1+\hat{\alpha}_{\mathrm{T}}}{\hat{S}}\bar{A}_s\bar{k}_0^2\right)\\
&\quad-\eta_1\bar{k}_0^2\left(1-\tfrac{1-\hat{\alpha}_{\mathrm{T}}}{\hat{S}}\bar{A}_sk_0^2\right),
\end{aligned} \qquad (7.133)
$$

and obtain the solvability condition

$$
\det\begin{pmatrix}1 & -1 & I_1\\ 0 & 1 & I_2\\ -\bar{k}_0 & -(1+\hat{\alpha}_{\mathrm{T}})\bar{k}_0 & I_3\end{pmatrix}=0, \qquad (7.134)
$$

$$
I_3+(1+\hat{\alpha}_{\mathrm{T}})\bar{k}_0I_2+\bar{k}_0I_1=0. \qquad (7.135)
$$

From (7.135), it follows that

$$\bar{k}_1\left[1 - \mathrm{i}\xi - \frac{3(1+\hat{\alpha}_{\mathrm{T}})\bar{A}_s\bar{k}_0^2}{\hat{S}}\right] - \sigma_1\hat{S}^2 - Q_0 - (1+\eta_0^2)$$
$$+\frac{\bar{k}_0(1+\hat{\alpha}_{\mathrm{T}})\bar{A}_s}{\hat{S}}\left(\mathrm{i}\bar{k}_0' + \frac{\mathrm{i}\xi\bar{k}_0}{\hat{S}^2} - \bar{A}_{s1}\mathcal{K}_0\bar{k}_0\right) = 0.$$

Noting that one may write

$$Q_0 = -\frac{2\mathrm{i}(1+\hat{\alpha}_{\mathrm{T}})\bar{A}_s\bar{k}_0\bar{k}_0'}{\hat{S}} + \mathrm{i}\xi\frac{(1+\hat{\alpha}_{\mathrm{T}})\bar{A}_s\bar{k}_0^2}{\hat{S}^3} - \frac{\eta_0^2}{2}\left(1 - \frac{1-\hat{\alpha}_{\mathrm{T}}^2}{\hat{S}}\bar{A}_s\bar{k}_0^2\right)$$
$$-\frac{\eta_0^2}{2}\left(1 - \frac{1+\hat{\alpha}_{\mathrm{T}}}{\hat{S}}\bar{A}_s\bar{k}_0^2\right)\left(1 - \frac{1+\hat{\alpha}_{\mathrm{T}}^2}{\hat{S}}\bar{A}_s\bar{k}_0^2\right),$$

we derive

$$k_1 = \frac{R_1(\xi)}{F(\xi)} - \mathrm{i}\frac{R_2(\xi)}{F(\xi)}\frac{\bar{k}_0'}{\bar{k}_0}, \tag{7.136}$$

where

$$\begin{aligned}
F(\xi) &= \left[1 - \mathrm{i}\xi - \frac{3(1+\hat{\alpha}_{\mathrm{T}})\bar{A}_s\bar{k}_0^2}{\hat{S}}\right]\\
R_1(\xi) &= S^2\sigma_1 + (1+\tfrac{1}{2}\eta_0^2) + \bar{A}_s\bar{A}_{s1}\mathcal{K}_0\frac{(1+\hat{\alpha}_{\mathrm{T}})\bar{k}_0^2}{\hat{S}}\\
&\quad+\frac{\eta_0^2}{2}\left[\left(\frac{1-\hat{\alpha}_{\mathrm{T}}^2}{\hat{S}}\bar{A}_s\bar{k}_0^2\right) - \left(1 - \frac{1+\hat{\alpha}_{\mathrm{T}}}{\hat{S}}\bar{A}_s\bar{k}_0^2\right)\left(1 - \frac{1+\hat{\alpha}_{\mathrm{T}}^2}{\hat{S}}\bar{A}_s\bar{k}_0^2\right)\right],\\
R_2(\xi) &= \frac{3(1+\hat{\alpha}_{\mathrm{T}})\bar{A}_s\bar{k}_0^2}{\hat{S}}.
\end{aligned}\tag{7.137}$$

Like the systems of dendritic growth discussed in the previous sections, the normal mode of the MVE solution (7.117) has a singular point $\zeta_c$ in the extended complex $\zeta = \xi + \mathrm{i}\xi_{\mathrm{I}}$-plane, which satisfies the equation

$$F(\zeta) = \left[1 - \mathrm{i}\zeta - \frac{3(1+\hat{\alpha}_{\mathrm{T}})\bar{A}_s\bar{k}_0^2}{\hat{S}}\right] = 0,$$

which yields

$$\frac{1+\hat{\alpha}_{\mathrm{T}}}{\hat{S}}\bar{A}_s\bar{k}_0^2 = \frac{1-\mathrm{i}\zeta_c}{3}, \tag{7.138}$$

and

$$\sigma_0 = \sqrt{\frac{4}{27(1+\hat{\alpha}_{\mathrm{T}})\bar{A}_s}}\left(\frac{1-\mathrm{i}\zeta_c}{\hat{S}}\right)^{3/2}, \tag{7.139}$$

or

$$\sigma_0 = e^{-i\frac{3\pi}{4}} \sqrt{\frac{4}{27(1+\hat{\alpha}_T)}} \frac{(\zeta_c + i)^{\frac{7}{4}} (\zeta_c - i)^{\frac{1}{4}}}{\sqrt{f(\zeta_c)}}. \tag{7.140}$$

The meromorphic function $\bar{k}_1(\zeta)$ can be expanded into a Laurent series near the isolated singular points $\zeta_c$, so that the condition

$$R_1(\zeta_c) = 0 \tag{7.141}$$

must hold. In terms of the formula (7.138), from the condition (7.141) we derive the solution $\sigma_1$ as a function of $\zeta_c$, namely,

$$\begin{aligned} \sigma_1 = & -\tfrac{1}{(1+\zeta_c^2)} \Big\{ 1 + \tfrac{1}{2}\eta_0^2 + \tilde{A}_{s1} \tfrac{1-i\zeta_c}{3(1+\zeta_c^2)^{3/2}} \\ & + \tfrac{\eta_0^2}{6} \Big[ (1-\hat{\alpha}_T)(1-i\zeta_c) - (2+i\zeta_c)\Big(1 - \tfrac{1+\hat{\alpha}_T^2}{1+\hat{\alpha}_T} \tfrac{1-i\zeta_c}{3}\Big)\Big]\Big\}. \end{aligned} \tag{7.142}$$

We note that this formula does not contain the function $\eta_1$. Again, it is seen that the higher-order approximations of the RPE do not affect the stability properties up to $\mathcal{O}(\varepsilon)$.

For special case of 2D dendritic growth with isotropic surface tension, by setting $\tilde{A}_{s1} = 0$ in (7.142), we obtain

$$\begin{aligned} \sigma_1 = & -\tfrac{1}{(1+\zeta_c^2)} \Big\{ 1 + \tfrac{1}{2}\eta_0^2 + \tfrac{\eta_0^2}{6} \Big[ (1-\hat{\alpha}_T)(1-i\zeta_c) \\ & -(2+i\zeta_c)\Big(1 - \tfrac{1+\hat{\alpha}_T^2}{1+\hat{\alpha}_T} \tfrac{1-i\zeta_c}{3}\Big)\Big]\Big\}. \end{aligned} \tag{7.143}$$

Furthermore, for the symmetric model $\hat{\alpha}_T = 1$, we obtain

$$\sigma_1 = -\tfrac{1}{(1+\zeta_c^2)} \Big[ 1 + \tfrac{1}{2}\eta_0^2 - \tfrac{\eta_0^2}{18}(2+i\zeta_c)^2 \Big]. \tag{7.144}$$

We note that the form of (7.144) is completely the same as (6.100), which was derived for axisymmetric dendritic growth in Chap. 6.

Up to this point, only $\sigma_0$ is a free parameter. So far, for the sake of clarity we have applied the notation $\zeta = \xi + i\xi_I$ to represent the extended complex $\xi$-plane. as mentioned before, for simplicity without confusion one may just use the same notation $\xi$, instead of using the notation $\zeta$, to represent the complex variable $\xi$, when the real variable $\xi$ is extended to the complex plane.

An important remark should be made here that the formula (7.144) has a singularity at $\zeta_c = \pm i$, which corresponds to $\sigma_0 = 0$. Later, we shall show that the system allows a real spectrum of eigenvalues with $|\sigma_0| \ll 1$ and $\zeta_c \approx -\mathrm{i}$, which yields the so-called LF instability mechanism. Therefore, the formula (7.144) is not applicable for the case of LF instability.

### 7.3.3 The Inner Equation near the Singular Point $\zeta_c$

Similar to the procedure adopted in Chap. 6, in the inner region, $|\zeta - \zeta_c| \ll 1$, $|\eta - 1| \ll 1$, one may introduce the inner variables

$$\zeta_* = \frac{\zeta - \zeta_c}{\varepsilon^\alpha}, \quad \eta_* = \frac{\eta - 1}{\varepsilon^\alpha}, \tag{7.145}$$

where $\alpha > 0$ is to be determined. In addition, we denote the interface shape function by

$$\eta_s(\zeta, t) = 1 + \frac{\tilde{h}}{\eta_0^2} = 1 + \varepsilon^\alpha \eta_{*s} \tag{7.146}$$

and

$$\eta_{*s} = \frac{\hat{h}(\zeta_*, t)}{\eta_0^2}, \tag{7.147}$$

so that with the inner variable, we have

$$\tilde{h}(\zeta, t) = \varepsilon^\alpha \hat{h}(\zeta_*, t). \tag{7.148}$$

Finally, we set

$$\begin{aligned} \tilde{T}(\zeta, \eta, t) &= \varepsilon^\alpha \hat{T}(\zeta_*, \eta_*, t), \\ \tilde{T}_S(\zeta, \eta, t) &= \varepsilon^\alpha \hat{T}_S(\zeta_*, \eta_*, t). \end{aligned} \tag{7.149}$$

We look for mode solutions and carry out the inner expansions:

$$\begin{aligned} \hat{T}(\zeta_*, \eta_*, t) &= \left\{ \nu_0(\varepsilon)\hat{T}_0(\zeta_*, \eta_*) + \nu_1(\varepsilon)\hat{T}_1(\zeta_*, \eta_*) + \dots \right\} e^{\frac{\sigma t}{\varepsilon \eta_0^2}}, \\ \hat{T}_S(\zeta_*, \eta_*, t) &= \left\{ \nu_0(\varepsilon)\hat{T}_{S0}(\zeta_*, \eta_*) + \nu_1(\varepsilon)\hat{T}_{S1}(\zeta_*, \eta_*) + \dots \right\} e^{\frac{\sigma t}{\varepsilon \eta_0^2}}, \\ \hat{h} &= \left\{ \nu_0(\varepsilon)\hat{h}_0 + \nu_1(\varepsilon)\hat{h}_1 + \cdots \right\} e^{\frac{\sigma t}{\varepsilon \eta_0^2}}. \end{aligned} \tag{7.150}$$

In terms of the inner variables, (7.84) becomes

$$\left(\frac{\partial^2}{\partial \zeta_*^2} + \frac{\partial^2}{\partial \eta_*^2}\right)\hat{T} = \varepsilon^{2\alpha-1}\sigma\eta_0^2(\zeta^2+\eta^2)\hat{T} + \varepsilon^{\alpha}\eta_0^2\left(\zeta\frac{\partial}{\partial \zeta_*} - \eta\frac{\partial}{\partial \eta_*}\right)\hat{T}, \tag{7.151}$$

and the interface boundary conditions (7.86)–(7.89) are transformed to the following: at $\eta_* = 0$,

1.

$$\hat{T} = \hat{T}_S + \hat{h} + (\text{h.o.t.}); \tag{7.152}$$

2.

$$\hat{T}_S = \frac{\varepsilon^{2-2\alpha}\bar{A}_s(\zeta)}{S(\zeta)}\frac{\partial^2 \hat{h}}{\partial \zeta_*^2} + (\text{h.o.t.}); \tag{7.153}$$

3.

$$\varepsilon^{1-\alpha}\frac{\partial}{\partial \eta_*}\left(\hat{T} - \hat{\alpha}_T\hat{T}_S\right) + \sigma S^2(\zeta)\hat{h} + \varepsilon^{1-\alpha}\zeta\frac{\partial \hat{h}}{\partial \zeta_*} = (\text{h.o.t.}). \tag{7.154}$$

Letting $\varepsilon \to 0$, for the leading-order approximation, the above inner system may yield a third-order complex ODE for the perturbed interface function $\hat{h}_0$:

$$\mathrm{i}\frac{\varepsilon^{3-3\alpha}(1+\hat{\alpha}_T)\bar{A}_s(\zeta)}{S}\frac{\partial^3 \hat{h}_0}{\partial \zeta_*^3} + \varepsilon^{1-\alpha}(\zeta+\mathrm{i})\frac{\partial \hat{h}_0}{\partial \zeta_*} + \sigma S^2\hat{h}_0 = 0. \tag{7.155}$$

The above inner equation can also be derived in the simpler way addressed in the previous chapter. Namely, one may start with the dispersion relationship (7.113) and rewrite it in the form

$$\left[\hat{S}^2\sigma_0 - \bar{k}_0(1-\mathrm{i}\zeta) + \frac{(1+\hat{\alpha}_T)\bar{A}_s(\zeta)\bar{k}_0^3}{\hat{S}}\right]\tilde{h}_0 = 0. \tag{7.156}$$

By making the following replacements in (7.156),

$$\bar{k}_0 \Longleftrightarrow -\mathrm{i}\varepsilon\frac{\mathrm{d}}{\mathrm{d}\zeta},$$

one may obtain the inner equation

$$\mathrm{i}\frac{\varepsilon^3(1+\hat{\alpha}_{\mathrm{T}})\bar{A}_s(\zeta)}{S}\frac{\partial^3\hat{h}_0}{\partial\zeta^3}+\varepsilon(\zeta+\mathrm{i})\frac{\partial\hat{h}_0}{\partial\zeta}+\sigma S^2\hat{h}_0=0. \tag{7.157}$$

Changing to the inner variable $\zeta_*$, the above equation can be rewritten in the form (7.155).

It should be pointed out that in order for the above approximate model equation to be applicable, the following condition must hold:

$$\varepsilon^{2\alpha-1}\sigma\to 0. \tag{7.158}$$

This condition ensures that all right-hand terms of (7.150) are negligible.

To solve Eq. (7.155), we use as before the transformation introduced in [3] and relate $\tilde{h}_0$ to a new unknown function $W_0(\xi)$, i.e.,

$$\tilde{h}_0=W_0(\zeta)\exp\left[\frac{\mathrm{i}}{\varepsilon}\int_{\zeta_c}^{\zeta}k_c(\zeta_1)\,\mathrm{d}\zeta_1\right], \tag{7.159}$$

where the reference wave number function $k_c(\zeta)$ is to be chosen later. It will be seen that

$$\mathrm{Re}\{k_3\}<\mathrm{Re}\{k_c\}<\mathrm{Re}\{k_1\}. \tag{7.160}$$

Therefore, in the outer region we have the fundamental solutions

$$\begin{aligned} H_1&=W_0^{(+)}(\zeta)\ \exp\left[\frac{\mathrm{i}}{\varepsilon}\int_{\zeta_c}^{\xi}k_c(\zeta_1)\,\mathrm{d}\zeta_1\right],\\ H_3&=W_0^{(-)}(\zeta)\ \exp\left[\frac{\mathrm{i}}{\varepsilon}\int_{\zeta_c}^{\zeta}k_c(\zeta_1)\,\mathrm{d}\zeta_1\right]. \end{aligned} \tag{7.161}$$

One can draw the same diagram of the relation between the $H$ waves and the $W$ waves as in Chap. 6:

| | | |
|---|---|---|
| $H_1$ wave | $\Longleftrightarrow$ | $W^{(+)}$ |
| (short-wavelength branch) | | (outgoing wave); |
| $H_3$ wave | $\Longleftrightarrow$ | $W^{(-)}$ |
| ( long-wavelength branch) | | (incoming wave). |

In accordance with the above, we set

$$\hat{h}_0 = \hat{W}_0(\zeta_*) \exp\left\{ \frac{\mathrm{i}}{\varepsilon} \int_{\zeta_c}^{\zeta} k_c(\zeta_1)\, \mathrm{d}\zeta_1 \right\} \tag{7.162}$$

in the inner region. Similar to what was done in Chap. 6, by setting

$$S^2 \left( \frac{\partial \Sigma}{\partial k_0} \right)_{k_0 = k_c} = -\frac{3(1+\hat{\alpha}_\mathrm{T})\bar{A}_s k_c^2}{S} + (1 - \mathrm{i}\zeta) = 0\,, \tag{7.163}$$

or

$$k_c(\zeta) = \sqrt{\frac{S}{3(1+\hat{\alpha}_\mathrm{T})\bar{A}_s}}\,\big(1 - \mathrm{i}\zeta\big) = \frac{\mathrm{e}^{-\mathrm{i}\frac{\pi}{4}}(\zeta+\mathrm{i})^{\frac{3}{4}}(\zeta-\mathrm{i})^{\frac{1}{4}}}{\sqrt{3(1+\hat{\alpha}_\mathrm{T})\bar{A}_s}}$$
$$= \frac{\mathrm{e}^{-\mathrm{i}\frac{\pi}{4}}(\zeta+\mathrm{i})^{\frac{7}{4}}(\zeta-\mathrm{i})^{\frac{5}{4}}}{\sqrt{3(1+\hat{\alpha}_\mathrm{T})f(\zeta)}} \tag{7.164}$$

$$\big(\mathrm{Re}\{k_c\} > 0\big)\,, \tag{7.165}$$

and neglecting all the small higher-order terms, Eq. (7.155) is changed to

$$\varepsilon^{3-3\alpha}\Omega_3 \frac{\mathrm{d}^3\hat{W}_0}{\mathrm{d}\zeta_*^3} + \mathrm{i}\varepsilon^{2-2\alpha}\Omega_2 \frac{\mathrm{d}^2\hat{W}_0}{\mathrm{d}\zeta_*^2} + \mathrm{i}S^2\Big[\sigma_0 - \Sigma_c(\zeta)\Big]\hat{W}_0 = 0, \tag{7.166}$$

where

$$\begin{aligned}
\Omega_3 &= -\tfrac{(1+\hat{\alpha}_\mathrm{T})\bar{A}_s(\zeta)}{S}, \\
\Omega_2 &= -\tfrac{3k_c(1+\hat{\alpha}_\mathrm{T})\bar{A}_s(\zeta)}{S} = -\mathrm{e}^{-\mathrm{i}\frac{\pi}{4}} \tfrac{\sqrt{3(1+\hat{\alpha}_\mathrm{T})f(\zeta)}}{(\zeta+\mathrm{i})^{\frac{3}{4}}(\zeta-\mathrm{i})^{\frac{5}{4}}}, \\
\Sigma_c(\zeta) &= \Sigma(k_c,\zeta) = \tfrac{k_c}{\hat{S}^2}(1-\mathrm{i}\zeta) - \tfrac{(1+\hat{\alpha}_\mathrm{T})\bar{A}_s(\zeta)}{\hat{S}^3}k_c^3 \\
&= \mathrm{e}^{-\mathrm{i}\frac{3\pi}{4}} \sqrt{\tfrac{4}{27(1+\hat{\alpha}_\mathrm{T})\bar{A}_s}} \left(\tfrac{\zeta+\mathrm{i}}{\zeta-\mathrm{i}}\right)^{\frac{3}{4}} \\
&= \mathrm{e}^{-\mathrm{i}\frac{3\pi}{4}} \sqrt{\tfrac{4}{27(1+\hat{\alpha}_\mathrm{T})}} \tfrac{(\zeta+\mathrm{i})^{\frac{7}{4}}(\zeta-\mathrm{i})^{\frac{1}{4}}}{\sqrt{f(\zeta)}}.
\end{aligned} \tag{7.167}$$

Equation (7.166) can be rewritten as

$$\varepsilon^{3-3\alpha}\frac{\mathrm{i}}{3k_c}\frac{\mathrm{d}^3\hat{W}_0}{\mathrm{d}\zeta_*^3} - \varepsilon^{2-2\alpha}\frac{\mathrm{d}^2\hat{W}_0}{\mathrm{d}\zeta_*^2} + A^2(\zeta)\Big[\sigma_0 - \Sigma_c(\zeta)\Big]\hat{W}_0 = 0, \tag{7.168}$$

where

$$A^2(\zeta) = -\frac{\hat{S}^2(\zeta)}{\Omega_2(\zeta)} = \frac{\mathrm{e}^{\frac{\mathrm{i}\pi}{2}}}{\sqrt{3(1+\hat{\alpha}_\mathrm{T})(1-\alpha_4)}}\frac{(\zeta+\mathrm{i})^{\frac{7}{4}}(\zeta-\mathrm{i})^{\frac{9}{4}}}{(\zeta^2+a_1^2)^{\frac{1}{2}}(\zeta^2+a_2^2)^{\frac{1}{2}}}. \tag{7.169}$$

Clearly, at $\zeta = \zeta_c$,

$$\sigma_0 - \Sigma_c(\zeta_c) = 0. \tag{7.170}$$

The variable $\zeta$ in the coefficient functions of (7.166) needs to be changed to the inner variable $\zeta_*$ using $\zeta = \zeta_c + \varepsilon^\alpha \zeta_*$. It is now clear that $\zeta = \zeta_c, \pm\mathrm{i}$ are the turning points, while $\zeta = \pm\mathrm{i}a_i$ $(i = 1, 2)$ are the singular points of the inner equation (7.166) at which $\bar{A}_s = 0$. The relative positions of these singular points compared to $\zeta_c$ are related to the values of the parameters $\sigma_0$ and $\alpha_4$, so that two cases need to be discussed separately.

#### 7.3.3.1 Case I: $|\sigma_0| = \mathcal{O}(1)$

In the case $|\sigma_0| = \mathcal{O}(1)$, all the singular points $\zeta = \pm\mathrm{i}a_i$ $(i = 1, 2)$ and the turning points $\zeta = \pm\mathrm{i}$ are far from $\zeta_c$. In the vicinity of the turning point $\zeta_c$, one may take Taylor expansions around $\zeta_c$ for the coefficient functions, so that

$$\sigma_0 - \Sigma_c(\zeta) = -\Sigma_c'(\zeta_c)(\zeta - \zeta_c) + \cdots . \tag{7.171}$$

By neglecting the small higher-order terms, one may rewrite the inner equation (7.166) in the form

$$\underset{(1)}{\varepsilon^{3-3\alpha}\frac{\mathrm{i}}{3k_c}\frac{\mathrm{d}^3\hat{W}_0}{\mathrm{d}\zeta_*^3}} - \underset{(2)}{\varepsilon^{2-2\alpha}\frac{\mathrm{d}^2\hat{W}_0}{\mathrm{d}\zeta_*^2}} + \underset{(3)}{A_0^2(\zeta - \zeta_c)\hat{W}_0} = 0\,, \tag{7.172}$$

where

$$A_0^2 = \frac{S^2(\zeta_c)}{\Omega_2(\zeta_c)}\Sigma_c'(\zeta_c) = \mathcal{F}(\alpha_4, \hat{\alpha}_{\mathrm{T}}, \zeta_c) = \mathcal{O}(1).$$

As in the isotropic case, we find that in the leading-order approximation, the left-hand side of (7.172) has the balance (2) $\approx$ (3) $\gg$ (1). This yields $\alpha = \frac{2}{3}$, which satisfies the condition (7.158). Moreover, in the leading-order approximation, the inner equation is reduced to the Airy equation:

$$\frac{\mathrm{d}^2\hat{W}_0}{\mathrm{d}\hat{\zeta}_*^2} + \hat{\zeta}_*\hat{W}_0 = 0, \tag{7.173}$$

where

$$\hat{\zeta}_* = \frac{B}{\varepsilon^{\frac{2}{3}}}(\zeta - \zeta_c), \quad B = A_0^{\frac{2}{3}}, \quad \frac{\pi}{3} < \arg(B) \le \frac{\pi}{2}. \tag{7.174}$$

The structure of the Stokes lines for this case is similar to that shown in Fig. 6.2.

#### 7.3.3.2 Case II: $|\sigma_0| \ll 1$

Note that as $\sigma_0 \to 0$, we have $\zeta_c \to -\mathrm{i}$, and as $\alpha_4 \to 0$, $a_i \to 1$ $(i = 1, 2)$. Hence in this case, the turning point $\zeta = -\mathrm{i}$ and the singular points $-\mathrm{i}a_i$ $(i = 1, 2)$ will be close to the turning point $\zeta_c$ and play an important role. To study this case, by neglecting the small higher-order terms, we first rewrite the inner equation (7.166) in the form

$$\underset{(1)}{\varepsilon^{3-3\alpha}\frac{\mathrm{i}}{3k_c}\frac{\mathrm{d}^3\hat{W}_0}{\mathrm{d}\zeta_*^3}} - \underset{(2)}{\varepsilon^{2-2\alpha}\frac{\mathrm{d}^2\hat{W}_0}{\mathrm{d}\zeta_*^2}} + \underset{(3)}{A^2(\zeta)\Big[\sigma_0 - \Sigma_c(\zeta)\Big]\hat{W}_0} = 0\,, \tag{7.175}$$

where

$$A^2(\zeta) = -\frac{\hat{S}^2(\zeta)}{\Omega_2(\zeta)} = \frac{\mathrm{e}^{\frac{\mathrm{i}\pi}{2}}}{\sqrt{3(1+\hat{\alpha}_{\mathrm{T}})(1-\alpha_4)}}\frac{(\zeta+\mathrm{i})^{\frac{7}{4}}(\zeta-\mathrm{i})^{\frac{9}{4}}}{(\zeta^2+a_1^2)^{\frac{1}{2}}(\zeta^2+a_2^2)^{\frac{1}{2}}}. \tag{7.176}$$

The leading-order approximation of the inner equation is closely related to the asymptotic behavior of $\sigma_0$, $\alpha_4$, and $\zeta_c$ as $\varepsilon \to 0$.

Let us assume that as $\varepsilon \to 0$,

$$\sigma_0 = \hat{\sigma}_0\varepsilon^{\nu_0}, \quad \alpha_4 = \hat{\alpha}_4\varepsilon^{\mu_0}, \quad \zeta_c = -\mathrm{i} + \hat{\delta}\varepsilon^{\beta_0}, \tag{7.177}$$

where $\nu_0, \mu_0, \beta_0 > 0$. We obtain

$$\begin{cases} \zeta_c + \mathrm{i} = \hat{\delta}\varepsilon^{\beta_0}, \\ \zeta + \mathrm{i} = \varepsilon^{\alpha}\zeta_* + \hat{\delta}\varepsilon^{\beta_0}\,, \end{cases} \tag{7.178}$$

and

$$\begin{cases} \zeta_c + \mathrm{i}a_i = \hat{\delta}\varepsilon^{\beta_0} + \mathrm{i}\hat{a}_i\varepsilon^{\frac{\mu_0}{2}}\,, \\ \zeta + \mathrm{i}a_i = \varepsilon^{\alpha}\zeta_* + \hat{\delta}\varepsilon^{\beta_0} + \mathrm{i}\hat{a}_i\varepsilon^{\frac{\mu_0}{2}} \\ \hat{a}_1 = -\sqrt{2\hat{\alpha}_4}\,, \quad \hat{a}_2 = \sqrt{2\hat{\alpha}_4}\,. \end{cases} \tag{7.179}$$

We assume that the singular points $\zeta = -\mathrm{i}, -\mathrm{i}a_1, -\mathrm{i}a_2$ are all in the inner region of $\zeta_c$. This implies that $|\zeta_c + \mathrm{i}| = \mathcal{O}(\varepsilon^{\alpha})$ and $|\zeta_c + \mathrm{i}a_i| = \mathcal{O}(\varepsilon^{\alpha})$ $(i = 1, 2)$. From (7.178)–(7.179), we deduce that

$$\beta_0 = \alpha = \frac{\mu_0}{2}. \tag{7.180}$$

The relationship of $\nu_0$ and $\beta_0$ can be determined via the formula (7.140). By applying (7.177) to (7.140), it follows that

$$\hat{\sigma}_0 \varepsilon^{\nu_0} \sim -\frac{\mathrm{e}^{-\mathrm{i}\frac{3\pi}{8}}}{2^{\frac{1}{4}} 3^{\frac{3}{2}}} \sqrt{\frac{(1+\hat{\alpha}_{\mathrm{T}})}{27(1-\alpha_4)}} \frac{\hat{\delta}^{\frac{7}{4}} \varepsilon^{\frac{7\beta_0}{4}}}{\left(\hat{\delta}^2 \varepsilon^{2\beta_0} + 2\hat{\alpha}_4 \varepsilon^{\mu_0}\right)^{\frac{1}{2}}} = \mathcal{O}(\varepsilon^{\frac{3\beta_0}{4}}). \tag{7.181}$$

On viewing (7.180), from (7.181) we obtain that

$$\nu_0 = \frac{3\beta_0}{4} = \frac{3\alpha}{4}, \tag{7.182}$$

$$\hat{\sigma}_0 = -\frac{\mathrm{e}^{-\mathrm{i}\frac{3\pi}{8}}}{2^{\frac{1}{4}} 3^{\frac{3}{2}}} \sqrt{\frac{(1+\hat{\alpha}_{\mathrm{T}})}{27(1-\alpha_4)}} \frac{\hat{\delta}^{\frac{7}{4}}}{\left(\hat{\delta}^2 + 2\hat{\alpha}_4\right)^{\frac{1}{2}}}. \tag{7.183}$$

Based on the above information, we now determine the orders of the magnitudes of each term on the left-hand side of (7.175). Note that

$$\begin{aligned}
\frac{\mathrm{i}}{3k_c} &\approx -\frac{\varepsilon^{-\frac{3}{4}\alpha}\sqrt{3(1+\hat{\alpha}_{\mathrm{T}})(1-\alpha_4)}}{3\times 2^{\frac{1}{4}}} \frac{(\zeta_*+\hat{\delta}+\mathrm{i}\hat{a}_1)(\zeta_*+\hat{\delta}+\mathrm{i}\hat{a}_1)}{(\zeta_*+\hat{\delta})^{\frac{7}{4}}} \\
A^2(\xi) &\approx \varepsilon^{\frac{3\alpha}{4}} \frac{\mathrm{e}^{\frac{\mathrm{i}\pi}{2}}}{\sqrt{3(1+\hat{\alpha}_{\mathrm{T}})(1-\alpha_4)}} \frac{(\zeta_*+\hat{\delta})^{\frac{7}{4}}}{(\zeta_*+\hat{\delta}+\mathrm{i})} \\
\left[\sigma_0 - \Sigma_c(\xi)\right] &= \mathrm{e}^{-\mathrm{i}\frac{3\pi}{4}} \sqrt{\frac{4}{27(1+\hat{\alpha}_{\mathrm{T}})}} \left[\frac{(\zeta_c+\mathrm{i})^{\frac{7}{4}}(\zeta_c-\mathrm{i})^{\frac{1}{4}}}{\sqrt{f(\zeta_c)}} - \frac{(\zeta+\mathrm{i})^{\frac{7}{4}}(\zeta-\mathrm{i})^{\frac{1}{4}}}{\sqrt{f(\zeta)}}\right] \\
&\approx -\varepsilon^{\frac{3\alpha}{4}} \frac{\mathrm{e}^{-\mathrm{i}\frac{3\pi}{2}}}{2^{\frac{3}{4}}} \sqrt{\frac{4}{27(1+\hat{\alpha}_{\mathrm{T}})(1-\alpha_4)}} \left[\frac{(\zeta_*+\hat{\delta})^{\frac{7}{4}}}{(\zeta_*+\hat{\delta}+\mathrm{i}\hat{a}_1)^{\frac{1}{2}}(\zeta_*+\hat{\delta}+\mathrm{i}\hat{a}_1)^{\frac{1}{2}}}\right. \\
&\qquad \left. - \frac{\hat{\delta}^{\frac{7}{4}}}{(\hat{\delta}+\mathrm{i}\hat{a}_1)^{\frac{1}{2}}(\hat{\delta}+\mathrm{i}\hat{a}_1)^{\frac{1}{2}}}\right].
\end{aligned}$$

We find that

$$(1) = \mathcal{O}(\varepsilon^{3-\frac{15\alpha}{4}}), \quad (2) = \mathcal{O}(\varepsilon^{2-2\alpha}) \quad \text{and} \quad (3) = \mathcal{O}(\varepsilon^{\frac{3\alpha}{2}}).$$

The balance on the left-hand side of (7.175) leads to

$$(1) \approx (2) \approx (3);$$

and

$$\alpha = \frac{4}{7}, \quad \nu_0 = \frac{3}{7} \quad \text{and} \quad \mu_0 = \frac{8}{7}. \tag{7.184}$$

In other words,

$$\sigma_0 = \mathcal{O}(\varepsilon^{\frac{3}{7}}), \quad \zeta_c + \mathrm{i} = \mathcal{O}(\varepsilon^{\frac{4}{7}}), \quad \text{and} \quad \alpha_4 = \mathcal{O}(\varepsilon^{\frac{8}{7}}), \tag{7.185}$$

which satisfies the condition (7.158). In the far field of the inner region, since $|\zeta_*| \gg 1$, the first term on the left-hand side of (7.175) is negligible, so that the inner equation (7.175) is further simplified as

$$\frac{\mathrm{d}\hat{W}_0}{\mathrm{d}\zeta_*^2} + A_*^2 \zeta_*^{\frac{3}{2}} \hat{W}_0 = 0 , \tag{7.186}$$

where

$$A_*^2 = \frac{2^{\frac{1}{4}}}{9(1+\hat{\alpha}_{\mathrm{T}})(1-\alpha_4)}, \tag{7.187}$$

$$\frac{\mathrm{d}\hat{W}_0}{\mathrm{d}\hat{\zeta}_*^2} + \hat{\zeta}_*^{\frac{3}{2}} \hat{W}_0 = 0 , \tag{7.188}$$

where

$$\hat{\zeta}_* = \frac{A_*^{\frac{4}{7}}}{\varepsilon^{\frac{4}{7}}}(\zeta - \zeta_c). \tag{7.189}$$

From the above results, we derive

$$\alpha_4^{\frac{7}{8}} = \mathcal{O}(\varepsilon) \quad \text{and} \quad \sigma_0^{\frac{11}{7}} \alpha_4^{\frac{2}{7}} = \mathcal{O}(\varepsilon). \tag{7.190}$$

It will be seen that this case is consistent with the quantization condition obtained later.

The structure of the Stokes lines emitted from $\zeta_* = 0$ for Eq. (7.188) is sketched in Fig. 7.2. The open angle between $(L_1)$ and $(L_2)$ is $\theta = \frac{4}{7}\pi$. The anti-Stokes line $(A_1)$ divides the whole complex plane into sectors $(S_1)$ and $(S_2)$.

#### 7.3.3.3 A Brief Summary

From the results derived above, one sees that for cases I and II, the inner equation in the far field of the inner region ($\zeta_* \gg 1$) can be approximately written in the

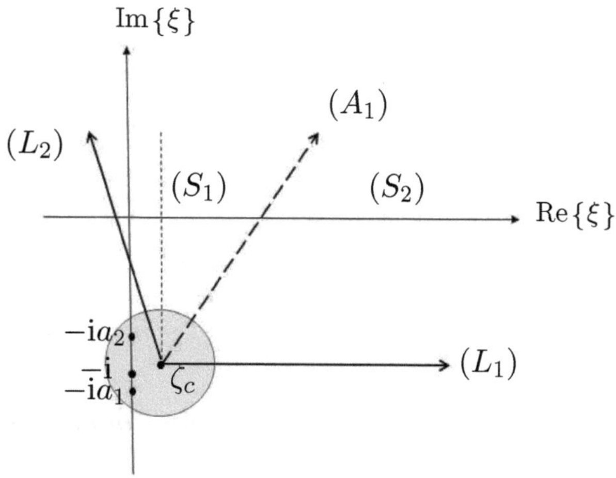


**Fig. 7.2** Sketch of the *Stokes lines* $(L_1)$, $(L_2)$, and the *anti-Stokes line* $(A_1)$ emanating from the turning point $\zeta_c$, in the case $|\sigma| \ll 1$

following unified form:

$$\frac{d^2\hat{W}_0}{d\hat{\xi}_*^2} + \hat{\xi}_*^{p_0}\hat{W}_0 = 0\,, \tag{7.191}$$

where

$$\hat{\xi}_* = \frac{B}{\varepsilon^\alpha}(\zeta - \zeta_c)\,. \tag{7.192}$$

For different cases, the constants $B$ and $p_0$ and the exponent number $\alpha$ are different. The general solution of (7.191) is

$$\hat{W}_0 = C_1\hat{\xi}_*^{\frac{1}{2}}H_\nu^{(1)}(z) + C_2\hat{\xi}_*^{\frac{1}{2}}H_\nu^{(2)}(z) \quad (z = 2\nu\hat{\xi}_*^{\frac{1}{2\nu}}), \tag{7.193}$$

where $C_1, C_2$ are arbitrary constants, $\nu = \frac{1}{p_0+2}$, and $H_\nu^{(1)}(z)$ and $H_\nu^{(2)}(z)$ are the $\nu$th-order Hankel functions of the first and second kinds, respectively.

The open angle at the turning point $\zeta_c$ between the two neighboring Stokes lines $(L_1)$ and $(L_2)$ is

$$\theta = \frac{2}{p_0 + 2}\pi\,. \tag{7.194}$$

The anti-Stokes line $(A_2)$ between $(L_1)$ and $(L_2)$ always intersects the real axis of $\zeta$ at a point $\zeta_c' > 0$ and divides the real axis into two parts. One part is in the sector $(S_1)$, while the other is in $(S_2)$.

The main results are summarized as follows:

- Case I : $\sigma_0 = \mathcal{O}(1)$, and

$$\alpha = \frac{2}{3}, \quad p_0 = 1, \quad \nu = \frac{1}{3}.$$

- Case II : $|\sigma_0| = \mathcal{O}(\varepsilon^{3/7}) \ll 1$ and $\alpha_4 \ll 1$,

$$\alpha = \frac{4}{7}, \quad p_0 = \frac{3}{2}, \quad \nu = \frac{2}{7}.$$

### 7.3.4 Matching Procedure and Connection Conditions

We now turn to matching the inner solution (7.193) with the outer solution (7.117) along different directions in each sector. Again, we apply the connection formula for the Hankel functions: When $\pi < \arg(z_1) < 2\pi$, we have

$$H_\nu^{(2)}(z_1) = H_\nu^{(2)}(z\mathrm{e}^{\mathrm{i}\pi}) = 2\cos(\nu\pi)H_\nu^{(2)}(z) + \mathrm{e}^{\mathrm{i}\nu\pi}H_\nu^{(1)}(z) \qquad (7.195)$$

$$(0 \leq \arg(z) \leq \pi),$$

and the asymptotic expansions, as $|z| \to \infty$,

$$\begin{aligned} H_\nu^{(1)}(z) &\sim \sqrt{\frac{2}{\pi z}}\mathrm{e}^{\mathrm{i}z-\mathrm{i}(\nu/2+1/4)\pi} \quad (-\pi < \arg(z) \leq 2\pi), \\ H_\nu^{(2)}(z) &\sim \sqrt{\frac{2}{\pi z}}\mathrm{e}^{-\mathrm{i}z+\mathrm{i}(\nu/2+1/4)\pi} \quad (-2\pi < \arg(z) \leq \pi). \end{aligned} \qquad (7.196)$$

We first match the inner solution (7.193) with the outer solution (7.117) as $\hat{\zeta}_* \to \infty$ in the sector $(S_2)$. In this sector, the outer solution satisfying the root condition (7.105) has $D_1' = 0$. Thus, to match with the outer solution, one must set $C_1 = 0$ and $C_2 = D = D_3' \times$ constant $\neq 0$ for the inner solution (7.193).

Furthermore, to match the inner solution with the outer solution in the sector $(S_1)$, one may apply the formula (7.196) and follow the procedure as presented in Chap. 6, which leads to the following connection condition:

$$\frac{D_1}{D_3} = \mathrm{i}2\cos(\nu\pi). \qquad (7.197)$$

So far, we have not applied the tip condition. When the tip condition (7.106) or (7.107) is applied, we obtain the quantization conditions, which determine the parameter $\sigma_0$ as a function of $\alpha_4$ and $\varepsilon$.

## 7.4 The Quantization Condition of Global Trapped-Wave Modes

Global wave instability occurs only in case I discussed in the last section. We define $\sigma_0 = \sigma_R - i\omega$ with $0 < \omega = \mathcal{O}(1)$. With the complex eigenvalue $\sigma_0$, the physical solution in the outer region is

$$\mathrm{Re}\left\{\tilde{h}_0(\xi, t)\right\} = \mathrm{Re}\left\{H(\xi)\mathrm{e}^{\frac{\sigma_0 t}{\varepsilon \eta_0^2}}\right\}, \tag{7.198}$$

where

$$H(\xi) = D_1 H_1 + D_3 H_3 . \tag{7.199}$$

We define

$$\begin{cases} d_1 = D_1 \mathrm{e}^{-i\chi_1} \\ d_3 = D_3 \mathrm{e}^{-i\chi_3} , \end{cases} \tag{7.200}$$

where

$$\begin{aligned} \chi_1 &= \frac{1}{\varepsilon}\int_0^{\xi_c} k_0^{(1)} \mathrm{d}\xi, \\ \chi_3 &= \frac{1}{\varepsilon}\int_0^{\xi_c} k_0^{(3)} \mathrm{d}\xi . \end{aligned} \tag{7.201}$$

Thus, it follows from the connection formula (7.197) that

$$\frac{d_1}{d_3} = i2\cos(\nu\pi)\mathrm{e}^{-i\chi}, \tag{7.202}$$

where

$$\chi = \frac{1}{\varepsilon}\int_0^{\xi_c} \left(k_0^{(1)} - k_0^{(3)}\right)\mathrm{d}\xi . \tag{7.203}$$

In this case, we have $\nu = \frac{1}{3}$. To satisfy the smooth tip conditions, the parameter $\sigma_0$ must be a proper function of $\varepsilon$ such that the tip $\xi = 0$ is located on the Stokes line $(L_3)$ and as $\varepsilon \to 0$, the following conditions hold:

1. For the symmetric S-modes,

$$d_1 k_0^{(1)}(0) + d_3 k_0^{(3)}(0) = 0 , \tag{7.204}$$

which gives

$$\mathrm{i}e^{-\mathrm{i}\chi} = -\frac{k_0^{(3)}(0)}{k_0^{(1)}(0)}\,. \tag{7.205}$$

2. For the antisymmetric A-modes,

$$d_1 + d_3 = 0\,, \tag{7.206}$$

which gives

$$\mathrm{i}e^{-\mathrm{i}\chi} = -1\,. \tag{7.207}$$

Thus, we obtain the following quantization conditions:

$$\chi = \frac{1}{\varepsilon}\int_0^{\xi_c}\left(k_0^{(1)} - k_0^{(3)}\right)\mathrm{d}\xi = \left(2n + 1 + \frac{1}{2} + \theta_0\right)\pi - \mathrm{i}\ln\alpha_0,$$
$$\left(n = 0, \pm 1, \pm 2, \ldots\right), \tag{7.208}$$

where

$$\begin{cases} \alpha_0\, e^{\mathrm{i}\theta_0\pi} = \dfrac{k_0^{(1)}(0)}{k_0^{(3)}(0)} & \text{for the S-modes;} \\ \alpha_0 = 1,\ \ \theta_0 = 0 & \text{for the A-modes.} \end{cases} \tag{7.209}$$

This spectrum contains two discrete sets of complex eigenvalues, $\sigma_{0n}$ $(n = 0, \pm 1, \pm 2, \ldots)$, given by the quantization conditions (7.208) as functions of $\varepsilon$ and $\alpha_4$, for the S-modes and A-modes, respectively. Note that the form of the quantization condition (7.208) for the 2D S-modes is the same as (6.182) for the 3D axisymmetric mode derived in the previous chapter.

For each $n$, the system allows a unique neutrally stable A-mode and a unique neutrally stable S-mode, whose corresponding critical numbers are denoted by $\varepsilon_{*n}^{(a)}$ and $\varepsilon_{*n}^{(s)}$, respectively. It can be shown that

$$\varepsilon_{*0}^{(a)} > \varepsilon_{*0}^{(s)} > \varepsilon_{*1}^{(a)} > \varepsilon_{*1}^{(s)} > \varepsilon_{*2}^{(a)} > \varepsilon_{*2}^{(s)} > \cdots\,.$$

Hence, the neutral A-mode ($n = 0$) is more stable than the neutral S-mode ($n = 0$).

In Fig. 7.3, we show the leading-order approximation of eigenvalues $\sigma_0$ of the A-modes ($n = 0, 1, 2,$) versus $\varepsilon$, for the two cases $\alpha_4 = 0$ and $0.1$. The smaller symbols describe the case $\alpha_4 = 0$, while the larger symbols describe the case $\alpha_4 = 0.1$. As we have seen for the case of a pure melt in Chap. 6, these eigenvalues for $n = 0, 1, 2, \ldots$ are all located on the same curve in the complex $\sigma$-plane, and they all

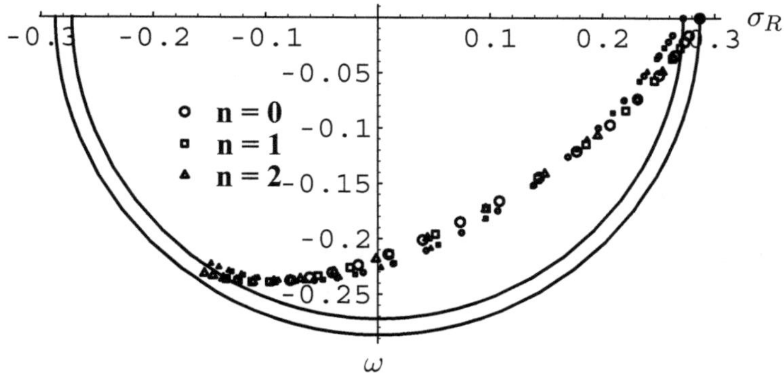


**Fig. 7.3** The variations with $\varepsilon$ of $\sigma_0 = \sigma_R - i\omega$ of GTW A-modes ($n = 0, 1, 2$) in the complex $\sigma_0$-plane. The smaller symbols describe the case $\alpha_4 = 0$, while the larger symbols correspond to $\alpha_4 = 0.1$

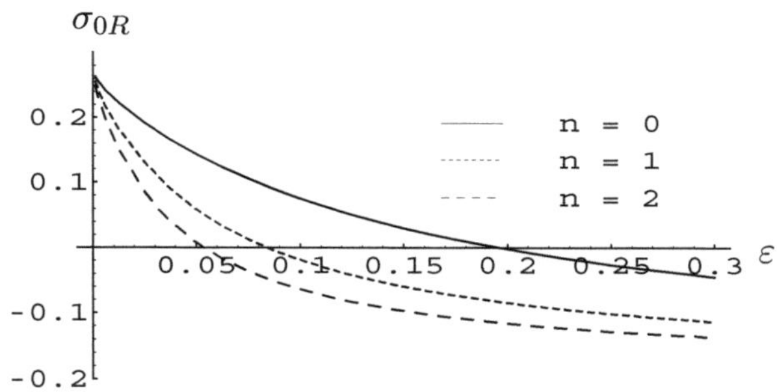


**Fig. 7.4** The variations of the real part of the zeroth-order approximation of eigenvalues, $\sigma_{0R}$, of 2D GTW A-modes ($n = 0, 1, 2$) with $\varepsilon$ for the case $\hat{\alpha}_T = 1$

tend to the point $\sigma_0 = \sigma_{max}$ as $\varepsilon \to 0$. As $\alpha_4 = 0$, we have $\hat{\alpha}_T = 1$, $\sigma_{max} = 0.2722$, while as $\alpha_4 = 0.1$, we have $\hat{\alpha}_T = 1$, $\sigma_{max} = 0.2869$.

In Figs. 7.4 and 7.5, we show for $\hat{\alpha}_T = 1$ the zeroth-order approximate values of $\sigma_{0R}$ and $\omega_0$ of GTW A-modes ($n = 0, 1, 2, 3$) versus $\varepsilon$. In Figs. 7.6 and 7.7, we show the zeroth-order approximate values of $\sigma_{0R}$ and $\omega_0$ of GTW S-modes ($n = 0, 1, 2, 3$) versus $\varepsilon$ for the same case $\hat{\alpha}_T = 1$.

It is seen that the system allows a unique neutrally stable A-mode ($n = 0$) and a unique neutrally stable S-mode ($n = 0$). The neutrally stable A-mode ($n = 0$) is more stable than the neutrally stable S-mode ($n = 0$).

In Fig. 7.8, we show the zeroth-order approximate critical numbers $\varepsilon_*^{(0)}$ corresponding to neutral GTW modes versus $\alpha_4$, while in Fig. 7.9 we show the eigenvalue $\omega_*^{(0)}$ of the neutral GTW modes versus $\alpha_4$. In these figures, the thin lines are used for the S-modes, the thick lines for the A-modes; the solid lines are for the case $\hat{\alpha}_T = 1$, and the dashed lines for the case $\hat{\alpha}_T = 0$.

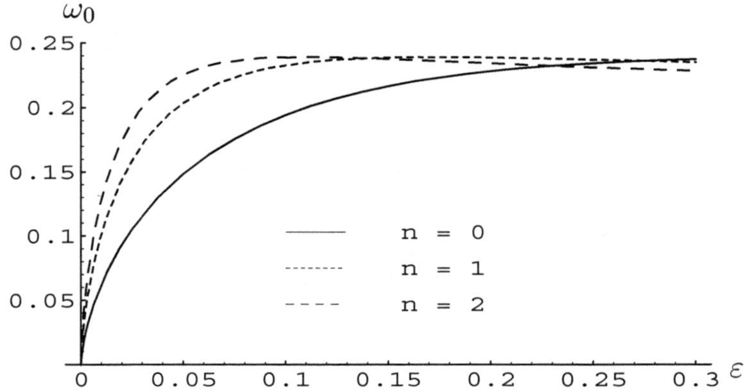


**Fig. 7.5** The variations of the imaginary part of the zeroth-order approximation of eigenvalues, $\omega_0$, of 2D GTW A-modes ($n = 0, 1, 2$) with $\varepsilon$ for the case $\hat{\alpha}_T = 1$

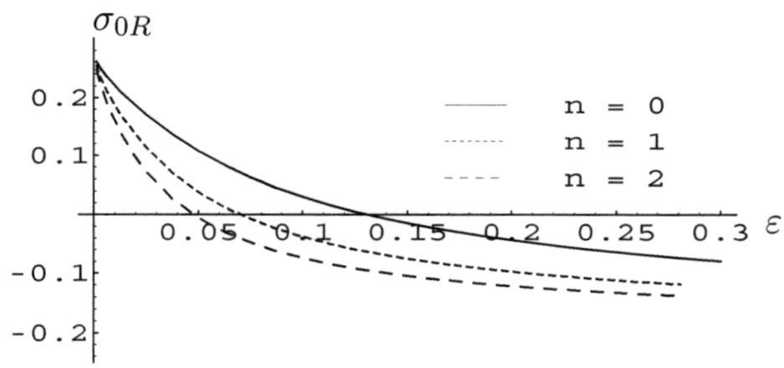


**Fig. 7.6** The variations of the real part of the zeroth-order approximation of eigenvalues, $\sigma_{0R}$, of 2D GTW S-modes ($n = 0, 1, 2$) with $\varepsilon$ for the case $\hat{\alpha}_T = 1$

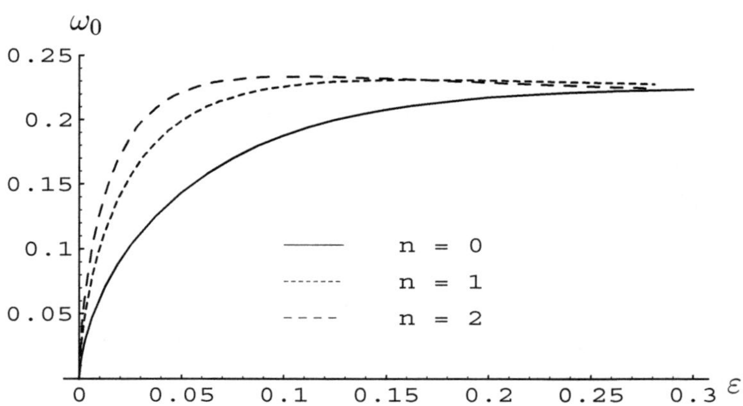


**Fig. 7.7** The variations of the imaginary part of the zeroth-order approximation of eigenvalues, $\omega_0$, of 2D GTW S-modes ($n = 0, 1, 2$) with $\varepsilon$ for the case $\hat{\alpha}_T = 1$

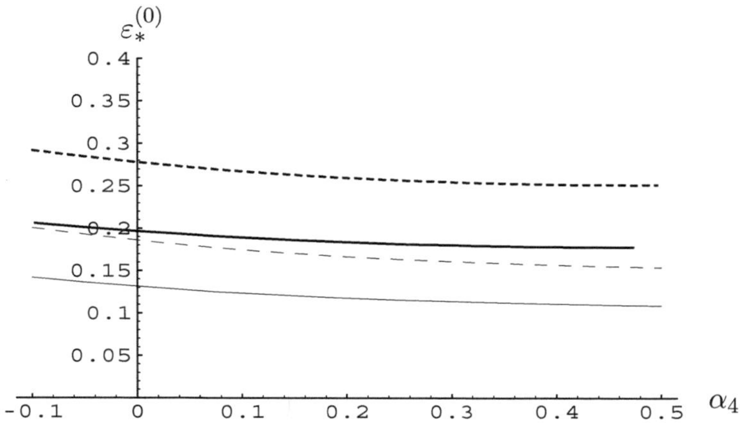


**Fig. 7.8** The variations of $\varepsilon_*^{(0)}$ of the GTW A-modes and S-modes ($n = 0$) with $\alpha_4$. The *thin lines* are used for the S-modes, the *thick lines* for the A-modes; the *solid lines* are for the case $\hat{\alpha}_T = 1$, and the *dashed lines* for $\hat{\alpha}_T = 0$

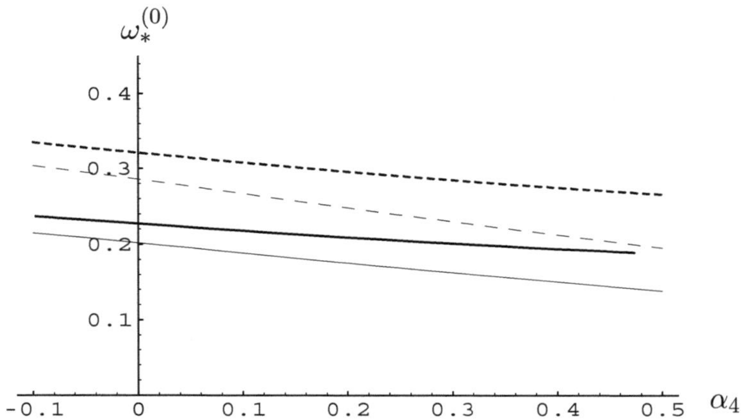


**Fig. 7.9** The variations of $\omega_*^{(0)}$ of the corresponding GTW A-modes and S-modes ($n = 0$) with $\alpha_4$

One can calculate the first-order approximate complex eigenvalues $\sigma = \sigma_0 + \varepsilon\sigma_1 = \sigma_R - i\omega$ for given $\varepsilon$ for the symmetric modes (S-modes) and antisymmetric modes (A-modes) using (7.142) and the quantization conditions (7.208) and (7.209). The first-order approximate critical number $\varepsilon_*^{(1)}$ can be calculated via the condition $\mathrm{Re}\{\sigma_0 + \varepsilon\sigma_1\} = 0$.

The results for the typical case ($\varepsilon = 0.1, \eta_0^2 = 0.01, \alpha_4 = 0$) are listed in Tables 7.1 and 7.2. The first-order approximate critical number $\varepsilon_*^{(1)}$, the corresponding frequencies $\omega_*^{(1)}$, and the phase velocity $V_p$ of the interfacial wave in the far field away from the tip, for the GTW neutral S-modes ($n = 0$) and the GTW neutral A-modes ($n = 0$) are computed with various Péclet numbers $\eta_0^2$. The numerical results

**Table 7.1** The eigenvalues of the 2D GTW S-modes ($\varepsilon = 0.1$; $\hat{\alpha}_T = 1$; $\eta_0^2 = 0.01$; $\alpha_4 = 0$)

| $n$ | $\sigma_0$ | $\sigma = \sigma_0 + \varepsilon\sigma_1$ | $V_p$ |
|---|---|---|---|
| 0 | (0.03028, −0.18770) | (−0.00284, −0.21147) | 1.01840 |
| 1 | (−0.03745, −0.22550) | (−0.05278, −0.23419) | 0.96369 |
| 2 | (−0.07369, −0.23310) | (−0.08319, −0.23741) | 0.91889 |
| 3 | (−0.09580, −0.23378) | (−0.10006, −0.23630) | 0.88604 |

**Table 7.2** The eigenvalues of the 2D GTW A-modes ($\varepsilon = 0.1$; $\hat{\alpha}_T = 1$; $\eta_0^2 = 0.01$; $\alpha_4 = 0$)

| $n$ | $\sigma_0$ | $\sigma = \sigma_0 + \varepsilon\sigma_1$ | $V_p$ |
|---|---|---|---|
| 0 | (0.07564, −0.19394) | (0.02688, −0.21228) | 1.04949 |
| 1 | (−0.01882, −0.23207) | (−0.03876, −0.24057) | 0.98563 |
| 2 | (−0.06396, −0.23832) | (−0.07537, −0.24255) | 0.93330 |
| 3 | (−0.08994, −0.23779) | (−0.09757, −0.24024) | 0.89592 |

**Table 7.3** The critical numbers $\varepsilon_*^{(1)}$ and the corresponding frequencies $\omega_*^{(1)}$ of the 2D neutral GTW S-modes ($n = 0$) for the case $\hat{\alpha}_T = 0$; $\alpha_4 = 0$

| $Pe_0$ | $T_\infty$ | $\varepsilon_*^{(1)}$ | $\omega_*^{(1)}$ | $V_p$ |
|---|---|---|---|---|
| 0.0005 | −0.02753 | 0.13777 | 0.29661 | 1.0209 |
| 0.0045 | −0.07976 | 0.13769 | 0.29653 | 1.0209 |
| 0.01 | −0.1159 | 0.13761 | 0.29643 | 1.0208 |
| 0.1 | −0.3133 | 0.13589 | 0.29451 | 1.0204 |
| 1.0 | −0.6557 | 0.12128 | 0.27667 | 1.0162 |
| 10.0 | −0.9208 | 0.06290 | 0.17152 | 1.0026 |

**Table 7.4** The critical numbers $\varepsilon_*^{(1)}$ and the corresponding frequencies $\omega_*^{(1)}$ of the 2D neutral GTW A-modes (n=0) for the case $\hat{\alpha}_T = 0$; $\alpha_4 = 0$

| $Pe_0$ | $T_\infty$ | $\varepsilon_*^{(1)}$ | $\omega_*^{(1)}$ | $V_p$ |
|---|---|---|---|---|
| 0.0005 | −0.02753 | 0.17941 | 0.32260 | 1.02819 |
| 0.0045 | −0.07976 | 0.17925 | 0.32247 | 1.02818 |
| 0.01 | −0.1159 | 0.17909 | 0.32231 | 1.02812 |
| 0.1 | −0.3133 | 0.17644 | 0.31979 | 1.02735 |
| 1.0 | −0.6557 | 0.15402 | 0.29660 | 1.02089 |
| 10.0 | −0.9208 | 0.07202 | 0.17497 | 1.00279 |

**Table 7.5** The critical numbers $\varepsilon_*^{(1)}$ and the corresponding frequencies $\omega_*^{(1)}$ of the 2D neutral GTW S-modes ($n = 0$) for the case $\hat{\alpha}_T = 1$; $\alpha_4 = 0$

| $Pe_0$ | $T_\infty$ | $\varepsilon_*^{(1)}$ | $\omega_*^{(1)}$ | $V_p$ |
|---|---|---|---|---|
| 0.0005 | −0.02753 | 0.09745 | 0.20976 | 1.02088 |
| 0.0045 | −0.07976 | 0.09737 | 0.20969 | 1.02088 |
| 0.01 | −0.1159 | 0.09730 | 0.20960 | 1.02085 |
| 0.1 | −0.3133 | 0.09612 | 0.20827 | 1.02034 |
| 1.0 | −0.6557 | 0.08573 | 0.19562 | 1.01622 |
| 10.0 | −0.9208 | 0.04448 | 0.12129 | 1.00254 |

are shown in Tables 7.3 and 7.4 for the case $\hat{\alpha}_T = 0$, while in Tables 7.5 and 7.6, results are given for the case $\hat{\alpha}_T = 1$. The variations of the critical number $\varepsilon_*^{(1)}$ with the anisotropy parameter $\alpha_4$ are shown in Tables 7.7, 7.8, 7.9, and 7.10.

**Table 7.6** The critical numbers $\varepsilon_*^{(1)}$ and the corresponding frequencies $\omega_*^{(1)}$ of the 2D neutral GTW A-modes (n=0) for the case $\hat{\alpha}_T = 1; \alpha_4 = 0$

| $Pe_0$ | $T_\infty$ | $\varepsilon_*^{(1)}$ | $\omega_*^{(1)}$ | $V_p$ |
|---|---|---|---|---|
| 0.0005 | −0.02753 | 0.12691 | 0.22813 | 1.02815 |
| 0.0045 | −0.07976 | 0.12675 | 0.22807 | 1.02817 |
| 0.01 | −0.1159 | 0.12667 | 0.22793 | 1.02809 |
| 0.1 | −0.3133 | 0.12480 | 0.22614 | 1.02732 |
| 1.0 | −0.6557 | 0.10894 | 0.20975 | 1.02086 |
| 10.0 | −0.9208 | 0.05093 | 0.12372 | 1.00278 |

**Table 7.7** The critical numbers $\varepsilon_*^{(1)}$ and the corresponding frequencies $\omega_*^{(1)}$ of the 2D neutral GTW A-modes ($n = 0$) for the case $\eta_0^2 = 0.001, \hat{\alpha}_T = 1$

| $\alpha_4$ | $\varepsilon_*^{(1)}$ | $\omega_*^{(1)}$ | $V_p$ |
|---|---|---|---|
| −0.050 | 0.1631 | 0.2524 | 1.0399 |
| −0.025 | 0.1366 | 0.2386 | 1.0329 |
| 0.000 | 0.1269 | 0.2281 | 1.0282 |
| 0.025 | 0.1210 | 0.2191 | 1.0244 |
| 0.050 | 0.1170 | 0.2110 | 1.0213 |
| 0.075 | 0.1139 | 0.2036 | 1.0187 |
| 0.100 | 0.1116 | 0.1967 | 1.0165 |
| 0.125 | 0.1097 | 0.1903 | 1.0146 |
| 0.150 | 0.1081 | 0.1842 | 1.0130 |
| 0.175 | 0.1069 | 0.1785 | 1.0115 |
| 0.200 | 0.1059 | 0.1730 | 1.0102 |
| 0.225 | 0.1051 | 0.1678 | 1.0091 |
| 0.250 | 0.1044 | 0.1628 | 1.0081 |
| 0.275 | 0.1039 | 0.1580 | 1.0072 |
| 0.300 | 0.1035 | 0.1534 | 1.0064 |

**Table 7.8** The critical numbers $\varepsilon_*^{(1)}$ and the corresponding frequencies $\omega_*^{(1)}$ of the 2D neutral GTW A-modes ($n = 0$) for the case $\eta_0^2 = 0.001, \hat{\alpha}_T = 0$

| $\alpha_4$ | $\varepsilon_*^{(1)}$ | $\omega_*^{(1)}$ | $V_p$ |
|---|---|---|---|
| −0.050 | 0.2306 | 0.3570 | 1.0399 |
| −0.025 | 0.1931 | 0.3374 | 1.0329 |
| 0.000 | 0.1794 | 0.3226 | 1.0282 |
| 0.025 | 0.1711 | 0.3098 | 1.0244 |
| 0.050 | 0.1654 | 0.2984 | 1.0213 |
| 0.075 | 0.1611 | 0.2879 | 1.0187 |
| 0.100 | 0.1577 | 0.2781 | 1.0165 |
| 0.125 | 0.1551 | 0.2691 | 1.0146 |
| 0.150 | 0.1529 | 0.2605 | 1.0194 |
| 0.175 | 0.1512 | 0.2524 | 1.0115 |
| 0.200 | 0.1498 | 0.2447 | 1.0102 |
| 0.225 | 0.1486 | 0.2373 | 1.0091 |
| 0.250 | 0.1477 | 0.2303 | 1.0081 |
| 0.275 | 0.1469 | 0.2235 | 1.0072 |
| 0.300 | 0.1463 | 0.2169 | 1.0064 |

**Table 7.9** The critical numbers $\varepsilon_*^{(1)}$ and the corresponding frequencies $\omega_*^{(1)}$ of the 2D neutral GTW S-modes ($n = 0$) for the case $\eta_0^2 = 0.001, \hat{\alpha}_T = 1$

| $\alpha_4$ | $\varepsilon_*^{(1)}$ | $\omega_*^{(1)}$ | $V_p$ |
|---|---|---|---|
| −0.050 | 0.1319 | 0.2435 | 1.0353 |
| −0.025 | 0.1057 | 0.2233 | 1.0262 |
| 0.000 | 0.0974 | 0.2097 | 1.0209 |
| 0.025 | 0.0926 | 0.1981 | 1.0170 |
| 0.050 | 0.0894 | 0.1878 | 1.0139 |
| 0.075 | 0.0870 | 0.1782 | 1.0114 |
| 0.100 | 0.0852 | 0.1692 | 1.0094 |
| 0.125 | 0.0838 | 0.1608 | 1.0077 |
| 0.150 | 0.0827 | 0.1526 | 1.0063 |
| 0.175 | 0.0818 | 0.1448 | 1.0051 |
| 0.200 | 0.0811 | 0.1372 | 1.0041 |
| 0.225 | 0.0805 | 0.1297 | 1.0034 |
| 0.250 | 0.0801 | 0.1224 | 1.0027 |
| 0.275 | 0.0798 | 0.1152 | 1.0021 |
| 0.300 | 0.0796 | 0.1080 | 1.0016 |

**Table 7.10** The critical numbers $\varepsilon_*^{(1)}$ and the corresponding frequencies $\omega_*^{(1)}$ of the 2D neutral GTW S-modes ($n = 0$) for the case $\eta_0^2 = 0.001, \hat{\alpha}_T = 0$

| $\alpha_4$ | $\varepsilon_*^{(1)}$ | $\omega_*^{(1)}$ | $V_p$ |
|---|---|---|---|
| −0.050 | 0.1863 | 0.3443 | 1.0352 |
| −0.025 | 0.1494 | 0.3159 | 1.0262 |
| 0.000 | 0.1378 | 0.2966 | 1.0209 |
| 0.025 | 0.1309 | 0.2802 | 1.0170 |
| 0.050 | 0.1263 | 0.2655 | 1.0139 |
| 0.075 | 0.1230 | 0.2520 | 1.0114 |
| 0.100 | 0.1205 | 0.2393 | 1.0094 |
| 0.125 | 0.1185 | 0.2273 | 1.0077 |
| 0.150 | 0.1169 | 0.2159 | 1.0063 |
| 0.175 | 0.1157 | 0.2048 | 1.0051 |
| 0.200 | 0.1147 | 0.1940 | 1.0041 |
| 0.225 | 0.1139 | 0.1835 | 1.0034 |
| 0.250 | 0.1133 | 0.1731 | 1.0027 |
| 0.275 | 0.1128 | 0.1629 | 1.0021 |
| 0.300 | 0.1125 | 0.1528 | 1.0016 |

The variations of the critical number $\varepsilon_*^{(1)}$ with the undercooling temperature $T_\infty$ are also shown inFigs. 7.10 and 7.11. It is noticed that in the small undercooling regime, the critical number $\varepsilon_*$ is insensitive to a change of undercooling, while in the large undercooling regime, as $\mathrm{Pe}_0 > 1.0$, the critical number $\varepsilon_*$ becomes a very sensitive function of the undercooling temperature. It decreases rapidly as the Péclet number $\mathrm{Pe}_0$ increases.

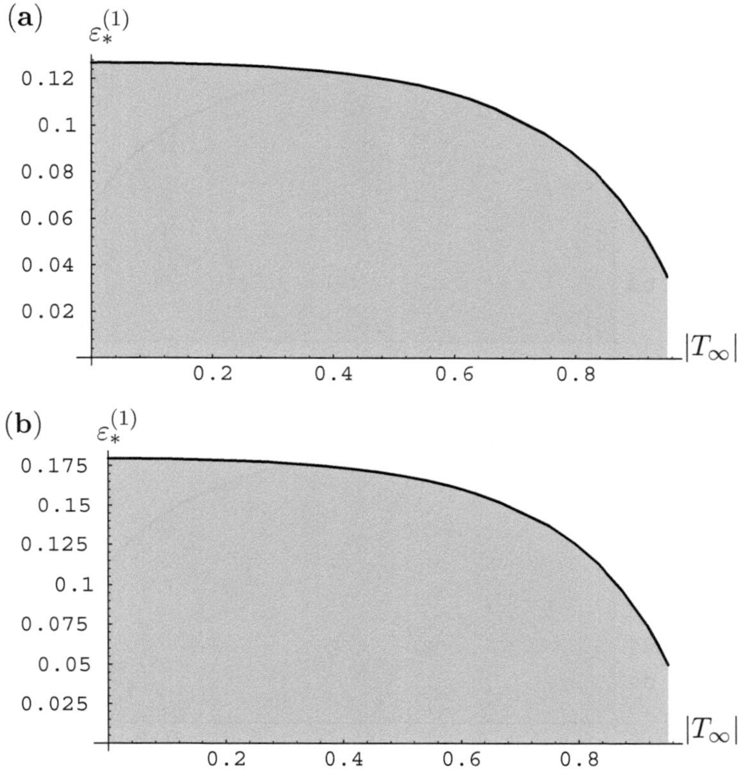


**Fig. 7.10** The variation of the critical number $\varepsilon_*^{(1)}$ for 2D neutral GTW A-modes (n=0) with the undercooling temperature $|T_\infty|$, for the cases (**a**) $\hat{\alpha}_T = 1$ (symmetric model); (**b**) $\hat{\alpha}_T = 0$ (one-sided model). The *shaded region* is the linearly unstable region of steady solutions

In Figs. 7.12 and 7.13, we show $\varepsilon_*^{(1)}$ and $\omega_*^{(1)}$ of GTW A-modes versus $\alpha_4$ for the cases $\eta_0^2 = 0.001, 1.0$ and $\hat{\alpha}_T = 0, 1$, respectively.

The numerical results for the cases $\eta_0^2 = 0.001$ and $\hat{\alpha}_T = 0, 1$ are also listed in Tables 7.4 and 7.6, respectively.

## 7.5 The Quantization Condition of Global Low-Frequency Modes

We turn to finding the quantization condition for the global modes with real eigenvalues $\sigma_0$. Note that in this case, the form of the tip smoothness condition is changed. Indeed, with the real spectrum of the eigenvalues, the physical solution

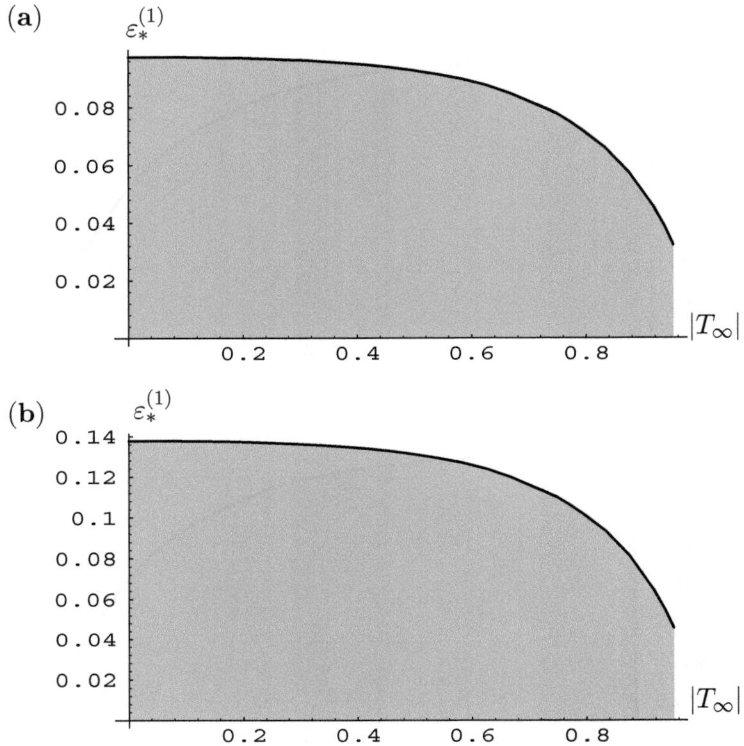


**Fig. 7.11** The variation of the critical number $\varepsilon_*^{(1)}$ for 2D neutral GTW S-modes (n=0) with the undercooling temperature $|T_\infty|$, for the cases (**a**) $\hat{\alpha}_{\mathrm{T}} = 1$ (symmetric model); (**b**) $\hat{\alpha}_{\mathrm{T}} = 0$ (one-sided model). The *shaded region* is the linearly unstable region of steady solutions

in the outer region is

$$\mathrm{Re}\{\tilde{h}(\xi,t)\} = \mathrm{Re}\{H(\xi)\}\exp\left\{\frac{\sigma_0 t}{\varepsilon\eta_0^2}\right\}. \tag{7.210}$$

As a consequence, the tip condition for symmetric modes becomes

$$\mathrm{Re}\{H'(0)\} = \mathrm{Re}\left\{\frac{id_1}{\varepsilon}k_0^{(1)}(0) + \frac{id_3}{\varepsilon}k_0^{(3)}(0)\right\} = 0. \tag{7.211}$$

Since the eigenmodes are determined to within an arbitrary multiplicative constant, without loss of generality, we can assume that $d_1$ is a positive real number and write

$$d_1 > 0, \quad d_3 = |d_3|\mathrm{e}^{\mathrm{i}\chi_0\pi}. \tag{7.212}$$

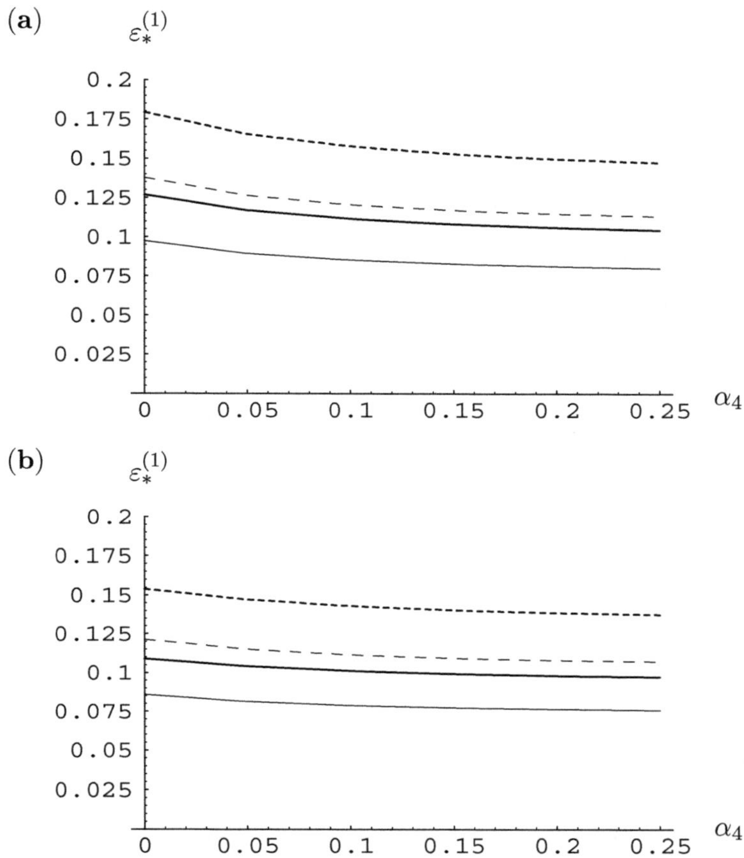


**Fig. 7.12** The variations of $\varepsilon_*^{(1)}$ of the neutral GTW A-modes ($n = 0$) in the first-order approximation with $\alpha_4$ for the cases (**a**) $\eta_0^2 = 0.001$, or $T_\infty = 0.03865$; (**b**) $\eta_0^2 = 1.0$, or $T_\infty = 0.6557$. The *thick lines* are for A-modes, while the *thin lines* are for S-modes. The *solid lines* are for the case $\hat{\alpha}_T = 1$, while the *dashed lines* are for $\hat{\alpha}_T = 0$

Noting that for a real eigenvalue $\sigma_0$, both $k_0^{(1)}(0)$ and $k_0^{(3)}(0)$ are real, the tip condition (7.211) yields $d_{3I} = 0$, so that the connection condition (7.202) can be written as

$$\frac{d_1}{d_3} = \left|\frac{d_1}{d_3}\right| \mathrm{e}^{-\mathrm{i}\chi_0\pi} = \mathrm{i}2\cos(\nu\pi)\mathrm{e}^{-\mathrm{i}\chi} \quad (\chi_0 = 0, 1), \tag{7.213}$$

or

$$\left|\frac{d_1}{d_3}\right| \mathrm{e}^{-\mathrm{i}\chi_0\pi} = 2\cos(\nu\pi)\mathrm{e}^{\mathrm{i}(2n+\frac{1}{2})\pi-\mathrm{i}\chi} \quad (\chi_0 = 0, 1). \tag{7.214}$$

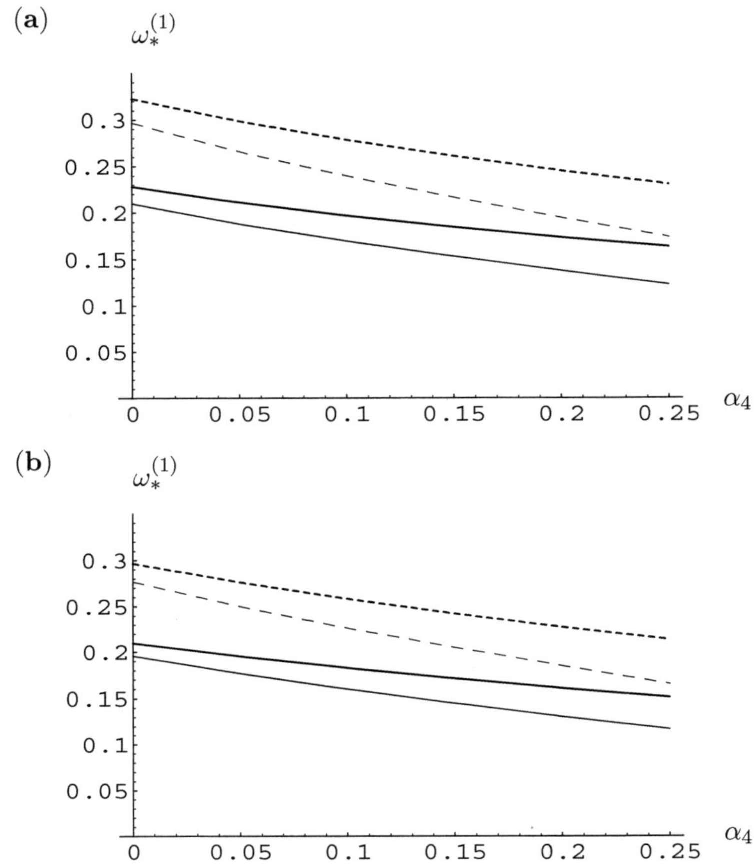


**Fig. 7.13** The variations of $\omega_*^{(1)}$ of the neutral GTW A-modes ($n = 0$) in the first-order approximation with $\alpha_4$ for the cases (**a**) $\eta_0^2 = 0.001$, or $T_\infty = 0.03865$; (**b**) $\eta_0^2 = 1.0$, or $T_\infty = 0.6557$. The *thick lines* are for A-modes, while the *thin lines* are for S-modes. The *solid lines* are for the case $\hat{\alpha}_T = 1$, while the *dashed lines* are for $\hat{\alpha}_T = 0$

In this case, $\nu = \frac{2}{7}$. It follows that

$$\mathrm{Re}\left[\frac{1}{\varepsilon}\int_0^{\xi_c}\left(k_0^{(1)} - k_0^{(3)}\right)\mathrm{d}\xi\right] = \left(2n + \frac{1}{2} + \chi_0\right)\pi \tag{7.215}$$

$$(n = 0, \pm 1, \pm 2, \ldots),$$

and

$$\left|\frac{d_1}{d_3}\right| = 2\cos\left(\frac{2\pi}{7}\right)\mathrm{e}^{\mathrm{Im}\{\chi\}}. \tag{7.216}$$

The quantization condition (7.215) determines the eigenvalues $\sigma_0^{(n)}$, while formula (7.216) determines the corresponding eigenfunctions.

For the antisymmetric modes, the tip condition is

$$\mathrm{Re}\{H(0)\} = \mathrm{Re}\{d_1 + d_3\} = 0, \tag{7.217}$$

which leads to

$$d_3 = -d_1 < 0. \tag{7.218}$$

In this case, the formula (7.202) leads to

$$\frac{d_1}{d_3} = -1 = \mathrm{i}2\cos\left(\frac{2\pi}{7}\right)\mathrm{e}^{-\mathrm{i}\chi}. \tag{7.219}$$

In summary, we obtain the quantization conditions

$$\begin{cases} \mathrm{Re}\left[\frac{1}{\varepsilon}\int_0^{\zeta_c}\left(k_0^{(1)} - k_0^{(3)}\right)\mathrm{d}\xi\right] = \left(2n + \frac{1}{2} + \chi_0\right)\pi \\ \qquad (n = 0, \pm 1, \pm 2, \ldots), \\ \left|\frac{d_1}{d_3}\right| = 2\cos\left(\frac{2\pi}{7}\right)\mathrm{e}^{\mathrm{Im}\{\chi\}}, \end{cases} \tag{7.220}$$

where

$$\chi_0 = \begin{cases} 0 \text{ or } 1, & \text{(for the S-modes);} \\ \cos(\chi_0\pi) = -2\cos\left(\frac{2\pi}{7}\right)\mathrm{e}^{\mathrm{Im}\{\chi\}}, & \text{(for the A-modes).} \end{cases} \tag{7.221}$$

The quantization condition (7.220) can be further simplified with the assumptions $|\sigma_0| \ll 1$ and $|\alpha_4| \ll 1$. For this purpose, we first derive an approximation of the singular point $\zeta_c$ as a function of $\sigma_0$ and $\alpha_4$. We start with the formula (7.183). Note that for the stability analysis, we are most interested in the quantization condition for the case of $\sigma_0 \to 0$ with fixed $\alpha_4$. Since in that case, $\zeta_c \to \mathrm{i}$, we may consider that $\hat{\delta}^2 \ll 2\hat{\alpha}_4 = \mathcal{O}(1)$. Hence the formula (7.183) can be simplified as

$$\hat{\sigma}_0 = -\frac{\mathrm{e}^{-\mathrm{i}\frac{11\pi}{8}}}{2^{\frac{3}{4}}3^{\frac{3}{2}}}\sqrt{\frac{(1+\hat{\alpha}_T)}{(1-\alpha_4)}}\frac{\hat{\delta}^{\frac{7}{4}}}{\left(\hat{\delta}^2 + 2\hat{\alpha}_4\right)^{\frac{1}{2}}} \approx \frac{\mathrm{e}^{\mathrm{i}\frac{3\pi}{8}}}{2^{\frac{5}{4}}3^{\frac{3}{2}}}\sqrt{\frac{(1+\hat{\alpha}_T)}{(1-\alpha_4)}}\frac{\hat{\delta}^{\frac{7}{4}}}{\hat{\alpha}_4^{\frac{1}{2}}},$$

or

$$\hat{\alpha}_4^{\frac{1}{2}}\hat{\sigma}_0 = \frac{\mathrm{e}^{\mathrm{i}\frac{3\pi}{8}}}{2^{\frac{5}{4}}3^{\frac{3}{2}}}\sqrt{1+\hat{\alpha}_T}\,\hat{\delta}^{\frac{7}{4}}.$$

This results in

$$\zeta_c + \mathrm{i} = \frac{2^{\frac{5}{7}} 3^{\frac{6}{7}}}{(1+\hat{\alpha}_{\mathrm{T}})^{\frac{2}{7}}} \mathrm{e}^{\mathrm{i}\frac{3\pi}{14}} \sigma_0^{\frac{4}{7}} \alpha_4^{\frac{2}{7}}. \tag{7.222}$$

Furthermore, from the local dispersion formula (7.113), one may derive the following expansion as $\sigma_0 \to 0$:

$$\begin{aligned} \bar{k}_0^{(3)} &= 0 + (1+\mathrm{i}\xi)\sigma_0 + \cdots, \\ \bar{k}_0^{(1)} &= \sqrt{\frac{\hat{S}}{(1+\hat{\alpha}_{\mathrm{T}})\bar{A}_s(\xi)}}(1-\mathrm{i}\xi) - \tfrac{1}{2}(1+\mathrm{i}\xi)\sigma_0 + \cdots, \end{aligned}$$

so that one has

$$\left(k_0^{(1)} - k_0^{(3)}\right) = \frac{1}{\sqrt{(1+\hat{\alpha}_{\mathrm{T}})}} \frac{(1-\mathrm{i}\xi)^{\frac{7}{4}}(1+\mathrm{i}\xi)^{\frac{5}{4}}}{f^{\frac{1}{2}}(\xi)} - \tfrac{3}{2}(1+\mathrm{i}\xi)\sigma_0 + \mathcal{O}(\sigma_0^2). \tag{7.223}$$

One may calculate the curve integral

$$\begin{aligned} \chi &= \frac{1}{\varepsilon}\int_0^{\zeta_c} \left(k_0^{(1)} - k_0^{(3)}\right)\mathrm{d}\xi \\ &= \left(\int_0^{-\mathrm{i}(a_1-0)} + \int_{(C_r)} + \int_{-\mathrm{i}(a_1+0)}^{-\mathrm{i}} + \int_{-\mathrm{i}}^{\zeta_c}\right)\left(k_0^{(1)} - k_0^{(3)}\right)\mathrm{d}\xi \\ &= \frac{1}{\varepsilon}\Big[(I) + (II) + (III) + (IV) + (V)\Big]. \end{aligned} \tag{7.224}$$

Here we have used the notation

$$\begin{aligned} &(I) + (II) + (III) + (IV) \\ &= \frac{1}{\sqrt{(1+\hat{\alpha}_{\mathrm{T}})}}\left(\int_0^{-\mathrm{i}(a_1-0)} + \int_{(C_r)} + \int_{-\mathrm{i}(a_1+0)}^{-\mathrm{i}} + \int_{-\mathrm{i}}^{\zeta_c}\right) Z(\xi)\mathrm{d}\xi, \end{aligned}$$

with integral taken along the path shown in Fig. 7.14. The line integral

$$(V) = -\int_0^{\zeta_c} \frac{3\sigma_0}{2}(1+\mathrm{i}\xi)\mathrm{d}\xi$$

is along the straight path in the complex $\zeta$-plane from $\zeta = 0$ to $\zeta = \zeta_c$ (Fig. 7.14). Furthermore, in the above we defined

$$Z(\xi) = \frac{(1-\mathrm{i}\xi)^{\frac{7}{4}}(1+\mathrm{i}\xi)^{\frac{5}{4}}}{(\xi^2+a_1^2)^{\frac{1}{2}}(\xi^2+a_2^2)^{\frac{1}{2}}} = \frac{(1-\mathrm{i}\xi)^{\frac{7}{4}}(1+\mathrm{i}\xi)^{\frac{5}{4}}}{(\xi+\mathrm{i}a_1)^{\frac{1}{2}}(\xi-\mathrm{i}a_1)^{\frac{1}{2}}(\xi+\mathrm{i}a_2)^{\frac{1}{2}}(\xi-\mathrm{i}a_2)^{\frac{1}{2}}}.$$

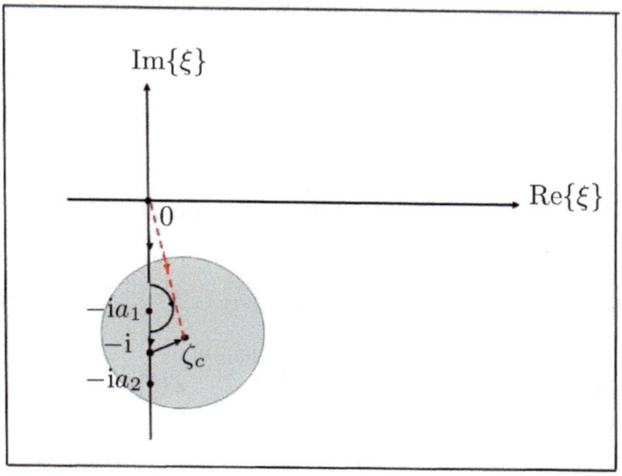


**Fig. 7.14** A sketch of the two contours used for the curve integrals in the complex $\zeta$-plane from $\zeta = 0$ to $\zeta = \zeta_c$

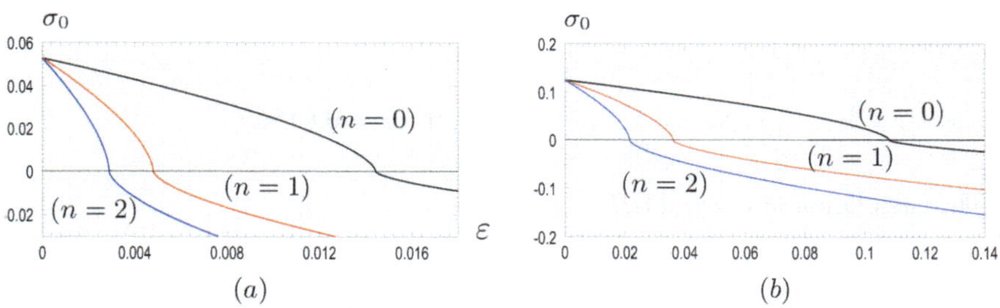


**Fig. 7.15** The real spectra of eigenvalues $\sigma_0$ corresponding to the LF instability mechanism versus $\varepsilon$ for the case $\hat{\alpha}_T = 1$, and (**a**) $\alpha_4 = 0.01$; (**b**) $\alpha_4 = 0.1$

The function $Z(\xi)$ has a branching point at $\xi = -\mathrm{i}a_1$.

To calculate (*I*) and (*III*), we let $\xi = -\mathrm{i}t$, and define

$$\hat{Z}(t) = \frac{(1-t)^{\frac{7}{4}}(1+t)^{\frac{5}{4}}}{(|a_1 - t|)^{\frac{1}{2}}(a_1+t)^{\frac{1}{2}}(a_2-t)^{\frac{1}{2}}(a_2+t)^{\frac{1}{2}}}.$$

Thus we may write

$$Z(\mathrm{i}t) = \begin{cases} \hat{Z}(t), & (0 \le t < a_1); \\ -\mathrm{i}\hat{Z}(t), & (a_1 < t \le 1). \end{cases}$$

The above-defined integrals are calculated as follows:

1. The calculation of the integral (*I*): Let $\xi = -\mathrm{i}t$,

$$\begin{aligned}(I) &= \frac{1}{\sqrt{(1+\hat{\alpha}_{\mathrm{T}})(1-\hat{\alpha}_4)}}\int_0^{-\mathrm{i}(a_1-0)} Z(\xi)\mathrm{d}\xi \\ &= -\frac{\mathrm{i}}{\sqrt{(1+\hat{\alpha}_{\mathrm{T}})(1-\hat{\alpha}_4)}}\int_0^{(a_1-0)} \hat{Z}(t)\mathrm{d}t = \mathrm{i}\mathcal{R}(\alpha_4).\end{aligned} \tag{7.225}$$

Noting that $\alpha_4 \ll 1$ and $a_1 = 1-\sqrt{2\alpha_4}, a_2 = 1+\sqrt{2\alpha_4}$, we derive

$$\begin{aligned}(I) = \mathrm{i}\mathcal{R}(\alpha_4) &= \frac{1}{\sqrt{(1+\hat{\alpha}_{\mathrm{T}})(1-\hat{\alpha}_4)}}\int_0^{-\mathrm{i}(a_1-0)} Z(\xi)\mathrm{d}\xi \\ &= -\frac{\mathrm{i}}{\sqrt{(1+\hat{\alpha}_{\mathrm{T}})}}\int_0^1 \frac{(1-t)^{\frac{7}{4}}(1+t)^{\frac{5}{4}}}{(|a_1-t|)^{\frac{1}{2}}(a_1+t)^{\frac{1}{2}}(a_2-t)^{\frac{1}{2}}(a_2+t)^{\frac{1}{2}}}\mathrm{d}t \\ &\approx \frac{\mathrm{i}}{\sqrt{(1+\hat{\alpha}_{\mathrm{T}})}}\int_0^1 (1-t)^{\frac{3}{4}}(1+t)^{\frac{1}{4}}\mathrm{d}t = \frac{\mathrm{i}\mathcal{R}_0}{\sqrt{(1+\hat{\alpha}_{\mathrm{T}})}},\end{aligned}$$

where

$$\mathcal{R}_0 = \int_0^1 (1-t)^{\frac{3}{4}}(1+t)^{\frac{1}{4}}\mathrm{d}t = 0.6156.$$

2. The calculation of integral (*III*):

$$\begin{aligned}(III) &= \frac{1}{\sqrt{(1+\hat{\alpha}_{\mathrm{T}})(1-\hat{\alpha}_4)}}\int_{(a_1+0)}^1 Z(\xi)\mathrm{d}\xi \\ &= -\frac{1}{\sqrt{(1+\hat{\alpha}_{\mathrm{T}})(1-\hat{\alpha}_4)}}\int_{(a_1+0)}^1 \hat{Z}(t)\mathrm{d}t.\end{aligned} \tag{7.226}$$

By setting $\hat{t} = 1-t$, we obtain

$$\hat{Z}(t) = \frac{(1-t)^{\frac{7}{4}}(1+t)^{\frac{5}{4}}}{(|a_1-t|)^{\frac{1}{2}}(a_1+t)^{\frac{1}{2}}(a_2-t)^{\frac{1}{2}}(a_2+t)^{\frac{1}{2}}},$$

$$\begin{aligned}(III) &= -\frac{1}{\sqrt{(1+\hat{\alpha}_{\mathrm{T}})(1-\hat{\alpha}_4)}}\int_{(a_1+0)}^1 \hat{Z}(t)\mathrm{d}t \\ &\approx -\frac{1}{\sqrt{(1+\hat{\alpha}_{\mathrm{T}})}}\int_0^{\sqrt{2\alpha_4}} \frac{\hat{t}^{\frac{7}{4}}(2-\hat{t})^{\frac{5}{4}}}{(\sqrt{2\alpha_4}-\hat{t})^{\frac{1}{2}}(\sqrt{2\alpha_4}+\hat{t})^{\frac{1}{2}}(2-\hat{t})}\mathrm{d}\hat{t} \\ &\approx -\int_0^{\sqrt{2\alpha_4}} \frac{\hat{t}^{\frac{7}{4}}(2)^{\frac{5}{4}}}{(2)(\sqrt{2\alpha_4}-\hat{t})^{\frac{1}{2}}(\sqrt{2\alpha_4}+\hat{t})^{\frac{1}{2}}}\mathrm{d}\hat{t}.\end{aligned} \tag{7.227}$$

Letting $s = \hat{t}/\sqrt{2\alpha_4}$, we obtain

$$\begin{aligned}
(III) &\approx -\frac{2^{\frac{1}{4}}}{\sqrt{(1+\hat{\alpha}_{\mathrm{T}})}}\int_0^{\sqrt{2\alpha_4}} \frac{\hat{t}^{\frac{7}{4}}}{(\sqrt{2\alpha_4}-\hat{t})^{\frac{1}{2}}(\sqrt{2\alpha_4}+\hat{t})^{\frac{1}{2}}}\mathrm{d}\hat{t} \\
&= -\frac{2^{\frac{1}{4}}}{\sqrt{(1+\hat{\alpha}_{\mathrm{T}})}}(2\alpha_4)^{\frac{7}{8}}\int_0^1 \frac{s^{\frac{7}{4}}}{\sqrt{1-s^2}}\mathrm{d}s \qquad (7.228)\\
&= -\frac{2^{\frac{9}{8}}\Gamma_0}{\sqrt{(1+\hat{\alpha}_{\mathrm{T}})}}\alpha_4^{\frac{7}{8}},
\end{aligned}$$

where

$$\Gamma_0 = \int_0^1 \frac{s^{\frac{7}{4}}}{\sqrt{1-s^2}}\mathrm{d}s = 0.8263.$$

3. The calculation of integral (*II*): Since $\zeta = -\mathrm{i}a_1$ is a branch point, we have

$$(II) = \int_{(C_r)} Z(\xi)\mathrm{d}\xi = 0.$$

4. The calculation of integral (*IV*) along the straight line from $\zeta = -\mathrm{i}$ to $\zeta = \zeta_c$:

$$(IV) = \int_{-\mathrm{i}}^{\zeta_c} \frac{(1-\mathrm{i}\xi)^{\frac{7}{4}}(1+\mathrm{i}\xi)^{\frac{5}{4}}}{(\xi+\mathrm{i}a_1)^{\frac{1}{2}}(\xi-\mathrm{i}a_1)^{\frac{1}{2}}(\xi+\mathrm{i}a_2)^{\frac{1}{2}}(\xi-\mathrm{i}a_2)^{\frac{1}{2}}}\mathrm{d}\xi$$

Let $\xi = -\mathrm{i} + t(\zeta_c + \mathrm{i})$. We have

$$\begin{aligned}
(IV) &\approx (\zeta_c+\mathrm{i})\int_0^1 \frac{\mathrm{e}^{-\mathrm{i}\frac{7\pi}{8}}t^{\frac{7}{4}}(\zeta_c+\mathrm{i})^{\frac{7}{4}}2^{\frac{5}{4}}}{\left[t^2(\zeta_c+\mathrm{i})^2+2\alpha_4\right]^{\frac{1}{2}}2\mathrm{i}}\mathrm{d}t \\
&= 2\mathrm{e}^{-\mathrm{i}\frac{11\pi}{8}}(\zeta_c+\mathrm{i})^{\frac{11}{4}}\int_0^1 \frac{t^{\frac{7}{4}}}{\left[t^2(\zeta_c+\mathrm{i})^2+2\alpha_4\right]^{\frac{1}{2}}}\mathrm{d}t.
\end{aligned}$$

In view of the assumption $\hat{\delta}^2 \ll 2\hat{\alpha}_4 = \mathcal{O}(1)$, one may write

$$(IV) \approx \sqrt{2}\mathrm{e}^{-\mathrm{i}\frac{11\pi}{8}}\frac{(\zeta_c+\mathrm{i})^{\frac{11}{4}}}{\alpha_4^{\frac{1}{2}}}\int_0^1 t^{\frac{7}{4}}\mathrm{d}t = -\frac{4\sqrt{2}\,\mathrm{e}^{-\mathrm{i}\frac{3\pi}{8}}}{11}\frac{(\zeta_c+\mathrm{i})^{11/4}}{\alpha_4^{1/2}}.$$

Moreover, using the formula (7.222), we obtain

$$(IV) \approx -\frac{144\times 2^{\frac{13}{28}}3^{\frac{5}{14}}}{11\times\left(1+\hat{\alpha}_{\mathrm{T}}\right)^{11/14}}\mathrm{e}^{\mathrm{i}\frac{27\pi}{28}}\sigma_0^{\frac{11}{14}}\alpha_4^{\frac{2}{7}}.$$

5. The calculation of integral $(V)$ along the straight line from $\zeta = 0$ to $\zeta = \zeta_c$: Set $\xi = t\zeta_c$. It follows that

$$
\begin{aligned}
(V) &= -\int_0^{\zeta_c} \frac{3\sigma_0}{2}(1 + \mathrm{i}\xi)\mathrm{d}\xi, \\
&= -\zeta_c \int_0^1 \frac{3\sigma_0}{2}(1 + \mathrm{i}t\zeta_c)\mathrm{d}t = -\zeta_c \frac{3\sigma_0}{2} \int_0^1 (1 + \mathrm{i}t\zeta_c)\mathrm{d}t \\
&= -\zeta_c \frac{3\sigma_0}{2}\left[1 + \frac{1}{2}\mathrm{i}\zeta_c\right].
\end{aligned}
$$

Using formula (7.222), we obtain

$$
\begin{aligned}
(V) &\approx \mathrm{i}\frac{9}{4}\sigma_0 - \frac{3}{2}A_c \mathrm{e}^{\mathrm{i}\frac{3\pi}{14}} \sigma_0^{\frac{11}{7}} \alpha_4^{\frac{2}{7}} \\
&= \mathrm{i}\frac{9}{4}\sigma_0 - \frac{3^{\frac{13}{7}}}{2^{\frac{2}{7}}(1 + \hat{\alpha}_{\mathrm{T}})^{\frac{2}{7}}} \mathrm{e}^{\mathrm{i}\frac{3\pi}{14}} \sigma_0^{\frac{11}{7}} \alpha_4^{\frac{2}{7}}.
\end{aligned}
\tag{7.229}
$$

We then have

$$
\begin{aligned}
\chi &= \frac{1}{\varepsilon} \int_0^{\zeta_c} \left(k_0^{(1)} - k_0^{(3)}\right)\mathrm{d}\xi \\
&= \frac{1}{\varepsilon}\Big[(I) + (II) + (III) + (IV) + (V)\Big] \\
&= \frac{1}{\varepsilon}\Bigg[\frac{\mathrm{i}}{\sqrt{(1 + \hat{\alpha}_{\mathrm{T}})}}\mathcal{R}_0 - \frac{2^{\frac{9}{8}}\Gamma_0}{\sqrt{(1 + \hat{\alpha}_{\mathrm{T}})}}\alpha_4^{\frac{7}{8}} \\
&\quad - \frac{144 \times 2^{\frac{13}{28}} 3^{\frac{5}{14}}}{11 \times \left(1 + \hat{\alpha}_{\mathrm{T}}\right)^{11/14}} \mathrm{e}^{\mathrm{i}\frac{27\pi}{28}} \sigma_0^{\frac{11}{14}} \alpha_4^{\frac{2}{7}} \\
&\quad + \mathrm{i}\frac{9}{4}\sigma_0 - \frac{3^{\frac{13}{7}}}{2^{\frac{2}{7}}(1 + \hat{\alpha}_{\mathrm{T}})^{\frac{2}{7}}} \mathrm{e}^{\mathrm{i}\frac{3\pi}{14}} \sigma_0^{\frac{11}{7}} \alpha_4^{\frac{2}{7}}\Bigg] + \text{(h.o.t.)}.
\end{aligned}
\tag{7.230}
$$

We obtain

$$
\begin{cases}
\mathrm{Re}\{\chi\} = \dfrac{1}{\varepsilon}\left[-\dfrac{2^{\frac{9}{8}}\Gamma_0}{\sqrt{(1 + \hat{\alpha}_{\mathrm{T}})}}\alpha_4^{\frac{7}{8}} + C_1\sigma_0^{\frac{11}{7}}\alpha_4^{\frac{2}{7}}\right] \\
\mathrm{Im}\{\chi\} = \dfrac{1}{\varepsilon}\left[\dfrac{1}{\sqrt{1 + \hat{\alpha}_{\mathrm{T}}}}\mathcal{R}_0 + \dfrac{9}{4}\sigma_0 + C_2\sigma_0^{\frac{11}{7}}\alpha_4^{\frac{2}{7}}\right],
\end{cases}
\tag{7.231}
$$

where

$$\begin{aligned}
\Gamma_0 &= \textstyle\int_0^1 \frac{s^{\frac{7}{4}}}{\sqrt{1-s^2}}\mathrm{d}s = 0.8263,\\
\mathcal{R}_0 &= \textstyle\int_0^1 (1-t)^{\frac{3}{4}}(1+t)^{\frac{1}{4}}\mathrm{d}t = 0.6156,\\
C_1 &= -\frac{144\times 2^{\frac{13}{28}}3^{\frac{5}{14}}}{11\times\left(1+\hat{\alpha}_{\mathrm{T}}\right)^{11/14}}\cos\left(\frac{27\pi}{28}\right) - \frac{3^{\frac{13}{7}}}{2^{\frac{2}{7}}(1+\hat{\alpha}_{\mathrm{T}})^{\frac{2}{7}}}\cos\left(\frac{3\pi}{14}\right) > 0,\\
C_2 &= -\frac{144\times 2^{\frac{13}{28}}3^{\frac{5}{14}}}{11\times\left(1+\hat{\alpha}_{\mathrm{T}}\right)^{11/14}}\sin\left(\frac{27\pi}{28}\right) - \frac{3^{\frac{13}{7}}}{2^{\frac{2}{7}}(1+\hat{\alpha}_{\mathrm{T}})^{\frac{2}{7}}}\sin\left(\frac{3\pi}{14}\right) < 0.
\end{aligned}$$

We calculate that for $\hat{\alpha}_{\mathrm{T}} = 0$,

$$C_1 = 18.27, \quad C_2 = -9.611,$$

while for $\hat{\alpha}_{\mathrm{T}} = 1$,

$$C_1 = 8.605, \quad C_2 = -7.164.s$$

The quantization condition (7.220) is therefore simplified as

$$\begin{cases}
-\dfrac{1.8021}{\sqrt{1+\hat{\alpha}_{\mathrm{T}}}}\alpha_4^{\frac{7}{8}} + C_1\sigma_0^{\frac{11}{7}}\alpha_4^{\frac{2}{7}} = \varepsilon\left(2n + \dfrac{1}{2} + \chi_0\right)\pi \\
\qquad (n = 0, \pm 1, \pm 2, \ldots), \\
\left|\dfrac{d_1}{d_3}\right| = 2\cos(2\pi/7)\mathrm{e}^{\mathrm{Im}\{\chi\}},
\end{cases} \tag{7.232}$$

where

$$\begin{cases}
\chi_0 = 0 \text{ or } 1, & \text{(for the S-modes);} \\
\cos(\chi_0\pi) = -2\cos(2\pi/7)\mathrm{e}^{\mathrm{Im}\{\chi\}}, & \text{(for the A-modes).}
\end{cases} \tag{7.233}$$

Since $\mathrm{Im}\{\chi\} > 0$ when $|\sigma_0| \ll 1$, we deduce from (7.233) that *the system allows only the LF S-modes; it does not permit LF A-modes.* For these S-modes, one may finally rewrite the quantization condition (7.232) in the form

$$\begin{cases}
C_1\sigma_0^{\frac{11}{7}}\alpha_4^{\frac{2}{7}} = \dfrac{1.8021}{\sqrt{1+\hat{\alpha}_{\mathrm{T}}}}\alpha_4^{\frac{7}{8}} - \varepsilon\left(n + \dfrac{1}{2}\right)\pi \\
\qquad (n = 0, 1, 2, \ldots), \\
\left|\dfrac{d_1}{d_3}\right| = 2\cos(2\pi/7)\mathrm{e}^{\mathrm{Im}\{\chi\}},
\end{cases} \tag{7.234}$$

where

$$\mathrm{sign}\left\{\frac{d_1}{d_3}\right\} = \begin{cases}
1 & \text{when } n \text{ is an even integer,} \\
-1 & \text{when } n \text{ is an odd integer.}
\end{cases} \tag{7.235}$$

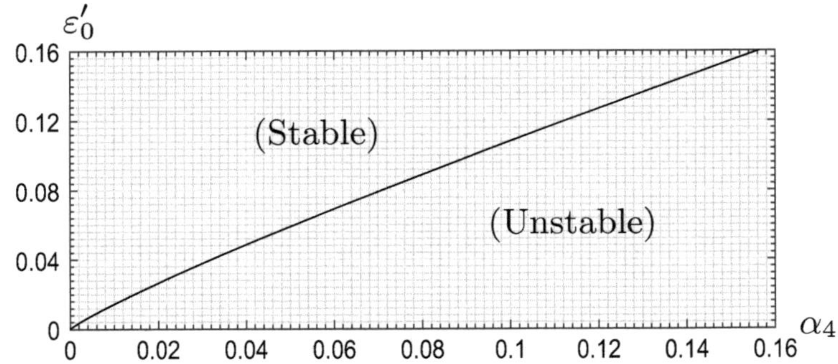


**Fig. 7.16** The critical number $\varepsilon_0'$ corresponding to the LF instability mechanism versus $\alpha_4$ for the case $\hat{\alpha}_T = 1$

The above quantization condition shows that as $\varepsilon \to 0$, $\alpha_4^{\frac{7}{8}} = \mathcal{O}(\varepsilon)$ and $\sigma_0^{\frac{11}{7}} \alpha_4^{\frac{2}{7}} = \mathcal{O}(\varepsilon)$. This is consistent with the asymptotic exponents (7.185) derived earlier.

For fixed $\varepsilon$ and $\alpha_4$, from the quantization condition (7.234), one can solve for a discrete set of eigenvalues $\{\sigma_n\}$ $(n = 0, 1, 2, 3, \ldots)$, as shown in Fig. 7.16. It is seen that there is a discrete set of neutral stable modes ($\sigma_n = 0$) corresponding to $\varepsilon_0' > \varepsilon_1' > \varepsilon_2' > \cdots > \varepsilon_n' > \cdots$. These neutral modes coincide with the steady needle crystal growth solutions predicted by the MSC theory.

For the real neutral mode $n = 0$, we obtain

$$\varepsilon = \varepsilon_s = \varepsilon_0' = \mathcal{B}_0 \alpha_4^{\frac{7}{8}}\,, \tag{7.236}$$

where $\mathcal{B}_0 = \frac{1.1473}{\sqrt{1+\hat{\alpha}_T}}$. This neutral mode ($n = 0$) is the most dangerous neutral mode of the LF mechanism.

There are $m+1$ purely growing modes and infinitely many decaying modes when $\varepsilon_{m+1} < \varepsilon < \varepsilon_m$. As $\varepsilon \to 0$, the first $n$ eigenvalues $\sigma_k$ $(k = 0, 1, 2, \ldots, n)$ all tend to the upper limit $\sigma'_{\max}$, which can be calculated as

$$\sigma'_{\max} = \alpha_4^{\frac{3}{8}} \left( \frac{1.8201}{C_1 \sqrt{1 + \hat{\alpha}_T}} \right)^{\frac{7}{11}}. \tag{7.237}$$

When $\varepsilon > \varepsilon_0'$, all modes are purely decaying. Thus the system will be stable, provided no other mechanism can cause the instability. As $\alpha_4 \to 0$, the low-frequency stability disappears as the upper limit $\sigma'_{\max}$ approaches 0.

The above low-frequency (LF) instability was first detected by Kessler and Levine numerically [4] and investigated by Bensimon et al. analytically [5] in 1986. It was concluded in the numerical work by Kessler and Levine that the steady needle solution with $\varepsilon = \varepsilon_0'$ was neutrally stable and the system did not allow a growing oscillatory mode (see [4]). This conclusion was subsequently accepted by many researchers as one of the key statements of the MSC theory. This conclusion, however, is incorrect. As indicated by Xu early in 1989–1991 [3, 6–9], the system is also subject to GTW instability. The steady needle solution with $\varepsilon = \varepsilon_0'$ is neutrally stable only under the LF instability mechanism; *it may become globally unstable due to the presence of GTW instability.*

The physical feature of LF and GTW instability are different and may be clarified as follows. Assume that a small initial perturbation of the interface shape is given. Due to the low-frequency instability, this initial perturbation of the interface shape will be purely growing or decaying without propagating and oscillating. As a consequence, the perturbed interface will diverge from or be restored to its original shape without oscillation, depending on whether it is unstable or stable. On the other hand, under the GTW instability mechanism, the initial perturbation will propagate and oscillate as a traveling wave with either growing or decaying amplitude. The perturbed interface will either depart from the original shape with oscillation if it is unstable under the GTW mechanism, or be restored to its original shape with oscillation if it is stable under the GTW mechanism.

## 7.6 The Selection Conditions for 2D Dendritic Growth

In the above, we have studied the stability properties of the nonclassical steady, or "nearly" steady, needle solutions. It is known that the system of dendritic growth with anisotropic surface tension has two coexisting instability mechanisms: the GTW and LF instabilities. Based on these instability mechanisms, the selection criterion for dendritic growth at the late stage of evolution can be derived naturally.

In Fig. 7.17, we plot the neutral curves $\{\gamma_0\}$ of the most dangerous GTW S-mode ($n = 0$) in the first-order approximation for cases of undercooling temperature $|T_\infty| = 0.01, 0.5$ and $\hat{\alpha}_T = 1.0$. Here, we assume that the system does not allow antisymmetric perturbations. In the same figure, we also plot the neutral curve $\{C_0\}$ of the most dangerous LF S-mode ($n = 0$) in the zeroth-order approximation. The shaded region below the two curves in Fig. 7.17 is the unstable region, whereas the remaining region is the globally stable region. The two types of neutral curve intersect each other at some critical number $\alpha_c$. It is calculated that $\alpha_c = 0.078$ as $|T_\infty| = 0.01$, and $\alpha_c = 0.075$ as $|T_\infty| = 0.5$. The variation of the critical number $\alpha_c$ with undercooling is also shown in Fig. 7.18a.

If the system allows both symmetric and antisymmetric perturbations, the neutral curves of the GTW S-mode ($n = 0$) shown in Fig. 7.17 should be replaced by the neutral curves of the GTW A-mode ($n = 0$). Then the corresponding critical numbers will be changed to $\alpha_c = 0.115$ for $|T_\infty| = 0.01$, and $\alpha_c = 0.108$ for $|T_\infty| = 0.5$. The variation of the critical number $\alpha_c$ with undercooling for such a system is shown in Fig. 7.18b. Thus, we can examine the stability properties for the basic state of the system in the $(\alpha_4, \varepsilon)$-plane.

For a given operation condition, $\alpha_4$ is fixed, whereas the parameter $\varepsilon$ associated with the tip velocity of the dendrite under investigation may slowly vary. Hence, the representative point of the basic state in the $(\alpha_4, \varepsilon)$-plane moves with time. In

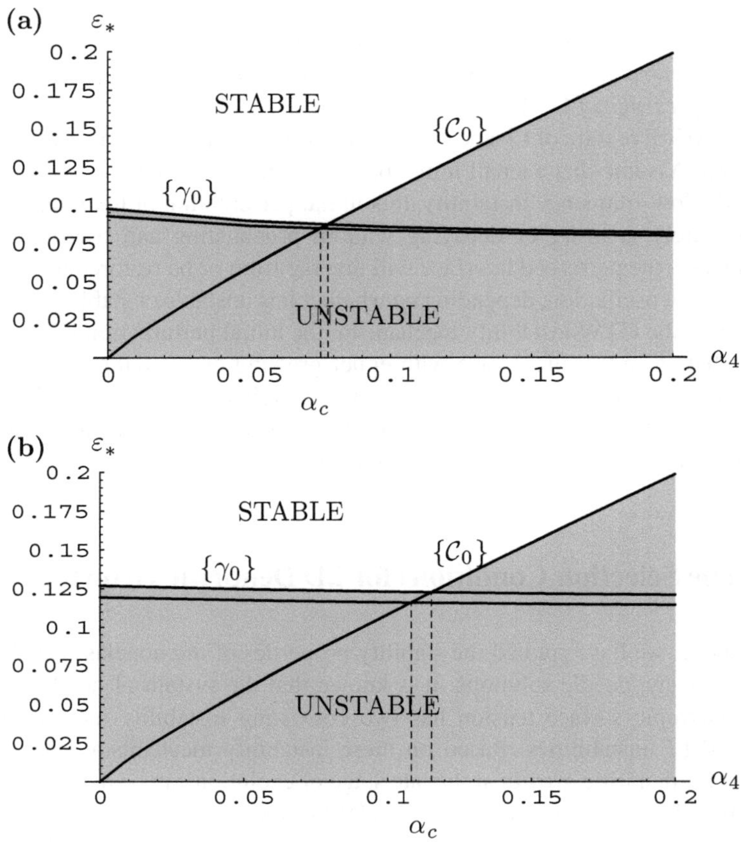


**Fig. 7.17** The neutral curves $\{\mathcal{C}_0\}$, $\{\gamma_0\}$ and the stability diagram of dendritic growth in the $(\alpha_4, \varepsilon)$-plane: (**a**) For the system prohibiting antisymmetric perturbations, which gives the critical numbers $\alpha_c = 0.075$ and 0.078, corresponding to $T_\infty = 0.01$ and 0.5. (**b**) For the system allowing both symmetric and antisymmetric perturbations, which gives the critical numbers $\alpha_c = 0.115$ and 0.108, corresponding to $T_\infty = 0.01$ and 0.5. The neutral curves of the GTW mechanism, $\{\gamma_0\}$, are given for the cases $\hat{\alpha}_T = 1$ and the undercooling $|T_\infty| = 0.01, 0.5$ from top to bottom. The *shaded region* is unstable, while the *unshaded region* is stable

general, as a dissipative dynamic system, as $t \to \infty$, dendritic growth will settle down in a low-dimensional subspace of the state space. In other words, it can be expected to exhibit one of the following three behaviors:

1. It may approach a steady solution, describing a smoothly growing needle and representing a fixed-point state in the state space.
2. It may approach a time-periodic solution describing an oscillatory growing dendrite, and representing a limit cycle in the state space.
3. It may have a chaotic limit state. The solution evolves with many short time scales and exhibits a chaotic pattern.

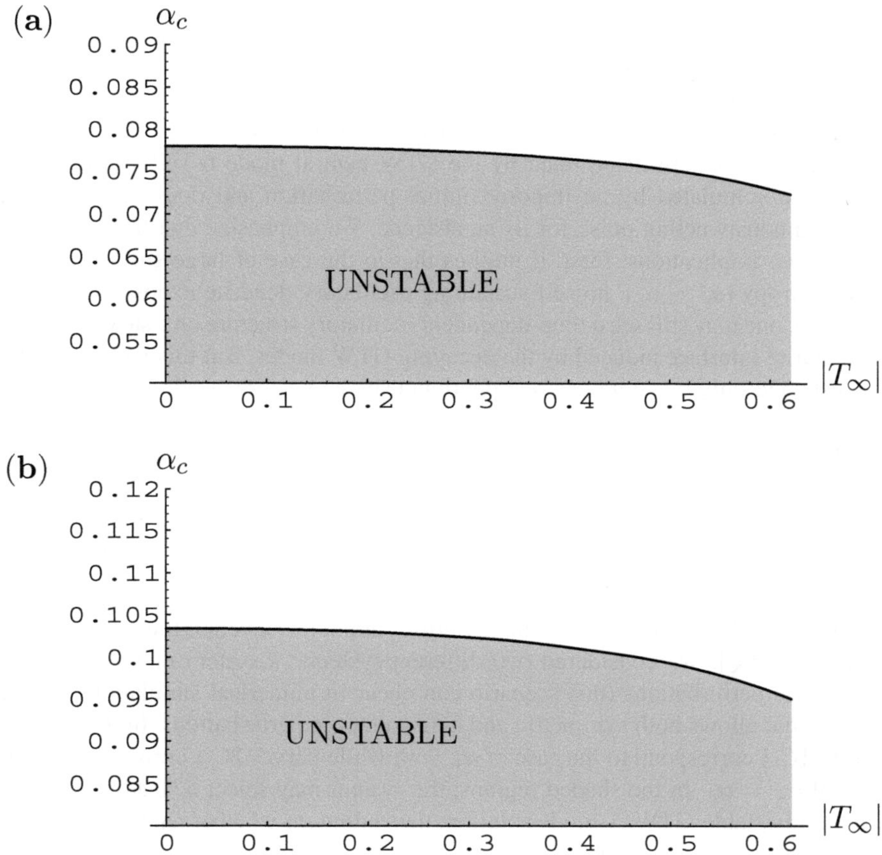


**Fig. 7.18** The variation of the critical number $\alpha_c$ with the undercooling temperature $T_\infty$: (**a**) For the system prohibiting antisymmetric perturbations; (**b**) For the system allowing both symmetric and antisymmetric perturbations

On the other hand, one can imagine that when the growth process in an experiment begins, the state point of the system must be located in the unstable region with a lower tip-growth rate. Otherwise, the state point will not move. Hence, if the system approaches a steady limit solution as $t \to \infty$, then this fixed point in the parameter plane must be on the neutral LF modes ($n = 0, 1, \ldots$) and must be on the neutral curve $\{\mathcal{C}_0\}$.

With the above consideration, we draw the following conclusions:

1. If the dendritic growth system exhibits a steady pattern as $t \to \infty$, then the corresponding steady limit solution must be on the neutral curve $\{\mathcal{C}_0\}$, and it occurs only when $\alpha_4 \geq \alpha_c$. In other words, for small anisotropy of surface tension, when $\alpha_4 < \alpha_c$, the steady needle solution is not observable due to

the existence of a number of growing oscillatory GTW modes. The selection criterion given by the MSC theory is apparently not applicable in this range.

2. If the dendrite exhibits a time-periodic oscillatory pattern as $t \to \infty$, then within a linear theory, the limit solution must correspond to a GTW neutral mode on the neutral curve $\{\gamma_0\}$. This occurs only when $0 \le \alpha_4 \le \alpha_c$. The unsteady oscillatory pattern determined by the GTW neutral mode is self-sustaining. It can be stimulated by an imposed initial perturbation and does not require a continuously acting noise for its persistence. We emphasize that this statement has two implications. First, it implies that in the case of large surface tension anisotropy ($\alpha_4 > \alpha_c$), no self-sustaining oscillatory dendrite is possible. In this range, one may still see a time-dependent oscillatory structure on a steady smooth dendrite interface induced by the decaying GTW modes. But this pattern can be sustained only by some external continuously acting force such as noise. Once these external forces cease, such an oscillatory structure disappears with time. On the other hand, when the criterion $\alpha_4 < \alpha_c$ is satisfied, the time-dependent oscillatory pattern may persist in the system described by the most dangerous neutral mode without the continuously acting external force or noises.
3. If the dendritic growth system exhibits a chaotic pattern as $t \to \infty$, the system evidently remains in the unstable region.

In Fig. 7.19 we show the selection criterion parameter $\varepsilon_*$ versus the undercooling parameter $|T_\infty|$. We considered two different systems: a system that allows only symmetric perturbations (this scenario can occur in numerical simulations) and a system that allows both symmetric and antisymmetric perturbations. In the figures, curves $\{C_1\}$ correspond to the case of $\alpha_4 = 0$, while curves $\{C_2\}$ correspond to the case of $\alpha_4 = \alpha_c$. In the shaded regions, the system may select a steady needle or a neutrally stable GTW A-mode solution, depending on whether $\alpha_4 > \alpha_c$ or $\alpha_4 < \alpha_c$. Hence, we call these narrow regions *coexistence belts*. Above the coexistence belt, only the steady solution can be selected, while below the coexistence belt, the system will exhibit a complex chaotic pattern, with no steady needle solution or oscillatory state solution.

The above conclusions appear to be in good agreement with experimental observations. So far, most experiments on dendritic growth from a pure melt are three-dimensional. These experimental results show little correlation between the selected values $\varepsilon = \varepsilon_*$ and the anisotropy [10]. These results, in agreement with the theory, suggest that the anisotropy for these materials lies in the range $0 \le \alpha_4 \le \alpha_c$ and that realistic dendritic growth is attracted to the neutral GTW mode as $t \to \infty$.

In recent years, several groups of researchers have performed numerical simulations of two-dimensional dendritic growth as an initial value problem [11, 12].The results of IFW theory are also in agreement with those numerical results.

Ihle and Müller-Krumbhaar (1994) have performed numerical simulations for the large undercooling case, $|T_\infty| = 0.5$, with various anisotropies [11]. Due to the numerical difficulty, they were unable to conduct a numerical simulation for the small undercooling case. They show that when $\alpha_4 = 0.15$, the numerical solutions are attracted to the LF neutral mode. At the later stages of evolution, the dendrite

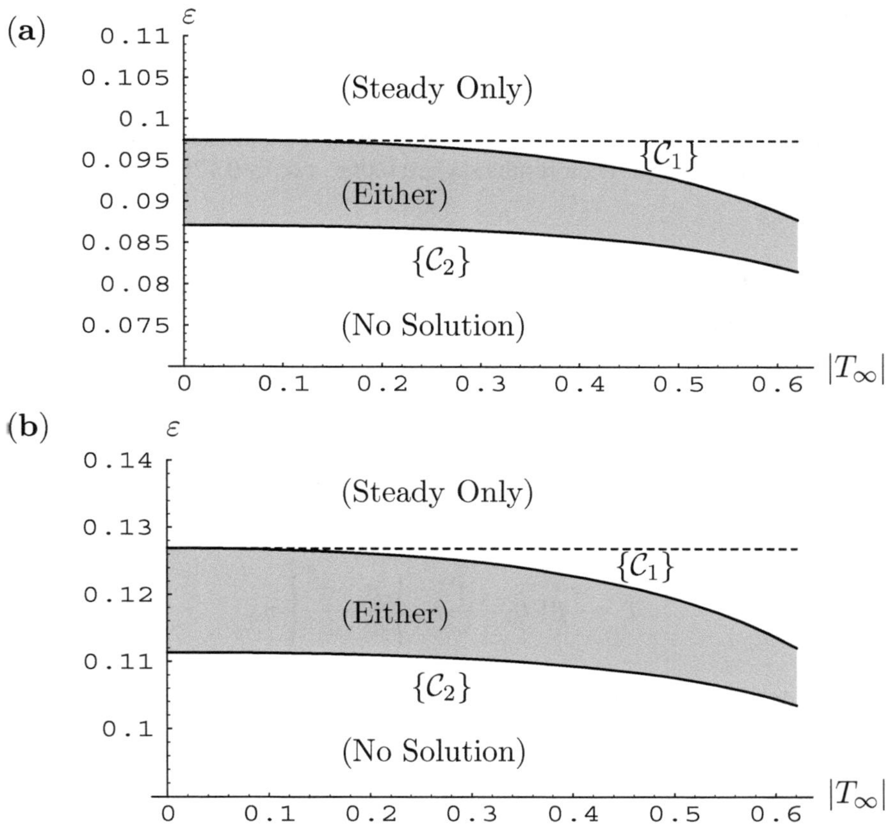


**Fig. 7.19** The selection criterion parameter $\varepsilon_*$ versus undercooling parameter $|T_\infty|$: (**a**) For the system prohibiting antisymmetric perturbations; (**b**) For the system allowing both symmetric and antisymmetric perturbations. In the figures, curves $\{C_1\}$ correspond to the case of $\alpha_4 = 0$, while curves $\{C_2\}$ correspond to the case of $\alpha_4 = \alpha_c$. The *shaded regions* are the coexistence belts, where the system may select a steady needle or a neutrally stable GTW A-mode solution, depending on whether $\alpha_4 > \alpha_c$ or $\alpha_4 < \alpha_c$. Above the coexistence belts, only a steady solution can be selected, while below the coexistence belt, no steady needle solution or oscillatory state solution can be selected

has a smooth interface with no side-branching. Then when $\alpha_4 = 0.1$, a strong time-oscillatory instability occurs, and the tip radius has up to 10% fluctuation with time. When $\alpha_4 = 0.05$, the initially steady needle solution undergoes "strongly irregular side-branching and large fluctuations in tip radius and velocity." At the end of the computation for this case, the tip velocity is still noticeably changing, and the numerical results on the tip radius $R_{\text{tip}}$ and $V_{\text{tip}}$ show the inconsistency with the scaling law (7.236). Although these numerical simulation results have been given various interpretations, they may be properly interpreted as evidence for the existence of the GTW instability mechanism, and they suggest that the system is dominated by this GTW mechanism when $\alpha_4$ is smaller than a critical number in the range $0.05 < \alpha_c \leq 0.1$. This critical number is in reasonably good agreement with the theoretical prediction $\alpha_c = 0.075$ for the case of large undercooling ($|T_\infty| = 0.5$). The numerical simulations [12] conducted by Brener, Müller-Krumbhaar, Saito, and Schiraishi (1994) in terms of the quasistatic approximation show the same scenario. Their results suggest $0.068 < \alpha_c \leq 0.125$.

## 7.7 The Effect of Kinetic Attachment at the Interface on Dendritic Growth

So far, we studied only the effect of anisotropic surface tension. It should be noted that in the system of solidification, the interface temperature is not affected only by the surface tension, it may be affected by the interface kinetics as well. With the inclusion of the effect of the interfacial kinetics, the Gibbs-Thomson condition is modified to the form

$$T = -\beta\Gamma U_{\mathrm{I}} - \frac{\Gamma}{\eta_0^2}\mathcal{K}_2\left\{\frac{\mathrm{d}}{\mathrm{d}\xi}, \frac{\mathrm{d}^2}{\mathrm{d}\xi^2}\right\}\eta_s, \tag{7.238}$$

where $\beta$ is the kinetic constant, while $U_{\mathrm{I}}$ is the local growth speed of the interface. It can be derived that in the parabolic coordinate system,

$$U_{\mathrm{I}} = \frac{1}{\sqrt{(\xi^2+\eta_s^2)(1+(\frac{\partial\eta_s}{\partial\xi})^2)}}\left[\frac{\partial(\xi\eta_s)}{\partial\xi} + \eta_0^2(\xi^2+\eta_s^2)\frac{\partial\eta_s}{\partial t}\right]. \tag{7.239}$$

As we have illustrated in the previous section, with the anisotropy, the coefficient of surface tension $\gamma$ might be described in the form

$$\begin{cases} \gamma = \hat{\gamma}\gamma_s(\hat{\theta}), \\ \gamma_s(\hat{\theta}) = 1 + \gamma_m\cos(m\hat{\theta}), \end{cases}$$

and $\Gamma = \hat{\Gamma}A_s(\hat{\theta})$. In the same manner, one may describe the kinetic constant $\beta$. We suppose that $\beta$ has not only an isotropic component, but also an anisotropic component. It can be written in the form

$$\beta = \hat{\beta}B_s(\hat{\theta}), \tag{7.240}$$

where

$$\hat{\beta} = \frac{\hat{\ell}_{\mathrm{k}}}{\hat{\ell}_{\mathrm{c}}}, \tag{7.241}$$

$\hat{\beta}$ is the isotropic kinetic constant, $\hat{\ell}_{\mathrm{k}}$ is the isotropic kinetic length, defined as

$$\hat{\ell}_{\mathrm{k}} = \frac{\hat{\beta}_0 \kappa_{\mathrm{T}} c_{\mathrm{p}} \rho}{\Delta H}, \tag{7.242}$$

and $\hat{\beta}_0$ is the *isotropic kinetic coefficient*. Moreover, similar to $A_s(\hat{\theta}) = \big[1 - \alpha_4 \cos(m\hat{\theta})\big]$, one may write

$$B_s(\hat{\theta}) = \big[1 - \beta_4 \cos(m\hat{\theta})\big], \tag{7.243}$$

where $\beta_4$ is the *anisotropic kinetic coefficient*. In terms the above notation, the Gibbs–Thomson condition may be written in the form

$$T = -\hat{\beta}\varepsilon^2 \eta_0^4 B_s(\hat{\theta}) U_{\mathrm{I}} - \varepsilon^2 \eta_0^2 A_s(\hat{\theta}) \mathcal{K}_2 \left\{ \frac{\mathrm{d}}{\mathrm{d}\xi}, \frac{\mathrm{d}^2}{\mathrm{d}\xi^2} \right\} \eta_{\mathrm{s}}. \tag{7.244}$$

We further define

$$\widehat{B} = \hat{\beta} \eta_0^2 \tag{7.245}$$

as the *isotropic kinetic parameter* to measure the effect of the isotropic kinetic attachment, and we suppose that $\widehat{B} \ll 1$ is a small parameter.

With the above definitions, the Gibbs–Thomson condition (7.244) is finally written as

$$T = -\varepsilon^2 \eta_0^2 \widehat{B} B_s(\hat{\theta}) U_{\mathrm{I}} - \varepsilon^2 \eta_0^2 A_s(\hat{\theta}) \mathcal{K}_2 \left\{ \frac{\mathrm{d}}{\mathrm{d}\xi}, \frac{\mathrm{d}^2}{\mathrm{d}\xi^2} \right\} \eta_{\mathrm{s}}. \tag{7.246}$$

In view of the formulas (7.8)–(7.13), we can write

$$B_s(\hat{\theta}) = \bar{B}_s(\xi) + \tilde{B}_{s0}(\xi)\tilde{\eta}_{\mathrm{s}}(\xi) + \tilde{B}_{s1}(\xi)\tilde{\eta}_{\mathrm{s}}'(\xi) \;, \tag{7.247}$$

where

$$\begin{aligned} \bar{B}_s(\xi) &= \frac{g(\xi)}{S^4}, \\ \tilde{B}_{s0}(\xi) &= -16\beta_4 \frac{\xi^2(1-\xi^2)}{S^6}, \\ \tilde{B}_{s1}(\xi) &= -16\beta_4 \frac{\xi(1-\xi^2)}{S^4}, \end{aligned} \tag{7.248}$$

$$\begin{aligned} g(\xi) &= (1-\beta_4)(1+\xi^2)^2 + 8\beta_4\xi^2 \\ &= (1-\beta_4)(\xi^2+b_1^2)(\xi^2+b_2^2), \end{aligned} \tag{7.249}$$

and

$$\begin{cases} b_1^2 = 1 + 3\beta_4 - 2\sqrt{2}\beta_4^{\frac{1}{2}}(1+\beta_4)^{\frac{1}{2}} \\ b_2^2 = 1 + 3\beta_4 + 2\sqrt{2}\beta_4^{\frac{1}{2}}(1+\beta_4)^{\frac{1}{2}}. \end{cases} \tag{7.250}$$

Especially, as $\beta_4 \ll 1$,

$$b_1 \approx 1 - \sqrt{2\beta_4}, \quad b_2 \approx 1 + \sqrt{2\beta_4}. \tag{7.251}$$

The investigation of the effect of kinetic attachment on instability mechanisms of dendritic growth is of both theoretical significance and practical interest. Up to now, experimental investigations on this topic have been quite limited, due to the fact that the quantitative values of the parameters related to the kinetics are very hard to measure. In accordance with this situation, the theoretical analyses on this topic have been quite preliminary [13].

### 7.7.1 Linear Perturbed System Around the Basic State

The basic states for the system with the effect of interfacial kinetics remain unchanged in the zeroth-order approximation. As before, one may separate the general unsteady solutions into two parts:

$$\begin{aligned} T &= T_B + \tilde{T}(\xi, \eta, t, \varepsilon), \\ T_S &= T_{SB} + \tilde{T}_S(\xi, \eta, t, \varepsilon), \\ \eta_s &= \eta_B + \tilde{h}(\xi, t, \varepsilon)/\eta_0^2, \end{aligned} \tag{7.252}$$

and linearize the general unsteady solution around the basic state. The problem is to determine the changes in the perturbed states caused by the kinetic effect.

We point out that the linearized perturbed system with the effect of the interface kinetics is the same as what was derived in the previous section, except for the Gibbs–Thomson interface condition. We list the set of interface conditions for the current system as follows: On the interface of the Ivantsov solution, $\eta = 1$,

$$(\tilde{T} - \tilde{T}_S) + \varepsilon^2 \eta_1 \frac{\partial(\tilde{T} - \tilde{T}_S)}{\partial \eta} = \tilde{h} + (\text{h.o.t}), \tag{7.253}$$

$$\begin{aligned} \tilde{T}_S + \varepsilon^2 \eta_1 \frac{\partial \tilde{T}_S}{\partial \eta} &= \varepsilon^2 \frac{\bar{A}_s(\xi)}{\hat{S}(\xi)} \left[ \frac{\partial^2 \tilde{h}}{\partial \xi^2} + \frac{\xi}{\hat{S}^2(\xi)} \frac{\partial \tilde{h}}{\partial \xi} - \frac{1}{\hat{S}^2(\xi)} \tilde{h} \right] \\ &= -\varepsilon^2 \frac{1}{\hat{S}^3} \left( \tilde{A}_{s0} \tilde{h} + \tilde{A}_{s1} \frac{\partial \tilde{h}}{\partial \xi} \right) - \varepsilon^2 \frac{\widehat{B}\bar{B}_s}{\hat{S}} \left[ \tilde{h} + \xi \frac{\partial \tilde{h}}{\partial \xi} + S^2 \frac{\partial \tilde{h}}{\partial t} \right] \\ &\quad + (\text{h.o.t}), \end{aligned} \tag{7.254}$$

and

$$\begin{aligned} &\frac{\partial}{\partial \eta} \left( \tilde{T} - \hat{\alpha}_T \tilde{T}_S \right) + \varepsilon^2 \eta_1 \frac{\partial^2}{\partial \eta^2} \left( \tilde{T} - \hat{\alpha}_T \tilde{T}_S \right) + \eta_0^2 \hat{S}^2(\xi) \frac{\partial \tilde{h}}{\partial t} + \xi \frac{\partial \tilde{h}}{\partial \xi} \\ &\quad + \varepsilon (1 + \eta_0^2) \tilde{h} + (\text{h.o.t}) = 0. \end{aligned} \tag{7.255}$$

One can see that the linear perturbed system now involves the parameters $\{\varepsilon, \widehat{B}, \alpha_4, \beta_4, T_\infty\}$. One may solve the related EVP problem by the MVE method as in the previous section. The converted perturbed system with the multiple variables $(\xi, \eta, \xi_+, \eta_+, t_+)$ is the same as that derived in the previous section, except for the interface conditions, which are modified to include the kinetics effect. The modified interface conditions are as follows: at $\eta_+ = 0$ or $\eta = 1$,

1. the thermodynamic equilibrium condition

$$\tilde{T} = \tilde{T}_S + \tilde{h} - \varepsilon \eta_1 \bar{k} \frac{\partial}{\partial \eta_+} (\tilde{T} - \tilde{T}_S) + \mathcal{O}(\varepsilon^2), \tag{7.256}$$

2. the Gibbs–Thomson condition

$$\begin{aligned} \tilde{T}_S = &\frac{\bar{A}_s}{\hat{S}(\xi)} \left\{ \left( \bar{k}^2 \frac{\partial^2}{\partial \xi_+^2} + 2\varepsilon \bar{k} \frac{\partial^2}{\partial \xi_+ \partial \xi} + \varepsilon \bar{k}' \frac{\partial}{\partial \xi_+} \right) + \varepsilon \frac{\xi \bar{k}}{\hat{S}^2(\xi)} \frac{\partial}{\partial \xi_+} \right\} \tilde{h} \\ &- \varepsilon \frac{\bar{k}}{\hat{S}^3} \tilde{A}_{s1} \frac{\partial \tilde{h}}{\partial \xi_+} - \varepsilon \frac{\widehat{B}\bar{B}_s}{\hat{S}} \left[ \varepsilon \tilde{h} + \xi \left( \bar{k} \frac{\partial \tilde{h}}{\partial \xi_+} + \varepsilon \frac{\partial \tilde{h}}{\partial \xi} \right) + \hat{S}^2 \frac{\partial \tilde{h}}{\partial t_+} \right] \\ &- \varepsilon \eta_1 \bar{k} \frac{\partial \tilde{T}_S}{\partial \eta_+} + \mathcal{O}(\varepsilon^2), \end{aligned} \tag{7.257}$$

3. the heat balance condition

$$\bar{k}\left(\frac{\partial \tilde{T}}{\partial \eta_+} - \hat{\alpha}_{\mathrm{T}}\frac{\partial \tilde{T}_{\mathrm{S}}}{\partial \eta_+}\right) + \varepsilon\eta_1\bar{k}^2\frac{\partial^2}{\partial \eta_+^2}(\tilde{T} - \hat{\alpha}_{\mathrm{T}}\tilde{T}_{\mathrm{S}}) + \varepsilon\left(\frac{\partial \tilde{T}}{\partial \eta} - \hat{\alpha}_{\mathrm{T}}\frac{\partial \tilde{T}_{\mathrm{S}}}{\partial \eta}\right)$$

$$+\hat{S}^2(\xi)\frac{\partial \tilde{h}}{\partial t_+} + \xi\left(k\frac{\partial \tilde{h}}{\partial \bar{\xi}_+} + \varepsilon\frac{\partial \tilde{h}}{\partial \xi}\right) + \varepsilon\Big(1 + \eta_0^2\Big)\tilde{h} + \mathcal{O}(\varepsilon^2) = 0 \quad (7.258)$$

To examine the kinetic effect, we let $\widehat{B}$ be fixed, and as before, we carry out the MVE for the perturbed state in the limit $\varepsilon \to 0$,

$$\begin{aligned}
\tilde{T} &= \tilde{\mu}_0(\varepsilon)\left\{\tilde{T}_0(\xi,\eta,\xi_+,\eta_+) + \varepsilon\tilde{T}_1(\xi,\eta,\xi_+,\eta_+) + \cdots\right\}\mathrm{e}^{\sigma t_+}\\
\tilde{T}_{\mathrm{S}} &= \tilde{\mu}_0(\varepsilon)\left\{\tilde{T}_{\mathrm{S}0}(\xi,\eta,\xi_+,\eta_+) + \varepsilon\tilde{T}_{\mathrm{S}1}(\xi,\eta,\xi_+,\eta_+) + \cdots\right\}\mathrm{e}^{\sigma t_+}\\
\tilde{h} &= \tilde{\mu}_0(\varepsilon)\left\{\tilde{h}_0(\xi,\hat{\xi}_+) + \varepsilon\tilde{h}_1(\xi,\bar{\xi}_+) + \cdots\right\}\mathrm{e}^{\sigma t_+},
\end{aligned} \quad (7.259)$$

and the asymptotic expansions for the wave number functions $(k, \bar{k})$ and the parameter $\sigma$:

$$\begin{aligned}
k(\xi,\eta,\varepsilon) &= k_0(\xi,\eta) + \varepsilon k_1(\xi,\eta) + \cdots,\\
\bar{k}(\xi,\varepsilon) &= \bar{k}_0(\xi) + \varepsilon\bar{k}_1(\xi) + \cdots,\\
\sigma &= \sigma_0 + \varepsilon\sigma_1 + \varepsilon^2\sigma_2 + \cdots.
\end{aligned} \quad (7.260)$$

It is important to note that the effect of kinetic attachment is described by the term with the factor $(\varepsilon\widehat{B}\bar{B}_s/\hat{S})$ in the interface condition (7.257). This factor has a singularity at the point $\xi = -\mathrm{i}$.

It is known that the global mode solutions do not depend only on the outer solution but also on the inner solution near the turning point $\zeta_c$. For the case $\sigma = \mathcal{O}(1)$, the system has a complex spectrum and a turning point $\xi = \zeta_c$ away from $-\mathrm{i}$. Hence, we deduce that for this case, as $\varepsilon \ll 1$, one can neglect the term of kinetic effect with $\left|\varepsilon\widehat{B}\bar{B}_s/\hat{S}\right| \ll 1$ in the regions in the complex $\xi$-plane, which includes the outer region and the inner region near $\zeta_c$. In other words, as the zeroth-order approximation, the system will be exactly the same as that discussed in the previous section with $\widehat{B} = 0$. The presence of interface kinetics affects only the first-order approximate solutions. On the other hand, for the case of real eigenvalues with $|\sigma_0| \ll 1$ and the corresponding turning point $\zeta_c \approx -\mathrm{i}$, the term $\left|\varepsilon\widehat{B}\bar{B}_s/\hat{S}\right|$ may become large in the inner region of $\zeta_c$. As a consequence, the effect of kinetic attachment is not negligible in the leading-order approximation for the global mode solutions.

In view of the above, one must distinguish the two cases and discuss them separately.

### 7.7.2 *The Complex Spectrum of Eigenvalues with* $|\sigma_0| = \mathcal{O}(1)$ *and GTW Instability*

For the complex spectrum of eigenvalues with $|\sigma_0| = \mathcal{O}(1)$, which leads the GTW instability, the system allows both S-modes and A-modes.

#### 7.7.2.1 Zeroth-Order Approximation Solutions for GTW Instability

In the zeroth-order approximation, since the kinetic effect can be neglected, the system reduces to that with $\widehat{B} = 0$. So, the eigenvalues corresponding to the GTW modes are determined by the quantization condition

$$\chi = \frac{1}{\varepsilon} \int_0^{\xi_c} \left(k_0^{(1)} - k_0^{(3)}\right) \mathrm{d}\xi = \left(2n + 1 + \frac{1}{2} + \theta_0\right) \pi - \mathrm{i} \ln \alpha_0,$$

$$\left(n = 0, \pm 1, \pm 2, \ldots\right), \tag{7.261}$$

where

$$\begin{cases} \alpha_0 \, \mathrm{e}^{\mathrm{i}\theta_0 \pi} = \dfrac{k_0^{(1)}(0)}{k_0^{(3)}(0)} & \text{for the S-modes;} \\ \alpha_0 = 1, \;\; \theta_0 = 0 & \text{for the A-modes.} \end{cases} \tag{7.262}$$

#### 7.7.2.2 First-Order Approximation Solutions for GTW Instability

We have the same equations as those derived in the previous section:

$$\begin{aligned} (k_0^2 + g_0^2) \left( \frac{\partial^2}{\partial \xi_+^2} + \frac{\partial^2}{\partial \eta_+^2} \right) \tilde{T}_1 &= a_0(\xi, \eta) \mathrm{e}^{\mathrm{i}\xi_+ - \eta_+} , \\ (k_0^2 + g_0^2) \left( \frac{\partial^2}{\partial \xi_+^2} + \frac{\partial^2}{\partial \eta_+^2} \right) \tilde{T}_{S1} &= a_{S0}(\xi, \eta) \mathrm{e}^{\mathrm{i}\xi_+ + \eta_+} . \end{aligned} \tag{7.263}$$

All the functions $\{a_0, a_{S0}, A_0, A_{S0}\}$ and $Q_0$ remain unchanged. With the solvability conditions (7.121), as before, the equations in the first-order approximation reduce to

$$\left( \frac{\partial^2}{\partial \xi_+^2} + \frac{\partial^2}{\partial \eta_+^2} \right) \tilde{T}_1 = 0, \quad \left( \frac{\partial^2}{\partial \xi_+^2} + \frac{\partial^2}{\partial \eta_+^2} \right) \tilde{T}_{S1} = 0. \tag{7.264}$$

The interface conditions are

(a)

$$\tilde{T}_1 = \tilde{T}_{S1} + \tilde{h}_1 - \eta_1 \bar{k}_0 \frac{\partial}{\partial \eta_+}(\tilde{T}_0 - \tilde{T}_{S0}), \tag{7.265}$$

(b)

$$\begin{aligned}\tilde{T}_{S1} = & \frac{\bar{A}_s(\xi)}{\hat{S}(\xi)}\left[\left(\bar{k}_0^2 \frac{\partial^2 \tilde{h}_1}{\partial \bar{\xi}_+^2} + 2\bar{k}_0 \bar{k}_1 \frac{\partial^2 \tilde{h}_0}{\partial \bar{\xi}_+^2} + \bar{k}_0' \frac{\partial \tilde{h}_0}{\partial \bar{\xi}_+}\right)\right. \\ & \left. + \frac{\xi \bar{k}_0}{\hat{S}^2(\xi)} \frac{\partial \tilde{h}_0}{\partial \bar{\xi}_+}\right] - \tilde{A}_{s1}(\xi) \frac{\bar{k}_0}{\hat{S}^3} \frac{\partial \tilde{h}_0}{\partial \bar{\xi}_+} \\ & - \frac{\widehat{B}\bar{B}_s}{\hat{S}}\left[\xi \bar{k}_0 \frac{\partial \tilde{h}_0}{\partial \xi_+} + \sigma_0 \hat{S}^2 \tilde{h}_0\right] - \eta_1 \bar{k}_0 \frac{\partial \tilde{T}_{S0}}{\partial \eta_+},\end{aligned} \tag{7.266}$$

(c)

$$\begin{aligned}& \bar{k}_0\left(\frac{\partial \tilde{T}_1}{\partial \eta_+} - \hat{\alpha}_T \frac{\partial \tilde{T}_{S1}}{\partial \eta_+}\right) + \bar{k}_1\left(\frac{\partial \tilde{T}_0}{\partial \eta_+} - \hat{\alpha}_T \frac{\partial \tilde{T}_{S0}}{\partial \eta_+}\right) + \eta_1 \bar{k}_0^2 \frac{\partial^2}{\partial \eta_+^2}(\tilde{T}_0 - \hat{\alpha}_T \tilde{T}_{S0}) \\ & + \left(\frac{\partial \tilde{T}_0}{\partial \eta} - \hat{\alpha}_T \frac{\partial \tilde{T}_{S0}}{\partial \eta}\right) + \hat{S}^2(\xi)(\sigma_0 \tilde{h}_1 + \sigma_1 \tilde{h}_0) \\ & + \xi\left(\bar{k}_0 \frac{\partial \tilde{h}_1}{\partial \bar{\xi}_+} + \bar{k}_1 \frac{\partial \tilde{h}_0}{\partial \xi_+}\right) + \left(1 + \eta_0^2\right)\tilde{h}_0 = 0.\end{aligned} \tag{7.267}$$

The only difference between the above system and that in the previous section is the interface condition (7.266). Consider the normal mode solutions (7.130) given in the previous section. One may derive the system

$$\begin{aligned}& \hat{A}_1 - \hat{A}_{S1} - \hat{D}_1 = I_1 \hat{D}_0, \\ & \hat{A}_{S1} + \tfrac{\bar{A}_s \hat{k}_0^2}{S} \hat{D}_1 = \tilde{I}_2 \hat{D}_0, \\ & -\hat{k}_0(\hat{A}_1 + \hat{\alpha}_T \hat{A}_{S1}) + (\sigma_0 S^2 + i\xi \hat{k}_0)\hat{D}_1 = I_3 \hat{D}_0.\end{aligned} \tag{7.268}$$

The above system is formally the same as (7.132), except the function $I_2$ in (7.132) is replaced by $\tilde{I}_2$, which is defined as

$$\begin{aligned}\tilde{I}_2 = & \tfrac{\bar{A}_s}{\hat{S}}\left[ik_0' + i\tfrac{\xi k_0}{\hat{S}^2} - 2k_0 k_1 - i\tilde{A}_{s1}(\xi)\tfrac{\bar{k}_0}{\hat{S}^3}\right] + \tfrac{\bar{A}_s \bar{k}_0^3}{\hat{S}} \eta_1 \\ & - \Delta I_2,\end{aligned}$$

where

$$\Delta I_2 = \frac{\widehat{B}\bar{B}_s}{\hat{S}}\left[\sigma_0\hat{S}^2 + \mathrm{i}\xi\bar{k}_0\left(1 - \frac{\bar{A}_s\bar{k}_0^2}{\hat{S}}\right)\right] \tag{7.269}$$

represents the effect of kinetic attachment. Furthermore, by replacing $I_2$ in the solvability condition (7.135), it follows that

$$\bar{k}_1\left[1 - \mathrm{i}\xi - \frac{3(1+\hat{\alpha}_{\mathrm{T}})\bar{A}_s\bar{k}_0^2}{S}\right] - \sigma_1\hat{S}^2 - Q_0 - (1+\eta_0^2) - (1+\hat{\alpha}_{\mathrm{T}})\bar{k}_0\Delta I_2$$
$$+\frac{\bar{k}_0(1+\hat{\alpha}_{\mathrm{T}})\bar{A}_s}{\hat{S}}\left(\mathrm{i}\bar{k}_0' + \frac{\mathrm{i}\xi\bar{k}_0}{\hat{S}^2} - \bar{A}_{s1}\frac{\hat{k}_0}{\hat{S}^3}\right) = 0.$$

We derive that

$$\bar{k}_1 = \frac{\widetilde{R}_1(\xi)}{F(\xi)} - \mathrm{i}\frac{R_2(\xi)}{F(\xi)}\frac{\bar{k}_0'}{\bar{k}_0}, \tag{7.270}$$

where

$$\widetilde{R}_1(\xi) = S^2\sigma_1 + (1+\tfrac{1}{2}\eta_0^2) + (1+\hat{\alpha}_{\mathrm{T}})\bar{k}_0\Delta I_2 + \bar{A}_s\tilde{A}_{s1}\frac{(1+\hat{\alpha}_{\mathrm{T}})\bar{k}_0^2}{\hat{S}^4}$$
$$+\frac{\eta_0^2}{2}\left[\left(\frac{1-\hat{\alpha}_{\mathrm{T}}^2}{S}\bar{A}_s\bar{k}_0^2\right) - \left(1 - \frac{1+\hat{\alpha}_{\mathrm{T}}}{S}\bar{A}_s\bar{k}_0 2\right)\left(1 - \frac{1+\hat{\alpha}_{\mathrm{T}}^2}{S}\bar{A}_s\bar{k}_0^2\right)\right]. \tag{7.271}$$

From the condition

$$\widetilde{R}_1(\zeta_c) = 0, \tag{7.272}$$

we obtain the solution $\sigma_1$ as a function of $\zeta_c$, namely,

$$\sigma_1 = -\frac{1}{(1+\zeta_c^2)}\left\{1 + \tfrac{1}{2}\eta_0^2 + (1+\hat{\alpha}_{\mathrm{T}})\bar{k}_0\Delta I_2 + \frac{\tilde{A}_{s1}}{\hat{S}^3}\frac{1-\mathrm{i}\zeta_c}{3}\right.$$
$$\left.+\frac{\eta_0^2}{6}\left[(1-\hat{\alpha}_{\mathrm{T}})(1-\mathrm{i}\zeta_c) - (2+\mathrm{i}\zeta_c)\left(1 - \frac{1+\hat{\alpha}_{\mathrm{T}}^2}{1+\hat{\alpha}_{\mathrm{T}}}\frac{1-\mathrm{i}\zeta_c}{3}\right)\right]\right\}. \tag{7.273}$$

In terms of the formula (7.138), one may write

$$\bar{k}_0^2 = \frac{\hat{S}(1-\mathrm{i}\zeta_c)}{3(1+\hat{\alpha}_{\mathrm{T}})\bar{A}_s}.$$

Then the term describing the effect of kinetic attachment can be written as

$$(1+\hat{\alpha}_{\mathrm{T}})\bar{k}_0\Delta I_2 = (1+\hat{\alpha}_{\mathrm{T}})\frac{\widehat{B}\bar{B}_s}{\hat{S}}\left[\sigma_0\bar{k}_0\hat{S}^2 + \mathrm{i}\xi\bar{k}_0^2\left(1 - \frac{\bar{A}_s\bar{k}_0^2}{\hat{S}}\right)\right]$$
$$= (1+\hat{\alpha}_{\mathrm{T}})\frac{\widehat{B}\bar{B}_s}{\hat{S}}\left[(1-\mathrm{i}\zeta)\bar{k}_0^2 - \frac{(1+\hat{\alpha}_{\mathrm{T}})\bar{A}_s\bar{k}_0^4}{\hat{S}} + \mathrm{i}\zeta\bar{k}_0^2\left(1 - \frac{\bar{A}_s\bar{k}_0^2}{\hat{S}}\right)\right]$$
$$= \widehat{B}\frac{\bar{B}_s}{\bar{A}_s}\frac{(1-\mathrm{i}\zeta_c)}{3}\left[1 + \tfrac{1}{3}(1+\hat{\alpha}_{\mathrm{T}}+\mathrm{i}\zeta)\left(\frac{1-\mathrm{i}\zeta_c}{1+\hat{\alpha}_{\mathrm{T}}}\right)\right].$$

### 7.7.3 The Real Spectrum of Eigenvalues with $|\sigma_0| \ll 1$ and LF Instability

For the case of a real spectrum of eigenvalues with $|\sigma_0| \ll 1$, as indicated before, one must include the effect of the term of kinetic attachment in the leading-order approximation. For this purpose, we define

$$\mathcal{M} = \varepsilon \widehat{B}, \tag{7.274}$$

and keep $\mathcal{M}$ fixed in the limit $\varepsilon \to 0$. Thus, in the leading-order approximation, we obtain the system

$$\left(\frac{\partial^2}{\partial \xi_+^2} + \frac{\partial^2}{\partial \eta_+^2}\right)\tilde{T}_0 = 0; \quad \left(\frac{\partial^2}{\partial \xi_+^2} + \frac{\partial^2}{\partial \eta_+^2}\right)\tilde{T}_{S0} = 0, \tag{7.275}$$

with the following boundary conditions:

1. As $\eta_+ \to \infty$,

$$\tilde{T}_0 \to 0\,. \tag{7.276}$$

2. As $\eta_+ \to -\infty$,

$$\tilde{T}_{S0} \to 0\,. \tag{7.277}$$

3. The interface conditions at $\eta_+ = 0$ and $\eta = 1$,

$$\tilde{T}_0 = \tilde{T}_{S0} + \tilde{h}_0\,, \tag{7.278}$$

$$\tilde{T}_{S0} = \frac{\bar{A}_s(\xi)\bar{k}_0^2}{\hat{\hat{S}}(\xi)}\frac{\partial^2 \tilde{h}_0}{\partial \bar{\xi}_+^2} - \frac{\mathcal{M}\bar{B}_s}{\hat{S}}\left[\xi \bar{k}_0 \frac{\partial \tilde{h}_0}{\partial \bar{\xi}_+} + \hat{S}^2 \sigma_0 \tilde{h}_0\right], \tag{7.279}$$

$$\frac{\partial}{\partial \eta_+}\left(\tilde{T}_0 - \hat{\alpha}_T \tilde{T}_{S0}\right) + \sigma_0 \hat{S}^2(\xi)\tilde{h}_0 + \bar{k}_0 \xi \frac{\partial \tilde{h}_0}{\partial \bar{\xi}_+} = 0\,. \tag{7.280}$$

4. The root conditions as $\xi$ and $\xi_+ \to \infty$,

$$\left\{\tilde{T}_0,\, \tilde{T}_{S0},\, \tilde{h}_0\right\} = 0. \tag{7.281}$$

5. The tip smoothness conditions at $\xi = \eta_+ = 0$,

$$\frac{\partial}{\partial \xi}\left\{\tilde{T}_0,\, \tilde{T}_{S0},\, \tilde{h}_0\right\} = 0 \quad \text{for the S-modes;} \tag{7.282}$$

$$\left\{\tilde{T}_0,\, \tilde{T}_{S0},\, \tilde{h}_0\right\} = 0 \quad \text{for the A-modes}\,. \tag{7.283}$$

For the normal mode form of the solutions

$$\begin{aligned}\tilde{T}_0 &= A_0(\xi,\eta)\exp\{i\xi_+ - \eta_+\}\\ \tilde{T}_{S0} &= A_{S0}(\xi,\eta_+)\exp\{i\xi_+ + \eta_+\}\\ \tilde{h}_0 &= \hat{D}_0\exp\{i\bar{\xi}_+\}\ ,\end{aligned} \tag{7.284}$$

we derive the local dispersion relationship

$$\begin{aligned}\sigma_0\hat{S}^2 = \hat{S}^2\Sigma\{\hat{k}_0,\xi\} &= -\frac{(1+\hat{\alpha}_T)\bar{A}_s}{\hat{S}}\bar{k}_0^3 - i\xi\frac{(1+\hat{\alpha}_T)\mathcal{M}\bar{B}_s}{\hat{S}}\bar{k}_0^2\\ &+\Big[1 - i\xi - (1+\hat{\alpha}_T)\mathcal{M}\bar{B}_s\hat{S}\sigma_0\Big]\bar{k}_0.\end{aligned} \tag{7.285}$$

For the special case $\mathcal{M}=0$, we regain

$$\sigma_0 = \frac{\bar{k}_0}{\hat{S}^2}\left[1 - \frac{(1+\hat{\alpha}_T)\bar{A}_s(\xi)\bar{k}_0^2}{\hat{S}}\right] - i\frac{\xi\bar{k}_0}{\hat{S}^2}\,.$$

As before, for given $\sigma_0$, from (7.285) one can solve for three roots $\{\bar{k}_0^{(1)},\bar{k}_0^{(2)},\bar{k}_0^{(3)}\}$. To derive the explicit formulas, we may introduce the notation

$$\begin{cases}\tilde{\sigma}_0 = \sigma_0 - \sigma_*(\sigma_0,\xi),\\ \tilde{k}_0 = \bar{k}_0 - k_*(\xi),\\ k_* = -i\xi\frac{\mathcal{M}\bar{B}_s}{3\bar{A}_s},\\ \sigma_*\hat{S}^2 = -\frac{(1+\hat{\alpha}_T)\bar{A}_s}{\hat{S}}k_*^3 - i\xi\frac{(1+\hat{\alpha}_T)\mathcal{M}\bar{B}_s}{\hat{S}}k_*^2 + \Big[1 - i\xi - (1+\hat{\alpha}_T)\mathcal{M}\bar{B}_s\hat{S}\sigma_0\Big]k_*\\ \quad = (1-i\xi)k_* + \mathcal{O}(\mathcal{M}^2) = -i\xi(1-i\xi)\frac{\mathcal{M}\bar{B}_s}{3\bar{A}_s} + \mathcal{O}(\mathcal{M}^2)\,.\end{cases} \tag{7.286}$$

We can rewrite the dispersion relationship (7.285) in the form

$$\tilde{\sigma}_0 = \widetilde{\Sigma}(\bar{k}_0,\xi) = \frac{\tilde{k}_0}{\hat{S}^2}\left[(1 - i\xi - \Lambda_0) - \frac{(1+\hat{\alpha}_T)\bar{A}_s(\xi)}{\hat{S}}\tilde{k}_0^2\right], \tag{7.287}$$

where

$$\begin{aligned}\Lambda_0 &= \frac{3(1+\hat{\alpha}_T)k_*^2\bar{A}_s}{\hat{S}} + i\frac{2\xi(1+\hat{\alpha}_T)\mathcal{M}\bar{B}_s k_*}{\hat{S}} + (1+\hat{\alpha}_T)\mathcal{M}\bar{B}_s\hat{S}\sigma_0\\ &= \frac{(1+\hat{\alpha}_T)}{3}\frac{\xi^2\mathcal{M}^2\bar{B}_s^2}{\hat{S}\bar{A}_s} + (1+\hat{\alpha}_T)\mathcal{M}\bar{B}_s\hat{S}\sigma_0\\ &= (1+\hat{\alpha}_T)\mathcal{M}\bar{B}_s\hat{S}\sigma_0 + \mathcal{O}(\mathcal{M}^2).\end{aligned} \tag{7.288}$$

The modified form of the local dispersion relationship (7.287) is formally the same as (7.113). From given $\sigma_0$, one may determine $\sigma_*$ and $\tilde{\sigma}$ from (7.286). Then with given $\tilde{\sigma}_0$, one can solve for the three roots from $\{\tilde{k}^{(i)}(\xi)\}$ $(i = 1, 2, 3)$ from the modified dispersion relationship (7.287):

$$\begin{cases} \tilde{k}_0^{(1)}(\xi) = \tilde{M}(\xi)\cos\left[\frac{1}{3}\cos^{-1}\left(\frac{\tilde{\sigma}_0}{\tilde{N}(\xi)}\right)\right] & \text{(short-wavelength branch)}, \\ \tilde{k}_0^{(2)}(\xi) = \tilde{M}(\xi)\cos\left[\frac{1}{3}\cos^{-1}\left(\frac{\tilde{\sigma}_0}{\tilde{N}(\xi)}\right) + \frac{2\pi}{3}\right] \\ \tilde{k}_0^{(3)}(\xi) = \tilde{M}(\xi)\cos\left[\frac{1}{3}\cos^{-1}\left(\frac{\tilde{\sigma}_0}{\tilde{N}(\xi)}\right) + \frac{4\pi}{3}\right] & \text{(long-wavelength branch)}, \end{cases} \tag{7.289}$$

where we define

$$\begin{cases} \tilde{M}(\xi) = \sqrt{\dfrac{4\Lambda_1\hat{S}(\xi)}{3}}(1 - \mathrm{i}\xi - \Lambda_0)^{\frac{1}{2}}, \\ \tilde{N}(\xi) = -\dfrac{\tilde{M}(\xi)}{3\hat{S}^2(\xi)}(1 - \mathrm{i}\xi - \Lambda_0). \end{cases} \tag{7.290}$$

Subsequently, one may derive the three wave number functions $\{\bar{k}^{(i)}(\xi)\}, (i = 1, 2, 3)$. The general outer solution of the $H$ wave in the outer region is then derived as

$$\tilde{h}_0 = D_1 H_1(\xi) + D_3 H_3(\xi) \tag{7.291}$$

$$= D_1 \exp\left[\frac{\mathrm{i}}{\varepsilon}\int_{\zeta_c}^{\xi} \bar{k}_0^{(1)} d\xi_1\right] + D_3 \exp\left[\frac{\mathrm{i}}{\varepsilon}\int_{\zeta_c}^{\xi} \bar{k}_0^{(3)} d\xi_1\right]. \tag{7.292}$$

The coefficients $\{D_1, D_3\}$ are arbitrary constants to be determined. As demonstrated before, the MVE solution obtained has a singularity at the point $\zeta_c$ in the complex $\zeta$-plane. Here we extend the real variable $\xi$ to the complex variable $\zeta = \xi + \mathrm{i}\xi_{\mathrm{I}}$. The singular point $\zeta_c$ is subject to the condition

$$\left(\frac{\partial \Sigma}{\partial \bar{k}_0}\right) = 0 \quad \Leftrightarrow \quad \left(\frac{\partial \widetilde{\Sigma}}{\partial \tilde{k}_0}\right) = 0 \tag{7.293}$$

and

$$\sigma_0 = \Sigma(\bar{k}_0, \zeta_c) \quad \Leftrightarrow \quad \tilde{\sigma}_0 = \widetilde{\Sigma}(\tilde{k}_0, \zeta_c). \tag{7.294}$$

It follows that

$$\tilde{\sigma}_0 = \sigma_0 - \sigma_* = \sqrt{\frac{4}{27(1 + \hat{\alpha}_{\mathrm{T}})\bar{A}_s}}\left[\frac{1 - \mathrm{i}\zeta_c - \Lambda_0}{\hat{S}}\right]^{3/2}. \tag{7.295}$$

For a given $\sigma_0$, one can solve for the singular point $\zeta_c$ from (7.295). In viewing $\mathcal{M} \ll 1$, we may make the expansion

$$\zeta_c = \zeta_{c0}(\sigma_0) + \mathcal{M}\zeta_{c1}(\sigma_0) + \cdots, \tag{7.296}$$

where $\zeta_{c0}$ can be solved from

$$\sigma_0 = \sqrt{\frac{4}{27(1+\hat{\alpha}_T)\bar{A}_s}}\left[\frac{1-i\zeta_{c0}}{\hat{S}}\right]^{3/2}. \tag{7.297}$$

The normal mode solutions with multiple variables obtained are only outer solutions in the region away from singular point $\zeta_c$. In other words, near the singular point $\zeta_c$, the solution cannot be expressed in the form of a normal mode. In order to find a more general form of inner solutions valid near the singular point, we utilize the replacement

$$\bar{k}_0 \Longleftrightarrow -i\varepsilon\frac{d}{d\zeta}$$

in the dispersion relationship (7.285), which leads to the following differential equation for the interface shape function:

$$-\varepsilon^3\frac{(1+\hat{\alpha}_T)\bar{A}_s}{\hat{S}}\frac{d^3h_0}{d\zeta^3} + \varepsilon^2\zeta\frac{(1+\hat{\alpha}_T)\mathcal{M}\bar{B}_s}{\hat{S}}\frac{d^2h_0}{d\zeta^2}$$
$$-\varepsilon\Big[(1-\Lambda_0) - i\zeta\Big]\frac{dh_0}{d\zeta} + i\sigma_0\hat{S}^2h_0 = 0. \tag{7.298}$$

With the inner variables in the inner region $|\zeta - \zeta_c| \ll 1$; $|\eta - 1| \ll 1$, we obtain

$$\zeta_* = \frac{\zeta - \zeta_c}{\varepsilon^\alpha}, \quad \zeta = \zeta_c + \varepsilon^\alpha\zeta_* \quad \eta_* = \frac{\eta - 1}{\varepsilon^\alpha}.$$

Equation (7.298) is transformed to the inner equation

$$-\varepsilon^{3-3\alpha}\frac{(1+\hat{\alpha}_T)\bar{A}_s}{\hat{S}}\frac{d^3\hat{h}_0}{d\zeta_*^3} + \varepsilon^{2-2\alpha}\zeta\frac{(1+\hat{\alpha}_T)\mathcal{M}\bar{B}_s}{\hat{S}}\frac{d^2\hat{h}_0}{d\zeta_*^2}$$
$$-\varepsilon^{1-\alpha}\Big[\zeta + i(1-\Lambda_0)\Big]\frac{d\hat{h}_0}{d\zeta_*} + i\sigma_0\hat{S}^2\hat{h}_0 = 0. \tag{7.299}$$

We use the $W$-wave representation, as before, for the solutions in the outer region and inner region, respectively, namely

$$\begin{aligned} h_0(\zeta) &= W_0(\zeta)\exp\left\{\tfrac{\mathrm{i}}{\varepsilon}\int_{\zeta_c}^{\zeta} k_c(\zeta_1)\,\mathrm{d}\zeta_1\right\}, \\ \hat{h}_0(\zeta_*) &= \hat{W}_0(\zeta_*)\exp\left\{\tfrac{\mathrm{i}}{\varepsilon}\int_{\zeta_c}^{\zeta} k_c(\zeta_1)\,\mathrm{d}\zeta_1\right\}. \end{aligned} \tag{7.300}$$

Equation (7.298) is then changed to

$$\varepsilon^3\Omega_3\frac{\mathrm{d}^3 W_0}{\mathrm{d}\zeta^3} + \mathrm{i}\varepsilon^2\Omega_2\frac{\mathrm{d}^2 W_0}{\mathrm{d}\zeta^2} - \varepsilon\Omega_1\frac{\mathrm{d}W_0}{\mathrm{d}\zeta} + \mathrm{i}\hat{S}^2\Big\{\sigma_0 - \Sigma(\bar{k}_0,\zeta)\Big\}W_0 = 0\,, \tag{7.301}$$

where

$$\begin{aligned} \hat{S}^2\Sigma\{\bar{k}_0,\zeta\} = &-\tfrac{(1+\hat{\alpha}_{\mathrm{T}})\bar{A}_s}{\hat{S}}\bar{k}_0^3 - \mathrm{i}\zeta\tfrac{(1+\hat{\alpha}_{\mathrm{T}})\mathcal{M}\bar{B}_s}{\hat{S}}\bar{k}_0^2 \\ &+(1-\mathrm{i}\zeta-\mathrm{i}\Lambda_0)\bar{k}_0, \end{aligned} \tag{7.302}$$

$$\begin{aligned} \Omega_3 &= \tfrac{1}{3!}\left[\tfrac{\partial^3(\hat{S}^2\Sigma)}{\partial\bar{k}_0^3}\right]_{\bar{k}_0=k_c} = -\tfrac{(1+\hat{\alpha}_{\mathrm{T}})\bar{A}_s}{\hat{S}}, \\ \Omega_2 &= \tfrac{1}{2!}\left[\tfrac{\partial^2(\hat{S}^2\Sigma)}{\partial\bar{k}_0^2}\right]_{\bar{k}_0=k_c} = -\tfrac{3\mathrm{i}(1+\hat{\alpha}_{\mathrm{T}})\bar{A}_s k_c+(1+\hat{\alpha}_{\mathrm{T}})\mathcal{M}\bar{B}_s}{\hat{S}}, \\ \Omega_1 &= \left[\tfrac{\partial(\hat{S}^2\Sigma)}{\partial\bar{k}_0}\right]_{\bar{k}_0=k_c} = -\tfrac{3(1+\hat{\alpha}_{\mathrm{T}})\bar{A}_s k_c^2+2\mathrm{i}\zeta(1+\hat{\alpha}_{\mathrm{T}})\mathcal{M}\bar{B}_s k_c}{\hat{S}} \\ &\qquad +(1-\mathrm{i}\zeta-\mathrm{i}\Lambda_0). \end{aligned} \tag{7.303}$$

We set

$$\begin{aligned} \Omega_1 = \left[\tfrac{\partial(\hat{S}^2\Sigma)}{\partial\bar{k}_0}\right]_{\bar{k}_0=k_c} &= -\tfrac{3(1+\hat{\alpha}_{\mathrm{T}})\bar{A}_s k_c^2+2\mathrm{i}\zeta(1+\hat{\alpha}_{\mathrm{T}})\mathcal{M}\bar{B}_s k_c}{\hat{S}} \\ &\quad +(1-\mathrm{i}\zeta-\mathrm{i}\Lambda_0) = 0. \end{aligned} \tag{7.304}$$

Then Eq. (7.301) is reduced to

$$\varepsilon^3\Omega_3\frac{\mathrm{d}^3 \hat{W}_0}{\mathrm{d}\zeta^3} + \mathrm{i}\varepsilon^2\Omega_2\frac{\mathrm{d}^2 \hat{W}_0}{\mathrm{d}\zeta^2} + \mathrm{i}\hat{S}^2\Big\{\sigma - \Sigma_c(\zeta)\Big\}\hat{W}_0 = 0\,, \tag{7.305}$$

and we have

$$\begin{aligned} \hat{S}^2\Sigma_c(\zeta) &= \hat{S}^2\Sigma(k_c,\zeta) \\ &= -\tfrac{(1+\hat{\alpha}_{\mathrm{T}})\bar{A}_s}{\hat{S}}k_c^3 + \mathrm{i}\zeta\tfrac{(1+\hat{\alpha}_{\mathrm{T}})\mathcal{M}\bar{B}_s}{\hat{S}}k_c^2 + (1-\mathrm{i}\zeta-\mathrm{i}\Lambda_0)k_c. \end{aligned} \tag{7.306}$$

Using the relationship $\zeta = \zeta_c + \varepsilon^{\alpha}\zeta^*$, from Eq. (7.305) we derive the inner equation for the interface shape function:

$$\varepsilon^{3-3\alpha}\Omega_3\frac{\mathrm{d}^3 \hat{W}_0}{\mathrm{d}\zeta_*^3} + \mathrm{i}\varepsilon^{2-2\alpha}\Omega_2\frac{\mathrm{d}^2 \hat{W}_0}{\mathrm{d}\zeta_*^2} + \mathrm{i}\hat{S}^2\Big[\sigma - \Sigma_c(\zeta)\Big]\hat{W}_0 = 0\,. \tag{7.307}$$

Equation (7.307) is formally exactly the same as (9.17) derived in the previous section. As matter of fact, as $\mathcal{M} \to 0$, (7.307) just reduces to (7.172). As a consequence, almost all the results on the inner solutions obtained in the previous section are applicable to this case. In what follows, we shall list these results without derivation.

The inner equation in the far field of the inner region ($|\zeta_*| \gg 1$) can be approximately written in the following unified form:

$$\frac{d^2\hat{W}_0}{d\hat{\xi}_*^2} + \hat{\zeta}_*^{p_0}\hat{W}_0 = 0\,, \tag{7.308}$$

where

$$\hat{\zeta}_* = \frac{B}{\varepsilon^\alpha}(\zeta - \zeta_c)\,. \tag{7.309}$$

The general solution of (7.308) is

$$\hat{W}_0 = C_1\hat{\zeta}_*^{\frac{1}{2}}H_\nu^{(1)}(z) + C_2\hat{\zeta}_*^{\frac{1}{2}}H_\nu^{(2)}(z) \quad (z = 2\nu\hat{\zeta}_*^{\frac{1}{2\nu}}), \tag{7.310}$$

where $C_1, C_2$ are arbitrary constants, $\nu = \frac{1}{p_0+2}$, and $H_\nu^{(1)}(z)$ and $H_\nu^{(2)}(z)$ are the $\nu$th-order Hankel functions of first and second kind, respectively. For the case $|\sigma_0| \ll 1$ under discussion, we have

$$\alpha = \frac{4}{7}\,, \quad p_0 = \frac{3}{2}\,, \quad \nu = \frac{2}{7}\,.$$

The pair of coefficients $\{D_1, D_3\}$ in (7.292) have different values in different sectors as shown in Fig. 7.2 and are subject to the following connection conditions:

- In sector $(S_2)$: $D_1' = 0$ and $D_3' = D' \neq 0$.
- In sector $(S_1)$: $\frac{D_1}{D_3} = \mathrm{i}2\cos(\nu\pi)$.

The quantization condition for LF modes will be formally the same as that derived in the previous section, namely,

$$\begin{cases} \mathrm{Re}\left[\frac{1}{\varepsilon}\int_0^{\xi_c}\left(k_0^{(1)} - k_0^{(3)}\right)d\xi\right] = \left(2n + \frac{1}{2} + \chi_0\right)\pi \\ \qquad (n = 0, \pm 1, \pm 2, \ldots)\,, \\ \left|\frac{d_1}{d_3}\right| = 2\cos\left(\frac{2\pi}{7}\right)e^{\mathrm{Im}\{\chi\}}, \end{cases} \tag{7.311}$$

where

$$\chi_0 = \begin{cases} 0 \text{ or } 1, & \text{(for the S-modes)}, \\ \cos(\chi_0\pi) = -2\cos\left(\frac{2\pi}{7}\right)e^{\mathrm{Im}\{\chi\}} & \text{(for the A-modes)}. \end{cases} \tag{7.312}$$

The quantization condition (7.311) can be simplified with the assumptions $\sigma_0 \ll 1$, $\mathcal{M} \ll 1$. To do so, we first approximate the expression of $\left(k_0^{(1)} - k_0^{(3)}\right)$ through the dispersion relationship (7.285). Let

$$k_0 = k_{00} + k_{01} + \mathcal{O}(\sigma_0, \mathcal{M}),$$

$$k_{00}^{(3)} = 0, \quad k_{00}^{(1)} = k_e,$$

where $|k_{01}| \ll |k_{00}|$. From (7.285), we derive that

$$\frac{(1+\hat{\alpha}_{\mathrm{T}})\bar{A}_s}{\hat{S}} k_e^2 - (1 - \mathrm{i}\zeta) = 0,$$

where

$$\begin{aligned} &k_e(\zeta) = \frac{e^{\mathrm{i}\frac{3\pi}{4}}}{\sqrt{(1+\hat{\alpha}_{\mathrm{T}})\bar{A}_s}} (\zeta - \mathrm{i})^{\frac{1}{4}} (\zeta + \mathrm{i})^{\frac{3}{4}}, \\ &\arg\{k_e(\zeta)\} > 0, \text{ and } \mathrm{Re}\{k_e(\zeta)\} > 0 \quad (\text{as } 0 \le \xi < \infty). \end{aligned} \tag{7.313}$$

Moreover, we derive

$$\begin{aligned} -2(1 - \mathrm{i}\zeta) k_{01}^{(1)} &= \mathrm{i}\tfrac{(1+\hat{\alpha}_{\mathrm{T}})\zeta \mathcal{M} \bar{B}_s}{S} k_e^2 + \sigma_0 S^2, \\ (1 - \mathrm{i}\zeta) k_{01}^{(3)} &= \sigma_0 \hat{S}^2. \end{aligned}$$

As a consequence, we obtain that

$$\begin{aligned} &k_0^{(1)}(\zeta, \sigma_0, \mathcal{M}) = k_e(\zeta, \alpha_4) - \tfrac{\sigma_0}{2}(1 + \mathrm{i}\zeta) - \tfrac{\mathrm{i}\zeta\mathcal{M}}{2}\tfrac{\bar{B}_s}{\bar{A}_s} + \mathcal{O}(\sigma_0, \mathcal{M}), \\ &k_0^{(3)}(\zeta, \sigma_0, \mathcal{M}) = \sigma_0(1 + \mathrm{i}\zeta) + \mathcal{O}(\sigma_0, \mathcal{M}), \end{aligned} \tag{7.314}$$

so that we have derived

$$\begin{aligned} \left(k_0^{(1)} - k_0^{(3)}\right) &\approx \frac{e^{\mathrm{i}\frac{3\pi}{4}}}{\sqrt{(1+\hat{\alpha}_{\mathrm{T}})\bar{A}_s}} (\zeta - \mathrm{i})^{\frac{1}{4}} (\zeta + \mathrm{i})^{\frac{3}{4}} - \frac{3\sigma_0}{2}(1 + \mathrm{i}\zeta) - \frac{\mathrm{i}\zeta\mathcal{M}}{2}\frac{\bar{B}_s}{\bar{A}_s} \\ &= \frac{1}{\sqrt{(1+\hat{\alpha}_{\mathrm{T}})}} \frac{(1 - \mathrm{i}\zeta^{\frac{7}{4}}(1 + \mathrm{i}\zeta)^{\frac{5}{4}}}{\sqrt{f(\zeta)}} - \frac{3\sigma_0}{2}(1 + \mathrm{i}\zeta) - \frac{\mathrm{i}\zeta\mathcal{M}}{2}\frac{\bar{B}_s}{\bar{A}_s} \\ &= \frac{1}{\sqrt{(1+\hat{\alpha}_{\mathrm{T}})}} \frac{(1 - \mathrm{i}\zeta)^{\frac{7}{4}}(1 + \mathrm{i}\zeta)^{\frac{5}{4}}}{(1-\alpha_4)^{\frac{1}{2}}(\zeta^2 + a_1^2)^{\frac{1}{2}}(\zeta^2 + a_2^2)^{\frac{1}{2}}} - \frac{3\sigma_0}{2}(1 + \mathrm{i}\zeta) \\ &\quad - \frac{\mathrm{i}\zeta\mathcal{M}}{2}\frac{\bar{B}_s}{\bar{A}_s}. \end{aligned}$$

The term $\frac{i\zeta\mathcal{M}}{2}\frac{\bar{B}_s}{\bar{A}_s}$ in the above represents the kinetic effect. Letting $\mathcal{M} = 0$, we regain the formula (7.223). Note that

$$\frac{\zeta\bar{B}_s}{\bar{A}_s} = \left(\frac{1-\beta_4}{1-\alpha_4}\right)\frac{\zeta(\zeta^2+b_1^2)(\zeta^2+b_2^2)}{(\zeta^2+a_1^2)(\zeta^2+a_2^2)}.$$

Thus, the kinetic effect term is given by

$$\frac{\zeta\bar{B}_s}{\bar{A}_s} = \zeta\frac{(1-\alpha_4)}{(1-\beta_4)}\frac{(\zeta^2+b_1^2)(\zeta^2+b_2^2)}{(\zeta^2+a_1^2)(\zeta^2+a_2^2)} \approx \zeta\frac{(\zeta^2+b_1^2)(\zeta^2+b_2^2)}{(\zeta^2+a_1^2)(\zeta^2+a_2^2)}.$$

We now calculate the integral:

$$-\frac{i\mathcal{M}}{2}\left(\int_0^{-i(a_1-0)} + \int_{(C_r)} + \int_{-i(a_1+0)}^{-i} + \int_{-i}^{\zeta_c}\right)\frac{\zeta\bar{B}_s}{\bar{A}_s}d\zeta$$
$$= -\tfrac{i\mathcal{M}}{2}\Big[(I') + (II') + (III') + (IV')\Big].$$

For this purpose, we carry out the following calculations:

1. The calculations of $(I') + (III')$:

$$(I') + (III') = \left(\int_0^{-i(a_1-0)} + \int_{-i(a_1+0)}^{-i}\right)\frac{\zeta\bar{B}_s}{\bar{A}_s}d\zeta$$

Letting $\zeta = -ia_1 t$, we derive that

$$\begin{aligned}
(I) + (III) &= \left(\int_0^{-i(a_1-0)} + \int_{-i(a_1+0)}^{-i}\right)\frac{\zeta\bar{B}_s}{\bar{A}_s}d\zeta \\
&= \frac{(1-\alpha_4)}{(1-\beta_4)}\left(\int_0^{-i(a_1-0)} + \int_{-i(a_1+0)}^{-i}\right)\frac{\zeta(\zeta^2+b_1^2)(\zeta^2+b_2^2)}{(\zeta^2+a_1^2)(\zeta^2+a_2^2)}d\zeta \\
&= -a_1^2\frac{(1-\alpha_4)}{(1-\beta_4)}\left(\int_0^{(1-0)} + \int_{(1+0)}^{1/a_1}\right)\frac{t(\hat{b}_1^2-t^2)(\hat{b}_2^2-t^2)}{(1-t^2)(\hat{a}_2^2-t^2)}dt \\
&= -\frac{a_1^2}{2}\frac{(1-\alpha_4)}{(1-\beta_4)}\left(\int_0^{(1-0)} + \int_{(1+0)}^{1/a_1^2}\right)\frac{(\hat{b}_1^2-\hat{t})(\hat{b}_2^2-\hat{t})}{(1-\hat{t})(\hat{a}_2^2-\hat{t})}d\hat{t},
\end{aligned}$$

where we have defined

$$\hat{b}_1^2 = b_1^2/a_1^2;\ \hat{b}_2^2 = b_2^2/a_1^2;\ \hat{a}_2^2 = a_2^2/a_1^2,$$

and we have used the variable $\hat{t} = t^2$.

Note that one may write

$$\frac{(\hat{b}_1^2-\hat{t})(\hat{b}_2^2-\hat{t})}{(1-\hat{t})(\hat{a}_2^2-\hat{t})}=1+\frac{(\hat{b}_1^2-1)(\hat{a}_2^2-1)}{\hat{b}_2^2-1}\frac{1}{1-\hat{t}}$$
$$+(\hat{b}_2^2-\hat{a}_2^2)\Big(1+\frac{\hat{b}_1^2-1}{\hat{b}_2^2-1}\Big)\frac{1}{\hat{a}_2^2-\hat{t}}$$
$$=1+\frac{(b_1^2-a_1^2)(a_2^2-a_1^2)}{b_2^2-a_1^2}\frac{1}{1-\hat{t}}$$
$$+(b_2^2-a_2^2)\Big(1+\frac{b_1^2-a_1^2}{b_2^2-a_1^2}\Big)\frac{1}{\hat{a}_2^2-\hat{t}}$$

and derive that

$$\Big(\int_0^{(1-0)}+\int_{(1+0)}^{1/a_1^2}\Big)\,1\mathrm{d}\hat{t}=\hat{t}\Big|_0^{1/a_1^2}=\tfrac{1}{a_1^2}\approx 1,$$

$$\Big(\int_0^{(1-0)}+\int_{(1+0)}^{1/a_1^2}\Big)\,\tfrac{1}{1-\hat{t}}\mathrm{d}\hat{t}=\ln|1-\hat{t}|\Big|_0^{1/a_1^2}$$
$$=\ln\Big|1-\tfrac{1}{a_1^2}\Big|\approx\tfrac{1}{2}\ln\alpha_4,$$

$$\Big(\int_0^{(1-0)}+\int_{(1+0)}^{1/a_1^2}\Big)\,\tfrac{1}{\hat{a}_2^2-\hat{t}}\mathrm{d}\hat{t}=\ln|\hat{a}_2^2-\hat{t}|\Big|_0^{1/a_1^2}$$
$$=\ln\Big|(1-\tfrac{1}{a_1^2\hat{a}_2^2}\Big|=\ln\Big|1-\tfrac{1}{a_2^2}\Big|\approx\tfrac{1}{2}\ln\alpha_4.$$

We then obtain

$$(I)'+(III)'=-\frac{a_1^2}{2}\left(\int_0^{(1-0)}+\int_{(1+0)}^{1/a_1^2}\right)\frac{(\hat{b}_1^2-\hat{t})(\hat{b}_2^2-\hat{t})}{(1-\hat{t})(\hat{a}_2^2-\hat{t})}\mathrm{d}\hat{t}$$
$$=-\frac{a_1^2}{2}+\mathcal{O}(\sqrt{\alpha_4}\ln\alpha_4)\approx-\frac{a_1^2}{2}.$$

2. The calculations of $(II')$: The point $\zeta=\mathrm{i}a_1$ is a simple pole, so that we have

$$\left(\int_{(C_r)}\right)\frac{1}{1-\hat{t}}\mathrm{d}\hat{t}=-\pi\mathrm{i}.$$

We obtain

$$(II)'=-\frac{a_1^2}{2}\left(\int_{(C_r)}\right)\frac{(\hat{b}_1^2-\hat{t})(\hat{b}_2^2-\hat{t})}{(1-\hat{t})(\hat{a}_2^2-\hat{t})}\mathrm{d}\hat{t}$$
$$=-\frac{a_1^2}{2}\left(\int_{(C_r)}\right)\frac{(\hat{b}_1^2-1)(\hat{a}_2^2-1)}{\hat{b}_2^2-1}\frac{1}{1-\hat{t}}\mathrm{d}\hat{t}=-\sqrt{2\alpha_4}\,\pi\mathrm{i}.$$

3. The calculations of $(IV')$ along the straight line from $\zeta = -\mathrm{i}$ to $\zeta = \zeta_c$: Letting $\zeta = -\mathrm{i} + s(\zeta_c + \mathrm{i})$, we obtain that

$$
\begin{aligned}
(IV)' &= \left(\int_{-\mathrm{i}}^{\zeta_c}\right) \frac{\zeta \mathcal{M} \bar{B}_s}{\bar{A}_s} \mathrm{d}\zeta = \frac{(1-\alpha_4)}{(1-\beta_4)} \left(\int_{-\mathrm{i}}^{\zeta_c}\right) \frac{\zeta(\zeta^2+b_1^2)(\zeta^2+b_2^2)}{(\zeta^2+a_1^2)(\zeta^2+a_2^2)} \mathrm{d}\zeta \\
&\approx -\mathrm{i}\frac{(b_1^2-1)(b_2^2-1)}{(a_1^2-1)(a_2^2-1)}(\zeta_c+\mathrm{i}) + \mathcal{O}((\zeta_c+\mathrm{i})^2) \\
&\approx -\mathrm{i}\frac{\beta_4}{\alpha_4}(\zeta_c+\mathrm{i}),
\end{aligned}
$$

since we have

$$
\frac{(b_1^2-1)(b_2^2-1)}{(a_1^2-1)(a_2^2-1)} \approx \frac{\beta_4}{\alpha_4}.
$$

Furthermore, from the formula (7.222 ) derived for the case $\mathcal{M} = 0$, we may write

$$
\zeta_c + \mathrm{i} = \frac{2^{\frac{5}{7}} 3^{\frac{6}{7}}}{(1+\hat{\alpha}_{\mathrm{T}})^{\frac{2}{7}}} \mathrm{e}^{\mathrm{i}\frac{3\pi}{14}} \sigma_0^{\frac{4}{7}} \alpha_4^{\frac{2}{7}} + \mathcal{O}(\mathcal{M}).
$$

We derive

$$
(IV)' \approx -\mathrm{i}\frac{\beta_4}{\alpha_4} \frac{2^{\frac{5}{7}} 3^{\frac{6}{7}}}{(1+\hat{\alpha}_{\mathrm{T}})^{\frac{2}{7}}} \mathrm{e}^{\mathrm{i}\frac{3\pi}{14}} \sigma_0^{\frac{4}{7}} \alpha_4^{\frac{2}{7}}.
$$

From the above, we derive that

$$
\begin{aligned}
&\frac{\mathrm{i}\mathcal{M}}{2} \left( \int_0^{-\mathrm{i}(a_1-0)} + \int_{(C_r)} + \int_{-\mathrm{i}(a_1+0)}^{-\mathrm{i}} + \int_{-\mathrm{i}}^{\zeta_c} \right) \frac{\zeta \bar{B}_s}{\bar{A}_s} \mathrm{d}\zeta \\
&\quad = -\frac{\mathrm{i}\mathcal{M}}{2} \left[ -\frac{1}{2} - \mathrm{i}\sqrt{2\alpha_4}\,\pi - \mathrm{i}\frac{\beta_4}{\alpha_4} \frac{2^{\frac{5}{7}} 3^{\frac{6}{7}}}{(1+\hat{\alpha}_{\mathrm{T}})^{\frac{2}{7}}} \mathrm{e}^{\mathrm{i}\frac{3\pi}{14}} \sigma_0^{\frac{4}{7}} \alpha_4^{\frac{2}{7}} \right].
\end{aligned}
$$

We now define

$$
\chi(\mathcal{M}) = \frac{1}{\varepsilon} \int_0^{\zeta_c} \left( k_0^{(1)} - k_0^{(3)} \right) \mathrm{d}\zeta = \chi(0) + \tilde{\chi}(\mathcal{M}), \tag{7.315}
$$

where

$$
\begin{aligned}
\tilde{\chi}(\mathcal{M}) &= -\frac{1}{\varepsilon}\frac{\mathrm{i}\mathcal{M}}{2} \left( \int_0^{-\mathrm{i}(a_1-0)} + \int_{(C_r)} + \int_{-\mathrm{i}(a_1+0)}^{-\mathrm{i}} + \int_{-\mathrm{i}}^{\zeta_c} \right) \frac{\zeta \bar{B}_s}{\bar{A}_s} \mathrm{d}\zeta \\
&= \frac{\mathrm{i}\mathcal{M}}{2\varepsilon} \left[ \frac{1}{2} + \mathrm{i}\sqrt{2\alpha_4}\,\pi + \mathrm{i}\frac{\beta_4}{\alpha_4} \frac{2^{\frac{5}{7}} 3^{\frac{6}{7}}}{(1+\hat{\alpha}_{\mathrm{T}})^{\frac{2}{7}}} \mathrm{e}^{\mathrm{i}\frac{3\pi}{14}} \sigma_0^{\frac{4}{7}} \alpha_4^{\frac{2}{7}} \right].
\end{aligned}
$$

If follows that

$$\begin{cases} \mathrm{Re}\{\tilde{\chi}(\mathcal{M})\} = -\dfrac{\mathcal{M}}{2\varepsilon}\left[\sqrt{2\alpha_4}\,\pi + \dfrac{\beta_4}{\alpha_4}\dfrac{2^{\frac{5}{7}}3^{\frac{6}{7}}}{(1+\hat{\alpha}_{\mathrm{T}})^{\frac{2}{7}}}\cos\left(\dfrac{3\pi}{14}\right)\sigma_0^{\frac{4}{7}}\alpha_4^{\frac{2}{7}}\right], \\ \mathrm{Im}\{\tilde{\chi}(\mathcal{M})\} = \dfrac{\mathcal{M}}{2\varepsilon}\left[\dfrac{1}{2} - \dfrac{\beta_4}{\alpha_4}\dfrac{2^{\frac{5}{7}}3^{\frac{6}{7}}}{(1+\hat{\alpha}_{\mathrm{T}})^{\frac{2}{7}}}\sin\left(\dfrac{3\pi}{14}\right)\sigma_0^{\frac{4}{7}}\alpha_4^{\frac{2}{7}}\right]. \end{cases} \tag{7.316}$$

Recall that

$$\begin{cases} \mathrm{Re}\{\chi(0)\} = \dfrac{1}{\varepsilon}\left[-\dfrac{2^{\frac{9}{8}}\Gamma_0}{\sqrt{1+\hat{\alpha}_{\mathrm{T}}}}\alpha_4^{\frac{7}{8}} + C_1\sigma_0^{\frac{11}{7}}\alpha_4^{\frac{2}{7}}\right], \\ \mathrm{Im}\{\chi(0)\} = \dfrac{1}{\varepsilon}\left[\dfrac{1}{\sqrt{1+\hat{\alpha}_{\mathrm{T}}}}\mathcal{R}_0 + \dfrac{9}{4}\sigma_0 + C_2\sigma_0^{\frac{11}{7}}\alpha_4^{\frac{2}{7}}\right]. \end{cases} \tag{7.317}$$

We finally obtain

$$\begin{cases} \mathrm{Re}\{\chi(\mathcal{M})\} = \frac{1}{\varepsilon}\Big[-\mathcal{M}\sqrt{\frac{\alpha_4}{2}}\,\pi - \frac{2^{\frac{9}{8}}\Gamma_0}{\sqrt{1+\hat{\alpha}_{\mathrm{T}}}}\alpha_4^{\frac{7}{8}} \\ \qquad + \left(C_1\sigma_0^{\frac{11}{7}} - \mathcal{M}\tilde{C}_1\sigma_0^{\frac{4}{7}}\right)\alpha_4^{\frac{2}{7}}\Big] \\ \mathrm{Im}\{\chi(\mathcal{M})\} = \frac{1}{\varepsilon}\left[\frac{1}{\sqrt{1+\hat{\alpha}_{\mathrm{T}}}}\mathcal{R}_0 + \frac{9}{4}\sigma_0 + \left(C_2\sigma_0^{\frac{11}{7}} - \mathcal{M}\tilde{C}_2\sigma_0^{\frac{4}{7}}\right)\alpha_4^{\frac{2}{7}}\right], \end{cases} \tag{7.318}$$

where

$$\tilde{C}_1 = \frac{\beta_4}{\alpha_4}\frac{3^{\frac{6}{7}}}{2^{\frac{2}{7}}(1+\hat{\alpha}_{\mathrm{T}})^{\frac{2}{7}}}\cos\left(\frac{3\pi}{14}\right), \quad \tilde{C}_2 = \frac{\beta_4}{\alpha_4}\frac{3^{\frac{6}{7}}}{2^{\frac{2}{7}}(1+\hat{\alpha}_{\mathrm{T}})^{\frac{2}{7}}}\sin\left(\frac{3\pi}{14}\right).$$

Noting that $\sigma_0^{\frac{11}{7}}\alpha_4^{\frac{2}{7}} = \mathcal{O}(\varepsilon)$ and $\mathcal{M}\sigma_0^{\frac{4}{7}}\alpha_4^{\frac{2}{7}} = \varepsilon\widehat{B}\sigma_0^{\frac{4}{7}}\alpha_4^{\frac{2}{7}} \ll \varepsilon$, the quantization condition for the LF modes can be rewritten as

$$C_1\sigma_0^{\frac{11}{7}}\alpha_4^{\frac{2}{7}} = \frac{1.8021}{\sqrt{1+\hat{\alpha}_{\mathrm{T}}}}\alpha_4^{\frac{7}{8}} - \varepsilon\left(n + \frac{1}{2} - \widehat{B}\sqrt{\alpha_4/2}\right)\pi\,. \tag{7.319}$$

For the neutral LF S-mode ($\sigma_0 = 0$) with index $n = 0$, we have

$$\frac{1.8021}{\sqrt{1+\hat{\alpha}_{\mathrm{T}}}}\alpha_4^{\frac{7}{8}} = \varepsilon\left(\frac{1}{2} - \widehat{B}\sqrt{\alpha_4/2}\right)\pi\,, \tag{7.320}$$

or

$$\varepsilon = \frac{1.1473}{\sqrt{1+\hat{\alpha}_{\mathrm{T}}}}\frac{\alpha_4^{\frac{7}{8}}}{\left(1 - \widehat{B}\sqrt{2\alpha_4}\right)}. \tag{7.321}$$

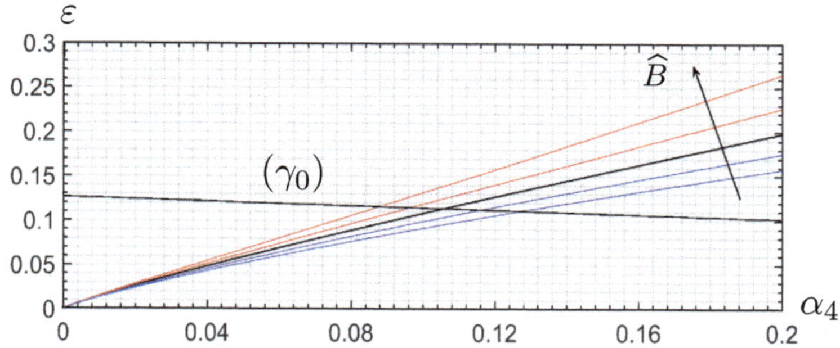


**Fig. 7.20** The *neutral curves* of the LF S-mode ($n = 0$) for the case $\hat{\alpha}_T = 1$ and $\widehat{B} = -0.2, -0.4, 0, 02, 04$. The *dashed line* ($\gamma_0$) is the *neutral curve* of the GTW A-mode ($n = 0$)

In Fig. 7.20 we plot the neutral curves of the LF S-mode ($n = 0$) with $\widehat{B} = -0.4, -0.2, 0, 0.2, 0.4$ for the case $\hat{\alpha}_T = 1$. In the same figure, we also plot the neutral curve of the GTW A-mode ($n = 0$). It is seen that the critical number $\alpha_c$ decreases with increasing kinetic attachment parameter $\widehat{B}$.

It is known that by adding a very small amount of some special species, such as adding an antifreeze protein to the undercooled liquid to a concentration of about $10^{-4}$, one might change the kinetic attachment property of the growth system at the interface remarkably, without much affecting other thermodynamic properties, including its interfacial energy and anisotropy of surface tension. Thus under a certain growth condition, with a decrease in the critical number $\alpha_c$, a growing dendrite with treelike interfacial pattern may be transformed into a smooth needle-like crystal without a side-branching structure. So far, no systematic investigation has ever been conducted on this interesting topic experimentally, though the needle-like crystal patterns with partially and totally suppressed side-branching structure on the interface have indeed been observed by some researchers in a system of ice crystal growth from undercooled water (Furukawa, Private communication, 2012).

## 7.8 Axially Symmetric Dendritic Growth with Anisotropy

For the case of 3D dendritic growth, the anisotropy of surface tension and kinetic effect are generally nonaxisymmetric. In the paraboloidal coordinate system $(\xi, \eta, \theta)$, the surface tension parameter $\Gamma$ and kinetic constant $\beta$ at the interface are functions of $(\xi, \theta)$, which may be written in the form $\Gamma = \hat{\Gamma}A(\xi, \theta)$ and $\beta = \hat{\beta}B(\xi, \theta)$, respectively. We carry out the Fourier series expansion

$$A(\xi, \theta) = \sum_{m=0}^{\infty} A_m(\xi) \cos m\theta, \quad B(\xi, \theta) = \sum_{m=0}^{\infty} B_m(\xi) \cos m\theta.$$

In the above expansions, the leading terms ($m = 0$) may be called the axial components of anisotropy, while all remaining terms ($m \geq 1$) are called the azimuthal components of anisotropy. If we take only the axial components of anisotropy into account while neglecting the effects of all azimuthal components, then these axial components may be assumed to have the same forms for the anisotropy as those for the 2D case. Namely,

$$A_0(\xi) = A_s(\xi), \quad B_0(\xi) = B_s(\xi). \tag{7.322}$$

Thus, along the Ivantsov paraboloid interface, the formula (7.3)–(7.12), the Gibbs–Thomson condition (7.20), and (7.246)–(7.249) are still applicable, except that the curvature operator $\mathcal{K}_2$ of the 2D case in the Gibbs–Thomson condition should be replaced by the curvature operator $\mathcal{K}$ of the axial case. As a consequence, the dispersion relationships (7.109) and (7.285), the formulas for $\sigma_1$ (7.273), and the quantization condition (7.208) for the axisymmetric GTW modes remain unchanged. Particularly, the low-frequency instability mechanism explored in the 2D case is still valid for the axisymmetric case; the quantization conditions (7.234)–(7.236) and (7.319)–(7.321) are still applicable.

## References


1. C. Herring, in *Structure and Properties of Solid Surfaces*, ed. by R. Gomer, C.S. Smith (University Chicago, Chicago, 1952)
2. X.J. Chen, Y.Q. Chen, J.P. Xu, J.J. Xu, Steady needle growth with 3D anisotropy of surface tension. Front. Phys. China **3**(4), 1–18 (2008)
3. J.J. Xu, Interfacial wave theory of solidification — dendritic pattern formation and selection of tip velocity. Phys. Rev. A 15 **432**, 930–947 (1991)
4. D.A. Kessler, H. Levine, Stability of dendritic crystals. Phys. Rev. Lett. **57**, 3069–3072 (1986)
5. D. Bensimon, P. Pelce, B.I. Shraiman, Dynamics of curved fronts and pattern selection. J. Phys. **48**, 2081–2087 (1987)
6. J.J. Xu, Interfacial wave theory for dendritic structure of a growing needle crystal (I): local instability mechanism. Phys. Rev. A **40**(3), 1599–1608 (1989)
7. J.J. Xu, Interfacial wave theory for dendritic structure of a growing needle crystal (II): wave-emission mechanism at the turning point. Phys. Rev. A **40**(3), 1609–1614 (1989)
8. J.J. Xu, Global neutral stable state and selection condition of tip growth velocity. J. Cryst. Growth **100**, 481–490 (1990)
9. J.J. Xu, Two-dimensional dendritic growth with anisotropy of surface tension. Physics (D) **51**, 579–595 (1991)
10. M.E. Glicksman, S.P. Marsh, The dendrite, in *Handbook of Crystal Growth, Volume 1: Fundamentals, Part B: Transport and Stability*, ed. by D.T.J. Hurle (Elsevier Science, North-Holland/Amsterdam, 1993)
11. T. Ihle, H. Müller-Krumbhaar, Fractal and compact growth morphologies in phase transitions with diffusion transport. Phys. Rev. E **49**, 2972–2991 (1994)
12. E. Brener, T. Ihle, H. Müller-Krumbhaar, Y. Saito, K. Shiraishi, Fluctuation effects on dendritic growth morphology. Physica A **204**, 96–110 (1994)
13. J.J. Xu, Stability and selection of dendritic growth with anisotropic kinetic attachment. J. Cryst. Growth **245**, 134–148 (2002)

# Chapter 8
# Three Dimensional Dendritic Growth from an Undercooled Binary Mixture

Up until now, we have considered dendritic growth only from a pure melt. Such a dendrite is also called a thermal dendrite. The dendritic structure on the interface may roughly describe the temperature profile in the melt. However, such a profile will not be traceable after the solidification is completed. A more practical dendritic growth system is the binary alloy system. The dendritic structure formed describes not only the interface shape but also the variation of impurity concentration. After the solidification is completed, the impurity distribution profile remains in the solid state and has a significant effect on the properties of the final materials. This chapter is devoted to this important subject.

We shall consider the case of a single free dendrite growing in an undercooled dilute binary mixture. The system is more complicated now, because it involves both the heat transport process and mass transport in the solute. As demonstrated in Chap. 1, in a binary mixture the concentration of the dilute solute undergoes a jump at the interface due to the different solubilities in the solid and liquid phases. This is measured by the segregation coefficient $\kappa$ defined by $C_S = \kappa C_L$, which relates the concentrations $C_S$ and $C_L$ on the solid and liquid sides of the interface, respectively. If $\kappa < 1$, the solute is rejected by the solid state.

Assume that the concentration of solute in the liquid state, away from the interface, is $C_\infty$, the solute diffusivity in the liquid state is $D$, while the diffusion of solute in the solid state is negligible. There are two different length scales in the present system: the thermal diffusion length $\ell_T$ and solute diffusion length $\ell_D$. Normally, $\ell_D \ll \ell_T$, whence as in Chap. 1, we utilize the solute diffusion length $\ell_D = D/U$ as the length scale instead of the thermal diffusion length $\ell_T = \kappa_T/U$ and set $\lambda = \ell_D/\ell_T$. Furthermore, we use $\ell_D/U$ as the time scale and $C_\infty$ as the scale for the concentration. As before, $\Delta H/(c_p\rho)$ is used as the scale for the temperature.



J.-J. Xu, *Interfacial Wave Theory of Pattern Formation in Solidification*, Springer Series in Synergetics, DOI 10.1007/978-3-319-52663-8_8

For simplicity, we restrict ourself in the case of the symmetric model with $\hat{\alpha}_T = 1$, and further assume that the surface tension is isotropic and that the system is free of convection in the liquid phase.

Note that the Gibbs–Thomson condition for this case involves a new parameter, the so-called morphological parameter $M$, defined as

$$M = -\frac{mC_\infty}{\Delta H/(c_p\rho)}, \tag{8.1}$$

where $m < 0$ is the slope of the liquidus in the phase diagram of the binary system as demonstrated in Chap. 1. Furthermore, the surface tension parameter $\Gamma$ is different from the one used for the pure melt system, for $\Gamma$ is now defined through the mass diffusion length as opposed to the thermal diffusion length. Namely,

$$\Gamma = \frac{\ell_c}{\ell_D}, \tag{8.2}$$

where $\ell_c$ is the capillary length proportional to the isotropic surface tension coefficient. Accordingly, the interfacial stability parameter $\varepsilon = \sqrt{\Gamma}/\eta_0^2$ is also different from the one used in Chap. 6 for the pure melt system.

It is seen that the present system contains six physical parameters: $\Gamma, \lambda, \kappa, M, T_\infty$, and $\eta_0^2$. It is obviously more complicated than that discussed in Chap. 6. Nevertheless, the ideas and approach developed so far can be readily applied to this system. The solutions of the present system have the same mathematical structure as those discussed before. For the case of zero surface tension, it allows an exact similarity solution of the paraboloidal needle crystal growth, like the Ivantsov solution. Therefore, one may adopt the paraboloidal coordinate system $(\xi, \eta, \theta)$, (3.2) introduced in Chap. 3. Moreover, for the general case $\varepsilon \neq 0$, one can define the basic state by the nonclassical steady needle or "nearly' steady" needle solutions addressed in the previous chapters.

## 8.1 Mathematical Formulation of the Problem

The moving paraboloidal coordinate system $(\xi, \eta, \theta)$ adopted is shown in Fig. 8.1. The unknown functions now consist of the temperature fields in the melt and solid states, $T(\xi, \eta, \theta, t)$, $T_S(\xi, \eta, \theta, t)$, respectively, the concentration field in the liquid, $C(\xi, \eta, \theta, t)$, and the interface shape $\eta_s(\xi, \theta, t)$. The unsteady growth process is subject to the heat conduction equation and mass diffusion equation, whose

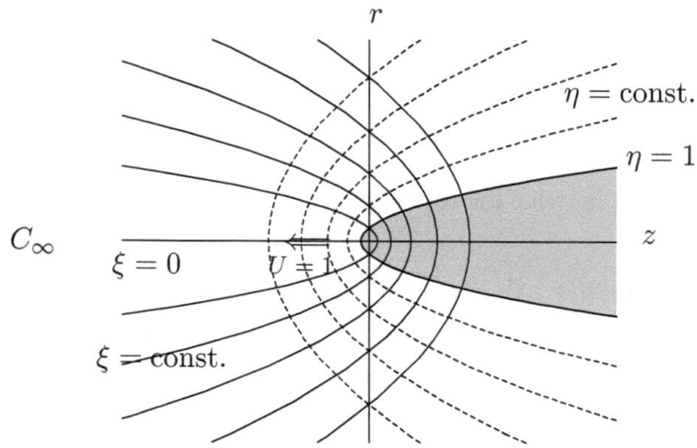


**Fig. 8.1** The paraboloidal coordinate system $(\xi, \eta, \theta)$ for three-dimensional dendritic growth

nondimensional forms are as follows:

$$\begin{aligned}
&\frac{\partial^2 T}{\partial \xi^2}+\frac{\partial^2 T}{\partial \eta^2}+\frac{1}{\xi}\frac{\partial T}{\partial \xi}+\frac{1}{\eta}\frac{\partial T}{\partial \eta}+\Big(\frac{1}{\xi^2}+\frac{1}{\eta^2}\Big)\frac{\partial^2 T}{\partial \theta^2}\\
&\quad=\lambda\eta_0^2\Big(\xi\frac{\partial T}{\partial \xi}-\eta\frac{\partial T}{\partial \eta}\Big)+\lambda\eta_0^4(\xi^2+\eta^2)\frac{\partial T}{\partial t},\\
&\frac{\partial^2 C}{\partial \xi^2}+\frac{\partial^2 C}{\partial \eta^2}+\frac{1}{\xi}\frac{\partial C}{\partial \xi}+\frac{1}{\eta}\frac{\partial C}{\partial \eta}+\Big(\frac{1}{\xi^2}+\frac{1}{\eta^2}\Big)\frac{\partial^2 C}{\partial \theta^2}\\
&\quad=\eta_0^2\Big(\xi\frac{\partial C}{\partial \xi}-\eta\frac{\partial C}{\partial \eta}\Big)+\eta_0^4(\xi^2+\eta^2)\frac{\partial C}{\partial t}.
\end{aligned}\tag{8.3}$$

Since we are considering the symmetric model with $\hat{\alpha}_T = 1$, the heat conduction equation for the solid phase is the same as the above for the liquid phase, provided $T$ is replaced by $T_S$.

The boundary condition are as follows:

1. The upstream condition: as $\eta \to \infty$,

$$T \to T_\infty, \ \ C \to 1. \tag{8.4}$$

2. The regularity condition in the solid state: as $\eta \to 0$,

$$T_S = \mathcal{O}(1). \tag{8.5}$$

3. The interface condition: at $\eta = \eta_s(\xi, \theta, t)$:

   (a) the thermodynamic equilibrium condition

$$T = T_S, \tag{8.6}$$

(b) the modified Gibbs–Thomson condition

$$T = -\frac{\Gamma}{\eta_0^2}\mathcal{K}_3\Big\{\eta_s(\xi,\theta,\varepsilon)\Big\} - MC\,, \tag{8.7}$$

where $\mathcal{K}_3/\eta_0^2$ is twice the local mean curvature of the interface $\eta_s(\xi,\theta,t)$ and

$$\begin{aligned}
\mathcal{K}_3\Big\{\eta_s(\xi,\theta,t,\varepsilon)\Big\} &= -\frac{1}{\left[\xi^2+\eta_s^2\right]^{\frac{3}{2}}\left[\xi^2\eta_s^2(1+\eta_s'^2)+\eta_{s,\theta}^2(\xi^2+\eta_s^2)\right]^{\frac{3}{2}}}\\
&\times\Bigg\{\Big[(1+\eta_s'^2)(\xi^2+\eta_s^2)\Big]\Big[-(\xi\eta_s)^2(\eta_s\eta_s'-\xi)+2\xi\eta_{s,\theta}^2(\xi^2+\eta_s^2)\\
&+\xi^2\eta_s\eta_{s,\theta\theta}(\eta_s\eta_s'-\xi)-\xi\eta_s(\xi\eta_s)'(\eta_{s,\theta}^2+\eta_s\eta_{s,\theta\theta})\Big]\\
&-2\Big[\eta_{s,\theta}\eta_s'(\xi^2+\eta_s^2)\Big]\Big[(\xi\eta_s)'\eta_{s,\theta}(\xi^2+\eta_s^2)+\xi\eta_s(\eta_s\eta_s'-\xi)(\xi\eta_{s,\theta})'\\
&-\xi\eta_s(\eta_s\eta_{s,\theta})'(\xi\eta_s)'\Big]+\xi\eta_s\Big[(\xi\eta_s)^2+\eta_{s,\theta}^2(\xi^2+\eta_s^2)\Big]\\
&\times\Big[(\xi\eta_s)''(\eta_s\eta_s'-\xi)+(\xi\eta_s)'(1-\eta_s'^2-\eta_s\eta_s'')\Big]\Bigg\}\,.
\end{aligned} \tag{8.8}$$

(c) the heat balance condition

$$\begin{aligned}
&\Big(\frac{\partial}{\partial\eta}-\eta_s'\frac{\partial}{\partial\xi}\Big)(T-T_S)-\eta_{s\theta}\Big(\frac{1}{\xi^2}+\frac{1}{\eta_s^2}\Big)\frac{\partial}{\partial\theta}(T-T_S)\\
&+\lambda\eta_0^4(\xi^2+\eta_s^2)\frac{\partial\eta_s}{\partial t}+\lambda\eta_0^2(\xi\eta_s)'=0\,.
\end{aligned} \tag{8.9}$$

(d) the mass balance condition for impurities

$$\begin{aligned}
&\frac{\partial C}{\partial\eta}-\eta_s'\frac{\partial C}{\partial\xi}-\eta_{s\theta}\Big(\frac{1}{\xi^2}+\frac{1}{\eta_s^2}\Big)\frac{\partial C}{\partial\theta}\\
&\qquad+(1-\kappa)C\Big[\eta_0^2(\xi\eta_s)'+\eta_0^4(\xi^2+\eta^2)\frac{\partial\eta_s}{\partial t}\Big]=0\,.
\end{aligned} \tag{8.10}$$

4. The tip smoothness condition: at $\xi=0$, $\eta=\eta_s(0,\theta,\varepsilon)$,

$$\frac{\partial}{\partial\xi}\{T,T_S,\eta_s\}=0\,. \tag{8.11}$$

The root conditions are treated as before.

## 8.2 Basic Steady-State Solution of the System

For given $(\xi, \eta, \theta)$, in a region behind of dendrite-tip and away from the root, we expand the basic state in the regular perturbation expansion (RPE) as $\varepsilon \to 0$:

$$\begin{aligned} T &= T_{\mathrm{B}}(\xi, \eta, \theta, \varepsilon) = T_*(\eta) + \varepsilon^2 \eta_0^2 T_1(\xi, \eta, \theta) + \cdots, \\ T_{\mathrm{S}} &= T_{\mathrm{BS}}(\xi, \eta, \theta, \varepsilon) = T_{\mathrm{S}*} + \varepsilon^2 \eta_0^2 T_{\mathrm{S}1}(\xi, \eta, \theta) + \cdots, \\ C &= C_{\mathrm{B}}(\xi, \eta, \theta, \varepsilon) = C_*(\eta) + \varepsilon^2 \eta_0^2 C_1(\xi, \eta, \theta) + \cdots, \\ \eta &= \eta_{\mathrm{B}}(\xi, \theta, \varepsilon) = 1 + \varepsilon^2 \eta_1(\xi, \theta) + \cdots. \end{aligned} \tag{8.12}$$

### 8.2.1 *The Zeroth-Order Approximation Solution*

For the zeroth-order case, it is easy to derive that $\varepsilon = 0$; the system (8.17)–(8.24) allows the following similarity solution:

$$\begin{aligned} T_* &= T_*(\eta) = T_\infty + \frac{\lambda \eta_0^2}{2} e^{\frac{\lambda \eta_0^2}{2}} E_1\Big(\frac{\lambda \eta_0^2 \eta^2}{2}\Big), \\ T_{\mathrm{S}*} &= T_*(1), \\ C_* &= C_*(\eta) = 1 + A_0\Big(\frac{\eta_0^2}{2}\Big) E_1\Big(\frac{\eta_0^2 \eta^2}{2}\Big), \\ \eta_* &= 1, \end{aligned} \tag{8.13}$$

where

$$\begin{aligned} A_0\Big(\frac{\eta_0^2}{2}\Big) &= \frac{\frac{1}{2}(1-\kappa)\eta_0^2}{e^{-\frac{\eta_0^2}{2}} - (1-\kappa)\frac{\eta_0^2}{2} E_1\big(\frac{\eta_0^2}{2}\big)}, \\ C_{\mathrm{S}*} &= \kappa C_*(1), \end{aligned} \tag{8.14}$$

$$T_\infty + \frac{\lambda \eta_0^2}{2} e^{\frac{\lambda \eta_0^2}{2}} E_1\Big(\frac{\lambda \eta_0^2}{2}\Big) = -M\Big[1 + A_0\Big(\frac{\eta_0^2}{2}\Big) E_1\Big(\frac{\eta_0^2}{2}\Big)\Big].$$

From the above, $\eta_0^2$ can be solved as a function of the undercooling temperature $T_\infty$, the ratio of diffusivities $\lambda$, and the morphological parameter $M$. The tip radius of the paraboloidal interface $\ell_{\mathrm{t}}$ is calculated to be

$$\ell_{\mathrm{t}} = \eta_0^2 \ell_{\mathrm{D}}. \tag{8.15}$$

If one defines the Péclet number of the binary mixture system as the ratio of the tip radius of the interface to the mass diffusion length, namely $\mathrm{Pe} = \frac{\ell_{\mathrm{t}}}{\ell_{\mathrm{D}}}$, then it follows that when $\varepsilon = 0$, we have $\mathrm{Pe}_0 = \eta_0^2$.

The jump in the concentration at the interface can be obtained from the interface condition (8.10) as

$$\Delta C = (1-\kappa)C_*(1) = -\frac{1}{\eta_0^2}\frac{\mathrm{d}C_*}{\mathrm{d}\eta}(1). \tag{8.16}$$

The first-order approximation solutions of the basic states may be nonaxisymmetric, due to the nonzero anisotropic surface tension, but they are of order $\mathcal{O}(\varepsilon^2)$. We shall not proceed further to solve the first-order approximation solution for the basic states, since we are interested in solving the eigenvalue problem (EVP) for the unsteady perturbed states around the basic steady state only with an accuracy of $\mathcal{O}(\varepsilon)$. So, we set

$$\begin{aligned}
&T_{\mathrm{B}}(\xi,\eta,\theta,\varepsilon) \approx T_*(\eta), \quad T_{\mathrm{BS}}(\xi,\eta,\theta,\varepsilon) \approx T_{\mathrm{S}*},\\
&C_{\mathrm{B}}(\xi,\eta,\theta,\varepsilon) \approx C_*(\eta), \quad C_{\mathrm{BS}}(\xi,\eta,\theta,\varepsilon) \approx C_{\mathrm{S}*},\\
&\eta_{\mathrm{B}}(\xi,\theta,\varepsilon) \approx 1.
\end{aligned}$$

## 8.3 Three-Dimensional Linear Perturbed States Around the Basic State

We write the general unsteady solutions in two parts:

$$\begin{aligned}
T &= T_{\mathrm{B}} + \tilde{T}(\xi,\eta,\theta,t,\varepsilon),\\
T_{\mathrm{S}} &= T_{\mathrm{SB}} + \tilde{T}_{\mathrm{S}}(\xi,\eta,\theta,t,\varepsilon),\\
\eta_{\mathrm{s}} &= \eta_{\mathrm{B}} + \tilde{h}(\xi,\theta,t,\varepsilon)/\eta_0^2\,.
\end{aligned} \tag{8.17}$$

The perturbed states may be nonaxisymmetric, even if some of the basic states are axisymmetric. The 3D linearized perturbed system is obtained as follows:

$$\begin{aligned}
&\tfrac{\partial^2\tilde{T}}{\partial\xi^2} + \tfrac{\partial^2\tilde{T}}{\partial\eta^2} + \left(\tfrac{1}{\xi} - \lambda\eta_0^2\xi\right)\tfrac{\partial\tilde{T}}{\partial\xi} + \left(\tfrac{1}{\eta} + \lambda\eta_0^2\eta\right)\tfrac{\partial\tilde{T}}{\partial\eta}\\
&\quad+\left(\tfrac{1}{\xi^2} + \tfrac{1}{\eta^2}\right)\tfrac{\partial^2\tilde{T}}{\partial\theta^2} = \lambda\eta_0^4(\xi^2+\eta^2)\tfrac{\partial\tilde{T}}{\partial t},\\
&\tfrac{\partial^2\tilde{C}}{\partial\xi^2} + \tfrac{\partial^2\tilde{C}}{\partial\eta^2} + \left(\tfrac{1}{\xi} - \eta_0^2\xi\right)\tfrac{\partial\tilde{C}}{\partial\xi} + \left(\tfrac{1}{\eta} + \eta_0^2\eta\right)\tfrac{\partial\tilde{C}}{\partial\eta}\\
&\quad+\left(\tfrac{1}{\xi^2} + \tfrac{1}{\eta^2}\right)\tfrac{\partial^2\tilde{C}}{\partial\theta^2} = \eta_0^4(\xi^2+\eta^2)\tfrac{\partial\tilde{C}}{\partial t},
\end{aligned} \tag{8.18}$$

with the following boundary conditions:

1. As $\eta \to \infty$,

$$\tilde{T} \to 0,\ \tilde{C} \to 0\,. \tag{8.19}$$

2. As $\eta \to 0$,

$$\{\tilde{T}_S, \tilde{C}_S\} \quad \text{are regular}\,. \tag{8.20}$$

3. The interface conditions: Since we neglect the effect of anisotropy of surface tension on the perturbed states, the interface conditions at the interface of the basic state are as follows: At $\eta = 1 + \bar{\eta}_s$,

   (a) the thermodynamic equilibrium condition

$$\tilde{T} = \tilde{T}_S + \lambda\tilde{h} + \mathcal{O}(\varepsilon^2)\,, \tag{8.21}$$

   (b) the Gibbs–Thomson condition

$$\tilde{T}_S = \frac{\varepsilon^2}{\hat{S}(\xi)}\left\{\frac{\partial^2\tilde{h}}{\partial\xi^2} + \left(1 + \frac{1}{\xi^2}\right)\frac{\partial^2\tilde{h}}{\partial\theta^2} + \frac{1+2\xi^2}{\xi\hat{S}^2(\xi)}\frac{\partial\tilde{h}}{\partial\xi} + \frac{\xi^2(2+\xi^2)+4}{\hat{S}^2(\xi)}\tilde{h}\right\} - M(\tilde{C} - \Delta C\tilde{h}) + \mathcal{O}(\varepsilon^2), \tag{8.22}$$

   where

$$\hat{S}(\xi) = \sqrt{1+\xi^2}\,, \tag{8.23}$$

   (c) the heat balance condition

$$\frac{\partial}{\partial\eta}\left(\tilde{T} - \tilde{T}_S\right) + \lambda S^2(\xi)\frac{\partial\tilde{h}}{\partial t} + \lambda\xi\frac{\partial\tilde{h}}{\partial\xi} + \lambda\left(2 + \lambda\eta_0^2\right)\tilde{h} = \mathcal{O}(\varepsilon^2), \tag{8.24}$$

   (d) the mass balance condition for impurity

$$\frac{\partial\tilde{C}}{\partial\eta} + \Delta C\left\{\eta_0^2 S^2(\xi)\frac{\partial\tilde{h}}{\partial t} + \xi\frac{\partial\tilde{h}}{\partial\xi}\right\} + \eta_0^2(1-\kappa)\tilde{C} + \tilde{h}\Delta C(2+\eta_0^2) = \mathcal{O}(\varepsilon^2). \tag{8.25}$$

   Note that the $\theta$-dependent terms in the above heat balance condition and mass balance condition are of order $\mathcal{O}(\varepsilon^2)$, which will not be taken into account in the later calculations.

4. The tip smoothness condition: at $\xi = 0,\ \eta = 1$,

$$\frac{\partial}{\partial\xi}\{\tilde{T}, \tilde{T}_S, \tilde{C}, \tilde{h}\} = 0\,. \tag{8.26}$$

5. The root condition: as $\xi \to \infty$,

$$\{\tilde{T}, \tilde{T}_S, \tilde{C}, \tilde{h}\} = 0\,. \tag{8.27}$$

## 8.4 Multiple Variables Expansion Solution in the Outer Region

In the outer region away from the tip and all other possible singular points, one may look for the asymptotic expansion of the solutions $\tilde{q}(\xi, \eta, \theta, t, \varepsilon)$ in terms of the multiple variables expansion (MVE) method. As before, we define the fast variables $\xi_+$, $\eta_+$, and $t_+$ as

$$\begin{aligned} \xi_+ &= \frac{\phi(\xi,\eta,\varepsilon)}{\varepsilon} = \frac{1}{\varepsilon}\int_0^{\xi} k(\xi_1,\eta,\varepsilon)\mathrm{d}\xi_1, \\ \eta_+ &= \frac{\psi(\xi,\eta,\varepsilon)}{\varepsilon} = \frac{1}{\varepsilon}\int_1^{\eta} k(\xi,\eta_1,\varepsilon)\mathrm{d}\eta_1, \\ t_+ &= \frac{t}{\eta_0^2\varepsilon}, \end{aligned} \tag{8.28}$$

and denote that at the interface $\eta = 1$, $\bar{\xi}_+(\xi,\varepsilon) = \xi_+(\xi, 1, \varepsilon)$, $\bar{\phi}(\xi,\varepsilon) = \phi(\xi,1,\varepsilon)$ and $\bar{k}(\xi,\varepsilon) = k(\xi,1,\varepsilon)$.

As in Chap. 6, the pair of phase functions $(\phi, \psi)$ are assumed to be subject to the following Cauchy–Riemann conditions:

$$\begin{aligned} &\phi_\xi = \psi_\eta = k(\xi,\eta,\varepsilon), \quad \phi_\eta = -\psi_\xi = g(\xi,\eta,\varepsilon), \\ &(\mathrm{Re}\{\phi_\xi > 0\}, \quad \mathrm{Re}\{\psi_\eta\} > 0). \end{aligned} \tag{8.29}$$

It follows that $\phi_\xi\psi_\xi + \phi_\eta\psi_\eta = 0$, and $\nabla^2\phi(\xi,\eta,\varepsilon) = \nabla^2\psi(\xi,\eta,\varepsilon) = 0$. Note that we did not introduce a fast variable $\theta_+$ in the $\theta$-direction, since it is assumed that the fast variations of the solutions occur only in the $\xi$- and $\eta$-directions, but not in the $\theta$-direction.

As was done in Chap. 6, by formally treating the fast and slow variables $(\xi, \eta, \theta, \xi_+, \eta_+, t_+)$ as independent variables, the system (8.18)–(8.27) can be converted to a system with multiple variables as follows: The converted equations in the liquid phase are

$$\begin{aligned} &(\phi_\xi^2 + \phi_\eta^2)\left(\frac{\partial^2}{\partial \xi_+^2} + \frac{\partial^2}{\partial \eta_+^2}\right)\tilde{T} + \varepsilon^2\left(\frac{1}{\xi^2} + \frac{1}{\eta^2}\right)\frac{\partial^2 \tilde{T}}{\partial \theta^2} \\ &= -\varepsilon\left(2\phi_\xi \frac{\partial^2 \tilde{T}}{\partial \xi_+ \partial \xi} + 2\psi_\xi \frac{\partial^2 \tilde{T}}{\partial \eta_+ \partial \xi}\right) - \varepsilon\left(2\psi_\eta \frac{\partial^2 \tilde{T}}{\partial \eta_+ \partial \eta} + 2\phi_\eta \frac{\partial^2 \tilde{T}}{\partial \xi_+ \partial \eta}\right) \end{aligned}$$

$$-\varepsilon\Big(\frac{1}{\xi}-\eta_0^2\xi\Big)\Big(\phi_\xi\frac{\partial\tilde{T}}{\partial\xi_+}+\psi_\xi\frac{\partial\tilde{T}}{\partial\eta_+}\Big)-\varepsilon\Big(\frac{1}{\eta}+\eta_0^2\eta\Big)\Big(\phi_\eta\frac{\partial\tilde{T}}{\partial\xi_+}+\psi_\eta\frac{\partial\tilde{T}}{\partial\eta_+}\Big)$$
$$+\varepsilon\eta_0^2(\xi^2+\eta^2)\frac{\partial\tilde{T}}{\partial t_+}+\mathcal{O}(\varepsilon^2), \tag{8.30}$$

and

$$(\phi_\xi^2+\phi_\eta^2)\Big(\frac{\partial^2}{\partial\xi_+^2}+\frac{\partial^2}{\partial\eta_+^2}\Big)\tilde{C}+\varepsilon^2\Big(\frac{1}{\xi^2}+\frac{1}{\eta^2}\Big)\frac{\partial^2\tilde{C}}{\partial\theta^2}$$
$$=-\varepsilon\left(2\phi_\xi\frac{\partial^2\tilde{C}}{\partial\xi_+\partial\xi}+2\psi_\xi\frac{\partial^2\tilde{C}}{\partial\eta_+\partial\xi}\right)-\varepsilon\left(2\psi_\eta\frac{\partial^2\tilde{C}}{\partial\eta_+\partial\eta}+2\phi_\eta\frac{\partial^2\tilde{C}}{\partial\xi_+\partial\eta}\right)$$
$$-\varepsilon\Big(\frac{1}{\xi}-\eta_0^2\xi\Big)\Big(\phi_\xi\frac{\partial\tilde{C}}{\partial\xi_+}+\psi_\xi\frac{\partial\tilde{T}}{\partial\eta_+}\Big)-\varepsilon\Big(\frac{1}{\eta}+\eta_0^2\eta\Big)\Big(\phi_\eta\frac{\partial\tilde{C}}{\partial\xi_+}+\psi_\eta\frac{\partial\tilde{C}}{\partial\eta_+}\Big)$$
$$+\varepsilon\eta_0^2(\xi^2+\eta^2)\frac{\partial\tilde{C}}{\partial t_+}+\mathcal{O}(\varepsilon^2). \tag{8.31}$$

From the boundary conditions at $\eta = 1$, one deduces $\psi(\xi,1,\varepsilon) = 0$ and $\phi_\eta(\xi,1,\varepsilon) = -\psi_\xi(\xi,1,\varepsilon) = g(\xi,1,\varepsilon) = 0$. By following the procedure developed in Chap. 6, one can solve the harmonic functions $\phi(\xi,\eta)$ and $\psi(\xi,\eta)$ in the whole $(\xi,\eta)$-plane once the function $k(\xi,1) = \bar{k}(\xi)$ is known. Namely, by letting the function $\bar{k}(\xi)$ of the real variable $\xi$ be extended to the function $\bar{k}(\zeta)$ in the complex plane $\zeta = \xi + \mathrm{i}\xi_{\mathrm{I}}$ by analytic continuation, one may obtain

$$\phi(\xi,\eta) = \frac{1}{2}\left\{\int_0^{\xi+\mathrm{i}(\eta-1)}\bar{k}(\zeta)\mathrm{d}\zeta+\int_0^{\xi-\mathrm{i}(\eta-1)}\bar{k}(\zeta)\mathrm{d}\zeta\right\},$$
$$\psi(\xi,\eta) = -\frac{\mathrm{i}}{2}\int_{\xi-\mathrm{i}(\eta-1)}^{\xi+\mathrm{i}(\eta-1)}\bar{k}(\zeta)\mathrm{d}\zeta.$$

The boundary conditions for the system with multiple variables are as follows:

1. As $\eta_+\to\infty$

$$\tilde{T}\to 0\,. \tag{8.32}$$

2. As $\eta_+\to-\infty$,

$$\tilde{T}_{\mathrm{S}}\to 0\,. \tag{8.33}$$

3. At the interface of the needle crystal, $\eta_+ = 0, \eta = 1$:
   (a) the thermodynamic equilibrium condition

$$\tilde{T} = \tilde{T}_S + \lambda\tilde{h} - \varepsilon\eta_1\bar{k}\frac{\partial}{\partial\eta_+}(\tilde{T} - \tilde{T}_S) + \mathcal{O}(\varepsilon^2)\,, \tag{8.34}$$

   (b) the Gibbs–Thomson condition

$$\begin{aligned}\tilde{T}_S + \varepsilon\eta_1\bar{k}\frac{\partial\tilde{T}_S}{\partial\eta_+} = \frac{1}{\hat{S}(\xi)}\Bigg\{&\left(\bar{k}^2\frac{\partial^2}{\partial\xi_+^2} + 2\varepsilon\bar{k}\frac{\partial^2}{\partial\xi_+\partial\xi} + \varepsilon\bar{k}'\frac{\partial}{\partial\xi_+}\right) \\ &+\varepsilon\left(\frac{1}{\xi} + \frac{\xi}{S^2(\xi)}\right)\left(\bar{k}\frac{\partial}{\partial\xi_+} + \varepsilon\frac{\partial}{\partial\xi}\right)\Bigg\}\tilde{h} + \frac{\varepsilon^2\hat{S}}{\xi^2}\frac{\partial^2\tilde{h}}{\partial\theta^2} \\ &-M\left(\tilde{C} + \varepsilon\eta_1\bar{k}\frac{\partial\tilde{C}}{\partial\eta_+} - \Delta C\tilde{h}\right) + \mathcal{O}(\varepsilon^2),\end{aligned} \tag{8.35}$$

   (c) the heat balance condition

$$\begin{aligned}&\bar{k}\left(\frac{\partial\tilde{T}}{\partial\eta_+} - \frac{\partial\tilde{T}_S}{\partial\eta_+}\right) + \varepsilon\eta_1\bar{k}^2\frac{\partial^2}{\partial\eta_+^2}(\tilde{T} - \tilde{T}_S) + \varepsilon\left(\frac{\partial\tilde{T}}{\partial\eta} - \frac{\partial\tilde{T}_S}{\partial\eta}\right) \\ &+\lambda S^2(\xi)\frac{\partial\tilde{h}}{\partial t_+} + \lambda\xi\left(k\frac{\partial\tilde{h}}{\partial\bar{\xi}_+} + \varepsilon\frac{\partial\tilde{h}}{\partial\xi}\right) + \varepsilon\lambda\left(2 + \lambda\eta_0^2\right)\tilde{h} \\ &= \mathcal{O}(\varepsilon^2),\end{aligned} \tag{8.36}$$

   (d) the mass balance condition

$$\begin{aligned}&k\frac{\partial\tilde{C}}{\partial\eta_+} + \varepsilon\frac{\partial\tilde{C}}{\partial\eta} + \varepsilon\eta_1\bar{k}^2\frac{\partial^2\tilde{C}}{\partial\eta_+^2} + \varepsilon\eta_0^2(1-\kappa)\tilde{C} \\ &+\Delta C\left\{S^2(\xi)\frac{\partial\tilde{h}}{\partial t_+} + \xi\left(k\frac{\partial\tilde{h}}{\partial\hat{\xi}_+} + \varepsilon\frac{\partial\tilde{h}}{\partial\xi}\right) + \varepsilon(2 + \eta_0^2)\tilde{h}\right\} \\ &+\mathcal{O}(\varepsilon^2) = 0\,.\end{aligned} \tag{8.37}$$

4. The tip smoothness condition: at $\xi = 0,\ \eta = 1$,

$$\frac{\partial}{\partial\xi}\{\tilde{T}, \tilde{T}_S, \tilde{C}, \tilde{h}\} = 0\,. \tag{8.38}$$

5. The root condition: as $\xi \to \infty$, one has

$$\{\tilde{T};\ \tilde{T}_S;\ \tilde{C};\ \tilde{h}\} = 0\,. \tag{8.39}$$

We now make the following multiple variables expansion (MVE) for the 3D perturbed states:

$$
\begin{aligned}
\tilde{T} &= \tilde{\mu}_0(\varepsilon)\left\{\tilde{T}_0(\xi,\eta,\xi_+,\eta_+) + \varepsilon\tilde{T}_1(\bullet) + \cdots\right\} e^{im\theta+\sigma t_+}, \\
\tilde{T}_S &= \tilde{\mu}_0(\varepsilon)\left\{\tilde{T}_{S0}(\xi,\eta,\xi_+,\eta_+) + \varepsilon\tilde{T}_{S1}(\bullet) + \cdots\right\} e^{im\theta+\sigma t_+}, \\
\tilde{C} &= \tilde{\mu}_0(\varepsilon)\left\{\tilde{C}_0(\xi,\eta,\xi_+,\eta_+) + \varepsilon\tilde{C}_1(\bullet) + \cdots\right\} e^{im\theta+\sigma t_+}, \\
\tilde{h} &= \tilde{\mu}_0(\varepsilon)\left\{\tilde{h}_0(\xi,\bar{\xi}_+) + \varepsilon\tilde{h}_1(\xi,\bar{\xi}_+) + \cdots\right\} e^{im\theta+\sigma t_+},
\end{aligned}
\tag{8.40}
$$

where $m = 0, 1, 2, \ldots$ and $\sigma = \sigma_R - i\omega \quad (\omega \geq 0)$ is generally a complex number.

In addition to the expansion form (8.40), we make the following asymptotic expansions for the functions $(\phi, \psi, \bar{k})$ and the parameter $\sigma$:

$$
\begin{aligned}
\phi(\xi,\eta,\varepsilon) &= \phi_0(\xi,\eta) + \varepsilon\phi_1(\xi,\eta) + \cdots, \\
\psi(\xi,\eta,\varepsilon) &= \psi_0(\xi,\eta) + \varepsilon\psi_1(\xi,\eta) + \cdots, \\
k(\xi,\eta,\varepsilon) &= k_0(\xi,\eta) + \varepsilon k_1(\xi,\eta) + \cdots, \\
\bar{k}(\xi,\varepsilon) &= \bar{k}_0(\xi) + \varepsilon\bar{k}_1(\xi) + \cdots, \\
g(\xi,\eta,\varepsilon) &= g_0(\xi,\eta) + \varepsilon g_1(\xi,\eta) + \cdots, \\
\bar{g}(\xi,\varepsilon) &= 0, \\
\sigma &= \sigma_0 + \varepsilon\sigma_1 + \varepsilon^2\sigma_2 + \cdots.
\end{aligned}
\tag{8.41}
$$

In the first step, we let $\sigma_0$ be an arbitrary constant. By substituting (8.40) and (8.41) into the system (8.30)–(8.39), one can successively derive each order of approximation.

## 8.5 The MVE Solutions in the Outer Region

### *8.5.1 The Zeroth-Order Approximation*

In the zeroth-order, we obtain

$$
\begin{aligned}
&\left(\frac{\partial^2}{\partial\xi_+^2} + \frac{\partial^2}{\partial\eta_+^2}\right)\tilde{T}_0 = 0, \\
&\left(\frac{\partial^2}{\partial\xi_+^2} + \frac{\partial^2}{\partial\eta_+^2}\right)\tilde{C}_0 = 0,
\end{aligned}
\tag{8.42}
$$

and the following boundary conditions:

1. As $\eta_+ \to \infty$,

$$\tilde{T}_0 \to 0 \quad \text{and} \quad \tilde{C}_0 \to 0 \,. \tag{8.43}$$

2. As $\eta_+ \to -\infty$,

$$\tilde{T}_{S0} \to 0 \,. \tag{8.44}$$

3. At $\eta_+ = 0, \eta = 1$,

$$\tilde{T}_0 = \tilde{T}_{S0} + \lambda \tilde{h}_0 \,, \tag{8.45}$$

$$\tilde{T}_{S0} = \frac{\bar{k}_0^2}{S(\xi)} \frac{\partial^2 \tilde{h}_0}{\partial \hat{\xi}_+^2} - M(\tilde{C}_0 - \Delta C \tilde{h}_0), \tag{8.46}$$

$$\bar{k}_0 \frac{\partial}{\partial \eta_+} \left( \tilde{T}_0 - \tilde{T}_{S0} \right) + \lambda \sigma_0 S^2(\xi) \tilde{h}_0 + \lambda \xi \bar{k}_0 \frac{\partial \tilde{h}_0}{\partial \hat{\xi}_+} = 0 \,, \tag{8.47}$$

$$\bar{k}_0 \frac{\partial \tilde{C}_0}{\partial \eta_+} + \Delta C \left[ \sigma_0 S^2(\xi) \tilde{h}_0 + \xi \bar{k}_0 \frac{\partial \tilde{h}_0}{\partial \hat{\xi}_+} \right] = 0 \,. \tag{8.48}$$

4. In the far field away from the tip, as $\xi \to \infty$,

$$\hat{h}_0 < \infty. \tag{8.49}$$

Hence, we have the normal mode solutions

$$\begin{aligned} \tilde{T}_0 &= A_0(\xi, \eta) \exp(\mathrm{i}\xi_+ - \eta_+), \\ \tilde{T}_{S0} &= A_{S0}(\xi, \eta) \exp(\mathrm{i}\xi_+ + \eta_+), \\ \tilde{C}_0 &= B_0(\xi, \eta) \exp(\mathrm{i}\xi_+ - \eta_+), \\ \tilde{h}_0 &= \hat{D}_0 \exp(\mathrm{i}\bar{\xi}_+) \,. \end{aligned} \tag{8.50}$$

We write

$$\hat{A}_0(\xi) = A_0(\xi, 1),\ \hat{A}_{S0}(\xi) = A_{S0}(\xi, 1),\ \hat{B}_0(\xi) = B_0(\xi, 1), \tag{8.51}$$

and then from (8.45)–(8.48), we derive

$$
\begin{aligned}
\hat{A}_0 - \hat{A}_{S0} - \lambda \hat{D}_0 &= 0,\\
\hat{A}_{S0} + \Big(\frac{\bar{k}_0^2}{S} - M\Delta C\Big)\hat{D}_0 + M\hat{B}_0 &= 0,\\
-\bar{k}_0(\hat{A}_0 + \hat{A}_{S0}) + \lambda(\sigma_0 S^2 + \mathrm{i}\xi\bar{k}_0)\hat{D}_0 &= 0\\
-\bar{k}_0\hat{B}_0 + \Delta C(\sigma_0 S^2 + \mathrm{i}\xi\bar{k}_0)\hat{D}_0 &= 0\,.
\end{aligned}
\tag{8.52}
$$

For a nontrivial solution, one must have

$$
\Delta = \det\begin{pmatrix}
1 & -1 & -\lambda & 0\\
0 & 1 & (\bar{k}_0^2/S - M\Delta C) & M\\
-\bar{k}_0 & -\bar{k}_0 & \lambda(\sigma_0 S^2 + \mathrm{i}\xi\bar{k}_0) & 0\\
0 & 0 & \Delta C(\sigma_0 S^2 + \mathrm{i}\xi\bar{k}_0) & -\bar{k}_0
\end{pmatrix} = 0\,. \tag{8.53}
$$

From this we derive the local dispersion relationship

$$
\sigma_0 = \frac{\bar{k}_0}{\hat{S}^2}\left(1 - \frac{2\bar{k}_0^2}{(\lambda + 2M\Delta C)\hat{S}}\right) - \mathrm{i}\frac{\xi\bar{k}_0}{S^2}, \tag{8.54}
$$

which may be written as

$$
\sigma_0 = \Sigma(\xi, \bar{k}_0) = \frac{\bar{k}_0}{\hat{S}^2}\left(1 - \frac{2\bar{k}_0^2}{\Lambda_c\hat{S}}\right) - \mathrm{i}\frac{\xi\bar{k}_0}{S^2}\,, \tag{8.55}
$$

where we have defined

$$
\Lambda_c = \lambda + 2M\Delta C\,. \tag{8.56}
$$

One may further derive that

$$
\begin{cases}
\hat{A}_{S0} = -\lambda\frac{\bar{k}_0^2}{\Lambda_c S}\hat{D}_0,\\
\hat{A}_0 + \hat{A}_{S0} = \lambda\Big(1 - \frac{2\bar{k}_0^2}{\Lambda_c\hat{S}}\Big)\hat{D}_0,\\
\hat{A}_0 - \hat{A}_{S0} = \lambda\hat{D}_0,\\
\hat{B}_0 = \Delta C\Big(1 - \frac{2\bar{k}_0^2}{\Lambda_c\hat{S}}\Big)\hat{D}_0,
\end{cases}
\tag{8.57}
$$

where $\hat{D}_0$ is an arbitrary constant. The above dispersion formula is a generalization of the dispersion relationship for dendritic growth from a pure melt. In fact, the dispersion relationship for dendritic growth from a pure melt coincides with the above formula for the special case

$$\Lambda_c = 1. \tag{8.58}$$

For a given constant $\sigma$, one can find three roots of (8.55) as functions of $\xi$ yielding results similar to those presented in Fig. 6.1, namely

$$\begin{cases} \bar{k}_0^{(1)}(\xi) = \hat{M}(\xi)\cos\left[\frac{1}{3}\cos^{-1}\left(\frac{\sigma_0}{\hat{N}(\xi)}\right)\right] & \text{(short-wave branch)}, \\ \bar{k}_0^{(2)}(\xi) = \hat{M}(\xi)\cos\left[\frac{1}{3}\cos^{-1}\left(\frac{\sigma_0}{\hat{N}(\xi)}\right) + \frac{2\pi}{3}\right], & \\ \bar{k}_0^{(3)}(\xi) = \hat{M}(\xi)\cos\left[\frac{1}{3}\cos^{-1}\left(\frac{\sigma_0}{\hat{N}(\xi)}\right) + \frac{4\pi}{3}\right] & \text{(long-wave branch)}, \end{cases} \tag{8.59}$$

where we define

$$\begin{cases} \hat{M}(\xi) = \sqrt{\dfrac{2\Lambda_c\hat{S}(\xi)}{3}}(1 - \mathrm{i}\xi)^{\frac{1}{2}}, \\ \hat{N}(\xi) = -\dfrac{\hat{M}(\xi)}{3\hat{S}^2(\xi)}(1 - \mathrm{i}\xi). \end{cases} \tag{8.60}$$

These wave number functions respectively yield three basic solutions. We call the solution $\tilde{h}_0 = H_1(\xi) = \mathrm{e}^{\sigma_0 t + \mathrm{i}\bar{\xi}_+^{(1)}}$ the $H_1$-wave; the solution $\tilde{h}_0 = H_2(\xi) = \mathrm{e}^{\sigma_0 t + \mathrm{i}\bar{\xi}_+^{(2)}}$ the $H_2$-wave; and the solution $\tilde{h}_0 = H_3(\xi) = \mathrm{e}^{\sigma_0 t + \mathrm{i}\bar{\xi}_+^{(3)}}$ the $H_3$-wave, in which

$$\bar{\xi}_+^{(i)} = \frac{1}{\xi}\int_{\xi_0}^{\xi} \bar{k}^{(i)}(\xi_1)\mathrm{d}\xi_1, \quad (i = 1, 2, 3).$$

Note that $\mathrm{Re}\{\bar{k}_0^{(1)}(\xi)\} > \mathrm{Re}\{\bar{k}_0^{(3)}(\xi)\} > 0 > \mathrm{Re}\{\bar{k}_0^{(2)}(\xi)\}$. Only the short-wave branch, the $H_1$-wave with the larger positive number $\mathrm{Re}\{\bar{k}_0^{(1)}\}$, and the long-wave branch, the $H_3$-wave, with the smaller positive number $\mathrm{Re}\{\bar{k}_0^{(3)}\}$, are physically meaningful. In order for the solutions to satisfy the boundary conditions (8.43) and (8.44), the $H_2$-wave with negative number $\mathrm{Re}\{\bar{k}_0^{(2)}\}$ must be ruled out. Thus, the general $H$-wave solution of the interface shape function has two components: $\tilde{h}_0(\xi) = H(\xi) = D_0^{(1)}H_1(\xi) + D_0^{(3)}H_3(\xi)$, and one may write the outer solution in the form

$$\begin{aligned} \tilde{h} = {} & D_0^{(1)}\exp\left[\frac{\sigma t}{\varepsilon\eta_0^2} + \frac{\mathrm{i}}{\varepsilon}\int_{\xi_c}^{\xi}\left(\bar{k}_0^{(1)} + \varepsilon\bar{k}_1^{(1)} + \cdots\right)\mathrm{d}\xi_1\right] \\ & + D_0^{(3)}\exp\left[\frac{\sigma t}{\varepsilon\eta_0^2} + \frac{\mathrm{i}}{\varepsilon}\int_{\xi_c}^{\xi}\left(\bar{k}_0^{(3)} + \varepsilon\bar{k}_1^{(3)} + \cdots\right)\mathrm{d}\xi_1\right] + \cdots. \end{aligned} \tag{8.61}$$

### 8.5.2 First-Order Approximation

The first-order approximation solution will fully determine the amplitude $\{A_0(\xi,\eta), A_{S0}(\xi,\eta)\}$, the functions $\bar{k}_1(\xi)$, and $\sigma_1$. The equation for the first-order approximation can be obtained from (8.30)–(8.31) as the following:

$$
\begin{aligned}
&(k_0^2+g_0^2)\left(\frac{\partial^2}{\partial\xi_+^2}+\frac{\partial^2}{\partial\eta_+^2}\right)\tilde{T}_1\\
&\quad=-\left(2k_0\frac{\partial^2\tilde{T}_0}{\partial\xi_+\partial\xi}-2g_0\frac{\partial^2\tilde{T}_0}{\partial\eta_+\partial\xi}\right)-\left(2k_0\frac{\partial^2\tilde{T}_0}{\partial\eta_+\partial\eta}+2g_0\frac{\partial^2\tilde{T}_0}{\partial\xi_+\partial\eta}\right)\\
&\quad\quad-\Big(\frac{1}{\xi}-\lambda\eta_0^2\xi\Big)\Big(k_0\frac{\partial\tilde{T}_0}{\partial\xi_+}-g_0\frac{\partial\tilde{T}_0}{\partial\eta_+}\Big)-\Big(\frac{1}{\eta}+\lambda\eta_0^2\eta\Big)\Big(g_0\frac{\partial\tilde{T}_0}{\partial\xi_+}+k_0\frac{\partial\tilde{T}_0}{\partial\eta_+}\Big)\\
&\quad\quad+\lambda\eta_0^2(\xi^2+\eta^2)\frac{\partial\tilde{T}_0}{\partial t_+},
\end{aligned}
\tag{8.62}
$$

and

$$
\begin{aligned}
&(k_0^2+g_0^2)\left(\frac{\partial^2}{\partial\xi_+^2}+\frac{\partial^2}{\partial\eta_+^2}\right)\tilde{C}_1\\
&\quad=-\left(2k_0\frac{\partial^2\tilde{C}_0}{\partial\xi_+\partial\xi}-2g_0\frac{\partial^2\tilde{C}_0}{\partial\eta_+\partial\xi}\right)-\left(2k_0\frac{\partial^2\tilde{C}_0}{\partial\eta_+\partial\eta}+2g_0\frac{\partial^2\tilde{C}_0}{\partial\xi_+\partial\eta}\right)\\
&\quad\quad-\Big(\frac{1}{\xi}-\eta_0^2\xi\Big)\Big(k_0\frac{\partial\tilde{C}_0}{\partial\xi_+}-g_0\frac{\partial\tilde{C}_0}{\partial\eta_+}\Big)-\Big(\frac{1}{\eta}+\eta_0^2\eta\Big)\Big(g_0\frac{\partial\tilde{C}_0}{\partial\xi_+}+k_0\frac{\partial\tilde{C}_0}{\partial\eta_+}\Big)\\
&\quad\quad+\eta_0^2(\xi^2+\eta^2)\frac{\partial\tilde{C}_0}{\partial t_+}.
\end{aligned}
\tag{8.63}
$$

The interface conditions are as follows: at the interface of the needle crystal $\eta_+=0$, $\eta=1$,

1. the thermodynamic equilibrium condition

$$\tilde{T}_1=\tilde{T}_{S1}+\lambda\tilde{h}_1-\eta_1\bar{k}_0\frac{\partial}{\partial\eta_+}(\tilde{T}_0-\tilde{T}_{S0}), \tag{8.64}$$

2. the Gibbs–Thomson condition

$$\tilde{T}_{S1} + \eta_1 \bar{k}_0 \frac{\partial \tilde{T}_{S0}}{\partial \eta_+} = \frac{1}{S(\xi)} \left\{ \bar{k}_0^2 \frac{\partial^2 \tilde{h}_1}{\partial \xi_+^2} + 2\bar{k}_0 \bar{k}_1 \frac{\partial^2 \tilde{h}_0}{\partial \xi_+^2} + \bar{k}_0' \frac{\partial \tilde{h}_0}{\partial \xi_+} \right.$$
$$\left. + \bar{k}_0 \left( \frac{1}{\xi} + \frac{\xi}{S^2(\xi)} \right) \frac{\partial \tilde{h}_0}{\partial \xi_+} \right\} - M \left( \tilde{C}_1 + \eta_1 \bar{k}_0 \frac{\partial \tilde{C}_0}{\partial \eta_+} + \eta_0^2 \Delta C \tilde{h}_1 \right), \tag{8.65}$$

3. the heat balance condition

$$\bar{k}_0 \left( \frac{\partial \tilde{T}_1}{\partial \eta_+} - \frac{\partial \tilde{T}_{S1}}{\partial \eta_+} \right) + \bar{k}_1 \left( \frac{\partial \tilde{T}_0}{\partial \eta_+} - \frac{\partial \tilde{T}_{S0}}{\partial \eta_+} \right) + \eta_1 \bar{k}_0^2 \frac{\partial^2}{\partial \eta_+^2} (\tilde{T}_0 - \tilde{T}_{S0})$$
$$+ \left( \frac{\partial \tilde{T}_0}{\partial \eta} - \frac{\partial \tilde{T}_{S0}}{\partial \eta} \right) + \lambda S^2(\xi)(\sigma_0 \tilde{h}_1 + \sigma_1 \tilde{h}_0)$$
$$+ \lambda \xi \left( \bar{k}_0 \frac{\partial \tilde{h}_1}{\partial \bar{\xi}_+} + \bar{k}_1 \frac{\partial \tilde{h}_0}{\partial \xi_+} \right) + \lambda \left( 2 + \lambda \eta_0^2 \right) \tilde{h}_0 = 0, \tag{8.66}$$

4. the mass balance condition

$$\bar{k}_0 \frac{\partial \tilde{C}_1}{\partial \eta_+} + \bar{k}_1 \frac{\partial \tilde{C}_0}{\partial \eta_+} + \eta_1 \bar{k}_0^2 \frac{\partial^2 \tilde{C}_0}{\partial \eta_+^2} + \frac{\partial \tilde{C}_0}{\partial \eta} + \eta_0^2 (1 - \kappa) \tilde{C}_0$$
$$+ \Delta C \left[ S^2(\xi)(\sigma_0 \tilde{h}_1 + \sigma_1 \tilde{h}_0) + \xi \left( \bar{k}_0 \frac{\partial \tilde{h}_1}{\partial \bar{\xi}_+} + \bar{k}_1 \frac{\partial \tilde{h}_0}{\partial \xi_+} \right) \right.$$
$$\left. + \left( 2 + \eta_0^2 \right) \tilde{h}_0 \right] = 0. \tag{8.67}$$

The first-order approximate solutions can be obtained in the form of normal modes:

$$\begin{aligned} \tilde{T}_1 &= A_1(\xi, \eta) \exp\{i\xi_+ - \eta_+\}, \\ \tilde{T}_{S1} &= A_{S1}(\xi, \eta) \exp\{i\xi_+ + \eta_+\}, \\ \tilde{C}_1 &= B_1(\xi, \eta) \exp\{i\xi_+ - \eta_+\}, \\ \tilde{h}_1 &= \hat{D}_1 \exp\{i\bar{\xi}_+\}. \end{aligned} \tag{8.68}$$

We then derive

$$(k_0^2+g_0^2)\left(\frac{\partial^2}{\partial\xi_+^2}+\frac{\partial^2}{\partial\eta_+^2}\right)\tilde{T}_1 = a_0(\xi,\eta)e^{i\xi_+-\eta_+}\,,$$
$$(k_0^2+g_0^2)\left(\frac{\partial^2}{\partial\xi_+^2}+\frac{\partial^2}{\partial\eta_+^2}\right)\tilde{T}_{S1} = a_{S0}(\xi,\eta)e^{i\xi_++\eta_+}\,, \tag{8.69}$$
$$(k_0^2+g_0^2)\left(\frac{\partial^2}{\partial\xi_+^2}+\frac{\partial^2}{\partial\eta_+^2}\right)\tilde{C}_1 = b_0(\xi,\eta)e^{i\xi_++\eta_+}\,,$$

where

$$\begin{aligned} a_0 = 2(k_0-ig_0)\left(\frac{\partial A_0}{\partial\eta}-i\frac{\partial A_0}{\partial\xi}\right)+A_0\Big\{\sigma_0\lambda\eta_0^2(\xi^2+\lambda\eta^2) \\ -(ik_0+g_0)\Big[\Big(\frac{1}{\xi}-\lambda\eta_0^2\xi\Big)+i\Big(\frac{1}{\eta}+\lambda\eta_0^2\eta\Big)\Big]\Big\} \end{aligned} \tag{8.70}$$

$$\begin{aligned} a_{S0} = -2(k_0+ig_0)\left(\frac{\partial A_{S0}}{\partial\eta}+i\frac{\partial A_{S0}}{\partial\xi}\right)+A_{S0}\Big\{\sigma_0\lambda\eta_0^2(\xi^2+\eta^2) \\ -(ik_0-g_0)\Big[\Big(\frac{1}{\xi}-\lambda\eta_0^2\xi\Big)-i\Big(\frac{1}{\eta}+\lambda\eta_0^2\eta\Big)\Big]\Big\}, \end{aligned} \tag{8.71}$$

and

$$\begin{aligned} b_0 = 2(k_0-ig_0)\left(\frac{\partial B_0}{\partial\eta}-i\frac{\partial B_0}{\partial\xi}\right)+B_0\Big\{\sigma_0\eta_0^2(\xi^2+\eta^2) \\ -(ik_0+g_0)\Big[\Big(\frac{1}{\xi}-\eta_0^2\xi\Big)+i\Big(\frac{1}{\eta}+\eta_0^2\eta\Big)\Big]\Big\}. \end{aligned} \tag{8.72}$$

To eliminate the secular terms on the right-hand side of (8.69), one must set

$$a_0 = a_{S0} = b_0 = 0\,. \tag{8.73}$$

Then from (8.73), it follows that

$$\begin{aligned} \left(\frac{\partial}{\partial\eta}-i\frac{\partial}{\partial\xi}\right)\ln A_0(\xi,\eta) &= \Phi_0(\xi,\eta) \\ \left(\frac{\partial}{\partial\eta}+i\frac{\partial}{\partial\xi}\right)\ln A_{S0}(\xi,\eta) &= \Phi_{S0}(\xi,\eta), \\ \left(\frac{\partial}{\partial\eta}+i\frac{\partial}{\partial\xi}\right)\ln B_0(\xi,\eta) &= \Psi_0(\xi,\eta), \end{aligned} \tag{8.74}$$

where

$$\begin{aligned}
\Phi_0(\xi,\eta) &= -\frac{\sigma_0\lambda\eta_0^2(\xi^2+\eta^2)}{2\bar{k}_0[\xi+\mathrm{i}(\eta-1)]}+\frac{\mathrm{i}}{2}\left[\left(\frac{1}{\xi}-\eta_0^2\xi\right)+\mathrm{i}\left(\frac{1}{\eta}+\lambda\eta_0^2\eta\right)\right]\\
\Phi_{S0}(\xi,\eta) &= \frac{\sigma_0\lambda\eta_0^2(\xi^2+\eta^2)}{2\bar{k}_0[\xi-\mathrm{i}(\eta-1)]}-\frac{\mathrm{i}}{2}\left[\left(\frac{1}{\xi}-\eta_0^2\xi\right)-\mathrm{i}\left(\frac{1}{\eta}+\lambda\eta_0^2\eta\right)\right],\\
\Psi_0(\xi,\eta) &= \frac{\sigma_0\eta_0^2(\xi^2+\eta^2)}{2\bar{k}_0[\xi-\mathrm{i}(\eta-1)]}-\frac{\mathrm{i}}{2}\left[\left(\frac{1}{\xi}-\eta_0^2\xi\right)-\mathrm{i}\left(\frac{1}{\eta}+\eta_0^2\eta\right)\right].
\end{aligned} \tag{8.75}$$

Equations (8.74) are of first-order hyperbolic type. The functions $A_0(\xi,\eta)$ and $A_{S0}(\xi,\eta)$ can be solved in the $(\xi,\eta)$-plane as an initial value problem given initial conditions on the line $\eta=1$:

$$A_0(\xi,1)=\hat{A}_0(\xi);\quad A_{S0}(\xi,1)=\hat{A}_{S0}(\xi);\quad B_0(\xi,1)=\hat{B}_0(\xi),$$

when the wave function $\bar{k}_0(\xi)$ is known.

Moreover, with the above results, we may derive that at $\eta=1$,

$$\begin{aligned}
\hat{D}_0Q_0 &= \frac{\partial}{\partial\eta}(A_0-A_{S0})\Big|_{\eta=1}=\mathrm{i}\frac{\partial}{\partial\xi}(\hat{A}_0+\hat{A}_{S0})+\hat{A}_0\Phi_0-\hat{A}_{S0}\Phi_{S0}\\
&= \mathrm{i}\frac{\partial}{\partial\xi}(\hat{A}_0+\hat{A}_{S0})-(\hat{A}_0+\hat{A}_{S0})\frac{\sigma_0\lambda\eta_0^2(1+\xi^2)}{2\bar{k}_0(\xi)}\\
&\quad+\frac{\mathrm{i}}{2}(\hat{A}_0+\hat{A}_{S0})\left(\frac{1}{\xi}-\lambda\eta_0^2\xi\right)-\frac{1}{2}(1+\lambda\eta_0^2)(\hat{A}_0-\hat{A}_{S0})
\end{aligned} \tag{8.76}$$

and

$$\begin{aligned}
\hat{D}_0Q_c &= \frac{\partial B_0}{\partial\eta}\Big|_{\eta=1}\\
&= \mathrm{i}\frac{\partial\hat{B}_0}{\partial\xi}-\hat{B}_0\left[\frac{\sigma_0\eta_0^2}{2\bar{k}_0}\hat{S}^2(\xi)+\frac{\eta_0^2}{2}(1+\mathrm{i}\xi)+\frac{1}{2}-\frac{\mathrm{i}}{2\xi}\right].
\end{aligned} \tag{8.77}$$

With formulas (8.57), we obtain

$$\begin{cases}
Q_0=-\mathrm{i}\frac{2\lambda}{\Lambda_c}\frac{d}{d\xi}\left(\frac{\bar{k}_0^2}{S}\right)-\lambda\left(1-\frac{2\bar{k}_0^2}{\Lambda_cS}\right)\left[\frac{\lambda\sigma_0\eta_0^2}{2\bar{k}_0}S^2(\xi)+\frac{i\lambda\eta_0^2}{2}\xi-\frac{\mathrm{i}}{2\xi}\right]\\
\quad-\frac{\lambda}{2}(1+\lambda\eta_0^2)\\
Q_c=-\mathrm{i}\frac{2\Delta C}{\Lambda_c}\frac{d}{d\xi}\left(\frac{\bar{k}_0^2}{S}\right)-\Delta C\left(1-\frac{2\bar{k}_0^2}{\Lambda_cS}\right)\left[\frac{\sigma_0\eta_0^2}{2\bar{k}_0}S^2(\xi)+\frac{\eta_0^2}{2}(1+\mathrm{i}\xi)\right.\\
\quad\left.+\frac{1}{2}-\frac{\mathrm{i}}{2\xi}\right].
\end{cases} \tag{8.78}$$

The above results will be needed later to solve for $\bar{k}_1(\xi)$.

We now turn to solving the equations under the conditions (8.73):

$$\begin{aligned}
\left(\frac{\partial^2}{\partial \xi_+^2}+\frac{\partial^2}{\partial \eta_+^2}\right)\tilde{T}_1 &= 0,\\
\left(\frac{\partial^2}{\partial \xi_+^2}+\frac{\partial^2}{\partial \eta_+^2}\right)\tilde{T}_{S1} &= 0\,,\\
\left(\frac{\partial^2}{\partial \xi_+^2}+\frac{\partial^2}{\partial \eta_+^2}\right)\tilde{C}_1 &= 0\,.
\end{aligned} \tag{8.79}$$

By setting

$$\begin{cases}
\hat{A}_1(\xi) = A_1(\xi,1),\\
\hat{A}_{S1}(\xi) = A_{S1}(\xi,1)\,,\\
\hat{B}_1(\xi) = B_1(\xi,1)\,,
\end{cases} \tag{8.80}$$

from the interface boundary conditions (8.64)–(8.67), we derive that

$$\begin{cases}
\hat{A}_1-\hat{A}_{S1}-\lambda\hat{D}_1 = I_1\hat{D}_0,\\
\hat{A}_{S1}+\left(\frac{\bar{k}_0^2}{S}-M\Delta C\right)\hat{D}_1+M\hat{B}_1 = I_2\hat{D}_0,\\
-\bar{k}_0(\hat{A}_1+\hat{A}_{S1})+\lambda(\sigma_0 S^2+\mathrm{i}\xi\bar{k}_0)\hat{D}_1 = I_3\hat{D}_0,\\
\bar{k}_0\hat{B}_1-\Delta C(\sigma_0 S^2+\mathrm{i}\xi\bar{k}_0)\hat{D}_1 = I_4\hat{D}_0,
\end{cases} \tag{8.81}$$

where

$$\begin{aligned}
I_1 &= \lambda\eta_1\bar{k}_0\left(1-\frac{2\bar{k}_0^2}{\Lambda_c\hat{S}}\right),\\
I_2 &= -\eta_1\bar{k}_0\hat{A}_{S0}+\frac{1}{S}\left[\mathrm{i}\bar{k}_0'+\mathrm{i}\bar{k}_0\left(\frac{1}{\xi}+\frac{\xi}{S^2}\right)-2\bar{k}_0\bar{k}_1\right]\\
&\quad+\eta_1\bar{k}_0M\Delta C\left(1-\frac{2\bar{k}_0^2}{\Lambda_c S}\right),\\
I_3 &= \lambda\bar{k}_1\left(1-\frac{2\bar{k}_0^2}{\Lambda_c\hat{S}}\right)-\lambda\eta_1\bar{k}_0^2-\lambda(\sigma_1 S^2+\mathrm{i}\xi\bar{k}_1)-Q_0\\
&\quad-\lambda(2+\lambda\eta_0^2),\\
I_4 &= Q_c+\Delta C\left[(\sigma_1 S^2+\mathrm{i}\xi\bar{k}_1+(2+\eta_0^2)\right]\\
&\quad+\Delta C\left(1-\frac{2\bar{k}_0^2}{\Lambda_c S}\right)\left[\eta_1\bar{k}_0^2-\bar{k}_1+\eta_0^2(1-\kappa)\right].
\end{aligned} \tag{8.82}$$

The determinant of the coefficient matrix of the above inhomogeneous system (8.81) is zero. For existence of the solution for $\{\hat{A}_1, \hat{A}_{S1}, \hat{B}_1, \hat{D}_1\}$, the solvability condition

$$\det\begin{pmatrix} 1 & -1 & I_1 & 0 \\ 0 & 1 & I_2 & M \\ -\bar{k}_0 & -\bar{k}_0 & I_3 & 0 \\ 0 & 0 & I_4 & \bar{k}_0 \end{pmatrix} = 0, \tag{8.83}$$

must be satisfied, which leads to

$$I_3 + 2\bar{k}_0 I_2 + \bar{k}_0 I_1 - 2MI_4 = 0. \tag{8.84}$$

The condition (8.84) may be rewritten in the form

$$\eta_1 P_1 + \bar{k}_1 P_2 + P_3 = 0, \tag{8.85}$$

where

$$P_1 = -\lambda\bar{k}_0^2 - 2\bar{k}_0^2\hat{A}_{S0} = 0, \tag{8.86}$$

$$\begin{aligned} P_2 &= \lambda\Big(1 - \tfrac{2\bar{k}_0^2}{\Lambda_c\hat{S}}\Big) - \mathrm{i}\lambda\xi - \tfrac{4\bar{k}_0^2}{\hat{S}} - 2M\Big[\mathrm{i}\Delta C\xi - \Delta C\Big(1 - \tfrac{2\bar{k}_0^2}{\Lambda_c S}\Big)\Big] \\ &= \Lambda_c\Big[1 - \mathrm{i}\xi - \tfrac{6\bar{k}_0^2}{\Lambda_c\hat{S}}\Big] = \Lambda_c\tfrac{\partial(\hat{S}^2\Sigma_0)}{\partial\bar{k}_0}, \end{aligned} \tag{8.87}$$

and

$$\begin{aligned} P_3 = &-(Q_0 + 2MQ_c) - (\lambda + 2M\Delta C)\sigma_1\hat{S}^2 - \lambda(2 + \lambda\eta_0^2) \\ &-2M\Delta C\Big[(2 + \eta_0^2) + \eta_0^2(1-\kappa)\Big(1 - \tfrac{2\bar{k}_0^2}{\Lambda_c\hat{S}}\Big)\Big] \\ &+\tfrac{2\bar{k}_0}{\hat{S}}\Big[\mathrm{i}\bar{k}_0' + \mathrm{i}\bar{k}_0\Big(\tfrac{1}{\xi} + \tfrac{\xi}{\hat{S}^2}\Big)\Big]. \end{aligned} \tag{8.88}$$

In the above,

$$\begin{aligned} (Q_0 + 2MQ_c) = &-2\mathrm{i}\left(\tfrac{2\bar{k}_0\bar{k}_0'}{\hat{S}} - \tfrac{\bar{k}_0^2\xi}{\hat{S}^3}\right) + \tfrac{\mathrm{i}\Lambda_c}{2\xi}\Big(1 - \tfrac{2\bar{k}_0^2}{\Lambda_c\hat{S}}\Big) \\ &-\tfrac{1}{2}\eta_0^2(\lambda^2 + 2M\Delta C)\Big(1 - \tfrac{2\bar{k}_0^2}{\Lambda_c\hat{S}}\Big)\Big[1 - \mathrm{i}\xi - \tfrac{2\bar{k}_0^2}{\Lambda_c\hat{S}}\Big] \\ &-\Big(1 - \tfrac{2\bar{k}_0^2}{\Lambda_c\hat{S}}\Big)\Big[\mathrm{i}\eta_0^2\xi(\tfrac{1}{2}\lambda^2 + M\Delta C) + M\Delta C(1 + \eta_0^2)\Big] \\ &-\tfrac{\lambda}{2}(1 + \lambda\eta_0^2). \end{aligned}$$

One may further split $P_3$ into the following four parts:

$$P_3 = -\Lambda_c\sigma_1\hat{S}^2 + P_{31}\frac{1}{\xi} + P_{32}\frac{\bar{k}_0'}{\bar{k}_0} + P_{33},$$

where

$$\begin{aligned}
P_{31} &= \tfrac{2i\bar{k}_0^2}{\hat{S}} - \tfrac{i\Lambda_c}{2}\Big(1 - \tfrac{2\bar{k}_0^2}{\Lambda_c\hat{S}}\Big) = -\tfrac{1}{2}iP_2 + \tfrac{1}{2}\Lambda_c\xi,\\
P_{32} &= 4i\tfrac{\bar{k}_0\bar{k}_0'}{\hat{S}} + \tfrac{2i\bar{k}_0\bar{k}_0'}{\hat{S}} = \tfrac{6i\bar{k}_0\bar{k}_0'}{\hat{S}},\\
P_{33} &= -\tfrac{\lambda}{2}(3 + \lambda\eta_0^2) - 2M\Delta C\Big[(2 + \eta_0^2) + \eta_0^2(1-\kappa)\Big(1 - \tfrac{2\bar{k}_0^2}{\Lambda_c\hat{S}}\Big)\Big]\\
&\quad + \tfrac{1}{2}\eta_0^2(\lambda^2 + 2M\Delta C)\Big(1 - \tfrac{2\bar{k}_0^2}{\Lambda_c\hat{S}}\Big)\Big[1 - i\xi - \tfrac{2\bar{k}_0^2}{\Lambda_c\hat{S}}\Big]\\
&\quad + \Big(1 - \tfrac{2\bar{k}_0^2}{\Lambda_c\hat{S}}\Big)\Big[i\eta_0^2\xi(\tfrac{1}{2}\lambda^2 + M\Delta C) + M\Delta C(1 + \eta_0^2)\Big].
\end{aligned}$$

From the solvability condition (8.85), we finally obtain

$$\bar{k}_1 = \frac{i}{2\xi} + \frac{R_1(\xi)}{F(\xi)} - \frac{R_2(\xi)}{F(\xi)}\frac{\bar{k}_0'}{\bar{k}_0}, \tag{8.89}$$

where we have used the notation

$$\begin{cases}
F(\xi) = \Lambda_c\left[1 - i\xi - \dfrac{6\bar{k}_0^2}{\Lambda_c\hat{S}}\right] = \Lambda_c\dfrac{\partial(\hat{S}^2\Sigma)}{\partial\bar{k}_0}\\
R_1(\xi) = \Lambda_c\sigma_1\hat{S}^2 - \Big[P_{33}(\xi) + \tfrac{1}{2}\Lambda_c\Big],\\
R_2(\xi) = \tfrac{6\bar{k}_0^2}{\hat{S}}.
\end{cases} \tag{8.90}$$

Just as was done before, one may extend all functions of the real variable $\xi$ to the corresponding functions of the complex variable $\zeta = \xi + i\xi_I$ by analytic continuation. It is seen that similar to the case of pure melt, the wave function $\bar{k}_1(\zeta)$ has a singularity at $\zeta = \zeta_c$, which is the complex root of $F(\xi)$ satisfying the equation

$$\frac{\partial(\hat{S}^2\Sigma)}{\partial\bar{k}_0} = \left[1 - i\xi - \frac{6\bar{k}_0^2}{\Lambda_c\hat{S}}\right] = 0. \tag{8.91}$$

We may also solve for $\zeta_c$ from the equation

$$\sigma_0 = \sqrt{\frac{2\Lambda_c}{27}}\left[\frac{1 - i\zeta_c}{\hat{S}(\zeta_c)}\right]^{\frac{3}{2}} = e^{i\frac{3\pi}{4}}\sqrt{\frac{2\Lambda_c}{27}}\left(\frac{\zeta_c + i}{\zeta_c - i}\right)^{\frac{3}{4}}. \tag{8.92}$$

Noting that at $\zeta = \zeta_c$,

$$\frac{1 - 2\bar{k}_0^2}{\Lambda_c \hat{S}} = \frac{2 + \mathrm{i}\zeta_c}{3},$$

one may rewrite this as

$$\begin{aligned} P_{33} = &-\Big[\tfrac{\lambda}{2}(3 + \lambda\eta_0^2) + 2M\Delta C(2 + \eta_0^2)\Big] \\ &+ M\Delta C\Big(\tfrac{2+\mathrm{i}\zeta_c}{3}\Big)\big[1 - \eta_0^2(1 - 2\kappa)\big] \\ &+ \tfrac{1}{2}\eta_0^2(\lambda^2 + 2M\Delta C)\Big(\tfrac{2+\mathrm{i}\zeta_c}{3}\Big)^2. \end{aligned} \tag{8.93}$$

From (8.89), it is seen that the solution $\bar{k}_1(\zeta)$ has an isolated singularity at the tip $\zeta = 0$ and

$$\bar{k}_1(\zeta) \sim \frac{\mathrm{i}}{2\zeta}, \tag{8.94}$$

as $\zeta \to 0$. Furthermore, as $\zeta \to \zeta_c$, it follows from (8.90) that

$$k_1 \sim \frac{R_1(\xi_c)}{A(\zeta - \zeta_c)^{\frac{1}{2}}} + O\left(\frac{1}{\zeta - \zeta_c}\right). \tag{8.95}$$

Since $\zeta_c$ is an isolated singular point of the meromorphic function $\bar{k}_1(\zeta)$, one deduces that $R_1(\zeta_c) = 0$, whence

$$\sigma_1 = \frac{1}{\Lambda_c \hat{S}^2(\zeta_c)}\left[\frac{1}{2}\Lambda_c + P_{33}(\zeta_c)\right]. \tag{8.96}$$

We see that just as for the system of dendritic growth from pure melt, the first-order approximation of the eigenvalue $\sigma_1$ for the binary system is also independent of the first-order approximation solution $\eta_1$ of the basic state. We see also that the normal mode solution (8.61) is inapplicable in the vicinity of $\zeta = 0$ and $\zeta_c$, since the function $\bar{k}_1(\zeta)$ has a singularity at these points.

To derive the global solution, we need to find the inner solutions near the singular point $\zeta_c$ and match the outer solutions with the inner solutions.

## 8.6 The Inner Solutions near the Singular Point $\zeta_c$

In the vicinity of $\zeta_c$ ($|\zeta - \zeta_c| \ll 1; |\eta - 1| \ll 1$), we introduce the inner variables

$$
\begin{aligned}
\zeta_* &= \frac{\zeta - \zeta_c}{\varepsilon^\alpha}, \\
\eta_* &= \frac{\eta - 1}{\varepsilon^\alpha},
\end{aligned}
\tag{8.97}
$$

and set

$$
\begin{aligned}
\tilde{T}(\zeta, \eta, t) &= \varepsilon^\alpha \hat{T}(\zeta_*, \eta_*, t) \\
\tilde{T}_S(\zeta, \eta, t) &= \varepsilon^\alpha \hat{T}_S(\zeta_*, \eta_*, t), \\
\tilde{C}(\zeta, \eta, t) &= \varepsilon^\alpha \hat{C}(\zeta_*, \eta_*, t), \\
\tilde{h}(\zeta, t) &= \varepsilon^\alpha \hat{h}(\zeta_*, t).
\end{aligned}
\tag{8.98}
$$

Then at the interface, we have

$$
\eta_{*s} = \frac{\hat{h}(\zeta_*, t)}{\eta_0^2}. \tag{8.99}
$$

In terms of these inner variables, the perturbed system (8.18) can be expressed in the form

$$
\begin{aligned}
\frac{\partial^2 \hat{T}}{\partial \zeta_*^2} + \frac{\partial^2 \hat{T}}{\partial \eta_*^2} &= \left\{ \varepsilon^{2\alpha-1} \lambda \sigma \eta_0^2 (\zeta^2 + \eta^2) - \varepsilon^\alpha \lambda \eta_0^2 \left( \zeta \frac{\partial}{\partial \zeta_*} - \eta \frac{\partial}{\partial \eta_*} \right) \right. \\
&\quad \left. - \varepsilon^\alpha \left( \frac{1}{\zeta} \frac{\partial}{\partial \zeta_*} + \frac{1}{\eta} \frac{\partial}{\partial \eta_*} \right) \right\} \hat{T} + (\text{h.o.t}), \\
\frac{\partial^2 \hat{C}}{\partial \zeta_*^2} + \frac{\partial^2 \hat{C}}{\partial \eta_*^2} &= \left\{ \varepsilon^{2\alpha-1} \sigma \eta_0^2 (\zeta^2 + \eta^2) - \varepsilon^\alpha \eta_0^2 \left( \zeta \frac{\partial}{\partial \zeta_*} - \eta \frac{\partial}{\partial \eta_*} \right) \right. \\
&\quad \left. - \varepsilon^\alpha \left( \frac{1}{\zeta} \frac{\partial}{\partial \zeta_*} + \frac{1}{\eta} \frac{\partial}{\partial \eta_*} \right) \right\} \hat{C} + (\text{h.o.t}),
\end{aligned}
\tag{8.100}
$$

where $\zeta = \zeta_c + \varepsilon^\alpha \zeta_*$, $\eta = 1 + \varepsilon^\alpha \eta_*$ are slowly varying variables.

The boundary conditions (8.21)–(8.27) are transformed to the following forms:

1. As $\eta_* \to \infty$,

$$
\hat{T} \to 0, \quad \hat{C} \to 0. \tag{8.101}
$$

2. As $\eta_* \to -\infty$,

$$\hat{T}_S \to 0 \, . \tag{8.102}$$

3. At the interface $\eta_* = 0$,

   (a)

$$\hat{T} = \hat{T}_S + \lambda \hat{h} + (\text{h.o.t}) \, , \tag{8.103}$$

   (b)

$$\hat{T}_S = \frac{\varepsilon^{2-2\alpha}}{S(\zeta)} \frac{\partial^2 \hat{h}}{\partial \zeta_*^2} - M(\hat{C} - \Delta C \hat{h}) + (\text{h.o.t}) \, , \tag{8.104}$$

   (c)

$$\varepsilon^{1-\alpha} \frac{\partial}{\partial \eta_*} \left( \hat{T} - \hat{T}_S \right) + \lambda \sigma S^2(\zeta) \hat{h} + \varepsilon^{1-\alpha} \lambda \zeta \frac{\partial \hat{h}}{\partial \zeta_*} = (\text{h.o.t}), \tag{8.105}$$

   (d)

$$\varepsilon^{1-\alpha} \frac{\partial \hat{C}}{\partial \eta_*} + \Delta C \left\{ \sigma S^2(\zeta) \hat{h} + \varepsilon^{1-\alpha} \zeta \frac{\partial \hat{h}}{\partial \zeta_*} \right\} = (\text{higher-order terms}) \, . \tag{8.106}$$

   (e) Matching conditions with the outer solution away from the singular point $\zeta_c$: as $\zeta_* \to \infty$,

$$\left\{ h(\zeta) \Leftrightarrow \hat{h}(\zeta_*) \right\}. \tag{8.107}$$

Assume that the inner expansion is in the following form:

$$\begin{aligned} \hat{T}(\zeta_*, \eta_*, t) &= \left\{ \nu_0(\varepsilon) \hat{T}_0(\zeta_*, \eta_*) + \nu_1(\varepsilon) \hat{T}_1(\bullet) + \dots \right\} e^{\frac{\sigma t}{\varepsilon \eta_0^2} + im\theta} \\ \hat{C}(\zeta_*, \eta_*, t) &= \left\{ \nu_0(\varepsilon) \hat{C}_0(\zeta_*, \eta_*) + \nu_1(\varepsilon) \hat{C}_1(\bullet) + \dots \right\} e^{\frac{\sigma t}{\varepsilon \eta_0^2} + im\theta} \\ \hat{h} &= \left\{ \nu_0(\varepsilon) \hat{h}_0 + \nu_1(\varepsilon) \hat{h}_1 + \cdots \right\} e^{\frac{\sigma t}{\varepsilon \eta_0^2} + im\theta} . \end{aligned} \tag{8.108}$$

### 8.6.1 Leading-Order Approximation

Letting $\varepsilon \to 0$, then if $\alpha > \frac{1}{2}$ in the leading-order approximation we have the inner system

$$\frac{\partial^2 \hat{T}_0}{\partial \zeta_*^2} + \frac{\partial^2 \hat{T}_0}{\partial \eta_*^2} = 0,$$
$$\frac{\partial^2 \hat{T}_{S0}}{\partial \zeta_*^2} + \frac{\partial^2 \hat{T}_{S0}}{\partial \eta_*^2} = 0,$$
$$\frac{\partial^2 \hat{C}_0}{\partial \zeta_*^2} + \frac{\partial^2 \hat{C}_0}{\partial \eta_*^2} \approx 0. \tag{8.109}$$

The interface conditions are as follows: At the interface $\eta_* = 0$, $\eta = 1$,

$$\hat{T}_0 = \hat{T}_{S0} + \lambda \hat{h}_0, \tag{8.110}$$

$$\hat{T}_{S0} = \frac{\varepsilon^{2-2\alpha}}{S(\zeta)} \frac{\partial^2 \hat{h}_0}{\partial \zeta_*^2} - M(\hat{C}_0 - \Delta C \hat{h}_0), \tag{8.111}$$

$$\varepsilon^{1-\alpha} \frac{\partial}{\partial \eta_*}\left(\hat{T}_0 - \hat{T}_{S0}\right) + \lambda \sigma_0 S^2(\zeta) \hat{h}_0 + \varepsilon^{1-\alpha} \lambda \zeta \frac{d\hat{h}_0}{d\zeta_*} = 0, \tag{8.112}$$

$$\varepsilon^{1-\alpha} \frac{\partial \hat{C}_0}{\partial \eta_*} + \Delta C \left\{ \sigma_0 S^2(\zeta) \hat{h}_0 + \varepsilon^{1-\alpha} \zeta \frac{\partial \hat{h}_0}{\partial \zeta_*} \right\} = 0. \tag{8.113}$$

The inner solution can be rewritten in the form

$$\left(\mathrm{i}\frac{\partial}{\partial \zeta_*} + \frac{\partial}{\partial \eta_*}\right)\left(\mathrm{i}\frac{\partial}{\partial \zeta_*} - \frac{\partial}{\partial \eta_*}\right)\left[\hat{T}_0, \hat{T}_{S0}, \hat{C}_0\right] = 0. \tag{8.114}$$

We see that

1. In the liquid phase region $\eta > 1$,

$$\left(\mathrm{i}\frac{\partial}{\partial \zeta_*} - \frac{\partial}{\partial \eta_*}\right)\left[\hat{T}_0, \hat{C}_0\right] = 0. \tag{8.115}$$

2. In the solid phase region $\eta < 1$,

$$\left(\mathrm{i}\frac{\partial}{\partial \zeta_*} + \frac{\partial}{\partial \eta_*}\right)\hat{T}_{S0} = 0. \tag{8.116}$$

As a consequence, at the interface we have

$$\frac{\partial}{\partial \eta_*}(\hat{T}_0 - \hat{T}_{S0}) = \mathrm{i}\tfrac{\partial}{\partial \zeta_*}(\hat{T}_0 + \hat{T}_{S0}),$$
$$\frac{\partial}{\partial \eta_*}\hat{C}_0 = \mathrm{i}\tfrac{\partial}{\partial \zeta_*}\hat{C}_0.$$

From (8.112), we derive

$$\mathrm{i}\varepsilon^{1-\alpha}\frac{\partial}{\partial \zeta_*}\Big(\hat{T}_0 + \hat{T}_{S0}\Big) + \lambda\sigma_0 S^2(\zeta)\hat{h}_0 + \varepsilon^{1-\alpha}\lambda\zeta\frac{\mathrm{d}\hat{h}_0}{\mathrm{d}\zeta_*} = 0. \tag{8.117}$$

From (8.113), we derive

$$\mathrm{i}\varepsilon^{1-\alpha}\frac{\partial \hat{C}_0}{\partial \zeta_*} + \Delta C\Big[\sigma_0 S^2(\zeta)\hat{h}_0 + \varepsilon^{1-\alpha}\zeta\frac{\mathrm{d}\hat{h}_0}{\mathrm{d}\zeta_*}\Big] = 0. \tag{8.118}$$

Moreover, using (8.110) and (8.111), we derive

$$\Big(\hat{T}_0 + \hat{T}_{S0}\Big) = 2\frac{\varepsilon^{2-2\alpha}}{S(\zeta)}\frac{\mathrm{d}^2\hat{h}_0}{\mathrm{d}\zeta_*^2} - 2M(\hat{C}_0 - \Delta C\hat{h}_0) + \lambda\hat{h}_0. \tag{8.119}$$

Hence,

$$\frac{\mathrm{d}}{\mathrm{d}\zeta_*}\Big(\hat{T}_0 + \hat{T}_{S0}\Big) = 2\frac{\varepsilon^{2-2\alpha}}{S(\zeta)}\frac{\mathrm{d}^3\hat{h}_0}{\mathrm{d}\zeta_*^3} - 2M\frac{\partial \hat{C}_0}{\partial \zeta_*}$$
$$+(\lambda - 2\eta_0^2 M\Delta C)\frac{\partial \hat{h}_0}{\partial \zeta_*}. \tag{8.120}$$

We derive

$$\mathrm{i}\varepsilon^{1-\alpha}\Bigg[2\frac{\varepsilon^{2-2\alpha}}{S(\zeta)}\frac{\mathrm{d}^3\hat{h}_0}{\mathrm{d}\zeta_*^3} - 2M\frac{\partial \hat{C}_0}{\partial \zeta_*} + (\lambda + 2M\Delta C)\frac{\partial \hat{h}_0}{\partial \zeta_*}\Bigg]$$
$$+\lambda\sigma_0 S^2(\zeta)\hat{h}_0 + \varepsilon^{1-\alpha}\lambda\zeta\frac{\mathrm{d}\hat{h}_0}{\mathrm{d}\zeta_*} = 0, \tag{8.121}$$

or

$$\mathrm{i}\varepsilon^{3-3\alpha}\frac{2}{S(\zeta)}\frac{\mathrm{d}^3\hat{h}_0}{\mathrm{d}\zeta_*^3} + \mathrm{i}\varepsilon^{1-\alpha}\Lambda_c(1 - \mathrm{i}\zeta)\frac{\partial \hat{h}_0}{\partial \zeta_*} + \Lambda_c\sigma_0 S^2(\zeta)\hat{h}_0 = 0. \tag{8.122}$$

We finally derive the third-order complex ODE for the unsteady interface perturbations by neglecting the small higher-order terms:

$$\left[ \mathrm{i}\varepsilon^{3-3\alpha} \frac{2}{\Lambda_c S(\zeta)} \frac{\mathrm{d}^3 \hat{h}_0}{\mathrm{d}\zeta_*^3} + \varepsilon^{1-\alpha}(\zeta + \mathrm{i}) \frac{\partial \hat{h}_0}{\partial \zeta_*} \right] + \sigma_0 S^2(\zeta) \hat{h}_0 = 0. \tag{8.123}$$

The inner equation (8.123) can also be derived in a simpler way starting with the dispersion relationship (8.55). Making the following replacements in (8.55),

$$\bar{k}_0 \Longleftrightarrow -\frac{\mathrm{i}}{\varepsilon} \frac{\mathrm{d}}{\mathrm{d}\xi},$$

one may obtain the inner equation,

$$\left[ \mathrm{i} \frac{2\varepsilon^3}{\Lambda_c S(\zeta)} \frac{\mathrm{d}^3 \hat{h}_0}{\mathrm{d}\zeta^3} + \varepsilon(\zeta + \mathrm{i}) \frac{\partial \hat{h}_0}{\partial \zeta} \right] + \sigma_0 S^2(\zeta) \hat{h}_0 = 0. \tag{8.124}$$

With the inner variables, Eq. (8.124) is changed to the form of (8.123).

It can be shown that the above equation has a simple turning point singularity at $\zeta = \zeta - \zeta_c$, and it reduces to the Airy equation as $\zeta \to \zeta_c$. To show this, we use the transformation

$$H(\zeta) = W(\zeta) \exp\left\{ \frac{\mathrm{i}}{\varepsilon} \int_{\zeta_c}^{\zeta} k_c(\zeta_1)\, \mathrm{d}\zeta_1 \right\} \tag{8.125}$$

to describe the solution for the interface shape $H(\zeta)$ using the function $W(\zeta)$, where the reference wave number function $k_c(\zeta)$ is to be chosen later:

$$\mathrm{Re}\{k_0^{(3)}\} < \mathrm{Re}\{k_c\} < \mathrm{Re}\{k_0^{(1)}\}. \tag{8.126}$$

Thus, these two zeroth-order approximate basic outer solutions in the outer region can be expressed as

$$\begin{aligned} H_1(\zeta) &= W_0^{(+)}(\zeta) \exp\left\{ \frac{\mathrm{i}}{\varepsilon} \int_{\zeta_c}^{\zeta} k_c(\zeta_1)\, \mathrm{d}\zeta_1 \right\}, \\ H_3(\zeta) &= W_0^{(-)}(\zeta) \exp\left\{ \frac{\mathrm{i}}{\varepsilon} \int_{\zeta_c}^{\zeta} k_c(\zeta_1)\, \mathrm{d}\zeta_1 \right\}. \end{aligned} \tag{8.127}$$

The general outer solutions $\tilde{h}_0(\zeta)$ then can be expressed in the form

$$W_0 = D_1 W_0^{(+)} + D_3 W_0^{(-)}. \tag{8.128}$$

As with the system discussed in Chap. 6, one can draw the following diagram of the relation between the $H$ waves and the $W$ waves:

$H_1$ wave (short-wave branch) $\Longleftrightarrow$ $W^{(+)}$ (outgoing wave);

$H_3$ wave (long-wave branch) $\Longleftrightarrow$ $W^{(-)}$ (incoming wave).

In accordance with the above, in the inner region for the inner solution we can set

$$\hat{h}_0(\zeta) = \hat{W}_0(\zeta) \exp\left\{ \frac{\mathrm{i}}{\varepsilon} \int_{\zeta_c}^{\zeta} k_c(\zeta_1)\, \mathrm{d}\zeta_1 \right\}. \tag{8.129}$$

Then Eq. (8.123) is changed to

$$\varepsilon^3 \Omega_3 \frac{\mathrm{d}^3 \hat{W}_0}{\mathrm{d}\zeta^3} + \mathrm{i}\varepsilon^2 \Omega_2 \frac{\mathrm{d}^2 \hat{W}_0}{\mathrm{d}\zeta^2} - \varepsilon \Omega_1 \frac{\mathrm{d}\hat{W}_0}{\mathrm{d}\zeta} + \mathrm{i}S^2 \Big\{ \sigma - \Sigma_c(\zeta) \Big\} \hat{W}_0 = 0\,, \tag{8.130}$$

where

$$\begin{aligned}
\Omega_3 &= \tfrac{1}{3!} \left[ \frac{\partial^3 (S^2 \Sigma)}{\partial \hat{k}_0^3} \right]_{\hat{k}_0 = k_c} = -\frac{1}{\Lambda_c S}, \\
\Omega_2 &= \tfrac{1}{2!} \left[ \frac{\partial^2 (S^2 \Sigma)}{\partial \hat{k}_0^2} \right]_{\hat{k}_0 = k_c} = -\frac{3k_c}{\Lambda_c S}, \\
\Omega_1 &= \left[ \frac{\partial (S^2 \Sigma)}{\partial \hat{k}_0} \right]_{\hat{k}_0 = k_c} = -\frac{3k_c^2}{\Lambda_c S} + (1 - \mathrm{i}\zeta), \\
\Sigma_c(\zeta) &= \tfrac{1}{S^2} \Sigma(k_c, \zeta) = \tfrac{1}{S^2} \left[ \hat{k}_c (1 - \mathrm{i}\zeta) - \frac{\hat{k}_c^3}{\Lambda_c S} \right].
\end{aligned} \tag{8.131}$$

If we set

$$\Omega_1 = -\frac{3k_c^2}{\Lambda_c S} + (1 - \mathrm{i}\zeta) = 0 \tag{8.132}$$

or

$$k_c(\zeta) = \mathrm{e}^{-\mathrm{i}\frac{\pi}{4}} \sqrt{\frac{\Lambda_c S}{3} (\zeta + \mathrm{i})} \qquad \big( \mathrm{Re}\{k_c\} > 0 \big)\,, \tag{8.133}$$

we derive

$$\Sigma_c(\zeta) = \Sigma(k_c, \zeta) = \mathrm{e}^{-\mathrm{i}\frac{3\pi}{4}} \sqrt{\frac{2\Lambda_c}{27}} \left( \frac{\zeta + \mathrm{i}}{\zeta - \mathrm{i}} \right)^{\frac{3}{4}}. \tag{8.134}$$

Then (8.130) can be changed to

$$\varepsilon^3\Omega_3\frac{\mathrm{d}^3\hat{W}_0}{\mathrm{d}\zeta^3}+\mathrm{i}\varepsilon^2\Omega_2\frac{\mathrm{d}^2\hat{W}_0}{\mathrm{d}\zeta^2}+\mathrm{i}S^2\Big\{\sigma-\Sigma_c(\zeta)\Big\}\hat{W}_0=0\,. \tag{8.135}$$

We now describe the inner solution with the inner variable $\zeta_*$ so that (8.135) is changed to

$$\varepsilon^{(3-\alpha^3)}\Omega_3\frac{\mathrm{d}^3\hat{W}_0}{\mathrm{d}\zeta_*^3}+\mathrm{i}\varepsilon^{(2-2\alpha)}\Omega_2\frac{\mathrm{d}^2\hat{W}_0}{\mathrm{d}\zeta_*^2}+\mathrm{i}S^2\Big\{\sigma-\Sigma_c(\zeta)\Big\}\hat{W}_0=0. \tag{8.136}$$

Up to this point, we have not taken into account the property of the singular point $\zeta_c$. The above equation is applicable in the vicinity of any point $\zeta_c$, as long as $\alpha>\frac{1}{2}$. We now consider that $\zeta=\zeta_c$ is a simple turning point singularity of Eq. (8.136), namely,

$$\sigma-\Sigma_c(\zeta_c)=-\Sigma_c'(\zeta_c)(\zeta-\zeta_c)+\cdots\,.$$

It is found that to balance the leading terms of Eq. (8.136) in the limit $\zeta_*\to 0$, one must set

$$\alpha=\frac{2}{3}, \tag{8.137}$$

whence $\zeta_*=\frac{(\zeta-\zeta_c)}{\varepsilon^{2/3}}$. After neglecting all small higher-order terms, the inner equation (8.136) for $\hat{W}_0(\zeta_*)$ reduces to the following Airy equation: [1]

$$\frac{\mathrm{d}^2\hat{W}_0}{\mathrm{d}\zeta_*^2}+A^2\zeta_*\hat{W}_0=0\,, \tag{8.138}$$

where

$$A=\left(\frac{\Lambda_c S^3}{3k_c}\frac{\mathrm{d}\Sigma_c}{\mathrm{d}\zeta}\right)^{\frac{1}{2}}_{\zeta=\zeta_c}=\mathrm{e}^{\mathrm{i}\pi/4}\sqrt{\frac{\Lambda_c}{3}}\left(\frac{\zeta_c+\mathrm{i}}{\zeta_c-\mathrm{i}}\right)^{\frac{1}{4}} \tag{8.139}$$
$$\left(\frac{-\pi}{4}<\arg\{A\}<0\right).$$

This can be written as the standard Airy equation

$$\frac{\mathrm{d}^2\hat{W}_0}{\mathrm{d}\hat{\xi}_*^2}+\hat{\xi}_*\hat{W}_0=0\,, \tag{8.140}$$

by introducing the new inner variable

$$\hat{\zeta}_* = \frac{A^{2/3}}{\varepsilon^{2/3}}(\zeta - \zeta_c)\,. \tag{8.141}$$

The general solution of the above Airy equation is

$$\hat{W}_0 = D_{*1}\hat{\zeta}_*^{\frac{1}{2}}H^{(1)}_{\frac{1}{3}}(z) + D_{*2}\hat{\zeta}_*^{\frac{1}{2}}H^{(2)}_{\frac{1}{3}}(z), \tag{8.142}$$
$$\left(z = \frac{2}{3}\hat{\zeta}_*^{\frac{3}{2}}\right),$$

where $H^{(1)}_\nu(z)$ is the $\nu$th-order Hankel function of the first kind, while $H^{(2)}_\nu(z)$ is the Hankel function of the second kind. In order to match with the outer solution, which satisfies the downstream far-field condition (8.49) as $\hat{\zeta}_* \to \infty$, the inner solution must be

$$\hat{W}_0 = D_*\hat{\zeta}_*^{\frac{1}{2}}H^{(2)}_{\frac{1}{3}}(z). \tag{8.143}$$

By matching the inner solution with the outer solution in the sector $(S_1)$ and $(S_2)$, we verify

$$k_1 \sim \frac{\mathrm{i}}{4(\zeta - \zeta_c)} + o(1), \quad R_1(\zeta_c) = 0, \tag{8.144}$$

and obtain the connection condition

$$\frac{D_0^{(1)}}{D_0^{(3)}} = \mathrm{e}^{\frac{\mathrm{i}\pi}{2}}, \quad D_0^{(3)} = D_3'. \tag{8.145}$$

This condition determines the coefficients $D_0^{(1)}$ and $D_0^{(3)}$ in the outer solutions (8.61) with a free parameter $D_3'$.

## 8.7 Tip Inner Solution in the Tip Region

As indicated in the previous section, the MVE normal mode solution is also invalid in a neighborhood of the point $(\zeta = 0, \eta = 1)$ in $(\zeta, \eta)$-space, which may be called the tip region. One needs, therefore, to find the new asymptotic form of the solution in this region. In doing so, we define the tip inner variables

$$\begin{aligned}
\hat{\xi} &= \tfrac{\hat{k}\xi}{\varepsilon}, \quad \hat{\zeta} = \tfrac{\hat{k}\zeta}{\varepsilon}, \\
\hat{\eta} &= \tfrac{\hat{k}(\eta-1)}{\varepsilon}, \\
\hat{t} &= \tfrac{t}{\eta_0^2\varepsilon},
\end{aligned} \tag{8.146}$$

where $|\zeta| \ll \varepsilon$ and $|\eta - 1| \ll 1$. The tip solution can be expressed as a function of these inner variables and expanded in the following asymptotic form as $\varepsilon \to 0$:

$$\begin{aligned}
\hat{T}(\hat{\xi},\hat{\eta},\theta,\hat{t}) &= \Big\{\hat{\mu}_0(\varepsilon)\hat{T}_0(\hat{\xi},\hat{\eta}) + \hat{\mu}_1(\varepsilon)\hat{T}_1(\bullet) + \cdots\Big\}e^{im\theta+\sigma\hat{t}}, \\
\hat{T}_S(\hat{\xi},\hat{\eta},\theta,\hat{t}) &= \Big\{\hat{\mu}_0(\varepsilon)\hat{T}_{S0}(\hat{\xi},\hat{\eta}) + \hat{\mu}_1(\varepsilon)\hat{T}_{S1}(\bullet) + \cdots\Big\}e^{im\theta+\sigma\hat{t}}, \\
\hat{C}(\hat{\xi},\hat{\eta},\theta,\hat{t}) &= \Big\{\hat{\mu}_0(\varepsilon)\hat{C}_0(\hat{\xi},\hat{\eta}) + \hat{\mu}_1(\varepsilon)\hat{C}_1(\bullet) + \cdots\Big\}e^{im\theta+\sigma\hat{t}}, \\
\hat{h}(\hat{\xi},\theta,\hat{t}) &= \Big\{\hat{\mu}_0(\varepsilon)\hat{h}_0(\hat{\xi}) + \hat{\mu}_1(\varepsilon)\hat{h}_1(\hat{\xi}) + \cdots\Big\}e^{im\theta+\sigma\hat{t}} \\
\hat{k} &= \hat{k}_0 + \varepsilon\hat{k}_1 + \cdots .
\end{aligned} \tag{8.147}$$

At the zeroth-order approximation in the tip region, the system (8.18)–(8.27) is reduced to

$$\begin{aligned}
\frac{\partial^2\hat{T}_0}{\partial\hat{\xi}^2} + \frac{\partial^2\hat{T}_0}{\partial\hat{\eta}^2} + \frac{\hat{k}_0^2}{\hat{\xi}}\frac{\partial\hat{T}_0}{\partial\hat{\xi}} - \frac{m^2\hat{k}_0^2}{\hat{\xi}^2}\hat{T}_0, &= 0 \\
\frac{\partial^2\hat{T}_{S0}}{\partial\hat{\xi}^2} + \frac{\partial^2\hat{T}_{S0}}{\partial\hat{\eta}^2} + \frac{\hat{k}_0^2}{\hat{\xi}}\frac{\partial\hat{T}_{S0}}{\partial\hat{\xi}} - \frac{m^2\hat{k}_0^2}{\hat{\xi}^2}\hat{T}_{S0} &= 0, \\
\frac{\partial^2\hat{C}_0}{\partial\hat{\xi}^2} + \frac{\partial^2\hat{C}_0}{\partial\hat{\eta}^2} + \frac{\hat{k}_0^2}{\hat{\xi}}\frac{\partial\hat{C}_0}{\partial\hat{\xi}} - \frac{m^2\hat{k}_0^2}{\hat{\xi}^2}\hat{C}_0 &= 0,
\end{aligned} \tag{8.148}$$

with the following boundary conditions at $\hat{\eta} = 0$:

$$\hat{T}_0 = \hat{T}_{S0} + \lambda\hat{h}_0, \tag{8.149}$$

$$\hat{T}_{S0} = \hat{k}_0^2\left(\frac{\partial^2\hat{h}_0}{\partial\hat{\xi}^2} + \frac{1}{\hat{\xi}}\frac{\partial\hat{h}_0}{\partial\hat{\xi}} - \frac{m^2}{\hat{\xi}^2}\hat{h}_0\right) - M(\tilde{C}_0 - \Delta C\tilde{h}_0), \tag{8.150}$$

$$\hat{k}_0\frac{\partial}{\partial\hat{\eta}}(\hat{T}_0 - \hat{T}_{S0}) + \lambda\sigma_0\hat{h}_0 = 0. \tag{8.151}$$

$$\hat{k}_0\frac{\partial\tilde{C}_0}{\partial\hat{\eta}} + \Delta C\sigma_0\tilde{h}_0 = 0\,. \tag{8.152}$$

This system admits the inner solutions

$$\begin{aligned}
\hat{T}_0 &= \hat{a}_0 H_m^{(1)}(\hat{\xi})\mathrm{e}^{-\hat{\eta}}, \\
\hat{T}_{S0} &= \hat{a}_{S0} H_m^{(1)}(\hat{\xi})\mathrm{e}^{\hat{\eta}}, \\
\hat{C}_0 &= \hat{b}_0 H_m^{(1)}(\hat{\xi})\mathrm{e}^{\hat{\eta}}, \\
\hat{h}_0 &= \hat{d}_0 H_m^{(1)}(\hat{\xi}),
\end{aligned} \tag{8.153}$$

where $H_m^{(1)}(\hat{\xi})$ is the $m$th-order Hankel function of the first kind. From the interface conditions (8.149)–(8.152), it follows that

$$\begin{aligned}
\hat{a}_0 - \hat{a}_{S0} - \lambda \hat{d}_0 &= 0, \\
\hat{a}_{S0} + \left(\hat{k}_0^2 + \eta_0 M \Delta C\right)\hat{d}_0 + M\hat{b}_0 &= 0, \\
-\hat{k}_0(\hat{a}_0 + \hat{a}_{S0}) + \lambda \sigma_0 \hat{d}_0 &= 0, \\
-\hat{k}_0 \hat{b}_0 + \Delta C \sigma_0 \hat{d}_0 &= 0,
\end{aligned} \tag{8.154}$$

which leads to the local dispersion relationship in the tip region:

$$\sigma_0 = \bar{k}_0 \left(1 - \frac{2\bar{k}_0^2}{\Lambda_c}\right). \tag{8.155}$$

For fixed $\sigma_0$, (8.155) gives three roots for $\hat{k}_0$. Comparing the local dispersion relationship in the tip region (8.155) with the local dispersion relationship in the outer region (8.55), one can evidently write

$$\begin{cases}
\hat{k}_0^{(1)} = \bar{k}_0^{(1)}(0), \\
\hat{k}_0^{(2)} = \bar{k}_0^{(2)}(0), \\
\hat{k}_0^{(3)} = \bar{k}_0^{(3)}(0).
\end{cases} \tag{8.156}$$

The root $\hat{k}_0^{(2)}$ must be ruled out, due to the fact that $\mathrm{Re}\{\hat{k}_0^{(2)}\} < 0$. Therefore, the general solution of $\hat{h}_0$ in the tip region is

$$\hat{h}_0 = \hat{d}_0^{(1)} H_m^{(1)}\left(\frac{\hat{k}_0^{(1)}\xi}{\varepsilon}\right) + \hat{d}_0^{(3)} H_m^{(1)}\left(\frac{\hat{k}_0^{(3)}\xi}{\varepsilon}\right). \tag{8.157}$$

The above tip inner solution must match the outer solution. Let us set

$$\hat{d}_0^{(1)} = d_0, \quad \hat{d}_0^{(3)} = -\hat{Q}_0 d_0, \tag{8.158}$$

where $\hat{Q}_0$ is to be determined. We can express the tip inner solution as

$$\hat{h}_0 = d_0 \left\{ H_m^{(1)}\left(\frac{\hat{k}_0^{(1)}\xi}{\varepsilon}\right) - \hat{Q}_0 H_m^{(1)}\left(\frac{\hat{k}_0^{(3)}\xi}{\varepsilon}\right) \right\}. \tag{8.159}$$

As $\hat{\xi} \to \infty$, one has

$$\hat{h}_0 \approx d_0 \sqrt{\frac{2}{\pi}} e^{-(m+\frac{1}{2})\frac{\pi}{2}} \times \left\{ \sqrt{\frac{\varepsilon}{k_0^{(1)}(0)\xi}}\, e^{\frac{ik_0^{(1)}(0)\xi}{\varepsilon}} - \hat{Q}_0 \sqrt{\frac{\varepsilon}{k_0^{(3)}(0)\xi}}\, e^{\frac{ik_0^{(3)}(0)\xi}{\varepsilon}} \right\}. \tag{8.160}$$

Now we need to match the tip inner solution (8.159) with the outer solution derived in the last section, namely,

$$\hat{\mu}_0(\varepsilon)\sqrt{\frac{2\varepsilon}{\pi}} d_0 \left\{ \sqrt{\frac{1}{\hat{k}_0^{(1)}\hat{\xi}}} e^{\frac{i\hat{k}_0^{(1)}\xi}{\varepsilon}} - \hat{Q}_0 \sqrt{\frac{1}{\hat{k}_0^{(3)}\hat{\xi}}} e^{\frac{i\hat{k}_0^{(3)}\xi}{\varepsilon}} \right\}$$

$$\Updownarrow$$

$$\tilde{\mu}_0(\varepsilon)\left( D_1 \exp\left\{\frac{i}{\varepsilon}\int_{\zeta_c}^{\xi} \bar{k}_0^{(1)} d\xi'\right\} + D_3 \exp\left\{\frac{i}{\varepsilon}\int_{\zeta_c}^{\xi} \bar{k}_0^{(3)} d\xi'\right\} \right)$$
$$= \tilde{\mu}_0(\varepsilon)\left[ D_1 e^{-i\chi_1} \exp\left\{\frac{i}{\varepsilon}\int_{0}^{\xi} \bar{k}_0^{(1)} d\xi'\right\} + D_3 e^{-i\chi_3} \exp\left\{\frac{i}{\varepsilon}\int_{0}^{\xi} \bar{k}_0^{(3)} d\xi'\right\} \right].$$

Here we have used the notation

$$\chi_1(\varepsilon) = \frac{1}{\varepsilon}\int_0^{\zeta_c} \bar{k}_0^{(1)}(\xi')d\xi', \quad \chi_3(\varepsilon) = \frac{1}{\varepsilon}\int_0^{\zeta_c} \bar{k}_0^{(3)}(\xi')d\xi'. \tag{8.161}$$

In order to satisfy the above matching condition, the parameter $\sigma_0$ must be a proper function of $\varepsilon$ such that as $\varepsilon \to 0$, the two functions $\exp\left\{-i\chi_1 + \frac{i}{\varepsilon}\int_0^{\xi} \bar{k}_0^{(1)} d\xi'\right\}$ and $\exp\left\{-i\chi_3 + \frac{i}{\varepsilon}\int_0^{\xi} \bar{k}_0^{(3)} d\xi'\right\}$ are of the same order of magnitude. It follows from the matching condition that

$$\hat{\mu}_0(\varepsilon)\varepsilon^{\frac{1}{2}} = \tilde{\mu}_0(\varepsilon) \tag{8.162}$$

and

$$\frac{D_3 e^{-i\chi_3}}{D_1 e^{-i\chi_1}} = -ie^{i(\chi_1-\chi_3)} = -\hat{Q}_0 \left(\frac{\bar{k}_0^{(1)}(0)}{\bar{k}_0^{(3)}(0)}\right)^{\frac{1}{2}}, \tag{8.163}$$

or

$$\hat{Q}_0 = \hat{Q}_0(\sigma_0, \varepsilon) = \mathrm{i}e^{\mathrm{i}(\chi_1-\chi_3)} \left( \frac{\bar{k}_0^{(3)}(0)}{\bar{k}_0^{(1)}(0)} \right)^{\frac{1}{2}} . \tag{8.164}$$

Up to this point, $\hat{Q}_0$ in the tip solution is free.

## 8.8 Global Trapped-Wave (GTW) Modes and Quantization Condition

We now turn to the second step: constructing the global eigenmodes and deriving the quantization condition for the eigenvalues by applying the tip smoothness condition

$$\frac{\partial}{\partial \hat{\xi}} \{\hat{T}, \hat{T}_S, \hat{h}\} = 0 . \tag{8.165}$$

To examine the tip smoothness condition $\hat{h}_0(0) = \mathcal{O}(1), \hat{h}_0'(0) = 0$, one needs to investigate the behavior of the function $\hat{h}_0(\hat{\xi})$ as $\hat{\xi} \to 0$ for each of the cases $m = 0, 1, 2, \ldots$, respectively. Note that as $\hat{\xi} \to 0$,

$$H_m^{(1)}\left(\frac{\hat{k}_0\xi}{\varepsilon}\right) \sim \begin{cases} -\mathrm{i}\frac{(m-1)!}{\pi}\left(\frac{2\varepsilon}{\hat{k}_0\xi}\right)^m + O\left(\frac{\hat{k}_0\xi}{2\varepsilon}\right)^m, & (m > 0), \\ \mathrm{i}\frac{2}{\pi}\ln\left(\frac{\hat{k}_0\xi}{\varepsilon}\right) + 1 + \mathrm{i}\frac{2}{\pi}(\gamma - \ln 2) + O\left(\frac{\hat{k}_0\xi}{\varepsilon}\right), & (m = 0) \end{cases}$$

(see p. 360 in [1]). It follows that

- For the case $m = 0$, we have

$$\hat{h}_0(\xi) \approx d_0 \left\{ \mathrm{i}\frac{2}{\pi} \ln\left(\frac{\hat{k}_0^{(1)}\xi}{\varepsilon}\right) - \mathrm{i}\hat{Q}_0 \frac{2}{\pi} \ln\left(\frac{\hat{k}_0^{(3)}\xi}{\varepsilon}\right) \right\} + \mathcal{O}(1).$$

  To ensure $\hat{h}_0(0) = \mathcal{O}(1)$, one must set $\hat{Q}_0 = 1$. In the meanwhile, the tip smoothness condition is automatically satisfied, since

$$\hat{h}_0'(0) = 0.$$

- For the case $m = 1$, we have

$$\hat{h}_0(\hat{\xi}) \approx d_0 \left\{ \left( \frac{\hat{k}_0^{(1)}\xi}{2\varepsilon} - \mathrm{i}\frac{2\varepsilon}{\pi \hat{k}_0^{(1)}\xi} \right) - \hat{Q}_0 \left( \frac{\hat{k}_0^{(3)}\xi}{2\varepsilon} - \mathrm{i}\frac{2\varepsilon}{\pi \hat{k}_0^{(3)}\xi} \right) \right\} .$$

To ensure $\hat{h}_0(0) = \mathcal{O}(1)$, one must set $\hat{Q}_0 = \frac{\hat{k}_0^{(3)}}{\hat{k}_0^{(1)}}$. However, in the meanwhile, one has

$$\hat{h}'_0(0) = \frac{d_0\hat{k}_0^{(3)}}{2}\left\{\frac{\hat{k}_0^{(1)}}{\hat{k}_0^{(3)}} - \frac{\hat{k}_0^{(3)}}{\hat{k}_0^{(1)}}\right\} \neq 0,$$

since $\hat{k}_0^{(1)} \neq \hat{k}_0^{(3)}$, which implies that the tip smoothness condition cannot be satisfied. In other words, the system does not allow the mode with $m = 1$.

- For the case $m > 1$, we have

$$\hat{h}_0(\hat{\xi}) \approx d_0\left\{\left[\frac{1}{m!}\left(\frac{\hat{k}_0^{(1)}\xi}{2\varepsilon}\right)^m - \mathrm{i}\frac{(m-1)!}{\pi}\left(\frac{2\varepsilon}{\hat{k}_0^{(1)}\xi}\right)^m\right]\right.$$
$$\left. -\hat{Q}_0\left[\frac{1}{m!}\left(\frac{\hat{k}_0^{(3)}\xi}{2\varepsilon}\right)^m - \mathrm{i}\frac{(m-1)!}{\pi}\left(\frac{2\varepsilon}{\hat{k}_0^{(3)}\xi}\right)^m\right]\right\}.$$

To ensure $\hat{h}_0(0) = \mathcal{O}(1)$, one must set $\hat{Q}_0 = \left(\frac{\hat{k}_0^{(3)}}{\hat{k}_0^{(1)}}\right)^m$. In the meanwhile, the tip smoothness condition

$$\hat{h}'_0(0) = 0$$

is automatically satisfied.

In summary, we derive that in order for the tip solution (8.159) to satisfy the tip condition, one must set

$$\hat{Q}_0 = \left(\frac{\hat{k}_0^{(3)}}{\hat{k}_0^{(1)}}\right)^m = \left(\frac{\bar{k}_0^{(3)}(0)}{\bar{k}_0^{(1)}(0)}\right)^m, \tag{8.166}$$

and only the modes $m = 0$ and $m = 2, 3, \ldots$ are allowable. The $m = 0$ mode gives an axisymmetric interfacial traveling wave, while the $m = 2, 3, \ldots$ modes give a multiarm spiral interfacial wave.

Formula (8.166) is the quantization condition for the eigenvalues. Combining with (8.164), one may rewrite (8.166) in the form

$$\frac{1}{\varepsilon}\int_0^{\xi_c}\left(\bar{k}_0^{(1)} - \bar{k}_0^{(3)}\right)\mathrm{d}\zeta = \left(2n + 1 + \frac{1}{2} - \frac{2m-1}{2}\theta_0\right)\pi$$
$$+\,\mathrm{i}\frac{2m-1}{2}\ln\alpha_0, \quad (m = 0, 2, 3, \ldots; n = 0, 1, 2, \ldots), \tag{8.167}$$

where

$$\alpha_0\, \mathrm{e}^{\mathrm{i}\theta_0\pi} = \frac{\bar{k}_0^{(1)}(0)}{\bar{k}_0^{(3)}(0)}\,. \tag{8.168}$$

This quantization condition gives a discrete set of complex eigenvalues

$$\sigma_0^{(m,n)}\ \big(m = 0, 2, 3, \ldots; \quad n = 0, 1, 2, \ldots\big)$$

as functions of the parameters $\varepsilon$. For the special axisymmetric case (m=0), we have

$$\frac{1}{\varepsilon}\int_0^{\zeta_c}\left(\bar{k}_0^{(1)} - \bar{k}_0^{(3)}\right)\mathrm{d}\zeta = \left(2n + 1 + \frac{1}{2} + \frac{1}{2}\theta_0\right)\pi - \mathrm{i}\,\frac{1}{2}\ln\alpha_0. \tag{8.169}$$

This quantization condition gives a discrete set of complex eigenvalues

$$\sigma_0^{(0,n)}\ \ (n = 0, 1, 2, \ldots)\,.$$

The shape of the dendritic interface corresponding to a global neutral mode can be written in the form

$$\eta_s(\xi, \theta, t) \approx 1 + \hat{h}_0(\xi)\mathrm{e}^{-\mathrm{i}\frac{\omega_* t}{\eta_0^2\varepsilon_*} + \mathrm{i}m\theta} \qquad (\text{as } \varepsilon \to 0). \tag{8.170}$$

The parameter $\sigma = -\mathrm{i}\omega_*$ is the eigenvalue corresponding to the critical number $\varepsilon_* = \varepsilon_*^{(m,n)}$.

The quantization condition (8.167) is formally the same as that derived in previous chapter for the pure melt system. The only difference from the case of a pure melt is that here, the wave number functions $k_0^{(1)}(\xi), k_0^{(3)}(\xi)$ are solved from the dispersion formula (8.55). Consequently, the eigenvalues $\sigma_0$ for the present system now depend on the parameters $\varepsilon$ and the thermodynamic parameters $(\lambda, \kappa, M, \Delta C)$ via the parameter $\Lambda_c$. Therefore, in view of pattern formation with one parameter $\Lambda_c$, in the leading-order approximation, the system of pure melt is actually equivalent to the system of binary mixture. For the special case $\Lambda_c = \lambda + 2M\Delta C = 1$, the dendritic patterns displayed in the binary system will be completely the same as those in the system of pure melt, since the dispersion formula (8.96) reduces to (6.242).

So far, there have been only very limited experimental reports in the literature on nonaxisymmetric spiral dendritic pattern formation in the binary system [2]. We leave further discussion on spiral dendritic pattern formation to the future after such patterns are observed experimentally under certain conditions. Thus, we focus on the special axisymmetric case ($m = 0$).

## 8.9 Axisymmetric Global Modes ($m = 0$)

The quantization condition for the axisymmetric case ($m = 0$) is

$$\frac{1}{\varepsilon}\int_0^{\xi_c}\left(k_0^{(1)} - k_0^{(3)}\right)\mathrm{d}\xi = \left(2n + 1 + \frac{1}{2} + \frac{\theta_0}{2}\right)\pi - \frac{\mathrm{i}}{2}\ln\alpha_0 \qquad (8.171)$$

$$\left(n = 0, \pm 1, \pm 2, \pm 3, \ldots\right),$$

where

$$\alpha_0\,\mathrm{e}^{\mathrm{i}\theta_0\pi} = \frac{k_0^{(1)}(0)}{k_0^{(3)}(0)}. \qquad (8.172)$$

The zeroth-order approximate eigenvalues $\sigma_0$ can be calculated from (8.172) as a function of $\varepsilon$ and $\lambda, \kappa, M, \Delta C$, and $T_\infty$. Then for given $\varepsilon, \lambda, M, \kappa$, and $T_\infty$, one may obtain a discrete set of eigenvalues $\sigma_n \sim \{\sigma_0 + \varepsilon\sigma_1\}_n$ ($n = 0, 1, 2, \ldots$), and a unique global neutrally stable solution with $\sigma_R = 0$, so that the critical number $\varepsilon_*$ is obtained as a function of $T_\infty$, $\lambda$, $\kappa$, and $M$.

Numerical computations for the selected neutrally stable modes have been carried out for various parameters. To present these results, it is appropriate to introduce the following definitions:

- the effective undercooling temperature $\Delta T$, which is defined as

$$\Delta T = -T_\infty - M\,; \qquad (8.173)$$

- the dimensionless tip velocity $U_{\mathrm{tip}}$, which is defined as

$$U_{\mathrm{tip}} = \frac{(U)_{\mathrm{D}}}{\kappa_{\mathrm{T}}/\ell_{\mathrm{c}}} = \lambda\varepsilon_*^2\eta_0^4\,; \qquad (8.174)$$

- the dimensionless tip radius $R_{\mathrm{tip}}$, which is defined as

$$R_{\mathrm{tip}} = \frac{\ell_{\mathrm{t}}}{\ell_{\mathrm{c}}} = \frac{1}{\varepsilon_*^2\eta_0^2}\,; \qquad (8.175)$$

- the dimensionless frequency of oscillation of the interface $\Omega_*$, which is defined as

$$\Omega_* = \frac{\omega_*}{\varepsilon_*\eta_0^2}. \qquad (8.176)$$

It should be remarked that for the special case $M = 0$, the problem in hand reduces to the problem of dendritic growth from pure melt. However, the quantities $\varepsilon, \eta_0^2$

used here for a binary system are different from what we used in Chap. 6 for the case of pure melt, since the length scales are chosen differently in the two cases. To distinguish them from each other, let us denote the parameters $\varepsilon, \varepsilon_*, \eta_0^2$, and other parameters used for the case $M = 0$, by adding the subscript "melt." Thus, in the case $M = 0$, we have defined

$$(\varepsilon)_{\text{melt}} = \frac{\ell_T/\ell_c}{(\eta_0^2)_{\text{melt}}}, \tag{8.177}$$

$$(U_{\text{tip}})_{\text{melt}} = \frac{(U)_D}{\ell_c/\kappa_T} = (\varepsilon_*^2)_{\text{melt}}(\eta_0^4)_{\text{melt}}, \tag{8.178}$$

$$(R_{\text{tip}})_{\text{melt}} = \frac{\ell_t}{\ell_c} = \frac{1}{(\varepsilon_*^2)_{\text{melt}}(\eta_0^2)_{\text{melt}}}, \tag{8.179}$$

and the frequency of oscillation of the interface

$$(\Omega_*)_{\text{melt}} = \frac{(\omega_*)_{\text{melt}}}{(\varepsilon_*)_{\text{melt}}(\eta_0^2)_{\text{melt}}}. \tag{8.180}$$

However, from (8.15), we can derive

$$(\eta_0^2)_{\text{melt}} = \lambda \lim_{M\to 0} \eta_0^2 . \tag{8.181}$$

Thus we obtain the following connection between the quantities used for the two cases: fixing $T_\infty$, $\lambda$, and $\kappa$, and letting $M \to 0$, we find that

$$\frac{\varepsilon_*^2}{\lambda} \to (\varepsilon_*^2)_{\text{melt}}, \tag{8.182}$$

$$\frac{\Omega_*}{\lambda} \to (\Omega_*)_{\text{melt}}, \tag{8.183}$$

$$U_{\text{tip}} \to (U_{\text{tip}})_{\text{melt}}, \tag{8.184}$$

$$R_{\text{tip}} \to (R_{\text{tip}})_{\text{melt}} . \tag{8.185}$$

The relationships (8.184) and (8.185) imply that for fixed $\Delta T$, as $M \to 0$, the limiting values of $U_{\text{tip}}$ and $R_{\text{tip}}$ will be irrelevant to the parameters $\lambda$ and $\kappa$. Our numerical results satisfy this property.

The zeroth-order approximation, $\varepsilon_*^{(0)}$, of the critical number $\varepsilon_*$ is calculated through the zeroth-order approximation of the eigenvalue $\sigma$, namely by the equation $\text{Re}\{\sigma_0(\varepsilon)\} = 0$. The first-order approximation, $\varepsilon_*^{(1)}$, is to be calculated through the equation $\text{Re}\{\sigma_0(\varepsilon) + \varepsilon\sigma_1(\varepsilon)\} = 0$, but $\sigma_1$ can be calculated only for the cases in which the effect of $\eta_1$ is negligible.

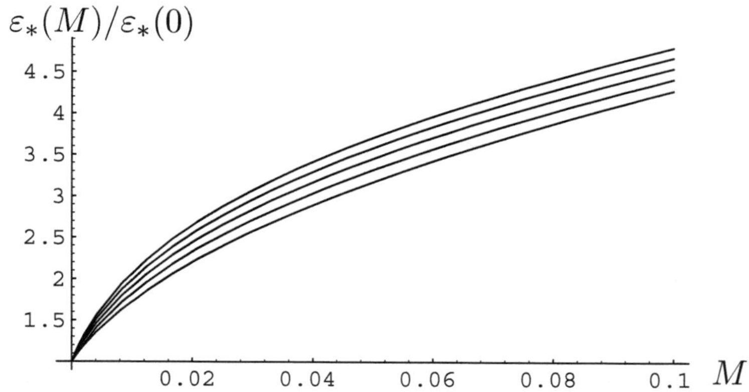


**Fig. 8.2** The variation of the ratio $\varepsilon_*(M)/\varepsilon_*(0)$ with $M$ for $\lambda = 0.01114$, $\kappa = 0.103$, and the five values $\Delta T = 0.01, 0.02, 0.03, 0.04, 0.05$ from *bottom* to *top*

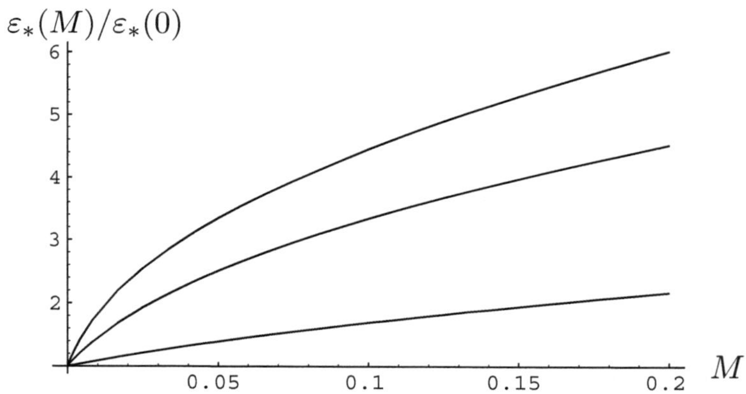


**Fig. 8.3** The variation of the ratio $\varepsilon_*(M)/\varepsilon_*(0)$ with $M$ for $\lambda = 0.01114$, $\Delta T = 0.021$, and the three values $\kappa = 0.1, 0.5, 0.9$ from *top* to *bottom*

In the following, we shall give mainly the numerical results of the zeroth-order approximation solutions.

The variation of the ratio of the critical numbers $\varepsilon_*(M)/\varepsilon_*(0)$ with the parameter $M$, for various values of the effective undercooling parameter $\Delta T$ and the parameters $\kappa$ and $\lambda$, is shown in Figs. 8.2, 8.3, and 8.4, respectively. Note that when $M = 0$, the zeroth-order critical number is given by $\varepsilon_*^{(0)}(0) = (\varepsilon_*^{(0)})_{\text{melt}}\sqrt{\lambda} = 0.1590\sqrt{\lambda}$.

The variation of the oscillation frequency of the interface, $\Omega_*$, with the parameter $M$, under different parameters $\Delta T$, $\kappa$, and $\lambda$, is shown in Figs. 8.5, 8.6, and 8.7, respectively. It is seen that the frequency $\Omega_*$ increases with $M$ and $\Delta T$, while it decreases as $\kappa$ increases.

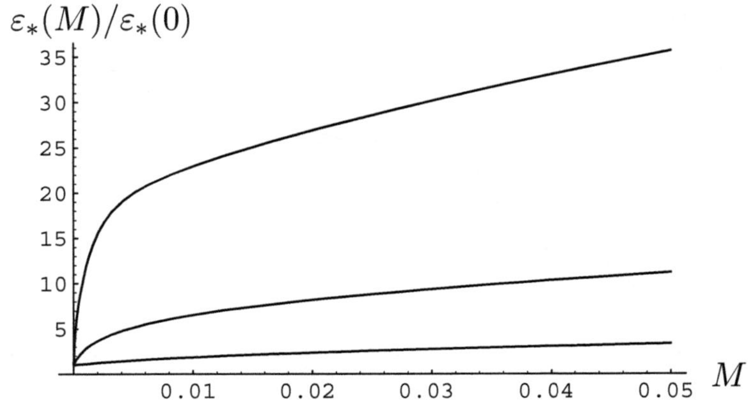


**Fig. 8.4** The variation of the ratio $\varepsilon_*(M)/\varepsilon_*(0)$ with $M$ for $\kappa = 0.103$, $\Delta T = 0.021$, and the three values $\lambda = 0.011, 0.001, 0.0001$ from *top* to *bottom*

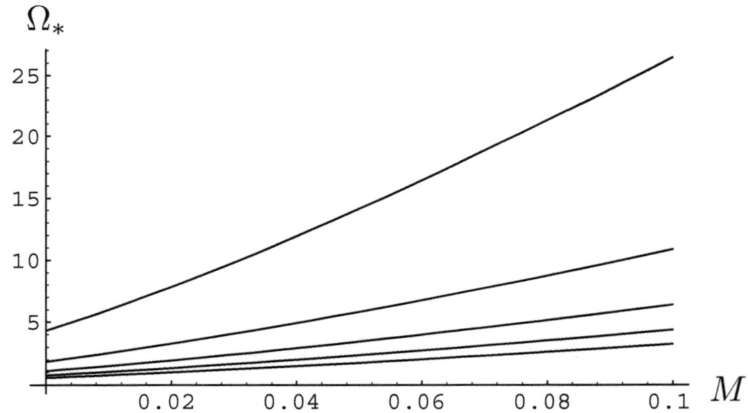


**Fig. 8.5** The variation of $\Omega_*$ with $M$ for $\lambda = 0.01114$, $\kappa = 0.103$, and the five values $\Delta T = 0.01, 0.02, 0.03, 0.04, 0.05$ from *top* to *bottom*

With the numerical results for $\varepsilon_*$, one can calculate the tip velocity $U_{\text{tip}}$ and the tip radius $R_{\text{tip}}$ of the dendrite. The variation of the tip velocity $U_{\text{tip}}$ with the parameter $M$, for different values of the parameters $\Delta T$, $\kappa$, and $\lambda$, is shown in Figs. 8.8, 8.9, and 8.10, respectively.

The variation of the tip radius $R_{\text{tip}}$ with the parameter $M$, for different values of the parameters $\Delta T$, $\kappa$, and $\lambda$, is shown in Figs. 8.11, 8.12, and 8.13, respectively.

It is seen that with other parameters fixed, as $M$ increases, the tip velocity $U_{\text{tip}}$ has a maximum value, while $R_{\text{tip}}$ has a minimum value.

To describe the interface pattern, in Fig. 8.14 we show the tips of dendrites for $\lambda = 0.01114, M = 0.05, \Delta T = 0.021$ with different values $\kappa = 0.1, 0.5, 0.9$. It is seen that the dendrite's shape is quite sensitive to the variation of $\kappa$.

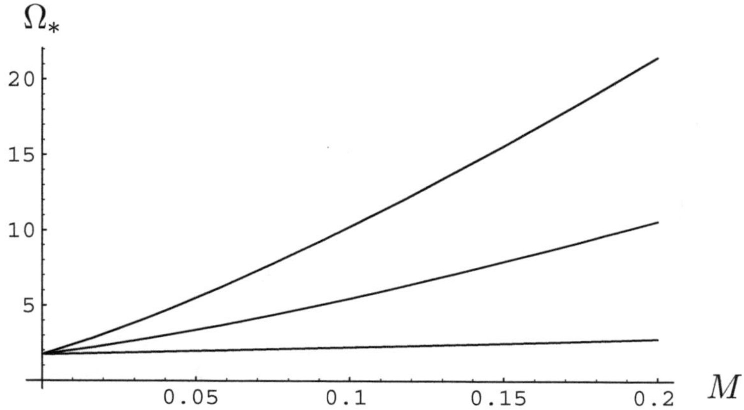


**Fig. 8.6** The variation of $\Omega_*$ with $M$ for $\lambda = 0.01114$, $\Delta T = 0.021$, and the three values $\kappa = 0.1, 0.5, 0.9$ from *top* to *bottom*

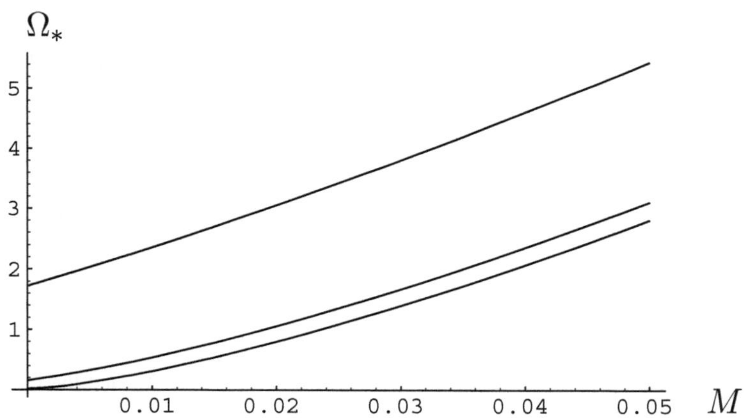


**Fig. 8.7** The variation of $\Omega_*$ with $M$ for $\kappa = 0.103$, $\Delta T = 0.021$, and the three values $\lambda = 0.011, 0.001, 0.0001$ from *top* to *bottom*

## 8.10 Comparisons of Theoretical Results with Experimental Data

So far, the most systematic experiments on dendritic growth from binary mixtures have been conducted by Lipton, Glicksman, and Kurz for the system SCN–acetone and SCN–argon in 1984 and 1987 [3–7]. The data in these various papers are not in complete agreement. More complete and accurate data exist for SCN–acetone, and this was given by Chopra, Glicksman, and Singh in 1988 [8].

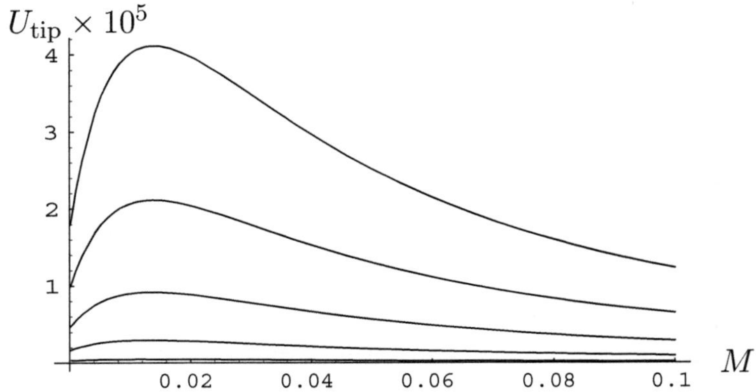


**Fig. 8.8** The variation of $U_{\text{tip}}$ with $M$ for $\lambda = 0.01114$, $\kappa = 0.103$, and the five values $\Delta T = 0.01, 0.02, 0.03, 0.04, 0.05$ from *bottom* to *top*

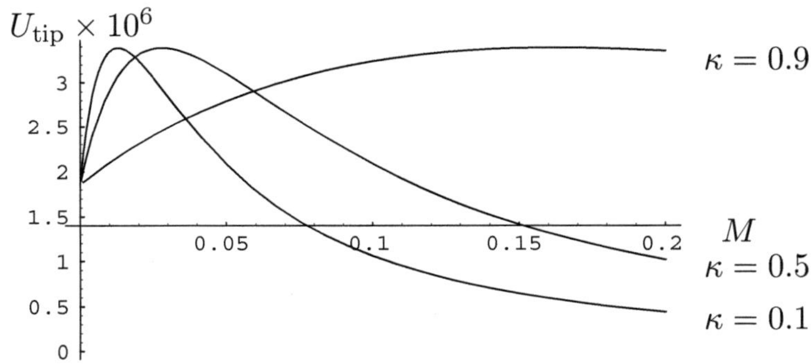


**Fig. 8.9** The variation of $U_{\text{tip}}$ with $M$ for $\lambda = 0.01114$, $\Delta T = 0.021$, and the three values $\kappa = 0.1, 0.5, 0.9$ from *bottom* to *top*

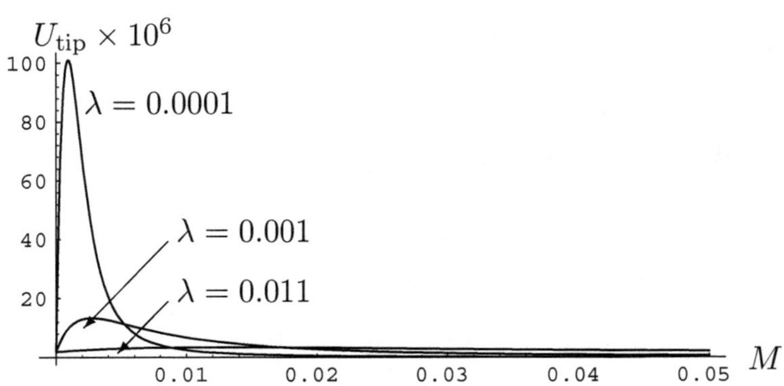


**Fig. 8.10** The variation of $U_{\text{tip}}$ with $M$ for $\kappa = 0.103$, $\Delta T = 0.021$, and the three values $\lambda = 0.011, 0.001, 0.0001$ from *bottom* to *top*

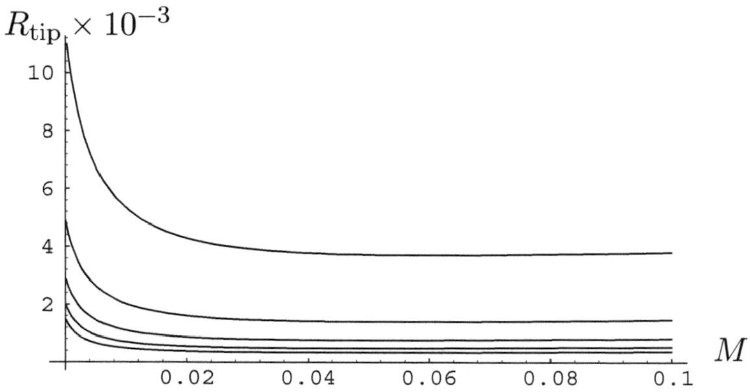


**Fig. 8.11** The variation of $R_{tip}$ with $M$ for $\lambda = 0.01114$, $\kappa = 0.103$, and the five values $\Delta T = 0.01, 0.02, 0.03, 0.04, 0.05$ from *top* to *bottom*

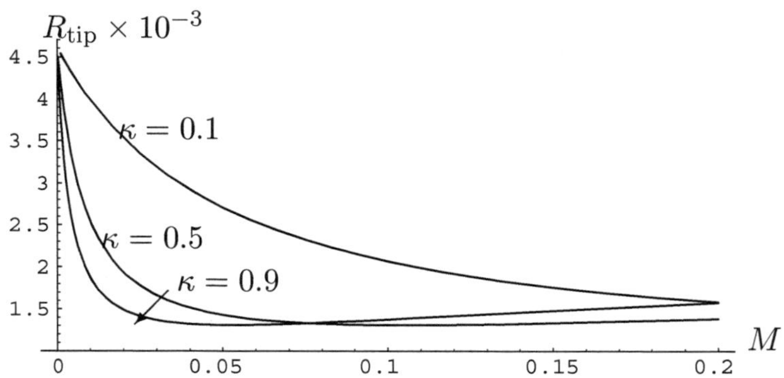


**Fig. 8.12** The variation of $R_{tip}$ with $M$ for $\lambda = 0.01114$, $\Delta T = 0.021$, and the three values $\kappa = 0.1, 0.5, 0.9$ from *top* to *bottom*

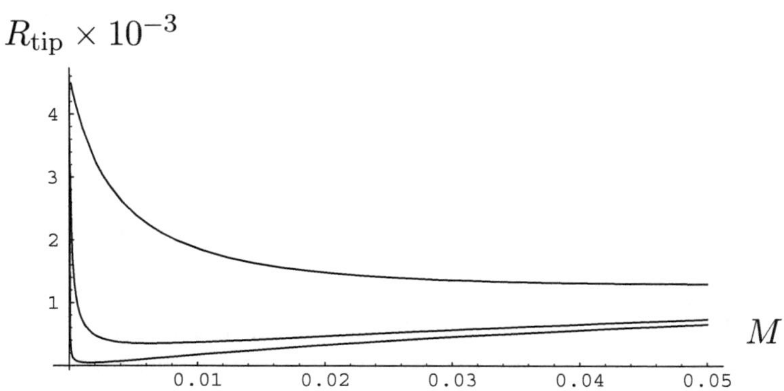


**Fig. 8.13** The variation of $R_{tip}$ with $M$ for $\kappa = 0.103$, $\Delta T = 0.021$, and the three values $\lambda = 0.011, 0.001, 0.0001$ from *top* to *bottom*

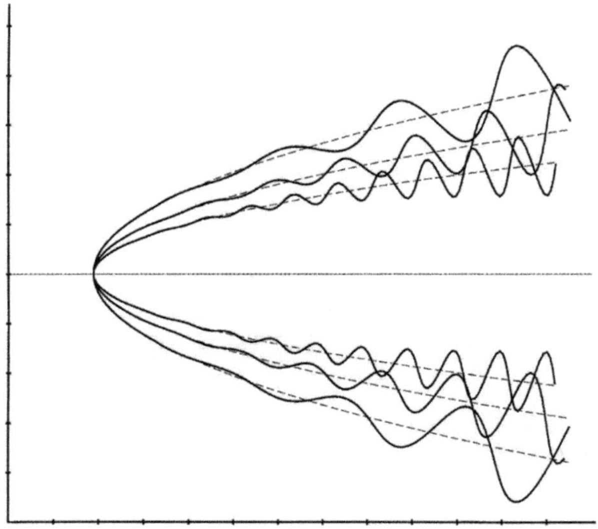

**Fig. 8.14** The tips of dendrites with the three values $\kappa = 0.1, 0.5, 0.9$ from *inside* to *outside*

**Table 8.1** The thermodynamic properties of SCN–acetone mixtures

| | |
|---|---|
| $\lambda(=\kappa_D/\kappa_T)$ | 0.01114 |
| $\kappa_D$ | $1.27 \times 10^{-9}\,\mathrm{m^2 s^{-1}}$ |
| $\Delta H$ | $46.26 \times 10^3\,\mathrm{J\,kg^{-1}}$ |
| $c_p$ | $1937.5\,\mathrm{J\,kg^{-1}\,K^{-1}}$ |
| $\kappa_T$ | $1.14 \times 10^{-7}\,\mathrm{m^2 s^{-1}}$ |
| $\gamma T_{M0}/\Delta H$ | $6.63 \times 10^{-8}\,\mathrm{K\,m}$ |
| $m$ | $-2.16\,\mathrm{K\,mol\%^{-1}}$ |
| $\kappa$ | $0.103\,\mathrm{mol\%\,mol\%^{-1}}$ |

In the following, we shall compare the theory with these experimental data.

The properties of SCN–acetone mixtures are listed in Table 8.1. Based on these data, we calculate that for this system $\ell_c = 2.7726 \times 10^{-9}$(m), the unit of temperature is $[T] = \frac{\Delta H}{(c_p\rho)} = 23.8761\,\mathrm{K}$, the unit of velocity is $[U] = \frac{\kappa_T}{\ell_c} = 41.1173$ (m/s), and the concentration of solute at the far field is $C_\infty = 11.0583\,M$, where $M$ is the morphological parameter. In terms of these data, we convert the experimental data given in [8] into our dimensionless quantities $U_{tip}, R_{tip}$, and $M$.

Comparisons between the zeroth-order approximation solutions and the experimental data for the variation of $U_{tip}$ and $R_{tip}$ versus $\Delta T$ for $\kappa = 0.103, \lambda = 0.01114$, and the four values $M = 0.00407, 0.00995, 0.02713, 0.03608$ are shown in Fig. 8.15a–d. Noting that there is no adjustable parameter in the theory, the agreement between the theoretical results and the experimental data is reasonably good. The discrepancy between the experimental data and the theoretical curves

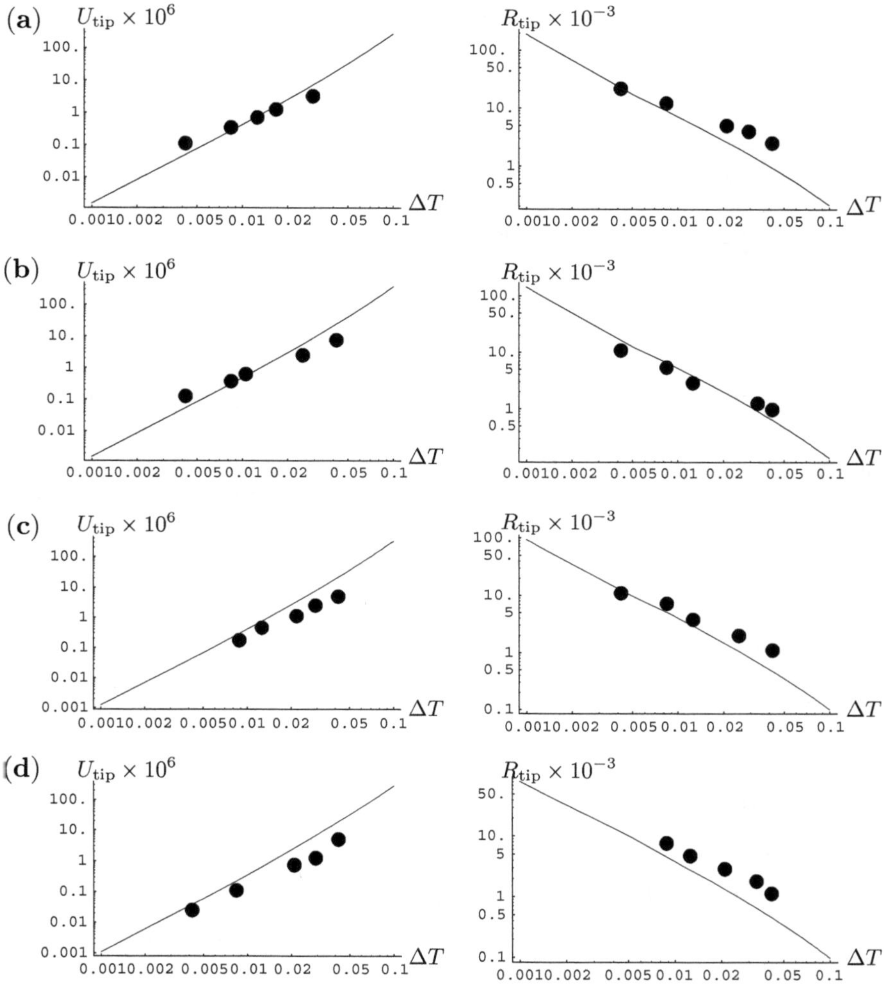


**Fig. 8.15** The variation of $U_{tip}$ and $R_{tip}$ with $\Delta T$ for $\kappa = 0.103, \lambda = 0.01114$, and **(a)** $M = 0.00407$; **(b)** $M = 0.00995$; **(c)** $M = 0.02713$; **(d)** $M = 0.03608$

might indicate the effects of some physical parameters neglected in our analysis, and/or the effect of higher-order contributions and nonlinearity. It is also due to the inaccuracy of the experimental measurements. In principle, when $M \to 0$, the results of dendritic growth from a binary system should approach the results of a pure substance system. However, one sees that the experimental data of a binary system are inconsistent with this principle. In contrast, the theoretical data obtained with the binary system are consistent with the results of a pure substance system.

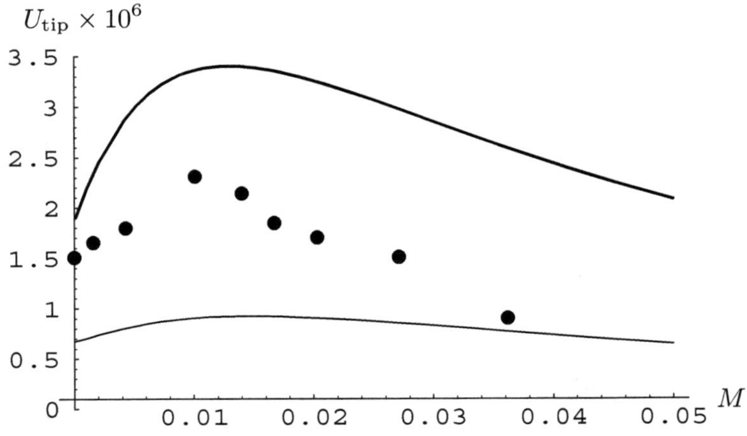


**Fig. 8.16** The variation of $U_{\text{tip}}$ with $M$ for the case $\Delta T = 0.021, \kappa = 0.103, \lambda = 0.01114$. The *bold line* shows the zeroth-order approximation solution, while the *thin line* is the first-order approximation solution

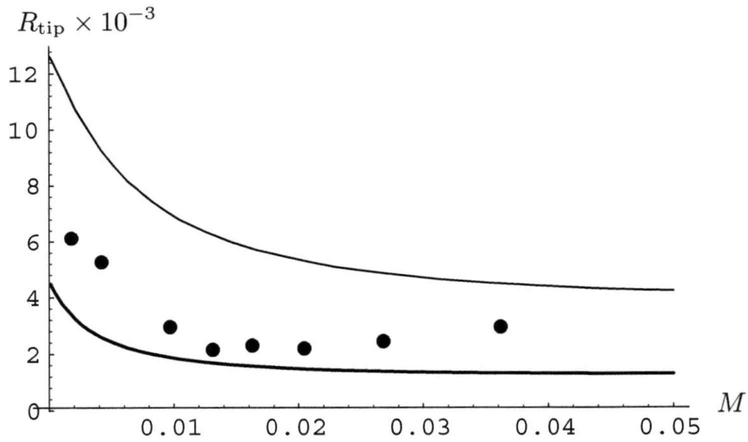


**Fig. 8.17** The variations of $R_{\text{tip}}$ with $M$ for the case $\Delta T = 0.021, \kappa = 0.103, \lambda = 0.01114$. The *bold line* shows the zeroth-order approximation solution, while the *thin line* is the first-order approximation solution

Comparisons of theoretical results and experimental data for $U_{\text{tip}}$ and $R_{\text{tip}}$ versus $M$ for $\Delta T = 0.021, \kappa = 0.103, \lambda = 0.01114$ are shown in Figs. 8.16 and 8.17, respectively. In these figures, the zeroth-order approximation solution is shown by the bold lines, while the first-order approximation solution is shown by the thin lines. The experimental data fall between the two lines. The overall agreement between the theoretical results and the experimental data is reasonably good. Particularly, in the regime of $M > 0.03$, the experimental data are quite close to the first-order approximation solution. Again, one sees that in the range of $M \ll 1$, the experimental data of the binary system are inconsistent with the data of a pure melt system, unlike our theoretical data. For $M = 0.000580$, with $\lambda = 0.01114$, $\Delta T = 0.021$, our numerical data show that $\varepsilon_* = 0.01191$, $U_{\text{tip}} \approx 0.8979 \times 10^{-6}$, and $R_{\text{tip}} \approx 9.355 \times 10^3$. Converting to the critical number for the system of pure melt, we have $(\varepsilon_*)_{\text{melt}} = \varepsilon_*/\lambda^{\frac{1}{2}} = 0.01128$, which is very close to the critical number 0.1108 that we derived for the system of pure melt with a small undercooling in Chap. 6. Since it has been seen that our results for pure melt agree with experimental data very well, we conclude that the above-listed results of $U_{\text{tip}}$ and $R_{\text{tip}}$ for $M = 0.000580$ should be correct, whereas the experimental data in this regime are not as accurate as the updated data obtained for the system of pure melt SCN.

## References


1. M. Abramovitz, I.A. Stegun, *Handbook of Mathematical Functions* (National Bureau of Standards, Washington, 1964)
2. S. Akamatsu, M. Perrut, S. Bottin-Rousseau, G. Faivre, Spiral two-phase dendrites. Phys. Rev. Lett. **104**, 056101 (2010)
3. J. Lipton, M.E. Glicksman, W. Kurz, Dendritic growth into undercooled alloy melts. Mater. Sci. Eng. **65**, 57–63 (1984)
4. J.S. Langer, *Lectures in the Theory of Pattern Formation, USMG NATO AS Les Houches Session XLVI 1986 — Le hasard et la matière / chance and matter*, ed. by J. Souletie, J. Vannimenus, R. Stora (Elsevier Science, Amsterdam, 1986)
5. D.A. Kessler, J. Koplik, H. Levine, Pattern formation far from equilibrium: the free space dendritic crystal growth, in *Patterns, Defects and Microstructures in Non-equilibrium Systems* (NATO A.R.W., Austin, 1986)
6. J.J. Xu, Global asymptotic solution for axi-symmetric dendrite growth with small undercooling, in *Structure and Dynamics of Partially Solidified System*, ed. by D.E. Loper. NATO ASI Series, vol. 125 (Springer, Dordrecht, 1987), pp. 97–109
7. J. Lipton, M. E. Glicksman, W. Kurz, Equiaxed dendrite growth in alloys at small supercooling. Metall. Trans. A **18**, 341–345 (1987)
8. M. Copra, M.E. Glicksman, N.B. Singh, Dendritic solidification in binary alloys. Metall. Trans. A **19**, 3087–3096 (1988)

# Chapter 9
# Viscous Fingering in a Hele–Shaw Cell

In this chapter, we study another interfacial phenomenon: the formation of viscous fingers in a Hele–Shaw cell. This phenomenon occurs in an entirely different physical system from dendritic growth, but it raises similar issues and can thus be treated using the same approach established in the previous chapters. We are interested in the study of this phenomenon in this book, not only because the subject itself occupies an important position in the field of pattern formation, but also because the resolution for this problem provides a keystone for analytically studying further pattern formation problems in solidification, such as cellular growth and eutectic growth.

## 9.1 Introduction

The formation of viscous fingers in porous media has, for almost half a century, been a phenomenon of considerable interest within a broad field of science and technology, including secondary oil recovery, fixed bed regeneration in chemical processing, and underground water engineering [1–6, 8–28]. This phenomenon can be demonstrated in systems such as those shown in Figs. 9.1 and 9.2 [8]. It was recognized early on that the formation of these fingers is associated with some kind of instability mechanism for the interface, which is caused either by the different viscosities and densities of fluids or by the surface tension at the interface.

The first scientific investigation and experimental observation of the fingering phenomenon in a vertical packed column driven by gravity and viscosity was made by Hill in 1952 [7]. Hill studied the displacement of sugar liquors by water from



J.-J. Xu, *Interfacial Wave Theory of Pattern Formation in Solidification*,
Springer Series in Synergetics, DOI 10.1007/978-3-319-52663-8_9

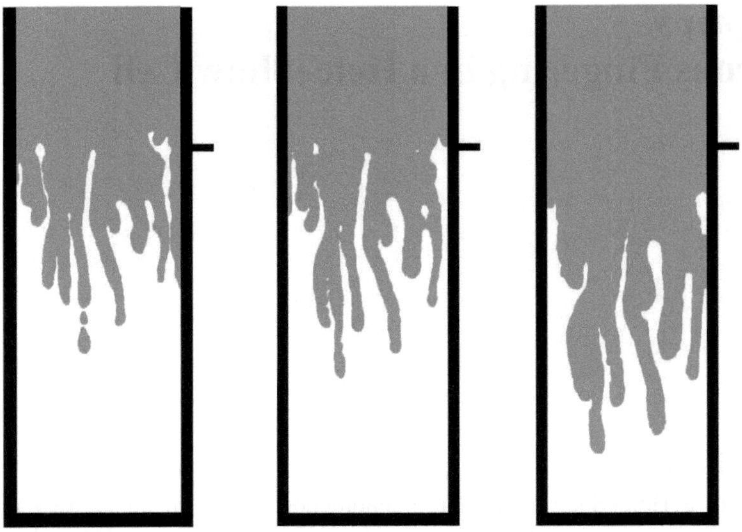

**Fig. 9.1** Viscous fingering formation in a vertical packed column

columns of granular bone charcoal, and carried out a stability analysis for the one-dimensional flat interface.

Fingering phenomena can also be driven by the surface tension on the interface and the difference in the viscosity of two immiscible fluids. Evidently, most fingerings that occur in a horizontal process belong to this category. These phenomena were first studied by Chouke et al. [4] and Saffman and Taylor [19] in the late 1950s. These investigators performed a rigorous linear stability analysis for a flat interface. This instability mechanism for a flat interface has become known as the Saffman–Taylor instability, but more precisely, it should be called the Chouke–Saffman–Taylor instability. To simulate the viscous fingering phenomenon in porous media, Saffman and Taylor [19] first used a Hele–Shaw cell as a simplified tool and made systematic experimental observations.

The device used by Saffman and Taylor is sketched in Fig. 9.3. The cell has a very thin gap $b$ and a width $2W$. To begin with, the cell is filled with a viscous fluid. Then a less viscous fluid is injected into the cell in order to displace the more viscous fluid. In the early stage of evolution, the flat interface between the two fluids will immediately develop into fingers with different sizes, due to the Chouke–Saffman–Taylor instability. However, at a later stage of evolution, experiments show that when the injection pressure is sufficiently small, or the surface tension fairly large, these small figures gradually disappear and the cell is eventually dominated by a single smooth, steady finger that occupies about one-half the cell width, as shown

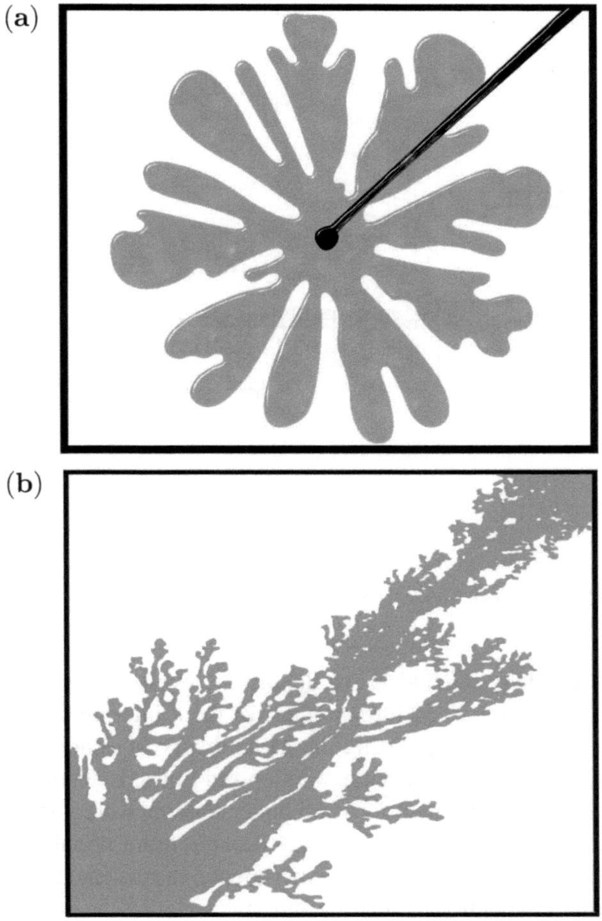


Fig. 9.2 Viscous fingering formation in some horizontal devices: (**a**) a liquid with lower viscosity is injected into the cell from the center of the top plate; (**b**) a liquid with lower viscosity is injected into the cell from the *lower-left corner* of the top plate and the fluid is drawn out from the *upper-right corner*

in Fig. 9.4a [21]. This kind of interfacial pattern evolution was observed in detail by Saffman and Taylor in 1958.

When the injection pressure increases or the surface tension decreases, the system may permit another type of stationary interfacial pattern, namely oscillatory, dendrite-like fingers that have a much narrower width. These oscillatory fingers were first discovered by Jacob, et al. in 1985, with a Hele–Shaw cell into one of whose plates a regular network of uniform grooves was etched (cf. [1]). The network of uniform grooves on the plate for the formation of dendritic finger in Hele–Shaw flow was considered by many researchers an analogy of the anisotropy of surface tension for dendritic growth in solidification. Based on the observation of Jacob

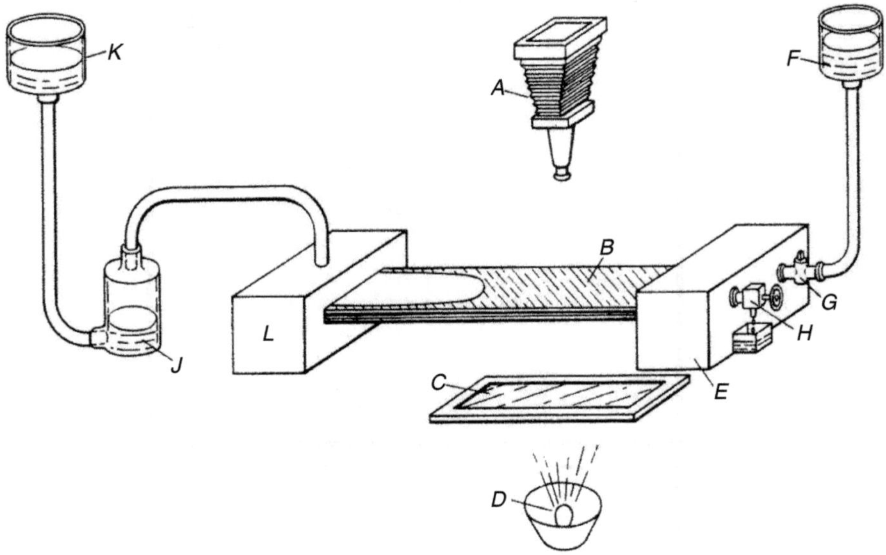


**Fig. 9.3** A sketch of the device used by Saffman and Taylor

et al., these researchers concluded that dendritic growth in solidification could not exist without anisotropy. Even further, some of them interpreted this discovery as experimental evidence of supporting the MSC theory of dendritic growth, which stated that the smooth steady needle solution did not exist without anisotropy of surface tension. These interpretations and understanding were widely disseminated throughout scientific community, and they became quite popular for a period of time. However, before long, some new experimental results came out, and it became clear that the above-mentioned analogy and interpretations were all inadequate. In 1986, the same type of oscillatory dendritic fingers as observed by Jacob et al. were found by Couder et al., later by Matsushita et al. and Yokoyama et al., by engraving a single regular groove in one of the Hele–Shaw plates, using a preexisting tiny bubble, or stretching a thin wire, which may be considered a rigid bubble, to guide the finger (cf. [5, 14] and [27]). Obviously, the single groove, bubble, and thin wire, which affect the finger's tip only locally, can hardly be interpreted as an analogy of the anisotropic surface tension in dendritic growth, which affects the entire interface. It was even more unexpected when in 1987, Kopf-Sill and Homsy claimed to have discovered that the oscillatory dendrite-like fingers might be formed in a Hele–Shaw cell with plates without engraving and without bubble and wire at the tip [13]. They reported that these fingers were normally very difficul to observe, but were still observed after the cell was carefully cleaned many times by them.

When the injection pressure becomes even larger, or the surface tension even smaller, the oscillatory fingers of all kinds observed by Couder et al. and by Jacob

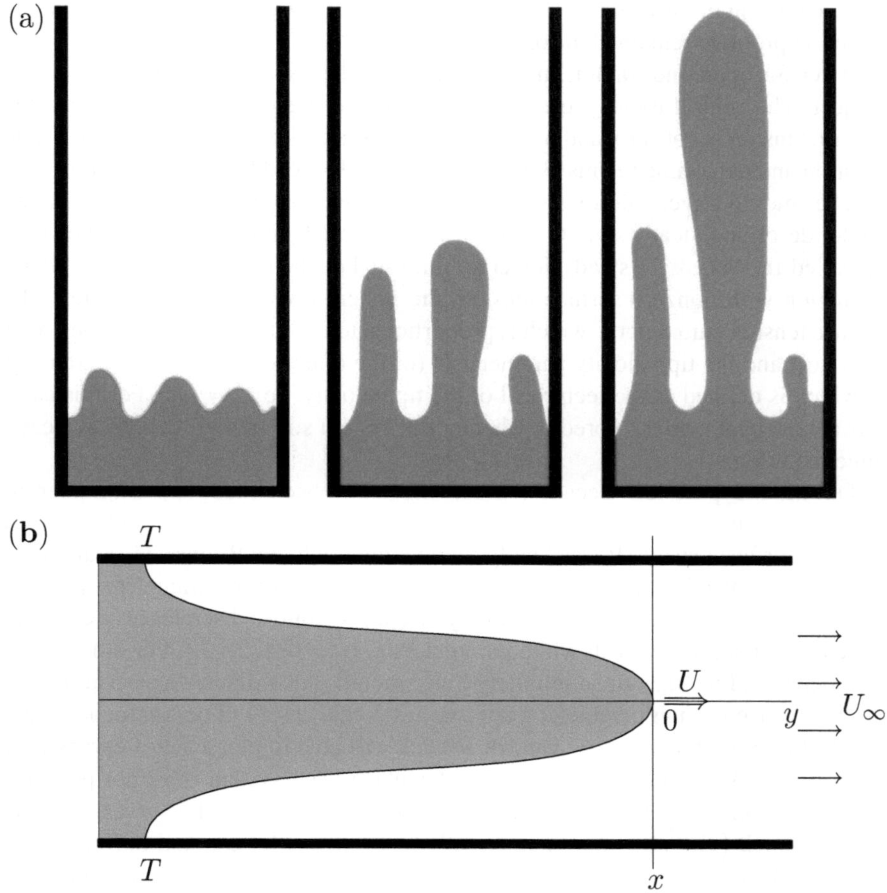


**Fig. 9.4** A typical process of smooth finger formation in a Hele–Shaw cell: (**a**) finger evolution at an early stage of the process; (**b**) sketch of a single dominant smooth finger formed in the final stage of the process

et al. are not sustained. These fingers will split and spread. Thus the viscous fingers appear as a chaotic pattern.

The smooth finger solution with zero surface tension was obtained by Saffman and Taylor in 1958. The solution contains an undetermined parameter that represents the tip velocity or the asymptotic width of the finger.

The most significant contribution made by Saffman and Taylor was their identification of the selection problem. On the basis of their experimental observations, Saffman and Taylor discovered that the tip velocity was actually uniquely determined by the operating conditions. They also initiated a linear stability analysis for the steady finger solution with zero surface tension and found that the solution was unstable. They did not find the steady smooth finger solutions with nonzero surface tension or perform a stability analysis for such a general case. As such, the selection problem remained unsolved.

After Saffman and Taylor, many other researchers devoted themselves to this subject. The central issues were how to specify the basic state solution when the surface tension is not zero, and how to perform a stability analysis for the basic state solution and determine the instability mechanisms induced by surface tension.

The smooth finger solutions with zero surface tension obtained by Saffman and Taylor describe a steady smooth finger with an infinitely long root, and so they may be called the "classical steady finger solutions." For the case of symmetric finger formation with nonzero surface tension, the system involves two parameters: the surface tension parameter $\varepsilon$, which is proportional to the surface tension at the finger interface, and the tip velocity parameter $U$ (or the effective finger width parameter $\lambda$, which is defined as the reciprocal of the tip velocity, i.e., $\lambda = \frac{1}{U}$). For this case, it is a nontrivial matter to predict whether the system still allows a classical steady finger solution.

During the past few decades, great effort has been made by a number of investigators to look for the classical steady finger solution to the problem with nonzero surface tension. It was finally found numerically by Romero [17], and later confirmed by Vanden-Broeck [23], that for the surface tension parameter $\varepsilon \neq 0$ and for every given operating condition, the system allows only a discrete set of classical steady finger solutions, with width parameters $\lambda_i$ $(i = 0, 1, 2, \ldots)$. For other width parameters $\lambda$, fingers with an infinitely long smooth root will have a cusp at the tip. As the surface tension parameter $\varepsilon$ approaches 0, this cusp will be transcendentally small. Based on these results, Vanden-Broeck was able to plot a set of curves $\{C_i\}$ $(i = 0, 1, 2, 3, \ldots)$ in the $(\lambda, \varepsilon)$ parameter plane. The system does not permit a classical finger solution for points not on these curves. Among the curves $\{C_i\}$, the lowest branch $\{C_0\}$ was found numerically in 1980 by McLean and Saffman. The results of Vanden-Broeck were soon verified by a number of investigators in terms of the microscopic solvability approach (see, for instance, [9]).

A number of researchers, such as Tanveer [22] and Bensimon et al. [3], have studied the stability around these classical steady finger solutions. Tanveer concluded that the solution with $n = 0$ is neutrally stable for all surface tension parameters $0 < \varepsilon \ll 1$. The same statement was made by Bensimon (1986) and other investigators.

In the meantime, numerical solutions for the initial value problem were obtained by a number of investigators such as DeGregoria and Schwartz [6], Kessler and Levine [11], and Bensimon [2]. The numerical results of DeGregoria and Schwartz showed that the slowly evolving fingers at the later stages of evolution spontaneously split when the surface tension is sufficiently small. As a consequence, the classical steady finger solutions will all be unstable when the surface tension parameter of fingering is less than a critical value.

The numerical results of DeGregoria and Schwartz apparently contradict the linear stability analysis by Tanveer and others. To reconcile the disagreement, Bensimon [2] speculated, on the basis of his numerical experiments, that the observed instability in the numerical simulation is a nonlinear mechanism where the threshold amplitude for destabilization decreases with the surface tension parameter of fingering. Bensimon did not perform a nonlinear bifurcation analysis to support his conjecture. So the contradiction between the numerical evidence and linear stability analysis by Tanveer and others was not reconciled until the global trapped-wave (GTW) instability for Hele–Shaw flow was discovered in 1991 [24].

The theoretical difficulty encountered in the fingering problem, due to the nonexistence of a classical steady finger solution with arbitrary tip velocity, is quite similar to what we have seen for the problem of dendritic growth. For points off the curves $\{C_i\}$ in the $(\varepsilon, \lambda)$ parameter plane, stability analysis for the classical steady state is meaningless, since these steady-state solutions do not exist. Consequently, one cannot divide the entire parameter plane into a stable region and an unstable region.

In order to overcome this difficulty, one needs to consider a broader class of basic state solutions, such as the "nearly" steady needle solution at the later stages of evolution ($t \gg 1$) and study their global stability properties. The results show that the fingering system also possesses two types of instability mechanism:

1. the low-frequency (LF) instability mechanism, which is caused by perturbations with low frequencies and is associated with the real spectrum of eigenvalues. The neutral modes of the LF mechanism coincide with the classical steady-state solutions obtained by Vanden-Broeck and others based on the microscopic solvability condition (MSC) theory.
2. the global trapped-wave (GTW) instability mechanism, which is induced by perturbations with a high frequency and is connected with a complex spectrum of eigenvalues.

When the surface tension parameter of fingering $\varepsilon$ is smaller than a critical value, the system is dominated by the GTW mechanism, and the steady finger solutions will be linearly unstable. Hence, Tanveer's conclusion is incorrect as $\varepsilon \to 0$.

In this chapter we shall restrict ourselves to the study of viscous finger formation in a rectangular Hele–Shaw cell and explore the finger selection mechanism in the later stages of evolution.

## 9.2 Mathematical Formulation of the Problem

Consider the evolution of a finger developing in a Hele–Shaw cell in the positive $y$ direction of a moving coordinate system $\{x, y\}$, that is fixed at the finger's tip. The flow velocity in the upstream far field is set to a constant $U_\infty$, but the tip velocity, in general, will change with time (see Fig. 9.4b).

For simplicity, we shall assume that the dynamic viscosity of the more viscous fluid is $\mu$, while the viscosity of the less viscous fluid is zero; all physical quantities have been averaged along the thickness of the cell, so that the problem is treated as two-dimensional. We use one-half of the cell width $W$ as the length unit and use the flow velocity in the upstream, $U_\infty$, as the velocity unit; the product $U_\infty W$ is used as the unit of the potential function $\phi(x, y, t)$ of the absolute velocity field

$\mathbf{u}(x, y, t)$. In terms of these units, $y_s(x, t)$ represents the interface's shape, while the dimensionless tip velocity is denoted by $U(t)$. The effective width of the finger is defined as $\lambda(t) = 1/U$. Moreover, we assume that for every finite time $t > 0$, the finger has a finite length with a triple point T ($x = 1, y = y_T$) downstream. This triple point is also a stagnation point of the flow field.

The general unsteady flow in a Hele–Shaw cell, in its dimensionless form, is therefore described by Laplace's equation

$$\nabla^2 \phi = \frac{\partial^2 \phi}{\partial x^2} + \frac{\partial^2 \phi}{\partial y^2} = 0 \tag{9.1}$$

with the following boundary and regularity conditions:

1. Upstream far-field condition: as $y \to \infty$,

$$\mathbf{u} = \mathbf{e}_y : \quad \frac{\partial \phi}{\partial y} = 1 , \quad \frac{\partial \phi}{\partial x} = 0 . \tag{9.2}$$

2. The slipping condition on the sidewalls: at $x = \pm 1$,

$$\frac{\partial \phi}{\partial x} = 0 . \tag{9.3}$$

3. At the interface: $\tilde{S}(x, y, t) = y - y_s(x, t) = 0$,

   (i) the dynamic condition: according to Darcy's law, the fluid velocity $\mathbf{u}$ is proportional to the gradient of the pressure. Namely, in the dimensional form, we have

$$\mathbf{u} = -\frac{b^2}{12\mu} \nabla p , \tag{9.4}$$

   where $b$ is the thickness of the cell. Thus, the potential field $\phi$ is proportional to the pressure field. So again in the dimensional form, we have

$$\phi = -\frac{b^2 p}{12\mu} . \tag{9.5}$$

   Moreover, one can derive from thermodynamics that due to the presence of the surface tension, the pressure will have a jump across the interface, which is proportional to the curvature of the interface, $\mathcal{K}$, and the surface tension coefficient, $\gamma$. Assuming that the pressure $p_0$ of the less viscous fluid inside the finger is constant and equal to zero, in the dimensionless form it is found that at the interface,

$$\phi = \varepsilon^2 \mathcal{K}\{y_s(x, t)\} ; \tag{9.6}$$

(ii) the kinematic condition: assuming that the two fluids are immiscible, the interface is a contact surface of two fluids with no transverse mass flux. Thus, we have

$$\frac{\partial \phi}{\partial n} = V_n = U(t)(\mathbf{e}_y \cdot \mathbf{n}) - \frac{\frac{\partial \tilde{S}}{\partial t}}{|\nabla \tilde{S}|}. \tag{9.7}$$

Here, $\{\mathbf{e}_x, \mathbf{e}_y\}$ are the unit coordinate vectors along the $x$- and $y$-axes, respectively; $\mathbf{n}$ is the outward normal vector of the interface, and the surface tension parameter of fingering is defined as

$$\varepsilon^2 = \frac{b^2 \gamma}{12 \mu U_\infty W^2}. \tag{9.8}$$

4. The downstream triple point condition: at $x = \pm 1, y = y_T(t)$,

$$\mathbf{u} = 0, \quad \text{or} \quad \frac{\partial \phi}{\partial x} = \frac{\partial \phi}{\partial y} = 0. \tag{9.9}$$

Since the triple point T is a stagnation point, from kinematic considerations it follows that

$$y = y_T(t) = -L - \int_0^{t-t_0} U(t_1) \mathrm{d}t_1, \tag{9.10}$$

where $L$ is the length of the finger at $t = t_0$.
5. The tip smoothness conditions.

In order to fully determine a time-evolving solution, one also needs to specify the initial conditions. The above PDE system is quite similar to that of two-dimensional dendritic growth described in Chap. 7. The dynamic condition (9.6) has the same form as the Gibbs–Thomson condition. Hence, it should not be surprising that the solutions for these two systems may display very similar mathematical structures. Note also that the mathematical formulation given above is different from the "classical Saffman–Taylor problem." The problem formulated by Saffman and Taylor is a special case of the above system with the triple point T set at infinity. Clearly, the system given here is more realistic and allows a wider class of solutions than the classical Saffman–Taylor problem.

## 9.3 The Smooth Finger Solution with Zero Surface Tension

For the case of zero surface tension ($\varepsilon = 0$), the system allows a special class of steady finger solutions with a constant tip velocity $U_0 = \frac{1}{\lambda_0}$ and the triple point at infinity. For this case, the parameter $0 < \lambda_0 < 1$ also describes the asymptotic width of the finger. This special steady solution is called the Saffman–Taylor (ST) solution. The ST solution plays the same role for the system of viscous fingering as the Ivantsov solution for dendritic growth. In our coordinate system, the ST solution is subject to the reduced system

$$\nabla^2 \phi_* = 0\,, \tag{9.11}$$

with the following boundary conditions:

1. As $y \to \infty, \;\; |x| < 1$,

$$\frac{\partial \phi_*}{\partial y} = 1\,, \quad \frac{\partial \phi_*}{\partial x} = 0\,. \tag{9.12}$$

2. As $y \to -\infty, \;\; \lambda_0 < |x| < 1$,

$$\frac{\partial \phi_*}{\partial x} = \frac{\partial \phi_*}{\partial y} = 0\,. \tag{9.13}$$

3. At $x = \pm 1$,

$$\frac{\partial \phi_*}{\partial x} = 0\,. \tag{9.14}$$

4. At the interface, $y = y_*(x)$,

$$\phi_* = 0\,, \tag{9.15}$$

$$\frac{\partial \phi_*}{\partial y} - y'_* \frac{\partial \phi_*}{\partial x} = \frac{1}{\lambda_0}. \tag{9.16}$$

Here it is assumed that $0 < \lambda_0 = \frac{1}{U_0} < 1$.

One can introduce the stream function by $\psi_*(x, y)$. The functions $\phi_*(x, y)$ and $\psi_*(x, y)$ are subject to the Cauchy–Riemann conditions

$$\frac{\partial \phi_*}{\partial x} = \frac{\partial \psi_*}{\partial y}; \quad \frac{\partial \psi_*}{\partial y} = -\frac{\partial \psi_*}{\partial x}. \tag{9.17}$$

Furthermore, let us introduce the notation

$$\eta = \phi_*(x, y), \quad \xi = -\psi_*(x, y). \tag{9.18}$$

Using the variables $(\xi, \eta)$ as the independent variables, (9.18) leads to the inverse functions

$$x = X(\xi, \eta), \quad y = Y(\xi, \eta). \tag{9.19}$$

We define the complex variables and functions

$$Z = x + \mathrm{i}y, \quad \widehat{\zeta} = \xi + \mathrm{i}\eta,$$

and

$$W = W(Z) = \phi_*(x, y) + \mathrm{i}\psi_*(x, y),$$

$$\widehat{\zeta} = \widehat{\zeta}(Z) = \xi(Z) + \mathrm{i}\eta(Z) = \mathrm{i}W(Z). \tag{9.20}$$

One may examine the conformal mapping of the Z-plane to the $\widehat{\zeta}$-plane made by the complex function $\widehat{\zeta} = \widehat{\zeta}(Z)$.

- From the interface condition (9.15) at $y = y_*(x)$, it follows that

$$\frac{\mathrm{d}}{\mathrm{d}x}\Big\{\psi_*(x, y_*(x))\Big\} = \frac{\partial\psi_*}{\partial x} + y_*'\frac{\partial\psi}{\partial y} = -\left(\frac{\partial\phi_*}{\partial y} - y_*'\frac{\partial\phi}{\partial x}\right)$$
$$= -\frac{1}{\lambda_0}. \tag{9.21}$$

  Let $\psi_*(0) = 0$. Hence on the interface, one has $\eta = \phi_* = 0$, $\xi = -\psi_* = \frac{x}{\lambda_0}$, which implies that the interface ($y = y_*(x), |x| \leq \lambda_0$) in the $(x, y)$-plane is mapped to the line segment ($\eta = 0, |\xi| \leq 1$) in the $(\xi, \eta)$-plane .
- From (9.12), one derives that as $y \to \infty$, $\xi = x$, $\eta \sim y$. Hence

$$\xi(\pm 1, \infty) = \pm 1, \quad \eta(x, \infty) = \infty,$$

  which implies that the line segment ($|x| \leq 1, y \gg 1$) in the $(x, y)$-plane is mapped to the line segment ($|\xi| \leq 1, \eta \gg 1$) in the $(\xi, \eta)$-plane.
- From the root condition (9.13), as $y \to -\infty$, $\lambda_0 \leq |x| \leq 1$, one derives $\xi(x, y) \sim$ const, $\eta(x, y) \sim$ const. Let the potential at the root be zero, $\eta(x, -\infty) = \phi_*(x, -\infty) = 0$. Hence, one has

$$\xi(x, -\infty) = \xi(\pm 1, -\infty), \quad \eta(x, -\infty) = \eta(\pm 1, -\infty) = 0.$$

  Moreover it will be seen below from the side-wall condition that

$$\xi(x, -\infty) = \xi(\pm 1, -\infty) = \pm 1.$$

  Therefore, one deduces that the line segment ($|x| \geq \lambda_0, y = -\infty$) in the $(x, y)$-plane is mapped to the singular point ($\xi = \pm 1, \eta = 0$) in the $(\xi, \eta)$-plane.
- Finally from side-wall condition (9.14), one derives that at $x = \pm 1$, $\frac{\partial\eta}{\partial x} = -\frac{\partial\xi}{\partial y} = 0$, so that $\xi = \xi(\pm 1, y) = \xi(\pm 1, \pm\infty) = \pm 1$, which implies that the side walls ($x = \pm 1, |y| < \infty$) are mapped to the semi-infinite lines ($\xi = \pm 1, 0 \leq \eta < \infty$) in the $(\xi, \eta)$-plane.

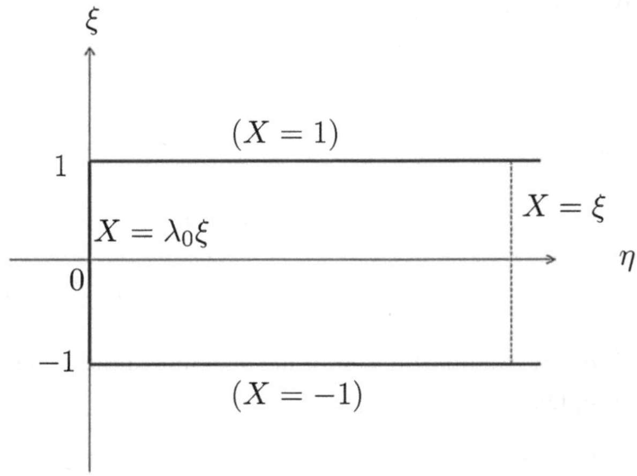


**Fig. 9.5** The boundary conditions imposed in the $(\xi, \eta)$-plane

The above mapping is sketched in Fig. 9.5.

One may transform the free boundary problem (FBP) (9.11)–(9.14) imposed in the $(x, y)$-plane for the harmonic function $\phi_*(x, y)$ into the following boundary value problem (BVP) imposed in the $(\xi, \eta)$-plane for the harmonic function $X(\xi, \eta)$. The governing equation is

$$\nabla^2 X(\xi, \eta) = 0\,. \tag{9.22}$$

The boundary conditions are as follows:

1. at $\eta = 0, |\xi| \leq 1$ (corresponding to the interface condition),

$$X = \lambda_0 \xi; \tag{9.23}$$

2. at $\xi = \pm 1, 0 < \eta < \infty$ (corresponding to the side-wall condition),

$$X = \pm 1; \tag{9.24}$$

3. as $\eta \to \infty$, $|\xi| \leq 1$ (corresponding to the far field condition),

$$X = \xi. \tag{9.25}$$

By the method of separation of variables, one can solve the above BVP (9.22)–(9.25). Let us expand $X(\xi, \eta)$ into the following Fourier sine series:

$$X = \xi + \sum_{n=1}^{n=\infty} a_n \mathrm{e}^{-n\pi\eta} \sin n\pi\xi. \tag{9.26}$$

The boundary conditions (9.24) and (9.25) are automatically satisfied. To satisfy the boundary condition (9.23), we set $\eta = 0$ in (9.26). It follows that

$$(\lambda_0 - 1)\xi = \sum_{n=1}^{n=\infty} a_n \sin n\pi\xi.$$

We obtain

$$a_n = (-1)^{n-1}\frac{2(1-\lambda_0)}{n\pi}, \tag{9.27}$$

so that we have

$$X = \xi - \frac{2(1-\lambda_0)}{\pi}\sum_{n=1}^{n=\infty}\frac{(-1)^{n-1}}{n}\mathrm{e}^{-n\pi\eta}\sin n\pi\xi. \tag{9.28}$$

On the other hand, from the Cauchy–Riemann conditions

$$\frac{\partial X}{\partial \xi} = \frac{\partial Y}{\partial \eta}, \quad \frac{\partial Y}{\partial \xi} = -\frac{\partial X}{\partial \eta},$$

we derive that

$$\frac{\partial}{\partial \eta}Y(\xi,\eta) = 1 + 2(1-\lambda_0)\sum_{n=1}^{n=\infty}(-1)^n\mathrm{e}^{-n\pi\eta}\cos n\pi\xi.$$

Therefore, we obtain

$$y(\xi,\eta) = \eta + \frac{2(1-\lambda_0)}{\pi}\sum_{n=1}^{n=\infty}\frac{(-1)^{n-1}\mathrm{e}^{-n\pi\eta}}{n}\cos n\pi\xi + C. \tag{9.29}$$

Noting that at the tip, $\psi_*(0,0) = 0$, we have $Y(0,0) = 0$. Hence we derive the integral constant

$$C = -\frac{2(1-\lambda_0)}{\pi}\sum_{n=1}^{n=\infty}\frac{(-1)^n}{n} = -\frac{2(1-\lambda_0)}{\pi}\ln 2.$$

We finally obtain that

$$Y(\xi,\eta) = \eta + \frac{2(1-\lambda_0)}{\pi}\sum_{n=1}^{n=\infty}\frac{(-1)^{n-1}\mathrm{e}^{-n\pi\eta}}{n}\cos n\pi\xi - \frac{2(1-\lambda_0)}{\pi}\ln 2. \tag{9.30}$$

It follows that

$$Z = X + \mathrm{i}Y = \widehat{\zeta} + \mathrm{i}\frac{2(1-\lambda_0)}{\pi}\sum_{n=1}^{\infty}\frac{(-1)^{n-1}}{n}\mathrm{e}^{-n\pi\eta}\mathrm{e}^{\mathrm{i}n\pi\xi}$$
$$-\mathrm{i}\frac{2(1-\lambda_0)}{\pi}\ln 2,$$

or

$$Z = X + \mathrm{i}Y = \widehat{\zeta} + \mathrm{i}\frac{2(1-\lambda_0)}{\pi}\ln(1+\mathrm{e}^{\mathrm{i}\pi\widehat{\zeta}}) - \mathrm{i}\frac{2(1-\lambda_0)}{\pi}\ln 2. \tag{9.31}$$

Noting that

$$\ln\left(1+\mathrm{e}^{\mathrm{i}n\widehat{\zeta}}\right) = \frac{\mathrm{i}\pi\widehat{\zeta}}{2} + \ln\cos\left(\frac{\pi\widehat{\zeta}}{2}\right) + \ln 2,$$

We finally obtain the Saffman–Taylor solution as

$$Z = x + \mathrm{i}y = Z(\widehat{\zeta}) = \lambda_0\widehat{\zeta} + \mathrm{i}\frac{2(1-\lambda_0)}{\pi}\ln\cos\left(\frac{\pi\widehat{\zeta}}{2}\right). \tag{9.32}$$

By setting $\eta = 0$ or $\widehat{\zeta} = \xi$ in the above, one derives that at the interface shape function,

$$Z = Z_*(\xi) = X_*(\xi) + \mathrm{i}Y_*(\xi) = \lambda_0\xi + \mathrm{i}\frac{2(1-\lambda_0)}{\pi}\ln\cos\left(\frac{\pi\xi}{2}\right),$$

which leads to the parametric form of the interface shape function

$$\begin{cases} x_* = X_*(\xi) = \lambda_0\xi, \\ y_* = Y_*(\xi) = \frac{2(1-\lambda_0)}{\pi}\ln\cos\left(\frac{\pi\xi}{2}\right), \end{cases}$$

and the function of the interface

$$y = y_*(x) = \frac{2(1-\lambda_0)}{\pi}\ln\cos\left(\frac{\pi x}{2\lambda_0}\right). \tag{9.33}$$

The ST-solution (9.32) and (9.33) is very important, playing a similar role in pattern formation theory to that of Ivantsov's solution in systems of dendritic growth.

Furthermore, Saffman obtained a set of analytical solutions for the initial value problem with zero surface tension ($\varepsilon = 0$) and a special class of initial conditions. Those time-evolving fingers have finite roots, and their tip velocities vary with time. Saffman showed that as $t \to \infty$ these solutions converge to the ST steady solutions [18].

## 9.4 Mathematical Formulation of the Problem with Zero Surface Tension

We formulate the problem of viscous fingering in a curvilinear coordinate based on the ST solution.

### 9.4.1 The System of Curvilinear Coordinates $(\xi, \eta)$

It is known that the variables $\{\xi = \xi(x, y); \eta = \eta(x, y)\}$ constitute a new orthogonal curvilinear coordinate system in the $(x, y)$-plane as shown in Fig. 9.6:

$$\begin{cases} x = X(\xi, \eta), \\ y = Y(\xi, \eta). \end{cases} \tag{9.34}$$

The Lamé coefficients $\mathcal{G}_1$ and $\mathcal{G}_2$ along the two coordinate curves under the transformation (9.34) are

$$\mathcal{G}_1 = \mathcal{G}_2 = \mathcal{G} = |Z'(\widehat{\zeta})| = \sqrt{X_\xi^2 + Y_\xi^2}\,. \tag{9.35}$$

From the ST solution, we have

$$\begin{cases} Z'(\widehat{\zeta}) = \lambda_0 - \mathrm{i}(1 - \lambda_0) \tan\left(\frac{\pi\widehat{\zeta}}{2}\right), \qquad Z'(0) = \lambda_0 = \frac{1}{U_0} \\ Z''(\widehat{\zeta}) = -\mathrm{i}\frac{(1-\lambda_0)}{2} \sec^2\left(\frac{\pi\widehat{\zeta}}{2}\right). \end{cases} \tag{9.36}$$

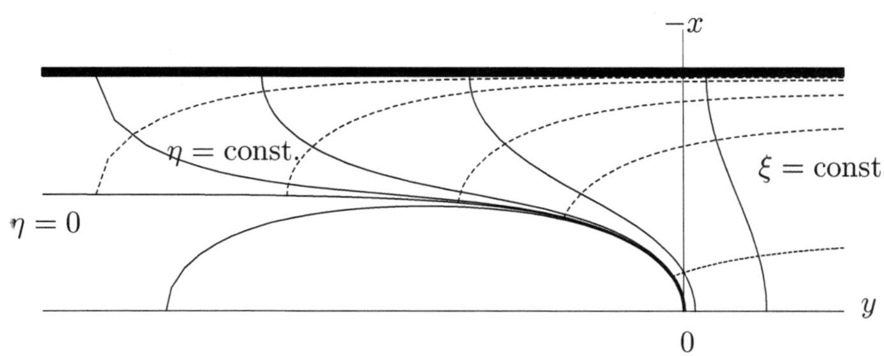


**Fig. 9.6** Sketch of the orthogonal curvilinear coordinate system $(\xi, \eta)$ based on the ST zero surface tension steady-state solutions

Furthermore, on the interface ($\eta = 0$), we have

$$X_\xi(\xi, 0) = Y_\eta(\xi, 0) = \lambda_0,$$

$$X_\eta(\xi, 0) = -Y_\xi(\xi, 0) = (1 - \lambda_0) \tan\left(\frac{\pi\xi}{2}\right),$$

$$X_{\xi\xi}(\xi, 0) = -X_{\eta\eta}(\xi, 0) = Y_{\xi\eta}(\xi, 0) = 0,$$

$$Y_{\xi\xi}(\xi, 0) = -Y_{\eta\eta}(\xi, 0) = -X_{\xi\eta}(\xi, 0) = -\frac{\pi(1 - \lambda_0)}{2} \sec^2\left(\frac{\pi\xi}{2}\right),$$

$$\mathcal{G}_0 = \mathcal{G}(\xi, 0) = \sqrt{\lambda_0^2 + (1 - \lambda_0)^2 \tan^2\left(\frac{\pi\xi}{2}\right)}. \tag{9.37}$$

With an arbitrary constant parameter $0 < \lambda_0 < 1$, the ST solution is an analytic function in the domain $-\infty < y < \infty$, $-1 < x < 1$. The interface of the solution has an infinitely long root, smoothly extending to $y = -\infty$, and has no singularity at the tip.

### 9.4.2 *Mathematical Formulation of the Problem in the* $(\xi, \eta)$ *Coordinate System*

To formulate the problem of general unsteady flow in a Hele–Shaw cell with $\varepsilon \neq 0$, we adopt a curvilinear coordinate system $(\xi, \eta)$. We denote the potential function of a general unsteady flow by $\phi = \phi(\xi, \eta, t)$ and the interface shape by $\eta = \eta_s(\xi, t)$. Obviously, in this system, the ST solution has the simple form

$$\phi_* = \eta, \quad \eta_* = 0. \tag{9.38}$$

For the general case $\varepsilon \neq 0$, the system (9.1)–(9.9) may be rewritten as

$$\frac{\partial^2\phi}{\partial\xi^2} + \frac{\partial^2\phi}{\partial\eta^2} = 0 \tag{9.39}$$

with the following boundary conditions:

1. As $\eta \to \infty$,

$$\frac{\partial\phi}{\partial\eta} = 1. \tag{9.40}$$

2. At $\xi = \pm 1$,

$$\frac{\partial\phi}{\partial\xi} = 0. \tag{9.41}$$

3. At $\eta = \eta_s(\xi, t)$,

$$\phi = \varepsilon^2 \mathcal{K}\Big\{\eta_s(\xi, t)\Big\} \tag{9.42}$$

and

$$\frac{\partial \phi}{\partial \eta} - \eta_s' \frac{\partial \phi}{\partial \xi} = U_0\big(Y_\eta - \eta_s' Y_\xi\big) + \mathcal{G}^2\Big(\frac{\partial \eta_s}{\partial t}\Big). \tag{9.43}$$

Here the prime represents the derivative with respect to $\xi$, $U_0 = \frac{1}{\lambda_0}$ is the tip velocity of the ST solution observed in the laboratory frame, and the curvature operator is

$$\mathcal{K}\Big[\eta_s(\xi, t)\Big] = -\frac{1}{\mathcal{G}(\xi, \eta_s)}\Bigg[\frac{\eta_s''}{\big(1 + \eta_s'^2\big)^{3/2}} + \frac{\Pi_0(\xi, \eta_s)}{\mathcal{G}^2(\xi, \eta_s)\big(1 + \eta_s'^2\big)^{1/2}} - \frac{\Pi_1(\xi, \eta_s)}{\mathcal{G}^2(\xi, \eta_s)\big(1 + \eta_s'^2\big)^{1/2}}\eta_s'\Bigg], \tag{9.44}$$

where

$$\begin{aligned} \Pi_0(\xi, \eta) &= \Big(Y_{\xi\xi} X_\xi - X_{\xi\xi} Y_\xi\Big), \\ \Pi_1(\xi, \eta) &= \Big(Y_{\xi\xi} Y_\xi + X_{\xi\xi} X_\xi\Big). \end{aligned} \tag{9.45}$$

4. At the triple point T, $\xi = \pm 1, \eta = \eta_T(t) \approx \mathrm{e}^{\frac{\pi y_T}{1-\lambda_0}}$,

$$\frac{\partial \phi}{\partial \eta} = \frac{\partial \phi}{\partial \xi} = 0. \tag{9.46}$$

This system is highly nonlinear due to the interface conditions (9.42). As before, we consider the "nearly" steady finger solutions when $t \gg 1$. Thus we employ the slow time variable defined as

$$\tau = \varepsilon(t - t_0) \quad (t_0 \gg 1), \tag{9.47}$$

and assume that the solutions $\{\phi_B, \eta_B\}$ are functions of the variables $(\xi, \eta, \tau)$. In terms of this slow variable $\tau$, the tip velocity $U(t)$ is written as

$$U(t) = U(\tau, \varepsilon), \tag{9.48}$$

and the location of the triple point T is expressed in the $(x, y)$-coordinate system as

$$x_T = \pm 1; \quad y_T(\tau) = -\frac{1}{\varepsilon}\left(\bar{L} + \int_0^\tau U(\tau, \varepsilon)\mathrm{d}\tau\right) = \frac{\bar{y}_T(\tau, \varepsilon)}{\varepsilon}, \tag{9.49}$$

where $\bar{L} = \mathcal{O}(1)$. In the $(\xi, \eta)$-coordinate system, the triple point is expressed as

$$\xi = \xi_{\mathrm{T}} = \pm 1; \qquad \eta = \bar{\eta}_{\mathrm{T}}(\tau) \approx \mathrm{e}^{\frac{\pi \bar{y}_{\mathrm{T}}}{\varepsilon(1-\lambda_0)}} \ll 1. \tag{9.50}$$

The downstream triple point condition (9.9) and the kinematic interface condition (9.7) can therefore be transformed into the following forms:

$$\frac{\partial \phi}{\partial \xi} = \frac{\partial \phi}{\partial \eta} = 0 \tag{9.51}$$

at $\xi = \pm 1,\ \eta = \bar{\eta}_{\mathrm{T}}(\tau)$ , and

$$\frac{\partial \phi_{\mathrm{B}}}{\partial \eta} - \eta_{\mathrm{B}}' \frac{\partial \phi_{\mathrm{B}}}{\partial \xi} = U_0 \big(Y_\eta - \eta_{\mathrm{s}}' Y_\xi\big) + \varepsilon \mathcal{G}^2 \frac{\partial \eta_{\mathrm{B}}}{\partial \tau} \tag{9.52}$$

at $\eta = \eta_{\mathrm{s}}(\xi, \tau)$.

### 9.4.3 *The Regular Perturbation Expansion Solution for the Basic State as $\varepsilon \to 0$*

Let the basic state of the system be given by the notation $q_{\mathrm{B}} =: \{\phi_{\mathrm{B}}, \eta_{\mathrm{B}}\}$. Since the ST solution is the solution for the above system at $\varepsilon = 0$, one can carry out the following regular perturbation expansion (RPE) as $\varepsilon \to 0$:

$$\phi_{\mathrm{B}}(\xi, \eta, \tau, \varepsilon) \sim \phi_* + \varepsilon^2 \phi_1(\xi, \eta) + \varepsilon^2 \phi_2(\xi, \eta) + \cdots , \tag{9.53}$$

$$\eta_{\mathrm{B}}(\xi, \tau, \varepsilon) \sim \eta_* + \varepsilon^2 \eta_1(\xi) + \varepsilon^2 \eta_2(\xi) + \cdots , \tag{9.54}$$

or

$$q_{\mathrm{B}}(\xi, \eta, \tau, \varepsilon) \sim q_* + \varepsilon^2 q_1(\xi, \eta) + \varepsilon^2 q_2(\xi, \eta) + \cdots , \tag{9.55}$$

where the first term $q_* =: \{\phi_*, \eta_* = 0\}$ is the ST solution. By substituting the expansion into the system (9.39)–(9.52), one can find terms $\phi_n, \eta_n$. These solutions are all time independent; namely, they do not contain the slow time variable $\tau$. Furthermore, at $\xi = \pm 1$, the solutions $\eta_n$ all vanish, i.e.,

$$\eta_n(\pm 1) = 0 \qquad (n = 1, 2, 3, \ldots). \tag{9.56}$$

Certainly, the solution $q_{\mathrm{B}}$ may have the generalized asymptotic expansion

$$q_{\mathrm{B}} = q_* + (\mathcal{R}_N) + (\mathcal{S}_N) , \tag{9.57}$$

where $\mathcal{R}_N$ is the partial sum of first $N$ terms in the RPE part, while $\mathcal{S}_N$ is the SPE part. The SPE part may depend on the slow time variable $\tau$. The time-dependent SPE part is smaller beyond all orders. In this sense, the solutions $q_B$ are called the "nearly" steady finger solutions, which we shall define as the basic state solutions for stability analysis.

The exact form of these basic state solutions $q_B$ is hard to find, but this is not important. Our goal is to study the stability of these solutions. Here what is the most important to us is that these basic state solutions exist and can be well approximated by the RPE part in the region away from the triple point at the later stages of evolution. Particularly, they are close to the ST solutions, so that

$$\begin{cases} \phi_B(\xi, \eta, \tau, \varepsilon) = \phi_*(\eta) + \mathcal{O}(\varepsilon^2), \\ \eta_B(\xi, \tau, \varepsilon) = \mathcal{O}(\varepsilon^2), \\ \bar{U}(\tau, \varepsilon) = U_0 + \mathcal{O}(\varepsilon^2), \\ \lambda(\tau, \varepsilon) = \lambda_0 + \mathcal{O}(\varepsilon^2). \end{cases} \tag{9.58}$$

The statement that the basic state solution can be approximated by an ST solution with high accuracy is well supported by the experimental evidence. For instance, the classic experiments conducted by Saffman and Taylor have shown that the fingers are very close to an ST solution, once the width $\lambda_0$ is properly selected (see Fig. 9.7 and refer to [19]). The effect of these higher-order small error terms in (9.58) on the calculations of the eigenvalues and eigenmodes of perturbed states that are to be carried out in the following sections are negligible.

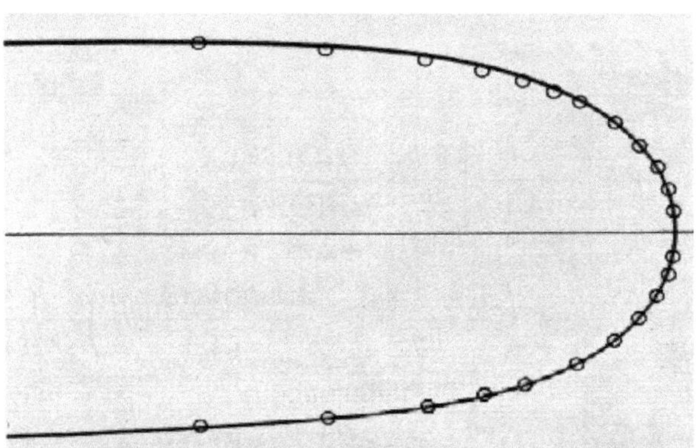

**Fig. 9.7** Comparison of a realistic smooth finger with the ST solution with a properly selected width $\lambda_0$. The *solid line* is the experimental curve, while the *dots* are calculated from the ST solution

## 9.5 The Linear Perturbed System and the Outer Solutions

### 9.5.1 The Linear Perturbed System and the Multiple Variables Expansions

We consider perturbations around the basic state induced by infinitesimal initial perturbations. The unsteady solution is written as

$$\begin{aligned} \phi(\xi,\eta,t) &= \phi_{\mathrm{B}} + \tilde{\phi}(\xi,\eta,t), \\ \eta_{\mathrm{s}}(\xi,t) &= \eta_{\mathrm{B}} + \tilde{h}(\xi,t)\,. \end{aligned} \tag{9.59}$$

The homogeneous linear perturbed system is derived as

$$\nabla^2\tilde{\phi} = \frac{\partial^2\tilde{\phi}}{\partial\xi^2} + \frac{\partial^2\tilde{\phi}}{\partial\eta^2} = 0, \tag{9.60}$$

with the following boundary conditions:

1. As $\eta \to \infty$,

$$\frac{\partial\tilde{\phi}}{\partial\eta} = \frac{\partial\tilde{\phi}}{\partial\xi} = 0\,. \tag{9.61}$$

2. At $\xi = \pm 1$,

$$\frac{\partial\tilde{\phi}}{\partial\xi} = 0\,. \tag{9.62}$$

3. On the interface, $\eta = \eta_{\mathrm{B}}$,

$$\begin{aligned} \tilde{\phi} + \tilde{h} = -\frac{\varepsilon^2}{\hat{\mathcal{G}}(\xi)}\Bigg[&\frac{\partial^2\tilde{h}}{\partial\xi^2} - \frac{\hat{\Pi}_1(\xi)}{\hat{\mathcal{G}}^2(\xi)}\frac{\partial\tilde{h}}{\partial\xi} \\ &+ \left(\frac{\partial\Pi_0(\xi,\eta_B)}{\partial\eta} - \frac{2\hat{\Pi}_0(\xi)\mathcal{G}_\eta(\xi,\eta_B)}{\hat{\mathcal{G}}(\xi)}\right)\frac{\tilde{h}}{\hat{\mathcal{G}}^2(\xi)}\Bigg], \end{aligned} \tag{9.63}$$

$$\frac{\partial\tilde{\phi}}{\partial\eta} + U_0\hat{Y}_\xi(\xi)\frac{\partial\tilde{h}}{\partial\xi} - \hat{\mathcal{G}}^2(\xi)\frac{\partial\tilde{h}}{\partial t} - U_0\hat{Y}_{\eta\eta}(\xi)\tilde{h} = 0\,, \tag{9.64}$$

where

$$\begin{cases} \hat{\mathcal{G}}(\xi) = \mathcal{G}(\xi,\eta_{\mathrm{B}}), \;\; \hat{Y}_\xi(\xi) = Y_\xi(\xi,\eta_{\mathrm{B}}), \;\; \hat{Y}_{\eta\eta}(\xi) = Y_{\eta\eta}(\xi,\eta_{\mathrm{B}}) \\ \hat{\Pi}_0(\xi) = \Pi_0(\xi,\eta_{\mathrm{B}}), \;\; \hat{\Pi}_1(\xi) = \Pi_1(\xi,\eta_{\mathrm{B}}). \end{cases} \tag{9.65}$$

The left-hand sides of (9.65) actually depend on the variables $\xi$, $\tau$, and $\varepsilon$ via $\eta_B$. We have dropped $\varepsilon$ for simplicity and neglected $\tau$ in $\eta_B$ and explicitly show only the variable $\xi$.

4. At the triple point, $\xi = \pm 1$, $\eta = \eta_T$,

$$\tilde{h} = \frac{\partial \tilde{\phi}}{\partial \xi} = \frac{\partial \tilde{\phi}}{\partial \eta} = 0\,. \tag{9.66}$$

5. At the tip, $\xi = 0$,

$$\frac{\partial}{\partial \xi}\{\tilde{\phi},\, \tilde{h}\} = 0 \quad \text{for a symmetric mode (S-mode);} \tag{9.67}$$

and

$$\{\tilde{\phi}, \tilde{h}\} = 0 \quad \text{for an antisymmetric mode (A-mode).} \tag{9.68}$$

The above system contains the two parameters $\varepsilon$ and $\lambda_0$. It leads to a linear eigenvalue problem, as one looks for the following type of solutions:

$$\tilde{\phi} = \hat{\phi}(\xi, \eta, \varepsilon)\mathrm{e}^{\sigma t},$$

$$\tilde{h} = \hat{h}(\xi, \varepsilon)\mathrm{e}^{\sigma t}. \tag{9.69}$$

The eigenvalues $\sigma = \sigma_R - \mathrm{i}\omega$ must be a function of $(\lambda_0, \varepsilon)$.

One needs to solve the problem only in the half-plane $\xi \geq 0$. The solution in the other half-plane $\xi < 0$ can then be derived from the symmetry or antisymmetry. As before, we shall solve this eigenvalue problem in two steps. As the first step, we solve the system (9.60)–(9.66) for given parameters $(\sigma, \lambda_0, \varepsilon)$. For this purpose, we shall look for the uniformly valid asymptotic solution to (9.60)–(9.66) with fixed parameters $(\sigma, \lambda_0)$ in the limit $\varepsilon \to 0$. The solutions obtained satisfy all the boundary conditions except the tip condition (9.67) or (9.68). Then, as the second step, we apply the tip conditions. This determines the eigenvalue $\sigma = \sigma(\lambda_0, \varepsilon)$.

To obtain a uniformly valid asymptotic solution in the outer region away from the finger-root, similar to those used for dendritic growth, we introduce the fast variables $\{\xi_+, \eta_+, t_+\}$:

$$\begin{aligned}
\xi_+ &= \frac{\hat{\phi}(\xi, \eta, \varepsilon)}{\varepsilon} = \frac{1}{\varepsilon}\int_{\xi_0}^{\xi} k(\xi_1, \eta, \varepsilon)\mathrm{d}\xi_1,\\
\eta_+ &= \frac{\hat{\psi}(\xi, \eta, \varepsilon)}{\varepsilon} = \frac{1}{\varepsilon}\int_{0}^{\eta} k(\xi, \eta_1, \varepsilon)\mathrm{d}\eta_1,\\
t_+ &= \frac{t}{\eta_0^2 \varepsilon},
\end{aligned} \tag{9.70}$$

where the pair of phase function $\hat{\phi}$ and $\hat{\psi}$ satisfies the Cauchy-Riemman conditions

$$\hat{\phi}_\xi = \hat{\psi}_\eta = k(\xi, \eta, \varepsilon); \ \hat{\phi}_\eta = -\hat{\psi}_\xi = g(\xi, \eta, \varepsilon).$$

Furthermore, we denote that, at the interface $\eta = 0$, $\bar{k}(\xi, \varepsilon) = k(\xi, 0, \varepsilon)$, $\bar{\phi}(\xi, \varepsilon) = \hat{\phi}(\xi, 0, \varepsilon)$, $\bar{\psi}(\xi, \varepsilon) = \hat{\psi}(\xi, 0, \varepsilon)$, and

$$\bar{\xi}_+(\xi, \varepsilon) = \xi_+(\xi, 0, \varepsilon) = \frac{\bar{\phi}(\xi, \varepsilon)}{\varepsilon} = \frac{1}{\varepsilon}\int_{\xi_0}^{\xi} \bar{k}(\xi_0, \varepsilon) d\xi_1.$$

As was derived for the dendritic growth systems, we may write

$$\begin{aligned}
\hat{\phi}(\xi, \eta) &= \frac{1}{2}\left\{\int_0^{\xi+i\eta} \bar{k}(t)dt + \int_0^{\xi-i\eta} \bar{k}(t)dt\right\}, \\
\hat{\psi}(\xi, \eta) &= -\frac{i}{2}\int_{\xi-i\eta}^{\xi+i\eta} \bar{k}(t)dt, \\
k(\xi, \eta) &= \frac{1}{2}\left\{\bar{k}(\xi + i\eta) + \bar{k}(\xi - i\eta)\right\}, \\
g(\xi, \eta) &= \frac{i}{2}\left\{\bar{k}(\xi + i\eta) - \bar{k}(\xi - i\eta)\right\}.
\end{aligned} \tag{9.71}$$

The converted system with the multiple variables $(\xi, \eta, \xi_+, \eta_+, \tau, t_+)$ is as follows:

$$\begin{aligned}
(k^2 + g^2)\left(\frac{\partial^2}{\partial \xi_+^2} + \frac{\partial^2}{\partial \eta_+^2}\right)\tilde{\phi} = &-\varepsilon\left(2k\frac{\partial^2 \tilde{\phi}}{\partial \xi_+ \partial \xi} + 2g\frac{\partial^2 \tilde{\phi}}{\partial \eta_+ \partial \xi}\right) \\
&-\varepsilon\left(2k\frac{\partial^2 \tilde{\phi}}{\partial \eta_+ \partial \eta} - 2g\frac{\partial^2 \tilde{\phi}}{\partial \xi_+ \partial \eta}\right),
\end{aligned} \tag{9.72}$$

$$\left(|\xi| \le 1, \ 0 \le \xi_+ < \infty, \ \eta \ge 0\right),$$

with the following boundary conditions:

1. In the upstream far field, as $\eta_+ \to \infty$ and $\eta \to \infty$,

$$\frac{\partial \tilde{\phi}}{\partial \eta_+} = \mathcal{O}(\varepsilon), \quad \frac{\partial \tilde{\phi}}{\partial \xi_+} = \mathcal{O}(\varepsilon). \tag{9.73}$$

2. At the sidewall, $\xi = 1$ and $\xi_+ \to \infty$,

$$\frac{\partial \tilde{\phi}}{\partial \xi_+} = \mathcal{O}(\varepsilon). \tag{9.74}$$

3. On the interface, $\eta_+ = \eta = 0$,

$$\tilde{\phi} + \tilde{h} = -\frac{1}{\hat{\mathcal{G}}_0(\xi)} \left\{ k^2 \frac{\partial^2 \tilde{h}}{\partial \bar{\xi}_+^2} + 2\varepsilon k \frac{\partial^2 \tilde{h}}{\partial \xi \partial \bar{\xi}_+} + \varepsilon k' \frac{\partial \tilde{h}}{\partial \bar{\xi}_+} \right.$$
$$\left. -\varepsilon k \frac{\hat{\Pi}_{1,0}(\xi)}{\hat{\mathcal{G}}_0^2(\xi)} \frac{\partial \tilde{h}}{\partial \bar{\xi}_+} \right\} + (\text{h.o.t}), \tag{9.75}$$

$$\left( \frac{\partial \tilde{\phi}}{\partial \eta_+} + \varepsilon \frac{\partial \tilde{\phi}}{\partial \eta} + \varepsilon \hat{K} \frac{\partial \tilde{\phi}}{\partial \bar{\xi}_+} \right) + U_0 \hat{Y}_{\xi,0}(\xi) \left( k \frac{\partial \tilde{h}}{\partial \bar{\xi}_+} + \varepsilon \frac{\partial \tilde{h}}{\partial \xi} \right) - \hat{\mathcal{G}}_0{}^2(\xi) \frac{\partial \tilde{h}}{\partial t_+}$$
$$-\varepsilon U_0 \hat{Y}_{\eta\eta,0}(\xi) \tilde{h} = (\text{h.o.t.}). \tag{9.76}$$

4. At the triple point, $\xi = 1$, $\xi_+ \to \infty$, and $\eta \to 0$,

$$\tilde{h} = 0, \quad \frac{\partial \tilde{\phi}}{\partial \xi_+} = \mathcal{O}(\varepsilon), \quad \frac{\partial \tilde{\phi}}{\partial \eta_+} = \mathcal{O}(\varepsilon). \tag{9.77}$$

5. At the tip, $\xi = \xi_+ = 0$,

$$\frac{\partial}{\partial \xi_+} \{\tilde{\phi}, \tilde{h}\} = 0 \quad \text{for a symmetric mode (S-mode);} \tag{9.78}$$

and

$$\{\tilde{\phi}, \tilde{h}\} = 0 \quad \text{for an antisymmetric mode (A-mode).} \tag{9.79}$$

One may make the following multiple variables expansion for the perturbed state:

$$\begin{aligned} \tilde{\phi} &= \{\tilde{\phi}_0(\xi, \eta, \tau, \xi_+, \eta_+) + \varepsilon \tilde{\phi}_1(\xi, \eta, \xi_+, \eta_+) + \cdots\} \mathrm{e}^{\sigma t_+}, \\ \tilde{h} &= \{\tilde{h}_0(\xi, \bar{\xi}_+) + \varepsilon \tilde{h}_1(\xi, \bar{\xi}_+) + \cdots\} \mathrm{e}^{\sigma t_+}, \\ k(\xi, \tau, \varepsilon) &= k_0 + \varepsilon \hat{k}_1 + \varepsilon^2 k_2 + \cdots, \\ \sigma(\tau, \varepsilon) &= \sigma_0 + \varepsilon \sigma_1 + \varepsilon^2 \sigma_2 \cdots. \end{aligned} \tag{9.80}$$

One can substitute (9.80) into the system (9.73)–(9.79) and successively derive each order of approximation, as we did for the problem of dendritic growth in the previous sections.

Later, for convenience, we shall employ a new variable $\rho$ defined by

$$\rho = -U_0 \hat{Y}_{\xi,0}(\xi), \tag{9.81}$$

rather than using $\xi$. The variable $\rho$ is connected with the coordinate $y$ along the interface of the basic state. It is also related to the arc length $\ell(\xi)$ measured along the interface of the ST solution starting from the tip and determined by

$$\ell = \int_0^{\xi} \hat{\mathcal{G}}_0(\xi_1)\mathrm{d}\xi_1 \,. \tag{9.82}$$

One can derive

$$\xi = \int_0^{\rho} \frac{\mathrm{d}\rho_1}{\hat{G}(\rho_1)} \tag{9.83}$$

and

$$\ell = \int_0^{\rho} \hat{F}(\rho_1)\mathrm{d}\rho_1 \,, \tag{9.84}$$

where

$$\hat{G}(\rho) = -U_0 \hat{Y}_{\xi\xi,0}(\xi), \quad \hat{F}(\rho) = -\frac{\hat{\mathcal{G}}_0(\xi)}{U_0 \hat{Y}_{\xi\xi,0}(\xi)}. \tag{9.85}$$

As $\xi$ goes from 0 to 1, $\ell$ goes from 0 to $\infty$, whereas $\rho$ goes from 0 to $\infty$.

The coefficients of the above linear perturbed equation are determined by the ST solution, and we have used the notation

$$\begin{aligned}
\hat{Y}_{\xi,0}(\xi) &= Y_{\xi}(\xi, 0), \\
\hat{\mathcal{G}}_0(\xi) &= \mathcal{G}(\xi, 0), \\
\hat{\Pi}_{0,0}(\xi) &= \Pi_0(\xi, 0), \\
\hat{\Pi}_{1,0}(\xi) &= \Pi_1(\xi, 0) \,.
\end{aligned} \tag{9.86}$$

### 9.5.2 The Zeroth-Order Approximation Solutions

In the zeroth-order approximation, we obtain

$$\left( \frac{\partial^2}{\partial \xi_+^2} + \frac{\partial^2}{\partial \eta_+^2} \right) \tilde{\phi}_0 = 0, \tag{9.87}$$

or

$$\mathrm{i}\frac{\partial \tilde{\phi}_0}{\partial \xi_+} = \frac{\partial \tilde{\phi}_0}{\partial \eta_+} \tag{9.88}$$

$$\big(0 < \xi < 1,\ 0 < \xi_+ < \infty,\ \eta_+ = \mathcal{O}(1)\big)$$

with the following boundary conditions:

1. As $\eta_+ \to \infty$,

$$\frac{\partial \tilde{\phi}_0}{\partial \eta_+} = \frac{\partial \tilde{\phi}_0}{\partial \xi_+} = 0\,. \tag{9.89}$$

2. At $\xi = 1$ and as $\xi_+ \to \infty$,

$$\frac{\partial \tilde{\phi}_0}{\partial \xi_+} = 0\,. \tag{9.90}$$

3. At $\eta = \eta_+ = 0$,

$$\tilde{\phi}_0 + \tilde{h}_0 = -\frac{\bar{k}_0^2}{\hat{\mathcal{G}}_0} \frac{\partial^2 \tilde{h}_0}{\partial \bar{\xi}_+^2}\,, \tag{9.91}$$

$$\bar{k}_0 \frac{\partial \tilde{\phi}_0}{\partial \eta_+} + U_0 \bar{k}_0 \hat{Y}_{\xi,0}(\xi) \frac{\partial \tilde{h}_0}{\partial \bar{\xi}_+} - \sigma_0 \hat{\mathcal{G}}_0^2(\xi) \tilde{h}_0 = 0\,. \tag{9.92}$$

4. At $\xi = 1, \eta = 0$, and as $\xi_+ \to \infty$,

$$\{\tilde{\phi}_0;\ \tilde{h}_0\} = 0\,. \tag{9.93}$$

We consider the mode solutions for the above system:

$$\begin{aligned} \tilde{\phi}_0 &= A_0(\xi, \eta)\ \exp\{\mathrm{i}\xi_+ - \eta_+,\} \\ \tilde{h}_0 &= \hat{D}_0\ \exp\{\mathrm{i}\bar{\xi}_+\}\,. \end{aligned} \tag{9.94}$$

The coefficient $\hat{D}_0$ in the zeroth-order approximation is a constant. To satisfy the interface conditions (9.91) and (9.92) at $\eta = 0$, the wave number function $k_0(\xi)$ is subject to the local dispersion relationship

$$\hat{\mathcal{G}}_0^2 \sigma_0 = \mathrm{i}\bar{k}_0 U_0 \hat{Y}_{\xi,0}(\xi) + \bar{k}_0 \left(1 - \frac{\bar{k}_0^2}{\hat{\mathcal{G}}_0}\right). \tag{9.95}$$

We now change from the variable $\xi$ to $\rho$. To leading order,

$$\rho = -\frac{\hat{Y}_{\xi,0}(\xi)}{\lambda_0} = \left(\frac{1-\lambda_0}{\lambda_0}\right) \tan\left(\frac{\pi\xi}{2}\right). \tag{9.96}$$

Accordingly,

$$\hat{\mathcal{G}}_0(\xi) = \lambda_0 S(\rho)$$
$$\hat{G}(\rho) = \frac{\pi}{2}\frac{\lambda_0}{1-\lambda_0}\left[\rho^2 + \left(\frac{1-\lambda_0}{\lambda_0}\right)^2\right], \tag{9.97}$$
$$\hat{F}(\rho) = \frac{\lambda_0 S(\rho)}{\hat{G}(\rho)},$$

where

$$S(\rho) = \sqrt{1+\rho^2}\,. \tag{9.98}$$

Moreover, in terms of $\rho$, the normal mode solution is expressed in the form

$$\tilde{h}_0 = \tilde{h}_0(\rho) = \hat{D}_0 \exp\left\{\frac{\mathrm{i}}{\varepsilon}\int_{\rho_0}^{\rho} \tilde{k}_0 \mathrm{d}\rho\right\}, \tag{9.99}$$

where $k_0(\xi) = \tilde{k}_0(\rho)G(\rho)$ and $\rho_0 = \rho(\xi_0)$. The corresponding local dispersion relationship is then transformed to

$$\sigma_0 = \Sigma_*(\rho, \tilde{k}_0) = \frac{\hat{G}(\rho)\tilde{k}_0}{\lambda_0^2 S^2(\rho)}\left[(1-\mathrm{i}\rho) - \frac{\hat{G}^2(\rho)\tilde{k}_0^2}{\lambda_0 S(\rho)}\right]. \tag{9.100}$$

The local dispersion relationship can be written in the form

$$\sigma_e = \Sigma(\rho, k_e) = \frac{k_e}{S^2(\rho)}\left[1 - \mathrm{i}\rho - \frac{k_e^2}{S(\rho)}\right], \tag{9.101}$$

provided one sets

$$k_e = \frac{\hat{G}(\rho)}{\lambda_0^{\frac{1}{2}}}\tilde{k}_0, \quad \text{and} \quad \sigma_e = \lambda_0^{\frac{3}{2}}\sigma_0\,. \tag{9.102}$$

Notice that this form is exactly like that for dendritic growth.

For a given constant $\sigma_0$, one can find three roots of (9.101) (see Fig. 9.8), namely

$$\begin{cases} k_e^{(1)}(\rho) = M(\rho)\cos\left\{\frac{1}{3}\cos^{-1}\left(\frac{\sigma_e}{N(\rho)}\right)\right\} & \text{(short-wave branch)}, \\ k_e^{(2)}(\rho) = M(\rho)\cos\left\{\frac{1}{3}\cos^{-1}\left(\frac{\sigma_e}{N(\rho)}\right) + \frac{2\pi}{3}\right\}, & \\ k_e^{(3)}(\rho) = M(\rho)\cos\left\{\frac{1}{3}\cos^{-1}\left(\frac{\sigma_e}{N(\rho)}\right) + \frac{4\pi}{3}\right\} & \text{(long-wave branch)}, \end{cases} \tag{9.103}$$

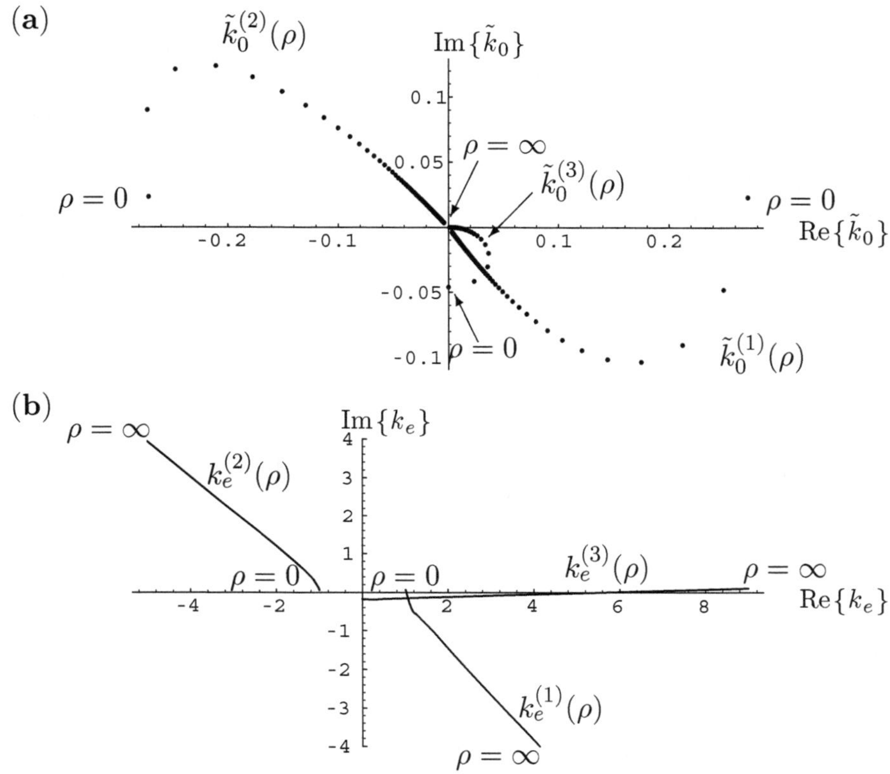


**Fig. 9.8** The distribution of curves of the wave number functions for a typical case, $\sigma_0 = (0, -0.7)$, $\lambda_0 = 0.4$: (**a**) the variation of $\{\tilde{k}_0^{(1)}, \tilde{k}_0^{(2)}, \tilde{k}_0^{(3)}\}$ with $\rho$ in the complex $\tilde{k}_0$-plane; (**b**) the variation of $\{k_e^{(1)}, k_e^{(2)}, k_e^{(3)}\}$ with $\rho$ in the complex $k_e$-plane

where $M(\rho) = \sqrt{\frac{4S(\rho)}{3}}\left(1 - \mathrm{i}\rho\right)^{\frac{1}{2}}$ and $N(\rho) = -\frac{M(\rho)}{3S^2(\rho)}\left(1 - \mathrm{i}\rho\right)$.

In order for the potential $\tilde{\phi}_0$ to satisfy the boundary condition (9.89), $\mathrm{Re}\{\tilde{k}_0\} > 0$ is needed. Consequently, only $\{k_e^{(1)}(\rho), k_e^{(3)}(\rho)\}$ in (9.103) are meaningful. Thus, the general solution in the outer region is

$$\begin{aligned}\tilde{h} &= D_1 H_1(\rho) + D_3 H_3(\rho)\\ &= D_1 \exp\left\{\frac{\mathrm{i}}{\varepsilon}\int_{\rho_0}^{\rho}\left(\tilde{k}_0^{(1)} + \varepsilon\tilde{k}_1^{(1)} + \cdots\right)\mathrm{d}\rho_1\right\}\\ &\quad + D_3 \exp\left\{\frac{\mathrm{i}}{\varepsilon}\int_{\rho_0}^{\rho}\left(\tilde{k}_0^{(3)} + \varepsilon\tilde{k}_1^{(3)} + \cdots\right)\mathrm{d}\rho_1\right\},\end{aligned} \tag{9.104}$$

where the constants $D_1$ and $D_3$ are to be determined. As we have seen in the case of dendritic growth, the MVE solution (9.104) is not uniformly valid in the entire complex $\rho$-plane (or $\xi$-plane). As usual, we may define complex variables $\zeta = \xi + \mathrm{i}\xi_I$ and $\varrho = \rho + \mathrm{i}\rho_I$, and extend all the analytic functions $q(\xi)$ (or $q(\rho)$) to $q(\zeta)$ (or $q(\varrho)$), respectively by analytic continuation.

There is a singularity for this solution at a point $\varrho_c$ (or $\zeta_c$). This singularity is not yet evident in the leading-order approximation (9.104), but it can be clearly seen in higher-order approximations. We shall not show the results of the first-order approximation solution. The derivation is very similar to that in Chap. 6. The singular point $\varrho_c$ satisfies the equation

$$\frac{\partial \Sigma_*(\varrho, \tilde{k}_0)}{\partial \tilde{k}_0} = 0 \quad \text{or} \quad \left(\tilde{k}_0^{(1)}(\zeta) - \tilde{k}_0^{(3)}(\zeta)\right) = 0\,. \tag{9.105}$$

Combining (9.105) and (9.100), one finds that $\varrho_c$ is also the root of

$$\sigma_0 = \mathrm{e}^{-\mathrm{i}\frac{3\pi}{4}} \sqrt{\frac{4}{27\lambda_0^3}} \left(\frac{\varrho_c + \mathrm{i}}{\varrho_c - \mathrm{i}}\right)^{\frac{3}{4}}\,. \tag{9.106}$$

As before, we choose $\varrho_c$ (corresponding to $\zeta_c$) as the lower limit of the integrals in (9.104).

For the finger system, we still have $\mathrm{Im}\{\tilde{k}_0^{(1)}\} < 0$ and $\mathrm{Im}\{\tilde{k}_0^{(3)}\} > 0$, for $\rho \gg 1$. So, as $\rho \to \infty$, $H_1$ increases exponentially, whereas $H_3$ decreases exponentially. To satisfy the root condition (9.93) in the far fields, the solution must be approximated by the subdominant function $H_3$. Namely,

$$\tilde{h}_0 \sim DH_3\,, \tag{9.107}$$

where the constant $D$ is proportional to the characteristic amplitude of the initial perturbation $\delta$.

As we demonstrated in Chap. 6, the asymptotic form (9.107) cannot be applied to the entire complex $\varrho$-plane, due to the Stokes phenomenon. The structure of the Stokes lines for the finger system will be shown later. It will be seen that similarly to what occurs in dendritic growth, one anti-Stokes line ($A_2$) in the complex $\varrho$-plane (or $\zeta$-plane) crosses the real axis at the point $\rho_c'$ and divides the whole complex plane into sectors ($S_1$) and ($S_2$). For a uniformly valid solution, the coefficients $\{D_1, D_3\}$ in (9.104) will, in general, be different in different sectors. We denote the coefficients of the solution (9.104) in ($S_1$) by $\{D_1, D_3'\}$, and in ($S_2$) by $\{D_1' = 0, D_3' = D\}$. To determine the connection condition for these two pairs of constants, one needs to derive the inner solution in the inner region near the singular point $\varrho_c$ and match the inner solution with the outer region.

## 9.6 The Inner Equation near the Singular Point $\zeta_c$

In order to obtain the asymptotic expansion in the inner region $|\zeta-\zeta_c| \ll 1, |\eta| \ll 1$, we introduce the inner variables

$$\begin{aligned} \zeta_* &= \frac{\zeta - \zeta_c}{\varepsilon^\alpha}, \\ \eta_* &= \frac{\eta}{\varepsilon^\alpha}, \end{aligned} \tag{9.108}$$

and denote the inner solution by

$$\begin{aligned} \tilde{h}(\zeta, t) &= \varepsilon^\alpha \hat{h}(\zeta_*, t), \\ \tilde{\phi}(\zeta, \eta, t) &= \varepsilon^\alpha \hat{\phi}(\zeta_*, \eta_*, t), \\ \tilde{\psi}(\zeta, \eta, t) &= \varepsilon^\alpha \hat{\psi}(\zeta_*, \eta_*, t), \end{aligned} \tag{9.109}$$

where $\alpha$ is to be determined. We make the following inner expansion:

$$\begin{aligned} \hat{\phi}(\zeta_*, \eta_*, t) &= \Big[\nu_0(\varepsilon)\hat{\phi}_0(\zeta_*, \eta_*) + \nu_1(\varepsilon)\hat{\phi}_1(\zeta_*, \eta_*) + \cdots\Big] e^{\frac{\sigma t}{\varepsilon}}, \\ \hat{\psi}(\zeta_*, \eta_*, t) &= \Big[\nu_0(\varepsilon)\hat{\psi}_0(\zeta_*, \eta_*) + \nu_1(\varepsilon)\hat{\psi}_1(\zeta_*, \eta_*) + \cdots\Big] e^{\frac{\sigma t}{\varepsilon}}, \\ \hat{h} &= \Big[\nu_0(\varepsilon)\hat{h}_0 + \nu_1(\varepsilon)\hat{h}_1 + \cdots\Big] e^{\frac{\sigma t}{\varepsilon}}. \end{aligned} \tag{9.110}$$

In terms of the inner variables, (9.108) is expressed in the form

$$\left(\frac{\partial^2}{\partial \zeta_*^2} + \frac{\partial^2}{\partial \eta_*^2}\right)\hat{\phi} = 0. \tag{9.111}$$

The boundary conditions (9.61), (9.63), and (9.64) can be written as

1. As $\eta_* \to \infty$,

$$\hat{\phi} \to 0. \tag{9.112}$$

2. At $\eta_* = 0$,

$$\hat{\phi} + \hat{h} = -\frac{\varepsilon^{2-2\alpha}}{\mathcal{G}_0}\frac{\partial^2 \tilde{h}}{\partial \zeta_*^2} + \text{(h.o.t.)}, \tag{9.113}$$

$$\varepsilon^{1-\alpha}\frac{\partial \hat{\phi}}{\partial \eta_*} - \sigma_0 \mathcal{G}_0^2(\zeta)\hat{h} + \varepsilon^{1-\alpha} U_0 Y_{\xi,0}(\zeta)\frac{\partial \hat{h}}{\partial \zeta_*} = \text{(h.o.t.)}. \tag{9.114}$$

Letting $\varepsilon \to 0$, by neglecting small higher-order terms, the above inner equations lead to the following third-order complex ordinary differential equation for the interface perturbation $\hat{h}$:

$$-\mathrm{i}\frac{\varepsilon^{3-3\alpha}}{\mathcal{G}_0}\frac{\mathrm{d}^3\hat{h}}{\mathrm{d}\zeta_*^3}+\varepsilon^{1-\alpha}\Big\{\mathrm{i}-U_0Y_{\xi,0}(\zeta)\Big\}\frac{\mathrm{d}\hat{h}}{\mathrm{d}\zeta_*}+\sigma_0\mathcal{G}_0^2\hat{h}=0. \tag{9.115}$$

As we have demonstrated in last chapter, the inner equation (9.115) can also be derived directly through the dispersion relationship (9.100).

With the new inner variable $\varrho_* = (\varrho-\varrho_c)/\varepsilon^{\alpha}$, (9.115) can be written as

$$\mathrm{i}\frac{\varepsilon^{3-3\alpha}G^3}{\lambda_0 S}\frac{\mathrm{d}^3\hat{h}}{\mathrm{d}\varrho_*^3}+\varepsilon^{1-\alpha}G(\rho)\big(\rho+\mathrm{i}\big)\frac{\mathrm{d}\hat{h}}{\mathrm{d}\varrho_*}+\lambda_0^2S^2\sigma_0\hat{h}=0. \tag{9.116}$$

As before, we use the transformation

$$\tilde{h}=W(\xi)\,\exp\left\{\frac{\mathrm{i}}{\varepsilon}\int_{\varrho_c}^{\varrho}\tilde{k}_c(\rho_1)\,\mathrm{d}\rho_1\right\}, \tag{9.117}$$

where the reference wave number function $\tilde{k}_c(\varrho)$ is chosen in the same way as in Chap. 6, i.e.,

$$\left(\frac{\partial S^2\Sigma_*}{\partial\tilde{k}_0}\right)_{\tilde{k}_0=\tilde{k}_c}=\frac{3G^2\tilde{k}_c^2}{\lambda_0 S}-(1-\mathrm{i}\varrho)=0\,, \tag{9.118}$$

or

$$\tilde{k}_c(\varrho)=\frac{\sqrt{\frac{\lambda_0 S}{3}\left(1-\mathrm{i}\varrho\right)}}{G(\varrho)}=\frac{\mathrm{e}^{-\frac{i\pi}{4}}\lambda_0^{\frac{1}{2}}(\varrho-i)^{\frac{1}{4}}(\varrho+i)^{\frac{3}{4}}}{3^{\frac{1}{2}}G(\varrho)} \tag{9.119}$$
$$\big(\mathrm{Re}\{\tilde{k}_c\}>0\big).$$

From this definition, we know that

$$\mathrm{Re}\{\tilde{k}_0^{(3)}\}<\mathrm{Re}\{\tilde{k}_c\}<\mathrm{Re}\{\tilde{k}_0^{(1)}\}. \tag{9.120}$$

Thus in the outer region, we have

$$\begin{aligned} H_1&=W_0^{(+)}(\rho)\exp\left\{\frac{\mathrm{i}}{\varepsilon}\int_{\varrho_c}^{\varrho}\tilde{k}_c(\rho_1)\,\mathrm{d}\rho_1\right\},\\ H_3&=W_0^{(-)}(\rho)\exp\left\{\frac{\mathrm{i}}{\varepsilon}\int_{\varrho_c}^{\varrho}\tilde{k}_c(\rho_1)\,\mathrm{d}\rho_1\right\}. \end{aligned} \tag{9.121}$$

One can draw the same diagram of the relationship between the $H$ waves and $W$ waves as that for dendritic growth:

$$\begin{array}{lll} H_1 \text{ wave} & \Longleftrightarrow & W^{(+)} \\ \text{(short-wave branch)} & & \text{(outgoing wave);} \\ H_3 \text{ wave} & \Longleftrightarrow & W^{(-)} \\ \text{(long-wave branch)} & & \text{(incoming wave).} \end{array}$$

In accordance with the above, in the inner region, we set

$$\hat{h} = \hat{W}(\varrho_*) \exp\left\{ \frac{\mathrm{i}}{\varepsilon} \int_{\varrho_c}^{\varrho} \tilde{k}_c(\rho_1)\, \mathrm{d}\rho_1 \right\}. \tag{9.122}$$

In terms of the transformation (9.117), Eq. (9.116) is transformed to

$$\underset{(1)}{\mathrm{i}\varepsilon^{3-3\alpha}\Omega_3 \frac{\mathrm{d}^3 \hat{W}}{\mathrm{d}\varrho_*^3}} - \underset{(2)}{\varepsilon^{2-2\alpha}\Omega_2 \frac{\mathrm{d}^2 \hat{W}}{\mathrm{d}\varrho_*^2}} - \underset{(3)}{\lambda_0^2 S^2 \Big\{ \sigma_0 - \Sigma_c(\varrho) \Big\} \hat{W}} = 0, \tag{9.123}$$

where

$$\begin{aligned} \Omega_3 &= -\frac{G^3}{\lambda_0 S} \\ \Omega_2 &= -\frac{3G^3 \tilde{k}_c}{\lambda_0 S} = -\mathrm{e}^{-\mathrm{i}\frac{\pi}{4}} \frac{3^{\frac{1}{2}} G^2}{\lambda_0^{\frac{1}{2}}} \left( \frac{\varrho + \mathrm{i}}{\varrho - \mathrm{i}} \right)^{\frac{1}{4}} \\ \Sigma_c(\varrho) &= \Sigma_*(\tilde{k}_c, \varrho) = \mathrm{e}^{-\mathrm{i}\frac{3\pi}{4}} \frac{2}{3^{\frac{3}{2}} \lambda_0^{\frac{3}{2}}} \left( \frac{\varrho + \mathrm{i}}{\varrho - \mathrm{i}} \right)^{\frac{3}{4}}. \end{aligned} \tag{9.124}$$

Clearly, we have

$$\sigma_0 - \Sigma_c(\varrho_c) = 0. \tag{9.125}$$

Therefore, $\varrho_c$ is a simple turning point of (9.123). Besides $\varrho_c$, the inner equation has two other turning points, $\varrho = \pm\mathrm{i}$, and two polar points, $\varrho = \pm ia$. The relative positions of these singular points with respect to $\varrho_c$ are related to the parameters $\sigma_0, \lambda_0$. For a single-valued analytical solution, we set the branch cut along the Stokes line $(L_1)$ (see Fig. 9.9). Thus, we must choose the singular point $\varrho_c$ with $\mathrm{Re}\{\varrho_c\} > 0$ and $\mathrm{Im}\{\varrho_c\} < 0$. The singular points $\varrho = \mathrm{i}, ia$ are far from the turning point $\varrho_c$, so their influence on the inner solution is negligible. However, as $\sigma_0 \to 0$, $\varrho_c \to -\mathrm{i}$, while as $\lambda_0 \to \frac{1}{2}$, $-ia \to -\mathrm{i}$. Hence, the singular points $\varrho = -\mathrm{i}, -ia$ may enter the inner region of $\varrho_c$ and consequently influence the behavior of the inner solution. Two cases are found to be significant, as detailed in the following two sections.

### 9.6.1 *Case I:* $|\sigma_0| = \mathcal{O}(1)$

It follows that $|\varrho_c + \mathrm{i}| \gg \mathcal{O}(\varepsilon^\alpha)$ and $|\varrho_c + \mathrm{i}a| \gg \mathcal{O}(\varepsilon^\alpha)$.

In this case, all the singular points $\varrho = \pm\mathrm{i}$ and $\varrho = \pm\mathrm{i}a$ are far from $\varrho_c$. By balancing the leading terms of (9.123), it is found that as $\varepsilon \to 0$, (1) $\ll$ (2) $\approx$ (3), and one must set $\alpha = \frac{2}{3}$, so that in the leading-order approximation, (9.123) is reduced to

$$\varepsilon^{2-2\alpha}\Omega_2 \frac{\mathrm{d}^2\hat{W}}{\mathrm{d}\varrho_*^2} + \lambda_0^2 S^2\Big[\sigma_0 - \Sigma_c(\varrho)\Big]\hat{W} = 0\,. \tag{9.126}$$

We now expand all coefficient functions of (9.126) in a Taylor series near $\varrho = \varrho_c$. Note that

$$\begin{aligned}
\frac{\lambda_0^2 S^2(\rho)}{\Omega_2(\varrho)}\Sigma_c'(\varrho_c) &= \frac{\lambda_0}{3G^2(\rho)}\frac{(\rho+\mathrm{i})^{\frac{3}{4}}}{(\varrho_c+\mathrm{i})^{\frac{1}{4}}}\frac{(\rho-\mathrm{i})^{5/4}}{(\varrho_c-\mathrm{i})^{7/4}},\\
G(\varrho) &= \frac{\pi}{2a}(\varrho-\mathrm{i}a)(\varrho+\mathrm{i}a),\\
a &= \frac{1-\lambda_0}{\lambda_0}
\end{aligned} \tag{9.127}$$

(9.126) can be rewritten in the form

$$\begin{aligned}
-\varepsilon^{2-2\alpha}\frac{\mathrm{d}^2\hat{W}}{\mathrm{d}\varrho_*^2} + \frac{A_0^2}{(\varrho_c-\mathrm{i})^{\frac{1}{2}}}\frac{(\varrho-\varrho_c)}{(\varrho-\mathrm{i}a)(\varrho+\mathrm{i}a)}\frac{(\varrho+\mathrm{i})^{\frac{3}{4}}}{(\varrho_c+\mathrm{i})^{\frac{1}{4}}}\hat{W}\\
= \text{(higher order terms)}\,,
\end{aligned} \tag{9.128}$$

where

$$A_0 = \frac{2a\lambda_0^{\frac{1}{2}}}{3^{\frac{1}{2}}\pi}\,. \tag{9.129}$$

To leading order, the inner equation (9.128) is finally reduced to Airy's equation

$$\frac{\mathrm{d}^2\hat{W}_0}{\mathrm{d}\hat{\varrho}_*^2} + \hat{\varrho}_*\hat{W}_0 = 0\,, \tag{9.130}$$

where

$$\begin{aligned} \hat{\varrho}_* &= \frac{B}{\varepsilon^{\frac{2}{3}}}(\rho - \varrho_c)\,, \\ B &= \hat{A}_1^{\frac{2}{3}} \\ \hat{A}_1 &= \frac{\mathrm{i}A_0}{(\varrho_c - \mathrm{i}a)^{\frac{1}{2}}(\varrho_c + \mathrm{i}a)^{\frac{1}{2}}}\left(\frac{\varrho_c + \mathrm{i}}{\varrho_c - \mathrm{i}}\right)^{\frac{1}{4}}, \end{aligned} \tag{9.131}$$

and

$$0 < \arg(\hat{A}_1) \leq \frac{3\pi}{4}, \quad \text{and} \quad 0 < \arg(B) \leq \frac{\pi}{2}\,. \tag{9.132}$$

The structure of the Stokes lines is sketched in Fig. 9.9a.

### 9.6.2 *Case II:* $|\sigma_0| \ll 1$

For this case, we set

$$\sigma_0 = \hat{\sigma}_0 \varepsilon^{\nu_0}, \qquad a = 1 + \hat{a}\varepsilon^{\mu_0}, \quad \text{or} \quad \lambda_0 = \frac{1}{2} - \frac{\hat{a}}{4}\varepsilon^{\mu_0}. \tag{9.133}$$

We then derive

$$\varrho_c = -\mathrm{i} + \mathrm{i}\hat{\delta}\varepsilon^{\frac{4\nu_0}{3}}\,, \quad \hat{\delta} = \frac{9}{2^{\frac{1}{3}}}\lambda_0^2 \hat{\sigma}_0^{\frac{4}{3}}\,, \tag{9.134}$$

and subsequently,

$$\begin{cases} \varrho + \mathrm{i}a = (\varrho - \varrho_c) + (\varrho_c + \mathrm{i}a) = \varepsilon^{\alpha}\varrho_* + \mathrm{i}\hat{\delta}\varepsilon^{\frac{4\nu_0}{3}} + \mathrm{i}\hat{a}\varepsilon^{\mu_0} \\ \varrho + \mathrm{i} = (\varrho - \varrho_c) + (\varrho_c + \mathrm{i}) \ = \varepsilon^{\alpha}\varrho_* + \mathrm{i}\hat{\delta}\varepsilon^{\frac{4\nu_0}{3}}. \end{cases} \tag{9.135}$$

We further assume that both the singular point $-\mathrm{i}a$ and the zero point $-\mathrm{i}$ are in the inner region, namely,

$$|\varrho_c + \mathrm{i}| = \mathcal{O}(\varepsilon^{\alpha}), \quad |\varrho_c + \mathrm{i}a| = \mathcal{O}(\varepsilon^{\alpha}). \tag{9.136}$$

It then follows that $\alpha = \frac{4\nu_0}{3} = \mu_0$. From the above results, one can derive that the left-hand side of (9.123) is dominated by the second and third terms, namely (1) $\ll$ (2) $\approx$ (3). By balancing the leading-order terms, it is found that

$$\alpha = \frac{4}{3}, \quad \nu_0 = 1, \quad \text{and} \quad \mu_0 = \frac{4}{3}\,. \tag{9.137}$$

Therefore, we have $(\lambda_0 - \frac{1}{2})^{\frac{3}{4}} = \mathcal{O}(\sigma_0)$. This relationship between the quantities $\sigma_0$ and $(\lambda_0 - \frac{1}{2})$ is consistent with the quantization conditions that we derive later.

In the far field, as $|\hat{\varrho}_*| \to \infty$, as the leading-order approximation, the inner equation is reduced to

$$\frac{d\hat{W}_0}{d\hat{\varrho}_*^2} + \hat{\varrho}_*^{-\frac{1}{4}}\hat{W}_0 = 0, \tag{9.138}$$

where

$$\begin{aligned} \hat{\varrho}_* &= \frac{B}{\varepsilon^{\frac{4}{3}}}(\rho - \varrho_c), \\ B &= \frac{\hat{A}_3^{8/7}}{(i\hat{\delta})^{1/7}}, \\ \hat{A}_3 &= \frac{iA_0}{(\varrho_c - ia)(\varrho_c - i)^{\frac{1}{4}}}. \end{aligned} \tag{9.139}$$

The structure of the Stokes lines when $\frac{1}{2} \leq \lambda_0 < 1$ (or $0 < a < 1$) is sketched in Fig. 9.9b.

We see that for both cases I and II, the inner equation in the far field of the inner region can be written in the following unified form:

$$\frac{d^2\hat{W}_0}{d\hat{\varrho}_*^2} + \hat{\varrho}_*^{p_0}\hat{W}_0 = 0\,, \tag{9.140}$$

where

$$\hat{\varrho}_* = \frac{B}{\varepsilon^{\alpha}}(\varrho - \varrho_c) \gg 1\,. \tag{9.141}$$

For the different cases, the constants $B$, $p_0$, and the exponent $\alpha$ are different. Moreover, the anti-Stokes line always intersects the real $\varrho$-axis at a point $\rho_c' > 0$.

The general solution of (9.140) is

$$\hat{W}_0 = D_1\hat{\varrho}_*^{\frac{1}{2}}H_\nu^{(1)}(z) + D_2\hat{\varrho}_*^{\frac{1}{2}}H_\nu^{(2)}(z) \qquad (z = 2\nu\hat{\varrho}_*^{\frac{1}{2\nu}}), \tag{9.142}$$

where $\nu = \frac{1}{p_0+2}$, and $H_\nu^{(1)}(z)$ and $H_\nu^{(2)}(z)$ are the $\nu$th-order Hankel functions of the first and second kind, respectively. For future use, we summarize the results below.

- Case I: $|\sigma_0| \geq \mathcal{O}(\varepsilon^{\frac{1}{2}})$,

$$\alpha = \frac{2}{3},\ p_0 = 1, \quad \text{and} \quad \nu = \frac{1}{3}\,;$$

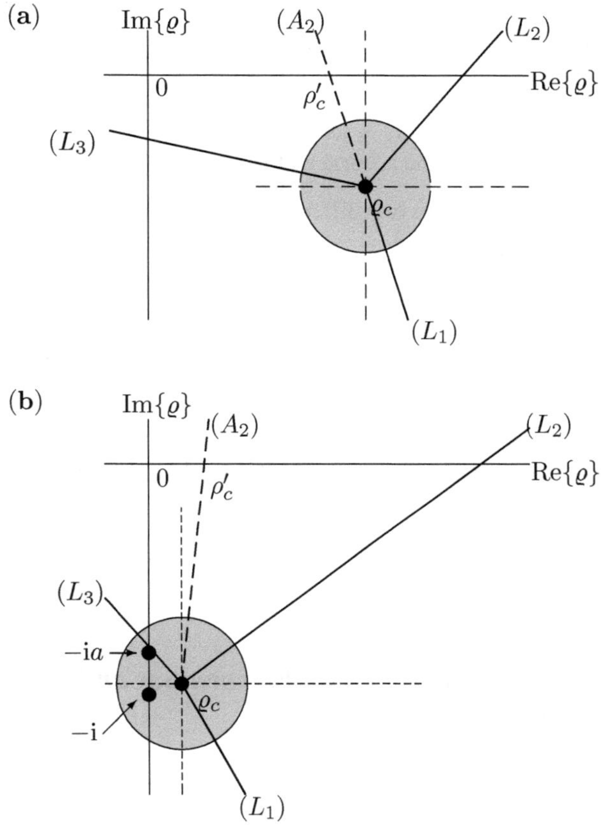


**Fig. 9.9** A sketch of the Stokes lines $(L_1)$, $(L_2)$, $(L_3)$ and the anti-Stokes line $(A_2)$: (**a**) for case I; (**b**) for case II

- Case II: $|\sigma_0| \leq \mathcal{O}(\varepsilon^{\frac{3}{5}})$, $(\lambda_0 - \frac{1}{2})^{\frac{3}{4}} = \mathcal{O}(\sigma_0)$,

$$\alpha = \frac{4}{3}, \; p_0 = -\frac{1}{4}, \quad \text{and} \quad \nu = \frac{4}{7}.$$

We now turn to matching the inner solution (9.142) with the outer solution (9.104) in the intermediate regions in each sector. Matching the inner solution with the outer solution in sector $(S_2)$ gives $D_1 = 0$ and $D_3 = D_3' \times \text{constant} \neq 0$. Furthermore, in sector $(S_1)$, the matching condition gives the connection formula

$$\frac{D_1}{D_3} = \mathrm{i}2\cos(\nu\pi). \tag{9.143}$$

So far, we have derived the asymptotic solution in the limit $\varepsilon \to 0$, for given parameters $\sigma_0, \lambda_0$. As the second step, we shall require the solution to satisfy the tip condition. Thus the parameter $\sigma_0$ must be chosen as a proper function of $\lambda_0$ and $\varepsilon$. As in the case of dendritic growth, we find two different types of spectra, namely

1. the complex spectrum $\sigma_0 = \sigma_R - i\omega$ $(\omega > 0)$;
2. the real spectrum (within the accuracy of the asymptotic solution).

These spectra correspond to two different instability mechanisms:

1. the global trapped-wave (GTW) instability, which is induced by perturbations with a high frequency ($|\omega| = \mathcal{O}(1)$);
2. the low-frequency (LF) instability mechanism, which is induced by perturbations with low frequency ($\omega \leq \mathcal{O}(\varepsilon)$).

## 9.7 Eigenvalue Spectra and Instability Mechanisms

The derivations that follow are almost completely the same as those for 2D dendritic growth with anisotropy.

### 9.7.1 *The Spectrum of Complex Eigenvalues and GTW Instability*

Consider $\sigma_0 = \sigma_R - i\omega$ with $\omega > 0$. With the complex eigenvalue $\sigma_0$, the physical solution in the outer region is

$$\mathrm{Re}\left\{\tilde{h}_0(\xi, t)\right\} = \mathrm{Re}\left\{H(\rho)\mathrm{e}^{\frac{\sigma_0 t}{\varepsilon}}\right\}, \tag{9.144}$$

where

$$H(\rho) = D_1 H_1 + D_3 H_3 \,. \tag{9.145}$$

We define

$$\begin{cases} d_1 = D_1 \mathrm{e}^{-\mathrm{i}\chi_1}, \\ d_3 = D_3 \mathrm{e}^{-\mathrm{i}\chi_3}, \end{cases} \tag{9.146}$$

where

$$\begin{aligned} \chi_1 &= \frac{1}{\varepsilon}\int_0^{\varrho_c} \tilde{k}_0^{(1)} \mathrm{d}\rho, \\ \chi_3 &= \frac{1}{\varepsilon}\int_0^{\varrho_c} \tilde{k}_0^{(3)} \mathrm{d}\rho \,. \end{aligned} \tag{9.147}$$

Thus, it follows from the connection formula (9.143) that

$$\frac{d_1}{d_3} = i2\cos(\nu\pi)\mathrm{e}^{-\mathrm{i}\chi}, \tag{9.148}$$

where

$$\chi = \frac{1}{\varepsilon}\int_0^{\rho_c}\left(\tilde{k}_0^{(1)} - \tilde{k}_0^{(3)}\right)\mathrm{d}\rho\,. \tag{9.149}$$

To satisfy the smooth tip conditions, the coefficients $d_1$ and $d_3$ must satisfy the following conditions:

(i) for the symmetric S-modes,

$$\frac{d_3}{d_1} = -\frac{\tilde{k}_0^{(1)}(0)}{\tilde{k}_0^{(3)}(0)}; \tag{9.150}$$

(ii) for the anti-symmetric A-modes,

$$d_1 = -d_3\,. \tag{9.151}$$

Combining (9.148) with (9.150) or (9.151), one obtains the following quantization conditions:

$$\frac{1}{\varepsilon}\int_0^{\rho_c}\left(\tilde{k}_0^{(1)} - \tilde{k}_0^{(3)}\right)\mathrm{d}\rho = \left(2n+1+\frac{1}{2}+\theta_0\right)\pi \\ -\mathrm{i}\ln\alpha_0 \quad \left(n = 0, \pm 1, \pm 2, \cdots\right), \tag{9.152}$$

where

$$\begin{cases} \alpha_0\,\mathrm{e}^{\mathrm{i}\theta_0\pi} = \dfrac{\tilde{k}_0^{(1)}(0)}{\tilde{k}_0^{(3)}(0)} & \text{(for the S-modes);} \\ \alpha_0 = 1,\ \theta_0 = 0 & \text{(for the A-modes)}\,. \end{cases} \tag{9.153}$$

The system has complex eigenvalues with $|\sigma_0| = \mathcal{O}(1)$, corresponding to case I discussed in the previous section. This spectrum contains two discrete sets of complex eigenvalues for S-modes and A-modes respectively,

$$\sigma_{0n}\ \ (n = 0, \pm 1, \pm 2, \ldots),$$

which are functions of $\varepsilon$ and $\lambda_0$. Given $\lambda_0$, the system allows a set of neutrally stable modes ($\sigma_R = 0$) when $\varepsilon = \varepsilon_{*n}$ ($n = 0, 1, 2, \ldots$) with $\varepsilon_* = \varepsilon_{*0} > \varepsilon_{*1} > \varepsilon_{*2} > \cdots$. Or given $\varepsilon$, the system allows a set of global neutrally stable modes with $\lambda_0 = \lambda_{0n}$ ($n = 0, 1, 2, \ldots$) and $\lambda_{00} < \lambda_{01} < \cdots$. These eigenmodes are all traveling waves

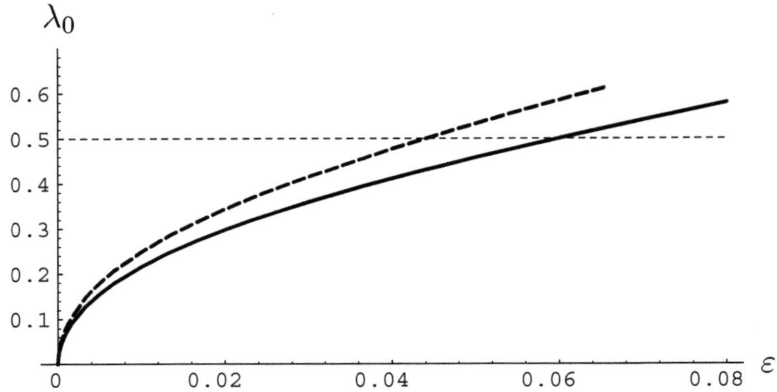


**Fig. 9.10** The neutral curves of GTW modes in the $(\lambda_0, \varepsilon)$-plane. The *dashed line* represents the S-mode, while the *solid line* is for the A-mode

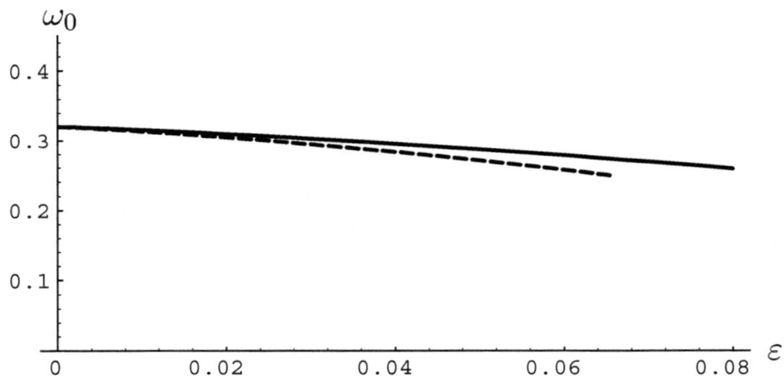


**Fig. 9.11** Variation of the frequency, $\omega_0$, of the neutral GTW modes with $\varepsilon$. The *dashed line* represents the S-mode, while the *solid line* is for the A-mode

propagating along the interface. The neutral curves for the S-mode and A-mode are shown in Fig. 9.10. It is found that the neutral A-modes are more stable than the neutral S-modes. Furthermore, in Fig. 9.11, we show the variation with $\varepsilon$ of the frequency $\omega_0$ of the corresponding neutral GTW modes.

As in Chap. 7, we can prove that the system does not allow other branches in the complex spectrum corresponding to case II. We can first simplify the quantization condition (9.152) with the assumption $|\sigma_0| \ll 1$. From the local dispersion formula (9.101),

$$\begin{cases} k_e^{(3)}(\varrho, \sigma_e) = (1 + \mathrm{i}\varrho)\sigma_e + \cdots, \\ k_e^{(1)}(\varrho, \sigma_e) = \mathrm{e}^{-\mathrm{i}\pi/4}(\varrho + \mathrm{i})^{\frac{3}{4}}(\varrho - \mathrm{i})^{\frac{1}{4}} - \frac{1}{2}(1 + \mathrm{i}\varrho)\sigma_e + \cdots, \end{cases} \tag{9.154}$$

as $\sigma_0 \to 0$. By definition

$$\begin{cases} \sigma_e = \lambda_0^{\frac{3}{2}} \sigma_0, \\ \tilde{k}_0 = \frac{\lambda_0^{\frac{1}{2}}}{G(\varrho)} k_e = \frac{2(1-\lambda_0)}{\pi \lambda_0^{\frac{1}{2}}} \frac{k_e}{(\varrho + \mathrm{i}a)(\varrho - \mathrm{i}a)} . \end{cases} \tag{9.155}$$

Then one has

$$\left(\tilde{k}_0^{(1)} - \tilde{k}_0^{(3)}\right) = \frac{2(1-\lambda_0)}{\pi \lambda_0^{\frac{1}{2}}} \times \left\{ \frac{(1-\mathrm{i}\varrho)^{\frac{3}{4}}(1+\mathrm{i}\varrho)^{\frac{1}{4}}}{(\varrho - \mathrm{i}a)(\varrho + \mathrm{i}a)} - \frac{3\sigma_0}{2} \lambda_0^{\frac{3}{2}} \frac{(1+\mathrm{i}\varrho)}{(\varrho - \mathrm{i}a)(\varrho + \mathrm{i}a)} \right\} + \mathcal{O}(\sigma_0^2). \tag{9.156}$$

To continue the calculation for $\chi$, one must consider two distinct ranges of $\lambda_0$.

#### 9.7.1.1 Range 1: $\frac{1}{2} \leq \lambda_0 < 1$ or $0 \leq a < 1$

For this case,

$$\begin{aligned} \int_0^{\varrho_c} \left(\tilde{k}_0^{(1)} - \tilde{k}_0^{(3)}\right) \mathrm{d}\rho &= \left(\int_0^{-\mathrm{i}(a-0)}\right) + \left(\int_{-\mathrm{i}(a+0)}^{-\mathrm{i}}\right) + \left(\int_{(C_r)}\right) + \left(\int_{-\mathrm{i}}^{\varrho_c}\right) \\ &= -\mathrm{i} \frac{2(1-\lambda_0)}{\pi \lambda_0^{\frac{1}{2}}} \int_0^1 \frac{(1-x)^{\frac{3}{4}}(1+x)^{\frac{1}{4}}}{(a-x)(a+x)} dx \\ &\quad + \mathrm{i}\sigma_0 \frac{3\lambda_0(1-\lambda_0)}{\pi} \int_0^1 \frac{(1+x)}{(a-x)(a+x)} dx \\ &\quad + \frac{(2\lambda_0 - 1)^{\frac{3}{4}}}{\lambda_0^{\frac{1}{2}}} - \frac{3\lambda_0}{2}\sigma_0 + \mathrm{i}\sigma_0 \frac{4 \cdot 3^{\frac{3}{2}} \lambda_0^{\frac{3}{4}}(1-\lambda_0)}{2\lambda_0 - 1} + \mathcal{O}(\sigma_0^2), \end{aligned} \tag{9.157}$$

where the integral path $(C_r)$ is a semicircle connecting the points $\rho = -\mathrm{i}(a-0)$ and $\rho = -\mathrm{i}(a+0)$ clockwise. It follows that

$$\chi = \frac{1}{\varepsilon} \int_0^{\varrho_c} \left(\tilde{k}_0^{(1)} - \tilde{k}_0^{(3)}\right) \mathrm{d}\rho \approx \frac{1}{\varepsilon} \left[ \frac{(2\lambda_0 - 1)^{\frac{3}{4}}}{\lambda_0^{\frac{1}{2}}} - \mathrm{i}\hat{B}_1 \frac{1-\lambda_0}{\lambda_0^{\frac{1}{2}}} \right] - \frac{\sigma_0}{\varepsilon} \left[ \frac{3\lambda_0}{2} - \mathrm{i}\lambda_0(1-\lambda_0)\hat{B}_2 \right], \tag{9.158}$$

where

$$\hat{B}_1 = \frac{2}{\pi}\int_0^1 \frac{(1-x)^{\frac{3}{4}}(1+x)^{\frac{1}{4}}}{(a-x)(a+x)}dx\,, \tag{9.159}$$

$$\hat{B}_2 = \frac{3}{\pi}\int_0^1 \frac{(1-x)}{(a-x)(a+x)}dx + \frac{4\cdot 3^{\frac{3}{2}}\lambda_0^2}{2\lambda_0 - 1}\,. \tag{9.160}$$

#### 9.7.1.2 Range 2: $0 < \lambda_0 < \frac{1}{2}$ or $a > 1$

For this case,

$$\int_0^{\varrho_c}\left(\tilde{k}_0^{(1)} - \tilde{k}_0^{(3)}\right)\mathrm{d}\rho = -\mathrm{i}\frac{2(1-\lambda_0)}{\pi\lambda_0^{\frac{1}{2}}}\int_0^1 \frac{(1-x)^{\frac{3}{4}}(1+x)^{\frac{1}{4}}}{(a-x)(a+x)}dx + \mathrm{i}\sigma_0\frac{3\lambda_0(1-\lambda_0)}{\pi}\int_0^1 \frac{(1-x)}{(a-x)(a+x)}dx + \mathcal{O}(\sigma_0^2)\,, \tag{9.161}$$

so,

$$\chi \approx \frac{\mathrm{i}}{\varepsilon}\left[\sigma_0\lambda_0(1-\lambda_0)\hat{B}_2 - \hat{B}_1\frac{1-\lambda_0}{\lambda_0^{\frac{1}{2}}}\right]. \tag{9.162}$$

By substituting $\chi$ into the quantization condition (9.152), we find that the eigenvalues are $\sigma_0 = \mathcal{O}(1)$ for both cases. This contradicts our original assumption that $|\sigma_0| \ll 1$. Therefore, we conclude that the system allows only the complex spectrum with $|\sigma_0| = \mathcal{O}(1)$.

### *9.7.2 The Spectrum of Real Eigenvalues and LF Instability*

We now consider the spectrum of real eigenvalues $\sigma_0$. The physical solution in the outer region for this case is

$$\mathrm{Re}\{\tilde{h}(\xi,t)\} = \mathrm{Re}\{H(\xi)\}\mathrm{e}^{\frac{\sigma_0 t}{\varepsilon}}\,. \tag{9.163}$$

The tip smoothness condition for symmetric modes is

$$\mathrm{Re}\{H(\xi)\}'(0) = \mathrm{Re}\{H'(0)\} = \mathrm{Re}\left\{\frac{id_1}{\varepsilon}\tilde{k}_0^{(1)}(0) + \frac{id_3}{\varepsilon}\tilde{k}_0^{(3)}(0)\right\} = 0\,. \tag{9.164}$$

Without loss of generality, we assume $d_1$ to be a positive real number and write

$$d_1 > 0, \quad \text{and} \quad d_3 = |d_3|e^{i\chi_0\pi}. \tag{9.165}$$

Since for a real eigenvalue $\sigma_0$, $k_0^{(1)}(0)$ and $k_0^{(3)}(0)$ are both real, the tip condition (9.78) for the S-mode becomes $d_{3I} = 0$. We therefore have

$$\frac{d_1}{d_3} = \left|\frac{d_1}{d_3}\right| e^{-i\chi_0\pi} = i2\cos(\nu\pi)e^{-i\chi} \quad (\chi_0 = 0, 1)\,. \tag{9.166}$$

This leads to the quantization condition

$$\mathrm{Re}\left\{\frac{1}{\varepsilon}\int_0^{\varrho_c}\left(\tilde{k}_0^{(1)} - \tilde{k}_0^{(3)}\right)\mathrm{d}\rho\right\} = \left(2n + \frac{1}{2} + \chi_0\right)\pi \tag{9.167}$$

$$(n = 0, \pm 1, \pm 2, \ldots)\,,$$

$$\left|\frac{d_1}{d_3}\right| = 2\cos(\nu\pi)e^{\mathrm{Im}\{\chi\}}\,. \tag{9.168}$$

For the antisymmetric modes, the tip condition is

$$\mathrm{Re}\{H(0)\} = \mathrm{Re}\{d_1 + d_3\} = 0\,, \tag{9.169}$$

which leads to

$$\left|\frac{d_1}{d_3}\right| = -\cos(\chi_0\pi) \leq 1\,. \tag{9.170}$$

We now have

$$\frac{d_1}{d_3} = \left|\frac{d_1}{d_3}\right| e^{-i\chi_0\pi} = -\cos(\chi_0\pi)e^{-i\chi_0\pi} = i2\cos(\nu\pi)e^{-i\chi}, \tag{9.171}$$

and this leads to the quantization condition

$$\mathrm{Re}\left\{\frac{1}{\varepsilon}\int_0^{\varrho_c}\left(\tilde{k}_0^{(1)} - \tilde{k}_0^{(3)}\right)\mathrm{d}\rho\right\} = \left(2n + \frac{1}{2} + \chi_0\right)\pi \tag{9.172}$$

$$(n = 0, \pm 1, \pm 2, \ldots),$$

$$\left|\frac{d_1}{d_3}\right| = -\cos(\chi_0\pi) = 2\cos(\nu\pi)e^{\mathrm{Im}\{\chi\}}\,. \tag{9.173}$$

The above two quantization conditions can be written in the same form as follows:

$$\mathrm{Re}\left\{\frac{1}{\varepsilon}\int_0^{\varrho_c}\left(\tilde{k}_0^{(1)}-\tilde{k}_0^{(3)}\right)\mathrm{d}\rho\right\} = \left(2n+\frac{1}{2}+\chi_0\right)\pi \tag{9.174}$$

$$(n = 0, \pm 1, \pm 2, \ldots),$$

$$\left|\frac{d_1}{d_3}\right| = 2\cos(\nu\pi)\mathrm{e}^{\mathrm{Im}\{\chi\}}\,. \tag{9.175}$$

and

$$\chi_0 = \begin{cases} 0 \text{ or } 1 & \text{(for S-modes)} \\ 1 + \dfrac{\cos^{-1}\left|\frac{d_1}{d_3}\right|}{\pi} & \text{(for A-modes)}\,. \end{cases} \tag{9.176}$$

The quantization condition (9.174) determines the eigenvalues $\sigma_{0n}$, while formula (9.175) determines the corresponding eigenfunction. We have found the real spectrum only for $\sigma_0 \ll 1$. In this case, the quantization condition (9.174) can be simplified in terms of formula (9.158) or (9.162).

When $0 < \lambda_0 < \frac{1}{2}$ or $a > 1$, from (9.162), it follows that $\mathrm{Re}\{\chi\} = 0$, and as a result, the system will have no eigenvalues. On the other hand, when $\frac{1}{2} \le \lambda_0 < 1$ or $0 \le a < 1$, in terms of (9.158) we obtain

$$\mathrm{Re}\{\chi\} = \frac{1}{\varepsilon}\left[\frac{(2\lambda_0-1)^{\frac{3}{4}}}{\lambda_0^{\frac{1}{2}}} - \frac{3\lambda_0}{2}\sigma_0\right], \tag{9.177}$$

$$\mathrm{Im}\{\chi\} = \frac{1}{\varepsilon}\left[-\hat{B}_1\frac{1-\lambda_0}{\lambda_0^{\frac{1}{2}}} + \sigma_0\hat{B}_2\lambda_0(1-\lambda_0)\right]. \tag{9.178}$$

Based on these results, we obtain the following simplified quantization conditions:

(1) For the S-modes,

$$\frac{3}{2}\lambda_0\sigma_0 = \frac{(2\lambda_0-1)^{\frac{3}{4}}}{\lambda_0^{\frac{1}{2}}} - \varepsilon\left(n+\frac{1}{2}\right)\pi\,, \quad \left(\frac{1}{2} \le \lambda_0 < 1\right). \tag{9.179}$$

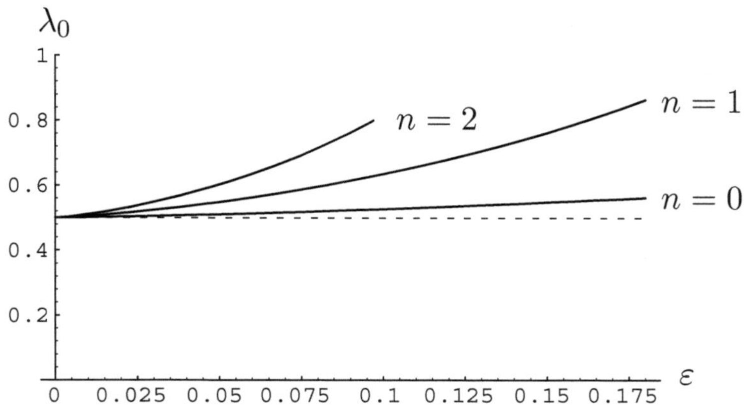


**Fig. 9.12** The neutral curves of the LF S-modes in the $(\lambda_0, \varepsilon)$-plane, with $n = 0, 1, 2$ from bottom to top

(2) For the A-modes,

$$\frac{3}{2}\lambda_0\sigma_0 = \frac{(2\lambda_0 - 1)^{\frac{3}{4}}}{\lambda_0^{\frac{1}{2}}} - \varepsilon\left(2n + \frac{1}{2} + \chi_0\right)\pi,$$

$$\cos(\chi_0\pi) = -2\cos\left(\frac{2\pi}{7}\right)e^{\mathrm{Im}\{\chi\}} \tag{9.180}$$

$$\left(\frac{1}{2} \leq \lambda_0 < 1\right).$$

It is seen from the quantization conditions (9.179) and (9.181) that one must have eigenvalues $\sigma_0 = \mathcal{O}(\varepsilon)$ and $(\lambda_0 - \frac{1}{2})^{\frac{3}{4}} = \mathcal{O}(\sigma_0)$. Case II, discussed in the previous section, is obviously consistent with this relationship. Hence $\nu = \frac{2}{7}$.

For fixed $\varepsilon$ and $\lambda_0$, from the above quantization condition (9.179) or (9.181) one can obtain a discrete set of eigenvalues $\{\sigma_{0n}\}$ $(n = 0, 1, 2, 3, \ldots)$. Given $\lambda_0$, the system allows a discrete set of neutral stable modes ($\sigma_{0n} = 0$) corresponding to $\varepsilon = \varepsilon'_n$ $(n = 1, 2, \ldots)$, where $\varepsilon'_0 > \varepsilon'_1 > \varepsilon'_2 > \cdots > \varepsilon'_n > \cdots$. On the other hand, given $\varepsilon > 0$, the system allows a discrete set of neutral modes with width parameters $\lambda_0 = \lambda_{0n}$, or tip velocity $U_0 = U_{0n}$. The neutral S-modes are determined by the formula

$$\frac{(2\lambda_{0n} - 1)^{\frac{3}{4}}}{\sqrt{\lambda_{0n}}} = \varepsilon\left(n + \frac{1}{2}\right)\pi, \quad \left(\frac{1}{2} \leq \lambda_0 < 1\right). \tag{9.181}$$

The neutral curves of the S-modes in the $(\lambda_0, \varepsilon)$-plane are shown in Fig. 9.12.

The neutral A-modes are determined by

$$\frac{(2\lambda_{0n}-1)^{\frac{3}{4}}}{\sqrt{\lambda_{0n}}} = \varepsilon\left(2n+\frac{1}{2}+\chi_0\right)\pi\,,$$
$$\cos(\chi_0\pi) = -2\cos\left(\frac{2\pi}{7}\right)\exp\left[-\frac{\hat{B}_1}{\varepsilon}\frac{(1-\lambda_{0n})}{\lambda_{0n}^{\frac{1}{2}}}\right]\,, \tag{9.182}$$

for $\frac{\pi}{2} < \chi_0 < \pi$. It is found that the neutral S-mode ($n = 0$) is more stable than the neutral A-mode ($n = 0$). These neutral S-modes are steady finger solutions that coincide with the classical steady finger solutions discovered numerically by Vanden-Broeck [23] and by some other investigators analytically in terms of the microscopic solvability condition (MSC) theory. The LF instability was first discovered numerically by Kessler and Levine [12]. It was also studied by Bensimon et al. using a different approach [3].

In Fig. 9.13, we plot the neutral curve $\{\gamma_0\}$ of the A-mode ($n = 0$) of the GTW mechanism and the neutral curve $\{C_0\}$ of the S-mode ($n = 0$) of the LF mechanism in the parameter plane $(\lambda, \varepsilon)$. These two neutral curves intersect each other at a critical number $\varepsilon_c = 0.0908$. The shaded region below these two curves shown in Fig. 9.13 is stable, while the region above either of the curves is unstable. It is evident that when the injection velocity $U_\infty$ is very large or the surface tension $\gamma$ very small such that $\varepsilon < 0.0908$, the steady finger solutions are all linearly unstable.

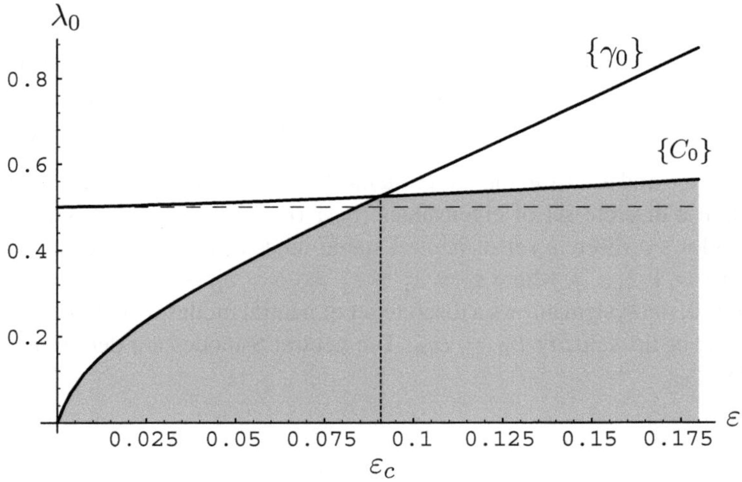


**Fig. 9.13** The stability diagram of viscous fingering in the system without a nose bubble. Here $\{C_0\}$ is the neutral curve of the S-mode ($n = 0$) of the LF mechanism, while $\{\gamma_0\}$ is the neutral curve of the A-mode ($n = 0$) of the GTW mechanism. The *shaded region* is stable, whereas the remaining part of plane is the unstable region

## 9.8 Fingering Flow with a Nose Bubble

Now we turn to studying the finger formation problem with a tiny nose bubble as sketched in Fig. 9.14.

### 9.8.1 *The Basic State of Finger Formation with a Nose Bubble and Its Linear Perturbation*

The presence of the bubble introduces an additional perturbation to the Saffman–Taylor solution. The bubble's effect can be simulated by modifying the interface condition at the finger tip. Assume that the effective bubble radius is $r_b$ and the interface shape in the coordinate system $(x, y)$ is denoted by the function $y_s(x, t)$. The entire interface of the finger consists of of the curves $T \to B$, $B \to O' \to B'$, and $B' \to T'$, which are joined at the triple points $B$ and $B'$.

For simplicity, we assume that the small nose bubble is solid, with shape and size unchanged during the whole process, and that the slip condition $u_n = 0$ is applicable. Hence, the bubble surface $BO'B'$ coincides with the streamline $\psi = 0$. Furthermore, several kinds of boundary conditions may be imposed at the triple points $B$ and $B'$ where the surface of the bubble intersects with the interface. Either the triple point location on the bubble's surface or the contact angle between the bubble and the interface of the finger are prescribed. Specifically, we may consider the following scenarios:

1. The position of the triple point on the bubble is fixed, but the contact angle is free. Namely,

$$y_s(r_b, t) = b\,. \tag{9.183}$$

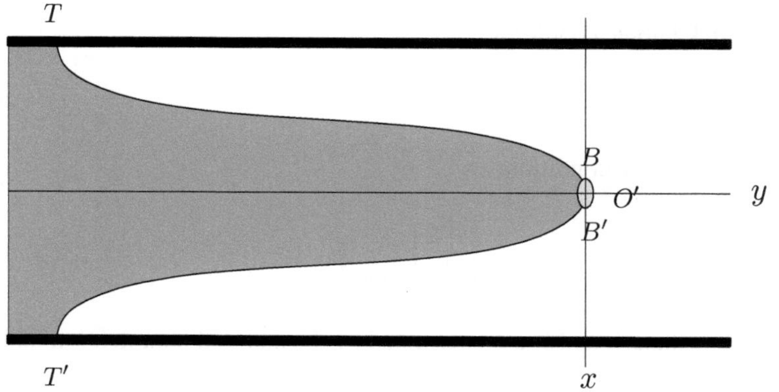


**Fig. 9.14** Sketch of a finger with a nose bubble

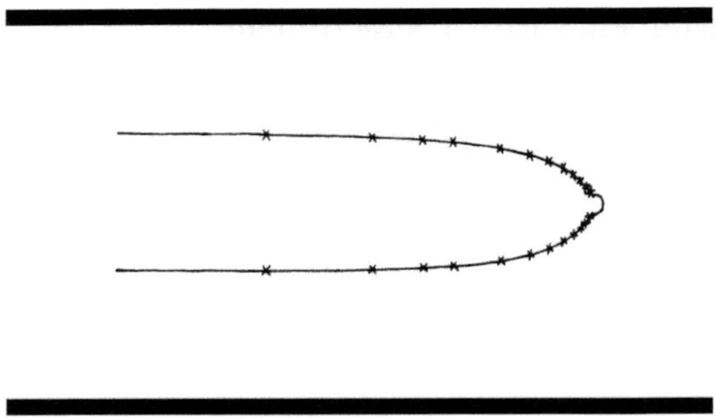

**Fig. 9.15** The comparison of an experimental record of a smooth finger with a nose bubble with a properly selected ST solution. The *solid line* is the experimental curve, while the cross points are calculated from the ST solution

We call this condition the *stick condition of the bubble.*

2. The contact angle of the finger surface with the tangential direction $\hat{\tau}$ along the bubble is fixed, but the position of the triple point can freely slip along the bubble. Namely,

$$\left\{\frac{dy_s}{d\hat{\tau}}\right\}_{\text{jump}}(r_b, t) = a\,. \tag{9.184}$$

We call this condition the *slip condition of the bubble.*

In the above, the constants $a$ and $b$ are material parameters depending on the physical properties of the bubble and the fluids.

The problem can be formulated in the $(\xi, \eta)$ coordinate system based on the ST solution. The triple point location on the nose bubble $\xi_b, \eta_b$ can be determined by the effective size parameter $r_b$. Accordingly, the triple point conditions at $B$ can be specified as follows:

(1)' The stick bubble condition,

$$\eta_s(\xi_b, t) = \tilde{b}\,. \tag{9.185}$$

(2)' The slip bubble condition,

$$\left\{\frac{d\eta_s}{d\xi}\right\}_{\text{jump}}(\xi_b, t) = \tilde{a}\,, \tag{9.186}$$

where the constants $\tilde{a}$ and $\tilde{b}$ are determined by the material parameters $a$ and $b$.

The exact form of the generalized steady-state solution for the present system is even harder to obtain than in the previous case without a nose bubble. However, since in practice, $\varepsilon$ and $r_b$ are both small, it is expected that these basic state solutions $q_B$ can be well approximated by the ST solution in the region away from the finger's root. This argument can be justified by the experimental evidence, as shown in Fig. 9.15 [5]. In the figure, the solid line shows the experimental curve of a smooth finger with a nose bubble, while the cross points represent the ST solution with the selected tip velocity through the experimental data. One sees an excellent agreement between the experimental curve and the ST solution data. This experimental curve is supposed to represent the exact solution $q_B$. It shows that the presence and the nature of the nose bubble has very little effect on the basic state in the region far from the bubble.

Next we consider infinitesimal perturbations around the basic state and perform a linear stability analysis. One can see that although the modification of the basic state induced by presence of the nose bubble is negligible, the effect of the bubble on the perturbed state is significant.

The linear perturbed system, in the leading approximation as $\varepsilon$ and $r_b \to 0$, remains the same as that discussed in previous sections except that the tip condition is now changed due to the presence of the bubble. We shall once again apply the new variable $\rho$ instead of $\xi$. At the triple point $B$, $\xi = \xi_b$ and $\rho = \rho_b$. We define the effective bubble size $r_b$ by the curve length $\ell = \ell_b$ measured from $\xi = 0$ to $\xi = \xi_b$ along the ST finger, namely

$$r_b = \ell_b \approx F(0)\rho_b = \frac{2}{\pi}\Big(\frac{\lambda_0^2}{1-\lambda_0}\Big)\rho_b\,. \tag{9.187}$$

Thus, at the bubble triple point, we have

(i) the stick condition:

$$\text{as} \quad \rho \to \rho_b + 0\,, \quad \tilde{h} = 0 \quad \text{and} \quad \frac{\partial \tilde{h}}{\partial \rho} < \infty\,, \tag{9.188}$$

or

(ii) the slip condition:

$$\text{as} \quad \rho \to \rho_b + 0\,, \quad \tilde{h} < \infty \quad \text{and} \quad \frac{\partial \tilde{h}}{\partial \rho} = 0\,. \tag{9.189}$$

## 9.8.2 The Quantization Conditions for the System with a Nose Bubble

The entire procedure for the stability analysis remains the same as in the previous sections for the system without a nose bubble. Both the GTW mechanism and LF mechanism remain valid for the present system. The respective quantization conditions in the leading-order approximation are almost the same as (9.152) and (9.174) or (9.179) and (9.181). The only change is in the lower limit of the integrations due to the bubble effect.

The results are as follows:

### 9.8.2.1 The GTW Mechanism

The branch of complex eigenvalues $\sigma_0 = \mathcal{O}(1)$ is determined by the following quantization condition:

(i) for the S-modes:

$$\frac{1}{\varepsilon}\int_{\rho_b}^{\varrho_c}\left(\tilde{k}_0^{(1)} - \tilde{k}_0^{(3)}\right)\mathrm{d}\rho = \left(2n + 1 + \frac{1}{2} + \theta_0\right)\pi - \mathrm{i}\ln\alpha_0 \tag{9.190}$$

$$\left(n = 0, \pm 1, \pm 2, \ldots\right),$$

where

$$\alpha_0\,\mathrm{e}^{\mathrm{i}\theta_0\pi} = \frac{\tilde{k}_0^{(1)}(\rho_b)}{\tilde{k}_0^{(3)}(\rho_b)} = \frac{k_0^{(1)}(\rho_b)}{k_0^{(3)}(\rho_b)}, \tag{9.191}$$

(ii) for the A-mode:

$$\frac{1}{\varepsilon}\int_{\rho_b}^{\varrho_c}\left(\tilde{k}_0^{(1)} - \tilde{k}_0^{(3)}\right)\mathrm{d}\rho = \left(2n + 1 + \frac{1}{2}\right)\pi \tag{9.192}$$

$$\left(n = 0, \pm 1, \pm 2, \ldots\right).$$

These quantization conditions give two discrete sets of complex eigenvalues

$$\sigma_{0n}\ \ (n = 0, \pm 1, \pm 2, \ldots),$$

which are functions of $\varepsilon$, $\lambda_0$, and $\rho_b$.

The global neutrally stable modes ($\sigma_R = 0$) correspond to $\varepsilon = \varepsilon_{*0} = \varepsilon_* > \varepsilon_{*1} > \varepsilon_{*2} > \cdots$, for given $\lambda_0, \ell_b$, or $\lambda_0 = \lambda_{00} < \lambda_{01} < \cdots$, for given $\varepsilon, \ell_b$. In Fig. 9.16, we show the neutral curves of A-modes ($n = 0$) in the $(\varepsilon, \lambda_0)$-plane for various bubble sizes $\ell_b$, and the variation of $\omega_*$ of the corresponding neutral modes with $\varepsilon$, respectively.

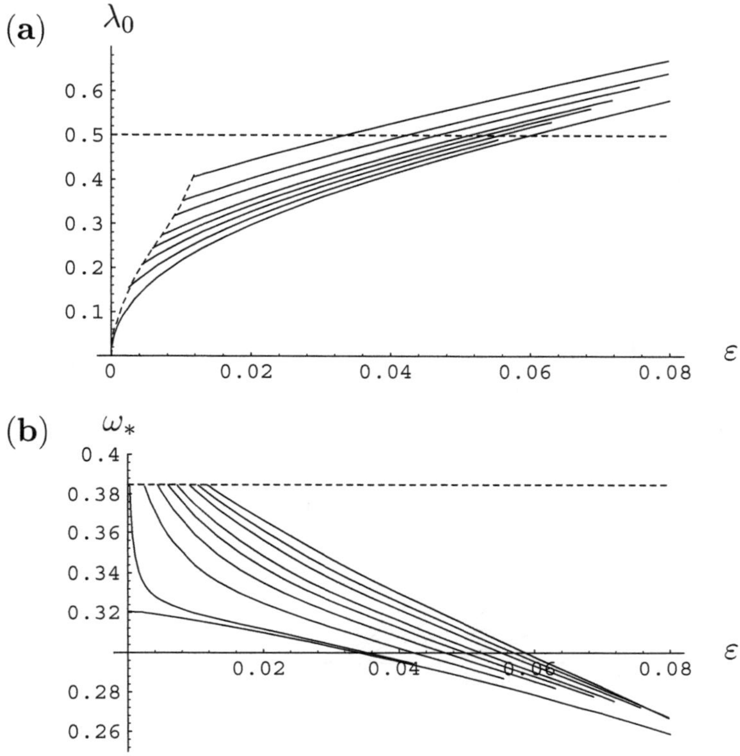


**Fig. 9.16** For a variety of values of the parameter $\ell_b$ (from top to bottom, $\ell_b = 0.12, 0.08, 0.06, 0.04, 0.03, 0.02, 0.01, 0.0$): (**a**) the variation of $\lambda_0$ of the antisymmetric neutral mode ($n = 0$) with $\varepsilon$; (**b**) the variation of $\omega_*$ of the antisymmetric neutral mode ($n = 0$) with $\varepsilon$

In Fig. 9.17, we show the neutral curves of S-modes with ($n = 0$) in the parameter $(\varepsilon, \lambda_0)$-plane for various of bubble sizes $\ell_b$, and the variations of $\omega_*$ of the corresponding neutral modes with $\varepsilon$, respectively.

We found that there exists a "prohibition zone" in the parameter plane where a single-valued oscillatory neutral mode is no longer permitted. The boundary of this zone is marked by the dashed lines in Figs. 9.16 and 9.17.

#### 9.8.2.2 The LF Mechanism

The branch of real eigenvalues $|\sigma_0| \ll 1$ is determined by the quantization condition

$$\mathrm{Re}\,\{\chi\} = \left(2n + \tfrac{1}{2} + \chi_0\right)\pi\,, \quad \left(\tfrac{1}{2} \le \lambda_0 < 1\right) \tag{9.193}$$

$$(n = 0, \pm 1, \pm 2, \ldots)\,, \tag{9.194}$$

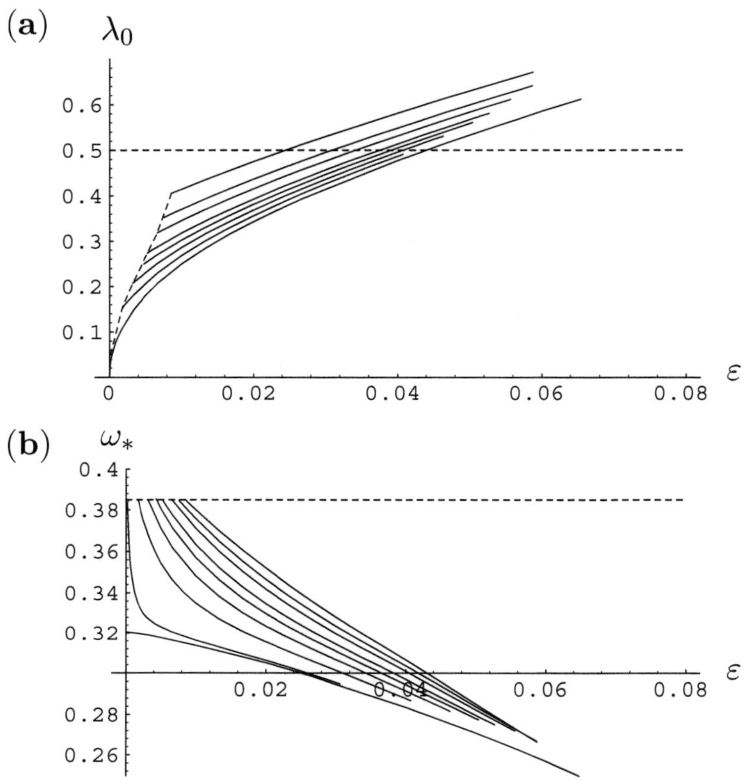


**Fig. 9.17** For a variety of values of the parameter $\ell_b$ (from top to bottom, $\ell_b = 0.12, 0.08, 0.06, 0.04, 0.03, 0.02, 0.01, 0.0$): (**a**) the variation of $\lambda_0$ of the symmetric neutral mode ($n = 0$) with $\varepsilon$; (**b**) the variation of $\omega_*$ of the symmetric neutral mode ($n = 0$) with $\varepsilon$

where

$$\begin{cases} \chi_0 = 0, \text{ or } 1 & \text{for S-modes,} \\ \cos(\chi_0 \pi) = -2\cos(\frac{2\pi}{7})e^{\mathrm{Im}\{\chi\}} & \text{for A-modes,} \end{cases} \tag{9.195}$$

and

$$\chi = \frac{1}{\varepsilon} \int_{\rho_b}^{\varrho_c} \left( \tilde{k}_0^{(1)} - \tilde{k}_0^{(3)} \right) d\rho \,. \tag{9.196}$$

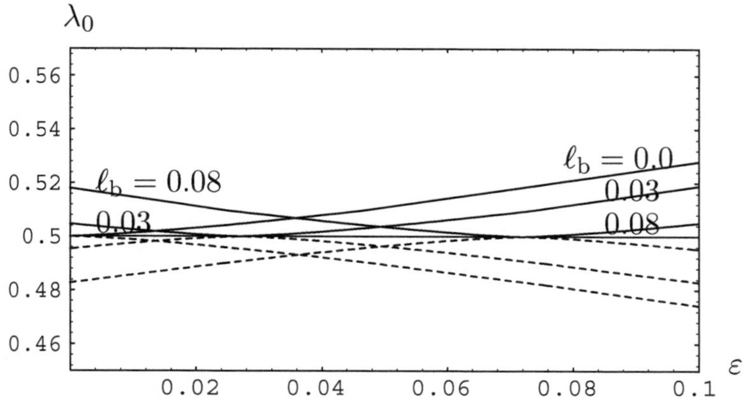


**Fig. 9.18** The neutral curves of the LF mode ($n = 0$) for the systems with a nose bubble: $\ell_b = 0.0, 0.03, 0.08$

Since $|\sigma_0| \ll 1$, the above quantization condition can be further simplified. For S-modes, it reduces to

$$\frac{3}{2}\lambda_0\sigma_0 = \frac{(2\lambda_0 - 1)^{\frac{3}{4}}}{\lambda_0^{\frac{1}{2}}} + \frac{\ell_b(1 - \frac{2}{3}\lambda_0^{\frac{3}{2}}\sigma_0)}{\lambda_0^{\frac{1}{2}}} - \varepsilon\left(n + \frac{1}{2}\right)\pi \qquad (9.197)$$

$$\left(\frac{1}{2} \le \lambda_0 < 1\right).$$

For fixed $\varepsilon$, we obtain a discrete set of neutral modes with width parameters $\lambda_0 = \lambda_{0n}$ determined by the formula

$$\frac{(2\lambda_{0n} - 1)^{\frac{3}{4}} + \ell_b}{\sqrt{\lambda_{0n}}} = \varepsilon\left(n + \frac{1}{2}\right)\pi\,, \quad \left(\frac{1}{2} \le \lambda_{0n} < 1\right). \qquad (9.198)$$

These neutral modes are steady smooth finger solutions with a tiny nose bubble. For a given operation condition with $\varepsilon$ fixed, the width of the steady smooth finger $\lambda_0$ is reduced, due to the effect of the nose bubble. The neutral curves of the S-modes ($n = 0$) for various bubble sizes are shown in Fig. 9.18. In this figure, the dashed lines should be disregarded, since they describe the branches with $\lambda_0 < 0.5$.

## 9.9 The Selection Criteria of Finger Solutions

The selection criteria for finger solutions can be derived based on our understanding of instability mechanisms obtained in the preceding sections. For this purpose, we can utilize the stability diagram in Fig. 9.13.

Setting $\lambda_0$ as the effective width of the finger, we can examine the basic state solution of the system in the $(\lambda_0, \varepsilon)$-plane. Noting that if the system has a steady limit solution as $t \to \infty$, that this fixed-point solution must be the classical steady finger solution with infinitely long stem, and that it must be an LF neutral mode, one can make the following statements:

1. If the finger system exhibits a steady pattern as $t \to \infty$, then the limit solution must be on the neutral curve $\{\mathcal{C}_0\}$, and it occurs only when $\varepsilon \geq \varepsilon_c$. In other words, for a small surface tension when $\varepsilon < \varepsilon_c$, the steady needle solution can never persist, due to the existence of a number of growing oscillatory GTW modes. Therefore, the conclusion drawn by Tanveer and others that the steady finger solution $(n = 0)$ is linearly stable for the entire range $0 < \varepsilon \ll 1$ is incorrect, and the selection criterion given by the MSC theory is not applicable in the range $0 \leq \varepsilon < \varepsilon_c$.
2. If the finger system exhibits a time-periodic oscillatory pattern as $t \to \infty$, then the limit solution, in the scope of a linear theory, must correspond to a GTW neutral mode on the neutral curve $\{\gamma_0\}$, and it occurs only when $0 \leq \varepsilon \leq \varepsilon_c$. As we have explained for the case of dendritic growth, the unsteady oscillatory pattern determined by the GTW neutral mode is self-sustaining. It does not need a continuously acting noise for its persistence. This statement also implies that for large surface tension, when $\varepsilon > \varepsilon_c$, no self-sustaining oscillatory finger is possible.
3. If the finger system exhibits a chaotic pattern as $t \to \infty$, it must be in the unstable region.

The above statements appear to be in good agreement with both experimental observations and numerical simulations of the initial value problem. Experimental observations show that the realistic finger system approaches a steady limit within a certain range of $\varepsilon$, as $t \to \infty$. Therefore, from the above statement (1), one may further conclude that the neutral modes of the LF mechanism will in fact be selected for $\varepsilon > \varepsilon_c$. This conclusion has been verified by many experimental observations, as well as in numerical simulations.

In what follows, we attempt to directly examine the theoretical statement (2) in terms of the available experimental evidence. Although the time-periodic oscillatory finger is hard to observe in experiments with a "pure" system, it is easily observed experimentally for systems with a nose bubble. Once a system has time-periodic limit solutions, according to statement (2), such a time-periodic oscillatory finger solution must correspond to the GTW neutral mode, and it will occur only in the range $0 < \varepsilon \leq \varepsilon_c$. Thus, one may use experimental data to check whether this conclusion is correct.

The dependence of the GTW mechanism on the nose bubble size has been derived in the preceding section. In Fig. 9.19, we compare the theory with the experimental data of Couder et al. [5].

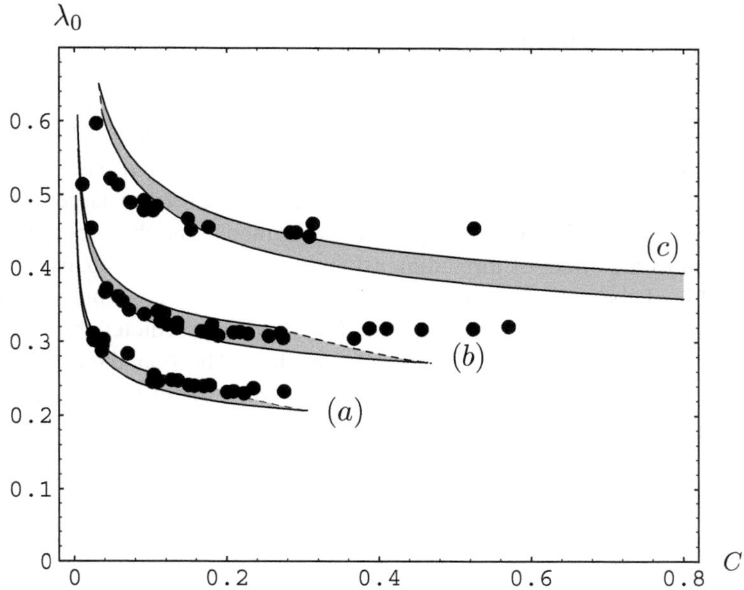


**Fig. 9.19** The theoretical predictions compared with experimental data of Couder et al. [5] for the cases (a) $W = 12\,\text{cm}$, $r_b = 0.24$–$0.30\,\text{cm}$, $\ell_b = 0.02$–$0.025$; (b) $W = 6\,\text{cm}$, $r_b = 0.24$–$0.30\,\text{cm}$, $\ell_b = 0.04$–$0.05$; (b) $W = 2\,\text{cm}$, $r_b = 0.12$–$0.18\,\text{cm}$, $\ell_b = 0.06$–$0.09$. The *shaded regions* show the neutrally stable states predicted by the theory, while the *dots* are the experimental data

In the experiments by Couder et al. [5], the average width of the fingers, $\lambda$, was measured. Three sets of experimental data provided by Couder et al. correspond to three different cell widths ($W = 12\,\text{cm}, 6\,\text{cm}$, and $2\,\text{cm}$). The thickness of the cell was $b = 0.1\,\text{cm}$. The bubble size for each run was not precisely recorded, but the authors estimated that the range of bubble diameters was (0.1–0.8 cm). Couder et al. used a stability parameter $C$ different from the $\varepsilon$ that we used. The relationship between the two is

$$C = \sqrt{\frac{\lambda_0}{\varepsilon}}\,\bar{b}\,, \tag{9.199}$$

where $\bar{b}$ is the dimensionless thickness parameter of the cell defined by $\bar{b} = b/W$, and $b$ is the thickness of the cell. The dimensional effective bubble size is $R_b = \ell_b W$. Thus for a given dimensional bubble size and cell width $W$, we can determine the bubble size parameter $\ell_b$ and plot the neutral curve in the same $(\lambda_0, C)$-plane as shown in Fig. 9.13. Thus, we can compare the theoretical results with the experimental data of Couder et al.

It is seen that the overall agreement between the two is surprisingly good. Specifically, the theory predicts that in case (a) ($W = 12\,\text{cm}$), the bubble size in their experiments is in the range $R_b = (0.24$–$0.30\,\text{cm})$, or $\ell_b = (0.020$–$0.025)$. The experimental data for this case are fully covered by the neutral curves with the values of $\ell_b$ within the above range. In case (b), the bubble size parameter is in the range $\ell_b = (0.04–0.05)$. The quantitative agreement between the selected width $\lambda$ predicted by our calculations and the measured data is still very good, but a small discrepancy is evident. For case (c) ($W = 2\,\text{cm}$), we assume that the dimensional bubble size in the experiments is in the range $R_b = (0.12–0.18\,\text{cm})$, so that ($\ell_b = 0.06–0.09$). The selected $\lambda$ predicted by our theory is in good qualitative agreement with the experimental data, but a larger numerical discrepancy appears. The discrepancy between the theory and the experiments may be attributed to the effect of three-dimensionality, since the discrepancy increases as the width of the cell decreases. Moreover, for a given nose bubble size, the existence of the prohibition zone in the $(\lambda_0, \varepsilon)$-plane predicted by the theory is also very well verified by the experimental data. The boundary of this zone is shown by the dashed lines in Fig. 9.19. It is very interesting to see that no experimental data for the values of $C$ for case (a) were reported beyond the dashed line. This implies that inside the prohibition zone, oscillatory fingers are indeed not observable.

## *Appendix: The Forms of Some Operators in the System of the Curvilinear Coordinate System* $(\xi, \eta)$

For the curvilinear coordinate system $(\xi, \eta)$, we have

$$\mathbf{r} = X(\xi, \eta)\mathbf{i} + Y(\xi, \eta)\mathbf{j}, \tag{9.200}$$

where $\mathbf{i}$ and $\mathbf{j}$ are the unit vectors along the $x$-axis and $y$-axis respectively. Let the vectors $\mathbf{e}_1, \mathbf{e}_2$ be the unit vectors along the $\xi, \eta$ directions respectively. One can write

$$\begin{aligned} \mathbf{e}_1 &= \frac{\partial \mathbf{r}}{\partial \xi} \Big/ \left| \frac{\partial \mathbf{r}}{\partial \xi} \right| = \frac{1}{\mathcal{G}} \left( X_\xi \mathbf{i} + Y_\xi \mathbf{j} \right), \\ \mathbf{e}_2 &= \frac{\partial \mathbf{r}}{\partial \eta} \Big/ \left| \frac{\partial \mathbf{r}}{\partial \eta} \right| = \frac{1}{\mathcal{G}} \left( X_\eta \mathbf{i} + Y_\eta \mathbf{j} \right). \end{aligned} \tag{9.201}$$

To derive the mathematical formulation of the problem in the coordinate system $\{\xi, \eta\}$, we denote the field function by $\Phi = \Phi(\xi, \eta)$ and the interface shape by $\eta = \eta_s(\xi, t)$. We have the formulas

$$\begin{aligned} \nabla \Phi &= \frac{\mathbf{e}_1}{\mathcal{G}} \frac{\partial \Phi}{\partial \xi} + \frac{\mathbf{e}_2}{\mathcal{G}} \frac{\partial \Phi}{\partial \eta}, \\ (\mathbf{j} \cdot \nabla)\Phi &= \frac{1}{\mathcal{G}^2} \left( Y_\xi \frac{\partial \Phi}{\partial \xi} + X_\xi \frac{\partial \Phi}{\partial \eta} \right), \end{aligned} \tag{9.202}$$

as well as

$$\nabla^2\Phi = \nabla\cdot(\nabla\Phi) = \frac{1}{\mathcal{G}^2}\left(\frac{\partial^2\Phi}{\partial\xi^2} + \frac{\partial^2\Phi}{\partial\eta^2}\right) \tag{9.203}$$

and

$$\frac{\partial\Phi}{\partial n} = \mathbf{n}\cdot\nabla\Phi = \frac{1}{\left(1+\eta_s'^2\right)^{1/2}\mathcal{G}}\left(\frac{\partial\Phi}{\partial\eta} - \eta_s'\frac{\partial\Phi}{\partial\xi}\right). \tag{9.204}$$

In the moving frame, the local growth speed at the interface is

$$u_I = -\frac{\frac{\partial s}{\partial t}}{|\nabla s|} = \mathcal{G}\frac{\frac{\partial\eta_s}{\partial t}}{\left(1+\eta_s'^2\right)^{1/2}}, \tag{9.205}$$

$$\mathbf{j}\cdot\mathbf{n} = \frac{Y_\eta - Y_\xi\eta_s'}{\mathcal{G}\left(1+\eta_s'^2\right)^{1/2}}, \tag{9.206}$$

$$\frac{\partial\Phi}{\partial n} = \mathbf{n}\cdot\nabla\Phi = \frac{1}{\left(1+\eta_s'^2\right)^{1/2}\mathcal{G}}\left(\frac{\partial\Phi}{\partial\eta} - \eta_s'\frac{\partial\Phi}{\partial\xi}\right). \tag{9.207}$$

The curvature operator can be derived as

$$\begin{aligned}\mathcal{K}\Big\{\eta_s(\xi,t)\Big\} = -\frac{1}{\mathcal{G}(\xi,\eta_s)}\Bigg\{&\frac{\eta_s''}{\left(1+\eta_s'^2\right)^{3/2}} + \frac{\Pi_0(\xi,\eta_s)}{\mathcal{G}^2(\xi,\eta_s)\left(1+\eta_s'^2\right)^{1/2}}\\ &+\frac{\Pi_1(\xi,\eta_s)}{\mathcal{G}^2(\xi,\eta_s)\left(1+\eta_s'^2\right)^{1/2}}\eta_s'\Bigg\},\end{aligned} \tag{9.208}$$

where

$$\begin{aligned}\Pi_0(\xi,\eta) &= \left(Y_{\xi\xi}X_\xi - X_{\xi\xi}Y_\xi\right),\\ \Pi_1(\xi,\eta) &= \left(X_{\xi\xi}X_\xi + Y_{\xi\xi}Y_\xi\right).\end{aligned} \tag{9.209}$$

Assume that as $\epsilon \to 0$,

$$\eta_B(\xi) = \epsilon h_1(\xi) + \cdots.$$

We derive that

$$\mathcal{K}\{\eta_B(\xi)\} = \mathcal{K}_0(\xi) + \epsilon\mathcal{K}_1(\xi) + \cdots, \tag{9.210}$$

where

$$\begin{cases} \mathcal{K}_0(\xi) = -\dfrac{\Pi_{0,0}}{\mathcal{G}_0^3}, \\ \mathcal{K}_1(\xi) = \dfrac{2\pi^2\lambda_0(1-\lambda_0)}{\mathcal{G}_0^3\mathcal{R}_0}\left[1 - \dfrac{2(1-\lambda_0)}{\lambda_0\mathcal{R}_0} + \dfrac{3\Pi_{0,0}}{\pi\mathcal{G}_0^2}\right]h_1(\xi) \\ \qquad - \dfrac{2\pi(1-\lambda_0)^2\tan(\frac{\pi\xi}{2})}{\mathcal{R}_0\mathcal{G}_0^3}h_1'(\xi) - \dfrac{1}{\mathcal{G}_0}h_1''(\xi). \end{cases} \tag{9.211}$$

At the root point T, $\xi = \pm 1$, $\eta = \eta_T(t)$: since $Y_\xi(\pm 1, \eta_s) = -X_\eta(\pm 1, \eta_s) = 0$, the slope of the interface shape $y_s$ can be calculated as

$$\frac{dy_s}{dx}\bigg|_{x=\pm W} = \frac{Y_\xi + Y_\eta \eta_s'}{X_\xi + X_\eta \eta_s'}\bigg|_{X=\pm 1} = \eta_s'(\pm 1). \tag{9.212}$$

## References


1. E. Ben-Jacob, R. Godbey, N.D. Goldenfield, J. Koplik, H. Levine, T. Mueller, L.M. Sander, Experimental demonstration of the role of anisotropy in interfacial pattern formation. Phys. Rev. Lett. **55**, 1315–1318 (1985)
2. D. Bensimon, Stability of viscous fingering. Phys. Rev. A **33**, 1302–1308 (1986)
3. D. Bensimon, P. Pelce, B.I. Shraiman, Dynamics of curved fronts and pattern selection. J. Phys. **48**, 2081–2087 (1987)
4. R.L. Chouke, P. Van Meurs, C. Van der Pol, The instability of slow immiscible viscous liquid-liquid displacements in permeable media. Trans. AIME **216**, 188–194 (1959)
5. Y. Couder, N. Gerard, M. Rabaud, Narrow fingers in the Saffman–Taylor instability. Phys. Rev. A **34**, 5175–5178 (1986)
6. A.J. DeGregoria, L.W. Schwartz, A boundary integral method for two-phase displacement in Hele–Shaw cell. J. Fluid Mech. **164**, 383–400 (1986)
7. S. Hill, Channeling in packed columns. Chem. Eng. Sci. **1**, 247–253 (1952)
8. G.M. Homsy, Viscous fingering in porous media. Annu. Rev. Fluid Mech. **19**, 271–311 (1987)
9. D.C. Hong, J.S. Langer, Analytic theory of the selection mechanism in the Saffman–Taylor problem. Phys. Rev. Lett. **56**, 2032–2035 (1986)
10. S.D. Howison, A.A. Lacey, J.R. Ockendon, Hele–Shaw free boundary problems with suction. Q. J. Mech. Appl. Math. **41**, 183–193 (1988)
11. D.A. Kessler, H. Levine, Stability of finger patterns in Hele–Shaw cells. Phys. Rev. A **32**, 1930–1933 (1985)
12. D.A. Kessler, H. Levine, Theory of the Saffman–Taylor finger pattern. Phys. Rev. A **33**, 2621–2633 (1986)
13. A.R. Kopf-Sill, G.M. Homsy, Narrow fingers in a Hele–Shaw cell. Phys. Fluids **30**(9), 2607–2609 (1987)
14. M. Matsushita, H. Yamada, Dendritic growth of single viscous finger under the influence of linear anisotropy. J. Cryst. Growth **99**, 161–165 (1990)
15. J. Mclean, P.G. Saffman, The effect of surface tension on the shape of fingers in a Hele–Shaw cell. J. Fluid Mech. **102**, 455–469 (1981)
16. E. Meiburg, G.M. Homsy, Nonlinear unstable viscous fingers in Hele–Shaw cell: numerical simulation. Phys. Fluids **31**(3), 429–439 (1988)

17. L.A. Romero, PhD. Thesis, California Institute of Technology (1981)
18. P.G. Saffman, Exact solution for the growth of fingers from a flat interface between two fluids. Q. J. Mech. Appl. Math. **12**, 146–150 (1959)
19. P.G. Saffman, G.I. Taylor, The penetration of a fluid into a porous medium or Hele–Shaw cell containing a more viscous liquid. Proc. R. Soc. Lond. Ser. A **245**, 312–329 (1958)
20. H. Segur, S. Tanveer, H. Levine (eds.), *Asymptotics Beyond All Orders*. NATO ASI Series, Series B: Physics, vol. 284 (Plenum, New York, 1991)
21. P. Tabling, G. Zocchi, A. Libchaber, An experimental study of the Saffman–Taylor instability. J. Fluid Mech. **177**, 67–82 (1987)
22. S. Tanveer, Analytic theory for the selection of a symmetric Saffman–Taylor instability. Phys. Fluids **30**(8), 1589–1605 (1987)
23. J.-M. Vanden-Broeck, Fingers in a Hele–Shaw cell with surface tension. Phys. Fluids **26**(8), 2033–2034 (1983)
24. J.J. Xu, Global instability of viscous fingering in a Hele–Shaw cell: formation of oscillatory fingers. Eur. J. Appl. Math. **2**, 105–132 (1991)
25. J.J. Xu, Interfacial wave theory for oscillatory finger's formation in a Hele–Shaw cell: a comparison with experiments. Eur. J. Appl. Math. **7**, 169–199 (1996)
26. J.J. Xu, Interfacial instabilities and fingering formation in Hele–Shaw flow. IMA J. Appl. Math. **57**, 101–135 (1996)
27. J. Yokoyama, Y. Kitagawa, H. Yamada M. Matsushita, Dendritic growth of viscous fingers under linear anisotropy. Phys. A **204**, 789–799 (1994)
28. P.A. Zhuravlev, On the motion of a fluid in channels. Zap. Leningr. Gorn. In-ta. **33**(3), 54–61 (1956)

# Chapter 10
# Spatially Periodic Deep-Cellular Growth

## 10.1 Introduction

In this chapter, we shall investigate interfacial spatially periodic pattern formation in directional solidification. The most prominent phenomenon of this kind is the so-called deep-cellular growth from a binary mixture with low concentration of the minor species. [1–4, 8, 18, 21, 22, 25–30, 33–42].

The typical experimental device used for investigating this kind of phenomenon is the Hele–Shaw cell sketched in Fig. 10.1, similar to that utilized in the previous chapter. The system consists of a thin sample material and two uniform temperature zones: a hot zone with a temperature $T_{\mathrm{H}}$ higher than the melting temperature $T_{\mathrm{M0}}$ of a flat interface, and a cold zone with a temperature $T_{\mathrm{C}}$ lower than the melting temperature. The sample is pulled at a constant speed $V$ along the direction from the hot zone to cold zone.

Experimental results show that when the pulling speed $V$ is sufficiently small, the interface will be flat, located somewhere between the two zones. When the pulling speed increases beyond a critical number $V_{\mathrm{c}}$, the flat interface becomes unstable due to the Mullins–Sekerka instability studied in Chap. 2. The pattern formation on a planar interface near the critical number $V_{\mathrm{c}}$ has been well studied and described by the so-called weakly nonlinear theory [1, 4] and [9].

When the pulling speed increases further, the interface pattern evolves continuously and eventually forms a steady deep-cellular structure. The fingers are aligned periodically with equal primary spacing $\ell_w$ (see Fig. 10.2). The cellular array may be transformed to a dendritic array when the pulling speed is larger than another critical number.

The whole pattern-formation process and its correlation with the operating conditions are very complicated in the regime $V \gg V_{\mathrm{c}}$, because the problem becomes strongly nonlinear. One of most important issues for applications in the industries of material processing is to determine the steady states of the system at



J.-J. Xu, *Interfacial Wave Theory of Pattern Formation in Solidification*, Springer Series in Synergetics, DOI 10.1007/978-3-319-52663-8_10

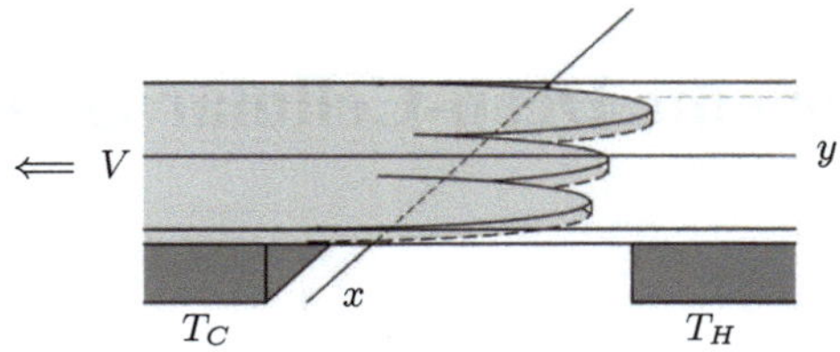


**Fig. 10.1** A sketch of directional solidification device, a Hele–Shaw cell

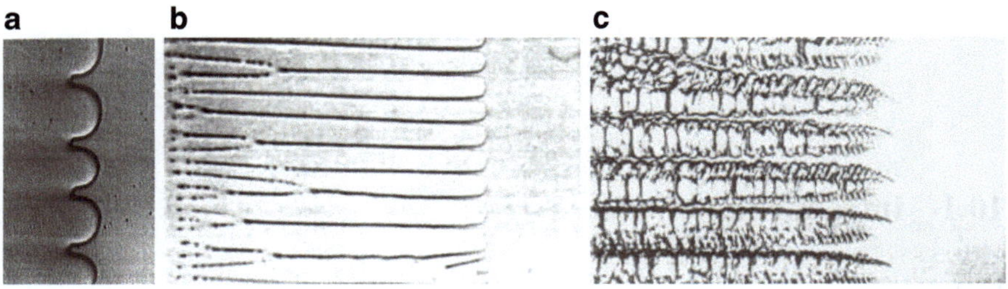


**Fig. 10.2** Typical experimental photos of cellular and dendritic array growth in directional solidification with increasing pulling velocity $V$: (**a**) shallow cellular, (**b**) deep-cellular array, (**c**) dendritic array (refer to [1])

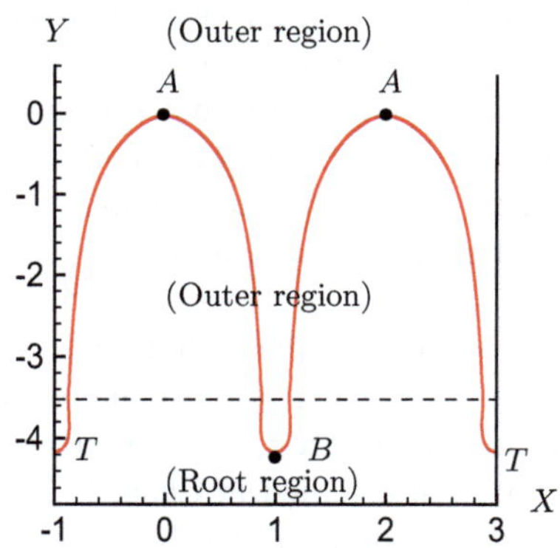


**Fig. 10.3** A sketch of the cellular-array interface. ($A$): The cell tip; ($B$): The bottom of the root

the late stage of growth. The prominent feature of the steady deep-cellular pattern is that it has a long groove with interface closure at the end, as sketched in Fig. 10.3. Hence, such topics as determining the primary spacing of the cellular structure, the underlying mechanism for the interface closure, and the location of the closed groove have elicited intense interest for decades.

The subject has been studied by a number of researchers analytically, such as Pelcé and Pumir [30], Weeks and coworkers [39, 40], Caroli et al. [4], Davis [8], Billia and Trivedi [1, 36], Karma [20, 24]. The previous analyses did not give the global solutions, since they failed to be valid in the root region. The desirable solutions should be able to determine the global interface shape of a cell as well as to provide sufficient information on the locations of the cell's tip and bottom of the root, the total length of the cell, the concentrations of the impurity at the cell's tip and bottom, and particularly the mathematical relationships between all these quantities and the operating conditions.

In addition, many numerical works (see, for instance, [6, 7, 11, 14–17, 32]) have appeared in recent years, which may show the global patterns of cellular arrays. However, the numerical simulations cannot sufficiently explore the singular properties of the system at the bottom of the root, nor can they well elucidate the underlying physical mechanisms of the phenomenon.

Global uniformly valid asymptotic solutions for steady states were derived by Xu and Chen in 2011–2015 by extending the IFW to the system of deep-arrayed cellular growth [5, 46, 47]. Xu and Chen explored the mechanism of the interface closure and found that the linear perturbed system has two spectra of eigenvalues: a real spectrum and a complex spectrum. Based on these findings, they determined that the system was subject to two global instability mechanisms: the oscillatory trapped-wave (GTW) instability mechanism and the low-frequency (LF) mechanism. In this chapter we shall systematically demonstrate these results.

To begin, let us adopt the two-dimensional model by neglecting the second principal curvature on the meniscus between the two planes of the Hele–Shaw cell. Consider the case of small Péclet number $\epsilon$ defined as the ratio of the radius of the cell's tip to mass diffusion length. The 2D model is quite practical, and the assumption of $\epsilon \ll 1$ can be well justified by experimental data $\epsilon = 0.02 - 0.14$ (refer to the experiments conducted by Somboonsuk et al. for the system succinonitrile–acetone in [36, 37]).

The first objective of the study is to obtain the global uniformly valid asymptotic solution for the steady state in the limit $\epsilon \to 0$. It will be seen that the system allows infinitely many global steady-state solutions of arrayed cellular growth with two free parameters: the Péclet number $\epsilon$ and the asymptotic width parameter $\lambda_0$ under given growth conditions and material properties. These two parameters are related to the geometry of the cellular pattern: tip radius and primary spacing. It will also be seen that the steady-state solutions have a complicated structure of three internal layers in the root region, which are subject to a quantization condition substantially affected by the surface tension. The eigenvalue determined by such a quantization condition yields the cell's full length of the corresponding global steady state.

## 10.2 Steady State of the System of Deep-Cellular Growth from a Binary Mixture

### *10.2.1 Mathematical Formulation of the Problem*

The system involves the temperature field and concentration field governed by the heat conduct equation and mass diffusion equation, respectively. The system can be simplified by the following considerations. Since the minor species in the binary mixture is dilute, one may consider it an impurity. In view of the fact that the solute diffusion in the solid phase is negligible, one may adopt the one-sided model by assuming that the mass diffusivity in the solid phase is zero. Since the cell is very thin, one may assume that the temperature distribution in the mixture is the same as prescribed along the external walls of the Hele–Shaw cell. Moreover, for the sake of simplicity, one may assume that the temperature field in the mixture in the region where the interfacial microstructure forms is linear with given gradient $(G)_{\mathrm{D}}$. As a consequence, one needs to solve only the concentration field governed by the mass diffusion equation.

We further assume that the thermodynamic properties other than the diffusivity are the same for both solid and liquid phases and that there is no convection in the system.

There are several characteristic lengths involved in the system as specified below. The solute diffusion length is defined as $\ell_{\mathrm{D}} = \kappa_{\mathrm{D}}/V$, where $\kappa_{\mathrm{D}}$ is the solute diffusivity. We use the tip radius $\ell_{\mathrm{t}}$ as the length scale and assume that $\ell_{\mathrm{t}} \ll \ell_{\mathrm{D}}$. The primary spacing of the cellular pattern is denoted by $\ell_w$. The pulling speed $V$ is used as the velocity scale, and $\ell_{\mathrm{t}}/V$ is used as the time scale. The scales of the temperature $T$ and concentration $C$ are set as $\Delta H/(c_{\mathrm{p}}\rho)$ and $C_\infty$, respectively. Here $\Delta H$ is the latent heat release per unit of volume of the solid phase, $c_{\mathrm{p}}$ is the specific heat, $\rho$ is the density of the melt, and $C_\infty$ is the impurity concentration in the far field away from the interface. The melting temperature $T_{\mathrm{M}}$ is used as the reference temperature. One may define the following dimensionless parameters:

- the Péclet number, $\epsilon = \ell_{\mathrm{t}}/\ell_{\mathrm{D}}$;
- the morphological parameter, $M = -\frac{mC_\infty}{\Delta H/(c_{\mathrm{p}}\rho)}$, where $m < 0$ is the slope of the liquidus in the phase diagram;
- the surface-tension parameter, $\Gamma = \frac{\ell_{\mathrm{c}}}{\ell_{\mathrm{t}}} = \frac{\ell_{\mathrm{c}}\ell_{\mathrm{D}}}{\ell_{\mathrm{t}}^2}\frac{\ell_{\mathrm{t}}}{\ell_{\mathrm{D}}}$, where $\ell_{\mathrm{c}}$ is the capillary length defined as $\ell_{\mathrm{c}} = \gamma c_{\mathrm{p}}\rho T_{\mathrm{M0}}/(\Delta H)^2$, and $\gamma$ is the surface-tension coefficient;
- the dimensionless gradient of the temperature, $G = \frac{\ell_{\mathrm{D}}}{\Delta H/(c_{\mathrm{p}}\rho)}(G)_{\mathrm{D}}$;
- the ratio of two length scales, $\lambda_{\mathrm{G}} = \ell_{\mathrm{D}}/\ell_{\mathrm{G}} = G/M$, where $\ell_{\mathrm{G}} = -mC_\infty/(G)_{\mathrm{D}}$;
- the primary spacing parameter, $W = \ell_w/\ell_{\mathrm{t}}$.

It should be noted that the surface tension parameter $\Gamma$ defined above for the system of directional solidification is different from the surface tension parameter $\Gamma$ defined for the systems of free dendritic growth system in Chaps. 6 and 8, because the length scale here is chosen differently. In most practical cases of directional solidification, experimental data has shown that the surface-tension parameter $\Gamma$ defined here is very small, so that one may assume that $\Gamma = \mathcal{O}(\epsilon^2)$ as $\epsilon \to 0$. Accordingly, one may set $\Gamma = \epsilon^2\hat{\Gamma}$, where $\hat{\Gamma} = \frac{1}{\epsilon}\big(\ell_{\mathrm{c}}\ell_{\mathrm{D}}/\ell_{\mathrm{t}}^2\big) = \mathcal{O}(1)$. This implies that the parameter $\big(\ell_{\mathrm{c}}\ell_{\mathrm{D}}/\ell_{\mathrm{t}}^2\big)$ is small and has the same order of magnitude as the Péclet number $\epsilon$. We define $\varepsilon = \sqrt{\Gamma}$ as the surface tension stability parameter. The relationship between the two parameters $\varepsilon$ and $\epsilon$ is as follows

$$\varepsilon = \epsilon\sqrt{\hat{\Gamma}}; \quad \epsilon = \varepsilon\hat{\mathrm{Pe}}, \tag{10.1}$$

where $\hat{\mathrm{Pe}} = 1/\sqrt{\hat{\Gamma}} = \mathcal{O}(1)$. Since $\varepsilon$ and $\epsilon$ have the same order of magnitude, one can use either of them as the basic small parameter in the asymptotic analysis. In what follows, let us use $\epsilon$ as the basic small parameter.

Due to the periodicity of the solution, one may consider only a single cell. For this case, the problem is equivalent to finger-like crystal growth in a channel with fixed side walls at $x = \pm W$. For the case of arrayed cellular growth in a Hele–Shaw cell with infinitely large width, of course, $W$ is a free parameter describing half of the primary spacing.

### 10.2.2 *Mathematical Formulation of the Problem in a Curvilinear Coordinate System* $(\xi, \eta)$

Before proceeding further, we nondimensionalize all the physical quantities with their scales defined above. We shall use the same notation to describe all dimensionless quantities as that used for dimensional quantities without risk of confusion.

Assume that the origin of rectangular coordinates $(x, y)$ is located at the cell tip (see Fig. 10.3); the temperature distribution in the growth zone is given by

$$T = \epsilon G(y - y_0) + \mathcal{O}(\epsilon),$$

where $y_0$ is a unknown number to be determined. Note that at $y = y_0$, we have $T = 0$, which corresponds to the melting temperature. Since the location of the melting temperature in the device can be measured, the solution for $y_0$ yields the location of the origin of the system or the tip of cell in the device; it may also yield the temperature at the tip.

The other unknowns of the problem are the concentration field $C$ and the interface shape. The concentration field is governed by the solute diffusion equation and coupled with the known temperature field by the Gibbs–Thomson condition imposed at the interface.

In the limiting case of $\epsilon = 0$, the concentration field under study is governed by the Laplace equation, which has an intrinsic similarity to the potential field in the system of viscous fingering studied in the previous chapter. As a consequence, one may formulate the problem by adopting the curvilinear coordinate system $(\xi, \eta)$ associated with the Saffman–Taylor (ST) solution. Similarly to what we did in the previous chapter, we define $\xi = -\Psi, \eta = \Phi, \widehat{\zeta} = \xi + i\eta$, where $\Psi(X, Y)$ and $\Phi(X, Y)$ are the stream and potential functions of Hele–Shaw flow, respectively. The Saffman–Taylor solution derived is

$$Z = X + iY = Z(\widehat{\zeta}) = \lambda_0 \widehat{\zeta} + i\frac{2(1-\lambda_0)}{\pi} \ln\cos\left(\frac{\pi\widehat{\zeta}}{2}\right), \tag{10.2}$$

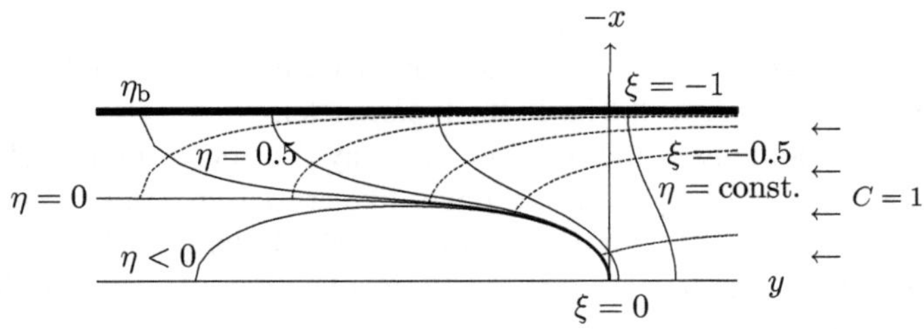


**Fig. 10.4** Sketch of the orthogonal curvilinear coordinate system $(\xi, \eta)$ based on the ST zero surface tension steady-state solutions

and

$$
\begin{aligned}
X(\xi, \eta) &= \xi - \tfrac{2(1-\lambda_0)}{\pi}\Theta, \\
Y(\xi, \eta) &= (2\lambda_0 - 1)\eta + \tfrac{(1-\lambda_0)}{\pi}\ln\left(\tfrac{\mathcal{R}}{4}\right),
\end{aligned}
\tag{10.3}
$$

where $\lambda_0$ is the asymptotic width parameter relative to the primary spacing parameter $W$, and

$$
\begin{aligned}
\Theta &= \tan^{-1}\Big[\sin\pi\xi/\big(\cos\pi\xi + e^{\pi\eta}\big)\Big], \\
\mathcal{R} &= \left(e^{\pi\eta} - 1\right)^2 + 2e^{\pi\eta}\left(1 + \cos\pi\xi\right).
\end{aligned}
$$

The variables $\{\xi = \xi(X, Y); \eta = \eta(X, Y)\}$ constitute a new orthogonal curvilinear coordinate system in the $(X, Y)$-plane as shown in Fig. 10.4, where $x = WX(\xi, \eta), y = WY(\xi, \eta)$. In the curvilinear coordinate system $(\xi, \eta)$, the linear distribution of the temperature field is described by $T_B = \epsilon G\big[WY(\xi, \eta) - y_0\big]$. The problem of directional solidification of a system of a binary mixture was generally formulated in Chap. 1. The mathematical formulation of the cellular growth in terms of the system of coordinates $(\xi, \eta)$ are given in Appendix 1.

### *10.2.3 The Basic Steady-State Solutions in the Far Field*

We begin by studying the basic state solution of cellular growth, denoted by $\{C_B(\xi, \eta), \eta_B(\xi)\}$. The basic steady-state solution is subject to Eq. (10.265):

$$
\frac{\partial^2 C_B}{\partial \xi^2} + \frac{\partial^2 C_B}{\partial \eta^2} + \epsilon W\left(Y_\xi \frac{\partial C_B}{\partial \xi} + X_\xi \frac{\partial C_B}{\partial \eta}\right) = 0, \tag{10.4}
$$

and it satisfies the boundary condition

$$C_B \to 1 \tag{10.5}$$

as $\eta \to \infty$. In the far field away from the tip, since $\eta \gg 1$, the microstructure at the interface on the concentration field is negligible. Hence, one may deduce that the concentration field is independent of $\xi$, and Eq. (10.4) reduces to

$$\frac{\partial^2 C_B}{\partial \eta^2} + \epsilon W X_\xi \frac{\partial C_B}{\partial \eta} = 0. \tag{10.6}$$

Furthermore, noting that as $\eta \to \infty$, $X_\xi \to 1$ exponentially, (10.6) reduces to the one-dimensional diffusion equation:

$$\frac{\partial^2 C_B}{\partial \eta^2} + \epsilon W \frac{\partial C_B}{\partial \eta} = 0. \tag{10.7}$$

The solution of (10.7) with the condition (10.5) is

$$C_B \sim 1 + Q_0(\epsilon) e^{-W\epsilon\eta}, \tag{10.8}$$

where $Q_0(\epsilon)$ is an arbitrary constant.

### 10.2.4 The Mathematical Formulation of the Problem in the Near Field

In the near field of the interface, $\eta = \mathcal{O}(1)$, the steady-state solution is subject to the system

$$\frac{\partial^2 C_B}{\partial \xi^2} + \frac{\partial^2 C_B}{\partial \eta^2} + \epsilon W \left( Y_\xi \frac{\partial C_B}{\partial \xi} + X_\xi \frac{\partial C_B}{\partial \eta} \right) = 0\,, \tag{10.9}$$

with the following boundary conditions:

1. In the up-stream far field: Away from the cell tips, the solution in the near field should match the solution in the far field. Hence, it is posited that as $\eta \to \infty$,

$$C_B \sim 1 + Q_0(\epsilon) e^{-\epsilon W \eta}. \tag{10.10}$$

2. At the side walls $\xi = \pm 1$,

$$\frac{\partial C_B}{\partial \xi} = 0. \tag{10.11}$$

3. At the interface $\eta = \eta_{\mathrm{B}}(\xi, \epsilon)$,

$$C_{\mathrm{B}} = y_* - \epsilon\lambda_{\mathrm{G}} WY(\xi, \eta) - \frac{\epsilon^2 \hat{\Gamma}}{MW}\mathcal{K}\{\eta_{\mathrm{B}}(\xi, \epsilon)\}, \tag{10.12}$$

$$\frac{\partial C_{\mathrm{B}}}{\partial \eta} - \eta_{\mathrm{B}}' \frac{\partial C_{\mathrm{B}}}{\partial \xi} - \epsilon W(1-\kappa) C_{\mathrm{B}}\left(Y_\xi \eta_{\mathrm{B}}' - Y_\eta\right) = 0, \tag{10.13}$$

where $\mathcal{K}\{\eta_{\mathrm{B}}(\xi, \epsilon)\}$ is the twice mean curvature operator, and as before, let us designate that $\mathcal{K} > 0$ *when an interfacial finger points to the liquid phase side*, and $\kappa$ is the segregation coefficient. We define $y_* = \epsilon\lambda_{\mathrm{G}} y_0$, and assume that $y_* = y_{*0} + \epsilon y_{*1} + \cdots$, as $\epsilon \to 0$.
4. At the cell tip, $\xi = \eta = 0$:

$$\frac{\partial \eta_{\mathrm{B}}}{\partial \xi}(0) = \eta_{\mathrm{B}}(0) = 0; \tag{10.14}$$

5. At the bottom of the root, $\xi = \pm 1$, $\eta = \eta_{\mathrm{b}}$:

$$\eta_{\mathrm{B}}(\pm 1) = \eta_{\mathrm{b}}; \ \frac{\partial \eta_{\mathrm{B}}}{\partial \xi}(\pm 1) = 0. \tag{10.15}$$

### *10.2.5 Generalized Asymptotic Solution in the Outer Region*

Now turn to solving the system (10.9)–(10.15) in the whole physical region $|\xi| \leq 1, |\eta| < \infty$ and first look for the outer solution. Assume that $|\eta_{\mathrm{B}}| = \mathcal{O}(\epsilon)$. We take the Taylor expansion for the interface conditions (10.12) and (10.13) around the interface $\eta = 0$ as follows:

$$\begin{aligned} C_{\mathrm{B}}(\xi, 0) + \eta_{\mathrm{B}} \frac{\partial C_{\mathrm{B}}}{\partial \eta}(\xi, 0) + \cdots &= y_* - \epsilon W \lambda_{\mathrm{G}}\big[Y_0(\xi) + Y_{\eta,0}(\xi)\eta_{\mathrm{B}}\big] \\ &\quad - \frac{\epsilon^2 \hat{\Gamma}}{MW}\left[\mathcal{K}_0(\xi) - \frac{1}{\mathcal{G}_0}\frac{\mathrm{d}^2 \eta_{\mathrm{B}}}{\mathrm{d}\xi^2}\right] + (\text{h.o.t.}), \end{aligned} \tag{10.16}$$

where

$$\mathcal{K}_0(\xi) = \frac{\pi\lambda_0(1-\lambda_0)\cos(\frac{\pi\xi}{2})}{2\big[\lambda_0^2 + (1-2\lambda_0)\sin^2(\frac{\pi\xi}{2})\big]^{\frac{3}{2}}},$$

and

$$\begin{aligned}
&\frac{\partial C_{\mathrm{B}}}{\partial \eta}(\xi,0)+\eta_{\mathrm{B}}\frac{\partial^2 C_{\mathrm{B}}}{\partial \eta^2}(\xi,0)-\frac{\partial \eta_B}{\partial \xi}\frac{\partial C_{\mathrm{B}}}{\partial \xi}(\xi,0)\\
&-\epsilon W(1-\kappa)\Big[C_{\mathrm{B}}(\xi,0)+\eta_{\mathrm{B}}\frac{\partial C_{\mathrm{B}}}{\partial \eta}(\xi,0)\Big]\\
&\times\Big[Y_{\xi,0}(\xi)\frac{\partial \eta_B}{\partial \xi}-Y_{\eta,0}(\xi)-Y_{\eta\eta,0}(\xi)\eta_{\mathrm{B}}\Big]\\
&+(\text{h.o.t})=0.
\end{aligned} \tag{10.17}$$

In the above, we have applied the asymptotic formulas in the limit $\eta \to 0$:

$$\begin{cases}
\mathcal{R}(\xi)\sim\mathcal{R}_0(\xi)=2\big[1+\cos(\pi\xi)\big],\\
Y(\xi,\eta)\sim Y_0(\xi)=\frac{(1-\lambda_0)}{\pi}\ln\left[\frac{1+\cos(\pi\xi)}{2}\right]\\
Y_\xi(\xi,\eta)\sim Y_{\xi,0}(\xi)=-(1-\lambda_0)\tan\left(\frac{\pi\xi}{2}\right),\\
Y_\eta(\xi,\eta)\sim Y_{\eta,0}(\xi)=\lambda_0,\quad Y_{\eta\eta}(\xi,\eta)\sim Y_{\eta\eta,0}(\xi)=\frac{2\pi(1-\lambda_0)}{\mathcal{R}_0},\\
\mathcal{G}(\xi,\eta)=\sqrt{Y_\xi^2+Y_\eta^2}\sim\mathcal{G}_0(\xi)=\sqrt{\lambda_0^2+(1-\lambda_0)^2\tan^2\left(\frac{\pi\xi}{2}\right)}.
\end{cases} \tag{10.18}$$

We point out that the functions $Y(\xi,\eta), Y_\xi(\xi,\eta)$ in the above have a singularity at $\xi=\pm1, \eta=0$. Hence the above boundary conditions are applicable only in the outer region far from the root $\xi=\pm1, \eta=0$. To find the outer solution in the outer region away from this singular point, we shall truncate the singularity at $\xi=\pm1, \eta=0$ by modifying the functions $Y(\xi,\eta), Y_\xi(\xi,\eta)$ in the following manner. Let $\xi_*=1-\delta_*$, $(0<\delta_*\ll 1)$; let $Q$ be the representative of the functions $Y(\xi,\eta), Y_\xi(\xi,\eta)$, namely, $Q\equiv\{Y, Y_\xi\}$. Denote the near field by $(D)=(0\le|\xi|<1, \eta=O(1))$, the region near the singular points by $(D_*)=(\xi_*\le|\xi|<1, \eta=0)$. Then the outer region can be considered to be $((D)-(D_*))$. We modify $Q(\xi,\eta)$ by $\widetilde{Q}(\xi,\eta)$ defined as

$$\widetilde{Q}(\xi,\eta)=\begin{cases} Q(\xi_*,0) & \text{as} \quad (\xi,\eta)\in(D_*);\\ Q(\xi,\eta) & \text{as} \quad (\xi,\eta)\in((D)-(D_*)).\end{cases}$$

It may be assumed that if $\delta_*$ is chosen sufficiently small, then in the outer region the solutions for the slightly modified system with the functions $\widetilde{Q}(\xi,\eta)$ will be very close to the outer solution for the original system with the functions $Q(\xi,\eta)$.

Equation (10.9) plus the upstream far-field conditions, side-wall condition, tip smoothness conditions, and interface conditions (10.16)–(10.17) together give an inhomogeneous system in the outer region. The general asymptotic solution of the system in the outer region consists of the following two parts:

$$C_B(\xi,\eta,\epsilon) = (\mathrm{I}) + (\mathrm{II}) = \bar{C}_B(\xi,\eta,\varepsilon) + \tilde{C}_B(\xi,\eta,\varepsilon),$$
$$\eta_B(\xi,\epsilon) = (\mathrm{I}) + (\mathrm{II}) = \bar{\eta}_B(\xi,\epsilon) + \tilde{\eta}_B(\xi,\epsilon). \tag{10.19}$$

Part I of the solution, $\bar{C}_B(\xi,\eta,\epsilon)$, is the particular solution of an inhomogeneous system, which may be expanded in the regular perturbation expansion (RPE) form; part II of the solution, $\tilde{C}_B(\xi,\eta,\epsilon)$, is the general solution of the associated homogeneous system, which may be expanded in the singular perturbation (SPE) form. In next section, we attempt to find part I of the solution first.

### *10.2.6 Regular Perturbation Expansion of the Solution in the Outer Region*

It has been found that in the far field $\eta \gg 1$, the solution (10.8) depends on the slow far-field variable, $\tilde{\eta} = \epsilon\eta$. Thus, in the near field $\eta = \mathcal{O}(1)$, one may deduce that the concentration field can be described as functions of multiple variables $(\xi,\eta,\tilde{\eta})$. Hence, we define $\bar{C}_B(\xi,\eta,\epsilon) = C(\xi,\eta,\tilde{\eta},\epsilon)$ and solve $C(\xi,\eta,\tilde{\eta},\epsilon)$ with the MVE expansion method. The converted system with multiple variables is derived as

$$\epsilon^2\frac{\partial^2 C}{\partial\tilde{\eta}^2} + \epsilon^2 W X_\xi \frac{\partial C}{\partial\tilde{\eta}} + \frac{\partial^2 C}{\partial\xi^2} + \frac{\partial^2 C}{\partial\eta^2} + 2\epsilon\frac{\partial^2 C}{\partial\eta\partial\tilde{\eta}}$$
$$+\epsilon W\Big(Y_\xi\frac{\partial C}{\partial\xi} + X_\xi\frac{\partial C}{\partial\eta}\Big) = 0 \tag{10.20}$$

with the following boundary conditions:

1. As $\eta \to \infty$,

$$C \sim 1 + Q_0(\epsilon)\mathrm{e}^{-W\tilde{\eta}}, \tag{10.21}$$

2. Symmetric condition: at the side walls $\xi = \pm 1$,

$$\frac{\partial C}{\partial\xi} = 0, \tag{10.22}$$

3. Interface conditions: at $\eta = \tilde{\eta} = 0$,

$$\mathcal{C}(\xi,0,0) + \eta_{\mathrm{B}}\frac{\partial \mathcal{C}}{\partial \eta}(\xi,0,0) + \epsilon\eta_{\mathrm{B}}\frac{\partial \mathcal{C}}{\partial \tilde{\eta}}(\xi,0,0) + \cdots$$
$$= y_* - \epsilon W\lambda_{\mathrm{G}}\Big[Y(\xi,0) + Y_\eta(\xi,0)\eta_{\mathrm{B}}\Big]$$
$$-\frac{\epsilon^2\hat{\Gamma}}{MW}\left[\mathcal{K}_0(\xi) - \frac{1}{\mathcal{G}_0}\frac{\mathrm{d}^2\eta_{\mathrm{B}}}{\mathrm{d}\xi^2}\right] + (\text{h.o.t}), \tag{10.23}$$

$$\left[\frac{\partial \mathcal{C}}{\partial \eta} + \eta_{\mathrm{B}}\frac{\partial^2 \mathcal{C}}{\partial \eta^2} + \epsilon\eta_{\mathrm{B}}\frac{\partial^2 \mathcal{C}}{\partial \eta \partial \tilde{\eta}} - \frac{\partial \eta_B}{\partial \xi}\frac{\partial \mathcal{C}}{\partial \xi}\right] + \left[\epsilon\frac{\partial \mathcal{C}}{\partial \tilde{\eta}} + \epsilon\eta_{\mathrm{B}}\frac{\partial^2 \mathcal{C}}{\partial \tilde{\eta}\partial \eta}\right.$$
$$\left. +\epsilon^2\eta_{\mathrm{B}}\frac{\partial^2 \mathcal{C}}{\partial \tilde{\eta}^2}\right] - \epsilon W(1-\kappa)\left[\mathcal{C} + \eta_{\mathrm{B}}\frac{\partial \mathcal{C}}{\partial \eta} + \epsilon\eta_{\mathrm{B}}\frac{\partial \mathcal{C}}{\partial \tilde{\eta}}\right]$$
$$\times\Big[Y_\xi(\xi,0)\frac{\partial \eta_B}{\partial \xi} - Y_\eta(\xi,0) - Y_{\eta\eta}(\xi,0)\eta_{\mathrm{B}}\Big]$$
$$+(\text{h.o.t}) = 0. \tag{10.24}$$

We expand the solution of the above system as $\epsilon \to 0$ in the following RPE form:

$$\begin{aligned} \mathcal{C}(\xi,\eta,\tilde{\eta},\epsilon) &= \mathcal{C}_0(\xi,\eta,\tilde{\eta}) + \epsilon\mathcal{C}_1(\xi,\eta,\tilde{\eta}) + \cdots \\ \tilde{\eta}_{\mathrm{B}}(\xi,\epsilon) &= \epsilon\Big[h_1(\xi) + \epsilon h_2(\xi) + \cdots\Big]. \end{aligned} \tag{10.25}$$

By substituting (10.25) into (10.20)–(10.24), one may successively derive the approximate solutions of every order.

#### 10.2.6.1 Zeroth-Order Approximation $\mathcal{O}(\epsilon^0)$

The solution for the zeroth-order approximation is trivial. Without giving the derivations, we write the solution as

$$\mathcal{C}_0(\xi,\eta,\tilde{\eta}) = B_0(\tilde{\eta}), \tag{10.26}$$

where $B_0(\tilde{\eta})$ is subject to the interface condition $B_0(0) = y_{*0}$ and the far- field condition

$$B_0(\tilde{\eta}) \sim 1 + \hat{Q}_0 \mathrm{e}^{-W\tilde{\eta}}, \quad (\tilde{\eta} \to \infty). \tag{10.27}$$

#### 10.2.6.2 First-Order Approximation $\mathcal{O}(\epsilon^1)$

In the first-order approximation, one has the equation

$$\frac{\partial^2 \mathcal{C}_1}{\partial \xi^2} + \frac{\partial^2 \mathcal{C}_1}{\partial \eta^2} = 0, \tag{10.28}$$

and the following boundary conditions:

1. As $\eta \to \infty$,

$$\mathcal{C}_1 \sim \text{const} \times e^{-W\tilde{\eta}}, \tag{10.29}$$

2. Symmetric condition: at the side walls $\xi = \pm 1$,

$$\frac{\partial \mathcal{C}_1}{\partial \xi} = 0, \tag{10.30}$$

3. and the interface conditions, at $\eta = 0, \tilde{\eta} = 0$:

$$\mathcal{C}_1 = y_{*1} - W\lambda_G Y(\xi, 0), \tag{10.31}$$

$$\frac{\partial \mathcal{C}_1}{\partial \eta} + \frac{\partial \mathcal{C}_0}{\partial \tilde{\eta}} W(1-\kappa)\mathcal{C}_0 Y_\eta(\xi, 0) = 0. \tag{10.32}$$

The solution is derived as

$$\mathcal{C}_1(\xi, \eta, \tilde{\eta}) = -W\lambda_G Y(\xi, \eta) + B_1(\tilde{\eta}) + D_1(\tilde{\eta})\eta. \tag{10.33}$$

Noting that as $\eta \to \infty$, $Y(\xi, \eta) \sim \eta + \beta_0$, where $\beta_0 = -\frac{2(1-\lambda_0)}{\pi} \ln 2$, one obtains that

$$\mathcal{C}_1(\xi, \eta, \tilde{\eta}) \sim B_1(\tilde{\eta}) - W\lambda_G\beta_0 + \left[D_1(\tilde{\eta}) - W\lambda_G\right]\eta, \ (\eta \to \infty).$$

To ensure the uniform validity of the multiple variables expansion (MVE) solution in the limit $\eta \to \infty$, one must eliminate the secular term $\left[D_1(\tilde{\eta}) - W\lambda_G\right]\eta$ on the right-hand side of the above expression by setting

$$D_1(\tilde{\eta}) = W\lambda_G. \tag{10.34}$$

Furthermore, from the interface conditions (10.31)–(10.32), we derive that

$$B_1(0) = y_{*1}, \quad B_1(\infty) = W\lambda_G\beta_0; \tag{10.35}$$

and

$$W\lambda_G = -B_0'(0) - W\Delta_0, \tag{10.36}$$

where we have defined

$$\Delta_0 = \lambda_0\Big[(1-\kappa)y_{*0} - \lambda_G\Big]. \tag{10.37}$$

From (10.36), we derive the following relationship between $\lambda_0$ and $y_{*0}$:

$$\lambda_0 = \frac{y_{*0} - 1 - \lambda_G}{(1-\kappa)y_{*0} - \lambda_G}. \tag{10.38}$$

Moreover, from (10.36), we derive

$$\lambda_0 = \frac{y_{*0} - 1 - \lambda_G}{(1-\kappa)y_{*0} - \lambda_G}. \tag{10.39}$$

For the special isothermal case $\lambda_G = 0$, the above formula reduces to that obtained by Pelcé and Pumir [30].

Note that in the zeroth-order approximation, the concentration at the tip of the interface is given by $c_{*0} = \mathcal{C}_0(0,0,0) = B_0(0) = y_{*0}$ and $0 < \lambda_0 < 1$. From (10.39), we deduce that $(1-\kappa)c_{*0} - \lambda_G > (c_{*0} - 1) - \lambda_G > 0$ always holds. Recall the constitutional instability criterion that given the gradient of temperature $G$, when $V = V_c$, $\lambda_G = \hat{\lambda}_G = \frac{(1-\kappa)}{\kappa}$, the flat interface becomes unstable. We then deduce that as a necessary condition, $0 < \lambda_G < \hat{\lambda}_G = \frac{(1-\kappa)}{\kappa}$ must hold for the system of cellular growth. Finally, since the concentration at the tip on the solid side is $c_{s,*0} = \kappa c_{*0} \le 1$, we deduce that $c_{*0} < \frac{1}{\kappa}$. Thus, one can determine the range of the parameters $(c_{*0}, \lambda_G)$ in the $(c_{*0}, \lambda_G)$-plane. From the above analysis, it is seen that during cellular growth, the parameters $(c_{*0}, \lambda_G)$ will always stay in the shaded triangular region shown in Fig. 10.5.

The exact form of the function $B_1(\tilde{\eta})$ in (10.33) remains to be determined.

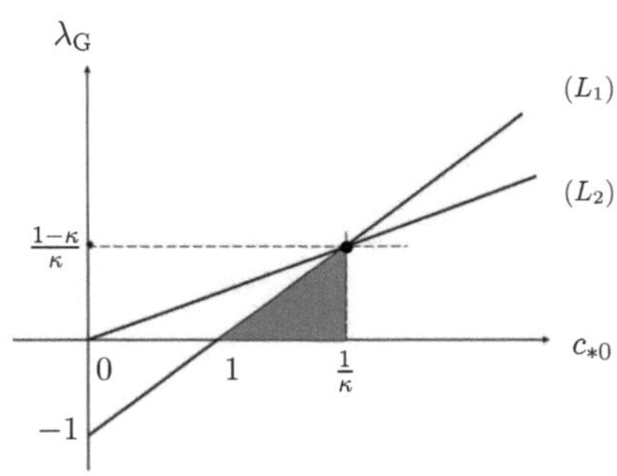


**Fig. 10.5** Sketch of the variation region of $(c_{*0}, \lambda_G)$, the *shaded region*: $(1-\kappa)c_{*0} - \lambda_G > (c_{*0} - 1) - \lambda_G > 0$. In the figure, $(L_1)$ represents the line $\lambda_G - c_{*0} + 1 = 0$, while $(L_2)$ represents the line $\lambda_G - (1-\kappa)c_{*0} = 0$

#### 10.2.6.3 The Second-Order Approximation $\mathcal{O}(\epsilon^2)$

One has the equation

$$\frac{\partial^2 \mathcal{C}_2}{\partial \xi^2} + \frac{\partial^2 \mathcal{C}_2}{\partial \eta^2} = \tilde{H}_2, \tag{10.40}$$

where

$$\tilde{H}_2 = -2\frac{\mathrm{d}D_1}{\mathrm{d}\tilde{\eta}} + W^2\lambda_G \mathcal{G}^2(\xi,\eta) - WD_1(\tilde{\eta})X_\xi(\xi,\eta) - \left[\frac{\mathrm{d}^2 B_0}{\mathrm{d}\tilde{\eta}^2} + WX_\xi(\xi,\eta)\frac{\mathrm{d}B_0}{\mathrm{d}\tilde{\eta}}\right], \tag{10.41}$$

and the following boundary conditions:

1. As $\eta \to \infty$,

$$\mathcal{C}_2 \sim \text{const.} \times \mathrm{e}^{-W\tilde{\eta}}, \tag{10.42}$$

2. Symmetric condition: at the side walls $\xi = \pm 1$,

$$\frac{\partial \mathcal{C}_2}{\partial \xi} = 0, \tag{10.43}$$

3. and the interface conditions, at $\eta = 0, \tilde{\eta} = 0$:

$$\mathcal{C}_2 = y_{*2} - \left(\frac{\partial \mathcal{C}_1}{\partial \eta} + \frac{\partial \mathcal{C}_0}{\partial \tilde{\eta}}\right)h_1 - W\lambda_\mathrm{G} Y_{\eta,0}(\xi)h_1 - \frac{\hat{\Gamma}}{MW}\mathcal{K}_0(\xi), \tag{10.44}$$

$$\frac{\partial \mathcal{C}_2}{\partial \eta} + B_1'(0) - W\frac{\Delta_0}{\lambda_0}\left[Y_{\xi,0}(\xi)h_1\right]' + W(1-\kappa)\mathcal{C}_1\lambda_0 = 0. \tag{10.45}$$

To eliminate the secular terms in the solution and ensure the uniform validity of the MVE solution, in the limit $\eta \to \infty$ it is necessary for the solution to satisfy the condition

$$\mathcal{H}_2(\tilde{\eta}) = \lim_{\eta \to \infty} \int_{-1}^{+1} \tilde{H}_2(\xi_1, \eta, \tilde{\eta})\mathrm{d}\xi_1 = 0, \tag{10.46}$$

which leads to

$$\frac{\mathrm{d}^2 B_0}{\mathrm{d}\tilde{\eta}^2} + W\frac{\mathrm{d}B_0}{\mathrm{d}\tilde{\eta}} = 0, \tag{10.47}$$

and consequently yields $\hat{Q}_0 = y_{*0} - 1$ and

$$\mathcal{C}_0 = B_0(\tilde{\eta}) = 1 + (y_{*0} - 1)\mathrm{e}^{-W\tilde{\eta}}. \tag{10.48}$$

One may solve the solution $\mathcal{C}_2$ and $h_1$ in terms of the Fourier series expansion. Before doing so, we point out that the inhomogeneous equation (10.40) has the particular solution

$$P_2(\xi,\eta,\tilde{\eta}) = -\frac{1}{2}\big[2D_1'(\tilde{\eta}) + B_0''(\tilde{\eta})\big]\eta^2 - \frac{1}{2}W\big[D_1(\tilde{\eta}) + B_0'(\tilde{\eta})\big]\eta Y(\xi,\eta)$$
$$+\frac{1}{2}W^2\lambda_G Y^2(\xi,\eta). \tag{10.49}$$

Moreover, at the tip $\xi = \eta = 0$, we have $P_2(0,0,0) = \frac{\partial}{\partial\eta}P_2(0,0,0) = 0$, while as $\eta \to \infty$,

$$P_2(\xi,\eta,\tilde{\eta}) \sim \tfrac{1}{2}\beta_0 W\big[W\lambda_{\mathrm{G}} - B_0'(\tilde{\eta})\big]\eta + \tfrac{1}{2}W^2\lambda_{\mathrm{G}}\beta_0^2.$$

Thus, one may write the solution $\mathcal{C}_2(\xi,\eta,\tilde{\eta})$ in the form

$$\mathcal{C}_2(\xi,\eta,\tilde{\eta}) = B_2(\tilde{\eta}) + D_2(\tilde{\eta})\eta + P_2(\xi,\eta,\tilde{\eta}) + \phi_2(\xi,\eta,\tilde{\eta}), \tag{10.50}$$

where $\phi_2(\xi,\eta,\tilde{\eta})$ satisfies the Laplace equation. We then deduce that in order for the solution $\mathcal{C}_2(\xi,\eta,\tilde{\eta})$ to satisfy the far-field condition in the limit $\eta \to \infty$, one must set

$$\begin{aligned} D_2(\tilde{\eta}) &= -\tfrac{1}{2}\beta_0 W\Big[W\lambda_{\mathrm{G}} - B_0'(\tilde{\eta})\Big], \\ B_2(\tilde{\eta}) &= -\tfrac{1}{2}W^2\lambda_{\mathrm{G}}\beta_0^2, \end{aligned} \tag{10.51}$$

and assume that $\phi_2(\xi,\eta,\tilde{\eta}) \sim \text{const.} \times \mathrm{e}^{-W\tilde{\eta}}$, as $\eta \to \infty$.

By applying the interface conditions (10.44) and (10.45) to the solution (10.50), it follows that at $\eta = \tilde{\eta} = 0$,

$$\begin{aligned} &-\tfrac{1}{2}W^2\lambda_{\mathrm{G}}\beta_0^2 + \phi_2(\xi,0,0) + P_2(\xi,0,0) \\ &\qquad = y_{*2} + W\Delta_0 h_1(\xi,0,0) - \tfrac{\hat{\Gamma}}{MW}\mathcal{K}_0(\xi), \end{aligned} \tag{10.52}$$

and

$$\frac{\partial\phi_2}{\partial\eta}(\xi,0,0) + \frac{\partial P_2}{\partial\eta}(\xi,0,0) + D_2(0) + B_1'(0)$$
$$+W(1-\kappa)\mathcal{C}_1(\xi,0,0)\lambda_0 = W\frac{\Delta_0}{\lambda_0}\Big[Y_{\xi,0}(\xi)h_1\Big]'. \tag{10.53}$$

In order to find the solution for $\mathcal{C}_2(\xi, \eta, \tilde{\eta})$, we first expand the functions $P_2(\xi, \eta, \tilde{\eta})$, $\phi_2(\xi, \eta, \tilde{\eta})$, and $h_1(\xi)$ in the following Fourier cosine series over the interval $(0 < \xi < 1)$:

$$\begin{cases} \phi_2(\xi, \eta, \tilde{\eta}) = \sum_{m=0}^{\infty} A_{2,m}(\tilde{\eta}) \cos(m\pi\xi) \mathrm{e}^{-m\pi\eta}, \\ P_2(\xi, \eta, \tilde{\eta}) = \sum_{m=0}^{\infty} E_{2,m}(\eta, \tilde{\eta}) \cos(m\pi\xi); \\ h_1(\xi) = \sum_{m=0}^{\infty} B_{1,m} \cos(m\pi\xi). \end{cases} \tag{10.54}$$

In the above, $E_{2,m}(\eta, \tilde{\eta})$ can be directly derived from the form of $P_2(\xi, \eta, \tilde{\eta})$, while the Fourier coefficients $\{A_{2,m}(\tilde{\eta}), B_{1,m}\}$ are unknown. Furthermore, to satisfy the far-field condition, one needs to require that $A_{2,0}(\tilde{\eta}) = A_{2,0}(0)\mathrm{e}^{-W\tilde{\eta}}$. Secondly, we let

$$\begin{aligned} \mathcal{K}_0(\xi) &= \sum_{m=0}^{\infty} \alpha_m \cos(m\pi\xi), \\ Y_0(\xi) &= \sum_{m=0}^{\infty} \beta_m \cos(m\pi\xi). \end{aligned} \tag{10.55}$$

The corresponding Fourier coefficients can be calculated as the integrals

$$\begin{aligned} \alpha_m &= \int_0^1 \mathcal{K}_0(\xi) \cos(m\pi\xi) \mathrm{d}\xi, \\ \beta_m &= \int_0^1 Y_0(\xi) \cos(m\pi\xi) \mathrm{d}\xi. \end{aligned}$$

Subsequently, one may derive that

$$\begin{aligned} \alpha_0 &= \tfrac{1}{\lambda_0}; \quad \beta_0 = -\tfrac{(2\ln 2)(1-\lambda_0)}{\pi}; \\ \beta_m &= \tfrac{2(-1)^{m-1}(1-\lambda_0)}{m\pi} \quad (m = 1, 2, \ldots). \end{aligned} \tag{10.56}$$

Finally, let us consider the function $\left[Y_{\xi,0}(\xi)h_1(\xi)\right]'$. Recall that the function $Y_\xi(\xi, \eta)$ has a singularity at $(\xi = 1, \eta = 0)$, ao that the function $\left[Y_{\xi,0}(\xi)h_1(\xi)\right]'$ cannot be expanded into a Fourier cosine series on the interval $(0, 1)$. However, as we stated before, to find the outer solution away from the root point, we may truncate the singularity of $Y_\xi(\xi, \eta)$ at $\xi = 1, \eta = 0$ and modify it with the function $\widetilde{Y}_\xi(\xi, \eta)$. In accordance with this, the function $Y_{\xi,0}(\xi) = Y_\xi(\xi, 0)$ is replaced by the function

$\widetilde{Y}_{\xi,0}(\xi) = \widetilde{Y}_{\xi}(\xi, 0)$. It is assumed that in the outer region away from the point ($\xi = 1, \eta = 0$), the outer solution of the system will not be affected by such a small modification.

It is evident that the function $[\widetilde{Y}_{\xi,0}(\xi)h_1(\xi)]'$ can be expanded into a Fourier cosine series on the interval $(0, 1)$. Let

$$\left[\widetilde{Y}_{\xi,0}(\xi)h_1(\xi)\right]' = \sum_{m=0}^{\infty} \mathcal{N}_m \cos(m\pi\xi). \tag{10.57}$$

We derive

$$\mathcal{N}_m = \int_0^1 \left[\widetilde{Y}_{\xi,0}(\xi)h_1(\xi)\right]' \cos(m\pi\xi)\mathrm{d}\xi = \cos(n\pi\xi)\widetilde{Y}_{\xi,0}(\xi)h_1(\xi)\Big|_0^1$$
$$+m\pi \int_0^1 \widetilde{Y}_{\xi,0}(\xi)h_1(\xi)\sin(m\pi\xi)\mathrm{d}\xi. \tag{10.58}$$

Noting that $\widetilde{Y}_{\xi,0}(0) = 0$, $h_1(1) = B_{1,0}$, and $\widetilde{Y}_{\xi,0}(1) = Y_{\xi,0}(\xi_*)$, one can write

$$\begin{aligned}\mathcal{N}_m &= \int_0^1 \left[\widetilde{Y}_{\xi,0}(\xi)h_1(\xi)\right]' \cos(m\pi\xi)\mathrm{d}\xi \\ &= (-1)^{n+1} Y_{\xi,0}(\xi_*)B_{1,0} + m\pi \int_0^1 \widetilde{Y}_{\xi,0}(\xi)h_1(\xi)\sin(m\pi\xi)\mathrm{d}\xi.\end{aligned}$$

Here one may write the approximations

$$\int_0^1 \widetilde{Y}_{\xi,0}(\xi)h_1(\xi)\sin(m\pi\xi)\mathrm{d}\xi \approx \int_0^1 Y_{\xi,0}(\xi)h_1(\xi)\sin(m\pi\xi)\mathrm{d}\xi.$$

It follows that

$$\begin{aligned}\mathcal{N}_m &= \int_0^1 \left[\widetilde{Y}_{\xi,0}(\xi)h_1(\xi)\right]' \cos(m\pi\xi)\mathrm{d}\xi \\ &\approx (-1)^{n+1} Y_{\xi,0}(\xi_*)B_{1,0} + m\pi \sum_{n=0}^{\infty} \mathcal{Q}_{m,n} B_{1,n},\end{aligned} \tag{10.59}$$

where

$$\begin{aligned}\mathcal{Q}_{m,n} &= 2\int_0^1 Y_{\xi,0}(\xi)\cos(n\pi\xi)\sin(m\pi\xi)\mathrm{d}\xi \\ &(m = 1, 2, \ldots; n = 0, 1, 2, \ldots).\end{aligned}$$

Let $\mathcal{Q}_{m,n} = -(1-\lambda_0)\widehat{\mathcal{Q}}_{m,n}$, with

$$\widehat{\mathcal{Q}}_{m,n} = \frac{1}{\pi}\sum_{i=1}^{4}\int_0^{\frac{\pi}{2}}(-1)^i\frac{\cos(K_i t)}{\cos t}\mathrm{d}t \qquad (K_{1,2} = 2m+1 \pm 2n; K_{3,4} = 2m-1 \pm 2n). \tag{10.60}$$

We find that $\widehat{\mathcal{Q}}_{m,n}$ yields the lower triangular matrix (see formula (2.539) in [16])

$$\widehat{\mathcal{Q}}_{m,n} = \begin{cases} 0 & (n > m), \\ (-1)^{m+n+1}2 & (n < m), \\ -1 & (n = m). \end{cases} \tag{10.61}$$

By substituting the above Fourier expansions into the interface conditions (10.52)–(10.53), one may respectively derive that

$$-\frac{1}{2}W^2\lambda_{\mathrm{G}}\beta_0^2 - y_{*2} + \sum_{m=0}^{\infty}A_{2,m}(0)\cos(m\pi\xi) + \sum_{m=0}^{\infty}E_{2,m}(0,0)\cos(m\pi\xi) = W\Delta_0\sum_{m=0}^{\infty}B_{1,m}\cos(m\pi\xi) - \frac{\hat{\Gamma}}{MW}\sum_{m=0}^{\infty}\alpha_m\cos(m\pi\xi), \tag{10.62}$$

and

$$-\sum_{m=0}^{\infty}m\pi A_{2,m}(0)\cos(m\pi\xi) + \sum_{m=0}^{\infty}\frac{\partial E_{2,m}}{\partial\eta}(0,0)\cos(m\pi\xi) + D_2(0) + B_1'(0) + W(1-\kappa)\lambda_0\Big[y_{*1} - W\lambda_{\mathrm{G}}\sum_{m=0}^{\infty}\beta_m\cos(m\pi\xi)\Big] = W\frac{\Delta_0}{\lambda_0}\sum_{m=0}^{\infty}\mathcal{N}_m\cos(m\pi\xi). \tag{10.63}$$

With (10.62) and (10.63), for $m = 0$ one obtains

$$A_{2,0}(0) = W\Delta_0 B_{1,0} + \frac{1}{2}W^2\lambda_{\mathrm{G}}\beta_0^2 + y_{*2} - \frac{\hat{\Gamma}}{MW}\alpha_0 - E_{2,0}(0,0), \tag{10.64}$$

and

$$\frac{\partial E_{2,0}}{\partial \eta}(0,0) + D_2(0) + B_1'(0)$$
$$= -W(1-\kappa)\lambda_0\Big[y_{*1} - W\lambda_G\beta_0\Big] + (-1)^{n+1}Y_{\xi,0}(\xi_*)B_{1,0}. \qquad (10.65)$$

We calculate that

$$E_{2,0}(0,0) = \left[\frac{1}{6} + \frac{2(\ln 2)^2}{\pi^2}\right]W^2\lambda_G(1-\lambda_0)^2,$$
$$\frac{\partial E_{2,0}}{\partial \eta}(0,0) = \frac{1}{2}\beta_0\lambda_0 W^2\Big[(1-\kappa)y_{*0} + \lambda_G\Big].$$

On the other hand, for $(m = 1, 2, \ldots)$, one obtains

$$A_{2,m}(0) + E_{2,m}(0,0) = W\Delta_0 B_{1,m} - \frac{\hat{\Gamma}}{MW}\alpha_m, \qquad (10.66)$$

and

$$-m\pi A_{2,m}(0) + \frac{\partial E_{2,m}}{\partial \eta}(0,0) - \lambda_G W^2(1-\kappa)\lambda_0\beta_m$$
$$= (-1)^{n+1}Y_{\xi,0}(\xi_*)B_{1,0} + m\pi W\frac{\Delta_0}{\lambda_0}\Big(\sum_{n=0}^{\infty} \mathcal{Q}_{m,n}B_{1,n}\Big). \qquad (10.67)$$

By eliminating $A_{2,m}(0)$ from (10.66) and (10.67), one obtains that

$$m\pi\frac{\hat{\Gamma}}{MW}\alpha_m + m\pi E_{2,m}(0,0) + \frac{\partial E_{2,m}}{\partial \eta}(0,0) - \lambda_G W^2(1-\kappa)\lambda_0\beta_m$$
$$= m\pi W\Delta_0\Big(\sum_{n=1}^{\infty} \delta_{m,n}B_{1,n}\Big) + m\pi W\frac{\Delta_0}{\lambda_0}\Big(\sum_{n=0}^{\infty} \mathcal{Q}_{m,n}B_{1,n}\Big),$$
$$+(-1)^{n+1}W\frac{\Delta_0}{\lambda_0}Y_{\xi,0}(\xi_*)B_{1,0} \quad (m = 1, 2, 3, \ldots). \qquad (10.68)$$

Equation (10.68) can be rewritten in the form

$$-\frac{1}{W}\frac{\lambda_0}{(1-\lambda_0)\Delta_0}$$
$$\times\left[\frac{\hat{\Gamma}}{MW}\alpha_m + E_{2,m}(0,0) + \frac{1}{m\pi}\frac{\partial E_{2,m}}{\partial \eta}(0,0) - \frac{1}{m\pi}\lambda_G W^2(1-\kappa)\lambda_0\beta_m\right]$$

$$= \sum_{n=1}^{\infty} \left[ \hat{\mathcal{Q}}_{m,n} - \frac{\lambda_0}{(1-\lambda_0)} \delta_{m,n} \right] B_{1,n}$$

$$- \left[ \frac{(-1)^{n+1}}{m\pi(1-\lambda_0)} Y_{\xi,0}(\xi_*) + 2(-1)^m \right] B_{1,0}$$

$$(m = 1, 2, 3, \ldots). \tag{10.69}$$

A necessary condition for convergence of the Fourier series of the solution is $B_{1,n} \to 0$ as $n \to \infty$. To satisfy this condition, one must set $B_{1,0} = 0$; hence we have $h_1(1) = 0$.

One then derives the following linear system of equations for $\{B_{1,m}\}$:

$$\sum_{n=1}^{\infty} \left[ \widehat{\mathcal{Q}}_{m,n} - \frac{\lambda_0}{(1-\lambda_0)} \delta_{m,n} \right] B_{1,n}$$

$$= \frac{\lambda_0}{W(1-\lambda_0)\Delta_0} \left[ (1-\kappa) W^2 \lambda_0 \lambda_G \frac{\beta_m}{m\pi} - \frac{\hat{\Gamma}}{MW} \alpha_m - E_{2,m}(0,0) - \frac{1}{m\pi} \frac{\partial E_{2,m}}{\partial \eta}(0,0) \right]$$

$$(m = 1, 2, 3, \ldots). \tag{10.70}$$

Once the system of Eq. (10.70) is solved, from (10.67) we obtain that

$$A_{2,m}(0) = W\Delta_0 B_{1,m} - \frac{\hat{\Gamma}}{MW} \alpha_m - E_{2,m}(0,0) \tag{10.71}$$

$$(m = 1, 2, 3, \ldots).$$

Thus, we have finally derived the solutions for the concentration field, $\mathcal{C}_2(\xi, \eta, \tilde{\eta})$, and the interface shape function $h_1(\xi)$.

The value of $h_1(0)$ at the tip can be calculated numerically as $h_1(0) = \sum_{m=1}^{\infty} B_{1,m}$. In Fig. 10.6a, we show the interface shape function $\eta = h_1(\xi)$ for the typical case $\lambda_0 = 0.6$. By fitting the numerical data in the vicinity of $\xi = -1$ on log-log graphs, we obtain that as $\xi \to -1$, $h_1(\xi) \sim \mathrm{a}_0(1+\xi)^{\alpha}$ $(0 < \alpha < 1)$, as shown in Fig. 10.6b. The numerical values of these numbers are calculated as $\mathrm{a}_0 \approx 2.6$, $\alpha \approx 0.88$. Hence, we deduce that $h_1'(-1) = \infty$.

To plot the interface shape in the physical plane $(X, Y)$, we apply the transformation (10.3) to calculate $X(\xi, \eta_s), Y(\xi, \eta_s)$. Therefore, approaching the root as $\xi \to \pm 1$ and $\eta_s \to 0$, we have $Y(\xi, \eta_s) \to -\infty$, and the width of the finger approaches the following constant:

$$X(\xi, \eta_s) = \xi - \tfrac{2(1-\lambda_0)}{\pi} \tan^{-1} \left[ \tfrac{\sin(\pi\xi)}{\cos(\pi\xi) + e^{\pi\eta_s}} \right]$$

$$\approx -1 + \frac{2(1-\lambda_0)}{\pi} \tan^{-1} \left[ \frac{1}{\eta_s'(-1)} \right] \approx -1.$$

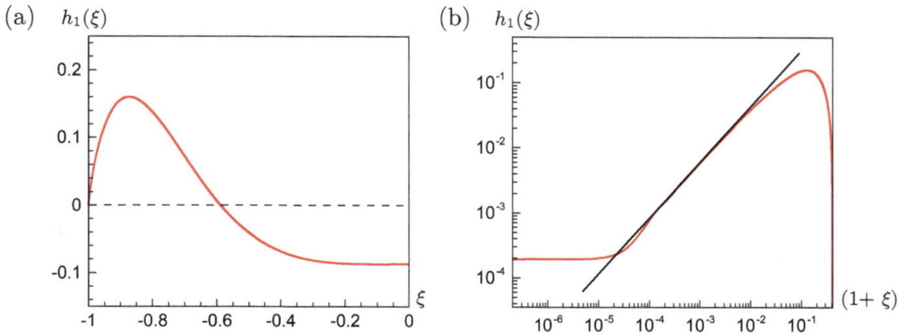


**Fig. 10.6** The characteristics of the solution $h_1(\xi)$ for the case $\epsilon = 0.1$, $M = 6.0$, $\hat{\Gamma} = 0.3$, $\kappa = 0.1$, and $\lambda_G = 2.0$, $\lambda_0 = 0.6$; (**a**) The graph of the function $h_1(\xi)$ in the outer region. (**b**) The behavior of the function $h_1(\xi)$ near $\xi = -1$; the *black line* is $h_1 = 2.63(1+\xi)^{0.88}$. Due to the unavoidable existence of numerical truncation error in the computation, the graph of the solution $h_1(\xi)$ shown in the figure does not vanish as $\xi \to -1$ as it should, theoretically speaking

Thus, we conclude that the root of the finger always extends to infinity with $Y(\xi, \eta_s) \to -\infty$, $X(\xi, \eta_s) \to -1$.

#### 10.2.6.4 The Third-Order Approximation $\mathcal{O}(\epsilon^3)$

Similar to what has been done for the system of second-order approximation, to eliminate the secular terms in the limit $\eta \to \infty$, one must apply the necessary condition to the inhomogeneous term $\tilde{H}_3(\xi, \eta, \tilde{\eta})$:

$$\mathcal{H}_3(\tilde{\eta}) = \lim_{\eta \to \infty} \int_{-1}^{+1} \tilde{H}_3(\xi_1, \eta, \tilde{\eta}) \mathrm{d}\xi_1 = 0. \tag{10.72}$$

Using the far-field conditions of the first-order approximation, we then derive that

$$B_1(\tilde{\eta}) = y_{*1} \mathrm{e}^{-W\tilde{\eta}} + W\lambda_G \beta_0 \big(1 - \mathrm{e}^{-W\tilde{\eta}}\big), \tag{10.73}$$

and we eventually obtain

$$\mathcal{C}_1 = W\lambda_G \eta - W\lambda_G Y(\xi, \eta) + \big[W\lambda_G \beta_0 + (y_{*1} - W\lambda_G \beta_0) \mathrm{e}^{-W\tilde{\eta}}\big], \tag{10.74}$$

where the constant $y_{*1}$ can be derived from (10.65) as follows:

$$y_{*1} = \frac{\beta_0 W}{2[(1-\kappa)\lambda_0 - 1]} \Big[ y_{*0} - \lambda_G (1 + 2\kappa\lambda_0) \Big]. \tag{10.75}$$

We do not attempt to proceed any further for higher-order approximate solutions. Here, we only point out that the RPE solution obtained possesses the following characteristics:

1. The RPE is a uniquely determined *particular solution* of the inhomogeneous system for $\{\bar{C}_{\mathrm{B}}(\xi, \eta, \varepsilon), \bar{\eta}_{\mathrm{B}}(\xi, \epsilon)\}$ in the outer region without any arbitrary constant.
2. The mass balance condition at the interface, in any $k$th-order approximation of RPE, always contains the following two terms:

$$\frac{\partial C_k}{\partial \eta} - W\frac{\Delta_0}{\lambda_0}\left[\widetilde{Y}_{\xi}(\xi, 0)h_k\right]' + \cdots = 0.$$

   One may similarly deduce that in order for the concentration function $C_k(\xi, \eta)$ to be expanded into a Fourier cosine series on the interval $(0, 1)$, the condition $h_k(1) = 0$ must hold. Therefore, we deduce that the interface shape given by the $N$th-order RPE approximate solution, $\eta = \bar{\eta}_{\mathrm{B}}(\xi, \epsilon) = \sum_{n=0}^{N} \epsilon^{n+1} h_n(\xi)$, will always vanish at $\xi = \pm 1$. This implies that the interface described by RPE approximate solutions will never have a closure at $\xi = \pm 1$, no matter how many terms are included.
3. The value of the above RPE solution at the tip does not vanish; namely $h_1(0) \neq 0$, which is inconsistent with the tip condition $\eta_{\mathrm{B}}(0) = 0$. Therefore, the RPE solution obtained is not valid in the region near the cell tip. In other words, it cannot yet be considered the complete outer solution.

To obtain the outer solution, one must include the *general solution* of the associated homogeneous system, $\{\tilde{C}_{\mathrm{B}}(\xi, \eta, \varepsilon); \tilde{\eta}_{\mathrm{B}}(\xi, \epsilon)\}$.

### *10.2.7 Singular Perturbation Expansion Part of the Solution in the Outer Region*

The general solution of the associated homogeneous system may have multiple length scales and can be expanded in SPE form with multiple variables: $\{\xi, \eta, \xi_+, \eta_+\}$, defined as

$$\xi_+ = \frac{1}{\sqrt{\epsilon \hat{\Gamma}}} \int_0^{\xi} k(\xi_1, \eta) d\xi_1,$$

$$\eta_+ = \frac{1}{\sqrt{\epsilon \hat{\Gamma}}} \int_0^{\eta} k(\xi, \eta_1) d\eta_1.$$

Let $\tilde{C}_{\mathrm{B}}(\xi,\eta,\varepsilon) = \tilde{C}(\xi,\eta,\xi_+,\eta_+)$ and $\tilde{\eta}_{\mathrm{B}}(\xi,\epsilon) = \tilde{h}(\xi,\xi_+)$. One may make the following singular perturbed expansions (SPE) for the solution:

$$\begin{aligned}
&\tilde{C}(\xi,\eta,\xi_+,\eta_+,\varepsilon) \sim \tilde{\mu}_0(\varepsilon)\tilde{C}_0(\xi,\eta,\xi_+,\eta_+) + \tilde{\mu}_1(\varepsilon)\tilde{C}_1(\xi,\eta,\xi_+,\eta_+) + \cdots, \\
&\tilde{\eta}_{\mathrm{s}}(\xi,\xi_+,\varepsilon) \sim \tilde{b}_0(\varepsilon)\tilde{h}_0(\xi,\xi_+) + \tilde{b}_1(\varepsilon)\tilde{h}_1(\xi,\xi_+) + \cdots, \\
&k(\xi,\eta,\varepsilon) \sim k_0(\xi,\eta) + \varepsilon k_1(\xi,\eta) + \cdots, \\
&g(\xi,\eta,\varepsilon) \sim g_0(\xi,\eta) + \varepsilon g_1(\xi,\eta) + \cdots.
\end{aligned} \tag{10.76}$$

It may be deduced from the interface condition that $\tilde{\mu}_0(\varepsilon) = \epsilon\tilde{b}_0(\epsilon)$. So in the leading-order approximation, we may write that $\tilde{C}_{\mathrm{B}}(\xi,\eta,\varepsilon) \approx \epsilon\tilde{b}_0(\epsilon)\tilde{C}_0(\xi,\eta,\xi_+,\eta_+)$, and $\tilde{\eta}_{\mathrm{B}}(\xi,\epsilon) \approx \tilde{b}_0(\epsilon)\tilde{h}_0(\xi,\xi_+)$, where the factor $\tilde{b}_0(\epsilon)$ is to be determined. The equation for the zeroth-order approximation is derived as

$$\frac{\partial^2\tilde{C}_0}{\partial\xi_+^2} + \frac{\partial^2\tilde{C}_0}{\partial\eta_+^2} = 0. \tag{10.77}$$

The boundary conditions are as follows:

1. At the side wall, $\xi = 1$,

$$\frac{\partial\tilde{C}_0}{\partial\xi_+} = 0\,. \tag{10.78}$$

2. On the interface $\eta = \eta_+ = 0$, letting $k_0(\xi,0) = \bar{k}_0(\xi)$, we have

$$\tilde{C}_0 = W\Delta_0\tilde{h}_0 + \frac{\bar{k}_0^2}{MW\mathcal{G}_0(\xi,0)}\frac{\partial^2\tilde{h}_0}{\partial\xi_+^2}, \tag{10.79}$$

$$\bar{k}_0\frac{\partial\tilde{C}_0}{\partial\eta_+} + WY_{\xi,0}(\xi,0)\bar{k}_0\Big[\lambda_{\mathrm{G}} - (1-\kappa)y_{*0}\Big]\frac{\partial\tilde{h}_0}{\partial\xi_+} = 0. \tag{10.80}$$

3. At the tip, $\xi = \xi_+ = 0$,

$$\frac{\partial\tilde{h}_0}{\partial\xi_+} = 0. \tag{10.81}$$

Consider the normal mode solutions

$$\tilde{C}_0 = \tilde{A}_0(\xi,\eta)\exp\{\mathrm{i}\xi_+ - \eta_+\}\,,\ \tilde{h}_0 = \bar{D}_0\exp\{\mathrm{i}\xi_+\}\,, \tag{10.82}$$

where the coefficient $\bar{D}_0$ is a constant. With the notation $\tilde{A}_0(\xi, 0) = \bar{A}_0(\xi)$, we derive from the interface conditions (10.79) and (10.80) that

$$\bar{A}_0(\xi) = \bar{D}_0\Big[W\Delta_0 - \tfrac{\bar{k}_0^2}{MW\mathcal{G}_0(\xi,0)}\Big],$$

and

$$\bar{k}_0^3 = \bar{k}_0 MW^2\Delta_0\mathcal{G}_0(\xi,0)\Big\{1 + \tfrac{\mathrm{i}}{\lambda_0}Y_{\xi,0}(\xi,0)\Big\}.$$

There are three roots for $\bar{k}_0$:

$$\bar{k}_0^{(1)}(\xi) = \bar{k}_s(\xi), \quad \bar{k}_0^{(2)}(\xi) = -\bar{k}_s(\xi), \quad \bar{k}_0^{(3)}(\xi) = 0, \tag{10.83}$$

where $\bar{k}_s(\xi) = W\sqrt{M\Delta_0\mathcal{G}_0(\xi,0)}\big[1 + \tfrac{\mathrm{i}}{\lambda_0}Y_{\xi,0}(\xi,0)\big]^{\frac{1}{2}}$, $(-1 < \xi \leq 0)$. Here, we have chosen that $\mathrm{Re}\{\bar{k}_s(\xi)\} > 0$. In accordance with the above, we obtain three fundamental solutions:

$$H_1(\xi) = \mathrm{e}^{\frac{\mathrm{i}}{\sqrt{\epsilon\Gamma}}\chi(\xi)}, \; H_2(\xi) = \mathrm{e}^{-\frac{\mathrm{i}}{\sqrt{\epsilon\Gamma}}\chi(\xi)}, \; H_3(\xi) = 1, \tag{10.84}$$

where we have defined

$$\chi(\xi) = \int_0^{\xi} \bar{k}_s(\xi_1)\mathrm{d}\xi_1 = \chi_{\mathrm{R}}(\xi) + \mathrm{i}\chi_{\mathrm{I}}(\xi). \tag{10.85}$$

Note that $\mathrm{Im}\{\bar{k}_s(\xi)\} > 0, \quad (-1 < \xi \leq 0)$. The solution $H_1(\xi)$ should be ruled out, because away from the interface, as $\eta$ increases, the corresponding concentration field tends to infinity exponentially, violating the far-field condition. As a consequence, the physically acceptable solution is a linear combination of two fundamental solutions $\{H_2(\xi), H_3(\xi)\}$. Note also that as $\xi \to -1$, we have

$$\mathcal{G}_0(\xi,0) \sim -(1-\lambda_0)\tan\left(\frac{\pi\xi}{2}\right), \tag{10.86}$$

$$Y_{\xi,0}(\xi,0) = -(1-\lambda_0)\tan\left(\frac{\pi\xi}{2}\right), \tag{10.87}$$

$$\bar{k}_s(\xi) \sim -\frac{1}{\sqrt{2}}(1+\mathrm{i})W(1-\lambda_0)\sqrt{\frac{M\Delta_0}{\lambda_0}}\tan\left(\frac{\pi\xi}{2}\right). \tag{10.88}$$

Consequently, $\mathrm{Re}\{\bar{k}_s(\xi)\} \to +\infty$, $\mathrm{Im}\{\bar{k}_s(\xi)\} \to +\infty$, as $\xi \to -1$. This implies that as $\xi \to -1$, the solution $|H_2(\xi)|$ approaches zero exponentially, which is subdominant compared with the solution $H_3(\xi) = 1$.

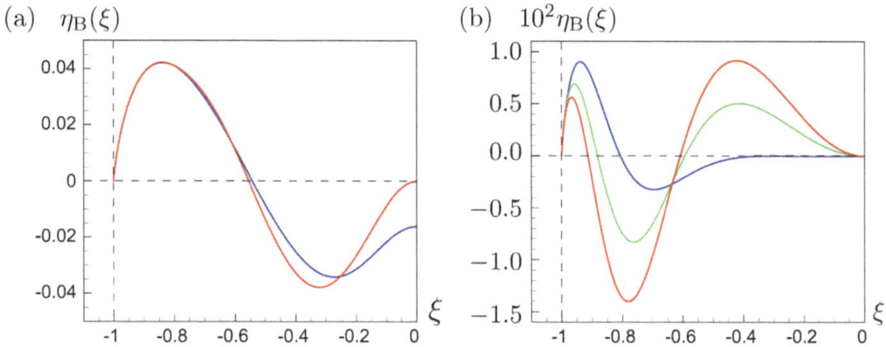


**Fig. 10.7** (**a**) The graph of the outer solution for the case $\epsilon = 0.1$, $\kappa = 0.1$, $\lambda_G = 0.8$, $\lambda_0 = 0.4$, $M = 1.0$, $\hat{\Gamma} = 2$. The top left-hand figure is the graph of the outer solution $\eta_B(\xi)$ (with inclusion of the SPE part) over $(-1 < \xi \leq 0)$, while the bottom right-hand figure is the graph of the RPE part of the outer solution $\epsilon h_1(\xi)$. (**b**) The graphs of the full solutions $\eta_B(\xi)$ in the $(\xi, \eta)$-plane for the typical cases: $\epsilon = 0.1$, $\kappa = 0.1$, $\lambda_G = 2.0$, $\lambda_0 = 0.6$, $M = 1.0$, and $\hat{\Gamma} = 1.0, 1.5, 2.0$ from bottom to top on the right-hand side of the figure

Up to this point, we have derived both the RPE and SPE of the basic state solution (10.19). By applying the tip condition $\eta_B(0, \epsilon) = 0$, we may finally obtain the outer solution for the interface shape as

$$\eta_B(\xi, \epsilon) = \epsilon \left[ h_1(\xi) - h_1(0) e^{\frac{\chi_I(\xi)}{\sqrt{\epsilon \hat{\Gamma}}}} \cos\left( \frac{\chi_R(\xi)}{\sqrt{\epsilon \hat{\Gamma}}} \right) \right] + \cdots . \tag{10.89}$$

The graph of the outer solution for the interface shape function $\eta_B(\xi)$ and the corresponding RPE part of the solution are shown in Fig. 10.7a for some typical cases, whereas the variation of the interface shape function of the outer solution with the parameter $\hat{\Gamma}$ is shown in Fig. 10.7b.

In Fig. 10.8, we show the interface shapes in the $(X, Y)$-plane for some typical cases (a)–(c) and compare the theoretical results with the experimental photos (A)–(C) given by Georgelin and Pocheau for the same cases with no adjustable parameter. In Fig. 10.9, we show a comparison of interface shapes between the theoretical result and the experimental data for a typical case: the pulling velocity $V = 12\,\mu\text{m/s}$, temperature gradient $G_D = 140\,\text{K/cm}$, primary spacing $\ell_w = 90\,\mu\text{m}$, and tip radius $\ell_t = 22.4\,\mu\text{m}$. It is seen that the theoretical results are in very good quantitative agreement with the experimental observations. Note that the theoretical model is 2D with no anisotropy of surface tension, while the experimental samples given by Georgelin and Pocheau are not perfectly two-dimensional and have a small amount of anisotropy. The results show that the effects of three-dimensionality and anisotropy of surface tension on the interface shapes in the region not close to the bottom of the root are insignificant, at least in the cases under study.

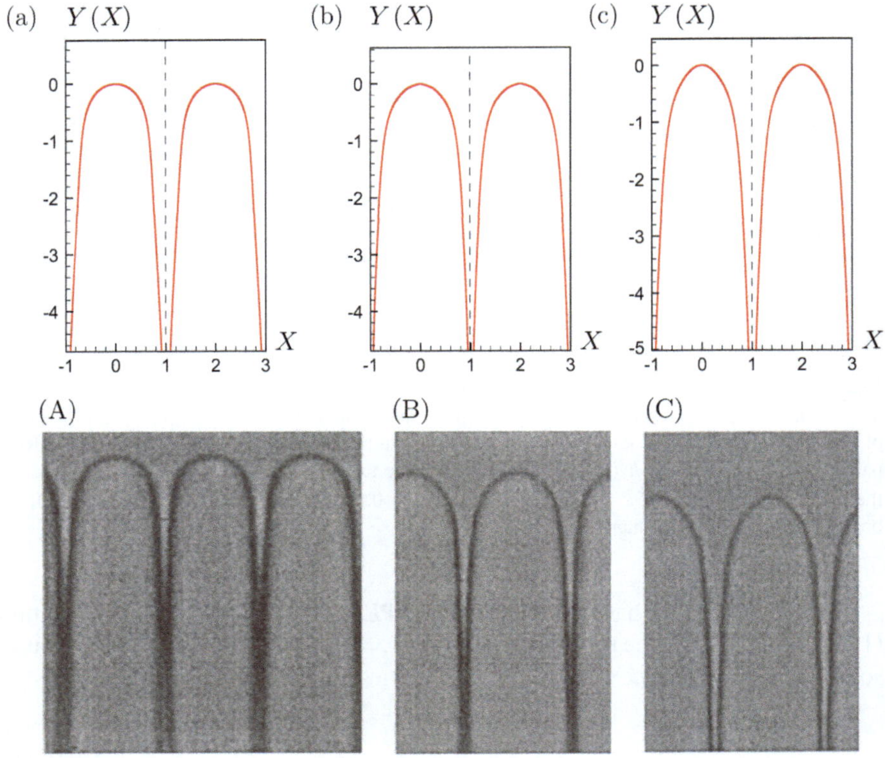


**Fig. 10.8** The interface shapes in the $(X, Y)$-plane described by the outer solutions for the typical cases experimentally observed by Georgelin and Pocheau, which yield the dimensionless parameters $\kappa = 0.29$, $M = 0.08432$ and (**a**) $\epsilon = 0.1289$, $W_0 = 1.725$; $\lambda_0 = 0.6020$, $\lambda_G = 0.7826$, $\hat{\Gamma} = 0.01164$, $\alpha = 0.88$, $a_0 = 1.9054$; (**b**) $\epsilon = 0.1991$, $W_0 = 2.007$; $\lambda_0 = 0.5760$, $\lambda_G = 0.7826$, $\hat{\Gamma} = 0.003157$, $\alpha = 0.88$, $a_0 = 2.2387$; (**c**) $\epsilon = 0.2667$, $W_0 = 3.000$; $\lambda_0 = 0.5077$, $\lambda_G = 0.3913$, $\hat{\Gamma} = 0.002629$, $\alpha = 0.88$. $a_0 = 1.7782$. The photos (**A**)–(**C**) are the interface shapes observed in the experiments by Georgelin and Pocheau in the same cases corresponding to (**a**)–(**c**), respectively. The dimensional experimental data for these cases are as follows: (**A**) Pulling velocity: $V = 12.0\,\mu$m/s, primary spacing $2\ell_w = 50.0\,\mu$m, gradient of temperature $G_D = 140.0 \times 10^{-4}$ K/($\mu$m) and tip radius $\ell_t = 14.5\,\mu$m; (**B**) Pulling velocity: $V = 12.0\,\mu$m/s, primary spacing $2\ell_w = 90.0\,\mu$m, gradient of temperature $G_D = 140.0 \times 10^{-4}$ K/($\mu$m) and tip radius $\ell_t = 22.4\,\mu$m; (**C**) Pulling velocity: $V = 24.0\,\mu$m/s, primary spacing $2\ell_w = 90.0\,\mu$m, gradient of temperature $G_D = 140.0 \times 10^{-4}$ K/($\mu$m) and tip radius $\ell_t = 15.0\,\mu$m

In terms of the above composite outer solution and the formulas $\bar{\ell}_t = 1 = \mathcal{K}\{\eta_B\}(0, \epsilon)/W$, one may finally derive that the primary spacing can be written in the form

$$W = W_0 + \epsilon W_1 + \cdots ,$$

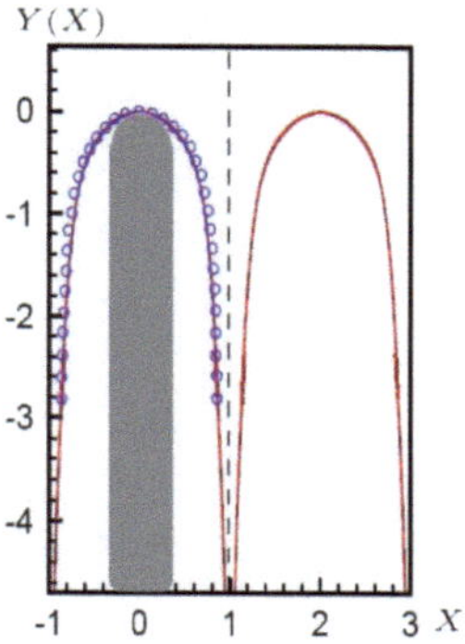


**Fig. 10.9** The interface shape in the $(X, Y)$-plane for the typical case $\kappa = 0.29$, $c_\infty = 1.2\,\text{mol}\%$, and the pulling velocity $V = 12\,\mu\text{m/s}$, the temperature gradient $G_D = 140.0 \times 10^{-4}\,\text{K}/(\mu\text{m})$, primary spacing $\ell_w = 90\,\mu\text{m}$, and the tip radius $\ell_t = 22.4\,\mu\text{m}$. The *gray area* describes the viscous finger given by the S-T solution. The *circles* represent the experimental data obtained by Pocheau et al. [14]

where

$$W_0 = \mathcal{K}_0(0) = \frac{\pi(1-\lambda_0)}{2\lambda_0^2}, \tag{10.90}$$

$$W_1 = \mathcal{K}_1(0) = -\frac{1}{\lambda_0}\eta_B''(0). \tag{10.91}$$

The outer solution obtained may also yield the concentration of impurity along the interface in the outer region:

$$\begin{aligned} \mathcal{C} &= y_{*0} + \epsilon\, \mathcal{C}_1(\xi, 0, 0) + \epsilon^2 \mathcal{C}_2(\xi, 0, 0) \\ &-\epsilon^2 h_1(0)\left[W\Delta_0 - \frac{\bar{k}_0^2(\xi)}{MW\mathcal{G}_0(\xi,0)}\right] e^{\frac{\chi_I(\xi)}{\sqrt{\epsilon\hat{\Gamma}}}} \cos\left(\frac{\chi_R(\xi)}{\sqrt{\epsilon\hat{\Gamma}}}\right) + \cdots \end{aligned} \tag{10.92}$$

as well as the location of the cell tip in the experimental system via the solution

$$y_* = y_{*0} + \epsilon y_{*1} + \cdots,$$

where

$$\begin{aligned} y_{*0} &= \frac{1 + \lambda_G(1-\lambda_0)}{1 - \lambda_0(1-\kappa)}; \\ y_{*1} &= \frac{\beta_0 W}{2[(1-\kappa)\lambda_0 - 1]}\Big[y_{*0} - \lambda_G(1 + 2\kappa\lambda_0)\Big]. \end{aligned} \tag{10.93}$$

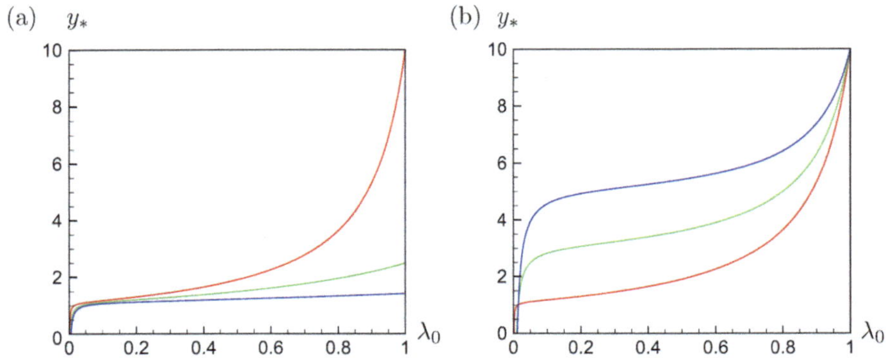


**Fig. 10.10** The variation of $y_*$ with $\lambda_0$ for the cases $\epsilon = 0.1$ and (**a**) $\lambda_G = 0.1$, and $\kappa = 0.1, 0.4, 0.7$ from top to bottom; (**b**) $\kappa = 0.1$, and $\lambda_G = 0.1, 2.0, 4.0$ from bottom to top

By neglecting the higher-order terms $\mathcal{O}(\epsilon^2)$, one may write the impurity concentration at the tip approximately as $\mathcal{C}_{\text{tip}} \approx y_*$. The variation of $y_*$ with $\lambda_0$ under different growth conditions and material properties is shown in Fig. 10.10a, b.

Moreover, the parameter $y_*$ determines the tip undercooling temperature as the function

$$T_{\text{tip}}(\lambda_G, \epsilon) = T_B(0,0) = \epsilon G y_0 = M y_* \approx M\big(y_{*0} + \epsilon y_{*1}\big), \tag{10.94}$$

depending on the parameters $\lambda_0, \kappa, \lambda_G$, and $\epsilon$. The theoretical results (10.93)–(10.94) can be compared with the experimental data.

Recall the constitutional instability criterion that in order for a flat interface to become unstable, the necessary condition $0 < \lambda_G < \hat{\lambda}_G = \frac{(1-\kappa)}{\kappa}$ must hold. On the other hand, one may write $\lambda_G = -\frac{\kappa_D}{mC_\infty}\frac{(G)_D}{V}$ and $\hat{\lambda}_G = -\frac{\kappa_D}{mC_\infty}\frac{(G)_D}{V_c}$, so that $\lambda_G = \hat{\lambda}_G \frac{V_c}{V}$.

Now let us take the limit $V \to \infty$, or $\lambda_G \to 0$ with fixed $C_\infty$ and temperature gradient $(G)_D$. Define

$$T_L(C_\infty) = \frac{M}{1 - \lambda_0(1-\kappa)} \approx \lim_{\lambda_G \to 0} T_{\text{tip}}(\lambda_G, \epsilon).$$

Thus, using (10.93), one can calculate $\Delta T_{\text{tip}} = |T_{\text{tip}}(\lambda_G, \epsilon) - T_L(C_\infty)|$ as

$$\begin{aligned} \Delta T_{\text{tip}} &= \big|T_{\text{tip}}(\lambda_G, \epsilon) - T_L(C_\infty)\big| \\ &= M\mathcal{A}\lambda_G = M\mathcal{A}\left(\frac{1-\kappa}{\kappa}\right)\frac{V_c}{V}, \end{aligned} \tag{10.95}$$

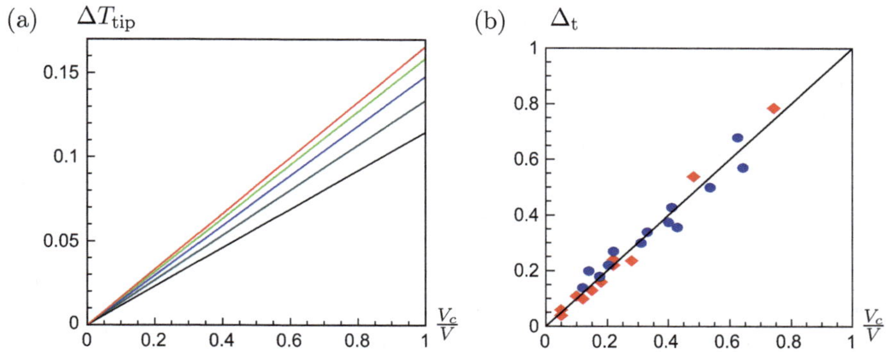


**Fig. 10.11** (**a**) The variation of $\Delta T_{\mathrm{tip}}$ with $\lambda_G$ for the cases $\epsilon = 0.1$, $\kappa = 0.3$, $M = 0.08432$, and $\lambda_0 = 0.3, 0.4, 0.5, 0.6, 0.7$ (or $W_0 = 12.2173, 5.8905, 3.1415, 1.7453, 0.9617$) from top to bottom; (**b**) The universal scaling law for a system of arrayed cellular growth. The experimental data are from Pocheau et al. for the directional solidification system of impure succinonitrile (SCN) [31]. The *solid full squares* are for the case $G_D = 78.0 \times 10^{-4}$ K/(μm), while the *full circles* are for the case $G_D = 140.0 \times 10^{-4}$ K/(μm)

where

$$\mathcal{A} = \frac{(1-\lambda_0)}{1-\lambda_0(1-\kappa)}\Big[1+\mathcal{O}(\epsilon)\Big]$$

is a function of the parameters $(\epsilon, \kappa, M)$ and $\lambda_0$ (or $W_0$).

In Fig. 10.11a, we show the variations of $\Delta T_{\mathrm{tip}}$ with parameter $V_c/V$ under different growth conditions and given material properties. It is seen that the graph of $\Delta T_{\mathrm{tip}}$ is a set of straight lines passing through the origin with different slopes. Furthermore, we take the limit $V \to V_c$ or $\lambda_G \to \hat{\lambda}_G$ with fixed $C_\infty$ and $(G)_D$ and define

$$\lim_{\lambda_G \to \hat{\lambda}_G} T_{\mathrm{tip}}(\lambda_G, \epsilon) = T_S(C_\infty).$$

Assume that the formula (10.95) can be approximately applied to the regime $V \approx V_c$ or $\lambda_G \approx \hat{\lambda}_G$. In terms of the formula (10.95), one derives

$$\Big|T_S(C_\infty) - T_L(C_\infty)\Big| = M\mathcal{A}\left(\frac{1-\kappa}{\kappa}\right).$$

Following the notation used by Pocheau et al. [31], we define the normalized tip temperature as $\Delta_t = |T_{\mathrm{tip}}(\lambda_G, \epsilon) - T_L(C_\infty)|/|T_S(\infty) - T_L(C_\infty)|$ and the reduced pulling speed as $\nu = \frac{V}{V_c}$. Then we derive that

$$\Delta_t = \lambda_G/\hat{\lambda}_G = V_c/V = \tfrac{1}{\nu}, \quad \text{or}, \quad \Delta_t \nu = 1, \qquad (10.96)$$

which yields a straight line with unit slope in the $(\Delta, 1/\nu)$-plane, and is a universal scaling law applicable to any arrayed-cellular growth system under any growth conditions.

It is known that under given control parameters $V, (G)_{\mathrm{D}}, M$, the primary spacing $W_0$ (or $\lambda_0$) varies in different experiments depending on the history of growth, so does the tip location parameter $y_*$. Thus, the scaling law (10.96) states that such a variation of primary spacing, practically, does not noticeably affect the value of $\Delta_{\mathrm{t}}$ in the whole extent of $W_0$.

The scaling law (10.96) is well supported by the experimental data measured by Pocheau et al. [31], who argued that the limit value of the tip temperature $T_{\mathrm{L}}(C_\infty)$ is very close to the liquidus temperature, while the limit value $T_{\mathrm{S}}(C_\infty)$ is close to the solidus temperature. Both temperatures are measurable with the flat interface [31]. They obtained quite accurate experimental data for $\Delta_{\mathrm{t}}$. In Fig. 10.11b, we show a comparison of the scaling law (10.96) with the experimental data under various growth conditions given by Pocheau et al. The best-fitting line to the experimental data claimed by Pocheau et al. has slope 0.998. It is seen that our theoretical result is in excellent agreement with their experimental observations.

The outer solution (10.89) obtained, however, is still not applicable in the region near the root. It fails to satisfy the root smoothness conditions: $\eta_{\mathrm{B}}(-1,\epsilon) = \eta_{\mathrm{T}}(-1,\epsilon) > 0$ and $\eta'_{\mathrm{B}}(-1,\epsilon) = 0$, yielding that the interface shape is not closed at $\xi = \pm 1$. Hence, it provides no information on the location of the bottom of the root, the total length of the cells, and the concentration of impurity at the bottom of the root. In order to find the global solution of cellular growth, we need to derive its inner solution in the root region.

## 10.3 The Inner Steady-State Solution in the Root Region and Interface Closure of Deep Cellular Growth

As shown in the previous section, the RPE solution of cellular growth has a finger-like interface, which extends to infinity without a closure, as shown in Fig. 10.9. Even with including the SPE part in the leading-order approximation, the full outer solution still shows no interface closure at the bottom of the root, since $\lim_{\xi\to -1} \tilde{\mathrm{h}}_0(\xi,\epsilon) = 0$, *decaying exponentially*. It can be proven that such a property of no interface closure remains in the approximations of the outer solution of all orders.

In reality, a remarkable feature of steady deep-cellular growth is that the system forms a long narrow groove, and the finger-like interface always has a smooth closure at the bottom. Thus, the outer solution obtained in the last section raises the following questions: What is the mechanism of the interface closure for steady deep-cellular growth? How can one determine the location of the closure and its dependence on the interfacial energy and other physical parameters?

In this section, we are going to explore the mechanism of interface closure. It is seen that the interface closure observed in experiments is a phenomenon determined by some transcendentally small factor with magnitude $\delta(\epsilon)$ "beyond all orders." The physical phenomena in which a transcendentally small factor "beyond all orders" plays crucial role have been previously encountered and extensively investigated in the areas of nonlinear science, fluid dynamics, and crystal growth. The Segur–Kruskal problem in needle crystal growth discussed in Chap. 5 is one prominent example on this subject, and the problem of interface closure in steady deep-cellular growth is another prominent example.

It is evident that the transcendentally small term missed by the RPE solution derived is very hard to "grasp" in terms of numerical simulations without analytical treatment. A typical example shows that as $\epsilon = 0.1$, one has $\delta(\epsilon) \approx 0.4 \times 10^{-8}$. Such "weak" information that we derive will be fully submerged by the "numerical noise" generated by the unavoidable rounding and truncation errors involved in the numerical simulations.

To demonstrate this argument more clearly, we carry out the numerical computations for the approximate outer solution with different truncation errors. The numerical results are shown in Fig. 10.12a, b. The solution was obtained in closed analytical form and it has been proven analytically that its interface extends to infinity with no interface closure. The numerical implementation of the this analytical solution, however, exhibits a false interface closure due to numerical truncation errors. Moreover, calculations with different truncation errors result in different locations of the interface closure.

The above demonstration casts serious doubt on the results of the previous numerical simulations in the literature. Though these numerical works showed interface closures in the system of deep-cellular growth in terms of different

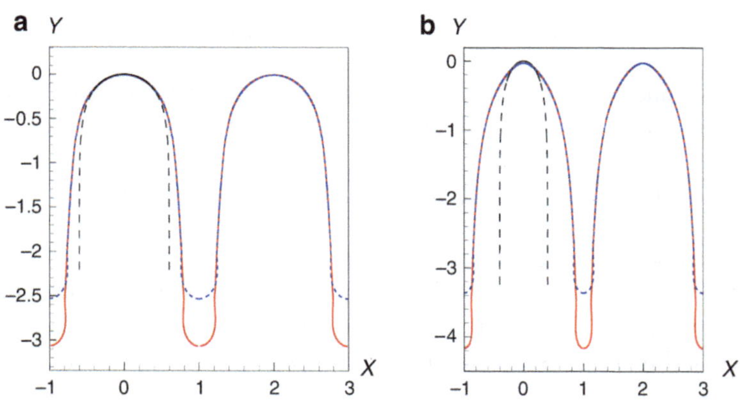


**Fig. 10.12** Numerical descriptions of the interface shapes computed using the analytical form of the first-order approximate outer solution obtained in [5]. The *blue dashed line* is calculated by the Fourier cosine series involved in the RPE part truncated at 6000 terms, while the *solid red line* is calculated by the Fourier cosine series involved truncated at 60,000 terms for the typical cases $\epsilon = 0.1, M = 6.0, hat\Gamma = 0.3, \kappa = 0.1$, and (**a**) $\lambda_G = 2.0, \lambda_0 = 0.6$; (**b**) $\lambda_G = 1.0, \lambda_0 = 0.4$. In both (**a**) and (**b**), the *black dashed line* is given by the S-T solution, corresponding to the case of $\epsilon = 0$

methods (see, for instance, [33, 38]), those interface closures may be actually caused by numerical errors.

It may be appropriate to conclude that in order to explore the mechanism of interface closure and its dependence on the physical parameters of the system, the analytical approach is not only preferable, but also necessary.

In the following subsections, we are going resolve this problem. For this purpose, we shall divide the whole physical region into the outer region away from the interface closure and the root region near the interface closure and study both the RPE and SPE forms of solutions in the root region. It will be seen that the root region has a very delicate structure comprising three inner subregions; the inner system in the root region has a simple turning point; its presence generates the so-called *trapped-waves mechanism*, which is responsible for the interface closure at the bottom of the root. The quantization condition derived from the trapped-waves mechanism yields a special eigenvalue problem (EVP), whose solution determines the location of the interface closure and its dependence on the interfacial energy and other physical parameters.

### 10.3.1 Mathematical Formulation of the Problem in the Root Region

To explore the mechanism of interface closure for steady deep-cellular growth, one must investigate the behavior of the solution with different length scales in the root region, $|\xi+1| \ll 1$, $|\eta-\eta_{\rm B}(\xi,\epsilon)| \ll 1$. We introduce the root variables

$$\hat{\xi} = \frac{1+\xi}{\delta(\epsilon)}, \quad \hat{\eta} = \frac{\eta-\eta_{\rm B}(\xi)}{\delta(\epsilon)} = \frac{\eta}{\delta(\epsilon)} - \eta_{\rm R}(\hat{\xi}). \tag{10.97}$$

Using these inner variables, the concentration field and interface shape function in the root region are respectively described in the following form:

$$C_{\rm B}(\xi,\eta,\epsilon) = C_{\rm R}(\hat{\xi},\hat{\eta},\epsilon), \quad \eta_{\rm B}(\xi,\epsilon) = \delta(\epsilon)\eta_{\rm R}(\hat{\xi},\epsilon), \tag{10.98}$$

where the subscript R refers to the root region.

We further assume that the interface shape function is analytic at the bottom of the root $\xi=-1$ or $\hat{\xi}=0$ and satisfies the root smoothness condition. Thus, one may carry out the Taylor series expansion $\eta_{\rm B}(\xi,\epsilon) = \delta\hat{\eta}_{\rm b} + \bar{\rm a}_2(1+\xi)^2 + \bar{\rm a}_3(1+\xi)^3 + \cdots$ and $\eta_{\rm R}(\hat{\xi},\epsilon) \approx \hat{P}_n(\hat{\xi}) = [\hat{\eta}_{\rm b} + {\rm a}_2\hat{\xi}^2 + {\rm a}_3\hat{\xi}^3 + \cdots + {\rm a}_n\hat{\xi}^n]$, where $n$ is any large integer.

Furthermore, in the far field of the root region, from the asymptotic behavior of the RPE part of the outer solution $\bar{h}_1(\xi) \sim {\rm a}_0(1+\xi)^{\alpha}$, as $(\xi \to -1)$, we deduce that $\eta_{\rm B}(\xi,\epsilon) \approx \delta(\epsilon)\eta_{\rm R}(\hat{\xi},\epsilon) \approx \epsilon\,{\rm a}_0(1+\xi)^{\alpha} = {\rm a}_0\epsilon\delta^{\alpha}(\epsilon)\hat{\xi}^{\alpha}$ holds as $\hat{\xi} \gg 1$. As a consequence, it we obtain $\delta(\epsilon) = \epsilon\delta^{\alpha}(\epsilon)$, or, $\delta(\epsilon) = \epsilon^{\frac{1}{1-\alpha}}$ and

$$\eta_{\rm R}(\hat{\xi},\epsilon) \approx {\rm a}_0\hat{\xi}^{\alpha}, \quad (\hat{\xi} \gg 1). \tag{10.99}$$

In view of the above, as an approximation, we may describe the interface shape function in the root region as

$$\eta_{\mathrm{R}}(\hat{\xi}, \epsilon) \approx \hat{\eta}_{\mathrm{T}}(\hat{\xi}) = \begin{cases} \hat{\eta}_{\mathrm{T}}^{(\mathrm{I})}(\hat{\xi}_*) = \mathrm{a}_0\hat{\xi}^{\alpha}, & (\mathrm{II}): \quad (\hat{\xi}_* < \hat{\xi} < \infty), \\ \hat{\eta}_{\mathrm{T}}^{(\mathrm{II})}(\hat{\xi}_*) = \hat{P}_n(\hat{\xi}), & (\mathrm{I}): \quad (0 \leq \hat{\xi} < \hat{\xi}_*), \end{cases} \tag{10.100}$$

satisfying the $(n-1)$th-order smoothness conditions

$$\hat{\eta}_{\mathrm{T}}^{(\mathrm{I})}(\hat{\xi}_*) = \hat{\eta}_{\mathrm{T}}^{(\mathrm{II})}(\hat{\xi}_*),$$
$$\tfrac{\mathrm{d}^k}{\mathrm{d}\hat{\xi}^k}\hat{\eta}_{\mathrm{T}}^{(\mathrm{I})}(\hat{\xi}_*) = \tfrac{\mathrm{d}^k}{\mathrm{d}\hat{\xi}^k}\hat{\eta}_{\mathrm{T}}^{(\mathrm{II})}(\hat{\xi}_*)\ (k = 1, 2, \ldots, n-1).$$

As a consequence, we derive that

$$\mathrm{a}_0\hat{\xi}_*^{\alpha} = \hat{\eta}_* = \hat{\eta}_{\mathrm{b}} + \mathrm{a}_2\hat{\xi}_*^2 + \mathrm{a}_3\hat{\xi}_*^3 + \cdots + \mathrm{a}_n\hat{\xi}_*^n,$$
$$\mathrm{a}_0\alpha\hat{\xi}_*^{\alpha-1} = 2\mathrm{a}_2\hat{\xi}_* + 3\mathrm{a}_3\hat{\xi}_*^2 + \cdots + n\mathrm{a}_n\hat{\xi}_*^{n-1},$$
$$\mathrm{a}_0\alpha(\alpha-1)\hat{\xi}_*^{\alpha-2} = 2\mathrm{a}_2 + (3\cdot 2)\mathrm{a}_3\hat{\xi}_* + \cdots + n(n-1)\mathrm{a}_n\hat{\xi}_*^{n-2},$$
$$\ldots$$

or in general,

$$\mathrm{a}_0(\alpha - i)!\hat{\xi}_*^{(\alpha-i)} = \sum_{i<m\leq n} \mathrm{a}_m \frac{m!\hat{\xi}_*^{m-i}}{(m-i)!} \quad (i = 0, 1, 2, \ldots, n-1). \tag{10.101}$$

The above-defined function $\hat{\eta}_{\mathrm{T}}(\hat{\xi})$ has up to $(n-1)$th order continuous derivatives in the entire root region $(0 \leq \hat{\xi} < \infty)$. With the $n$ conditions (10.101), the $n$ coefficients of $\hat{P}_n(\hat{\xi})$, $\{\hat{\eta}_{\mathrm{b}}, \mathrm{a}_2, \ldots, \mathrm{a}_n\}$, are fully determined as functions of the junction point $\hat{\xi}_*$. In the leading-order approximation, we set $n = 3$. Accordingly, the form of the interface shape is described by the function

$$\eta_{\mathrm{R}}(\hat{\xi}, \epsilon) \approx \hat{\eta}_{\mathrm{T}}(\hat{\xi}) = \begin{cases} \mathrm{a}_0\hat{\xi}^{\alpha}, & (\mathrm{II}):\ (\hat{\xi}_* < \hat{\xi} < \infty), \\ \hat{\eta}_{\mathrm{b}} + \mathrm{a}_2\hat{\xi}^2 + \mathrm{a}_3\hat{\xi}^3, & (\mathrm{I}):\ (0 \leq \hat{\xi} < \hat{\xi}_*), \end{cases} \tag{10.102}$$

as sketched in Fig. 10.13. In this case, the function $\hat{\eta}_{\mathrm{T}}(\hat{\xi})$ has continuous derivatives up to second order in the entire root region $(0 \leq \hat{\xi} < \infty)$, and the coefficients $\hat{\eta}_*$, $\hat{\eta}_{\mathrm{b}}$, $\mathrm{a}_2$, and $\mathrm{a}_3$ are derived as the following functions of $\hat{\xi}_*$:

$$\hat{\eta}_* = \mathrm{a}_0\hat{\xi}_*^{\alpha}, \quad \hat{\eta}_{\mathrm{b}} = \left[1 - \tfrac{\alpha(5-\alpha)}{6}\right]\hat{\eta}_*,$$
$$\mathrm{a}_2 = \tfrac{\alpha(3-\alpha)\hat{\eta}_*}{2\hat{\xi}_*^2}, \quad \mathrm{a}_3 = -\tfrac{\alpha(2-\alpha)\hat{\eta}_*}{3\hat{\xi}_*^3}.$$

It will be seen later that the parameter $\hat{\xi}_*$ can be determined as the eigenvalue of an eigenvalue problem.

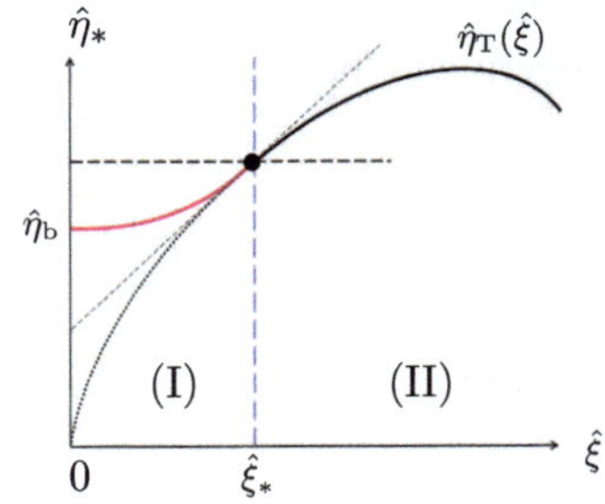


**Fig. 10.13** Sketch of the function $\hat{\eta}_{\rm T}(\hat{\xi})$ in the root region

With the approximation, one may rewrite the root variables $\hat{\xi}$ and $\hat{\eta}$ (10.97) as

$$\hat{\xi} = \frac{1+\xi}{\delta(\epsilon)}, \quad \hat{\eta} = \frac{\eta - \delta(\epsilon)\hat{\eta}_{\rm T}(\hat{\xi})}{\delta(\epsilon)} = \frac{\eta}{\delta(\epsilon)} - \hat{\eta}_{\rm T}(\hat{\xi}). \tag{10.103}$$

From (10.103), we derive that

$$\begin{aligned}
\frac{\partial}{\partial \eta} &= \frac{1}{\delta(\epsilon)}\frac{\partial}{\partial \hat{\eta}}, \quad \frac{\partial^2}{\partial \eta} = \frac{1}{\delta^2(\epsilon)}\frac{\partial^2}{\partial \hat{\eta}^2}, \\
\frac{\partial}{\partial \xi} &= \frac{1}{\delta(\epsilon)}\Big[\frac{\partial}{\partial \hat{\xi}} - \hat{\eta}_{\rm T}'(\hat{\xi})\frac{\partial}{\partial \hat{\eta}}\Big], \\
\frac{\partial^2}{\partial \xi^2} &= \frac{1}{\delta^2(\epsilon)}\Big[\frac{\partial^2}{\partial \hat{\xi}^2} + \hat{\eta}_{\rm T}'^2(\hat{\xi})\frac{\partial^2}{\partial \hat{\eta}^2} - 2\hat{\eta}_{\rm T}'(\hat{\xi})\frac{\partial^2}{\partial \hat{\eta}\partial \hat{\xi}} - \hat{\eta}_{\rm T}''(\hat{\xi})\frac{\partial}{\partial \hat{\eta}}\Big].
\end{aligned} \tag{10.104}$$

The concentration field in the root region may be derived as a function of the inner variables $\{\hat{\xi}, \hat{\eta}\}$ as follows:

$$\begin{aligned}
C_{\rm B}(\xi, \eta, \epsilon) &= C_{\rm B}\Big[-1 + \delta(\epsilon)\hat{\xi},\ \delta(\epsilon)(\hat{\eta} + \hat{\eta}_{\rm T}), \epsilon\delta(\epsilon)(\hat{\eta} + \hat{\eta}_{\rm T})\Big] \\
&= C_{\rm R}(\hat{\xi}, \hat{\eta}, \epsilon).
\end{aligned} \tag{10.105}$$

As before, we write the root solution in two parts:

$$\begin{aligned}
C_{\rm R}(\hat{\xi}, \hat{\eta}, \epsilon) &= \bar{C}_{\rm R}(\hat{\xi}, \hat{\eta}, \epsilon) + \tilde{C}_{\rm R}(\hat{\xi}, \hat{\eta}, \epsilon), \\
\eta_{\rm R}(\xi, \epsilon) &= \bar{\eta}_{\rm R}(\hat{\xi}, \epsilon) + \tilde{\eta}_{\rm R}(\hat{\xi}, \epsilon),
\end{aligned} \tag{10.106}$$

where $\bar{C}_{\rm R}(\hat{\xi}, \hat{\eta}, \epsilon)$ and $\bar{\eta}_{\rm R}(\hat{\xi}, \epsilon)$ are the RPE part of the root solution, while $\tilde{C}_{\rm R}(\hat{\xi}, \hat{\eta}, \epsilon)$ and $\tilde{\eta}_{\rm R}(\hat{\xi}, \epsilon)$ are the SPE part of the root solution. As is known from the previous derivations, the RPE part of the root solution will be uniquely determined with no arbitrary constant, and it matches the RPE part of the outer solution. One may write that $\bar{\eta}_{\rm R}(\hat{\xi}, \epsilon) = \hat{\eta}_{\rm T}(\hat{\xi}) + (\text{h.o.t.})$, and one can easily find that

$$\bar{C}_{\rm R}(\hat{\xi}, \hat{\eta}, \epsilon) = y_{*0} - \epsilon \ln \delta(\epsilon) W\lambda_G \hat{Y}_* + \epsilon\Big[y_{*1} - W\lambda_G \hat{Y}_0(\hat{\xi}, \hat{\eta})\Big] + (\text{h.o.t.}), \tag{10.107}$$

which matches the outer solution, $\bar{C}_{\rm B}(\xi, \eta, \epsilon)$.

Our major task is to find the SPE part of the root solution, $\{\tilde{C}_R(\hat{\xi}, \hat{\eta}, \epsilon), \tilde{\eta}_R(\hat{\xi}, \epsilon)\}$, which is transcendentally small as $\epsilon \to 0$ and is to be matched with the SPE part of the outer solution.

### 10.3.2 The SPE Part of the Root Solution in the Root Region

The inner systems governing the SPE-part of the solution $\{\tilde{C}_R(\hat{\xi}, \hat{\eta}, \epsilon), \tilde{\eta}_R(\hat{\xi}, \epsilon)\}$ can be obtained from (10.9)–(10.13). In doing so, one needs to change the outer variables $(\xi, \eta)$ to the root variables $(\hat{\xi}, \hat{\eta})$ and expand all the coordinate functions into their asymptotic forms, as $\epsilon \to 0$ with fixed $(\hat{\xi}, \hat{\eta})$. For convenience, we list the results below:

$$
\begin{aligned}
\mathcal{R}(\xi, \eta) &= \pi^2\delta^2(\epsilon)(\hat{\eta} + \hat{\eta}_T)^2 + \pi^2\delta^2(\epsilon)\hat{\xi}^2 + \cdots, \\
\Theta(\xi, \eta) &= -\arctan\left(\tfrac{\hat{\xi}}{\hat{\eta}+\hat{\eta}_T}\right) + \cdots;
\end{aligned}
$$

$$
\begin{aligned}
\hat{X}(\hat{\xi}, \hat{\eta}, \epsilon) &\sim -1 + \tfrac{2(1-\lambda_0)}{\pi} \arctan\left(\tfrac{\hat{\xi}}{\hat{\eta}+\hat{\eta}_T}\right) + \cdots \\
\hat{Y}(\hat{\xi}, \hat{\eta}, \epsilon) &\sim \ln\delta(\epsilon)\hat{Y}_* + \hat{Y}_0(\hat{\xi}, \hat{\eta}) + \delta(\epsilon)\hat{Y}_1(\hat{\xi}, \hat{\eta}) + \cdots, \\
Y_\eta(\xi, \eta) &= \tfrac{1}{\delta(\epsilon)}\hat{Y}_{\hat{\eta}}(\hat{\xi}, \hat{\eta}, \epsilon) \sim \tfrac{1}{\delta(\epsilon)}\hat{Y}_{\hat{\eta},0}(\hat{\xi}, \hat{\eta}) + \hat{Y}_{\hat{\eta},1}(\hat{\xi}, \hat{\eta}) + \cdots, \\
Y_\xi(\xi, \eta) &= \tfrac{1}{\delta(\epsilon)}\hat{Y}_{\hat{\xi}}(\hat{\xi}, \hat{\eta}, \epsilon) \sim \tfrac{1}{\delta(\epsilon)}\hat{Y}_{\hat{\xi},0}(\hat{\xi}, \hat{\eta}) + \cdots,
\end{aligned}
\tag{10.108}
$$

where

$$
\begin{aligned}
&\hat{Y}_* = \tfrac{2(1-\lambda_0)}{\pi}, \quad \hat{Y}_0 = \tfrac{(1-\lambda_0)}{\pi} \ln \tfrac{\pi^2\left[(\hat{\eta}+\hat{\eta}_T)^2+\hat{\xi}^2\right]}{4}, \quad \hat{Y}_1 = \lambda_0(\hat{\eta} + \hat{\eta}_T), \\
&\hat{Y}_{\hat{\eta},0}(\hat{\xi}, \hat{\eta}) = \tfrac{2(1-\lambda_0)}{\pi} \tfrac{(\hat{\eta}+\hat{\eta}_T)}{\hat{\xi}^2+(\hat{\eta}+\hat{\eta}_T)^2}, \quad \hat{Y}_{\hat{\eta},1}(\hat{\xi}, \hat{\eta}) = \lambda_0, \\
&\hat{Y}_{\hat{\xi},0}(\hat{\xi}, \hat{\eta}) = \tfrac{2(1-\lambda_0)}{\pi} \tfrac{\hat{\xi}}{\hat{\xi}^2+(\hat{\eta}+\hat{\eta}_T)^2},
\end{aligned}
\tag{10.109}
$$

and

$$
\begin{aligned}
\mathcal{G}(\xi, \eta) &= \sqrt{Y_\xi^2 + Y_\eta^2} = \tfrac{1}{\delta(\epsilon)}\hat{\mathcal{G}}(\hat{\xi}, \hat{\eta}, \epsilon) \sim \tfrac{1}{\delta(\epsilon)}\hat{\mathcal{G}}_0(\hat{\xi}, \hat{\eta}) + \cdots, \\
\hat{\mathcal{G}}_0(\hat{\xi}, \hat{\eta}) &= \tfrac{2(1-\lambda_0)}{\pi} \tfrac{1}{\sqrt{\hat{\xi}^2+(\hat{\eta}+\hat{\eta}_T)^2}}.
\end{aligned}
\tag{10.110}
$$

Moreover, one may write

$$
\begin{aligned}
\mathcal{K}\{\delta(\epsilon)\eta_R\} &= -\tfrac{1}{\hat{\mathcal{G}}_0(\hat{\xi},0)} \tfrac{\partial^2\tilde{\eta}_R}{\partial\hat{\xi}^2} + \hat{\mathcal{K}}_0(\hat{\xi}, 0) + o(\tilde{\eta}_R), \\
\hat{\mathcal{K}}_0(\hat{\xi}, 0) &= -\tfrac{\pi}{2(1-\lambda_0)} \tfrac{\hat{\eta}_T}{\sqrt{\hat{\xi}^2+\hat{\eta}_T^2}}.
\end{aligned}
\tag{10.111}
$$

Only explicitly showing the leading terms, one may write the inner equation as

$$\frac{\partial^2 \tilde{C}_R}{\partial \hat{\xi}^2} + \left[1 + \hat{\eta}_T'^2(\hat{\xi})\right]\frac{\partial^2 \tilde{C}_R}{\partial \hat{\eta}^2} - 2\hat{\eta}_T'(\hat{\xi})\frac{\partial^2 \tilde{C}_R}{\partial \hat{\eta}\partial \hat{\xi}} = \mathcal{O}(\text{h.o.t.}). \qquad (10.112)$$

The interface conditions in the root region can be linearized along $\hat{\eta} = 0$, and it follows that at $\hat{\eta} = 0$,

$$\tilde{C}_R + \frac{\partial \bar{C}_R}{\partial \hat{\eta}}\tilde{\eta}_R = -\epsilon W \lambda_G \frac{\partial \hat{Y}_0}{\partial \hat{\eta}}\tilde{\eta}_R + \frac{\epsilon^2 \hat{\Gamma}}{MW\hat{\mathcal{G}}_0(\hat{\xi}, 0)}\frac{\partial^2 \tilde{\eta}_R}{\partial \hat{\xi}^2} + \mathcal{O}(\text{h.o.t.}), \qquad (10.113)$$

and

$$\frac{\partial \tilde{C}_R}{\partial \hat{\eta}} + \frac{\partial^2 \bar{C}_R}{\partial \hat{\eta}^2}\tilde{\eta}_R - \frac{\partial \tilde{\eta}_R}{\partial \hat{\xi}}\frac{\partial \bar{C}_R}{\partial \hat{\xi}} - \epsilon W(1-\kappa)\bar{C}_R \hat{Y}_{\hat{\xi},0}\frac{\partial \tilde{\eta}_R}{\partial \hat{\xi}}$$
$$+ \epsilon W(1-\kappa)\tilde{C}_R \hat{Y}_{\hat{\eta},0} + \epsilon W(1-\kappa)\hat{Y}_{\hat{\eta},0}\frac{\partial \bar{C}_R}{\partial \hat{\eta}}\tilde{\eta}_R$$
$$+ \epsilon W(1-\kappa)\bar{C}_R \hat{Y}_{\hat{\eta}\hat{\eta},0}\tilde{\eta}_R = \mathcal{O}(\text{h.o.t.}). \qquad (10.114)$$

The system (10.112)–(10.114) is a linear homogeneous system of PDEs, which can be solved using the multiple-variables expansion method. Assuming that the SPE part of the root solutions contains multiple length scales, we introduce the following fast root variables:

$$\begin{cases} \hat{\xi}_+ = \dfrac{\hat{\phi}(\hat{\xi}, \hat{\eta})}{\hat{\beta}(\epsilon)} = \dfrac{1}{\hat{\beta}(\epsilon)}\displaystyle\int_0^{\hat{\xi}} k_*(\hat{\xi}_1, \hat{\eta})d\hat{\xi}_1, \\ \hat{\eta}_+ = \dfrac{\hat{\psi}(\hat{\xi}, \hat{\eta})}{\hat{\beta}(\epsilon)} = \dfrac{1}{\hat{\beta}(\epsilon)}\displaystyle\int_0^{\hat{\eta}} \dfrac{k_*(\hat{\xi}, \hat{\eta}_1)}{\sqrt{1+\hat{\eta}_T'^2}}d\hat{\eta}_1, \end{cases} \qquad (10.115)$$

where $\hat{\beta}$ is to be determined. In the above we have assumed that

$$\hat{\phi}_{\hat{\xi}} = \sqrt{1+\hat{\eta}_T'^2}\,\hat{\psi}_{\hat{\eta}} = k_*(\hat{\xi}, \hat{\eta}). \qquad (10.116)$$

Furthermore, we define

$$\hat{\psi}_{\hat{\xi}} = -\sqrt{1+\hat{\eta}_T'^2}\,\hat{\phi}_{\hat{\eta}} = g_*(\hat{\xi}, \hat{\eta}). \qquad (10.117)$$

In terms of these multiple variables $(\hat{\xi}, \hat{\eta}, \hat{\xi}_+, \hat{\eta}_+)$, we express the SPE-part of the root solution as

$$\tilde{C}_R(\hat{\xi}, \hat{\eta}, \epsilon) = \hat{C}_*(\hat{\xi}, \hat{\eta}, \hat{\xi}_+, \hat{\eta}_+, \epsilon), \quad \tilde{\eta}_R(\hat{\xi}, \epsilon) = \hat{\mathrm{h}}_*(\hat{\xi}, \hat{\xi}_+, \epsilon),$$

where $(\hat{\xi}, \hat{\eta}, \hat{\xi}_+, \hat{\eta}_+)$ are formally treated as the independent variables. Then the inner system (10.112)–(10.114) can be converted into a form with multiple variables $(\hat{\xi}, \hat{\eta}, \hat{\xi}_+, \hat{\eta}_+)$. We consider only the leading-order approximation in the limit of $\epsilon \to 0$:

$$\begin{aligned} \hat{C}_*(\hat{\xi}, \hat{\eta}, \hat{\xi}_+, \hat{\eta}_+, \epsilon) &\sim \epsilon \hat{b}_{*0}(\epsilon) \hat{C}_{*0}(\hat{\xi}, \hat{\eta}, \hat{\xi}_+, \hat{\eta}_+), \\ \hat{\mathrm{h}}_*(\hat{\xi}, \hat{\xi}_+, \epsilon) &\sim \hat{b}_{*0}(\epsilon) \hat{\mathrm{h}}_{*0}(\hat{\xi}, \hat{\xi}_+), \\ k_*(\hat{\xi}, \hat{\eta}, \epsilon) &\sim k_{*0}(\hat{\xi}, \hat{\eta}), \\ g_*(\hat{\xi}, \hat{\eta}, \epsilon) &\sim g_{*0}(\hat{\xi}, \hat{\eta}), \end{aligned} \tag{10.118}$$

where $|\hat{b}_{*0}(\epsilon)| \ll 1$ is an asymptotic factor.

### 10.3.3 The Zeroth-Order Approximation $\mathcal{O}(1)$

By balancing both sides of the interface conditions (10.112)–(10.114), we derive that

$$\hat{\beta} = \sqrt{\epsilon \hat{\Gamma}}, \tag{10.119}$$

and leading order equation as

$$\begin{aligned} &(k_{*0}^2 + g_{*0}^2)\left(\frac{\partial^2 \hat{C}_{*0}}{\partial \hat{\xi}_+^2} + \frac{\partial^2 \hat{C}_{*0}}{\partial \hat{\eta}_+^2}\right) + 2\hat{\eta}_T' \left[\frac{k_{*0} g_{*0}}{\mathcal{G}_T}\left(\frac{\partial^2 \hat{C}_{*0}}{\partial \hat{\xi}_+^2} - \frac{\partial^2 \hat{C}_{*0}}{\partial \hat{\eta}_+^2}\right)\right. \\ &\left. - \frac{k_{*0}^2 - g_{*0}^2}{\mathcal{G}_T} \frac{\partial^2 \hat{C}_{*0}}{\partial \hat{\xi}_+ \partial \hat{\eta}_+}\right] = 0. \end{aligned} \tag{10.120}$$

We now set $g_{*0}(\hat{\xi}, \hat{\eta}) = k_{*0}(\hat{\xi}, \hat{\eta})$. It follows from (10.116)–(10.117) that

$$k_{*0} = \int_0^{\hat{\eta}} \frac{\partial}{\partial \hat{\xi}} \left[\frac{k_{*0}}{(1 + \hat{\eta}_T'^2)^{1/2}}\right] d\hat{\eta},$$

so that

$$\frac{\partial k_{*0}}{\partial \hat{\eta}} - \frac{\partial}{\partial \hat{\xi}} \left[\frac{k_{*0}}{\sqrt{1 + \hat{\eta}_T'^2}}\right] = 0. \tag{10.121}$$

This is a governing equation of $k_{*0}(\hat{\xi},\hat{\eta})$, which can be solved for the function $k_{*0}(\hat{\xi},\hat{\eta})$, provided the boundary value $\hat{k}_0(\hat{\xi}) = k_{*0}(\hat{\xi},0)$ is determined. Furthermore, in this case, Eq. (10.120) reduces to

$$\left(1+\frac{\hat{\eta}'_{\mathrm{T}}}{\mathcal{G}_{\mathrm{T}}}\right)\frac{\partial^2\hat{C}_{*0}}{\partial\hat{\xi}_+^2}+\left(1-\frac{\hat{\eta}'_{\mathrm{T}}}{\mathcal{G}_{\mathrm{T}}}\right)\frac{\partial^2\hat{C}_{*0}}{\partial\hat{\eta}_+^2}=0, \tag{10.122}$$

where we have defined

$$\mathcal{G}_{\mathrm{T}}=\sqrt{1+\hat{\eta}'^2_{\mathrm{T}}(\hat{\xi})}. \tag{10.123}$$

The interface conditions in the leading-order approximation are derived in terms of the formula (10.107) as well as (10.113)–(10.114) as follows: at the interface $\hat{\eta}=\hat{\eta}_+=0$,

$$\hat{C}_{*0}=\frac{\hat{k}_0^2}{MW\hat{\mathcal{G}}_0(\hat{\xi},0)}\frac{\partial^2\hat{\mathrm{h}}_{*0}}{\partial\hat{\xi}_+^2}, \tag{10.124}$$

and

$$\frac{\hat{k}_0}{\mathcal{G}_{\mathrm{T}}}\left(\frac{\partial\hat{C}_{*0}}{\partial\hat{\eta}_+}-\frac{\partial\hat{C}_{*0}}{\partial\hat{\xi}_+}\right)-\hat{k}_0W\big[(1-\kappa)y_{*0}-\lambda_{\mathrm{G}}\big]\times\hat{Y}_{\xi,0}(\hat{\xi},0)\frac{\partial\hat{\mathrm{h}}_{*0}}{\partial\tilde{\xi}_+}=0. \tag{10.125}$$

The system allows the following normal mode solutions:

$$\begin{aligned}\hat{C}_{*0}&=\tilde{A}_{*0}(\hat{\xi},\hat{\eta})\exp\left\{\mathrm{i}\hat{\xi}_+-\lambda_c\hat{\eta}_+\right\},\\ \hat{\mathrm{h}}_{*0}&=\tilde{D}_{*0}\exp\left\{\mathrm{i}\hat{\xi}_+\right\},\end{aligned} \tag{10.126}$$

where the coefficient $\tilde{D}_{*0}$ is a free constant and

$$\lambda_c=[(\mathcal{G}_{\mathrm{T}}+\hat{\eta}'_{\mathrm{T}})/(\mathcal{G}_{\mathrm{T}}-\hat{\eta}'_{\mathrm{T}})]^{1/2}=\mathcal{G}_{\mathrm{T}}+\hat{\eta}'_{\mathrm{T}}. \tag{10.127}$$

From (10.124), we derive

$$\hat{A}_{*0}(\hat{\xi})=\tilde{A}_{*0}(\hat{\xi},0)=-\tilde{D}_{*0}\frac{\hat{k}_0^2}{MW\hat{\mathcal{G}}_0(\hat{\xi},0)}.$$

From (10.125), we derive

$$-\frac{\hat{A}_{*0}(\hat{\xi})}{\sqrt{1+\hat{\eta}'^2_{\mathrm{T}}}}\hat{k}_0(\lambda_c+\mathrm{i})-\mathrm{i}\hat{k}_0W\big[(1-\kappa)y_{*0}-\lambda_{\mathrm{G}}\big]\hat{Y}_{\xi,0}(\hat{\xi},0)\tilde{D}_{*0}=0,$$

or

$$\frac{\hat{A}_{*0}(\hat{\xi})}{\sqrt{1+\hat{\eta}_T'^2}}\,\hat{k}_0(\lambda_c+\mathrm{i})+\mathrm{i}\hat{k}_0 W\big[(1-\kappa)y_{*0}-\lambda_G\big]\hat{Y}_{\hat{\xi},0}(\hat{\xi},0)\tilde{D}_{*0}=0.$$

In order to satisfy the interface conditions (10.124) and (10.125), the function $\hat{k}_0(\hat{\xi})$ must satisfy the formula

$$-\frac{(\lambda_c+\mathrm{i})\hat{k}_0^3}{MW\hat{\mathcal{G}}_0(\hat{\xi},0)\sqrt{1+\hat{\eta}_T'^2}}+\mathrm{i}\hat{k}_0 W\hat{Y}_{\hat{\xi},0}(\hat{\xi},0)\big[(1-\kappa)y_{*0}-\lambda_G\big]=0. \quad (10.128)$$

There are three roots for $\hat{k}_0$:

$$\begin{aligned}
\hat{k}_0^{(1)}(\hat{\xi})&=\hat{k}_s(\hat{\xi})=\left(\frac{\mathrm{i}}{\lambda_c+\mathrm{i}}\right)^{\frac{1}{2}}\hat{\mathrm{m}}\sqrt{\hat{\mathcal{G}}_0(\hat{\xi},0)\hat{Y}_{\hat{\xi},0}(\hat{\xi},0)}\big(1+\hat{\eta}_T'^2\big)^{\frac{1}{4}},\\
\hat{k}_0^{(2)}(\hat{\xi})&=-\hat{k}_s(\hat{\xi})=-\left(\frac{\mathrm{i}}{\lambda_c+\mathrm{i}}\right)^{\frac{1}{2}}\hat{\mathrm{m}}\sqrt{\hat{\mathcal{G}}_0(\hat{\xi},0)\hat{Y}_{\hat{\xi},0}(\hat{\xi},0)}\big(1+\hat{\eta}_T'^2\big)^{\frac{1}{4}},\\
\hat{k}_0^{(3)}(\hat{\xi})&=0,
\end{aligned} \quad (10.129)$$

where we have defined

$$\hat{\mathrm{m}}=W\sqrt{M\big[(1-\kappa)y_{*0}-\lambda_G\big]}. \quad (10.130)$$

Recalling the formulas

$$\begin{cases}
\hat{\mathcal{G}}_0(\hat{\xi},0) &= \frac{2(1-\lambda_0)}{\pi}\frac{1}{\sqrt{\hat{\xi}^2+\hat{\eta}_T^2(\hat{\xi})}},\\
\hat{Y}_{\hat{\xi},0}(\hat{\xi},0) &= \frac{2(1-\lambda_0)}{\pi}\frac{\hat{\xi}}{\hat{\xi}^2+\hat{\eta}_T^2(\hat{\xi})}>0, \quad (0\le\hat{\xi}<\infty),
\end{cases} \quad (10.131)$$

one may rewrite

$$\hat{k}_s(\hat{\xi})=\left(\frac{\mathrm{i}}{\lambda_c+\mathrm{i}}\right)^{\frac{1}{2}}\frac{2(1-\lambda_0)\hat{\mathrm{m}}}{\pi}\frac{\hat{\xi}^{\frac{1}{2}}}{\big[\hat{\xi}^2+\hat{\eta}_T^2(\hat{\xi})\big]^{\frac{3}{4}}}\big(1+\hat{\eta}_T'^2\big)^{\frac{1}{4}}. \quad (10.132)$$

In accordance with the above, we obtain three fundamental solutions:

$$\begin{aligned}
\hat{H}_1(\hat{\xi})&=\mathrm{e}^{\frac{\mathrm{i}}{\sqrt{\epsilon\hat{\Gamma}}}\hat{\chi}(\hat{\xi})}=\mathrm{e}^{-\frac{\hat{\chi}_I(\hat{\xi})}{\sqrt{\epsilon\hat{\Gamma}}}}\left[\cos\left(\frac{\hat{\chi}_R(\hat{\xi})}{\sqrt{\epsilon\hat{\Gamma}}}\right)+\mathrm{i}\sin\left(\frac{\hat{\chi}_R(\hat{\xi})}{\sqrt{\epsilon\hat{\Gamma}}}\right)\right],\\
\hat{H}_2(\hat{\xi})&=\mathrm{e}^{-\frac{\mathrm{i}}{\sqrt{\epsilon\hat{\Gamma}}}\hat{\chi}(\hat{\xi})}=\mathrm{e}^{\frac{\hat{\chi}_I(\hat{\xi})}{\sqrt{\epsilon\hat{\Gamma}}}}\left[\cos\left(\frac{\hat{\chi}_R(\hat{\xi})}{\sqrt{\epsilon\hat{\Gamma}}}\right)-\mathrm{i}\sin\left(\frac{\hat{\chi}_R(\hat{\xi})}{\sqrt{\epsilon\hat{\Gamma}}}\right)\right],\\
\hat{H}_3(\hat{\xi})&=1,
\end{aligned} \quad (10.133)$$

where we have defined

$$\hat{\chi}(\hat{\xi}) = \int_a^{\hat{\xi}} \hat{k}_s(\hat{\xi}_1)\mathrm{d}\hat{\xi}_1 = \hat{\chi}_{\mathrm{R}}(\hat{\xi}) + \mathrm{i}\hat{\chi}_{\mathrm{I}}(\hat{\xi})$$

$$= \frac{2(1-\lambda_0)\hat{\mathrm{m}}}{\pi} \int_a^{\hat{\xi}} \left[ \frac{\mathrm{i}}{\lambda_c(\hat{\xi}_1)+\mathrm{i}} \right]^{\frac{1}{2}} \frac{\hat{\xi}_1^{\frac{1}{2}}(1+\hat{\eta}_{\mathrm{T}}'^2)^{\frac{1}{4}}}{[\hat{\xi}_1^2 + \hat{\eta}_{\mathrm{T}}^2(\hat{\xi}_1)]^{\frac{3}{4}}} \mathrm{d}\hat{\xi}_1. \quad (10.134)$$

Here the lower limit $a$ can be chosen arbitrarily. The fundamental solutions $\hat{H}_1(\hat{\xi}), \hat{H}_2(\hat{\xi}), \hat{H}_3(\hat{\xi})$ can be called $H$-waves. One may write the general inner solution in the root region as

$$\hat{\mathrm{h}}_{*0}(\hat{\xi}, \hat{\xi}_+) = \tilde{d}_1\hat{H}_1(\hat{\xi}) + \tilde{d}_2\hat{H}_2(\hat{\xi}) + \tilde{d}_3\hat{H}_3(\hat{\xi}), \quad (10.135)$$

where $(\tilde{d}_1, \tilde{d}_2, \tilde{d}_3)$ are arbitrary constants. To match with the outer solution, the constant solution $\hat{H}_3(\hat{\xi})$ for the interface shape function should be ruled out. Hence, one may write the general solution in the root inner region as

$$\hat{\mathrm{h}}_{*0}(\hat{\xi}, \hat{\xi}_+) = \tilde{d}_1\hat{H}_1(\hat{\xi}) + \tilde{d}_2\hat{H}_2(\hat{\xi}). \quad (10.136)$$

The pair of coefficients $(\tilde{d}_1, \tilde{d}_2)$ may have different values in different subregions, which will be determined in the following sections.

### *10.3.4 The Root Solution in Subregion (II)*

Subregion (II) contains the far field of the root region $\hat{\xi} \gg 1$, where the root solution must match the outer solution. It may be deduced that the root solution contains only one wave-type basic solution $\hat{H}_2(\hat{\xi})$ in subregion (II). As matter of fact, due to the asymptotic behavior of the outer solution that

$$\eta_{\mathrm{B}}(\xi, \epsilon) = \epsilon\Big[\bar{h}_1(\xi) + \bar{\mathrm{h}}_0(\xi, \epsilon) + \cdots\Big], \quad (10.137)$$

where

$$\bar{\mathrm{h}}_0(\xi, \epsilon) = -\mathrm{Re}\{\bar{h}_1(0)H_2(\xi)\} = -\bar{h}_1(0)\mathrm{e}^{\frac{\chi_{\mathrm{I}}(\xi)}{\sqrt{\epsilon\bar{\Gamma}}}} \cos\left(\frac{\chi_{\mathrm{R}}(\xi)}{\sqrt{\epsilon\hat{\bar{\Gamma}}}}\right) \to 0$$

decays exponentially as $\xi \to -1$, in order to match the outer solution, the root solutions of the interface shape function must have the form

$$\hat{\eta}_{\mathrm{R}}(\hat{\xi}, \epsilon) = \delta(\epsilon)\Big[\hat{\eta}_{\mathrm{T}}(\hat{\xi}) + \hat{b}_{*0}(\epsilon)\hat{d}_2\hat{H}_2(\hat{\xi}) + \cdots\Big], \quad (10.138)$$

where $\hat{H}_2(\hat{\xi})$ increases exponentially as $\hat{\xi} \to \infty$. The first terms on the right-hand sides of (10.137) and (10.138) are obviously matched. To match the SPE part of the root solution, $\hat{b}_{*0}(\epsilon)\hat{\mathrm{h}}_{*0}(\hat{\xi}, \hat{\xi}_+) = \hat{b}_{*0}(\epsilon)\hat{d}_2\hat{H}_2(\hat{\xi})$, with the SPE part of the outer solution, $\epsilon\bar{\mathrm{h}}_0(\xi, \epsilon)$, we set $\hat{\xi}_0 = \frac{1}{\delta(\epsilon)}$. Obviously, the point $\hat{\xi} = \hat{\xi}_0$ of the root variable corresponds to the point $\xi = 0$ of the outer variable. By choosing the lower limit $a = 0$, it follows from (10.134) that

$$\hat{\chi}(\hat{\xi}_0) = \frac{2(1-\lambda_0)\hat{\mathrm{m}}}{\pi} \int_0^{\hat{\xi}_0} \left[\frac{\mathrm{i}}{\lambda_c(\hat{\xi}_1) + \mathrm{i}}\right]^{\frac{1}{2}} \frac{\hat{\xi}_1^{\frac{1}{2}}\left[1 + \hat{\eta}_\mathrm{T}'^2(\hat{\xi}_1)\right]^{\frac{1}{4}}}{[\hat{\xi}_1^2 + \hat{\eta}_\mathrm{T}^2(\hat{\xi}_1)]^{\frac{3}{4}}} \mathrm{d}\hat{\xi}_1, \tag{10.139}$$

and

$$\hat{\chi}(\hat{\xi}) = \frac{2(1-\lambda_0)\hat{\mathrm{m}}}{\pi}\left\{ \int_0^{\hat{\xi}_0} \left[\frac{\mathrm{i}}{\lambda_c(\hat{\xi}_1) + \mathrm{i}}\right]^{\frac{1}{2}} \frac{\hat{\xi}_1^{\frac{1}{2}}\left[1 + \hat{\eta}_\mathrm{T}'^2(\hat{\xi}_1)\right]^{\frac{1}{4}}}{[\hat{\xi}_1^2 + \hat{\eta}_\mathrm{T}^2(\hat{\xi}_1)]^{\frac{3}{4}}} \mathrm{d}\hat{\xi}_1 \right.$$
$$\left. + \int_{\hat{\xi}_0}^{\hat{\xi}} \left[\frac{\mathrm{i}}{\lambda_c(\hat{\xi}_1) + \mathrm{i}}\right]^{\frac{1}{2}} \frac{\hat{\xi}_1^{\frac{1}{2}}\left[1 + \hat{\eta}_\mathrm{T}'^2(\hat{\xi}_1)\right]^{\frac{1}{4}}}{[\hat{\xi}_1^2 + \hat{\eta}_\mathrm{T}^2(\hat{\xi}_1)]^{\frac{3}{4}}} \mathrm{d}\hat{\xi}_1 \right\}. \tag{10.140}$$

On the other hand, in the integral

$$\chi(\xi) = \int_0^{\xi} W\sqrt{M\Delta_0\mathcal{G}_0(\xi_1, 0)}\left[1 + \frac{\mathrm{i}}{\lambda_0} Y_{\xi,0}(\xi_1, 0)\right]^{\frac{1}{2}} \mathrm{d}\xi_1, \tag{10.141}$$

with the root variable, the interval $(0, \xi)$ is changed to the interval $(\hat{\xi}_0, \hat{\xi})$. Noting the property of $\hat{\eta}_\mathrm{T}(\hat{\xi})$ shown by (10.102), one can consider that $\chi(\xi)$ matches the second part of $\hat{\chi}(\hat{\xi})$, namely

$$\int_0^{\xi} W\sqrt{M\Delta_0\mathcal{G}_0(\xi_1, 0)}\left[1 + \frac{\mathrm{i}}{\lambda_0} Y_{\xi,0}(\xi_1, 0)\right]^{\frac{1}{2}} \mathrm{d}\xi_1$$
$$\Updownarrow$$
$$\mathrm{e}^{\frac{\mathrm{i}\pi}{4}} \frac{2(1-\lambda_0)\hat{\mathrm{m}}}{\pi} \int_{\hat{\xi}_0}^{\hat{\xi}} \frac{\hat{\xi}_1^{\frac{1}{2}}\left[1 + \hat{\eta}_\mathrm{T}'^2(\hat{\xi}_1)\right]^{\frac{1}{4}}}{[\hat{\xi}_1^2 + \hat{\eta}_\mathrm{T}^2(\hat{\xi}_1)]^{\frac{3}{4}}} \mathrm{d}\hat{\xi}_1.$$

Hence, from the matching conditions of the outer solution with the inner solution, it follows that

$$\delta(\epsilon)\hat{b}_{*0}(\epsilon)\exp\left\{ \frac{-\mathrm{i}}{\sqrt{\epsilon\bar{\Gamma}}} \frac{2(1-\lambda_0)\hat{\mathrm{m}}}{\pi} \int_0^{\frac{1}{\delta}} \left[\frac{\mathrm{i}}{\lambda_c(\hat{\xi}_1)+\mathrm{i}}\right]^{\frac{1}{2}} \frac{\hat{\xi}_1^{\frac{1}{2}}\left[1+\hat{\eta}_\mathrm{T}'^2(\hat{\xi}_1)\right]^{\frac{1}{4}}}{[\hat{\xi}_1^2+\hat{\eta}_\mathrm{T}^2(\hat{\xi}_1)]^{\frac{3}{4}}} \mathrm{d}\hat{\xi}_1 \right\} = \epsilon,$$
$$\hat{d}_2 = -\bar{h}_1(0).$$

We then derive that in the far field of the root region, the root solution gives

$$\delta(\epsilon)\hat{\mathrm{h}}_{*0} = -\epsilon\Delta(\epsilon)\bar{h}_1(0)\hat{H}_2(\hat{\xi}) + \cdots, \tag{10.142}$$

where

$$\Delta(\epsilon) = \exp\left\{-\frac{1}{\sqrt{\epsilon\hat{\Gamma}}}\frac{2(1-\lambda_0)\hat{\mathrm{m}}}{\pi}\int_0^{\frac{1}{8}}\left[\frac{\mathrm{i}}{\lambda_c(\hat{\xi}_1)+\mathrm{i}}\right]^{\frac{1}{2}}\frac{\hat{\xi}_1^{\frac{1}{2}}\left[1+\hat{\eta}_{\mathrm{T}}'^2(\hat{\xi}_1)\right]^{\frac{1}{4}}}{[\hat{\xi}_1^2+\hat{\eta}_{\mathrm{T}}^2(\hat{\xi}_1)]^{\frac{3}{4}}}d\hat{\xi}_1\right\}.$$

As $\epsilon \to 0$, the factor $\Delta(\epsilon) \to 0$ decays exponentially.

### *10.3.5 Root Solution in Subregion (I)*

In subregion (I), which contains the bottom of the root region ($\hat{\xi} = \hat{\eta} = 0$), the physical solution $\mathrm{Re}\{\hat{\mathrm{h}}_{*0}(\hat{\xi})\}$ in the root region must satisfy the smoothness conditions at the bottom ($\hat{\xi} = 0, \hat{\eta} = \hat{\eta}_{\mathrm{b}}$):

$$\mathrm{Re}\{\hat{\mathrm{h}}_{*0}(0)\} = \mathrm{Re}\{\hat{\mathrm{h}}'_{*0}(0)\} = 0. \tag{10.143}$$

As a consequence, the root solution must have the form

$$\hat{\mathrm{h}}_{*0}(\hat{\xi}) = \tilde{d}_1\hat{H}_1(\hat{\xi}) + \tilde{d}_2\hat{H}_2(\hat{\xi}), \tag{10.144}$$

with $\tilde{d}_1$ and $\tilde{d}_2$ both nonzero. Now the same issues that one encountered in the previous chapters appear. How does the asymptotic solution (10.144), which consists of two waves in subregion (I) near the bottom of the root, connect with the asymptotic solution (10.142), which consists solely of one wave in the far field of the root region, subregion (II)? How does the root solution behave in the transition region? The key to settling these issues is that there exists a *simple turning point singularity* in the above-obtained asymptotic solutions. To explore such a singularity, we need to study the problem in the complex $\xi$-plane, similarly to how we proceeded in the previous chapters, and extend all the functions depending on the real variable $\hat{\xi}$, such as the wave number function $\hat{k}_s$, $\hat{\mathrm{h}}_{*0}$, $\hat{C}_{*0}$, to the corresponding complex functions by analytic continuation. In what follows, for simplicity, we shall use the same notation $\hat{\xi}$ to represent the complex variable $\hat{\xi}$ without risk of confusion.

We note from (10.132) that the wave number function has a complex root $\hat{k}_s(\hat{\xi}_c) = 0$. Indeed, from the property of $\hat{\eta}_{\mathrm{T}}(\hat{\xi})$ shown by (10.102), we may write $\hat{\eta}_{\mathrm{T}}(\hat{\xi}) = \hat{P}_3(\hat{\xi})$ approximately in the root region $\hat{\xi} = \mathcal{O}(1)$. By setting

$$\left[1+\hat{\eta}_{\mathrm{T}}'^2(\hat{\xi}_c)\right] = 1 + (2\mathrm{a}_2\hat{\xi}_c + 3\mathrm{a}_3\hat{\xi}_c^2)^2 = 0,$$

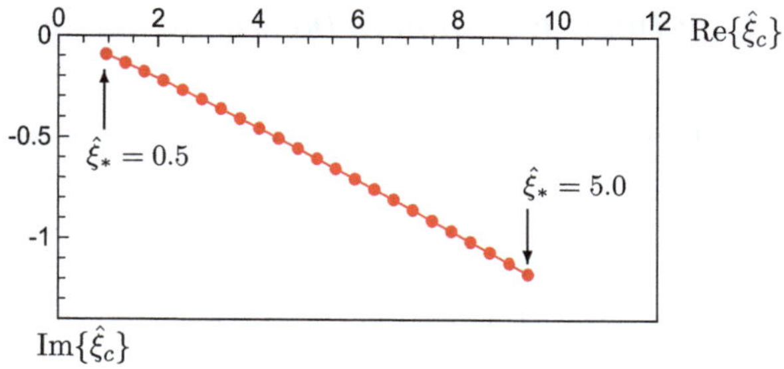


**Fig. 10.14** The variation of the turning point $\hat{\xi}_c$ in the complex $\hat{\xi}_c$-plane with the parameter $\hat{\xi}_*$ for the typical case $\alpha = 0.88, \mathrm{a} = 2.63, \kappa = 0.1, \lambda_0 = 0.6, \lambda_G = 2, M = 1.0$. The quantity $\mathrm{Re}\{\hat{\xi}_c\}$ increases as $\hat{\xi}_*$ increases from $0.5 \to 5.0$ with step size $\Delta\hat{\xi}_* = 0.2$

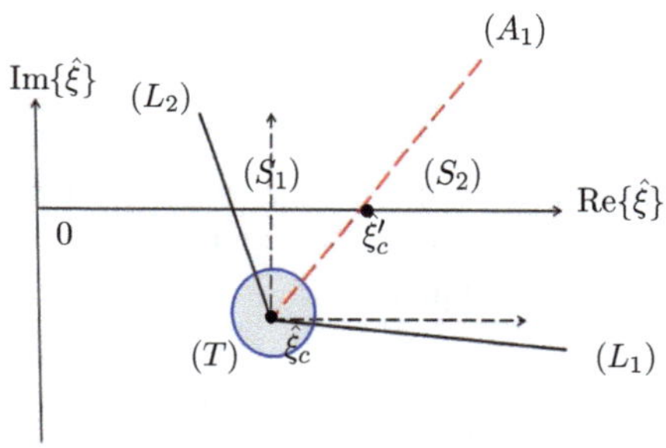


**Fig. 10.15** Sketch of the structure of Stokes and anti-Stokes lines emitted from the turning point $\hat{\xi}_c$ in the root region, where (T) is the vicinity of the turning point $\hat{\xi}_c$, $(L_1)$ and $(L_2)$ are the Stokes lines, while $(A_1)$ is the anti-Stokes line

we derive the complex turning points

$$\hat{\xi}_c = \frac{-2\mathrm{a}_2 \pm \sqrt{4\mathrm{a}_2^2 + 12\,\mathrm{i}\mathrm{a}_3}}{6\mathrm{a}_3}, \tag{10.145}$$

which are functions of the junction point of $\hat{\xi}_*$. Among the above two roots, we take the one with the minus sign; such a $\hat{\xi}_c$ is located in the fourth quadrant of the complex $\hat{\xi}$-plane with $\mathrm{Re}\{\hat{\xi}_c\} > 0, \mathrm{Im}\{\hat{\xi}_c\} < 0$, as shown in Figs. 10.14 and 10.15.

The presence of the turning point $\hat{\xi}_c$ plays a crucial role for the interface closure. The complex $\hat{\xi}$-plane will be divided by the Stokes lines and anti-Stokes lines emitted from $\hat{\xi}_c$ into several sectors, as shown in Fig. 10.15. To find the global uniformly valid root solution, one needs to study the properties of the turning point singularity and derive and solve the root inner equation in the vicinity of $(\hat{\xi}_c, 0)$ in the extended complex plane $(\hat{\xi}, \hat{\eta})$, and then match the root inner solution from both sides of the root inner region with the root solution in the root subregions (I) and (II), respectively.

### 10.3.6 The Root Inner Equation in the Vicinity of $(\hat{\xi}_c, 0)$

We introduce the following root inner variables in the vicinity of $(\hat{\xi}_c, 0)$, which we call subregion (T):

$$\tilde{\xi}_* = \frac{(\hat{\xi} - \hat{\xi}_c)}{(\epsilon\hat{\Gamma})^{\hat{\alpha}}}, \quad \tilde{\eta}_* = \frac{\hat{\eta}}{(\epsilon\hat{\Gamma})^{\hat{\alpha}}}. \tag{10.146}$$

It is evident that in subregion (T), the singular perturbation expansion part of the root solution does not have multiple length scales and cannot be expressed in the form of multiple variables. To derive the system valid in subregion (T), one needs to begin with the system (10.112)–(10.114). Now, denote the SPE-part of the root solution in subregion (T) by

$$\tilde{C}_*(\tilde{\xi}_*, \tilde{\eta}_*, \epsilon) = \hat{C}_S(\hat{\xi}, \hat{\eta}, \epsilon), \quad \tilde{h}_*(\tilde{\xi}_*, \epsilon) = \hat{\eta}_S(\hat{\xi}, \epsilon). \tag{10.147}$$

By neglecting the higher-order terms, as the leading-order approximation in the vicinity of the turning point, $\{\tilde{C}_*(\tilde{\xi}_*, \tilde{\eta}_*, \epsilon) \approx \tilde{C}_{*0}(\tilde{\xi}_*, \tilde{\eta}_*), \tilde{h}_*(\tilde{\xi}_*, \epsilon) \approx \tilde{h}_{*0}(\tilde{\xi}_*)\}$ satisfies the following root inner equation:

$$\frac{\partial^2 \tilde{C}_{*0}}{\partial \tilde{\xi}_*^2} + \left[1 + \hat{\eta}_T'^2(\hat{\xi})\right]\frac{\partial^2 \tilde{C}_{*0}}{\partial \tilde{\eta}_*^2} - 2\hat{\eta}_T'(\hat{\xi})\frac{\partial^2 \tilde{C}_{*0}}{\partial \tilde{\eta}_* \partial \tilde{\xi}_*} = 0. \tag{10.148}$$

We consider the following type of solutions for $\tilde{C}_{*0}$:

$$\tilde{C}_{*0} = \tilde{A}_{*0} \exp\left\{\mathrm{i}\tilde{\xi}_* - \tilde{\lambda}_c \tilde{\eta}_*\right\}. \tag{10.149}$$

By substituting (10.149) into (10.148), we obtain

$$-1 + \tilde{\lambda}_c^2(1 + \hat{\eta}_T'^2) + \mathrm{i}2\hat{\eta}_T'(\hat{\xi})\tilde{\lambda}_c = 0,$$

and

$$\tilde{\lambda}_c = \frac{1}{1 + \mathrm{i}\hat{\eta}_T'(\hat{\xi})}, \quad (\mathrm{Re}\{\tilde{\lambda}_c\} > 0). \tag{10.150}$$

From (10.149), we deduce that

$$\frac{\partial \tilde{C}_{*0}}{\partial \tilde{\eta}_*} = \mathrm{i}\tilde{\lambda}_c \frac{\partial \tilde{C}_{*0}}{\partial \tilde{\xi}_*}. \tag{10.151}$$

Furthermore, in the leading-order approximation we derive that $\tilde{C}_{*0} = \mathcal{O}(\epsilon \tilde{h}_{*0})$, and the interface conditions at the interface at $\tilde{\eta}_* = 0$ are

$$\tilde{C}_{*0} = \frac{1}{MW\hat{\mathcal{G}}_0(\hat{\xi},0)} \frac{\partial^2 \tilde{h}_{*0}}{\partial \tilde{\xi}_*^2}, \tag{10.152}$$

$$\frac{\partial \tilde{C}_{*0}}{\partial \tilde{\eta}_*} - W(1-\kappa) y_{*0} \hat{Y}_{0,\hat{\xi}}(\hat{\xi},0) \frac{\partial \tilde{h}_{*0}}{\partial \tilde{\xi}_*} = 0. \tag{10.153}$$

By applying the relationship (10.151), from (10.153) we derive that

$$\mathrm{i}\tilde{\lambda}_c \frac{\partial \tilde{C}_{*0}}{\partial \tilde{\xi}_*} - W\big[(1-\kappa)y_{*0} - \lambda_{\mathrm{G}}\big] \hat{Y}_{\hat{\xi},0}(\hat{\xi},0) \frac{\partial \tilde{h}_{*0}}{\partial \tilde{\xi}_*} = 0. \tag{10.154}$$

By combining (10.154) with (10.152), we derive

$$\mathrm{i}\tilde{\lambda}_c \frac{1}{MW\hat{\mathcal{G}}_0(\hat{\xi},0)} \frac{\partial^3 \tilde{h}_{*0}}{\partial \tilde{\xi}_*^3} - W\big[(1-\kappa)y_{*0} - \lambda_{\mathrm{G}}\big] \hat{Y}_{\hat{\xi},0}(\hat{\xi},0) \frac{\partial \tilde{h}_{*0}}{\partial \tilde{\xi}_*} = 0,$$

which can be rewritten as

$$\frac{\partial^3 \tilde{h}_{*0}}{\partial \tilde{\xi}_*^3} + \tilde{k}_s^2(\hat{\xi}) \frac{\partial \tilde{h}_{*0}}{\partial \tilde{\xi}_*} = 0. \tag{10.155}$$

In (10.155), we have defined

$$\tilde{k}_s^2(\hat{\xi}) = \mathrm{i}\hat{\mathrm{m}}^2 \hat{\mathcal{G}}_0(\hat{\xi},0) \hat{Y}_{\hat{\xi},0}(\hat{\xi},0)\big[1 + \mathrm{i}\hat{\eta}'_{\mathrm{T}}(\hat{\xi})\big],$$

so that

$$\tilde{k}_s(\hat{\xi}) = \mathrm{e}^{\frac{\mathrm{i}\pi}{4}} \frac{2(1-\lambda_0)\hat{\mathrm{m}}}{\pi} \frac{\hat{\xi}^{\frac{1}{2}}}{\big[\hat{\xi}^2 + \hat{\eta}_{\mathrm{T}}^2(\hat{\xi})\big]^{\frac{3}{4}}} \big[1 + \mathrm{i}\hat{\eta}'_{\mathrm{T}}(\hat{\xi})\big]^{\frac{1}{2}}. \tag{10.156}$$

It is seen that at $\hat{\xi} = \hat{\xi}_c$, we have $\tilde{k}_s(\hat{\xi}_c) = 0$, since $\big[1 + \mathrm{i}\hat{\eta}'_{\mathrm{T}}(\hat{\xi}_c)\big] = 0$. Moreover, one can write that as $\hat{\xi} \to \hat{\xi}_c$,

$$\tilde{k}_s^2(\hat{\xi}) = \mathcal{A}^2(\hat{\xi} - \hat{\xi}_c) + \mathcal{O}(\text{h.o.t.}),$$

where

$$\begin{aligned} \mathcal{A}^2 &= \mathrm{i}\frac{4(1-\lambda_0)^2\hat{\mathrm{m}}^2}{\pi^2} \frac{\hat{\xi}_c}{\big[\hat{\xi}_c^2 + \hat{\eta}_{\mathrm{T}}^2(\hat{\xi}_c)\big]^{\frac{3}{2}}} \big[\mathrm{i}\hat{\eta}''_{\mathrm{T}}(\hat{\xi}_c)\big] \\ &= -\frac{4(1-\lambda_0)^2\hat{\mathrm{m}}^2}{\pi^2} \frac{\hat{\xi}_c}{\big[\hat{\xi}_c^2 + \hat{\eta}_{\mathrm{T}}^2(\hat{\xi}_c)\big]^{\frac{3}{2}}} (2\mathrm{a}_2 + 6\mathrm{a}_3\hat{\xi}_c), \end{aligned} \tag{10.157}$$

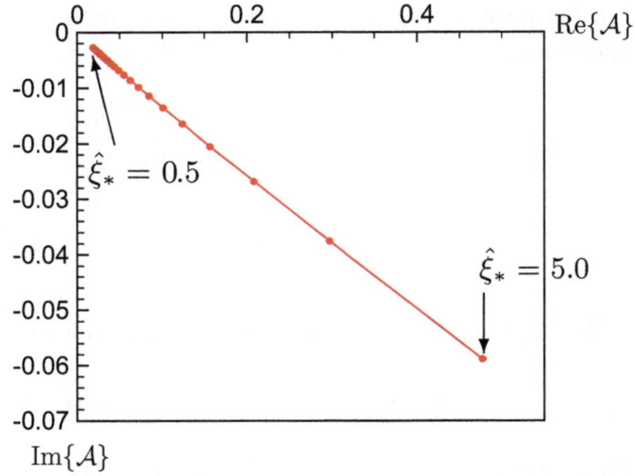


**Fig. 10.16** The variation of the coefficient $\mathcal{A}$ in the complex $\mathcal{A}$-plane with the parameter $\hat{\xi}_*$ for the typical case $\alpha = 0.88$, a $= 2.63$, $\kappa = 0.1$, $\lambda_0 = 0.6$, $\lambda_G = 2$, $M = 1$. The quantity Re$\{\mathcal{A}\}$ decreases as $\hat{\xi}_*$ increases from $0.5 \to 5.0$ with step size $\Delta\hat{\xi}_* = 0.2$. The argument arg$\{\mathcal{A}\}$ is approximately equal to $-0.04\pi$

which is a complex function of $\hat{\xi}_*$, as shown in Fig. 10.16. We derive

$$\tilde{k}_s^2(\hat{\xi}) \approx \mathcal{A}^2(\epsilon\hat{\Gamma})^{\hat{\alpha}}\tilde{\xi}_*.$$

As a consequence, in the leading-order approximation, (10.155) is reduced to

$$(\epsilon\hat{\Gamma})^{1-2\hat{\alpha}}\frac{\mathrm{d}^2 W_{*0}}{\mathrm{d}\tilde{\xi}_*^2} + (\epsilon\hat{\Gamma})^{\hat{\alpha}}\mathcal{A}^2\tilde{\xi}_* W_{*0} = 0. \tag{10.158}$$

Here we have defined

$$W_{*0}(\tilde{\xi}_*) = \frac{\mathrm{d}\hat{\mathrm{h}}_{*0}}{\mathrm{d}\tilde{\xi}_*}(\tilde{\xi}_*). \tag{10.159}$$

By setting $\hat{\alpha} = \frac{1}{3}$, we derive the Airy equation

$$\frac{\mathrm{d}^2 W_{*0}}{\mathrm{d}\tilde{\xi}_*^2} + \mathcal{A}^2\tilde{\xi}_* W_{*0} = 0. \tag{10.160}$$

Furthermore, by letting

$$\bar{\xi}_* = \mathcal{A}^{\frac{2}{3}}\tilde{\xi}_* = \mathcal{A}^{\frac{2}{3}}\frac{(\hat{\xi} - \hat{\xi}_c)}{(\epsilon\hat{\Gamma})^{\frac{1}{3}}},$$

one may further change (10.160) into the standard form of the Airy equation:

$$\frac{\mathrm{d}^2 W_{*0}}{\mathrm{d}\bar{\xi}_*^2} + \bar{\xi}_* W_{*0} = 0. \tag{10.161}$$

It is now seen clearly that the system has a simple turning point singularity at $\hat{\xi} = \hat{\xi}_c$, or $\bar{\xi}_* = 0$. In Fig. 10.15, we have sketched some Stokes lines and anti-Stokes lines emitted from $\hat{\xi}_c$ in the complex $\hat{\xi}$-plane. The anti-Stokes line $(A_1)$ divides the complex $\hat{\xi}$-plane into two sectors, $(S_1)$ and $(S_2)$, and intersects the real $\hat{\xi}$-axis at $\hat{\xi}'_c$. The intersection point $\hat{\xi}'_c$ divides the real $\hat{\xi}$-axis into two subintervals $(0, \hat{\xi}'_c)$ and $(\hat{\xi}'_c, \infty)$. The pair of constants $(\tilde{d}_1, \tilde{d}_2)$ in (10.136) will have different values in the sectors $(S_1)$ and $(S_2)$, and accordingly in the subintervals $(0, \hat{\xi}'_c)$ and $(\hat{\xi}'_c, \infty)$.

The open angle between two neighboring Stokes lines and two neighboring anti-Stokes lines emitted from the turning point $\hat{\xi}_c$ is $\theta = \frac{2}{1+2}\pi = \frac{2}{3}\pi$. The general solution of (10.161) is

$$W_{*0}(\overline{\xi}_*) = D_{*1}\overline{\xi}_*^{\frac{1}{2}} H^{(1)}_{\frac{1}{3}}(z) + D_{*2}\overline{\xi}_*^{\frac{1}{2}} H^{(2)}_{\frac{1}{3}}(z), \quad (z = \tfrac{2}{3}\overline{\xi}_*^{\frac{3}{2}}), \tag{10.162}$$

where $H^{(1)}_{\frac{1}{3}}(z)$ is the the Hankel function of the first kind of order $\frac{1}{3}$, while $H^{(2)}_{\frac{1}{3}}(z)$ is the Hankel function of the second kind of order $\frac{1}{3}$.

### 10.3.7 Matching

Away from the turning point $\hat{\xi} = \hat{\xi}_c$, the root inner solution in subregion (T) must match the root solution in the sector $(S_2)$, as well as the root solution in the sector $(S_2)$. The matching conditions will determine the unknown coefficients in the root solutions, which leads to a quantization condition for the eigenvalue $\hat{\xi}_*$.

### 10.3.8 The Root Inner Solutions in Subregion (T) Matching the Root Solution in the Sector $(S_2)$

We now let $\overline{\xi}_* \to \infty$ starting from the turning point $\overline{\xi}_* = 0$ in the sector $(S_2)$. From the solution (10.162), we derive that in the domain $(S_2)\bigcap(T)$, we have $D_{*1} = 0$, and the root inner solution can be written as

$$W_{*0}(\overline{\xi}_*) = D_{*2}\overline{\xi}_*^{\frac{1}{2}} H^{(2)}_{\frac{1}{3}}(z). \tag{10.163}$$

Hence, as $\overline{\xi}_* \to \infty$, we have

$$W_{*0}(\overline{\xi}_*) = W^{(T)}_{*0} = D_{*2}\overline{\xi}_*^{\frac{1}{2}} \mathrm{e}^{\mathrm{i}(5/12)\pi} \sqrt{\frac{2}{\pi z}}\mathrm{e}^{-\mathrm{i}z},$$

and

$$D_{*2}\mathrm{e}^{\mathrm{i}(5/12)\pi} = -\epsilon\Delta(\epsilon)h_1(0), \tag{10.164}$$

which may be considered an incoming W-wave in the sector $(S_2)$. The root solution in section (II) can finally be written as

$$\begin{aligned}\eta_{\mathrm{R}}(\hat{\xi},\epsilon) &= \delta(\epsilon)\hat{\eta}_{\mathrm{T}}(\hat{\xi}) - \epsilon\Delta(\epsilon)h_1(0)\hat{H}_2(\hat{\xi}) + \cdots \\ &= \delta(\epsilon)\hat{\eta}_{\mathrm{T}}(\hat{\xi}) - \epsilon h_1(0)\exp\left\{-\frac{1}{\sqrt{\epsilon\hat{\Gamma}}}\frac{2(1-\lambda_0)\hat{\mathrm{m}}}{\pi}\int_{\hat{\xi}}^{\frac{1}{\delta}}\left[\frac{\mathrm{i}}{\lambda_c+\mathrm{i}}\right]^{\frac{1}{2}}\right. \\ &\quad \left.\times\frac{\hat{\xi}_1^{\frac{1}{2}}\left[1+\hat{\eta}_{\mathrm{T}}'^2(\hat{\xi}_1)\right]^{\frac{1}{4}}}{[\hat{\xi}_1^2+\hat{\eta}_{\mathrm{T}}^2(\hat{\xi}_1)]^{\frac{3}{4}}}\mathrm{d}\hat{\xi}_1\right\}\hat{H}_2(\hat{\xi}) + \cdots .\end{aligned} \tag{10.165}$$

## 10.3.9 *The Root Inner Solutions in Subregion (T) Matching the Root Solution in the Sector $(S_1)$*

In sector $(S_1)$, according to the theory of Hankel functions, one has the formula

$$H^{(2)}_{\frac{1}{3}}(z\mathrm{e}^{\mathrm{i}\pi}) = H^{(2)}_{\frac{1}{3}}(z) + \mathrm{e}^{\mathrm{i}\frac{1}{3}\pi}H^{(1)}_{\frac{1}{3}}(z) \quad (-\pi < \arg(z) \leq \pi).$$

Hence, we deduce that the root inner solution (10.163) in $(S_1)\bigcap(T)$ can be expressed in the following form:

$$W_{*0}(\overline{\xi}_*) = D_{*2}\overline{\xi}_*^{\frac{1}{2}}\left[H^{(2)}_{\frac{1}{3}}(z) + \mathrm{e}^{\mathrm{i}\pi/3}H^{(1)}_{\frac{1}{3}}(z)\right]. \tag{10.166}$$

Thus, letting $\overline{\xi}_* \to \infty$ in sector $(S_1)$, we obtain

$$\begin{aligned}W_{*0}(\overline{\xi}_*) &= D_{*2}\overline{\xi}_*^{\frac{1}{2}}\left[\sqrt{\frac{2}{\pi z}}\mathrm{e}^{-\mathrm{i}z+\mathrm{i}(5/12)\pi} + \mathrm{e}^{\mathrm{i}\pi/3}\sqrt{\frac{2}{\pi z}}\mathrm{e}^{\mathrm{i}z-\mathrm{i}(5/12)\pi}\right] \\ &= -\epsilon\Delta(\epsilon)h_1(0)\overline{\xi}_*^{\frac{1}{2}}\sqrt{\frac{2}{\pi z}}\left[\mathrm{e}^{-\mathrm{i}z} - \mathrm{i}\mathrm{e}^{\mathrm{i}z}\right] \\ &= \tilde{b}_2W^{(-)}_{*0}(\hat{\xi}) + \tilde{b}_1W^{(+)}_{*0}(\hat{\xi}),\end{aligned} \tag{10.167}$$

where $\tilde{b}_2/\tilde{b}_1 = \mathrm{i}$.

On the other hand, for the root inner solution (10.144) in sector $(S_1)$,

$$\hat{\mathrm{h}}_{*0}(\hat{\xi}) = \tilde{d}_1\hat{H}_1(\hat{\xi}) + \tilde{d}_2\hat{H}_2(\hat{\xi}), \tag{10.168}$$

where the fundamental solutions $\hat{H}_1(\hat{\xi})$ and $\hat{H}_2(\hat{\xi})$ can be specified as

$$\hat{H}_1(\hat{\xi}) = \exp\left\{\frac{\mathrm{i}}{\sqrt{\epsilon\hat{\Gamma}}}\int_{\hat{\xi}_c}^{\hat{\xi}}\hat{k}_s(\hat{\xi}_1)\mathrm{d}\hat{\xi}_1\right\}, \quad \hat{H}_2(\hat{\xi}) = \exp\left\{-\frac{\mathrm{i}}{\sqrt{\epsilon\hat{\Gamma}}}\int_{\hat{\xi}_c}^{\hat{\xi}}\hat{k}_s(\hat{\xi}_1)\mathrm{d}\hat{\xi}_1\right\},$$

by setting the lower limit as $a = \hat{\xi}_c$. Moreover, one may express it in terms of the $W$-wave representation as

$$\hat{W}_{*0}(\hat{\xi}) = \tilde{d}_1 \hat{W}_{*0}^{(+)}(\hat{\xi}) + \tilde{d}_1 \hat{W}_{*0}^{(-)}(\hat{\xi}), \tag{10.169}$$

where, according to the definition, one may write

$$\hat{W}_{*0}(\hat{\xi}) \propto \hat{\mathrm{h}}'_{*0}(\hat{\xi}), \quad \hat{W}_{*0}^{(+)}(\hat{\xi}) \propto \hat{H}'_1(\hat{\xi}), \quad \hat{W}_{*0}^{(-)}(\hat{\xi}) \propto \hat{H}'_2(\hat{\xi})$$

with the same proportional constant. By comparing (10.167) with (10.169) and noting that

$$W_{*0}^{(+)}(\bar{\xi}_*) \quad \text{outgoing W-wave with } \tilde{k}_s$$
$$\Updownarrow$$
$$\hat{W}_{*0}^{(+)}(\hat{\xi}) \quad \text{W-wave with } \hat{k}_s,$$

and

$$W_{*0}^{(-)}(\bar{\xi}_*) \quad \text{incoming W-wave with } -\tilde{k}_s$$
$$\Updownarrow$$
$$\hat{W}_{*0}^{(-)}(\hat{\xi}) \quad \text{W-wave with } -\hat{k}_s,$$

we deduce that in order for the inner solution in the vicinity of the turning point (T) to match the root solution in the sector $(S_1)$, the connection conditions

$$\tilde{d}_2 = -\epsilon\Delta(\epsilon)h_1(0); \quad \tilde{d}_2/\tilde{d}_1 = \tilde{b}_2/\tilde{b}_1 = \mathrm{i}, \quad \tilde{d}_1 = -\mathrm{i}$$

must be satisfied. As a consequence, the root solution in the sector $(S_1)$ (10.144) becomes

$$\hat{\mathrm{h}}_{*0}(\hat{\xi}) = \tilde{d}_2\Big[\hat{H}_2(\hat{\xi}) - \mathrm{i}\hat{H}_1(\hat{\xi})\Big]. \tag{10.170}$$

### 10.3.10 *The Quantization Condition for the Eigenvalue $\hat{\xi}_*$*

From (10.170), the corresponding physical solution in the sector $(S_1)$ $(0 \le \hat{\xi} \le \hat{\xi}'_c)$ is derived as

$$\hat{\mathrm{h}}_{*0}(\hat{\xi}) = \tilde{d}_2\mathrm{Re}\Big\{\hat{H}_2(\hat{\xi}) - \mathrm{i}\hat{H}_1(\hat{\xi})\Big\}, \tag{10.171}$$

which should satisfy the root conditions at the bottom of the root region ($\hat{\xi} = 0, \hat{\eta} = \hat{\eta}_b$):

$$\hat{\mathrm{h}}_{*0}(0) = 0, \quad \hat{\mathrm{h}}'_{*0}(0) = 0. \tag{10.172}$$

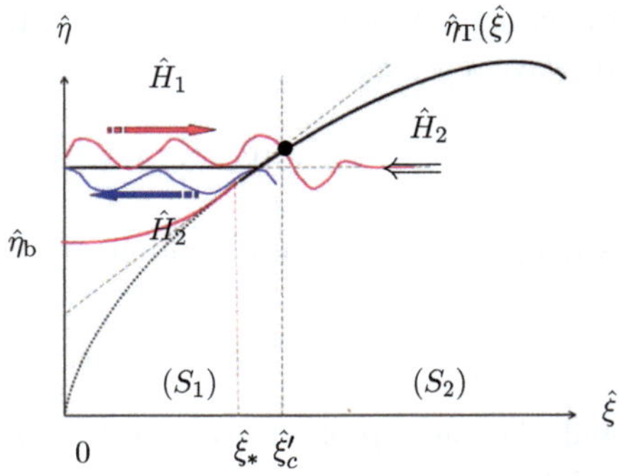


**Fig. 10.17** Sketch of wave diagram with $W$-wave representation in the root region

The behavior of the solution over the interval $(0 \leq \hat{\xi} < \hat{\xi}'_c)$ has a similarity to the Schrödinger waves in quantum mechanics trapped in a finite potential well $\hat{\eta}_{\mathrm{T}}(\hat{\xi})$, as sketched in Fig. 10.17. The waves in the subregion $(S_1)$ are trapped between the bottom of the root $\hat{\xi} = 0$ and the reflection point $\xi'_c$. Such a wave diagram may be called the *trapped-waves mechanism*, which was first discovered in the system of dendritic growth.

Due to $H'_1(0) = H'_3(0) = 0$, the smoothness condition $\hat{\mathrm{h}}'_{*0}(0) = 0$ is always satisfied. To satisfy the condition $\hat{\mathrm{h}}_{*0}(0) = 0$, we require

$$\mathrm{Re}\Big[\hat{H}_2(0) - \mathrm{i}\hat{H}_1(0)\Big] = 0. \tag{10.173}$$

Recalling that

$$\begin{aligned}\hat{\chi}(\hat{\xi}) &= \int_{\hat{\xi}_c}^{\hat{\xi}} \hat{k}_s(\hat{\xi}_1)\mathrm{d}\hat{\xi}_1 = \hat{\chi}_{\mathrm{R}}(\hat{\xi}) + \mathrm{i}\hat{\chi}_{\mathrm{I}}(\hat{\xi}) \\ &= \frac{2(1-\lambda_0)\hat{\mathrm{m}}}{\pi}\int_{\hat{\xi}_c}^{\hat{\xi}} \left[\frac{\mathrm{i}}{\lambda_c(\hat{\xi}_1)+\mathrm{i}}\right]^{\frac{1}{2}} \frac{\hat{\xi}_1^{\frac{1}{2}}(1+\hat{\eta}'^2_{\mathrm{T}})^{\frac{1}{4}}}{[\hat{\xi}_1^2+\hat{\eta}^2_{\mathrm{T}}(\hat{\xi}_1)]^{\frac{3}{4}}}\mathrm{d}\hat{\xi}_1,\end{aligned} \tag{10.174}$$

we define

$$\begin{aligned}\hat{Q}_{\mathrm{R}} &= -\frac{\hat{\chi}_{\mathrm{R}}(0)}{\sqrt{\epsilon\hat{\Gamma}}} \\ &= \frac{2}{\sqrt{\epsilon\hat{\Gamma}}}\frac{(1-\lambda_0)\hat{\mathrm{m}}}{\pi}\mathrm{Re}\left\{\int_0^{\hat{\xi}_c}\left[\frac{\mathrm{i}}{\lambda_c(\hat{\xi}_1)+\mathrm{i}}\right]^{\frac{1}{2}}\frac{\hat{\xi}_1^{\frac{1}{2}}[1+\hat{\eta}'^2_{\mathrm{T}}(\hat{\xi}_1)]^{\frac{1}{4}}}{[\hat{\xi}_1^2+\hat{\eta}^2_{\mathrm{T}}(\hat{\xi}_1)]^{\frac{3}{4}}}\mathrm{d}\hat{\xi}_1\right\}, \\ \hat{Q}_{\mathrm{I}} &= -\frac{\hat{\chi}_{\mathrm{I}}(0)}{\sqrt{\epsilon\hat{\Gamma}}} \\ &= \frac{2}{\sqrt{\epsilon\hat{\Gamma}}}\frac{(1-\lambda_0)\hat{\mathrm{m}}}{\pi}\mathrm{Im}\left\{\int_0^{\hat{\xi}_c}\left[\frac{\mathrm{i}}{\lambda_c(\hat{\xi}_1)+\mathrm{i}}\right]^{\frac{1}{2}}\frac{\hat{\xi}_1^{\frac{1}{2}}[1+\hat{\eta}'^2_{\mathrm{T}}(\hat{\xi}_1)]^{\frac{1}{4}}}{[\hat{\xi}_1^2+\hat{\eta}^2_{\mathrm{T}}(\hat{\xi}_1)]^{\frac{3}{4}}}\mathrm{d}\hat{\xi}_1\right\}.\end{aligned} \tag{10.175}$$

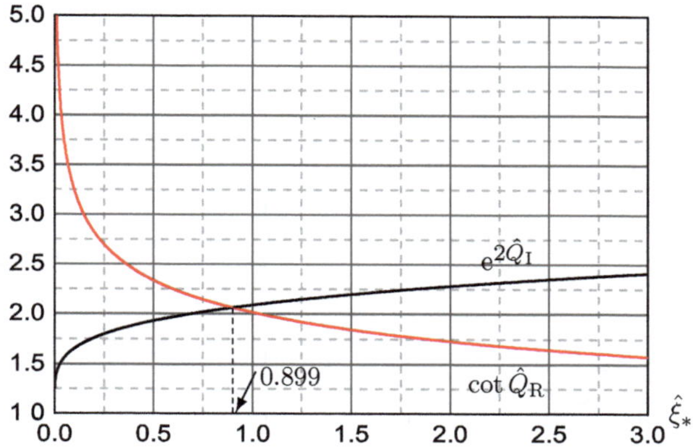


**Fig. 10.18** The graphs of the functions $Q_R(\hat{\xi}_*)$ and $Q_I(\hat{\xi}_*)$ for the typical case $\epsilon = 0.1, \kappa = 0.1, \lambda_0 = 0.4, \lambda_G = 1.0, M = 1.0$, and $\hat{\Gamma} = 1.5$. The two curves intersect at $\hat{\xi}_* = 0.899$

The bottom smoothness condition (10.173) finally yields the quantization condition

$$e^{-\hat{Q}_I} \cos \hat{Q}_R - e^{\hat{Q}_I} \sin \hat{Q}_R = 0,$$

or

$$e^{2\hat{Q}_I} = \cot \hat{Q}_R. \tag{10.176}$$

Note that $\hat{Q} = \hat{Q}_R + iQ_I$ is a function of the turning point $\hat{\xi}_c$, while $\xi_c$ is a complex function of the junction point $\hat{\xi}_*$, so that $\hat{Q}$ is a composite function of $\hat{\xi}_*$, since $\xi_c$ is a complex function of $\hat{\xi}_*$. Therefore, the quantization condition (10.176) yields the eigenvalue $\hat{\xi}_*$ as a function of the Péclet number $\epsilon$, the thermodynamic properties of the system, as well as the growth conditions. The graphs of $\hat{Q}_R$ and $Q_I$ as functions of $\hat{\xi}_*$ for a typical case are shown in Fig. 10.18. The curves intersect at $\hat{\xi}_* = 1.02$.

The variations of eigenvalue $\hat{\xi}_*$ with $\epsilon$ for the typical cases $\hat{\Gamma} = 1.0, 1.5, 2.0$ are shown in Fig. 10.19a. The variations of the corresponding $\hat{\xi}_c$ with $\epsilon$ for the case $\hat{\Gamma} = 1.0$ are shown in Fig. 10.19b. The outer solutions for these cases yield $\alpha \approx 0.88$ and $a_0 \approx 2.63$. These two parameters are rather insensitive to the parameters $\kappa$ and $\hat{\Gamma}$. It is seen that the value of $\hat{\xi}_*$ increases with increasing surface tension parameter $\hat{\Gamma}$ and Péclet number $\epsilon$.

Once $\hat{\xi}_*$ is determined, the parameters $\hat{\eta}_*$ and $\hat{\eta}_b$ are determined, then the inner solution in the whole root region $(0 \leq \hat{\xi} < \infty)$ is determined, which may be written as

$$\eta_R(\hat{\xi}, \epsilon) = \delta(\epsilon)\hat{\eta}_T(\hat{\xi}) - \epsilon\Delta(\epsilon)\bar{h}_1(0)e^{\frac{\hat{\chi}_I(\hat{\xi})}{\sqrt{\epsilon\hat{\Gamma}}}} \cos\left(\frac{\hat{\chi}_R(\hat{\xi})}{\sqrt{\epsilon\hat{\Gamma}}}\right) + \cdots. \tag{10.177}$$

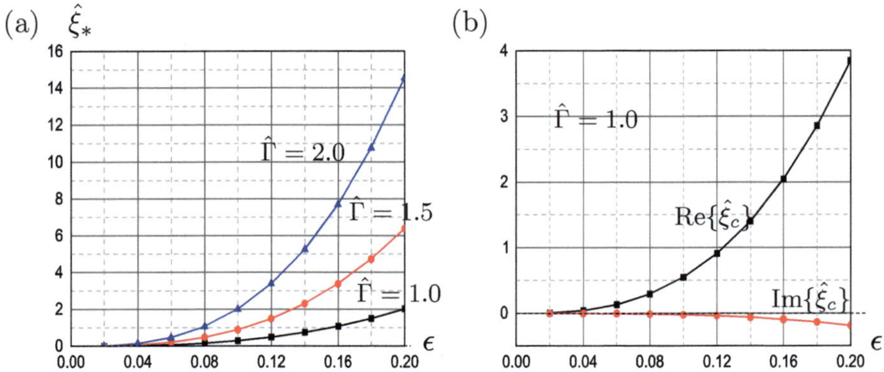


**Fig. 10.19** For the cases $\kappa = 0.1$, $\lambda_0 = 0.4$, $M = 1.0$, $\lambda_G = 1.0$, (**a**) the variations of $\hat{\xi}_*$ with the parameter $\epsilon$; (**b**) the variations of $\mathrm{Re}\{\hat{\xi}_c\}$ and $\mathrm{Im}\{\hat{\xi}_c\}$ with the parameter $\epsilon$. It is calculated from the outer solutions for these cases that $\alpha = 0.88$ and $a_0 = 2.63$

Together with the outer solution

$$\eta_B(\xi, \epsilon) = \epsilon\Big[\bar{h}_1(\xi) - \bar{h}_1(0)e^{\frac{\chi_I(\xi)}{\sqrt{\epsilon\hat{\Gamma}}}}\cos\Big(\frac{\chi_R(\xi)}{\sqrt{\epsilon\hat{\Gamma}}}\Big)\Big] + \cdots, \tag{10.178}$$

the composite solution in the whole physical region, including the outer region and inner region, is derived as

$$\begin{aligned}\eta_B(\xi, \epsilon) &= \delta(\epsilon)\hat{\eta}_T(\hat{\xi}) - \epsilon a_0(1+\xi)^\alpha \\ &+\epsilon\Big[\bar{h}_1(\xi) - \bar{h}_1(0)e^{\frac{\chi_I(\xi)}{\sqrt{\epsilon\hat{\Gamma}}}}\cos\Big(\frac{\chi_R(\xi)}{\sqrt{\epsilon\hat{\Gamma}}}\Big)\Big] + \cdots.\end{aligned} \tag{10.179}$$

The composite solutions $\eta_B(\xi)$ for the cases of $\lambda_0 = 0.4$, $\hat{\Gamma} = 1.0, 1.5, 2.0$ are shown in Fig. 10.20. The global interface shapes in the $(X, Y)$-plane for these cases are shown in Fig. 10.21. In the figure, for comparison we also show the Saffman–Taylor solution with a black dashed line and the outer solution with a solid red line. It is seen that in the root region, the difference between the Saffman–Taylor solution and the solution obtained is substantial. It is also seen that the effect of the surface tension parameter on the pattern formation in the root region is very important. The smaller the value of $\hat{\Gamma}$, the longer the total length of the finger.

In terms of the global asymptotic solution obtained, one can determine the properties of steady arrayed-cellular growth in the root region and the their dependence on the operating parameters, which include the total length of the finger, $Y_b$; the mean curvature of the bottom of the finger, $\mathcal{K}_b$; the concentration of impurity at the bottom of the root, $C_{root}$. The results are qualitatively consistent with those shown in [5], but there are some quantitative discrepancies between them, due to the correction of the quantization condition. It is anticipated that more precise

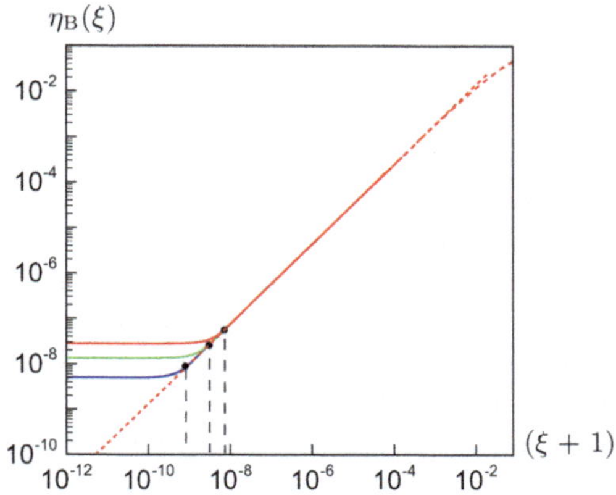


**Fig. 10.20** Graph of the composite solution $\eta_B(\xi)$ in the $(\xi, \eta)$-plane for the typical case $\epsilon = 0.1$, $\kappa = 0.1$, $\lambda_G = 1.0$, $\lambda_0 = 0.4$, $M = 1.0$ near $\xi = -1$ for $\hat{\Gamma} = 1.0, 1.5, 2.0$ from bottom to top on the left-hand side of the figure. The *bullets* show the junction points $\hat{\xi}_*$ in the root region. It is calculated from the outer solutions for these cases that $\alpha = 0.88$, $a_0 = 2.63$, and $\delta(\epsilon) = 0.4642 \times 10^{-8}$

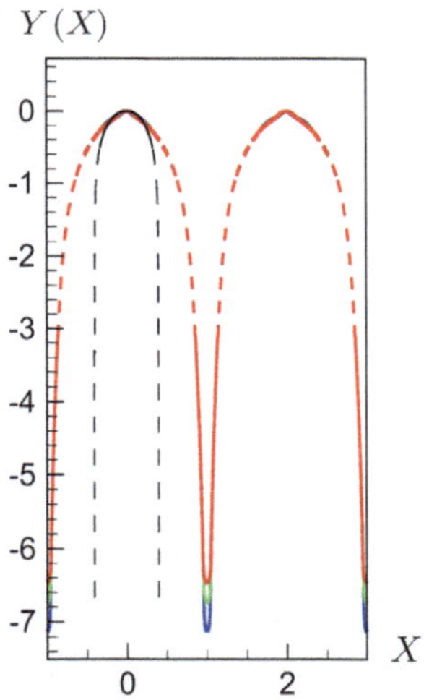


**Fig. 10.21** Interface shapes drawn in the $(X, Y)$-plane for the typical case $\epsilon = 0.1$, $\kappa = 0.1$, $M = 1.0$, and $\lambda_0 = 0.4$, $\lambda_G = 1.0$, $\hat{\Gamma} = 1.0, 1.5, 2.0$. It is seen that the total length of the finger increases with decreasing values of $\hat{\Gamma}$. In the figures, the *black dashed line* is given by the Saffman–Taylor solution, the red dashed line is given by the outer solution, which is not valid in the root region, while the *solid red line* is given by the root solution

experiments of deep-cellular growth will be performed in the future and that the experimental data on these quantities will become available. Thus, one may further examine the theoretical predictions in the light of new experimental data.

## 10.4 Global Instabilities, Origin and Essence of Side Branches

Let us now turn to investigating the global instability mechanisms of arrayed-cellular growth. As demonstrated before, the experimental results performed in Hele–Shaw cells show that for a system of cellular growth in a Hele–Shaw cell, when the pulling velocity $V$ is sufficiently small, the interface will be flat, located somewhere between the two zones. When the pulling velocity increases beyond a critical number $V_c$, the flat interface becomes unstable, due to the Mullins–Sekerka instability (see [28, 29]). Beyond $V_c$, the interface will continuously evolve, and after a complicated dynamical process, it eventually forms a steady arrayed-cellular structure with the tip located at, say, $y = 0$. These cells are aligned periodically with equal primary spacing $\ell_w$ and the relative width of cell $\lambda_0$ defined as the ratio of the asymptotic width of the cell to the primary spacing $\ell_w$ (see Fig. 10.22). It is further observed that when the pulling velocity is larger than a second critical number, the cellular array transforms to a dendritic array. The basic issues for the above phenomenon include determining the limiting steady state of arrayed-cellular growth as the time $t$ approaches $\infty$, whether the selected limiting state is unique, the stability properties of the steady state, the origin and essence of the side-branching appearing on the interface, and the mechanism for the transition from cellular array to dendritic array. Experimental results show that during the arrayed-dendritic growth, the frequency of oscillation at the interface is a constant independent of the distance behind of the tip, and the power spectrum of the interfacial oscillation contains a sequence of peaks with progressively decreasing heights. The following questions arise: What is the nature of the constant frequency observed? Are there special physical implications of these peaks and the distances between each peak in

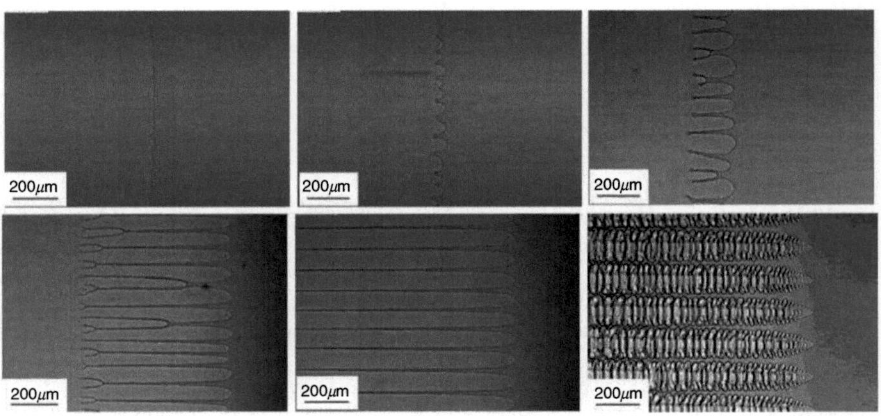


**Fig. 10.22** Some typical arrayed cellular and dendritic interface patterns observed in experiments of directional solidification, arranged in order of increasing pulling speed from left to right and from top to bottom. (Courtesy of W.D. Huang, Xin Lin)

the power spectrum obtained? Are there any intrinsic connections among all these data, and if so, what are they?

To answer all these questions, a profound understanding of the stabilities of a system of cellular growth is the critical element. This section will focus on this topic. It will explore the global stability mechanisms of the steady-state solutions in terms of the two-dimensional model of a single finger-like (or needle-like) crystal growth with isotropic surface tension in a channel. It will be shown that the system under study allows two types of global instability mechanisms: low-frequency (LF) instability and global trapped-wave (GTW) instability, which were first identified in Chaps. 6 and 7 for systems of free dendritic growth. These two stability mechanisms yield two branches of neutral curves: the finger-like branch, which yields a smooth steady or nearly steady arrayed-cellular interfacial pattern, and the tree-like branch, which yields a self-sustaining, oscillatory arrayed-dendritic interfacial pattern.

It is proposed that for the system of a finger-like crystal growth in a channel, as $t \to \infty$ the system selects the *most dangerous neutrally stable mode* uniquely as the limiting state. As a consequence, the oscillatory pattern observed is described by the *most dangerous neutrally stable GTW mode*. This linear neutrally stable GTW mode is expected to evolve to the limit cycles in the state space in the nonlinear regime. However, for the system of deep-cellular growth in the Hele–shaw cell with infinitely large width, the global neutral curve gives only the relationship between the tip radius $\ell_t$ and the primary spacing $\ell_w$, given the growth conditions and the material properties. In other words, in such a case, the limiting state and the selected primary spacing are not unique.

Moreover, it is found that the nature of the side-branching structure observed in cellular growth in directional solidification turns out to be the same as what is observed in free single dendritic growth, and the essence of side-branching formation is due to the presence of a discrete set of oscillatory GTW modes in the system and is generated by the so-called *trapped-wave instability mechanism* involving the reflections and amplification of interfacial waves. The side-branching structure observed is self-sustaining and can persist without noise continuously added to the system.

These findings agree with the experimental results by Georgelin and Pocheau [32] and the numerical simulations by Echebarria et al. [11] that were published in recent years.

### *10.4.1 Mathematical Formulation of General Unsteady Cellular Growth*

So far, we have been using the Péclet number $\epsilon$ as the basic small parameter. However, as indicated before, one may use the stability parameter $\varepsilon = \sqrt{\Gamma}$ as the basic parameter instead, since $\varepsilon$ and $\epsilon$ have the same order of magnitude. In this section, for convenience we shall use for the stability analysis the surface tension stability $\varepsilon$ as the basic small parameter. In doing so, we shall convert every $\epsilon$ appearing in the mathematical formulas to $\varepsilon$ in terms of the relationship $\epsilon = \varepsilon\hat{\text{Pe}}$.

As in the previous sections, we begin with the study of a finger-like crystal growth in the channel with fixed side walls at $x = \pm W$.

In the curvilinear coordinate system $(\xi, \eta)$, the general unsteady cellular growth is subject to the governing equation

$$\frac{\partial^2 C}{\partial \xi^2} + \frac{\partial^2 C}{\partial \eta^2} = \varepsilon W\hat{\text{Pe}}\left(\mathcal{G}^2\frac{\partial C}{\partial t} - Y_\xi\frac{\partial C}{\partial \xi} - X_\xi\frac{\partial C}{\partial \eta}\right) \tag{10.180}$$

and the following boundary conditions:

1. In the far field, as $\eta \to \infty$, $C \to 1$.
2. At the side walls, $\xi = \pm 1$: $\frac{\partial C}{\partial \xi} = 0$.
3. At the interface $\eta = \eta_s(\xi, t, \varepsilon)$,
   (a) the Gibbs–Thomson condition:
   $$C = y_* - \varepsilon W\hat{\text{Pe}}\lambda_G Y(\xi, \eta_s) - \frac{\varepsilon^2}{MW}\mathcal{K}\{\eta_s\},$$
   where $\mathcal{K}$ is the twice mean curvature, $y_* = \varepsilon\hat{\text{Pe}}\lambda_G y_0$, and
   (b) the mass conservation condition:
   $$\frac{\partial C}{\partial \eta} - \eta_s'\frac{\partial C}{\partial \xi} + \varepsilon W\hat{\text{Pe}}(1-\kappa)C\big(\mathcal{G}^2\frac{\partial \eta_s}{\partial t} - Y_\xi\eta_s' + Y_\eta\big) = 0,$$
   where $\kappa$ is the segregation coefficient.

In addition to the above boundary conditions, there are the smoothness conditions at the cell's tip and the properly imposed conditions at the bottom of the root.

## 10.4.2 The Basic Steady-State Solutions

In previous section, we obtained the global solution of steady cellular growth, whose interface shape in a typical case is shown in Fig. 10.21. This solution contains two free parameters, the tip radius $\ell_t$ and the primary spacing $\ell_w$, which will be used as the basic state. The global steady-state solution as $\varepsilon \to 0$ is written as

$$\begin{aligned} C_B(\xi, \eta, \varepsilon) &= \left[1 + (y_{*0} - 1)e^{-\varepsilon\hat{\text{Pe}}W\eta}\right] \\ &\quad + \varepsilon\hat{\text{Pe}}\Big\{y_{*1}e^{-\varepsilon\hat{\text{Pe}}W\eta} + W\lambda_G\big[\eta - Y(\xi, \eta)\big]\Big\} + \cdots \\ \eta_B(\xi, \varepsilon) &= \varepsilon\hat{\text{Pe}}h_1(\xi) + (\text{h.o.t.}). \end{aligned}$$

In the subregion near the interface, $\eta = \mathcal{O}(1)$, the steady solution for the concentration field can be simplified as

$$C_{\rm B} = y_{*0} + \varepsilon\hat{\rm Pe}\Big\{y_{*1} - W\lambda_G Y(\xi,\eta) + W\big[\lambda_{\rm G} - (y_{*0} - 1)\big]\eta\Big\} \\ + \cdots, \tag{10.181}$$

where $y_{*0} = \frac{1+\lambda_{\rm G}(1-\lambda_0)}{1-\lambda_0(1-\kappa)}$, $y_{*1} = \frac{W(1-\kappa)\lambda_0\lambda_{\rm G}\beta_0}{(1-\kappa)\lambda_0-1}$, and $\beta_0 = -\frac{2(1-\lambda_0)}{\pi}\ln 2$; $\lambda_0 = \frac{y_{*0}-1-\lambda_{\rm G}}{(1-\kappa)y_{*0}-\lambda_{\rm G}}$. Here $\lambda_0$ is the asymptotic width parameter, which is connected with the primary spacing $W$ via the formula $W = \frac{\pi(1-\lambda_0)}{2\lambda_0^2} + \cdots$, $(\varepsilon \to 0)$.

### 10.4.3 Linear Perturbed System

We split the unsteady solutions into two parts:

$$\begin{aligned} C(\xi,\eta,t,\varepsilon) &= C_{\rm B}(\xi,\eta,\varepsilon) + \widehat{C}(\xi,\eta,t,\varepsilon), \\ \eta_{\rm s}(\xi,t,\varepsilon) &= \eta_{\rm B}(\xi,\varepsilon) + \widehat{\eta}_{\rm s}(\xi,t,\varepsilon). \end{aligned} \tag{10.182}$$

The first part is the steady basic state, while the second part is the perturbed states. We assume that $\lambda_0$ is a given constant, and thus the primary spacing $W_0$ and cell's tip location $y_*$ are also undisturbed.

Substituting (10.182) into the above system, after linearization one obtains the linear perturbed system, whose coefficients are functions of the variables $(\xi,\eta)$ depending on the basic state $C_{\rm B}$ and the small parameter $\varepsilon$. Up to the first-order approximation $\mathcal{O}(\varepsilon)$ in the region near the interface, from (10.180) we derive the following formulas:

$$\begin{aligned} \frac{\partial C_{\rm B}}{\partial \eta}(\xi,\eta_{\rm B},\varepsilon) &\approx \varepsilon W\hat{\rm Pe}(\lambda_{\rm G} - y_{*0} + 1) - \varepsilon W\hat{\rm Pe}\lambda_G Y_\eta(\xi,0), \\ \frac{\partial^2 C_{\rm B}}{\partial \eta^2}(\xi,\eta_{\rm B},\varepsilon) &\approx -\varepsilon W\hat{\rm Pe}\lambda_G Y_{\eta\eta}(\xi,0), \\ \frac{\partial C_{\rm B}}{\partial \xi}(\xi,\eta_{\rm B},\varepsilon) &\approx -\varepsilon W\hat{\rm Pe}\lambda_G Y_\xi(\xi,0), \\ \frac{\partial^2 C_{\rm B}}{\partial \xi\partial \eta}(\xi,\eta_{\rm B},\varepsilon) &\approx -\varepsilon W\hat{\rm Pe}\lambda_G Y_{\xi\eta}(\xi,0). \end{aligned} \tag{10.183}$$

The perturbed system can then be written as follows:

$$\frac{\partial^2 \widehat{C}}{\partial \xi^2} + \frac{\partial^2 \widehat{C}}{\partial \eta^2} = \varepsilon W\hat{\rm Pe}\left(\mathcal{G}^2\frac{\partial \widehat{C}}{\partial t} - Y_\xi\frac{\partial \widehat{C}}{\partial \xi} - X_\xi\frac{\partial \widehat{C}}{\partial \eta}\right) \tag{10.184}$$

with the following boundary conditions:

1. In the far field, as $\eta \to \infty$, $\widehat{C} \to 0$.
2. At the side walls, $\xi = \pm 1$, $\frac{\partial \widehat{C}}{\partial \xi} = 0$.
3. At the interface $\eta = \eta_{\mathrm{B}}$,

$$\widehat{C} + \widehat{\eta}_{\mathrm{s}} C_{\mathrm{B},\eta}(\xi, 0) = -\varepsilon W \hat{\mathrm{Pe}} \lambda_{\mathrm{G}} Y_{\eta}(\xi, 0) \widehat{\eta}_{\mathrm{s}} + \frac{\varepsilon^2}{MW\mathcal{G}(\xi, 0)} \frac{\partial^2 \widehat{\eta}_{\mathrm{s}}}{\partial \xi^2} + (\text{h.o.t.}) \qquad (10.185)$$

and

$$\frac{\partial \widehat{C}}{\partial \eta} + \frac{\partial^2 C_{\mathrm{B}}}{\partial \eta^2} \widehat{\eta}_{\mathrm{s}} - \eta_{\mathrm{B},\xi} \frac{\partial \widehat{C}}{\partial \xi} - C_{\mathrm{B},\xi}(\xi, 0) \widehat{\eta}_{\mathrm{s},\xi}$$
$$+ \varepsilon W \hat{\mathrm{Pe}} (1 - \kappa) C_{\mathrm{B}}(\xi, 0) \big[ \mathcal{G}^2(\xi, 0) \widehat{\eta}_{\mathrm{s},t} - Y_{\xi} \widehat{\eta}_{\mathrm{s},\xi} \big]$$
$$+ \varepsilon W \hat{\mathrm{Pe}} (1 - \kappa) Y_{\eta}(\xi, 0) \widehat{C}$$
$$+ \varepsilon W \hat{\mathrm{Pe}} (1 - \kappa) C_{\mathrm{B}}(\xi, 0) Y_{\eta\eta}(\xi, 0) \widehat{\eta}_{\mathrm{s}} + (\text{h.o.t.}) = 0. \qquad (10.186)$$

### *10.4.4 Multiple-Variable Expansion (MVE) Solutions of Perturbed States in the Outer Region*

To solve the perturbed system, as before one may apply the multiple-variable expansion method with the fast variables defined as

$$\begin{cases} \xi_{+} = \dfrac{\phi(\xi, \eta)}{\beta(\varepsilon)} = \dfrac{1}{\beta(\varepsilon)} \displaystyle\int_{\xi_0}^{\xi} k(\xi_1, \eta) \mathrm{d}\xi_1, \\ \eta_{+} = \dfrac{\psi(\xi, \eta)}{\beta(\varepsilon)} = \dfrac{1}{\beta(\varepsilon)} \displaystyle\int_{0}^{\eta} k(\xi, \eta_1) \mathrm{d}\eta_1, \\ t_{+} = \dfrac{t}{\beta(\varepsilon)}, \end{cases} \qquad (10.187)$$

where the lower limit $\xi_0$ is to be specified later. It is seen that $\phi_{\xi} = \psi_{\eta} = k(\xi, \eta), \phi(\xi_0, \eta, \varepsilon) = 0, \psi(\xi, 0, \varepsilon) = 0$. Furthermore, we assume that the pair of functions $(\phi, \psi)$ is subject to the Cauchy–Riemann conditions $\phi_{\eta} = -\psi_{\xi} = g(\xi, \eta), g(\xi, 0) = g(\xi_0, \eta) = 0$, so that $g_{\xi} = k_{\eta}, g_{\eta} = -k_{\xi}$ (refer to Chaps. 6–9).

With the multiple variables $(\xi, \eta, \xi_+, \eta_+, t_+)$, we define

$$\widehat{C}(\xi, \eta, t, \varepsilon) = \tilde{C}(\xi, \eta, \xi_+, \eta_+, t_+, \varepsilon),$$
$$\widehat{\eta}_{\mathrm{s}}(\xi, t, \varepsilon) = \tilde{\eta}_{\mathrm{s}}(\xi, \xi_+, t_+, \varepsilon),$$

and write

$$\begin{aligned} C(\xi, \eta, t, \varepsilon) &= C_{\mathrm{B}}(\xi, \eta, \varepsilon) + \tilde{C}(\xi, \eta, \xi_+, \eta_+, t_+, \varepsilon), \\ \eta_{\mathrm{s}} &= \eta_{\mathrm{B}}(\xi, \varepsilon) + \tilde{\eta}_{\mathrm{s}}(\xi, \xi_+, t_+, \varepsilon). \end{aligned} \tag{10.188}$$

The system of Eq. (10.184) with the far-field condition, side-wall condition, and interface conditions (10.185)–(10.186) can be converted into the form of a multiple-variable system as follows:

$$\begin{aligned} &(k^2 + g^2)\Big(\frac{\partial^2 \tilde{C}}{\partial \xi_+^2} + \frac{\partial^2 \tilde{C}}{\partial \eta_+^2}\Big) + 2\beta\Big(k\frac{\partial^2}{\partial \xi_+ \partial \xi} - g\frac{\partial^2}{\partial \eta_+ \partial \xi}\Big)\tilde{C} \\ &+2\beta\Big(g\frac{\partial^2}{\partial \xi_+ \partial \eta} + k\frac{\partial^2}{\partial \eta_+ \partial \eta}\Big)\tilde{C} \\ &= \varepsilon\beta W\hat{\mathrm{Pe}}\left[\mathcal{G}^2\frac{\partial}{\partial t_+} - Y_\xi\left(k\frac{\partial}{\partial \xi_+} - g\frac{\partial}{\partial \eta_+} + \beta\frac{\partial}{\partial \xi}\right)\right. \\ &\left.-X_\xi\left(g\frac{\partial}{\partial \xi_+} + k\frac{\partial}{\partial \eta_+} + \beta\frac{\partial}{\partial \eta}\right)\right]\tilde{C} \\ &-\beta^2\Big(\frac{\partial^2}{\partial \xi^2} + \frac{\partial^2}{\partial \eta^2}\Big)\tilde{C} \\ &(-1 < \xi < 0, -\infty < \xi_+ < 0; 0 \le \eta, \eta_+ < \infty), \end{aligned} \tag{10.189}$$

with the following boundary conditions:

1. Away from the interface, $\eta_+ \to +\infty$,

$$\tilde{C} \to 0\,. \tag{10.190}$$

2. At the side wall, $\xi = -1$ and $\xi_+ \to -\infty$,

$$\frac{\partial \tilde{C}}{\partial \xi_+} = 0\,. \tag{10.191}$$

3. At the interface, we further take the Taylor expansion for the solution $(\tilde{C}, \tilde{\eta}_s)$ around $\eta = \eta_+ = 0$. It follows that at $\eta = \eta_+ = 0$,

$$\tilde{C} + \frac{\varepsilon h_1 \bar{k}}{\beta} \frac{\partial \tilde{C}}{\partial \eta_+} = -\Big[C_{B,\eta}(\xi, 0) + \varepsilon W\hat{Pe}\lambda_G Y_\eta(\xi, 0)\Big]\tilde{\eta}_s + \frac{\varepsilon^2 \bar{k}^2}{\beta^2(\varepsilon) MW\mathcal{G}_0(\xi)} \frac{\partial^2 \tilde{\eta}_s}{\partial \xi_+^2} + (\text{h.o.t}) \,; \tag{10.192}$$

and

$$\begin{aligned}
&\Big(\beta\frac{\partial \tilde{C}}{\partial \eta} + \bar{k}\frac{\partial \tilde{C}}{\partial \eta_+}\Big) - \eta_{B,\xi}\Big(\beta\frac{\partial \tilde{C}}{\partial \xi} + \bar{k}\frac{\partial \tilde{C}}{\partial \xi_+}\Big) + \beta\frac{\partial^2 C_B}{\partial \eta^2}\tilde{\eta}_s \\
&-C_{B,\xi}(\xi, 0)\Big(\beta\frac{\partial \tilde{\eta}_s}{\partial \xi} + \bar{k}\frac{\partial \tilde{\eta}_s}{\partial \xi_+}\Big) \\
&+\varepsilon W\hat{Pe}(1-\kappa)C_B(\xi, 0)\Big[\mathcal{G}_0^2(\xi)\Big(\beta\frac{\partial \tilde{\eta}_s}{\partial t} + \frac{\partial \tilde{\eta}_s}{\partial t_+}\Big) \\
&-Y_{\xi,0}(\xi)\Big(\beta\frac{\partial \tilde{\eta}_s}{\partial \xi} + \bar{k}\frac{\partial \tilde{\eta}_s}{\partial \xi_+}\Big)\Big] \\
&+\varepsilon\beta W\hat{Pe}(1-\kappa)Y_\eta(\xi, 0)\tilde{C} \\
&+\varepsilon\beta W\hat{Pe}(1-\kappa)C_B(\xi, 0)Y_{\eta\eta}(\xi, 0)\tilde{\eta}_s \\
&+(\text{h.o.t.}) = 0.
\end{aligned} \tag{10.193}$$

In the above, we have defined $\bar{k} = k(\xi, 0)$ and defined the functions $\mathcal{G}_0(\xi)$ and $Y_{\xi,0}(\xi)$ as

$$\begin{cases} \mathcal{G}_0(\xi) = \mathcal{G}(\xi, 0) = \sqrt{\lambda_0^2 + (1-\lambda_0)^2 \tan^2\left(\frac{\pi\xi}{2}\right)}, \\ Y_{\xi,0}(\xi) = Y_\xi(\xi, 0) = -(1-\lambda_0)\tan\left(\frac{\pi\xi}{2}\right). \end{cases} \tag{10.194}$$

4. At the tip, $\xi = \xi_+ = 0$,

$$\tilde{\eta}_s = 0; \quad \beta\frac{\partial \tilde{\eta}_s}{\partial \xi} + k\frac{\partial \tilde{\eta}_s}{\partial \xi_+} = 0. \tag{10.195}$$

5. Root condition: Assume that away from the tip, as $\xi_+ \to \infty$, the perturbed state cannot be growing **exponentially**. Under this assumption, it can be further shown that as $\xi_+ \to \infty$, the solution must be vanishing, namely

$$\tilde{\eta}_s \to 0. \tag{10.196}$$

Make the following singular perturbed expansion (SPE) for the solution:

$$\begin{aligned}
\tilde{C}(\xi, \xi_+, \eta, \eta_+, t_+, \varepsilon) &\sim \varepsilon\Big[\tilde{\mu}_0(\varepsilon)\tilde{C}_0(\xi, \eta, \xi_+, \eta_+) \\
&\quad + \tilde{\mu}_1(\varepsilon)\tilde{C}_1(\xi, \eta, \xi_+, \eta_+)\Big]e^{\sigma t_+} + \cdots \\
\tilde{\eta}_s(\xi, \xi_+, \varepsilon) &\sim \Big[\tilde{b}_0(\varepsilon)\tilde{h}_0(\xi, \xi_+) + \tilde{b}_1(\varepsilon)\tilde{h}_1(\xi, \xi_+) \\
&\quad + \cdots\Big]e^{\sigma t_+} \\
k(\xi, \eta, \varepsilon) &\sim k_0(\xi, \eta) + \varepsilon k_1(\xi, \eta) + \cdots, \\
\bar{k}(\xi, \varepsilon) &\sim \bar{k}_0(\xi) + \varepsilon \bar{k}_1(\xi) + \cdots, \\
\sigma(\varepsilon) &\sim \sigma_0 + \varepsilon\sigma_1 + \cdots.
\end{aligned} \tag{10.197}$$

#### 10.4.4.1 Leading-Order SPE Solution $\mathcal{O}(\varepsilon\tilde{\mu}_0(\varepsilon))$

In the leading-order approximation, we obtain $\tilde{b}_0(\varepsilon) = \tilde{\mu}_0(\varepsilon)$ and $\beta(\varepsilon) = \varepsilon^{1/2}$ and the equation

$$\frac{\partial^2 \tilde{C}_0}{\partial \xi_+^2} + \frac{\partial^2 \tilde{C}_0}{\partial \eta_+^2} = 0. \tag{10.198}$$

The boundary conditions are as follows:

1. In the far field away from the interface, $\eta_+ \to +\infty$,

$$\tilde{C}_0 \to 0.$$

2. At the side wall, $\xi = 1$,

$$\frac{\partial \tilde{C}_0}{\partial \xi_+} = 0.$$

3. At the interface $\eta = \eta_+ = 0$, we have

$$\tilde{C}_0 = -W\hat{\mathrm{P}}\mathrm{e}(\lambda_G - y_{*0} + 1)\tilde{h}_0 + \frac{\bar{k}_0^2}{MW\mathcal{G}_0(\xi)}\frac{\partial^2 \tilde{h}_0}{\partial \xi_+^2}, \tag{10.199}$$

and

$$\begin{aligned}
&\bar{k}_0\frac{\partial \tilde{C}_0}{\partial \eta_+} + \sigma_0 W\hat{\mathrm{P}}\mathrm{e}(1-\kappa)y_{*0}\bar{\mathcal{G}}_0^2(\xi)\tilde{h}_0 \\
&\quad + \bar{k}_0 W\hat{\mathrm{P}}\mathrm{e}\Big[\lambda_G - (1-\kappa)y_{*0}\Big]Y_{\xi,0}(\xi)\frac{\partial \tilde{h}_0}{\partial \xi_+} = 0,
\end{aligned} \tag{10.200}$$

where we have defined $\bar{k}_0 = k_0(\xi, 0)$.

4. At the tip, $\xi = \xi_+ = 0$,

$$\frac{\partial \tilde{h}_0}{\partial \xi_+} = \tilde{h}_0 = 0. \tag{10.201}$$

5. At the bottom of root, as $\xi_+ \to \infty$,

$$\tilde{h}_0 \to 0. \tag{10.202}$$

Consider the normal mode solutions

$$\tilde{C}_0 = \tilde{A}_0(\xi, \eta) \exp\{i\xi_+ - \eta_+\}, \quad \tilde{h}_0 = \hat{D}_0 \exp\{i\bar{\xi}_+\}, \tag{10.203}$$

where the coefficient $\hat{D}_0$ is an arbitrary complex constant. The far-field condition is automatically satisfied, provided $\mathrm{Re}\{k_0\} > 0$. Let $\hat{A}_0(\xi) = \tilde{A}_0(\xi, 0)$. By substituting (10.203) into the interface conditions (10.199) and (10.200), it follows that

$$\hat{A}_0(\xi) = -\Big[W\hat{\mathrm{Pe}}(\lambda_G - y_{*0} + 1) + \frac{\bar{k}_0^2}{MW\mathcal{G}_0(\xi)}\Big]\hat{D}_0,$$

and we have the dispersion formula

$$\sigma_0 = \Sigma(\bar{k}_0, \xi) = \frac{\bar{k}_0}{\bar{\Lambda}_0 \mathcal{G}_0^2(\xi)} \left\{ \Big[\lambda_0 + iY_{\xi,0}(\xi)\Big] - \frac{\bar{k}_0^2}{\bar{\Lambda}_1 \mathcal{G}_0(\xi)} \right\}, \tag{10.204}$$

where we have set

$$\bar{\Lambda}_0 = \frac{(1-\kappa)y_{*0}}{[(1-\kappa)y_{*0} - \lambda_G]} > 0, \tag{10.205}$$

$$\bar{\Lambda}_1 = MW^2\hat{\mathrm{Pe}}\big[(1-\kappa)y_{*0} - \lambda_G\big] > 0. \tag{10.206}$$

The dispersion formula (10.204) is a cubic polynomial in $\bar{k}_0$. It is similar to the one we derived for the system of dendritic growth in Chap. 6. For given complex parameter $\sigma_0$, one can solve for three complex wave numbers as functions of $\xi$: $\left\{\bar{k}_0^{(1)}(\xi), \bar{k}_0^{(2)}(\xi), \bar{k}_0^{(3)}(\xi)\right\}$.

#### 10.4.4.2 Transformation of Variables

For further analysis of the dispersion formula (10.204), we employ the new variable $\rho$ to replace $\xi$, which was first introduced in Chap. 9 for the investigation of the system of viscous fingering. Let

$$\rho = -\frac{Y_{\xi,0}(\xi)}{\lambda_0} = \left(\frac{1-\lambda_0}{\lambda_0}\right) \tan\left(\frac{\pi\xi}{2}\right). \tag{10.207}$$

It is seen that as $\xi \to 1^-$, $\rho \to \infty$. Thus, one may write $\mathcal{G}_0(\xi) = \lambda_0 S(\rho)$, where $S(\rho) = \sqrt{1+\rho^2}$. Moreover, one has

$$\begin{aligned}
\mathrm{d}\rho &= -\frac{Y_{\xi\xi,0}}{\lambda_0}\mathrm{d}\xi = \frac{\pi}{2}\frac{1-\lambda_0}{\lambda_0}\left[1+\tan^2\left(\frac{\pi\xi}{2}\right)\right]\mathrm{d}\xi,\\
\mathrm{d}\xi &= \frac{1}{G(\rho)}\mathrm{d}\rho,\\
G(\rho) &= -\frac{Y_{\xi\xi,0}(\xi)}{\lambda_0} = \frac{\pi}{2}\left(\frac{\lambda_0}{1-\lambda_0}\right)\left[\rho^2+\left(\frac{1-\lambda_0}{\lambda_0}\right)^2\right]\\
&= \frac{\pi}{2a}(\rho+\mathrm{i}a)(\rho-\mathrm{i}a),
\end{aligned} \tag{10.208}$$

where

$$a = \frac{1-\lambda_0}{\lambda_0}.$$

As demonstrated in Chap. 9, the new variable $\rho$ is associated with the arc length $\ell(\xi)$ measured along the interface of the basic state starting from the tip $\xi = 0$. As a matter of fact, one may write

$$\ell(\xi) = \int_0^{\xi} \mathcal{G}_0(\xi_1)\mathrm{d}\xi_1 = \int_0^{\rho} \mathcal{F}(\rho)\mathrm{d}\rho,$$

where

$$\mathcal{F}(\rho) = \frac{\mathcal{G}_0(\xi)}{G(\rho)} = -\frac{\lambda_0\mathcal{G}_0(\xi)}{Y_{\xi\xi,0}(\xi)}.$$

With the new variable $\rho$, one may express the perturbed solutions in the form

$$\begin{aligned}
\tilde{h}_0(\xi) &= \hat{D}_0 \exp\left\{\frac{\mathrm{i}}{\varepsilon^{\frac{1}{2}}}\int_{\xi_0}^{\xi}\bar{k}_0(\xi_1)\mathrm{d}\xi_1\right\}\\
&= \hat{D}_0 \exp\left\{\frac{\mathrm{i}}{\varepsilon^{\frac{1}{2}}}\int_{\rho_0}^{\rho}\tilde{k}_0(\rho_1)\mathrm{d}\rho_1\right\} = \tilde{h}_0(\rho),
\end{aligned} \tag{10.209}$$

where $\bar{k}_0(\xi) = \tilde{k}_0(\rho)G(\rho)$, the lower limit $\rho_0 = \rho(\xi_0)$.

Furthermore, in terms of the wave number function $\tilde{k}_0(\rho)$ and variable $\rho$, the form of dispersion formula becomes

$$\sigma_0 = \frac{G(\rho)\tilde{k}_0}{\bar{\Lambda}_0\lambda_0 S^2(\rho)}\left[(1-\mathrm{i}\rho) - \frac{G^2(\rho)\tilde{k}_0^2}{\bar{\Lambda}_1\lambda_0^2 S(\rho)}\right]. \tag{10.210}$$

It can be further rewritten as

$$\sigma_e = \Sigma(k_e, \rho) = \frac{k_e}{S^2(\rho)}\left[(1 - i\rho) - \frac{k_e^2}{S(\rho)}\right], \tag{10.211}$$

provided that one sets

$$k_e(\rho) = \tilde{k}_0(\rho)\frac{G(\rho)}{\lambda_0\sqrt{\bar{\Lambda}_1}}; \quad \sigma_e = \frac{\bar{\Lambda}_0}{\sqrt{\bar{\Lambda}_1}}\sigma_0.$$

The form of (10.211) is exactly the same as the dispersion formulas that we encountered in the system of dendritic growth with the one-sided model, as well as in the system of viscous fingering. This result indicates that the stability mechanisms of cellular growth have a profound similarity with those of dendritic growth and viscous fingering.

For a given complex number $\sigma_0$, one may find three roots of (10.211), namely,

$$\begin{cases} k_e^{(1)}(\rho) = M(\rho)\cos\left\{\frac{1}{3}\cos^{-1}\left(\frac{\sigma_e}{N(\rho)}\right)\right\} & \text{(SW-branch)}, \\ k_e^{(2)}(\rho) = M(\rho)\cos\left\{\frac{1}{3}\cos^{-1}\left(\frac{\sigma_e}{N(\rho)}\right) + \frac{2\pi}{3}\right\}, & \\ k_e^{(3)}(\rho) = M(\rho)\cos\left\{\frac{1}{3}\cos^{-1}\left(\frac{\sigma_e}{N(\rho)}\right) + \frac{4\pi}{3}\right\} & \text{(LW-branch)}, \end{cases} \tag{10.212}$$

where

$$M(\rho) = \sqrt{\frac{4S(\rho)}{3}}(1 - i\rho)^{\frac{1}{2}}, \quad N(\rho) = -\frac{M(\rho)}{3S^2(\rho)}(1 - i\rho).$$

The dispersion curves of $k_e^{(i)}(\rho)$ $(i = 1, 2, 3)$ in the complex plane are shown in Fig. 10.23a for a typical case. It is seen that as $\rho \to \infty$, we have $\mathrm{Im}\{k_e^{(1)}(\rho)\} < 0$ and $\mathrm{Im}\{k_e^{(3)}\}(\rho) > 0$. To display the feature of curves of $\tilde{k}_0^{(i)}(\rho)$ $(i = 1, 2, 3)$, we define $\hat{k}^{(i)}(\rho) = \frac{\tilde{k}_0^{(i)}(\rho)}{\lambda_0\sqrt{\bar{\Lambda}_1}}$. The curves of $\hat{k}^{(i)}(\rho)$ $(i = 1, 2, 3)$ are shown in Fig. 10.23b. Evidently, the sets of curves of $\tilde{k}_0^{(i)}(\rho)$ $(i = 1, 2, 3)$ and $\hat{k}^{(i)}(\rho)$ $(i = 1, 2, 3)$ have the same features in the complex plane. As $\rho \to \infty$, all these curves converge to the origin $\hat{k} = 0$, or $\tilde{k}_0 = 0$, from different directions.

The concentration field $\tilde{C}_0$ of the perturbed state given by the solution $H_2(\rho)$ having the wave function $k_e^{(2)}(\rho)$ cannot satisfy the far-field condition in the region away from the interface, due to $\mathrm{Re}\{\tilde{k}_0^{(2)}\} < 0$. As a consequence, the solution $H_2(\rho)$ must be ruled out. Only solutions $H_1(\rho)$ and $H_3(\rho)$ that have wave number function $k_e^{(1)}(\rho)$, and $k_e^{(3)}(\rho)$ in (10.212), respectively, are meaningful. Here, $H_3$, corresponding to the wave function $k_e^{(3)}(\rho)$ of the LW branch, is called as the long interfacial wave, whereas $H_1$, corresponding to the wave function $k_e^{(1)}(\rho)$ of the SW

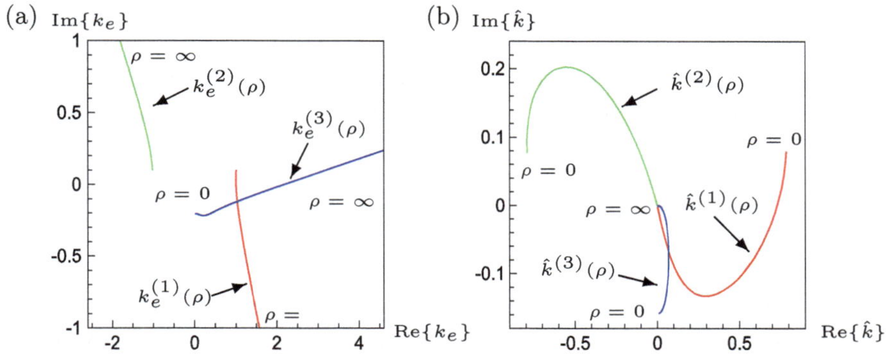


**Fig. 10.23** Typical dispersion curves for the case $M = 0.4, \kappa = 0.2, \lambda_G = 3.748, \lambda_0 = 0.55$, and $\sigma_0 = (3.187 \times 10^{-4}, -5.882 \times 10^{-3}), \varrho_c = (1.0, -0.9847)$: **(a)** The variations of the wave number function $k_e^{(i)}(\varrho)$. **(b)** The variations of the wave number function $\hat{k}^{(i)}(\varrho) = \frac{k_e^{(i)}(\varrho)}{G(\varrho)} = \frac{\tilde{k}_0^{(i)}(\varrho)}{\lambda_0\sqrt{\bar{\Lambda}_1}}$

branch, is called as the short interfacial wave. Thus, the general solution in the outer region is described as a combination of two fundamental interfacial traveling $H$-waves:

$$\begin{aligned}\tilde{h}_0(\rho) &= D_1 H_1(\rho) + D_3 H_3(\rho) \\ &= D_1 \exp\left\{\frac{\mathrm{i}}{\sqrt{\varepsilon}} \int_{\rho_0}^{\rho} \left[\tilde{k}_0^{(1)} + \varepsilon \tilde{k}_1^{(1)} + \cdots\right] \mathrm{d}\rho_1\right\} \\ &\quad + D_3 \exp\left\{\frac{\mathrm{i}}{\sqrt{\varepsilon}} \int_{\rho_0}^{\rho} \left[\tilde{k}_0^{(3)} + \varepsilon \tilde{k}_1^{(3)} + \cdots\right] \mathrm{d}\rho_1\right\}.\end{aligned} \tag{10.213}$$

The constants $D_1$ and $D_3$ are to be determined. Note that $\mathrm{Im}\{\tilde{k}_0^{(1)}\} < 0$ and $\mathrm{Im}\{\tilde{k}_0^{(3)}\} > 0$, as $\rho \gg 1$. As a consequence, as $\rho \to \infty$ along the real axis, $H_1(\rho)$ is dominant, increasing exponentially, whereas $H_3(\rho)$ is subdominant, decreasing exponentially.

So far, we have not yet applied the tip smoothness condition at $\rho = 0$ and the finiteness condition at the root, $\lim_{\rho\to\infty} |\tilde{h}_0(\rho)| < \infty$ for the perturbed states. It is seen that with a fixed pair of constant coefficients $\{D_1, D_3\}$, the solution (10.213) can never satisfy these two conditions simultaneously. To satisfy the root condition, the MVE solution can have only the subdominant component $H_3$, while to satisfy the tip condition, it must have both components $H_1$ and $H_3$. Hence we have the same scene and dilemma as encountered in studying the systems of dendritic growth and viscous fingering formation. The two interfacial waves $\{H_1, H_3\}$ appearing near the tip region must merge into a single interfacial wave $\{H_3\}$ somewhere before arriving at the root region. Hence as before, in order to resolve the dilemma, one must identify the location of the singular point of the MVE normal mode solutions, explore the behavior of the inner solution near the singular point, and determine the matching condition of the MVE normal mode solutions with the inner solution. It will be seen in the following sections that the MVE solution indeed has a turning-point singularity in the extended complex $\rho$-plane (or $\xi$-plane), as expected. Associated with this singular point and due to the Stokes phenomenon, the pair of coefficients $\{D_1, D_3\}$ are piecewise constant along the real axis of $\rho$ and in the complex $\rho$-plane.

#### 10.4.4.3 The First-Order Approximation and Singularity of the MVE Solution

The first-order approximation solution will fully determine the amplitude $A_0(\xi,\eta)$, the functions $\bar{k}_1(\xi)$, and $\sigma_1$. The results can be obtained by closely following the approach developed in the previous chapters. In order to keep the main body of the presentation clear, we move the detailed mathematical derivations to the appendices at the end of the chapter. There, it is shown that in order to derive the global uniformly valid asymptotic expansion solution, one must extend all the functions such as $\bar{k}_0(\xi)$, $Y_{\xi,0}(\xi)$, $\mathcal{G}_0(\xi)$ by analytic continuation to the corresponding complex functions $\bar{k}_0(\zeta)$, $Y_{\zeta,0}(\zeta)$, $\mathcal{G}_0(\zeta)$, etc. in the complex plane $\zeta = \xi + \mathrm{i}\xi_I$, and the problem is to be analyzed in the extended space $(\zeta,\eta)$. As a consequence, the wave-number function $\bar{k}_0(\zeta) \equiv \{\bar{k}_0^{(1)}(\zeta), \bar{k}_0^{(2)}(\zeta), \bar{k}_0^{(3)}(\zeta)\}$ is subject to the extended dispersion formula

$$\sigma_0 = \Sigma(\bar{k}_0,\zeta) = \frac{\bar{k}_0}{\bar{\Lambda}_0 \mathcal{G}_0^2(\zeta)}\left[\lambda_0 + \mathrm{i}Y_{\zeta,0}(\zeta) - \frac{\bar{k}_0^2}{\bar{\Lambda}_1 \mathcal{G}_0(\zeta)}\right]. \tag{10.214}$$

In accordance with the above, the function $\bar{k}_0(\rho)$ is extended by analytic continuation to the complex function $k_e(\varrho)$ in the complex plane $\varrho = \rho + \mathrm{i}\rho_I$ subject to the extended dispersion formula

$$\sigma_e = \Sigma(k_e,\varrho) = \frac{k_e}{S^2(\varrho)}\left[(1-\mathrm{i}\varrho) - \frac{k_e^2}{S(\varrho)}\right]. \tag{10.215}$$

It is found that the MVE solution has a singular point $(\zeta_c, 0)$ in the extended plane $(\zeta,\eta)$, or a singular point $(\varrho_c, 0)$ in the extended plane $(\varrho,\eta)$. The parameter $\zeta_c$ is subject to the equation

$$\frac{\partial \Sigma(\bar{k}_0,\zeta_c)}{\partial \bar{k}_0} = 0, \quad \text{and} \quad \left[\bar{k}_0^{(1)}(\zeta_c) - \bar{k}_0^{(3)}(\zeta_c)\right] = 0, \tag{10.216}$$

whereas the parameter $\varrho_c$ is subject to the equation

$$\frac{2}{\sqrt{27}}\left(\frac{1-\mathrm{i}\varrho_c}{1+\mathrm{i}\varrho_c}\right)^{\frac{3}{4}} = \frac{\bar{\Lambda}_0}{\sqrt{\bar{\Lambda}_1}}\sigma_0. \tag{10.217}$$

We obtain

$$\sigma_1 = \frac{1}{W\hat{\text{Pe}}(1-\kappa)y_{*0}\mathcal{G}_0^2(\zeta_c)}\Bigg[W\hat{\text{Pe}}\big[\lambda_{\text{G}} - (1-\kappa)y_{*0}\big]$$

$$\times Y_{\eta\eta}(\zeta_c, 0) - \frac{\mathrm{i}F_0(\zeta_c)\mathcal{G}_0'(\zeta_c)}{3\mathcal{G}_0(\zeta_c)}\Bigg], \tag{10.218}$$

where

$$F_0(\zeta) = -W\hat{\text{Pe}}(\lambda_{\text{G}} - y_{*0} + 1) + \mathrm{i}W\hat{\text{Pe}}$$
$$\times\big[(1-\kappa)y_{*0} - \lambda_{\text{G}}\big]Y_{\xi,0}(\zeta). \tag{10.219}$$

The formula (10.218) shows that once $\sigma_0$ is found, $\sigma_1$ follows. Up to this point, the parameter $\sigma_0$ still remains free. In the next section, we shall study the behavior of the solution near the singular point $\zeta_c$ and derive the parameter $\sigma_0$ and the global wave modes solutions in the extended $(\zeta, \eta)$-space.

### *10.4.5 The Inner Solution in the Vicinity of the Singular Point $(\zeta_c, 0)$*

The fact that the MVE solution is not valid at the point $(\zeta_c, 0)$ in the extended $(\zeta, \eta)$-space implies that the solutions in the vicinity of the point $(\zeta_c, 0)$ in the $(\zeta, \eta)$-space, $|\zeta - \zeta_c| \ll 1$, $|\eta| \ll 1$, no longer have multiple length scales and the form of normal modes. Thus, to derive the inner solution near the point $(\zeta_c, 0)$, one must start with the associated homogeneous system (10.184)–(10.186) and seek a different asymptotic form of the solution. In doing so, we introduce the inner variables

$$\zeta_* = \frac{\zeta - \zeta_c}{\varepsilon^\alpha}, \quad \eta_* = \frac{\eta}{\varepsilon^\alpha}, \tag{10.220}$$

where the exponent $\alpha$ is to be determined later. With the inner variables, the interface shape function is expressed as

$$\tilde{\eta}_\text{s}(\zeta, t, \varepsilon) = \varepsilon^\alpha \tilde{\eta}_{*\text{s}}(\zeta_*, t_+, \varepsilon),$$

while the concentration of impurity is expressed as

$$\tilde{C}(\zeta, \eta, t, \varepsilon) = \varepsilon^\alpha \tilde{C}_*(\zeta_*, \eta_*, t_+, \varepsilon).$$

The inner system is derived as

$$\left(\frac{\partial^2}{\partial\zeta_*^2}+\frac{\partial^2}{\partial\eta_*^2}\right)\tilde{C}_*+\varepsilon^{\alpha+1}W\hat{\mathrm{Pe}}\Big(Y_\xi\frac{\partial\tilde{C}_*}{\partial\zeta_*}+X_\xi\frac{\partial\tilde{C}_*}{\partial\eta_*}$$
$$+\frac{\mathcal{G}^2}{\beta(\varepsilon)}\frac{\partial\tilde{C}_*}{\partial t_+}\Big)=0, \tag{10.221}$$

with the following interface conditions: at $\eta_*=0$,

$$\tilde{C}_*=-\varepsilon W\hat{\mathrm{Pe}}\big[\lambda_{\mathrm{G}}-(y_{*0}-1)\big]\tilde{\eta}_{*\mathrm{s}}+\frac{\varepsilon^{2(1-\alpha)}}{MW\mathcal{G}_0(\zeta)}\frac{\partial^2\tilde{\eta}_{*\mathrm{s}}}{\partial\zeta_*^2}$$
$$+(\text{h.o.t.})\,; \tag{10.222}$$

and

$$\frac{\partial\tilde{C}_*}{\partial\eta_*}+\varepsilon W\hat{\mathrm{Pe}}Y_{\xi,0}(\zeta)\Big[\lambda_{\mathrm{G}}-(1-\kappa)y_{*0}\Big]\frac{\partial\tilde{\eta}_{*\mathrm{s}}}{\partial\zeta_*}$$
$$+\varepsilon^{1+\alpha}W\hat{\mathrm{Pe}}(1-\kappa)y_{*0}\mathcal{G}_0^2(\zeta)\frac{\sigma}{\beta(\varepsilon)}\frac{\partial\tilde{\eta}_{*\mathrm{s}}}{\partial t_+}$$
$$+(\text{h.o.t.})=0. \tag{10.223}$$

We carry out the following inner expansions for the solution:

$$\tilde{C}_*(\zeta_*,\eta_*,t_+,\varepsilon)=\varepsilon\Big[\tilde{\nu}_0(\varepsilon)\tilde{Y}_{*0}(\zeta_*,\eta_*)+\tilde{\nu}_1(\varepsilon)\tilde{Y}_{*1}(\zeta_*,\eta_*)+\ldots\Big]\mathrm{e}^{\sigma t_+},$$
$$\tilde{\eta}_{*\mathrm{s}}(\zeta_*,t_+,\varepsilon)=\Big[\tilde{\nu}_0(\varepsilon)\tilde{h}_{*0}(\zeta_*)+\tilde{\nu}_1(\varepsilon)\tilde{h}_{*1}(\zeta_*)+\cdots\Big]\mathrm{e}^{\sigma t_+}. \tag{10.224}$$

### *10.4.6 Leading-Order Inner Solution in the Vicinity of $\zeta_c$*

By dropping the small higher-order terms contained in the system (10.222)–(10.223), it follows that

$$\frac{\partial^2\tilde{C}_{*0}}{\partial\zeta_*^2}+\frac{\partial^2\tilde{C}_{*0}}{\partial\eta_*^2}=0. \tag{10.225}$$

With the boundary conditions at the interface $\eta_*=0$, we have

$$\tilde{C}_{*0}=-W\hat{\mathrm{Pe}}\big[\lambda_{\mathrm{G}}-(y_{*0}-1)\big]\tilde{h}_{*0}+\frac{\varepsilon^{1-2\alpha}}{MW\mathcal{G}_0(\zeta)}\frac{\partial^2\tilde{h}_{*0}}{\partial\zeta_*^2}, \tag{10.226}$$

and

$$\frac{\partial \tilde{C}_{*0}}{\partial \eta_*} + W\hat{\text{Pe}}\big[\lambda_G - (1-\kappa)y_{*0}\big]Y_{\xi,0}(\zeta)\frac{\partial \tilde{h}_{*0}}{\partial \zeta_*}$$
$$+\varepsilon^{\alpha-\frac{1}{2}}\sigma_0 W\hat{\text{Pe}}(1-\kappa)y_{*0}\mathcal{G}_0^2(\zeta)\tilde{h}_{*0} = 0. \tag{10.227}$$

Some additional small higher-order terms will be dropped later. Due to the analyticity of the solution, one has $\frac{\partial \tilde{C}_{*0}}{\partial \eta_*} = \mathrm{i}\frac{\partial \tilde{C}_{*0}}{\partial \zeta_*}$. Applying this formula at the interface $\eta = 0^+$, from (10.226)–(10.227) we derive

$$\varepsilon^{1-2\alpha}\frac{\mathrm{d}^3\tilde{h}_{*0}}{\mathrm{d}\zeta_*^3} + \bar{\Lambda}_1\mathcal{G}_0(\zeta)\Big[\lambda_0 + \mathrm{i}Y_{\xi,0}(\zeta)\Big]\frac{\mathrm{d}\tilde{h}_{*0}}{\mathrm{d}\zeta_*}$$
$$-\mathrm{i}\varepsilon^{\alpha-\frac{1}{2}}\sigma_0 MW^2\hat{\text{Pe}}(1-\kappa)y_{*0}\mathcal{G}_0^3(\zeta)\tilde{h}_{*0}$$
$$+(\text{h.o.t.}) = 0. \tag{10.228}$$

With the variable $\varrho$ and the corresponding inner variable $\varrho_* = \frac{\varrho-\varrho_c}{\varepsilon^\alpha}$, one may rewrite (10.227) in the form

$$\varepsilon^{1-2\alpha}G^3(\varrho)\frac{\mathrm{d}^3\tilde{h}_{*0}}{\mathrm{d}\varrho_*^3} + \lambda_0^2\bar{\Lambda}_1 G(\varrho)S(\varrho)(1-\mathrm{i}\varrho)\frac{\mathrm{d}\tilde{h}_{*0}}{\mathrm{d}\varrho_*}$$
$$-\mathrm{i}\varepsilon^{\alpha-\frac{1}{2}}\sigma_0 MW^2\hat{\text{Pe}}(1-\kappa)y_{*0}\lambda_0^3 S^3(\varrho)\tilde{h}_{*0}$$
$$+(\text{h.o.t.}) = 0. \tag{10.229}$$

In order to solve the above third-order ODE, we follow the approach developed in [45]. We first apply the following transformation for the solution in the outer region and the solution in the inner region, respectively. Namely,

$$\tilde{h}_0(\varrho) = \tilde{W}_0(\varrho)\ \exp\left[\frac{\mathrm{i}}{\sqrt{\varepsilon}}\int_{\varrho_c}^{\varrho}\tilde{k}_c(\varrho_1)\,\mathrm{d}\varrho_1\right],$$
$$\tilde{h}_{*0}(\varrho_*) = \tilde{W}_{*0}(\varrho_*)\exp\left[\frac{\mathrm{i}}{\sqrt{\varepsilon}}\int_{\varrho_c}^{\varrho}\tilde{k}_c(\varrho_1)\,\mathrm{d}\varrho_1\right]. \tag{10.230}$$

The reference wave number function $\tilde{k}_c(\varrho)$ is to be given later; it satisfies the inequalities $\text{Re}\{\tilde{k}_0^{(3)}\} < \text{Re}\{\tilde{k}_c\} < \text{Re}\{\tilde{k}_0^{(1)}\}$. With the above transformation, in the outer region the solution $\tilde{h}_0$ can be considered a combination of two $W$-waves instead of two $H$-waves. Namely, we may write

$$\tilde{W}_0(\varrho) = D_1\tilde{W}_0^{(+)}(\varrho) + D_3\tilde{W}_0^{(-)}(\varrho), \tag{10.231}$$

where

$$\tilde{W}_0^{(+)}(\varrho) = H_1(\varrho)\exp\left\{-\frac{\mathrm{i}}{\varepsilon^{\frac{1}{2}}}\int_{\varrho_c}^{\varrho}\tilde{k}_c(\varrho_1)\,\mathrm{d}\varrho_1\right\}, \tag{10.232}$$

$$\tilde{W}_0^{(-)}(\varrho) = H_3(\varrho)\exp\left\{-\frac{\mathrm{i}}{\varepsilon^{\frac{1}{2}}}\int_{\varrho_c}^{\varrho}\tilde{k}_c(\varrho_1)\,\mathrm{d}\varrho_1\right\}. \tag{10.233}$$

Here $\tilde{W}^{(+)}$, corresponding to the short $H$-wave $H_1(\varrho)$, is now an outgoing $W$-wave, while $\tilde{W}^{(-)}$, corresponding to the long $H$-wave $H_3(\varrho)$, is now an incoming $W$-wave.

We choose the reference wave number function $\tilde{k}_c$ from the condition $\left(\frac{\partial\Sigma}{\partial\tilde{k}_0}\right)_{\tilde{k}_0=\tilde{k}_c} = 0$, such that

$$\begin{aligned}\tilde{k}_c(\varrho) &= \frac{\sqrt{\lambda_0^2\bar{\Lambda}_1 S(1-\mathrm{i}\varrho)}}{3^{\frac{1}{2}}G(\varrho)}\\ &= \frac{\mathrm{e}^{-\frac{i\pi}{4}}\bar{\Lambda}_1^{\frac{1}{2}}\lambda_0(\varrho-\mathrm{i})^{\frac{1}{4}}(\varrho+\mathrm{i})^{\frac{1}{4}}(\varrho+\mathrm{i})^{\frac{1}{2}}}{3^{\frac{1}{2}}G(\varrho)}.\end{aligned} \tag{10.234}$$

From (10.228), by neglecting the small higher-order terms, one may derive the governing equation for the function $\tilde{W}_{*0}(\varrho_*)$ as follows:

$$\begin{aligned}&\varepsilon^{1-2\alpha}G^3(\varrho)\frac{\mathrm{d}^3\tilde{W}_{*0}(\varrho_*)}{\mathrm{d}\varrho_*^3} + \mathrm{i}\varepsilon^{\frac{1}{2}-\alpha}\lambda_0 G^2(\varrho)\sqrt{3\bar{\Lambda}_1 S(\varrho)(1-\mathrm{i}\varrho)}\frac{\mathrm{d}^2\tilde{W}_{*0}(\varrho_*)}{\mathrm{d}\varrho_*^2}\\ &-\mathrm{i}\varepsilon^{\alpha-\frac{1}{2}}MW^2\hat{\mathrm{P}}\mathrm{e}(1-\kappa)y_{*0}\lambda_0^3 S^3(\varrho)\big[\sigma_0-\Omega_0(\varrho)\big]\tilde{W}_{*0}(\varrho_*) = 0,\end{aligned} \tag{10.235}$$

where

$$\Omega_0(\varrho) = \frac{2}{\sqrt{27}}\frac{\sqrt{\bar{\Lambda}_1}}{\bar{\Lambda}_0}\frac{(1-\mathrm{i}\varrho)^{\frac{3}{2}}}{S^{\frac{3}{2}}(\varrho)} = \Sigma(\tilde{k}_c,\varrho). \tag{10.236}$$

It is now clearly seen that Eq. (10.234) has five isolated singular points: $\varrho = \pm\mathrm{i}, \pm ia$, and $\varrho = \varrho_c$, where $\varrho_c$ is a simple turning point of (10.234) in the complex $\varrho$-plane, and $\varrho = \pm\mathrm{i}, \pm ia$ are roots of the functions of $S(\varrho)$ and $G(\varrho)$, respectively. In view of the formula

$$\sigma_0 = \Omega_0(\varrho_c) = \frac{2}{\sqrt{27}}\frac{\sqrt{\bar{\Lambda}_1}}{\bar{\Lambda}_0}\left(\frac{1-\mathrm{i}\varrho_c}{1+\mathrm{i}\varrho_c}\right)^{\frac{3}{4}}, \tag{10.237}$$

to study the behavior of the inner solution in the vicinity of the turning point $\varrho_c$, one needs to distinguish two cases: case (I): $|\sigma_0| = \mathcal{O}(1)$; case (II): $|\sigma_0| \ll 1$. The location of the turning point $\varrho_c$ and the corresponding Stokes lines and anti-Stokes lines emitted from $\varrho_c$ are sketched in Fig. 10.24 for each case. We see in particular

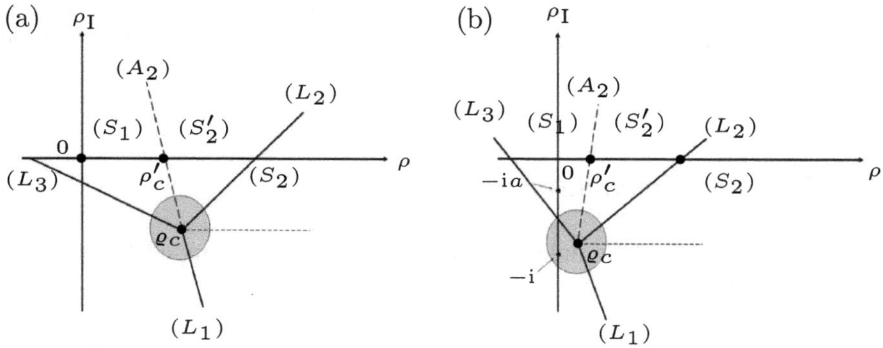


**Fig. 10.24** Sketch of the Stokes and anti-Stokes lines emanating from $\zeta_c$, (**a**) for the case (I): $\sigma_0 = \mathcal{O}(1)$; (**b**) for the case (II): $|\sigma_0| \ll 1$

that the cell's tip $\xi = 0$ and the bottom of the root $\xi = 1$ are located in different sectors: $(S_1)$ and $(S_2')$, separated by the anti-Stokes line $(A_2)$.

We derive that in the far field of the vicinity of $\varrho_c$, the inner equation can be reduced to the general Airy equation

$$\frac{\mathrm{d}^2 \tilde{W}_{*0}(\hat{\varrho}_*)}{\mathrm{d}\hat{\varrho}_*^2} + \hat{\varrho}_*^{p_0} \tilde{W}_{*0}(\hat{\varrho}_*) = 0, \qquad (10.238)$$

and the leading-order approximation of the inner solution has the form $\tilde{W}_{*0}(\hat{\varrho}_*) = D_* \hat{\varrho}_*^{\frac{1}{2}} H_\nu^{(2)}(z)(z = 2\nu \hat{\varrho}_*^{\frac{1}{2\nu}};\ \nu = \frac{1}{p_0+2})$. Here, the new variable $\hat{\varrho}_* = \mathcal{A}\varrho_*$ is used with the constant $\mathcal{A} = \mathcal{O}(1)$. Moreover, we derive that

- for case (I) ($|\sigma_0| = \mathcal{O}(1)$), we have $p_0 = 1, \theta_L = \frac{2}{3}\pi, \alpha = \frac{1}{3}, \nu = \frac{1}{3}$;
- for case (II) ($|\sigma_0| \ll 1$), we have $p_0 = \frac{7}{4}, \theta_L = \frac{8}{15}\pi, \alpha = \frac{2}{7}, \nu = \frac{4}{15}$.

In the above, $\theta_L$ is the open angle between two Stokes lines and anti-Stokes lines. By matching the inner solution with the outer solution in the sector $(S_1)$, the connection conditions for both cases are finally obtained in the following unified form:

1. In the domain $\varrho \in (S_1); \rho \in (0, \rho_c')$, we have

$$\frac{D_1}{D_3} = 2\mathrm{e}^{\mathrm{i}\frac{\pi}{2}} \cos(\nu\pi), \quad D_3 = D_3'. \qquad (10.239)$$

2. In the domain $\varrho \in (S_2)' \cup (S_2); \varrho \in (\rho_c', \infty)$, we have

$$D_1 = 0, \quad D_3 = D_3', \qquad (10.240)$$

where $\rho_c'$ is the intersection point of the anti-Stokes line $(A_2)$ with the real axis of $\varrho$.

We have not yet applied the tip smoothness condition. When the tip smoothness conditions are applied, the eigenvalue $\sigma_0$ will be determined as a function of the stability parameter $\varepsilon$ and other physical parameters.

## 10.4.7 Global Instability Mechanisms

### 10.4.7.1 Spectrum of Complex Eigenvalues and Global Trapped-Wave (GTW) Instability

Let us consider case (I) first. Case (I) leads to a spectrum of complex eigenvalues $\sigma_0 = \sigma_{0,R} - i\omega_0$ with $\omega_0 > 0$. The physical solution in the interval $(S_1)$ of the outer region is

$$\mathrm{Re}\{\tilde{h}_0(\varrho,t)\} = \mathrm{Re}\{H(\varrho)\mathrm{e}^{\frac{\sigma_0 t}{\sqrt{\varepsilon}}}\},$$

where

$$H(\varrho) = D_1 H_1 + D_3 H_3.$$

Let

$$d_1 = D_1 \mathrm{e}^{-\mathrm{i}\chi_1}, \quad d_3 = D_3 \mathrm{e}^{-\mathrm{i}\chi_3}, \tag{10.241}$$

where

$$\chi_1 = \frac{1}{\sqrt{\varepsilon}} \int_0^{\varrho_c} \tilde{k}_0^{(1)} \mathrm{d}\varrho, \quad \chi_3 = \frac{1}{\sqrt{\varepsilon}} \int_0^{\varrho_c} \tilde{k}_0^{(3)} \mathrm{d}\varrho. \tag{10.242}$$

Thus, it follows from the connection condition (10.239) that

$$\frac{d_1}{d_3} = \mathrm{i}\mathrm{e}^{-\mathrm{i}\chi}, \tag{10.243}$$

where

$$\chi = \frac{1}{\sqrt{\varepsilon}} \int_0^{\varrho_c} \left(\tilde{k}_0^{(1)} - \tilde{k}_0^{(3)}\right) \mathrm{d}\varrho.$$

To satisfy the tip's smoothness conditions, the constants $d_1$ and $d_3$ must satisy the conditions

$$\begin{cases} \dfrac{d_3}{d_1} = -\dfrac{\tilde{k}_0^{(1)}(0)}{\tilde{k}_0^{(3)}(0)}, & \text{(for symmetric S-modes);} \\[2ex] d_1 = -d_3 & \text{(for antisymmetric A-modes).} \end{cases} \tag{10.244}$$

Combining (10.243) with (10.244), one obtains the following quantization condition:

$$\frac{1}{\sqrt{\varepsilon}}\int_0^{\varrho_c}\left(\tilde{k}_0^{(1)}-\tilde{k}_0^{(3)}\right)\mathrm{d}\varrho=\left(2n+1+\frac{1}{2}+\theta_0\right)\pi$$
$$-\mathrm{i}\ln\alpha_0,\quad \left(n=0,1,2,\ldots\right),\qquad (10.245)$$

where

$$\begin{cases}\alpha_0\,\mathrm{e}^{\mathrm{i}\theta_0\pi}=\dfrac{\tilde{k}_0^{(1)}(0)}{\tilde{k}_0^{(3)}(0)} \text{ (for S-modes);}\\[2ex] \alpha_0=1,\ \theta_0=0 \text{ (for A-modes)}.\end{cases}\qquad (10.246)$$

The quantization condition (10.245) determines the spectrum of complex eigenvalues with $|\sigma_0|=\mathcal{O}(1)$, $\{\sigma_{0,n}\}$ $(n=0,1,2,\ldots)$, as functions of the parameters $\lambda_0$ and $\varepsilon$. Such a spectrum of complex eigenvalues yields two discrete sets of eigenmodes: S-modes and A-modes. Given $\lambda_0, \hat{\mathrm{Pe}}$ and other physical parameters, the system allows a set of neutrally stable modes ($\sigma_R=0$) when $\varepsilon=\varepsilon_{*,n}$ $(n=0,1,2,\ldots)$ with $\varepsilon_*=\varepsilon_{*,0}>\varepsilon_{*,1}>\varepsilon_{*,2}>\cdots$. For $\varepsilon$ in the range $\varepsilon_{*,k}<\varepsilon_*<\varepsilon_{*,k-1}$ where $k=1,2,\ldots$, the system allows $k$ unstable GTW modes with the indices $k=0,1,\ldots,k-1$, respectively. When $\varepsilon=\varepsilon_{*,k}$, the system allows a neutral stable mode with the index $n=k$, whose interface is oscillatory with the constant eigenfrequency $\omega_{*,k}$.

These eigenmodes are very similar to the *global traveling wave* (GTW) modes previously derived for dendritic growth, describing traveling waves propagating along the interface (refer to [10, 43, 45]). Therefore, we may also call this global oscillatory instability mechanism the *global traveling wave* (GTW) mechanism.

It should be noted that under the current notation, the parameters $\varepsilon, \lambda_G$ depend on both operation conditions: the pulling velocity $V$ and the temperature gradient $(G)_D$. In order to examine the effect of each of the operation conditions on the eigenvalue $\sigma_0$ separately, it is better to use the new control parameters:

- $\hat{v}=\sqrt{\ell_c V/\kappa_D}$, which measures the pulling velocity $V$, and
- $\hat{\beta}=\frac{\ell_c}{\ell_G}$, which measures the gradient of temperature.

Thus, one may write

$$\lambda_G=\hat{\beta}/\hat{v}^2;\quad \hat{\mathrm{Pe}}=\hat{v}^2/\varepsilon^3;\quad \bar{\Lambda}_1=MW^2\left[(1-\kappa)y_{*0}-\frac{\hat{\beta}}{\hat{v}^2}\right]\frac{\hat{v}^2}{\varepsilon^3}.$$

The system is now controlled separately by three independent operation parameters: $\{\hat{v},\hat{\beta},M\}$.

In Fig. 10.25a, b, we show the variations of $\sigma_{0,R}=\mathrm{Re}\{\sigma_0\}$ and $\omega_0=|\mathrm{Im}\{\sigma_0\}|$ of the S-modes with $\varepsilon$ for $n=0,1,2$ under a typical operation condition in the zeroth-order approximation. It is seen that given $\varepsilon$, the growth rate of the S-mode with index

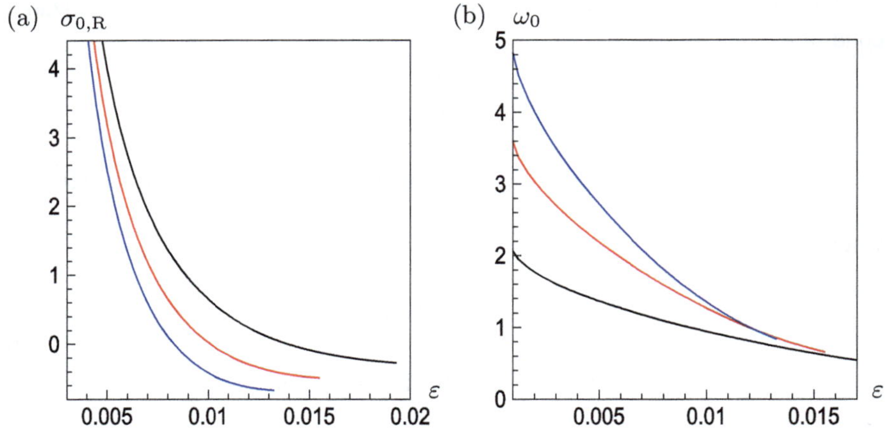


**Fig. 10.25** The eigenvalues of GTW modes calculated in terms of the quantization condition (10.245) for the case $\hat{v} = 0.01, M = 0.08, \kappa = 0.3, \hat{\beta} = 2.0 \times 10^{-5}, \lambda_G = 0.2, \lambda_0 = 0.45$, and $W_0 = 2.1332$. (**a**) The variations of the rear part of the eigenvalue $\sigma_{0,R} = \text{Re}\{\sigma_0\}$ of GTW S-modes ($n = 0, 1, 2$) with $\varepsilon$ from top to bottom. (**b**) The variation of the imaginary part of the eigenvalue $\omega_0 = |\text{Im}\{\sigma_0\}|$ of GTW S-modes ($n = 0, 1, 2$) with $\varepsilon$ from bottom to top

$n_i$ is greater than that of the S-mode with index $n_j$ if $n_i < n_j$. In this sense, we call the S-mode ($n = n_i$) more dangerous than the S-mode ($n = n_j$). We have also calculated the quantization condition for A-modes and found that with the same index $n$, the S-mode is more dangerous than the A-mode. Therefore, we conclude that in the GTW mechanism, the most dangerous mode is the S-mode with $n = 0$. Using the condition $\text{Re}\{\sigma_0\} = 0$, one may determine the zeroth-order approximate value of the critical surface tension parameter $\varepsilon_* = \varepsilon_*^{(0)}$ and the associated eigenfrequency $\omega_*^{(0)}$. In the first-order approximation, one may set the eigenvalue $\sigma \approx \sigma_0 + \varepsilon\sigma_1$ and calculate the first-order approximate critical stability parameter $\varepsilon_* = \varepsilon_*^{(1)}$ and the associated eigenfrequency $\omega_*^{(1)}$ via the condition $\text{Re}\{\sigma_0 + \varepsilon\sigma_1\} = 0$.

In Fig. 10.26a, we compare the zeroth-order and first-order approximations of the neutral curves of GTW S-modes ($n = 0$) in the $(\varepsilon, \lambda_0)$-plane for a typical case, whereas in Fig. 10.27a, we compare the zeroth-order and first-order approximations of the associated eigenfrequency $\omega_*^{(0)}$ and $\omega_*^{(1)}$ with $\lambda_0$. We see that with a fixed $\lambda_0$, we have $\varepsilon_*^{(1)} < \varepsilon_*^{(0)}$ and $\omega_*^{(1)} > \omega_*^{(0)}$.

In Fig. 10.28, we show the corresponding interface shape of the basic state added with the GTW S-neutral mode ($n = 0$). Since the interface shape of the GTW S-neutral mode has a self-sustaining side-branching structure and the associated basic state has a needle-like tip with relative width normally in the range ($0 < \lambda_0 < 0.5$), one may call the GTW neutral curve the needle-like branch.

Note that the parameter $\varepsilon$ is determined by the tip radius $\ell_t$, while the relative width parameter $\lambda_0$ depends on both the tip radius $\ell_t$ and the primary spacing $\ell_w$. To compare with the experimental data, it is better to express the neutral curves in the dimensionless control parameters' $(\hat{\ell}_t, \hat{\ell}_w)$-plane, where we define $\hat{\ell}_t = \ell_t/\ell_c$ and

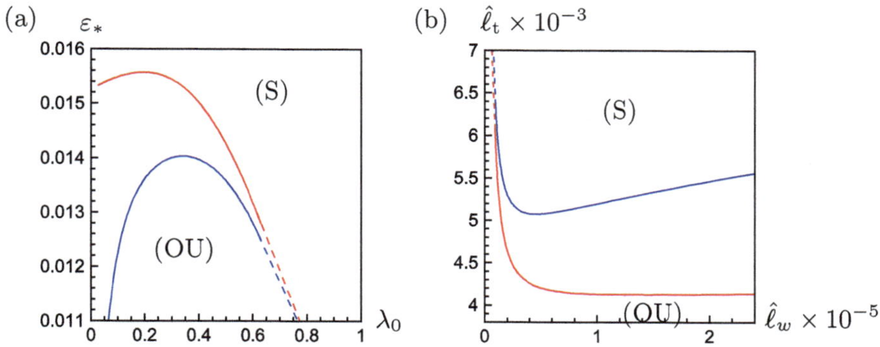


**Fig. 10.26** Under the operation condition, $\hat{v} = 0.5388 \times 10^{-2}, M = 0.09552, \kappa = 0.29, \hat{\beta} = 0.11485 \times 10^{-4}, \lambda_G = 0.4989$, which corresponds to $V = 16\,\mu\text{m/s}, G_D = 140^{-4}\,\text{K}/\mu\text{m}$, in the zeroth-order and first-order approximations from top to bottom. (**a**) The neutral curves of the GTW S-modes ($n = 0$) in the $(\lambda_0, \varepsilon)$-plane. (**b**) The neutral curves of the GTW S-modes ($n = 0$) in the $(\hat{\ell}_w, \hat{\ell}_t)$-plane

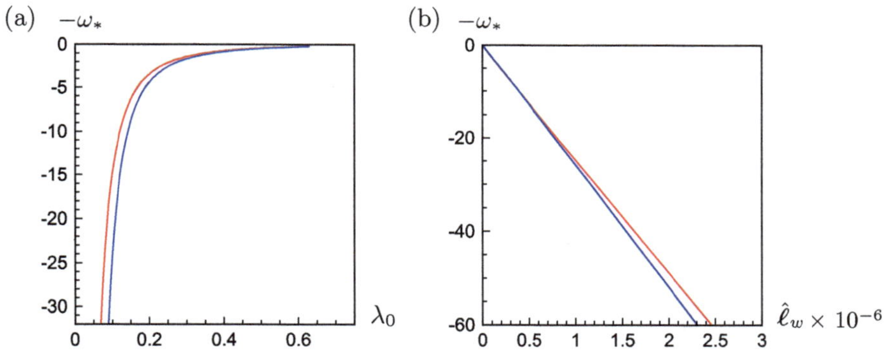


**Fig. 10.27** For the case $\hat{v} = 0.5388\times10^{-2}, M = 0.09552, \kappa = 0.29, \hat{\beta} = 0.11485\times10^{-4}, \lambda_G = 0.4989$, in the zeroth- and first-order approximations from top to bottom, (**a**) the variations of $\text{Im}\{\sigma_0\} = -\omega_0$ for the GTW S-neutral modes ($n = 0$) with $\lambda_0$; (**b**) the variations of $\text{Im}\{\sigma_0\} = -\omega_0$ for the GTW S- neutral modes ($n = 0$) with $\hat{\ell}_w$

$\hat{\ell}_w = \ell_w/\ell_c$. The neutral curves given in Fig. 10.26a are shown in the $(\hat{\ell}_w, \hat{\ell}_t)$-plane in Fig. 10.26b. The curves of $\omega_*$ shown in Fig. 10.27a are shown in the $(\hat{\ell}_w, \omega_*)$-plane in Fig. 10.27b.

The variation of the neutral curve of GTW S-mode ($n = 0$) in the first-order approximation with the temperature gradient $G_D$ and pulling velocity $V$, $M$, and $\kappa$ are shown in the $(\varepsilon, \lambda_0)$-plane in Fig. 10.29a, b and Fig. 10.30a, b, respectively. For the same cases, these neutral curves are also shown in the $(\hat{\ell}_w, \hat{\ell}_t)$-plane in Fig. 10.31a, b and Fig. 10.32a, b, respectively. It is seen that the smaller the temperature gradient $G_D$, the bigger the pulling velocity $V$, and the smaller the parameter $\kappa$, and the bigger the parameter $M$, the more unstable the system.

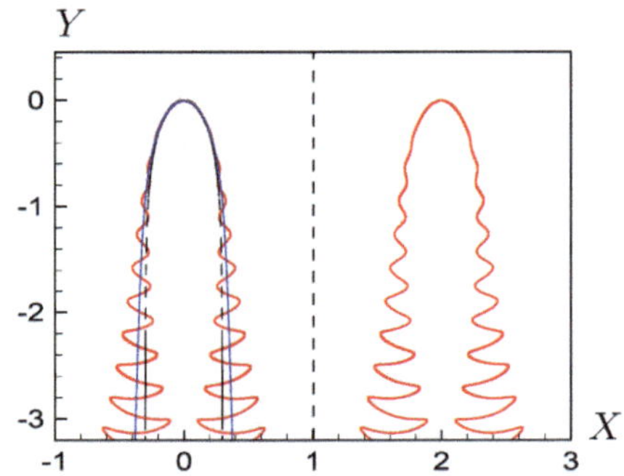


**Fig. 10.28** The arrayed-dendritic interface shape of the neutral GTW S-modes ($n = 0$) calculated using the formulas (10.209), (10.231), and (10.239) for the case $\hat{v} = 0.119 \times 10^{-2}, \kappa = 0.1, \hat{\beta} = 6.9 \times 10^{-7}, \lambda_G = 4.872 \times 10^{-3}, M = 0.50$, and $\lambda_0 = 0.3$. The *dashed line* closer to the centerline of the dendrite is the Saffman–Taylor solution; the *solid line* outside of the Saffman–Taylor solution is the first-order approximation of the basic state solution

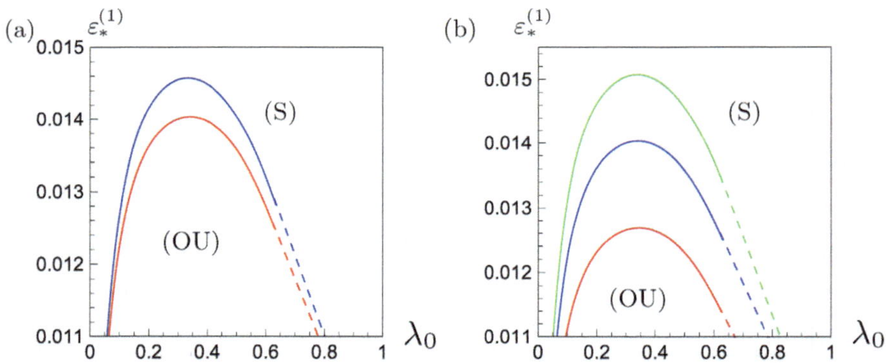


**Fig. 10.29** The neutral curves of the GTW S-modes ($n = 0$) in the first-order approximation in the $(\varepsilon, \lambda_0)$-plane calculated using the formulas (10.218) and (10.245) for the cases $C_\infty = 1.2$mol%, $M = 0.9552, \kappa = 0.29$, and (**a**) $V = 16\,\mu$m/s, $G_D = 78 \times 10^{-4}$ K/$\mu$m, $140 \times 10^{-4}$ K/$\mu$m, which corresponds to $\hat{v} = 0.5388 \times 10^{-2}$, and $\hat{\beta} = 0.8070 \times 10^{-5}, 0.1448 \times 10^{-4}, \lambda_G = 0.2780, 0.4989$, from top to bottom; (**b**) $G_D = 140 \times 10^{-4}$ K/$\mu$m, and $V = 12\,\mu$m/s, $16\,\mu$m/s, $20\,\mu$m/s, which correspond to $\hat{\beta} = 0.14485 \times 10^{-4}$ and $\hat{v} = 0.4666 \times 10^{-2}, 0.5388 \times 10^{-2}, 0.6024 \times 10^{-2}$; $\lambda_G = 0.6652, 0.4989, 0.3991$, from bottom to top

Furthermore, the variations of the eigenfrequency $\omega_*^{(1)}$ with $G_D$ and $V, \kappa$ are shown in Fig. 10.33a,,b, respectively. It is seen that the smaller the temperature gradient $G_D$ and the bigger the pulling velocity $V$, the bigger the eigenfrequency $\omega_*^{(1)}$.

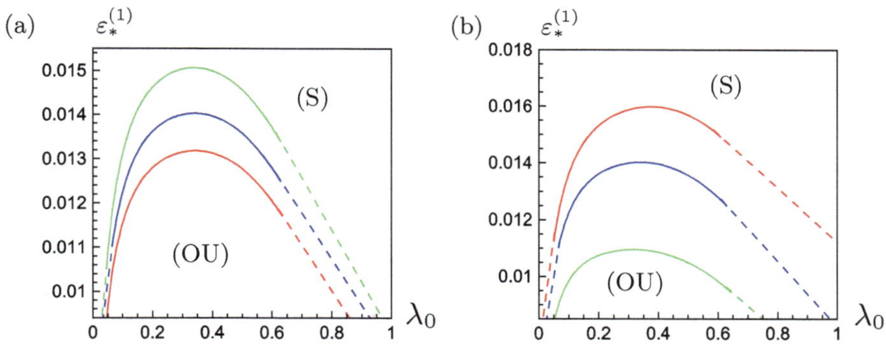


**Fig. 10.30** The neutral curves of the GTW S-modes ($n = 0$) in the first-order approximation in the $(\varepsilon, \lambda_0)$-plane for the cases $G_D = 140 \times 10^{-4}\,\mathrm{K}/\mu\mathrm{m}$, $V = 16\,\mu\mathrm{m/s}$, and (**a**) $\kappa = 0.29$, $M = 0.0796, 0.09552, 0.1194$, which correspond to $C_\infty = 1.0, 1.2, 1.5$ mol%, from bottom to top; (b) $M = 0.0955$, $C_\infty = 1.2$ mol%, $\kappa = 0.5, 0.29, 0.1$ from bottom to top

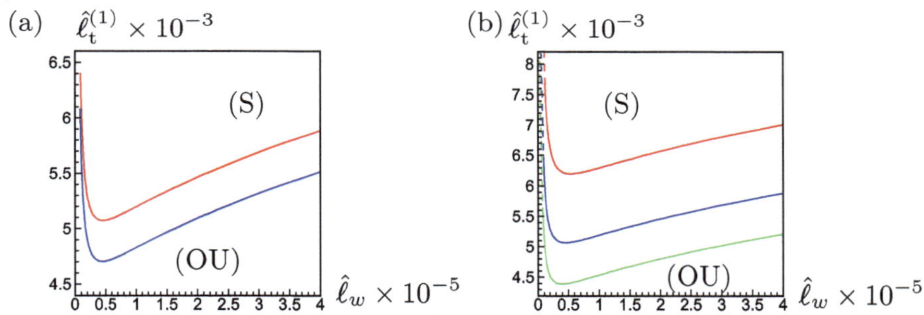


**Fig. 10.31** The neutral curves of the GTW S-modes ($n = 0$) in the first-order approximation in the $(\hat{\ell}_t, \hat{\ell}_w)$-plane for the cases $C_\infty = 1.2$mol%, $M = 0.9552$, $\kappa = 0.29$, and (**a**) $V = 16\,\mu\mathrm{m/s}$, $G_D = 78 \times 10^{-4}\,\mathrm{K}/\mu\mathrm{m}, 140 \times 10^{-4}\,\mathrm{K}/\mu\mathrm{m}$, which corresponds to the cases $\hat{v} = 0.5388 \times 10^{-2}$, and $\hat{\beta} = 0.8070 \times 10^{-5}, 0.1448 \times 10^{-4}$, $\lambda_G = 0.2780, 0.4989$ from top to bottom; (**b**) $G_D = 140 \times 10^{-4}\,\mathrm{K}/\mu\mathrm{m}$, and $V = 12\,\mu\mathrm{m/s}, 16\,\mu\mathrm{m/s}, 20\,\mu\mathrm{m/s}$, which correspond to the cases and $\hat{\beta} = 0.14485 \times 10^{-4}$, and $\hat{v} = 0.4666 \times 10^{-2}, 0.5388 \times 10^{-2}, 0.6024 \times 10^{-2}$; $\lambda_G = 0.6652, 0.4989, 0.3991$, from bottom to top

#### 10.4.7.2 Oscillatory Stability Criterion

From the above analysis, with given growth condition described by the three parameters $\hat{v}$, $\hat{\beta}$, and $M$, which, respectively, measure the pulling velocity $V$, gradient of temperature $G_D$, and concentration of impurity in the far field $C_\infty$, the neutral curve of the GTW S-mode ($n = 0$) divides the parameters $(\varepsilon, \lambda_0)$, or the $(\hat{\ell}_t, \hat{\ell}_w)$-plane, into two regions: the (S) region and the (Os.U) region. Accordingly,

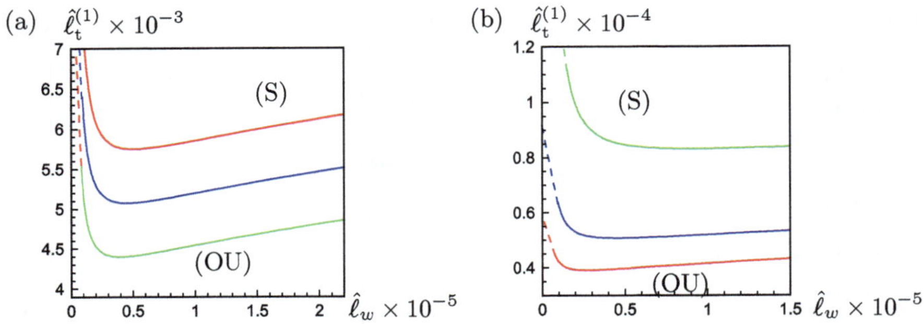


**Fig. 10.32** The neutral curves of the GTW S-modes ($n = 0$) in the first-order approximation in the $(\hat{\ell}_t, \hat{\ell}_w)$-plane for the cases $G_D = 140 \times 10^{-4}$ K/ μm, $V = 16$ μm/s, and (**a**) $\kappa = 0.29$, $M = 0.0796, 0.09552, 0.1194$, which corresponding to $C_\infty = 1.0, 1.2, 1.5$mol%, from top to bottom; (**b**) $M = 0.0955$, $C_\infty = 1.2$ mol%, $\kappa = 0.5, 0.29, 0.1$ from top to bottom

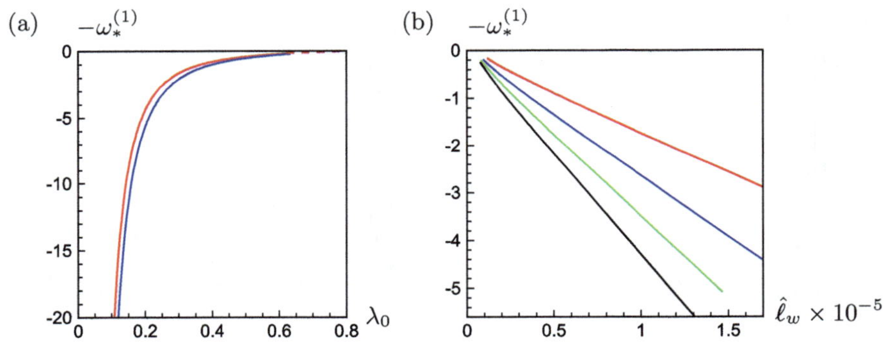


**Fig. 10.33** The variations of eigenfrequency $\omega_*^{(1)}$ in the first-order approximation with $\hat{\ell}_w$ for the cases (**a**) $V = 16$ μm/s and $G_D = 78 \times 10^{-4}$ K/ μm, $140 \times 10^{-4}$ K/ μm from bottom to top. (**b**) $G_D = 140 \times 10^{-4}$ K/ μm and $V = 12$ μm/s, 16 μm/s, 20 μm/s, 24 μm/s, from top to bottom

the basic state of the system is subject to the oscillatory stability criterion

$$\begin{cases} (\varepsilon, \lambda_0);\ (\hat{\ell}_t, \hat{\ell}_w) \in (\mathrm{S}): & \text{stable;} \\ (\varepsilon, \lambda_0);\ (\hat{\ell}_t, \hat{\ell}_w) \in (\mathrm{Os.U}): & \text{oscillating unstable.} \end{cases} \tag{10.247}$$

#### 10.4.7.3 Spectrum of Real Eigenvalues and Low-Frequency (LF) Instability

Case (II) leads to a spectrum of real eigenvalues with $|\sigma_0| \ll 1$. For this case, the physical solution in the outer region is given as $\mathrm{Re}\{\tilde{h}(\varrho, t)\} = \mathrm{Re}\{H(\varrho)\}\mathrm{e}^{\frac{\sigma_0 t}{\sqrt{\varepsilon}}}$. The system allows two types of global LF modes.

**1. Symmetric modes:** The tip's smoothness condition for these modes leads to

$$\begin{aligned} \mathrm{Re}\big\{H(\varrho)\big\}'(0) &= \mathrm{Re}\big\{H'(0)\big\} \\ &= \mathrm{Re}\left\{\frac{id_1}{\sqrt{\varepsilon}}\tilde{k}_0^{(1)}(0) + \frac{id_3}{\sqrt{\varepsilon}}\tilde{k}_0^{(3)}(0)\right\} = 0. \end{aligned} \tag{10.248}$$

Without loss of generality, we assume $d_1$ to be a positive real number and write

$$d_1 > 0, \quad \text{and} \quad d_3 = |d_3|\mathrm{e}^{\mathrm{i}\chi_0\pi}.$$

Since for a real eigenvalue $\sigma_0$, $k_0^{(1)}(0)$ and $k_0^{(3)}(0)$ are both real, the tip's smoothness condition (10.247) for the S-mode becomes $\mathrm{Im}\{d_3\} = 0$. The connection condition for this case leads to

$$\frac{d_1}{d_3} = \left|\frac{d_1}{d_3}\right|\mathrm{e}^{-\mathrm{i}\chi_0\pi}, \ (\chi_0 = 0, 1).$$

As a consequence, we obtain the quantization condition

$$\mathrm{Re}\Big\{\frac{1}{\sqrt{\varepsilon}}\int_0^{\varrho_c}\Big(\tilde{k}_0^{(1)} - \tilde{k}_0^{(3)}\Big)\mathrm{d}\varrho\Big\} = \Big(2n \pm \frac{1}{2}\Big)\pi (n = 0, 1, 2, \ldots). \tag{10.249}$$

**2. Antisymmetric modes:** The tip's smoothness condition for the antisymmetric modes is $\mathrm{Re}\big\{H(0)\big\} = \mathrm{Re}\big\{d_1 + d_3\big\} = 0$, which leads to

$$\left|\frac{d_1}{d_3}\right| = -\cos(\chi_0\pi) \leq 1.$$

Hence we have

$$\begin{aligned} \frac{d_1}{d_3} &= \left|\frac{d_1}{d_3}\right|\mathrm{e}^{-\mathrm{i}\chi_0\pi} = -\cos(\chi_0\pi)\mathrm{e}^{-\mathrm{i}\chi_0\pi}, \\ \chi_0 &= 1 + \frac{1}{\pi}\cos^{-1}\left|\frac{d_1}{d_3}\right|. \end{aligned}$$

Thus, we obtain the quantization condition

$$\mathrm{Re}\Big\{\frac{1}{\sqrt{\varepsilon}}\int_0^{\varrho_c}\Big(\tilde{k}_0^{(1)} - \tilde{k}_0^{(3)}\Big)\mathrm{d}\varrho\Big\} = \left(2n + \frac{1}{2} + \chi_0\right)\pi, (n = 0, 1, 2, \ldots). \tag{10.250}$$

The quantization condition (10.248) and (10.250) determine the eigenvalues $\sigma_{0n}$ and the corresponding eigenfunction.

It is noted that due to $|\sigma_0| \ll 1$, one may derive an approximate form for the integrals involved in (10.248) and (10.250). From the local dispersion relationship (10.210) and (10.211), one may derive the following asymptotic expansions for the wave number functions $k_e$ in the limit $\sigma_0 \to 0$:

$$\begin{cases} k_e^{(3)} = \mathrm{i}(\varrho - \mathrm{i})\sigma_e + \cdots, \\ k_e^{(1)} = \mathrm{e}^{-\frac{\mathrm{i}\pi}{4}}(\varrho + \mathrm{i})^{\frac{3}{4}}(\varrho - \mathrm{i})^{\frac{1}{4}} - \frac{\mathrm{i}}{2}(\varrho - \mathrm{i})\sigma_e + \cdots. \end{cases} \tag{10.251}$$

Recalling that

$$k_e(\varrho) = \tilde{k}_0(\varrho)\frac{G(\varrho)}{\lambda_0\sqrt{\bar{\Lambda}_1}},$$

we derive

$$\begin{aligned} \tilde{k}_0^{(1)}(\varrho) - \tilde{k}_0^{(3)}(\varrho) = \frac{2(1-\lambda_0)\sqrt{\bar{\Lambda}_1}}{\pi}\Bigg[ & \mathrm{e}^{-\frac{\mathrm{i}\pi}{4}}\frac{(\varrho + \mathrm{i})^{\frac{3}{4}}(\varrho - \mathrm{i})^{\frac{1}{4}}}{(\varrho + ia)(\varrho - ia)} \\ & - \frac{3\mathrm{i}}{2}\frac{(\varrho - \mathrm{i})}{(\varrho + ia)(\varrho - ia)}\sigma_e + \cdots \Bigg]. \end{aligned}$$

To calculate the integral along the path $(\mathcal{C}) : \varrho = 0 \to -\mathrm{i}$ contained in (10.249) and (10.250), one needs to separate two cases: (A) $(0 < \lambda_0 < \frac{1}{2})$ and (B) $(\frac{1}{2} < \lambda_0 < 1)$. We find that the system does not have an eigenvalue for case (A). Only case (B) is meaningful.

For the case $(\frac{1}{2} < \lambda_0 < 1)$, we have $0 \leq a < 1$, and the singular point $\varrho = -ia$ will be located in the middle of the path $(\mathcal{C})$. In this case, one can calculate the integral as follows:

$$\begin{aligned} (\mathrm{I}) &= \int_0^{\varrho_c} \left[\tilde{k}_0^{(1)}(\varrho) - \tilde{k}_0^{(3)}(\varrho)\right] \mathrm{d}\varrho \\ &= \left( \int_0^{-\mathrm{i}(a-0)} + \int_{C_r} + \int_{-\mathrm{i}(a+0)}^{-\mathrm{i}} + \int_{-\mathrm{i}}^{\varrho_c} \right) \left[\tilde{k}_0^{(1)}(\varrho) - \tilde{k}_0^{(3)}(\varrho)\right] \mathrm{d}\varrho, \end{aligned}$$

where $(\mathcal{C}_r)$ is a semicircle centered at $\varrho = -\mathrm{i}a$ connecting the points $\varrho = -\mathrm{i}(a-0)$ and $\varrho = -\mathrm{i}(a+0)$ in the clockwise direction. As a result, we have

$$\begin{aligned}
\mathrm{Re}\{\chi\} &= \frac{1}{\sqrt{\varepsilon}}\frac{(1-\lambda_0)\sqrt{\bar{\Lambda}_1}}{a}\left[(1-a)^{\frac{3}{4}}(1+a)^{\frac{1}{4}} - \frac{3}{2}(1+a)\sigma_e\right] \\
&= \frac{\sqrt{\bar{\Lambda}_1}}{\sqrt{\varepsilon}}\left[(2\lambda_0-1)^{\frac{3}{4}} - \frac{3}{2}\sigma_e\right] \\
\mathrm{Im}\{\chi\} &= \frac{2}{\sqrt{\varepsilon}}\frac{(1-\lambda_0)\sqrt{\bar{\Lambda}_1}}{a}\left[-\int_0^1 \frac{(1-x)^{\frac{3}{4}}(1+x)^{\frac{1}{4}}}{(a-x)(a+x)}\mathrm{d}x\right. \\
&\left.+\frac{3}{2}\int_0^1 \frac{(1+x)}{(a-x)(a+x)}\mathrm{d}x\right].
\end{aligned} \tag{10.252}$$

Based on these results, we derive the following simplified quantization conditions:

1. For the S-modes,

$$\frac{3}{2}\sigma_e = \frac{3}{2}\frac{\bar{\Lambda}_0}{\sqrt{\bar{\Lambda}_1}}\sigma_0 = (2\lambda_0-1)^{\frac{3}{4}} - \sqrt{\frac{\varepsilon}{\bar{\Lambda}_1}}\left(n \pm \frac{1}{2}\right)\pi(n = 0, 1, 2, \ldots); \tag{10.253}$$

2. For the A-modes,

$$\frac{3}{2}\sigma_e = \frac{3}{2}\frac{\bar{\Lambda}_0}{\sqrt{\bar{\Lambda}_1}}\sigma_0 = (2\lambda_0-1)^{\frac{3}{4}} - \sqrt{\frac{\varepsilon}{\bar{\Lambda}_1}}\left(2n + \frac{1}{2} + \chi_0\right)\pi(n = 0, 1, 2, \ldots). \tag{10.254}$$

For fixed $\varepsilon$ and $\lambda_0$, from (10.253) or (10.254) one can obtain a discrete set of eigenvalues $\{\sigma_{0n}\}$ $(n = 0, 1, 2, 3, \ldots)$. As a consequence, given $\lambda_0$, the system allows a discrete set of neutral stable modes ($\sigma_{0n} = 0$) corresponding to $\varepsilon = \varepsilon_n'$ $(n = 1, 2, \ldots)$, where $\varepsilon_0' > \varepsilon_1' > \varepsilon_2' > \cdots > \varepsilon_n' > \cdots$. On the other hand, given $\varepsilon > 0$, the system allows a discrete set of neutral modes with width parameters $\lambda_0 = \lambda_{0n}$.

We find that the neutral LF A-mode is more stable than the neutral LF S-mode with the same index $n$. The neutral LF S-modes are determined by the formula

$$\sqrt{\bar{\Lambda}_1}(2\lambda_0-1)^{\frac{3}{4}} = \sqrt{\varepsilon_*}\left(n+\frac{1}{2}\right)\pi, \quad \left(\frac{1}{2} \leq \lambda_0 < 1\right)(n = 0, 1, 2, \ldots). \tag{10.255}$$

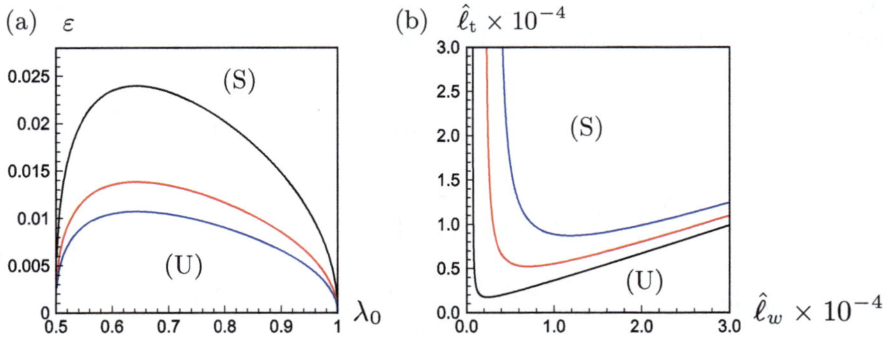


**Fig. 10.34** The neutral curves of the LF S-modes ($n = 0, 1, 2$) calculated using the formula (10.253) in the corresponding parameter plane for the case $\hat{v} = 0.5388 \times 10^{-2}, M = 0.9552 \times 10^{-1}, \kappa = 0.29, \hat{\beta} = 0.14485 \times 10^{-4}, \lambda_G = 0.4989$, (**a**) in the $(\lambda_0, \varepsilon)$-plane from top to bottom; (**b**) in the $(\hat{\ell}_w, \hat{\ell}_t)$-plane from bottom to top

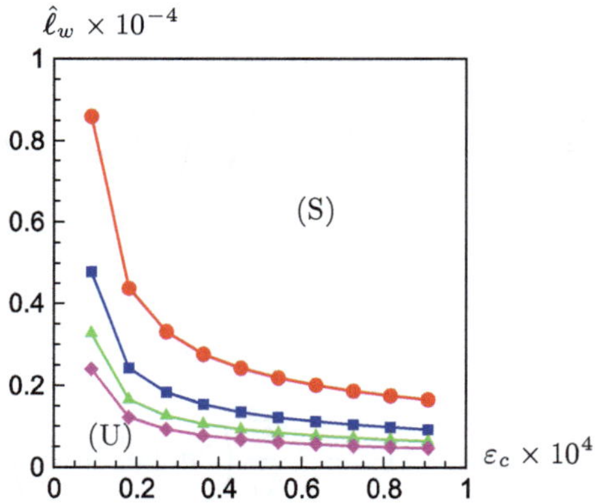


**Fig. 10.35** The neutral curves of the LF S-modes ($n = 0$) in the parameter $(\hat{\ell}_w, \varepsilon_c)$-plane for the cases $G_D = 140 \times 10^{-4}\,\mathrm{K}/\,\mu\mathrm{m}$, $M = 0.09552, \kappa = 0.29, \hat{\beta} = 0.1448 \times 10^{-4}$ and $\lambda_0 = 0.6, 0.7, 0.8, 0.9$ from top to bottom

The form of (10.255) is again very similar to what we obtained for the system of viscous fingering in Chap. 9. The neutral modes ($n = 0, 1, 2, \ldots$) of LF instability yield a discrete set of steady arrayed cellular growth solutions.

We have calculated (10.253) and (10.255) under some given operation parameters: $\{\hat{v}, \hat{\beta}, M\}$. The neutral curves for LF S-modes ($n = 0, 1, 2$) in the $(\lambda_0, \varepsilon)$-plane and $(\hat{\ell}_w, \hat{\ell}_t)$-plane for the typical case $\hat{v} = 0.5388 \times 10^{-2}, M = 0.9552 \times 10^{-1}, \kappa = 0.29, \hat{\beta} = 0.14485 \times 10^{-4}, \lambda_G = 0.4989$ are shown in Fig. 10.34a, b, respectively. One can see that in the LF mechanism, the most dangerous mode is the LF S-mode with ($n = 0$).

In Fig. 10.35, we show the neural curve of the LF S-mode ($n = 0$) in the parameter $(\hat{\ell}_w, \varepsilon_c)$-plane for various $\lambda_0$. One can see that as $\varepsilon_c$ (or the pulling velocity $V$) increases, the primary spacing $\hat{\ell}_w$ decreases with a fixed $\lambda_0$. This is in agreement with the experimental observations.

It should be pointed that although the leading-order approximation of eigenvalues $\sigma_0$ are real numbers, in the higher-order approximation, $\sigma$ may be complex with a small imaginary part. Hence, the unstable LF modes may be slowly oscillating with a low frequency.

### *10.4.8 Global Stability Diagram and Selection Principle of Arrayed-Cellular Growth*

The neutral curve of the LF S-mode ($n = 0$) shown in Fig. 10.34a may be called a finger-like branch, since it has a range of relative width ($0.5 < \lambda_0 < 1$) and a smooth interface shape like the Saffman–Taylor fingers. Accordingly, the LF S-modes may be called finger-like modes. The LF stability criterion, or say the stability criterion of the finger-like state, can be described as follows [see (10.34)]:

$$\begin{cases} (\varepsilon, \lambda_0);\ (\hat{\ell}_t, \hat{\ell}_w) \in (S): & \text{stable;} \\ (\varepsilon, \lambda_0);\ (\hat{\ell}_t, \hat{\ell}_w) \in (U): & \text{unstable, purely growing.} \end{cases} \tag{10.256}$$

On the other hand, the neutral curve of GTW S-mode ($n = 0$) may be called a needle-like branch. One may put the two neutral curves LF S-mode ($n = 0$) and GTW S-mode ($n = 0$) together in the $(\varepsilon, \lambda_0)$ and $(\hat{\ell}_w, \hat{\ell}_t)$ parameter planes as shown in Fig. 10.36a, b, respectively. Thus, we obtain the global stability diagrams of the system. The global stability region is bounded by the global neutral curves in the parameter $(\varepsilon, \lambda_0)$-plane, or $(\hat{\ell}_t, \hat{\ell}_w)$-plane. We propose that the selected limiting state

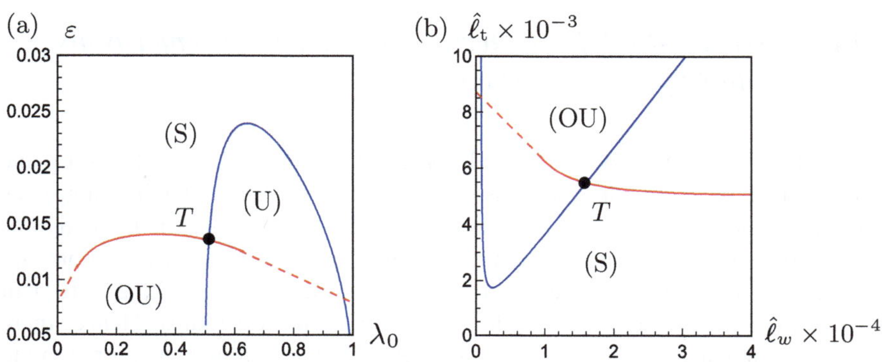


**Fig. 10.36** The two neutral curves of S-modes ($n = 0$), which correspond to LF stability and GTW stability, respectively for the case $\hat{v} = 0.5388 \times 10^{-2}$, $M = 0.9552 \times 10^{-1}$, $\kappa = 0.29$, $\hat{\beta} = 0.14485 \times 10^{-4}$, $\lambda_G = 0.4989$; (**a**) in the $(\lambda_0, \varepsilon)$ plane; (**b**) in the $(\hat{\ell}_w, \hat{\ell}_t)$ plane

must be on the global neutral curve, or rather that the selected state must be the most dangerous global neutrally stable mode, which may have a steady, or nearly steady, periodic arrayed-cellular pattern described by the neutral LF S-mode ($n = 0$) or the oscillatory periodic arrayed-dendritic interface structure described by the GTW S-mode ($n = 0$). This selection condition is called the *global neutrally stable mode* (GNSM) selection principle.

For the model problem of crystal growth in the channel with the fixed width $\ell_w$, the dimensionless primary spacing $\hat{\ell}_w$ is a given constant. Thus, with the GNSM selection principle, the tip radius $\hat{\ell}_t$ and the whole basic state are uniquely determined.

However, for the problem of deep arrayed-cellular growth in directional solidification, the imaginary side wall in the model problem is not a realistic solid wall. Hence, the primary spacing $\ell_w$ and the parameter $\hat{\ell}_w$ are free. Thus, the GNSM selection principle leads to a family of limiting states of a system of deep cellular growth with various primary spacing parameters $\hat{\ell}_w$ under given growth conditions $(\varepsilon_c, G_c, M)$. In other words, the GNSM selection condition for the system of cellular growth under the current setting does not determine the primary spacing parameter $\hat{\ell}_w$. It yields only a relationship between the tip radius and the primary spacing. Our theoretical result is consistent with the experimental observations by Pocheau et al. and Wang and his coworkers [9, 10, 12–14, 17, 23, 31, 48], which claimed that the value of the primary spacing parameter $\hat{\ell}_w$ is not unique and that it varies continuously within the available cell range, depending on the history of growth. It was based on such experimental observations that Pocheau et al. treated the primary spacing $\ell_w$ as another independent control parameter for cellular-array formation [14]. When the primary spacing $\ell_w$ is treated as an independent control parameter, like the width of the channel in the above model problem, the selected steady periodic deep-cellular array at the late stage of growth is, of course, fully determined.

### *10.4.9 Arrayed-Cellular Pattern Formation and Comparisons with Experimental Data*

In Fig. 10.37a–d we show the neutral curves of LF modes in the $(\hat{\ell}_w, \hat{\ell}_t)$-plane under some typical growth conditions. The experimental data for the limiting states for these cases measured by Pocheau et al. are also included in these figures for comparison. The thermodynamic parameters adopted in the calculations are shown in Table 10.1. Note that the thickness parameter $\delta/\ell_c = 2.12 \times 10^4\,\mu\text{m}$ is four to five times the size of the parameter of the tip radius. Hence, there exists a significant 3D effect in the experiments. The tip radius $\ell_t$ measured in the experiments is only one principal tip radius in the $(x, y)$-plane; the second principal radius along the thickness is omitted. As a consequence, with the experimental data the tip mean curvature reflected in the Gibbs–Thomson formula is systematically

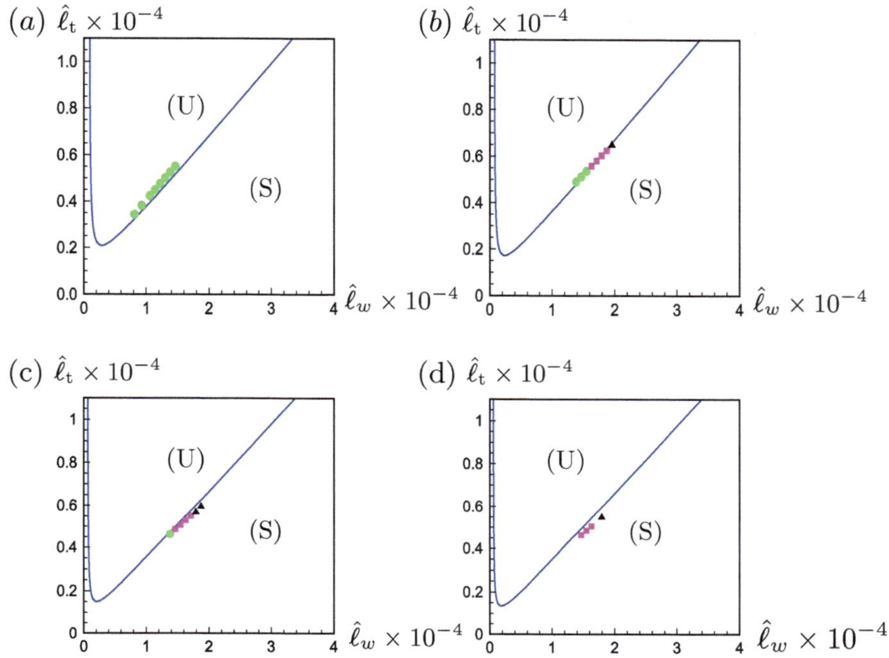


**Fig. 10.37** Selected cellular array growth for the case $G_D = 140.0 \times 10^{-4}\,\mathrm{K}/(\mu\mathrm{m})$, $\kappa = 0.29$, $M = 0.8432 \times 10^{-1}$, and (**a**) $V = 12\,\mu\mathrm{m/s}$; (**b**) $V = 16\,\mu\mathrm{m/s}$; (**c**) $V = 20\,\mu\mathrm{m/s}$; (**d**) $V = 24\,\mu\mathrm{m/s}$. Here, to eliminate the 3D effect occurring in the experiments, we have chosen $\widetilde{\ell_c} = 1.3\ell_c$ as the length scaling for the experimental data

**Table 10.1** The thermodynamic data used in the calculations for the nominally pure SCN system used by Pocheau et al. (2006, 2010) in the cell of growth with thickness $\delta = 50\,\mu\mathrm{m}$, width 45 mm, and length 150 mm

| | |
|---|---|
| $\kappa_D$ | $1.30 \times 10^{-5}\,\mathrm{cm^2\,s^{-1}}$ |
| $\kappa$ | 0.29 |
| $m$ | $1.9\,\mathrm{K\,mol\%^{-1}}$ |
| $c_\infty$ | 1.2 mol% |
| $\delta/\ell_c$ | $2.12 \times 10^4\,\mu\mathrm{m}$ |
| $\ell_c$ | $0.2359 \times 10^{-2}\,\mu\mathrm{m}$ |
| $c_p$ | $0.4630\,\mathrm{cal\,g^{-1}\,K^{-1}}$ |
| mol | 80.092 g/mol |
| $\Delta H$ | 11.051 cal/mol |

underestimated. To convert the 3D experimental system to an equivalent 2D model system for comparison with the 2D theory, one must compensate such an underestimation by setting a larger effective 2D surface tension coefficient with a factor $f_c > 1$ for the 2D model of the experimental system. Here, we set $f_c = 1.3$ in the calculations. Pocheau et al. classify the growth states observed in their experiments in three categories based on the amplitudes of oscillation of their interface shape: steady states, weakly dendritic states, and moderate dendritic

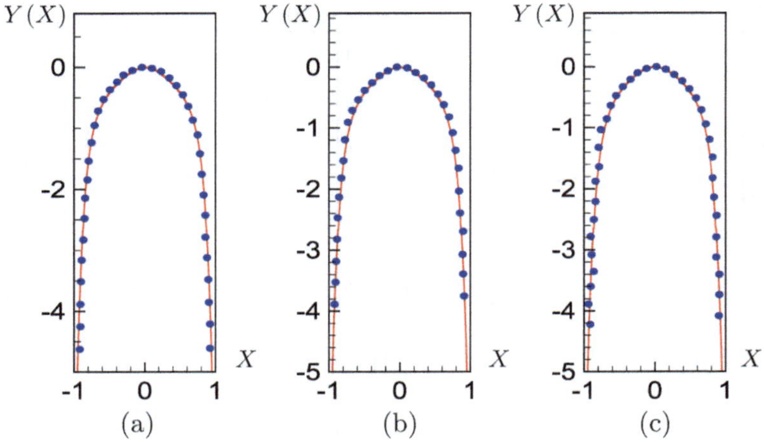


**Fig. 10.38** Interface shapes selected at the late stage of growth for the typical case $G_D = 140.0 \times 10^{-4}\,\mathrm{K}/(\mu\mathrm{m})$, $V = 20\,\mu\mathrm{m/s}$, $\kappa = 0.29$, $M = 0.8432 \times 10^{-1}$ and (**a**) $\ell_w = 65.0\,\mu\mathrm{m}$; (**b**) $\ell_w = 62.5\,\mu\mathrm{m}$; (**c**) $\ell_w = 60.0\,\mu\mathrm{m}$. The *dots* are experimental data given by Pocheau et al. [14]

states, but none for strong dendritic states. We marked these data in the figures with different symbols: full circles, squares, and triangles, respectively.

It is seen from Fig. 10.37a–d that the experimental points are quite close to the LF neutral curves. The discrepancies between theory and experiment are systematic, and they fall in the range of experimental measurement error, 10–20%. In Fig. 10.38a–c, we show the interface shapes predicted by the theory and compared with the experimental data under the same growth conditions. The quantitative agreements between both are quite good.

Hence, we conclude that the theoretical results are in very good quantitative agreement with these experimental data. The systematic discrepancy may be induced by physical effects, including the effects of 3D geometry, the anisotropy of surface tension, and interface kinetics that in the theoretical model are neglected.

### *10.4.10 Transition from Cellular Array to Dendritic Array*

We now put the experimental data by Pocheau et al. obtained in 2006 in the stability diagram shown in Fig. 10.39a–d. The stability diagram shows two branches of the global neutral curves: the finger-like branch and the needle-like branch, which cross each other at the transition point $T$. The steady smooth cellular array will lose its stability when the primary spacing $\hat{\ell}_w$ is larger than its critical number corresponding to the transition point $T$. It is seen that all the experimental data representing steady smooth cellular arrays are on the finger-like branch outside of the oscillatory unstable (Os.U) region. The experimental data that fall in the oscillatory unstable (Os.U) region all display some interfacial dynamical activities,

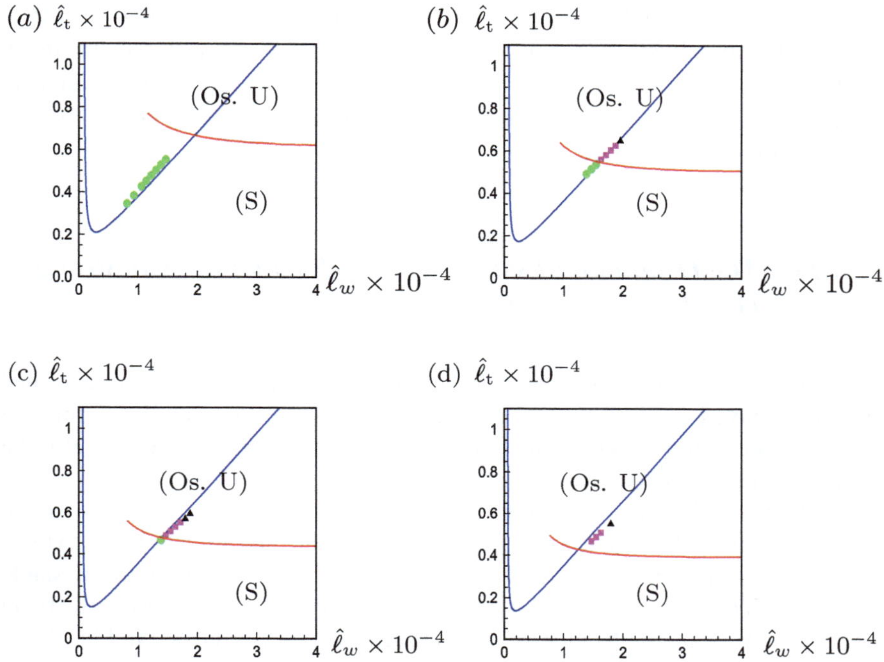


**Fig. 10.39** Selected cellular array growth for the case $G_D = 140.0 \times 10^{-4}\,\mathrm{K}/(\mu\mathrm{m})$, $\kappa = 0.29$, $M = 0.8432 \times 10^{-1}$, and (**a**) $V = 12\,\mu\mathrm{m/s}$; (**b**) $V = 16\,\mu\mathrm{m/s}$; (**c**) $V = 20\,\mu\mathrm{m/s}$; (**d**) $V = 24\,\mu\mathrm{m/s}$. In the experimental data, the full circles represent the steady cellular array, the *squares* represent weakly oscillatory cellular array, and the *triangles* represent the moderate oscillatory cellular array

such as weak oscillations and bursts of side-branching over the interface. This situation confirms that the system cannot maintain the steady smooth cellular array on the finger-like branch in the (Os.U) region, due to the presence of the unstable GTW modes in the system.

The experimental data representing the weak and moderate interfacial dynamical activities are slightly off the LF neutral curve and remain near it, but not close to the dendrite-like branch. This situation implies that the systems in these experiments were still in the transient process. The unstable steady smooth cellular array for these cases was still at the early stage of evolution. It is expected that if the experimental time lasted sufficiently long, the interface shapes would eventually evolve to a periodic deep-dendritic array. According to the GNSM selection principle, the data describing the strong arrayed-dendritic states would stay at the lower needle-like branch of the global neutral curve shown in Fig. 10.39a–d.

The experimental data pairs of $(\hat{\ell}_t, \hat{\ell}_w)$ obtained by Pocheau et al. (2006) do not contain the results for strong arrayed-dendritic patterns that can be used to verify the needle-like branch of the neutral curve in Fig. 10.39. It is expected that more complete and accurate experiments will be available in the future, showing that for the strong arrayed dendrites, as the value of $\hat{\ell}_w$ increases away from its critical number, the tip radius $\hat{\ell}_t$ of the dendrites on the needle-like branch remains much smaller than the tip radius $\hat{\ell}_t$ of the corresponding fingers on the finger-like branch. Moreover, the tip radius $\hat{\ell}_t$ of the strong arrayed dendrites approaches $\hat{\ell}_t$ of single dendritic growth as $\ell_w \to \infty$.

### 10.4.11 The Origin and Essence of Side-Branching in a Strong Oscillatory Dendritic Array

The conclusion drawn from our analysis is that the side-branching structure formed in strong arrayed-dendritic growth in directional solidification and that formed in single free dendritic pattern formation from undercooled melt are essentially the same type of phenomenon.

As described by the IFW theory in Chap. 6 for the system of a free single dendritic growth, the origin and essence of side-branching in cellular growth in directional solidification is also the presence of a discrete set of the unstable oscillatory GTW wave modes that are generated by the wave diagram involving the reflections and amplification of the interfacial waves as sketched in Fig. 10.40. This stability diagram was called the *global trapped-wave* (GTW) mechanism and interpreted in terms of wave reflection and interaction. The same interpretation can be applied to the current system: under the action of the incoming wave from the root $\tilde{W}_0^{(T)}(\varrho)$, an outgoing wave $\tilde{W}_0^{(+)}(\varrho)$ is reflected at the point $\rho_c'$ with the reflection rate $R = |D_1/D_3|$ and becomes the incoming wave $\tilde{W}_0^{(-)}(\varrho)$. The incoming wave $\tilde{W}_0^{(-)}(\varrho)$ is subsequently fully reflected at the cell's tip and becomes the outgoing wave $\tilde{W}_0^{(+)}(\varrho)$. During this course, the perturbed state is amplified. The amplification rate is the real part of the eigenvalue, $\sigma_R$, determined by the quantization condition (10.245). It is only under the condition $\sigma_R = 0$ that a stationary pattern persists.

As indicated before, in the global stable region in the $(\hat{\ell}_w, \hat{\ell}_t)$-plane shown in Fig. 10.39, all the GTW modes are decaying, and the system will remain at the steady arrayed-cellular interface. Of course, in that region the side-branches can be "artificially" generated by continuously applying sufficiently strong external perturbations or noise. The characteristics of these artificial side-branches are

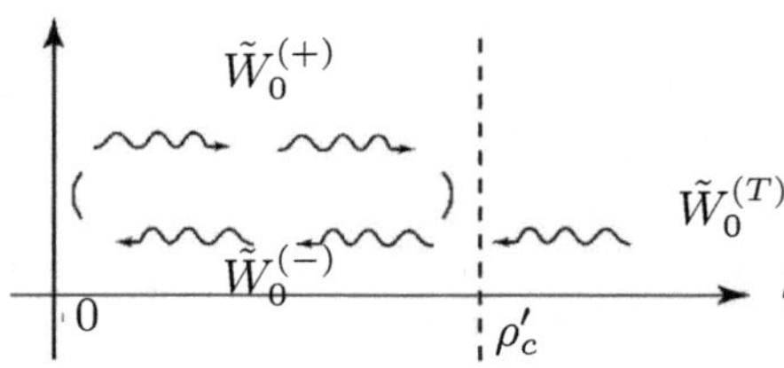


**Fig. 10.40** Wave diagram of the GTW mechanism

qualitatively different from those self-sustaining side-branches. The characteristic amplitude of the artificial side-branch depends on and is proportional to the strength of the external perturbations. The related dendritic patterns will disappear with time exponentially once the external forces stop. These features were observed in some numerical simulations, such as the numerical simulations by Echebarria et al. [11].

The IFW theory predicts that oscillatory arrayed-dendritic growth is described by the neutrally stable GTW S-mode ($n = 0$) located on the needle-like branch of the neutral curve. The side-branching on its interface is self-persistent with no need of any external enforced stimulations or noise. Its whole interface including the tips and side-branching oscillates with the same frequency, the eigenfrequency $\omega_*$ corresponding to the GTW S-mode ($n = 0$).

These conclusions are verified by experimental observations as well as numerical simulations. As a matter of fact, both the experiments by Georgelin and Pocheau [32] and the numerical simulation by Echebarria, Karma, and Gurevich [11] show that the frequency of the observed oscillation at the interface with side-branching is independent of the distance behind the tip. This is consistent with the conclusion of the IFW theory that the frequency of oscillation is an eigenfrequency of the system. Moreover, the experiments by Georgelin and Pocheau show that the amplitudes of oscillation at different points of the interface have cross-correlation. This is consistent with the IFW theory that the distribution of the oscillatory amplitude of the interface is mainly described by the eigenfunction of the GTW S-mode ($n = 0$).

In particular, in the numerical simulations, Echebarria et al. calculated the power spectrum of side-branching oscillation for a typical case of arrayed-dendritic growth. They found that there was a sequence of peaks in the graph of the power spectrum (refer to Fig. 16(b) in [11]). The physical implication of these peaks can be easily interpreted by comparing their locations with the period of primary harmonic resonance $T_* = 2\pi/\omega_*$, the periods of superharmonic resonance $2T_*, 3T_*, \ldots$, and the periods of subharmonic resonance $T_*/2, T_*/3, \ldots$. Here $\omega_*$ is the eigenfrequency $\omega_*$ of the GTW S-mode ($n = 0$), which has been calculated theoretically.

The numerical simulation of Echebarria et al. was conducted for the case $V = 32\,\mu\text{m/s}$, $G_D = 140\,\text{K/cm}$, the primary spacing $\ell_w = 138.9\,\mu\text{m}$. The system adopted is impure succinonitrile (SCN) alloy with anisotropy of surface tension $\epsilon_4 = 0.007$. The peak is found at the period $T = T_0 = 0.94\text{s}$. The stability parameter $\varepsilon_*$ and the eigenfrequency $\omega_*$ neutral GTW S-mode ($n = 0$) derived from the present theory are dimensionless numbers. Hence, for comparison, one needs to convert these dimensionless numbers to dimensional numbers in the given case.

Note that the numerical simulations by Echebarria et al. were based on the phase field model, the setting of whose parameters involves some uncertainties. For instance, with their notation, they set the ratio parameter $\epsilon = W/D_0$ in the range 36–72 associated with the thickness of the interface layer and the coupling constant $\lambda$ in the range 32–64. The choices of these constants, of course, may affect the numerical value of $T_0$. Moreover, as estimated by Echebarria et al., their numerical results have a 10% error. On the other hand, the numerical values of

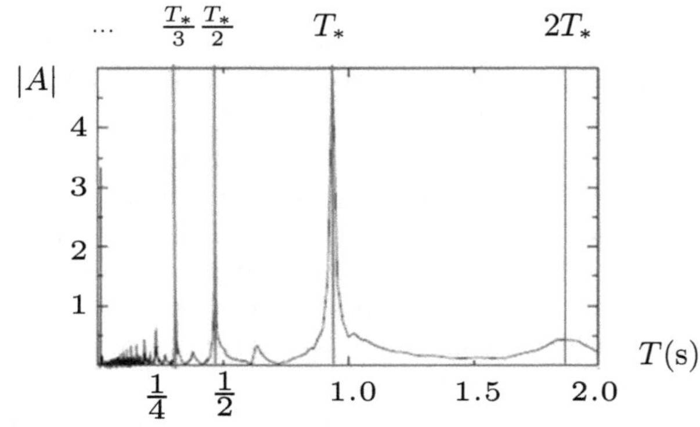


**Fig. 10.41** Comparison between the power spectrum and the periods of resonance for a typical case. The *curve* shows the power spectrum given in the numerical simulation by Echebarria et al. for a typical case (refer to Fig. 16(b) in [11]); the marks $2T_*, T_*, T_*/2, T_*/3, \ldots$ with *solid lines* are predicted by the IFW theory

$\varepsilon_*$ and $\omega_*$ predicted by analytical theory also involved a certain degree of error. For example, the theory neglected the effect of anisotropy of surface tension. As a consequence, it can be assumed that there is some systematic discrepancy between the numerical simulations and analytical theory. By taking these facts into account, in the comparison one may introduce an adjustment multiplier $f_c$. In this comparison, we set $f_c = 1.305$ to adjust the theoretical value of $\varepsilon_*$ given by the zeroth-order approximate solution of the neural GTW S-mode ($n = 0$). This leads to the eigenperiod $T_* = 0.940$s, the same value as $T_0$ that was provided by Echebarria et al.'s numerical simulation.

We mark all the periods of resonances $2T_*, T_*, T_*/2, T_*/3, \ldots$ with solid lines in the figure of the power spectrum given by Echebarria et al., as shown in Fig. 10.41. It is seen that all the peaks of the power spectrum curve are located at the places very close to the resonance lines. The highest peak is close to the period of the primary resonance, while the lower peaks are close to the periods of subharmonic and superharmonic resonances. The results are in full agreement with the assessment made by the theory.

Some researchers had adopted the so-called selected noise amplification (SNA) mechanism for explaining the origin and essence of side-branchings on the interface in cellular growth. As demonstrated in Chap. 6, it was before the presence of GTW instability was discovered that the SNA mechanism was proposed, attempting to reconcile the contradiction of the microscopic solvability condition (MSC) theory with the experimental observations on free dendritic growth. According to the MSC theory, the selected dendritic growth with the presence of anisotropy of surface tension is a needle-like crystal growth with no side-branching, whereas a growing dendrite is always observed to have side-branching. It is to offset such a defect of the MSC theory that "the selected noise amplification (SNA) mechanism" as an addition to the MSC theory was early proposed by some researchers. However, as has been indicated, the continuously acting noise can generate only artificial side-branching, not the experimentally observed self-sustaining side-breaching. The SNA mechanism does not elucidate the nature and numerical value of that specially "selected frequency," nor does it answer such questions as why only noise with such a frequency can be amplified, why such a special selected frequency is independent of the distance from the tip, and what is the implication of the discrete set of peaks appearing in the power spectrum. It appears that the SNA mechanism has created more unsolved issues after it attempted to fix one issue.

## 10.5 A Brief Summary

This chapter investigates steady deep arrayed-cellular growth in directional solidification and derives the analytical form of basic steady solutions. It also explores the interface-closure mechanism in the root region and determines the total length of the global cellular interface by solving an EVP. It also investigates the global instability mechanisms of the system and discusses the selection criterion. The eigenvalue problem for the linear perturbed states is solved in terms of a multiple-variable expansion method, and the associated global-modes solutions and the quantization conditions for the corresponding eigenvalues $\sigma$ are obtained. It is found that the stability mechanisms of the system of directional solidification have profound similarities with the systems of free dendritic growth and viscous fingering that are described in Chaps. 6 and 9, respectively, [1, 6–9, 12–17, 19, 20, 23, 24, 31, 32, 41, 44, 45, 48]. On the basis of the global stability mechanisms, we proposed the global neutral stable (GNSM) mode selection principle for the limiting state of growth, and demonstrated the origin and essence of side-branching. The theoretical predictions are well supported by the available experimental works by Pocheau et al. [14–16, 32], and the numerical simulations by Gurevich et al. and Echebarria et al. [11].

The main results are outlined as follows:

1. The global asymptotic solution for the basic steady state consists of the two parts RPE and SPE. The asymptotic factors of the RPE part are algebraic functions of $\epsilon$, whereas the asymptotic factors of the SPE part are exponential functions of $\epsilon$. Although at any fixed point other than the origin, each term of the SPE part is exponentially small as $\epsilon \to 0$, it is indispensable for describing the global features of the basic steady state.
2. The root region consists of three inner sublayers; the root inner system has a complex turning point $\hat{\xi}_c$, whose presence leads to the so-called *trapped-waves mechanism* and plays a crucial role for the interface-closure.
3. The quantization condition derived from the trapped-waves mechanism yields the eigenvalue $\hat{\xi}_*$ as a function of the parameter $\epsilon$ and other parameters of the system, which determines the total length of the cell, as well as the global features of the interfacial pattern.

4. The perturbed system has two discrete spectra of eigenvalues corresponding to the quantum numbers ($n = 0, 1, 2, \ldots$): the complex spectrum with $|\sigma| = \mathcal{O}(1)$ and the nearly real spectrum with $|\sigma| \ll 1$. Accordingly, the system involves two types of instability mechanisms: the global traveling wave (GTW) instability, whose neutral modes display deep, oscillatory arrayed-dendritic patterns, and the low frequency (LF) instability, whose neutral modes display steady smooth, or weakly oscillatory, arrayed-cellular patterns. Among all GTW modes, the symmetric GTW S-mode with $n = 0$ is the most dangerous oscillatory mode, while among the all LF modes, the symmetric LF S-mode with $n = 0$ is the most dangerous nearly steady or weakly oscillatory mode.
5. The neutral curves of the GTW S-mode ($n = 0$) and LF S-mode ($n = 0$) intersect with each other at the point $T$, as shown in Fig. 10.36. They divide the parameter $(\varepsilon, \lambda_0)$-plane (and the $(\hat{\ell}_w, \hat{\ell}_t)$-plane) into two subdomains: a stable domain and an unstable domain. Hence, the global neutral curve consists of finger-like branches that are part of the neutral curve of the LF S-mode $n = 0$, and the needle-like branch that is part of the neutral curve of the GTW S-mode ($n = 0$).
6. The *global neutrally stable mode* (GNSM) selection principle suggests that under given growth conditions, the fully developed limiting solution will be located on the global neutral curve.
7. The origin and essence of the side-branching in directional solidification are the same as those in free dendritic growth, due to the GTW mechanism that involves wave reflection and amplification along the interface. The side-branching structure on the interface is self-sustaining and can persist without continuous application of external perturbations or noise.
8. For the problem of finger-like crystal growth in a channel, the limiting state selected by the GNSM principle is unique. However, for 2D cellular growth in a Hele–Shaw cell, the limiting state selected by the GNSM criterion is not unique. The GNSM selection criterion yields only a deterministic relationship between the tip radius $\ell_t$ and primary spacing $\ell_w$.

## Appendix 1: The System of Cellular Growth Under the Curvilinear Coordinate System $(\xi, \eta)$

As given in Sect. 9.1, under the curvilinear coordinate system $(\xi, \eta)$, one may write

$$\mathbf{r} = X(\xi, \eta)\mathbf{i} + Y(\xi, \eta)\mathbf{j}, \tag{10.257}$$

where $\mathbf{i}$ and $\mathbf{j}$ are the unit vectors along the $x$-axis and $y$-axis respectively. Let the vectors $\mathbf{e}_1, \mathbf{e}_2$ be the unit vectors along the $\xi, \eta$ directions respectively. One can write

$$\begin{aligned}\mathbf{e}_1 &= \frac{\partial \mathbf{r}}{\partial \xi} \Big/ \left|\frac{\partial \mathbf{r}}{\partial \xi}\right| = \frac{1}{\mathcal{G}} \left(X_\xi \mathbf{i} + Y_\xi \mathbf{j}\right), \\ \mathbf{e}_2 &= \frac{\partial \mathbf{r}}{\partial \eta} \Big/ \left|\frac{\partial \mathbf{r}}{\partial \eta}\right| = \frac{1}{\mathcal{G}} \left(X_\eta \mathbf{i} + Y_\eta \mathbf{j}\right) .\end{aligned} \tag{10.258}$$

Denote the general unsteady concentration distribution by $C = C(\xi, \eta, t, \epsilon)$ and the interface shape by $\eta = \eta_s(\xi, t)$. We shall use the following formulas:

$$\begin{aligned}\nabla C &= \frac{\mathbf{e}_1}{\mathcal{G}} \frac{\partial C}{\partial \xi} + \frac{\mathbf{e}_2}{\mathcal{G}} \frac{\partial C}{\partial \eta}, \\ (\mathbf{j} \cdot \nabla) C &= \frac{1}{\mathcal{G}^2} \left( Y_\xi \frac{\partial C}{\partial \xi} + X_\xi \frac{\partial C}{\partial \eta} \right),\end{aligned} \tag{10.259}$$

and

$$\nabla^2 C = \nabla \cdot (\nabla C) = \frac{1}{\mathcal{G}^2} \left( \frac{\partial^2 C}{\partial \xi^2} + \frac{\partial^2 C}{\partial \eta^2} \right), \tag{10.260}$$

$$\frac{\partial C}{\partial n} = \mathbf{n} \cdot \nabla C = \frac{1}{\left(1 + \eta_s'^2\right)^{1/2} \mathcal{G}} \left( \frac{\partial C}{\partial \eta} - \eta_s' \frac{\partial C}{\partial \xi} \right). \tag{10.261}$$

In the moving frame, the local growth speed at the interface is

$$u_{\mathrm{I}} = -\frac{\frac{\partial s}{\partial t}}{|\nabla s|} = \mathcal{G} \frac{\frac{\partial \eta_s}{\partial t}}{\left(1 + \eta_s'^2\right)^{1/2}}, \tag{10.262}$$

$$\mathbf{j} \cdot \mathbf{n} = \frac{Y_\eta - Y_\xi \eta_s'}{\mathcal{G}\left(1 + \eta_s'^2\right)^{1/2}} . \tag{10.263}$$

The linear distribution of temperature is given by

$$\frac{T_{\mathrm{S}}(WY)}{M} = -c_* + \epsilon \lambda_{\mathrm{G}} WY(\xi, \eta) + o(\epsilon^2). \tag{10.264}$$

Using the above formulas, one may derive the mass diffusion equation

$$\frac{\partial^2 C}{\partial \xi^2} + \frac{\partial^2 C}{\partial \eta^2} = W\mathrm{Pe} \left( \mathcal{G}^2 \frac{\partial C}{\partial t} - Y_\xi \frac{\partial C}{\partial \xi} - X_\xi \frac{\partial C}{\partial \eta} \right) \tag{10.265}$$

and the interface conditions at $\eta = \eta_s(\xi, t)$:

1. The Gibbs–Thomson condition:

$$C = c_* - \epsilon\lambda_G WY(\xi, \eta) - \frac{\epsilon^2\hat{\Gamma}}{MW}\mathcal{K}\{\eta_s\} + o(\epsilon^2). \tag{10.266}$$

2. The mass conservation condition:

$$\frac{\partial C}{\partial \eta} - \eta_s'\frac{\partial C}{\partial \xi} + \epsilon W(1-\kappa)C\left(\mathcal{G}^2\frac{\partial \eta_s}{\partial t} - Y_\xi\eta_s' + Y_\eta\right) = 0. \tag{10.267}$$

In the above, the prime represents the derivative with respect to $\xi$. The expressions of the curvature operator $\mathcal{K}\Big\{\eta_s(\xi, t)\Big\}$ and the root condition are the same as given in Sect. 9.1. For the convenience of the reader, we have listed them again here:

$$\mathcal{K}\Big\{\eta_s(\xi,t)\Big\} = -\frac{1}{\mathcal{G}(\xi,\eta_s)}\Bigg\{\frac{\eta_s''}{\left(1+\eta_s'^2\right)^{3/2}} + \frac{\Pi_0(\xi,\eta_s)}{\mathcal{G}^2(\xi,\eta_s)\left(1+\eta_s'^2\right)^{1/2}}$$
$$+\frac{\Pi_1(\xi,\eta_s)}{\mathcal{G}^2(\xi,\eta_s)\left(1+\eta_s'^2\right)^{1/2}}\eta_s'\Bigg\}, \tag{10.268}$$

where

$$\begin{aligned}\Pi_0(\xi,\eta) &= \left(Y_{\xi\xi}X_\xi - X_{\xi\xi}Y_\xi\right),\\ \Pi_1(\xi,\eta) &= \left(X_{\xi\xi}X_\xi + Y_{\xi\xi}Y_\xi\right).\end{aligned} \tag{10.269}$$

Assume that as $\epsilon \to 0$,

$$\eta_B(\xi) = \epsilon h_1(\xi) + \cdots.$$

We derive that

$$\mathcal{K}\{\eta_B(\xi)\} = \mathcal{K}_0(\xi) + \epsilon\mathcal{K}_1(\xi) + \cdots, \tag{10.270}$$

where

$$\begin{cases}\mathcal{K}_0(\xi) = -\dfrac{\Pi_{0,0}}{\mathcal{G}_0^3},\\[2ex] \mathcal{K}_1(\xi) = \dfrac{2\pi^2\lambda_0(1-\lambda_0)}{\mathcal{G}_0^3\mathcal{R}_0}\left[1 - \dfrac{2(1-\lambda_0)}{\lambda_0\mathcal{R}_0} + \dfrac{3\Pi_{0,0}}{\pi\mathcal{G}_0^2}\right]h_1(\xi),\\[2ex] \qquad -\dfrac{2\pi(1-\lambda_0)^2\tan(\frac{\pi\xi}{2})}{\mathcal{R}_0\mathcal{G}_0^3}h_1'(\xi) - \dfrac{1}{\mathcal{G}_0}h_1''(\xi).\end{cases} \tag{10.271}$$

Particularly, at the tip of cell, one has

$$\mathcal{K}_0(0) = \frac{\pi(1-\lambda_0)}{2\lambda_0^2}, \quad \mathcal{K}_0'(0) = 0. \tag{10.272}$$

At the root point T, $\xi = \pm 1$, $\eta = \eta_T(t)$: since $Y_\xi(\pm 1, \eta_s) = -X_\eta(\pm 1, \eta_s) = 0$, the slope of the interface shape $y_s$ can be calculated as

$$\frac{dy_s}{dx}\Big|_{x=\pm W} = \frac{Y_\xi + Y_\eta \eta_s'}{X_\xi + X_\eta \eta_s'}\Big|_{X=\pm 1} = \eta_s'(\pm 1). \tag{10.273}$$

Thus, the root conditions are given as

$$\eta_s(\pm 1) = \eta_T, \quad \frac{\partial \eta_s}{\partial \xi}(\pm 1) = 0. \tag{10.274}$$

## Appendix 2: First-Order Approximate Solution

### *The Determination of the Wave Number Function $k_1(\xi)$*

The equation for the first-order approximation can be obtained from (10.189) as the following:

$$(k_0^2 + g_0^2)\left(\frac{\partial^2}{\partial \xi_+^2} + \frac{\partial^2}{\partial \eta_+^2}\right)\tilde{C}_1 = a_0(\xi, \eta)e^{i\xi_+ - \eta_+}, \tag{10.275}$$

where

$$a_0 = 2(k_0 - ig_0)\left(\frac{\partial A_0}{\partial \eta} - i\frac{\partial A_0}{\partial \xi}\right). \tag{10.276}$$

To eliminate the secular terms on the right-hand side of (10.275), one must set $a_0 = 0$. Then from (10.276), it follows that

$$\left(\frac{\partial}{\partial \eta} - i\frac{\partial}{\partial \xi}\right)A_0(\xi, \eta) = 0. \tag{10.277}$$

We note that (10.277) is a first-order hyperbolic equation. One can solve $A_0(\xi, \eta)$ in the $(\xi, \eta)$-plane as an initial value problem given the initial conditions on the line $\eta = 0$:

$$A_0(\xi, 0) = \hat{A}_0(\xi) = -\Big[W\hat{\text{Pe}}(\lambda_G - y_{*0} + 1) + \frac{\bar{k}_0^2}{MW\mathcal{G}_0(\xi)}\Big]\hat{D}_0.$$

We then derive that at $\eta = 0$,

$$\hat{D}_0 Q_0 = \frac{\partial}{\partial \eta} A_0(\xi, \eta)\Big|_{\eta=0} = -\mathrm{i}\hat{D}_0 \frac{\partial}{\partial \xi}\left[\frac{\bar{k}_0^2}{MW\mathcal{G}_0(\xi)}\right]. \tag{10.278}$$

These formulas will be needed later to solve for $k_1(\xi)$.

We now turn to solving the governing equation of $\tilde{C}_1$:

$$\Big(\frac{\partial^2}{\partial \xi_+^2} + \frac{\partial^2}{\partial \eta_+^2}\Big)\tilde{C}_1 = 0. \tag{10.279}$$

The corresponding interface conditions are as follows: at the interface $\eta_+ = 0, \eta = 0$,

$$\begin{aligned} &\tilde{C}_1 + W\hat{\mathrm{P}}\mathrm{e}(\lambda_\mathrm{G} - y_{*0} + 1)\tilde{h}_1 - \frac{\bar{k}_0^2}{MW\mathcal{G}_0(\xi)}\frac{\partial^2 \tilde{h}_1}{\partial \xi_+^2} \\ &\quad = \frac{2\bar{k}_0\bar{k}_1}{MW\mathcal{G}_0(\xi)}\frac{\partial^2 \tilde{h}_0}{\partial \xi_+^2} - h_1\bar{k}_0\frac{\partial \tilde{C}_0}{\partial \eta_+}, \end{aligned} \tag{10.280}$$

and

$$\begin{aligned} &\bar{k}_0\frac{\partial \tilde{C}_1}{\partial \eta_+} + \sigma_0 W\hat{\mathrm{P}}\mathrm{e}(1-\kappa)y_{*0}\mathcal{G}_0^2(\xi)\tilde{h}_1 + \bar{k}_0 W\hat{\mathrm{P}}\mathrm{e} \\ &\times\big[\lambda_\mathrm{G} - (1-\kappa)y_{*0}\big]Y_{\xi,0}(\xi)\frac{\partial \tilde{h}_1}{\partial \xi_+} \\ &= -\frac{\partial \tilde{C}_0}{\partial \eta} - h_1\bar{k}_0^2\frac{\partial^2 \tilde{C}_0}{\partial \eta_+^2} - \bar{k}_1\frac{\partial \tilde{C}_0}{\partial \eta_+} - W\hat{\mathrm{P}}\mathrm{e}\lambda_G Y_{\xi,0}(\xi)\bar{k}_1\frac{\partial \tilde{h}_0}{\partial \xi_+} \\ &\quad + W\hat{\mathrm{P}}\mathrm{e}\big[\lambda_\mathrm{G} - (1-\kappa)y_{*0})\big]Y_{\eta\eta}(\xi, 0)\tilde{h}_0 \\ &\quad - \sigma_1 W\hat{\mathrm{P}}\mathrm{e}(1-\kappa)y_{*0}\mathcal{G}_0^2(\xi)\tilde{h}_0 \\ &\quad + \bar{k}_1 W\hat{\mathrm{P}}\mathrm{e}(1-\kappa)y_{*0}Y_{\xi,0}(\xi)\frac{\partial \tilde{h}_0}{\partial \xi_+}. \end{aligned} \tag{10.281}$$

This first-order approximate system is inhomogeneous, which allows the form of normal modes solutions:

$$\tilde{C}_1 = A_1(\xi, \eta)\exp\{\mathrm{i}\xi_+ - \eta_+\}, \tag{10.282}$$

$$\tilde{h}_1 = \hat{D}_1 \exp\{\mathrm{i}\xi_+\} \tag{10.283}$$

and is subject to the solvability condition

$$I_2 + \bar{k}_0 I_1 = 0, \tag{10.284}$$

where

$$I_1 = -\frac{2\bar{k}_0\bar{k}_1}{MW\mathcal{G}_0(\xi)} + h_1(\xi)\bar{k}_0\Big[-W\hat{\text{Pe}}(\lambda_{\text{G}} - y_{*0} + 1) - \frac{\bar{k}_0^2}{MW\mathcal{G}_0(\xi)}\Big], \tag{10.285}$$

and

$$\begin{aligned} I_2 = &-Q_0 + h_1(\xi)\bar{k}_0^2\Big[W\hat{\text{Pe}}(\lambda_{\text{G}} - y_{*0} + 1) + \frac{\bar{k}_0^2}{MW\mathcal{G}_0(\xi)}\Big] \\ &-\bar{k}_1\Big[W\hat{\text{Pe}}(\lambda_{\text{G}} - y_{*0} + 1) + \frac{\bar{k}_0^2}{MW\mathcal{G}_0(\xi)} + \text{i}W\hat{\text{Pe}}\lambda_G Y_{\xi,0}(\xi)\Big] \\ &+\text{i}\bar{k}_1 W\hat{\text{Pe}}(1-\kappa)y_{*0}Y_{\xi,0}(\xi) + W\hat{\text{Pe}}\Big[\lambda_{\text{G}} - (1-\kappa)y_{*0}\Big]Y_{\eta\eta}(\xi,0) \\ &-\sigma_1 W\hat{\text{Pe}}(1-\kappa)y_{*0}\mathcal{G}_0^2(\xi). \end{aligned} \tag{10.286}$$

The formula (10.284) leads to

$$\begin{aligned} \bar{k}_1 = \frac{1}{F(\xi)}\Big\{ &-W\hat{\text{Pe}}\Big[\lambda_{\text{G}} - (1-\kappa)y_{*0}\Big]Y_{\eta\eta}(\xi,0) \\ &-\text{i}\frac{\partial}{\partial\xi}\Big[\frac{\bar{k}_0^2}{MW\mathcal{G}_0(\xi)}\Big] + \sigma_1 W\hat{\text{Pe}}(1-\kappa)y_{*0}\mathcal{G}_0^2(\xi)\Big\}, \end{aligned} \tag{10.287}$$

where we have defined

$$\begin{aligned} F(\xi) = &-W\hat{\text{Pe}}(\lambda_{\text{G}} - y_{*0} + 1) + \text{i}W\hat{\text{Pe}} \\ &\times\Big[(1-\kappa)y_{*0} - \lambda_{\text{G}}\Big]Y_{\xi,0}(\xi) - \frac{3\bar{k}_0^2}{MW\mathcal{G}_0(\xi)}. \end{aligned} \tag{10.288}$$

The formula (10.286) can be written in the following form:

$$\bar{k}_1(\xi) = \frac{R_1(\xi)}{F(\xi)} - \text{i}\frac{R_2(\xi)}{F(\xi)}\frac{\bar{k}_0'(\xi)}{\bar{k}_0(\xi)}, \tag{10.289}$$

where we have defined

$$\begin{cases} R_1(\xi) = -W\hat{\text{Pe}}\Big[\lambda_G - (1-\kappa)y_{*0}\Big]Y_{\eta\eta}(\xi,0) \\ \qquad -\dfrac{i\bar{k}_0^2}{MW}\dfrac{\partial}{\partial\xi}\left[\dfrac{1}{\mathcal{G}_0(\xi)}\right] + \sigma_1 W\hat{\text{Pe}}(1-\kappa)y_{*0}\mathcal{G}_0^2(\xi), \\ R_2(\xi) = \dfrac{2\bar{k}_0^2}{MW}\left[\dfrac{1}{\mathcal{G}_0(\xi)}\right]. \end{cases}$$

It is seen from (10.286) that the first-order approximate solution $\bar{k}_1(\xi)$ has a singularity at the complex root of the function $F(\xi)$.

## *The Location of the Singularity of MVE Solutions in the Extended Complex $\zeta$-Plane*

We now determine the location of the singular point. For this purpose, we extend all the functions such as $F(\xi)$, $\bar{k}_0(\xi)$, $Y_{\xi,0}(\xi)$, $\mathcal{G}_0(\xi)$, by analytic continuation, to the corresponding complex functions $F(\zeta)$, $\bar{k}_0(\zeta)$, $Y_{\zeta,0}(\zeta)$, $\mathcal{G}_0(\zeta)$ in the complex plane $\zeta = \xi + i\xi_I$. In the meanwhile, we extend all the associated functions defined on the real $\rho$-axis to the complex function $\tilde{k}_0(\varrho)$ in the complex plane $\varrho = \rho + i\rho_I$ by analytic continuation. We deduce that from the analytic continuation, the wave-number function $\bar{k}_0(\zeta) \equiv \{\bar{k}_0^{(1)}(\zeta), \bar{k}_0^{(2)}(\zeta), \bar{k}_0^{(3)}(\zeta)\}$ is subject to the following extended dispersion formula:

$$\sigma_0 = \Sigma(\bar{k}_0,\zeta) = \frac{\bar{k}_0}{\bar{\Lambda}_0\mathcal{G}_0^2(\zeta)}\left[\lambda_0 + iY_{\zeta,0}(\zeta) - \frac{\bar{k}_0^2}{\bar{\Lambda}_1\mathcal{G}_0(\zeta)}\right]. \tag{10.290}$$

Accordingly, the analytic function $\tilde{k}_0(\varrho) \equiv \left\{\tilde{k}_0^{(1)}(\varrho), \tilde{k}_0^{(2)}(\varrho), \tilde{k}_0^{(3)}(\varrho)\right\}$ is subject to the extended dispersion formula

$$\sigma_0 = \frac{G(\varrho)\tilde{k}_0}{\bar{\Lambda}_0\lambda_0 S^2(\varrho)}\left[(1-i\varrho) - \frac{G^2(\varrho)\tilde{k}_0^2}{\bar{\Lambda}_1\lambda_0^2 S(\varrho)}\right], \tag{10.291}$$

or

$$\sigma_e = \Sigma(k_e,\varrho) = \frac{k_e}{S^2(\varrho)}\left[(1-i\varrho) - \frac{k_e^2}{S(\varrho)}\right]. \tag{10.292}$$

It may be derived from the extended dispersion formula (10.290) that

$$\frac{\partial\Sigma(\bar{k}_0,\zeta)}{\partial\bar{k}_0} = \frac{F(\zeta)}{W\hat{\text{Pe}}(1-\kappa)y_{*0}\mathcal{G}_0^2(\zeta)}. \tag{10.293}$$

It follows from the above that at the singular point $F(\zeta_c) = 0$, we have

$$\frac{\partial \Sigma(\bar{k}_0, \zeta_c)}{\partial \bar{k}_0} = 0, \quad \text{and} \quad \left[\bar{k}_0^{(1)}(\zeta_c) - \bar{k}_0^{(3)}(\zeta_c)\right] = 0. \tag{10.294}$$

Combining (10.294) and (10.290), one derives that $\zeta_c$ is subject to the equation

$$\sigma_0 = \frac{2}{\sqrt{27}} \frac{\sqrt{\bar{\Lambda}_1}}{\bar{\Lambda}_0} \left[\frac{\lambda_0 + \mathrm{i}Y_{\xi,0}(\zeta_c)}{\mathcal{G}_0(\zeta_c)}\right]^{\frac{3}{2}}. \tag{10.295}$$

In accordance with the above, in the representation of the variable $\varrho$, the location of the singular point $\varrho = \varrho_c$ in the complex $\varrho$-plane must satisfy the equation

$$\frac{\partial \Sigma(\tilde{k}_0, \varrho)}{\partial \tilde{k}_0} = 0 \quad \text{or} \quad \left[\tilde{k}_0^{(1)}(\varrho) - \tilde{k}_0^{(3)}(\varrho)\right] = 0\,. \tag{10.296}$$

Combining (10.296) and (10.291), one finds that $\varrho_c$ is subject to the equation

$$\sigma_{\mathrm{e}} = \frac{2}{\sqrt{27}} \left(\frac{1 - \mathrm{i}\varrho_c}{1 + \mathrm{i}\varrho_c}\right)^{\frac{3}{4}} = \frac{\bar{\Lambda}_0}{\sqrt{\bar{\Lambda}_1}} \sigma_0. \tag{10.297}$$

Three complex singular points can be obtained from (10.297). Their locations depend on the eigenvalue $\sigma_0$, as well as on the parameters $\{\bar{\Lambda}_0, \bar{\Lambda}_1\}$. We are interested in the *singular point $\varrho_c$, which is located in the fourth quadrant of the complex $\varrho$-plane*.

As $\sigma_0 \to 0$, the singular point $\varrho_c$ approaches $-\mathrm{i}$. On the other hand, since $\sigma_0$ is a complex number with $|\sigma_0| = \mathcal{O}(1)$ and $0 \leq \mathrm{Re}\{\sigma_0\} \ll 1$, one of the turning points $\varrho_c$ in which we are interested is located in the fourth quadrant of the complex $\varrho$-plane and away from the point $\varrho = -\mathrm{i}$.

## *The First-Order Approximation of the Wave Number Function $\bar{k}_1(\zeta)$ and Eigenvalue $\sigma_1$*

We first determine the asymptotic behavior of $F(\zeta)$ as $\zeta \to \zeta_c$. By taking the complete derivative with respect to $\zeta$ on both sides of (10.290) for fixed $\sigma_0$, we derive that

$$\begin{aligned} 0 &= \frac{\partial \Sigma}{\partial \bar{k}_0} \bar{k}_0'(\zeta) + \frac{\partial \Sigma}{\partial \zeta} \\ &= \left[\frac{F(\zeta)}{\mathrm{W}\hat{\mathrm{P}}\mathrm{e}(1-\kappa) y_{*0} \mathcal{G}_0^2(\zeta)}\right] \bar{k}_0'(\zeta) + \frac{\partial \Sigma}{\partial \zeta}. \end{aligned} \tag{10.298}$$

Furthermore, by taking the partial derivative with respect to $\zeta$ of (10.290), we obtain

$$\begin{aligned}\frac{\partial\Sigma}{\partial\zeta} &= -\frac{\bar{k}_0}{\bar{\Lambda}_0\bar{\Lambda}_1\mathcal{G}_0^3}\left\{\frac{3\mathcal{G}_0'}{\mathcal{G}_0}\Big[(\lambda_0+\mathrm{i}Y_{\zeta,0})\bar{\Lambda}_1\mathcal{G}_0-\bar{k}_0^2\Big]\right.\\ &\quad\left.-\Big[(\lambda_0+\mathrm{i}Y_{\zeta,0})\bar{\Lambda}_1\mathcal{G}_0\Big]'\right\}\\ &= \frac{\bar{k}_0}{MW^2\hat{\mathrm{P}}\mathrm{e}(1-\kappa)y_{*0}\mathcal{G}_0^3(\zeta)}\left\{\Big[(\lambda_0+\mathrm{i}Y_{\zeta,0})\bar{\Lambda}_1\mathcal{G}_0\Big]'\right.\\ &\quad\left.-\frac{3\mathcal{G}_0'}{\mathcal{G}_0}\Big[(\lambda_0+\mathrm{i}Y_{\zeta,0})\bar{\Lambda}_1\mathcal{G}_0-\bar{k}_0^2\Big]\right\}.\end{aligned} \tag{10.299}$$

Finally, by combining the formula (10.298) with (10.297), we obtain

$$\begin{aligned}\frac{\bar{k}_0'(\zeta)}{\bar{k}_0(\zeta)} &= \frac{R_0(\zeta)}{F(\zeta)},\\ R_0(\zeta) &= \frac{1}{MW\mathcal{G}_0(\zeta)}\left\{\frac{3\mathcal{G}_0'}{\mathcal{G}_0}\Big[(\lambda_0+\mathrm{i}Y_{\xi,0})\bar{\Lambda}_1\mathcal{G}_0-\bar{k}_0^2\Big]\right.\\ &\quad\left.-\Big[(\lambda_0+\mathrm{i}Y_{\xi,0})\bar{\Lambda}_1\mathcal{G}_0\Big]'\right\}.\end{aligned} \tag{10.300}$$

It is seen that as $\zeta \to \zeta_c$, we have

$$\frac{\bar{k}_0'(\zeta)}{\bar{k}_0(\zeta)} \sim \frac{R_0(\zeta_c)}{F(\zeta)}, \quad \text{or} \quad \frac{\bar{k}_0'(\zeta)}{\bar{k}_0(\zeta)}F(\zeta) \sim R_0(\zeta_c).$$

On the other hand, from (10.288), by analytic continuation, we obtain

$$\begin{aligned}F(\zeta) &= -W\hat{\mathrm{P}}\mathrm{e}(\lambda_{\mathrm{G}}-y_{*0}+1)+\mathrm{i}W\hat{\mathrm{P}}\mathrm{e}\\ &\quad\times\Big[(1-\kappa)y_{*0}-\lambda_{\mathrm{G}}\Big]Y_{\zeta,0}(\zeta)-\frac{3\bar{k}_0^2}{MW\mathcal{G}_0(\zeta)}.\end{aligned} \tag{10.301}$$

It follows that

$$\begin{aligned}F'(\zeta) &= \mathrm{i}W\hat{\mathrm{P}}\mathrm{e}\Big[(1-\kappa)y_{*0}-\lambda_{\mathrm{G}}\Big]Y_{\zeta\zeta,0}(\zeta)-\frac{6\bar{k}_0^2\bar{k}_0'}{MW\mathcal{G}_0(\zeta)}\\ &\quad-\frac{3\bar{k}_0^2}{MW}\left[\frac{1}{\mathcal{G}_0(\zeta)}\right]'.\end{aligned} \tag{10.302}$$

From (10.302), we then obtain that as $\zeta \to \zeta_c$,

$$F(\zeta_c)F'(\zeta_c) \sim -\frac{6\bar{k}_0^3(\zeta_c)}{MW\mathcal{G}_0(\zeta_c)} = \mathcal{O}(1),$$

or

$$F(\zeta) \propto (\zeta - \zeta_c)^{\frac{1}{2}}. \tag{10.303}$$

Note that in the extended $\zeta$-plane, the formula (10.289) can be written in the following form:

$$\bar{k}_1(\zeta) = \frac{R_1(\zeta)}{F(\zeta)} - \mathrm{i}\frac{R_2(\zeta)}{F(\zeta)}\frac{\bar{k}_0'(\zeta)}{\bar{k}_0(\zeta)}, \tag{10.304}$$

where

$$\begin{cases} R_1(\zeta) = -W\hat{\mathrm{Pe}}\Big[\lambda_{\mathrm{G}} - (1-\kappa)y_{*0})\Big]Y_{\eta\eta}(\zeta,0) \\ \qquad -\dfrac{\mathrm{i}\bar{k}_0^2}{MW}\dfrac{\partial}{\partial\zeta}\left[\dfrac{1}{\mathcal{G}_0(\zeta)}\right] + \sigma_1 W\hat{\mathrm{Pe}}(1-\kappa)y_{*0}\mathcal{G}_0^2(\zeta), \\ R_2(\zeta) = \frac{2\bar{k}_0^2}{MW}\left[\frac{1}{\mathcal{G}_0(\zeta)}\right]. \end{cases}$$

It is clear from (10.304) that the solution $\bar{k}_1(\zeta)$ has an isolated singularity at $\zeta_c$, and as $\zeta \to \zeta_c$,

$$\bar{k}_1(\zeta) \sim \frac{m_1 R_1(\zeta_c)}{(\zeta-\zeta_c)^{\frac{1}{2}}} + \frac{m_2}{(\zeta-\zeta_c)} + \mathcal{O}(1), \tag{10.305}$$

where $m_1, m_2 = \mathcal{O}(1)$. Since around the isolated singular point $\zeta = \zeta_c$, the analytic function $\bar{k}_1(\zeta)$ can be expanded into a Laurent series, we deduce from (10.305) that

$$R_1(\zeta_c) = 0 \tag{10.306}$$

must hold. The condition (10.306) yields

$$\sigma_1 = \frac{1}{W\hat{\mathrm{Pe}}(1-\kappa)y_{*0}\mathcal{G}_0^2(\zeta_c)}\Bigg\{W\hat{\mathrm{Pe}}\Big[\lambda_{\mathrm{G}} - (1-\kappa)y_{*0}\Big] \times Y_{\eta\eta}(\zeta_c,0) - \frac{\mathrm{i}F_0(\zeta_c)\mathcal{G}_0'(\zeta_c)}{3\mathcal{G}_0(\zeta_c)}\Bigg\}. \tag{10.307}$$

## Appendix 3: Major Nomenclature

For the reader's convenience, we have appended here a list of the major notation used in this chapter:

### *Parameters*

- $\mathrm{Pe} = \frac{\ell_t}{\ell_D}$: the Péclet number, where $\ell_t$ is the tip radius.
- $\lambda_T = \frac{\ell_D}{\ell_T}$: the ratio of the two diffusion lengths $\ell_D$ and $\ell_T$.
- $G = \frac{\ell_D}{\Delta H/(c_p\rho)}(G)_D$: the dimensionless gradient of the temperature, where $(G)_D = \frac{T_H - T_C}{(L)_D}$.
- $M = -\frac{mC_\infty}{\Delta H/(c_p\rho)}$: the morphological parameter, where $m = (m_\alpha, m_\beta)$ is the slope of the liquidus.
- $\Gamma = \frac{\ell_c}{\ell_w} = \epsilon^2\hat{\Gamma}$: the interfacial stability parameter, $\ell_c = \frac{\gamma c_p \rho T_{M0}}{(\Delta H)^2}$.
- $\varepsilon = \sqrt{\Gamma}$, $\epsilon = \varepsilon\hat{\mathrm{Pe}}$, $\hat{\mathrm{Pe}} = 1/\sqrt{\hat{\Gamma}}$.
- $\lambda_G = \frac{\ell_D}{\ell_G} = -\frac{\kappa_D}{V}\frac{(G)_D}{mC_\infty} = \frac{G}{M}$.
- $y_* = \epsilon\lambda_G y_0 : \{y_{*0}; \epsilon y_{*1}; \cdots\}$.
- $\hat{v} = \sqrt{\ell_c V/\kappa_D}$: the parameter that is directly related to the pulling velocity $V$ and
- $\hat{\beta} = \frac{\ell_c}{\ell_G}$: the parameter that is directly related to the gradient of temperature.

### *Unknowns*

1. The basic steady solution in the outer region (with the slow variable $\tilde{\eta}$):
   $C_B(\xi,\eta,\epsilon) = \bar{C}_B(\xi,\eta,\epsilon) + \tilde{C}_B(\xi,\eta,\epsilon)$; $\eta_B(\xi,\epsilon) = \bar{\eta}_B(\xi,\epsilon) + \tilde{\eta}_B(\xi,\epsilon)$
   - RPE part:
     $\bar{C}_B(\xi,\eta,\epsilon) = C(\xi,\eta,\tilde{\eta},\epsilon)$: $\{C_0(\xi,\eta,\tilde{\eta}); C_1(\xi,\eta,\tilde{\eta}); C_2(\xi,\eta,\tilde{\eta}); \cdots\}$.
     $\bar{\eta}_B(\xi,\epsilon) : \{\bar{h}_0(\xi), \bar{h}_1(\xi), \ldots\}$
   - SPE part:
     Fast variables: $\{\xi_+, \eta_+\}$ with the factor $(\epsilon\hat{\Gamma})^{1/2}$
     $\tilde{C}_B(\xi,\eta,\epsilon) = \tilde{C}(\xi,\eta,\xi_+,\eta_+,\epsilon)$: $\{\tilde{C}_0(\xi,\eta,\xi_+,\eta_+); \tilde{C}_1(\xi,\eta,\xi_+,\eta_+); \cdots\}$
     $\tilde{\eta}_B(\xi,\epsilon) = \tilde{h}(\xi,\xi_+,\epsilon) : \{\tilde{h}_0(\xi,\xi_+), \tilde{h}_1(\xi,\xi_+), \ldots\}$
     wave number: $k(\xi,\eta) : \{k_0(\xi,\eta); k_1(\xi,\eta), \ldots\}$; $\bar{k}_0(\xi) = k_0(\xi, 0)$.
2. The basic steady solution in the root region (with inner variables $(\hat{\xi},\hat{\eta})$):
   $C_B(\xi,\eta,\epsilon) = \bar{C}_R(\hat{\xi},\hat{\eta},\epsilon) + \tilde{C}_R(\hat{\xi},\hat{\eta},\epsilon)$; $\eta_R(\xi,\epsilon) = \bar{\eta}_R(\xi,\epsilon) + \tilde{\eta}_R(\xi,\epsilon)$.

- RPE part:
  Fast variables: $\{\hat{\xi}_+, \hat{\eta}_+\}$ with the factor $(\epsilon\hat{\Gamma})^{1/2}$
  $C_{\rm B}(\xi, \eta, \epsilon) \approx C_{\rm R0}(\hat{\xi}, \hat{\eta}, \epsilon)$
  $\bar{\eta}_{\rm R}(\xi, \epsilon) \approx \hat{\eta}_{\rm T}(\hat{\xi})$
- SPE part:
  $\tilde{C}_{\rm R}(\xi, \eta, \epsilon) = \hat{C}_*(\hat{\xi}, \hat{\eta}, \hat{\xi}_+, \hat{\eta}_+, \epsilon)$: $\{\hat{C}_{*0}(\hat{\xi}, \hat{\eta}, \hat{\xi}_+, \hat{\eta}_+); \cdots\}$
  $\tilde{\eta}_{\rm R}(\xi, \epsilon) = \hat{\rm h}_*(\hat{\xi}, \hat{\xi}_+, \epsilon) : \{\hat{\rm h}_{*0}(\xi, \xi_+), \ldots\}$
  wave number: $k_*(\hat{\xi}, \hat{\eta}) : \{k_{*0}(\hat{\xi}, \hat{\eta}); \cdots\}$; $\hat{k}_0(\hat{\xi}) = k_{*0}(\hat{\xi}, 0)$.
- The turning point: $\hat{\xi}_c$,
- The junction point: $\hat{\xi}_*$.

3. The general unsteady solutions: (set $\epsilon = \varepsilon\hat{\rm P}{\rm e}, \varepsilon = \sqrt{\Gamma}$).
   $C(\xi, \eta, t, \varepsilon) + C_{\rm B}(\xi, \eta, \varepsilon) + \widehat{C}(\xi, \eta, t, \varepsilon)$;
   $\eta_{\rm s}(\xi, t, \varepsilon) + \eta_{\rm B}(\xi, \varepsilon) + \widehat{\eta}_{\rm s}(\xi, t, \varepsilon)$.
   The solutions for perturbed states:
   $\{\widehat{C}(\xi, \eta, t, \varepsilon); \widehat{\eta}_{\rm s}(\xi, t, \varepsilon)\}$.

- Fast variables: $\{\xi_+, \eta_+, t_+\}$ with the factor $\varepsilon^{1/2}$
- $\widehat{C}(\xi, \eta, t, \varepsilon) = \tilde{C}(\xi, \eta, \xi_+, \eta_+, t_+\varepsilon)$: $\{\tilde{C}_0(\xi, \eta, \xi_+, \eta_+); \cdots\}{\rm e}^{\sigma t_+}$
  $\widehat{\eta}_{\rm s}(\xi, t, \varepsilon) = \tilde{\eta}_{\rm s}(\xi, \xi_+, t_+, \varepsilon)$: $\{\tilde{h}_0(\xi, \xi_+); \cdots\}{\rm e}^{\sigma t_+}$
- $k(\xi, \eta, \varepsilon)$: $\{k_0(\xi, \eta); k_1(\xi, \eta); \cdots\}$: the wave number functions.
  $\bar{k}(\xi, \varepsilon) = k(\xi, 0, \varepsilon)$: $\{\bar{k}_0(\xi); \bar{k}_1(\xi); \cdots\}$.
- $\sigma$: $\{\sigma_0; \sigma_1; \cdots\}$: the eigenvalue.

# References


1. B. Billia, R. Trivedi, Pattern formation in crystal growth, in *Handbook of Crystal Growth, Volume 1: Fundamentals, Part B: Transport and Stability*, ed. by D.T.J. Hurle (Elsevier Science Publishers, Amsterdam, London, New York, Tokyo, 1993), pp. 899–1073
2. T.F. Bower, H.D. Brody, M.C. Flemings, Measurements of solute redistribution in dendritic solidification. Trans. Metall. Soc. AIME **236**, 624–633 (1966)
3. K. Brattkus, Capillary instabilities in deep cells during directional solidification. J. Phys. Fr. **50**, 2999–3006 (1989)
4. B. Caroli, C. Caroli, B. Roulet, Instability of planar solidification fronts, in *Solids Far from Equilibrium*, ed. by C. Godreche (Cambridge University Press, Cambridge, 1991), pp. 155–296
5. Y.Q. Chen, J.J. Xu, Global theory of steady deep-cellular growth in directional solidification. Phys. Rev. E **83**, 041603 (2011)
6. Y.Q. Chen, X.X. Tang, J.J. Xu, 3D interfacial wave theory of dendritic growth: (I). Multiple variables expansion solutions. Chin. Phys. B **18**, 671–685 (2009)
7. Y.Q. Chen, X.X. Tang, J.J. Xu, 3D interfacial wave theory of dendritic growth: (II). Non-axisymmetric global wave modes and selection of pattern formation. Chin. Phys. B **18**, 686–698 (2009)

8. S.H. Davis, Effect of flow on morphological stability, in *Handbook of Crystal Growth, Volume 1: Fundamentals, Part B: Transport and Stability*, ed. by D.T.J. Hurle (Elsevier Science Publishers, Amsterdam, London, New York, Tokyo, 1993), pp. 859–897
9. S.H. Davis, *Theory of Solidification* (Cambridge University Press, Cambridge, 2001)
10. G.L. Ding, W.D. Huang, X. Lin, Y.K. Zhou, Prediction of average spacing for constrained cellular/dendritic growth. J. Crys. Growth **177**(3/4), 281–288 (1997)
11. B. Echebarria, A. Karma, S. Gurevich, Onset of sidebranching in directional solidification. Phys. Rev. E **81**, 021608 (2010)
12. M. Georgelin, A. Pocheau, Onset of sidebranching in directional solidification. Phys. Rev. E **57**, 011604, 3190–3203 (1998)
13. M. Georgelin, A. Pocheau, Characterization of cell tip curvature in directional solidification. J. Cryst. Growth **268**, 272–283 (2004)
14. A. Pocheau, M. Gergelin, Shape of growth cells in directional solidification. Phys. Rev. E **73**, 011604 (2006)
15. M. Georgelin, S. Bodea, A. Pocheau, Coherence of dendritic sidebranching in directional solidification. Europhys. Lett. **77**, 46001 (2007)
16. I.S. Gradshteyn, I.M. Ryzhik, *Table of Integrals, Seriers and Products*, 7th edn. (Academic Press is an imprint of Elsevier, New York, 2007)
17. J. Kageyama, Y. Sasajima, M. Ichimura, Phase field simulation on directional solidification of succinonitrile (SCN)–acetone organic model alloy. Mater. Trans. **46**(9), 2003–2010 (2005)
18. A. Karma, P. Pelcé, Effective interface formalism: application to global instability in viscous fingering. Phys. Rev. A **41**, 4507 (1990)
19. J. Kevorkian, J.D. Cole, *Multiple Scale and Singular Perturbation Methods*. Applied Mathematical Sciences, vol. 114 (Springer, New York, Heidelberg, Berlin, 1996)
20. P. Kopczyncki, W.-J. Rappel, A. Karma, Phase-field method for computationally efficient modeling of solidification with arbitrary interface kinetics. Phys. Rev. Lett. **77**, 3387 (1996)
21. M. Kruskal, H. Segur, Asymptotics beyond all orders in a model of crystal growth. Stud. Appl. Math. **85**, 129–181 (1991)
22. P. Kurowski, S. de Cheveign, G. Faivre, C. Guthmann, Cusp instability in cellular growth. J. Phys. Fr. **50**, 3007–3019 (1989)
23. J.C. LaCombe, M.B. Koss, C. Giummarra, J.E. Frei, A.O. Lupulecscu, M.E. Glicksman, Evidence for eigenfrequencies in dendritic growth dynamics, in *Coherent Structures in Comples System*, ed. by D. Reguera, L.L. Bonilla, J.M. Rabi. Lecture Notes in Physics, vol. 567 (Springer, Berlin, Heideberg, New York, 2001), pp. 283–297
24. M. Losert, D.A. Stillman, H.Z. Cummins, P. Kopczyncki, W.-J. Rappel, A. Karma, Selection of doublet cellular patterns in directional solidification through spatially periodic perturbations. Phys. Rev. E **58**, 7492 (1998)
25. M. Mashaal, M. Ben Amar, V. Hakim, Directional solidification cells at low velocity. Phys. Rev. A **41**, 4421–4432 (1990)
26. J. Mclean, P.G. Saffman, The effect of surface tension on the shape of fingers in a Hele-Shaw cell. J. Fluid Mech. **102**, 455–469 (1981)
27. P. Molho, A.J. Simon, A. Libchaber, Péclet number and crystal growth in channel. Phys. Rev. A **42**, 904–910 (1990)
28. W.W. Mullins, R.F. Sekerka, Morphological stability of a particle growing by diffusion or heat flow. J. Appl. Phys. **34**, 323–329 (1963)
29. W.W. Mullins, R.F. Sekerka, Stability of a planar interface during solidification of a dilute binary alloy. J. Appl. Phys. **35**, 444–451 (1964)
30. P. Pelcé, A. Pumir, Cell shape in directional solidification in the small Péclet number limit. J. Cryst. Growth **73**, 337–342 (1985)
31. A. Pocheau, M. Georgelin, Cell tip undercooling in directional solidification. J. Cryst. Growth **206**, 215–229 (1999)
32. A. Pocheau, S. Bodea, M. Georgelin, Self-organized dendritic sidebranching in directional solidification: sidebranch coherence within uncorrelated bursts. Phys. Rev. E **80**, 031601 (2009)

33. N. Ramprasad, R.A. Brown, Ph.D. Thesis. MIT, Department of Chemical Engineering (1991)
34. J.W. Rutter, B. Chalmers, A prismatic substructure formed during solidification of metals. Can. J. Phys. **31**, 15–39 (1953)
35. P.G. Saffman, G.I. Taylor, The penetration of a fluid into a porous medium or Hele-Shaw cell containing a more viscous liquid. Proc. R. Soc. Lond. Ser. A **245**, 312–329 (1958)
36. K. Somboonsuk, J.T. Mason, R. Trivedi, Interdendritc spacing: part (I). Experimental studies. Metall. Trans. A **15A**, 967–975 (1984)
37. R. Trivedi, Interdendritc spacing: part (II). A comparison of theory and experiment. Metall. Trans. A **15A**, 977–982 (1984)
38. L.H. Ungar, R.A. Brown, Cellular interface morphologies in directional solidification: the one-sideed model. Phys. Rev. B **29**, 1367–1380 (1984). Cellular interface morphologies in directional solidification: IV. The formation of deep cell. Phys. Rev. B **31**, 5931–5940 (1985)
39. J.D. Weeks, W. van-Saarloons, Directional solidification cells with grooves for a small partition coeffcient. Phys. Rev. A **39**, 2772–2775 (1989)
40. J.D. Weeks, W. van-Saarloons, M. Grant, Stability and shape of cellular profiles in directional solidification: expansion and matching methods. J. Cryst. Growth **112**, 244–282 (1991)
41. J.J. Xu, Global instability of viscous fingering in a Hele–Shaw cell formation of oscillatory fingers. Eur. J. Appl. Math. **2**, 105–132 (1991)
42. J.J. Xu, Interfacial wave theory of solidification–dendritic pattern formation and selection of tip velocity. Phys. Rev. A15 **43**, 930–947 (1991)
43. J.J. Xu, Generalized needle solutions, interfacial instabilities and pattern formation. Phys. Rev. E **53**, 5031–5062 (1996)
44. J.J. Xu, Interfacial instabilities and fingering formation in Hele–Shaw flow. IMA J. Appl. Math. **57**, 101–135 (1996)
45. J.J. Xu, *Interfacial Wave Theory of Pattern Formation: Selection of Dendrite Growth and Viscous Fingering in a Hele–Shaw Flow* (Springer, Berlin, Heidelberg, New York, 1998)
46. J.J. Xu, Y.Q. Chen, Global stabilities, selection of steady cellular growth, and origin of side branches in directional solidification. Phys. Rev. E **83**, 061605 (2011)
47. J.J. Xu, Y.Q. Chen, Interface closure in the root region of steady deep-cellular growth in directional solidification. Eur. J. Appl. Math. **26**(3), 355–382 (2015)
48. J.J. Xu, D.S. Yu, Selection and resonance of dendritic growth with interference of oscillatory external sources. J. Cryst. Growth **226**, 378–392 (2001)

# Chapter 11
# Lamellar Eutectic Growth

## 11.1 Introduction

Eutectic growth in a binary mixture system is another fundamental subject in condensed matter physics and materials science. Unlike systems of cellular growth, the concentration of the two species (A) and (B) in a binary mixture of eutectic growth are moderate and close to the eutectic concentration $C_e$ (see Fig. 11.1).

The typical experimental device used for studying such phenomena is also the Hele–Shaw cell, similar to the one used in previous chapter and shown in Fig. 10.1. The coordinate system $(x, y)$ shown in Fig. 10.1 will be adopted for the analysis in this chapter.

The system consists of a thin sample material and two uniform temperature zones separated by a distance $(L)_D$: the hot zone with a temperature $T_H$ higher than the eutectic temperature $T_e$, and the cold zone with a temperature $T_C$ lower than the eutectic temperature $T_e$. The sample is pulled at a constant speed $V$ along the direction from the hot zone to the cold zone. We denote the concentration of species $(B)$ in the liquid phase of the mixture by $C$.

According to the phase diagram (see Fig. 11.1), due to the phase transition, a liquid state of the mixture is separated into two different solid phases as shown in Fig. 11.2: (1) the $\alpha$-phase, in which the species $(A)$ is the major component, while the species $(B)$ is the minor component; (2) the $\beta$-phase, in which the species $(B)$ is the major component, while $(A)$ is the minor component. One special feature of the eutectic interfacial pattern is that it has a triple point, where the angles between the interfaces depend on the thermodynamic properties of the system. Furthermore, experimental observations show that the system of steady eutectic growth displays spatially periodic interfacial patterns, which may be untilted or tilted, as seen in Fig. 11.2a or b, respectively.

The early analytical theory of steady eutectic growth was established by Jackson and Hunt (J–H) [7]. In the J–H theory, the steady state is described by a solution



J.-J. Xu, *Interfacial Wave Theory of Pattern Formation in Solidification*,
Springer Series in Synergetics, DOI 10.1007/978-3-319-52663-8_11

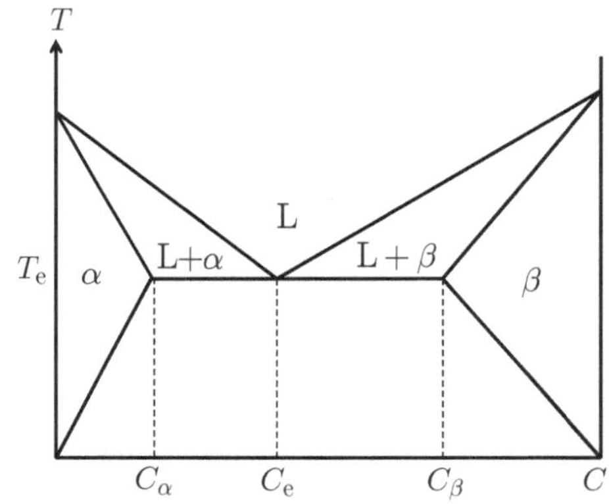


**Fig. 11.1** Sketch of a phase diagram of eutectic growth

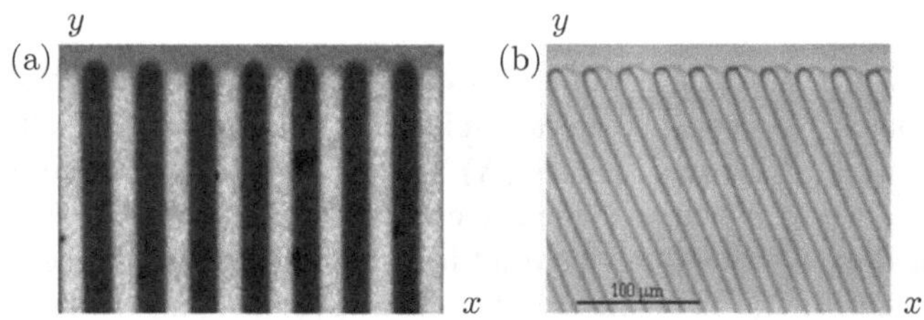


**Fig. 11.2** (**a**) Typical steady eutectic growth with $\alpha$–$\beta$ interface parallel to the $y$-axis [2]. (**b**) Typical steady tilted eutectic growth [6]

with **one free parameter** depending on the interlamellar spacing, and the interfaces are assumed to be flat. As a consequence, in the J–H solution, the effect of the triple point is totally neglected. It has long been recognized that the J–H solution was not self-consistent, since it treated the related free boundary problem as a boundary value problem. The first work treating eutectic growth as a free boundary problem was done by Nash in 1977 [11]. Nash derived a set of nonlinear integrodifferential equations for the system and solved it numerically for a simplified version that assumed that the interface was isothermal. In 1981, Datye et al. performed a linear stability analysis [4] for a system of untilted eutectic growth. However, as the basic state, Datye et al. adopted the steady solution with flat interface obtained by J–H. Again the presence of the triple point and its effect on the stability mechanisms of the system were ignored.

Later analytical work on this subject was done by Chen and Davis [3, 5] in terms of the multiple scales asymptotic approach. Chen and Davis assumed that in the leading-order approximation, the interface shape could be considered flat, as in the J–H solution. They attempted to find higher-order approximate solutions for the problem to improve on the J–H solution. The solution obtained by Chen and

Davis is applicable to the case that the *contact angles at the triple point are small.* In reality, however, the contact angles are normally not small: both the slope and curvature of the liquid–solid interface near the triple point may be quite large. As a consequence, although in some cases, the liquid–solid interface may be nearly flat in a local region away from the triple point, it can never be nearly flat in the global region including the triple point. The significant effect of the triple point on the behavior of the steady-state solution and its stability properties needs careful investigation.

Global uniformly valid asymptotic solutions for eutectic growth were derived by Xu, Chen, and Li in 2013–2014 [14, 15]. Their work took into account the triple point singularity and discussed the case of the tilt angle $\varphi \neq 0$, as shown in Fig. 11.2b. Their results will be presented in this chapter.

## 11.2 Mathematical Formulation of Eutectic Growth from a Binary Mixture

### *11.2.1 Scales and Dimensionless Parameters*

As usual, we neglect the solute diffusion in the solid phases and assume that all the thermodynamic properties except the diffusivity coefficients are the same for both the solid phase and the liquid phase. We also neglect the effect of convection in the liquid phase. Thus the system will not be affected by the fluid dynamics. We adopt the coordinate system $(x, y)$ moving along the tilted direction $\mathbf{U} = V(-\tan\varphi\,\mathbf{i} + \mathbf{j})$, where $\mathbf{i}$ and $\mathbf{j}$ are the unit coordinate vectors along the $x$-axis and $y$-axis, respectively. The origin of the coordinate system is located at the tip B of the $\alpha$-interface, as shown in Figs. 11.3 and 11.4.

Two important lengths of the system are the solute diffusion length, $\ell_D = \frac{\kappa_D}{V}$, and the thermal diffusion length, $\ell_T = \frac{\kappa_T}{V}$, where $\kappa_D$ and $\kappa_T$ are the solute diffusivity

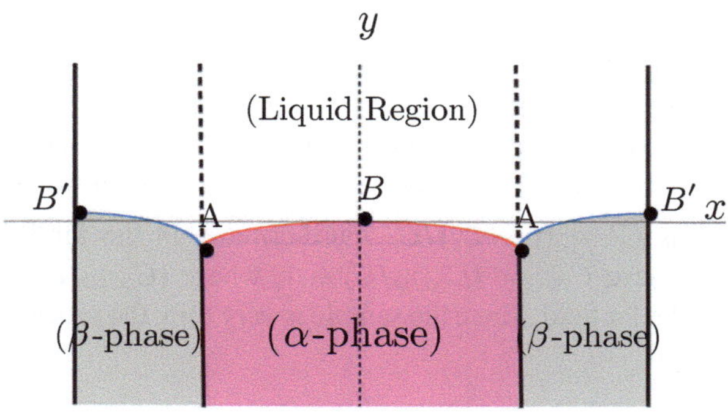


**Fig. 11.3** Sketch of the interface shape of steady untilted eutectic growth, where (A) is the triple point, (B) is the tip of $\alpha$-interface, and (B′) is the tip of the $\beta$-interface

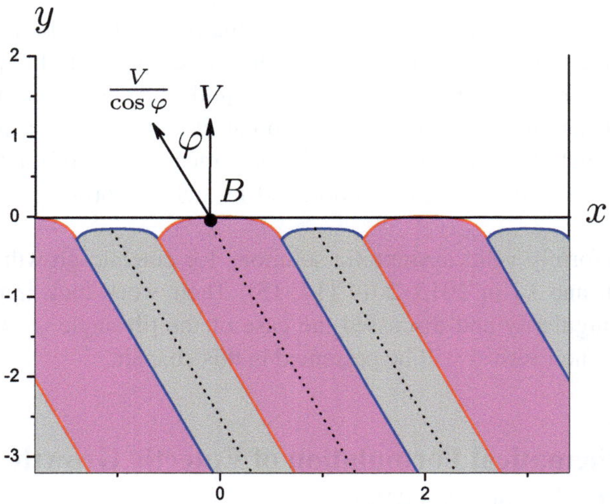


**Fig. 11.4** A sample of the interface shape of steady tilted eutectic growth with the tilt angle $\varphi$ predicted by the theoretical solution

and thermal diffusivity, respectively. Note that in contrast to the system of cellular growth, in that for **eutectic growth, the tip radius cannot be considered the characteristic length of the system**. This is because there are two different solid phases, and accordingly, there are two interface-tips, and the radii of curvature at these two interface-tips may have different orders of magnitudes. Moreover, the radius of each of the interface-tips may vary within quite a large range under nearly the same growth conditions. As a consequence, we use instead half the interlamellar spacing, $\ell_w$, as the length scale. We further assume that $\ell_w$ is much less than the solute diffusion length $\ell_D$.

The pulling velocity $V$ is used as the velocity scale, and $\frac{\ell_w}{V}$ is used as the time scale. The scales of the temperature $T$ and concentration $C$ are set as $\Delta H/(c_p\rho)$ and $C_e$, respectively. Here $\Delta H$ is latent heat release per unit of volume of the solid phase, $c_p$ is the specific heat, $\rho$ is the density of the melt, and $C_e$ is the eutectic concentration of species ($A$) in the phase diagram (see Fig. 11.1).

The eutectic temperature is used as the reference temperature, while the eutectic concentration is used as the reference concentration. Thus, the nondimensional temperature field is defined as $\bar{T} = \frac{T-T_e}{\Delta H/(c_p\rho)}$, while the nondimensional concentration field is defined as $\bar{C} = (C - C_e)/C_e$. Furthermore, in the far field away from the interface, we have $\bar{C}_\infty = (C_\infty)_D/C_e - 1$, where $(C_\infty)_D$ is the dimensional concentration in the far field. Since the sample is very thin, the process can be treated as two-dimensional.

The system involves the following dimensionless parameters:

- $\mathrm{Pe} = \epsilon = \frac{\ell_w}{\ell_D}$: the Péclet number;
- $\lambda_T = \frac{\ell_D}{\ell_T}$: the ratio of the two lengths, $\ell_D$ and $\ell_T$;
- $M = -\frac{mC_e}{\Delta H/(c_p\rho)}$: the morphological parameter, where $m = (m_\alpha, m_\beta)$ is the slope of the liquidus in the phase diagram;
- $\Gamma = \frac{\ell_c}{\ell_w}$: the surface tension parameter, where the capillary length is defined as $\ell_c = \frac{\gamma c_p \rho T_e}{(\Delta H)^2}$, the surface tension coefficient as $\gamma = (\gamma_\alpha, \gamma_\beta)$, and accordingly, $\Gamma = (\Gamma_\alpha, \Gamma_\beta)$; we write $R_c = \Gamma_\alpha/\Gamma_\beta$;
- $\Gamma_c = \frac{\ell_c}{\ell_D} = \Gamma\mathrm{Pe}$: the dimensionless surface tension coefficient, which is a thermodynamic property of the system;
- $G = \frac{\ell_D}{\Delta H/(c_p\rho)}(G)_D$: the dimensionless gradient of the temperature, where $(G)_D = \frac{T_H - T_C}{(L)_D}$, where $(G)_D = (T_H - T_C)/(L)_D$ is the dimensional gradient;
- $W_c = \frac{w_c}{\ell_w}$: the width parameter of the $\alpha$-phase.

Here, the parameters $M, \Gamma, m$ and the segregation coefficient $\kappa$ are piecewise constant functions having different values in different portions of the interface. There are more piecewise smooth functions to be specified later, such as the interface shape function and wave number functions, which have different expressions in different portions of the interface. We shall indicate such functions with the subscripts $\alpha$ and $\beta$ in the following manner. Let $q$ be the representative of these quantities. We write $q = q_\alpha$ for $x \in (0, w_0)$, and $q = q_\beta$ for $x \in (w_0, 1)$. Here $w_0$ is the coordinate of the triple point along the $x$-axis.

In practice (refer to the experiments by Mergy et al. [10]), all the length scales involved in the present system have order of magnitude

$$\ell_T \gg \ell_D \gg \ell_w \gg \ell_c.$$

Typically, one has $\ell_c = \mathcal{O}(10^{-6}\,\mathrm{mm})$ and $\ell_D = (1\text{–}2)$ mm and $\epsilon \approx 0.01$. Therefore, we may assume that $\mathrm{Pe} = \epsilon \ll 1$ and $\Gamma = \epsilon^2\bar{\Gamma}$, where $\bar{\Gamma} = \mathcal{O}(1)$.

Note that if $\kappa = 1$, the concentration field will be continuous at the liquid–solid interface. The binary system will behave like a system of a single species. Furthermore, if the Péclet number equals zero, the liquid–solid interface will become very flat in the local region away from the triple point. With this in mind, as a model system we consider the case that the Péclet number satisfies $\epsilon = \mathrm{Pe} \ll 1$ and that the segregation coefficients $\kappa$, in both the $\alpha$-phase and the $\beta$-phase, are close to unity. As a consequence, one may introduce the notation $\varpi = (1 - \kappa_\alpha) \ll 1$ and write that $|1 - \kappa_\beta| = \mathcal{O}(\varpi)$, $1 - \kappa_\beta = -\hat{\kappa}\varpi$, where $0 < \hat{\kappa} = \mathcal{O}(1)$.

Consider both the tilt angle $\varphi$ and the spacing parameter $\epsilon$ as the free parameters. The interfacial patterns of the basic steady states may be untilted or tilted, as sketched in Figs. 11.3 and 11.4, respectively.

We choose $\mathrm{Pe} = \epsilon \ll 1$ as a basic small parameter, and all other small parameters will be related to $\epsilon$ with certain orders of magnitudes. Only by properly setting the relationships between the orders of magnitude of these parameters in accordance with the intrinsic properties of the system is it possible to guarantee the existence of some form of global asymptotic solution in the limit $\epsilon \to 0$.

Similar to the system of cellular growth studied in the last chapter, the temperature field in the current system can be decoupled from the concentration field and expressed in the asymptotic form

$$T = T_S \sim T_e + (G)_D(y - y_*) + \mathcal{O}(\epsilon^2).$$

Here $y_*$ is the distance between the tip B of the $\alpha$-interface and the place where the temperature equals the eutectic temperature $T = T_e$. In other words, the parameter $y_*$ describes the location of the $\alpha$-interface's tip in the laboratory frame, or as well the temperature at the $\alpha$-interface's tip.

The unknowns for the general unsteady problem are now the concentration field $C(x, y, t)$, the shape function $y = h(x, t)$ of the liquid–solid interface, the location of the triple point $W_c$, and the parameters $\ell_w$ and $y_*$. The concentration field is governed by the solute diffusion equation and is subject to the Gibbs–Thomson condition and mass conservation condition imposed at the liquid–solid interface.

Moreover, due to the periodicity of the lamellar eutectic growth solution, one may consider a single lamella, as shown in Fig. 11.3. In that case, the problem is equivalent to eutectic growth in a channel with fixed side-walls $x = \pm\ell_w$. For eutectic growth in a Hele–Shaw cell with infinitely large width, of course, $\ell_w$ is a free parameter, representing half of the interlamellar spacing.

## 11.2.2 Formulation of the Problem in the Liquid Phase

Hereinafter, we use the same notation to describe all dimensionless quantities as that used for dimensional quantities.

We first write the system for steady eutectic growth in the liquid phase. Denote the basic steady-state solution with the subscript B. The steady equation for the concentration field is

$$\nabla^2 C_B + \epsilon \frac{\partial C_B}{\partial y} = 0. \tag{11.1}$$

The boundary conditions are as follows:

1. As $y \to \infty$, $C_B \to C_\infty$.
2. At the symmetric line $x = 0$ and side-wall $x = 1$,

$$\frac{\partial C_B}{\partial x} = \frac{\partial h_B}{\partial x} = 0. \tag{11.2}$$

3. At the interface, $y = h_B(x, \epsilon)$,

(a) the Gibbs–Thomson condition:

$$C_{\mathrm{B}} = \epsilon \frac{G}{M} y_* - \epsilon \frac{\mathrm{G}}{M} h_{\mathrm{B}} - \frac{\Gamma}{M} \mathcal{K}\{h_{\mathrm{B}}\} + \mathcal{O}(\text{h.o.t.}), \tag{11.3}$$

where

$$\mathcal{K}\{h\} = -h_{\mathrm{B}xx}/(1 + h_{\mathrm{B}x}^2)^{\frac{3}{2}}; \tag{11.4}$$

(b) the mass balance condition:

$$\frac{\partial C_{\mathrm{B}}}{\partial y} - h_{\mathrm{B}x} \frac{\partial C_{\mathrm{B}}}{\partial x} + \epsilon(1-\kappa)(1+C_{\mathrm{B}})(1+\tan\varphi\, h_{\mathrm{B}x}) = 0. \tag{11.5}$$

4. At the triple point, $x = \pm W_c$, $y = h_{\mathrm{B}}(x) = y_c$: Let $\gamma_\alpha$, $\gamma_\beta$, and $\gamma_{\alpha\beta}$ be the surface tensions on the $\alpha$-interface, $\beta$-interface, and $\alpha$–$\beta$-interface, respectively, while $\theta_\alpha$, $\theta_\beta$, $\theta_\ell$ are the contact angles over the $\alpha$, $\beta$, and liquid phases, respectively (see Fig. 11.5).

Assume that at the triple point, the $\alpha$–$\beta$ interface, as well as the liquid–solid interface, is in local thermodynamic equilibrium. From the mechanical balance, it can be derived that

$$\gamma_\alpha \cos\theta_\alpha + \gamma_\beta \cos\theta_\beta = -\gamma_{\alpha\beta}, \quad \gamma_\alpha \sin\theta_\alpha - \gamma_\beta \sin\theta_\beta = 0, \tag{11.6}$$

and $\theta_\ell = 2\pi - (\theta_\alpha + \theta_\beta)$. Hence, all the quantities $\gamma_\alpha$, $\gamma_\beta$, $\gamma_{\alpha\beta}$ and $\theta_\alpha$, $\theta_\beta$, $\theta_\ell$ can be considered known constants describing the thermodynamic properties of the system. For the general case of tilted growth, one may write that

$$\theta^- = \theta_\alpha - \frac{\pi}{2} - \varphi, \quad \theta^+ = \theta_\beta - \frac{\pi}{2} + \varphi, \tag{11.7}$$

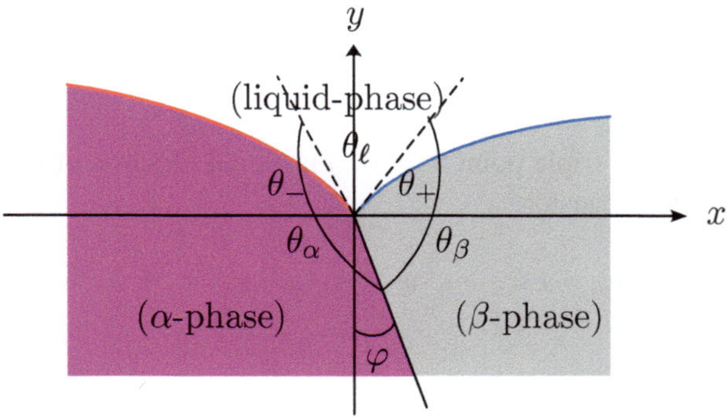


**Fig. 11.5** Sketch of the interface shapes near the triple point. Here the contact angles $\theta_\alpha$, $\theta_\beta$ are thermodynamic constants. The slope angles $\theta_\pm$ and tilt angle $\varphi$ may be time-dependent for the unsteady state. Their values for the basic state are denoted by the constants $\bar{\theta}_\pm$ and $\bar{\varphi}$, respectively

where $\varphi$ is the tilt angle defined as the counterclockwise rotation angle from the $y$-axis as shown in Fig. 11.5. Moreover, due to the steadiness of growth and the fact that the mass diffusivity in the solid state is negligibly small, it is deduced that the whole $\alpha$–$\beta$ interface in the solid state must not be in thermodynamic equilibrium; it is described as the straight line tilted from the $y$-axis with tilt angle $\varphi$ along which direction the $\alpha$–$\beta$-interface moves with speed $(V/\cos\varphi)$. In the meanwhile, the triple point drifts along the $x$-axis with speed $(V\tan\varphi)$, as sketched in Fig. 11.4.

### 11.2.3 The Formulation of the Problem in the Solid Phase

As indicated in the previous section, the interface between the $\alpha$ and $\beta$ phases is the tilted straight line tilted from the $y$-axis, moving with speed $V/\cos\varphi$. As a consequence, the concentration at the point $(x_*, y_*)$ in the solid state must be the same as that at the point on the interface $x_0 = x_* - (y_0 - y_*)\tan\varphi, y_0 = h_{\mathrm{B}}(x_0)$.

One may write the basic state in the solid phases as follows:

- In the $\alpha$-phase region, $|x| < W_c$, $h_{\mathrm{B}}(x) - y > 0$:

$$C_{\mathrm{S}}\big[x + (h_{\mathrm{B}}(x) - y)\tan\varphi, y\big] = C_{\mathrm{B}}\big[x, h_{\mathrm{B}}(x)\big]\kappa_\alpha. \tag{11.8}$$

- In the $\beta$-phase region, $W_c < |x| < 1$, $h_{\mathrm{B}}(x) - y > 0$:

$$C_{\mathrm{S}}\big[x + (h_{\mathrm{B}}(x) - y)\tan\varphi, y\big] = C_{\mathrm{B}}\big[x, h_{\mathrm{B}}(x)\big]\kappa_\beta. \tag{11.9}$$

Furthermore, one may write the total mass conservation condition as

$$\begin{aligned} C_\alpha W_c + C_\beta(1 - W_c) + \int_0^{W_c} C_{\mathrm{B}}\big[x, h_{\mathrm{B}}(x)\big]\kappa_\alpha \mathrm{d}x \\ + \int_{W_c}^{1} C_{\mathrm{B}}\big[x, h_{\mathrm{B}}(x)\big]\kappa_\beta \mathrm{d}x = C_\infty. \end{aligned} \tag{11.10}$$

The location of the triple point $x = W_c$ must be a function of $\epsilon$. Assume that it has the asymptotic expansion

$$W_c(\epsilon) = w_0 + \epsilon w_1 + \epsilon^2 w_2 + \cdots, \quad \text{as} \quad \epsilon \to 0. \tag{11.11}$$

### 11.2.4 The Formulation of the Problem in the Far Field

The leading-order approximation of the solution in the far field is the same as that derived in the previous chapter. A more accurate expression of the far-field solution is given below. As mentioned before, for $y \gg 1$, one may neglect the effect of the detailed interface shape and consider the interface to be flat. Thus, the role of the interface is like a line of mass sources and sinks periodically distributed along the $x$-axis, which may be described by the function of the source strength within one period:

$$Q(x) = \begin{cases} Q_\alpha(x) > 0, & |x| < W_c; \\ Q_\beta(x) < 0, & W_c < |x| < 1. \end{cases} \tag{11.12}$$

Hence the system may reduce to the one-dimensional diffusion equation

$$\frac{\partial^2 C_{\mathrm{B}}}{\partial y^2} + \epsilon \frac{\partial C_{\mathrm{B}}}{\partial y} = 0 \tag{11.13}$$

with the following boundary conditions:

1. At $y = 0, |x| < 1$:

$$\frac{\partial C_{\mathrm{B}}}{\partial y}(x, 0) = -Q(x).$$

2. As the far-field condition, we set that as $y \to \infty$,

$$C_{\mathrm{B}} = C_\infty.$$

The solution of the above system can be easily derived in terms of a Fourier series expansion over the interval $0 < x < 1$ (the detailed derivations are given in Section 11.14). Thus, one may derive the following asymptotic form of the solution in the far field: as $y \to \infty$,

$$\begin{aligned} C_{\mathrm{B}}(x, y) &= C_\infty + \hat{Q}(\epsilon, \varpi)\mathrm{e}^{-\epsilon y} \\ &\quad + \sum_{n=1}^{\infty} \hat{A}_n(\epsilon, \varpi) \left[ \cos(n\pi x) \exp\left( - n\pi y - \frac{1}{2}\epsilon y + \cdots \right) \right] \\ &= C_\infty + \hat{Q}(\epsilon, \varpi)\mathrm{e}^{-\epsilon y} \\ &\quad + \mathrm{e}^{-\frac{1}{2}\epsilon y} \sum_{n=1}^{\infty} \left[ \hat{A}_n(\epsilon, \varpi) \cos(n\pi x)\mathrm{e}^{-n\pi y} + \mathcal{O}(\text{h.o.t.}) \right], \end{aligned} \tag{11.14}$$

where $\hat{Q}, \hat{A}_n$ are undetermined constants. The form of the solution (11.14) can be considered a function of $(y, \tilde{y} = \epsilon y)$ and will be used later as the far-field condition for the solutions in the near field.

## 11.3 Formulation of the Problem with Multiple Variables in the Near Field

In viewing the form of the solution at the far field, we may assume that in the near field $y = \mathcal{O}(1)$, the steady-state solution, being a periodic function of the variable $x$ with unit period, must contain the group variable $\epsilon y$. Hence, to describe the solution in the near field, one may introduce the slow variable $\tilde{y} = \epsilon y$ and assume that the concentration field may be described in the multiple variables form $C_{\mathrm{B}}(x, y, \epsilon) = \bar{C}(x, y, \tilde{y}, \epsilon)$. With the following replacements in the system,

$$\frac{\partial}{\partial y} \Longleftrightarrow \frac{\partial}{\partial y} + \epsilon \frac{\partial}{\partial \tilde{y}}; \quad \frac{\partial^2}{\partial y^2} \Longleftrightarrow \frac{\partial^2}{\partial y^2} + 2\epsilon \frac{\partial^2}{\partial y \partial \tilde{y}} + \epsilon^2 \frac{\partial^2}{\partial \tilde{y}^2},$$

the system for steady eutectic growth (11.1)–(11.6) can be converted into the following system with multiple variables,

$$\frac{\partial^2 \bar{C}}{\partial x^2} + \frac{\partial^2 \bar{C}}{\partial y^2} + 2\epsilon \frac{\partial^2 \bar{C}}{\partial y \partial \tilde{y}} + \epsilon^2 \frac{\partial^2 \bar{C}}{\partial \tilde{y}^2} + \epsilon \frac{\partial \bar{C}}{\partial y} + \epsilon^2 \frac{\partial \bar{C}}{\partial \tilde{y}} = 0, \tag{11.15}$$

with the following boundary conditions:

1. As $y \to \infty$ and $\tilde{y} \to \infty$,

$$\bar{C} \sim C_\infty + \hat{Q}(\epsilon) \mathrm{e}^{-\tilde{y}} + \cdots . \tag{11.16}$$

2. The symmetry conditions at $x = 0, 1$,

$$\frac{\partial}{\partial x} \bar{C} = \frac{\partial}{\partial x} h_{\mathrm{B}} = 0, \quad h_{\mathrm{B}}(0) = 0. \tag{11.17}$$

3. At the interface, $y = h_{\mathrm{B}}(x), \tilde{y} = \epsilon h_{\mathrm{B}}(x)$, we have

   (a) the Gibbs–Thomson condition,

$$\bar{C} = \epsilon \frac{\mathrm{G}}{M} (y_* - h_{\mathrm{B}}) - \frac{\Gamma}{M} \mathcal{K}\{h_{\mathrm{B}}\} + (\text{h.o.t.}), \tag{11.18}$$

(b) the mass balance condition,

$$\frac{\partial \bar{C}}{\partial y} + \epsilon \frac{\partial \bar{C}}{\partial \tilde{y}} - h_{Bx} \frac{\partial \bar{C}}{\partial x} + \epsilon(1-\kappa)(1+\tan\varphi\, h_{Bx})\left(1+\bar{C}\right)$$
$$= 0. \tag{11.19}$$

4. At the triple point, $x = \pm W_c$, $y = h_B(x) = y_c$:

$$h'_B(W_c^-) = -\tan\theta^-; \quad h'_B(W_c^+) = \tan\theta^+, \tag{11.20}$$

where $\theta^-$ and $\theta^+$ are slope angles given by (11.103).

In general, as $\epsilon \to 0$, as the leading order approximation, the liquid–solid interface shape function might not approach zero. Namely, one may have $h_B(x) \sim h_0(x) = \mathcal{O}(1)$. However, with the additional assumptions $\varpi = 1 - \kappa_\alpha \ll 1$ and $|1-\kappa_\beta| = \mathcal{O}(\varpi)$, one may expect that the interface shape function satisfies the conditions $|h_B(x)| \ll 1$ and $|h'_B(x)| \ll 1$, except near the triple point $x = w_0$. Therefore, in the region far from $y = w_0$, one may linearize the interface conditions (11.18) and (11.19) around the flat interface $y = h_B = 0$ as follows:

1. the Gibbs–Thomson condition at $y = \tilde{y} = 0$,

$$\bar{C} + \frac{\partial \bar{C}}{\partial y} h_B + \frac{\partial \bar{C}}{\partial \tilde{y}}(\epsilon h_B) + \cdots = \epsilon \frac{G}{M}(y_{*0} + \epsilon y_{*1} + \cdots)$$
$$-\epsilon \frac{G}{M} h_B - \frac{\Gamma}{M}\mathcal{K}\{h_B\} + (\text{h.o.t.}), \tag{11.21}$$

2. the mass balance condition at $y = \tilde{y} = 0$,

$$\frac{\partial \bar{C}}{\partial y} + \frac{\partial^2 \bar{C}}{\partial y^2} h_B + \frac{\partial^2 \bar{C}}{\partial y \partial \tilde{y}}(\epsilon h_B) + \cdots$$
$$+\epsilon\left[\frac{\partial \bar{C}}{\partial \tilde{y}} + \frac{\partial^2 \bar{C}}{\partial \tilde{y}\partial y} h_B + \frac{\partial^2 \bar{C}}{\partial \tilde{y}^2}(\epsilon h_B) + \cdots\right]$$
$$-h_{Bx}\left[\frac{\partial \bar{C}}{\partial x} + \frac{\partial^2 \bar{C}}{\partial x \partial y} h_B + \frac{\partial^2 \bar{C}}{\partial x \partial \tilde{y}}(\epsilon h_B) + \cdots\right]$$
$$+\epsilon(1-\kappa)(1+\tan\varphi\, h_{Bx})\left[1 + \bar{C} + \frac{\partial \bar{C}}{\partial y} h_B + \frac{\partial \bar{C}}{\partial \tilde{y}}(\epsilon h_B) + \cdots\right] = 0. \tag{11.22}$$

The system now involves two small parameters, $(\epsilon, \varpi)$. At this point, we have required only that $\varpi \to 0$, as $\epsilon \to 0$. We have not yet specified the ratio of these two small parameters. For the time being, we may formally treat $\varpi$ and $\epsilon$ as if they were two independent small parameters, so that one may write the following form of regular perturbation expansion (RPE) solution in the limit of $\epsilon \to 0$, $\varpi \to 0$:

$$\begin{aligned}
\bar{C}_{\mathrm{B}}(x, y, \tilde{y}, \epsilon, \varpi) &= \bar{C}_{00}(x, y, \tilde{y}) + \varpi \bar{C}_{01}(x, y, \tilde{y}) + \epsilon \bar{C}_{10}(x, y, \tilde{y}) \\
&\quad + \epsilon\varpi \bar{C}_{11}(x, y, \tilde{y}) + \epsilon^2 \bar{C}_{20}(x, y, \tilde{y}) + \epsilon^2 \varpi \bar{C}_{21}(x, y, \tilde{y}) + \cdots \\
h_{\mathrm{B}}(x, \epsilon, \varpi) &= \varpi \bar{h}_{01}(x) + \epsilon \bar{h}_{10}(x) + \epsilon\varpi \bar{h}_{11}(x) + \cdots, \\
y_*(\epsilon, \varpi) &= \varpi y_{01} + \epsilon y_{10} + \epsilon\varpi y_{11} + \cdots.
\end{aligned} \tag{11.23}$$

Accordingly, we have

$$W_c(\epsilon, \varpi) = w_0 + \varpi w_{01} + \epsilon\varpi w_{11} + \cdots. \tag{11.24}$$

Later, we shall show that to balance both sides of the Gibbs–Thomson condition in the leading-order approximation, the two small parameters $(\epsilon, \varpi)$ must be related to each other such that $\varpi = \mathcal{O}(\epsilon^{1/2})$. Thus, one may set $\varpi = \hat{\varpi}\epsilon^{\frac{1}{2}}$, $\hat{\varpi} = \mathcal{O}(1)$.

## 11.4 The Steady-State Solution in the Liquid Phase

We turn to solving the MVE form of the solution for the steady state in the near field. By substituting (11.23) and (11.24) into the system (11.15)–(11.17), (11.21), and (11.22), one can derive each order of approximation successively.

### *11.4.1 The Solutions for the Concentration Field in Approximations of Orders* $(0, 0)$ *and* $(1, 0)$

The trivial solutions in the approximations of orders $(0, 0)$ and $(1, 0)$ can be easily derived as

$$\bar{C}_{00} = 0, \quad \bar{C}_{10} = 0.$$

### *11.4.2 The Solutions for the Concentration Field in the Approximation of Order* $(0, 1)$

The solution for the approximation of order $(0, 1)$ is subject to the equation

$$\left(\frac{\partial^2}{\partial x^2} + \frac{\partial^2}{\partial y^2}\right) \bar{C}_{01}(x, y, \tilde{y}) = 0. \tag{11.25}$$

The far-field condition as $y \to \infty$ is

$$\bar{C}_{01} \to \hat{C}_{\infty}.$$

Here we have defined $C_{\infty} = \varpi \hat{C}_{\infty}$ with $\hat{C}_{\infty} = \mathcal{O}(1)$. The interface conditions at $y = \tilde{y} = 0$ are as follows:

- the Gibbs–Thomson condition,

$$\bar{C}_{01} = 0; \tag{11.26}$$

- the mass balance condition,

$$\frac{\partial \bar{C}_{01}}{\partial y} = 0. \tag{11.27}$$

One may derive $\bar{C}_{01} = B_{01}(\tilde{y})$. From the far-field condition and Gibbs–Thomson condition, it follows that

$$B_{01}(0) = 0, \quad B_{01}(\infty) = \hat{C}_{\infty}. \tag{11.28}$$

The function $B_{01}(\tilde{y})$ is to be determined in the approximation of order $(2, 1)$ later.

### *11.4.3 The Solutions for the Triple Point Location in the Approximation of Order* (1, 0)

In the approximation of order $(1, 0)$, one can determine the location of the triple point in terms of the total mass conservation condition (11.10). In fact, since $h_{\mathrm{B}} \approx 0$ and at the interface we have $C_{\mathrm{B}}\big[x, h_{\mathrm{B}}(x)\big] \approx \bar{C}_{\mathrm{B}}\big[x, 0, 0\big] \approx 0$ and $W_c \approx w_0$, the condition (11.10) reduces to

$$C_{\infty} = C_{\alpha} W_c + C_{\beta}(1 - W_c) + \int_0^{W_c} C_{\mathrm{B}}\big[x, h_{\mathrm{B}}(x)\big]\kappa_{\alpha} \mathrm{d}x$$
$$+ \int_{W_c}^{1} C_{\mathrm{B}}\big[x, h_{\mathrm{B}}(x)\big]\kappa_{\beta} \mathrm{d}x \approx C_{\alpha} w_0 + C_{\beta}(1 - w_0). \tag{11.29}$$

Noting that $C_{\alpha} = \kappa_{\alpha} - 1, C_{\beta} = \kappa_{\beta} - 1$, it follows that

$$\kappa_{\alpha} w_0 + \kappa_{\beta}(1 - w_0) = 1 + C_{\infty}.$$

From the above, we derive that $\kappa_{\alpha} < (1 + C_{\infty}) < \kappa_{\beta}$, which is consistent with the assumption $C_{\infty} = \mathcal{O}(\varpi)$, and $C_{\infty} = \varpi \hat{C}_{\infty}$. Furthermore, we obtain

$$w_0 = (\hat{\kappa} - \hat{C}_{\infty})/(1 + \hat{\kappa}). \tag{11.30}$$

### 11.4.4 Solutions for the Concentration Field in the Approximation of Order (1, 1)

The equation for the approximation of order $(1,1)$ is

$$\left(\frac{\partial^2}{\partial x^2}+\frac{\partial^2}{\partial y^2}\right)\bar{C}_{11}=0. \tag{11.31}$$

In this approximation, one deduces that the interface is curved. The solution $\bar{h}_{01}(x)$ for the interface shape and solution $\bar{C}_{11}(x,y,\tilde{y})$ for concentration are coupled by the Gibbs–Thomson interface condition. To resolve the problem, we first find the solution $\bar{C}_{11}(x,y,\tilde{y})$ from (11.31) that satisfies the far-field condition $\bar{C}_{11}=0$ as $y\to\infty$, the symmetry condition $\frac{\partial\bar{C}_{11}}{\partial x}=0$ at $x=0,1$, and the mass balance condition at the interface. The solution obtained may still contain some unknown constants. Then we apply the remaining Gibbs–Thomson interface condition to find the complete form of $\bar{C}_{11}$, as well as the solution of interface shape $\bar{h}_{01}(x)$.

The mass balance condition (11.22) at $y=\tilde{y}=0$ can be written as

$$\frac{\partial\bar{C}_{11}}{\partial y}=-\frac{\partial\bar{C}_{01}}{\partial\tilde{y}}-\hat{\Theta}(x)=-B'_{01}(0)-\hat{\Theta}(x). \tag{11.32}$$

Here the piecewise constant function $\hat{\Theta}(x)$ is defined as

$$\hat{\Theta}(x)=\begin{cases}\hat{\Theta}_\alpha(x)=1, & (0\le x<w_0),\\ \hat{\Theta}_\beta(x)=-\hat{\kappa}, & (w_0\le x\le 1).\end{cases} \tag{11.33}$$

It is easy to verify that $1-\kappa(x)=\varpi\hat{\Theta}(x)$. The Fourier cosine expansion for $\hat{\Theta}(x)$ can be written as

$$\hat{\Theta}(x)=\frac{1}{2}\omega_0+\sum_{n=1}^{\infty}\omega_n\cos(n\pi x)\quad(0<x<1). \tag{11.34}$$

The corresponding Fourier coefficients are calculated as

$$\omega_0=2\int_0^1\hat{\Theta}(x)\mathrm{d}x=2(1+\hat{\kappa})w_0-2\hat{\kappa},$$

$$\omega_n=2\int_0^1\hat{\Theta}(x)\cos(n\pi x)\mathrm{d}x=\frac{2(1+\hat{\kappa})\sin(n\pi w_0)}{n\pi},$$

$$(n=1,2,\cdots). \tag{11.35}$$

Therefore, the condition (11.32) can be rewritten as at $y = 0$,

$$\frac{\partial \bar{C}_{11}}{\partial y} = -\left[B'_{01}(0) + \frac{1}{2}\omega_0\right] - \sum_{n=1}^{\infty} \omega_n \cos(n\pi x). \tag{11.36}$$

It will be proved later that

$$\left[B'_{01}(0) + \frac{1}{2}\omega_0\right] = 0. \tag{11.37}$$

Note that the harmonic function $\Omega_{11}(x, y)$ satisfying the Neumann condition

$$\frac{\partial \Omega_{11}}{\partial y}(x, 0) = -\sum_{n=1}^{\infty} \omega_n \cos(n\pi x)$$

is unique up to an arbitrary constant $d_{11,0}$; it can be expressed in the form of a Fourier cosine series:

$$\Omega_{11}(x, y) = \frac{1}{2}d_{11,0} + \bar{\Omega}(x, y) \quad (0 < x < 1), \tag{11.38}$$

where

$$\begin{aligned} \bar{\Omega}(x, y) &= \sum_{n=1}^{\infty} d_{11,n} \cos(n\pi x) e^{-n\pi y}, \\ d_{11,n} &= \frac{\omega_n}{n\pi} = \frac{2(1+\hat{\kappa})\sin(n\pi w_0)}{n^2\pi^2}, \quad (n = 1, 2, \cdots). \end{aligned} \tag{11.39}$$

It follows that

$$P_{11}(x) = \Omega_{11}(x, 0) = \frac{1}{2}d_{11,0} + \sum_{n=1}^{\infty} d_{11,n} \cos(n\pi x). \tag{11.40}$$

Furthermore, using the formula (see (11.194) in Section 11.15)

$$\begin{aligned} S_2(y) &= \sum_{n=1}^{\infty} \frac{\sin ny}{n^2} = -\frac{1}{2}\left[y \ln 2 + \int_0^y \ln(1 - \cos x) \mathrm{d}x\right] \\ &(0 < y < 2\pi), \end{aligned} \tag{11.41}$$

one may rewrite the Fourier series in (11.40) in the integral form

$$\begin{aligned} \widehat{P}_{11}(x) &= \sum_{n=1}^{\infty} d_{11,n} \cos(n\pi x) = -(1+\hat{\kappa})\frac{w_0}{\pi} \ln 2 \\ &- \frac{(1+\hat{\kappa})}{2\pi} \int_{(x-w_0)}^{(x+w_0)} \ln(1 - \cos \pi t) \mathrm{d}t. \end{aligned} \tag{11.42}$$

This formula will be useful in later derivations.

Since $\bar{C}_{11}(x, y, \tilde{y})$ and $\Omega_{11}(x, y)$ are both harmonic functions and satisfy the same Neumann condition, it follows that

$$\bar{C}_{11}(x, y, \tilde{y}) = \frac{1}{2} d_{11,0} B_{11}(\tilde{y}) + D_{11}(\tilde{y}) \bar{\Omega}_{11}(x, y), \tag{11.43}$$

where $B_{11}(\tilde{y})$ and $D_{11}(\tilde{y})$ are arbitrary functions depending on the slow variable $\tilde{y}$. By applying the boundary value at $y = \tilde{y} = 0$, we derive that

$$\begin{aligned}\bar{C}_{11}(x, 0, 0) &= \frac{1}{2} d_{11,0} B_{11}(0) + D_{11}(0) \bar{\Omega}_{11}(x, 0) = \frac{1}{2} d_{11,0} + \bar{\Omega}(x, 0) \\ &= \frac{1}{2} d_{11,0} + \sum_{n=1}^{\infty} d_{11,n} \cos(n\pi x),\end{aligned} \tag{11.44}$$

so that

$$B_{11}(0) = 1, \quad D_{11}(0) = 1. \tag{11.45}$$

Furthermore, by applying the far-field condition, we deduce that

$$B_{11}(\infty) = 0. \tag{11.46}$$

## *11.4.5 Approximations of Higher Order*

The complete forms of $D_{11}(\tilde{y})$ and $B_{01}(\tilde{y})$ can be determined in the approximation of order $(2, 1)$, while the function $B_{11}(\tilde{y})$ can be determined in the approximation of order $(3, 1)$, in terms of the uniformly valid requirements of the solution. As a matter of fact, the equation of the approximation of order $(2, 1)$ is

$$\left(\frac{\partial^2}{\partial x^2} + \frac{\partial^2}{\partial y^2}\right) \bar{C}_{21} = H_{21}, \tag{11.47}$$

where

$$H_{21} = -\Big(2\frac{\partial^2}{\partial \tilde{y} \partial y} + \frac{\partial}{\partial y}\Big) \bar{C}_{11} - \Big(\frac{\partial^2}{\partial \tilde{y}^2} + \frac{\partial}{\partial \tilde{y}}\Big) \bar{C}_{01} \tag{11.48}$$

$$= -(2D'_{11} + D_{11}) \frac{\partial}{\partial y} \Omega_{11}(x, y) + (B''_{01} + B'_{01}). \tag{11.49}$$

In order to eliminate the secular term as $y \to \infty$, we derive the solvability conditions $(B''_{01} + B'_{01}) = 0$; $(2D'_{11} + D_{11}) = 0$. It follows that

$$B_{01}(\tilde{y}) = \hat{C}_\infty(1 - \mathrm{e}^{-\tilde{y}}), \quad B'_{01}(0) = \hat{C}_\infty; \quad D_{11}(\tilde{y}) = \mathrm{e}^{-\frac{1}{2}\tilde{y}}. \tag{11.50}$$

Noting that $w_0 = (\hat{\kappa} - \hat{C}_\infty)/(1 + \hat{\kappa})$, we derive

$$\hat{C}_\infty = -\frac{1}{2}\omega_0 = -(1 + \hat{\kappa})w_0 + \hat{\kappa}, \tag{11.51}$$

which verifies formula (11.37). Furthermore, the equation of the approximation of order $(3, 1)$ is

$$\left(\frac{\partial^2}{\partial x^2} + \frac{\partial^2}{\partial y^2}\right)\bar{C}_{31} = H_{31}, \tag{11.52}$$

where

$$\begin{aligned} H_{31} &= -\Big(2\frac{\partial^2}{\partial\tilde{y}\partial y} + \frac{\partial}{\partial y}\Big)\bar{C}_{21} - \Big(\frac{\partial^2}{\partial\tilde{y}^2} + \frac{\partial}{\partial\tilde{y}}\Big)\bar{C}_{11} \\ &= -\Big(2\frac{\partial^2}{\partial\tilde{y}\partial y} + \frac{\partial}{\partial y}\Big)\bar{C}_{21} + \frac{1}{2}d_{11,0}(B''_{11} + B'_{11}) \\ &\quad -\frac{1}{4}\mathrm{e}^{-\frac{1}{2}\tilde{y}}\bar{\Omega}_{11}(x, y). \end{aligned} \tag{11.53}$$

In order to eliminate the secular term as $y \to \infty$, we derive that $(B''_{11} + B'_{11}) = 0$. Using the conditions $B_{11}(0) = 1$ and $B_{11}(\infty) = 0$, we derive that

$$B_{11}(\tilde{y}) = \mathrm{e}^{-\tilde{y}}. \tag{11.54}$$

We finally obtain the solution

$$\begin{aligned} \bar{C}(x, y, \tilde{y}, \epsilon, \varpi) &= \varpi\hat{C}_\infty(1 - \mathrm{e}^{-\tilde{y}}) \\ &\quad + \epsilon\varpi\left[\frac{1}{2}d_{11,0}\mathrm{e}^{-\tilde{y}} + \mathrm{e}^{-\frac{1}{2}\tilde{y}}\bar{\Omega}_{11}(x, y)\right] + \cdots. \end{aligned} \tag{11.55}$$

The solution for the concentration field (11.55) is uniformly valid in the whole physical region $(0 < y < \infty)$ and matches the far-field solution (11.14) as $y \to \infty$.

The term with the multiplier $d_{11,0}$ in (11.55) reflects the interaction between the interface shape and the concentration field. This term was omitted in the J–H theory.

The solution $\bar{C}_{11}(x, y, \tilde{y})$ has not yet been completed, since the constant $d_{11,0}$ is still unknown. To determine the constant $d_{11,0}$ and find the solution for the interface shape $\bar{h}_{01}(x)$, one needs to apply the Gibbs–Thomson interface condition.

## 11.5 The Solution for the Interface Shape Between Liquid and Solid

Let us now turn to deriving the solution for the interface shape between the liquid and solid phases in terms of the Gibbs–Thomson condition (11.21).

### *11.5.1 The Outer Solution for the Interface Shape in the Outer Region Away from the Triple Point*

In the approximation of order of $\mathcal{O}(1, 1)$, the Gibbs–Thomson condition at $y = \tilde{y} = 0$ reduces to

$$\bar{C}_{11} + \left( \frac{\partial \bar{C}_{00}}{\partial \tilde{y}} + \frac{\partial \bar{C}_{10}}{\partial y} \right) \bar{h}_{01}(x) = \frac{\mathrm{G}}{M} \Big[ y_{01} - \bar{h}_{01}(x) \Big]. \tag{11.56}$$

Since

$$\left( \frac{\partial \bar{C}_{00}}{\partial \tilde{y}} + \frac{\partial \bar{C}_{10}}{\partial y} \right) = 0,$$

from (11.56) one derives that

$$\bar{C}_{11}(x, 0) = P_{11}(x) = \frac{\mathrm{G}}{M} \Big[ y_{01} - \bar{h}_{01}(x) \Big]. \tag{11.57}$$

It follows that

$$\bar{h}_{01}(x) = y_{01} - \frac{M}{\mathrm{G}} \left[ \frac{1}{2} d_{11,0} + \widehat{P}_{11}(x) \right]. \tag{11.58}$$

Recall that

$$\begin{aligned} P_{11}(x) &= \frac{1}{2} d_{11,0} + \widehat{P}_{11}(x); \\ \widehat{P}_{11}(x) &= -\frac{w_0}{\pi}(1 + \hat{\kappa}) \ln 2 - \frac{1 + \hat{\kappa}}{2\pi} \int_{(x-w_0)}^{(x+w_0)} \ln(1 - \cos \pi t) \mathrm{d}t. \end{aligned} \tag{11.59}$$

One may derive that

$$\begin{aligned} \bar{h}'_{01}(x) &= -\frac{M}{\mathrm{G}} \widehat{P}'_{11}(x) = -\frac{1 + \hat{\kappa}}{2\pi} \frac{M}{\mathrm{G}} \ln \left[ \frac{1 - \cos \pi (x - w_0)}{1 - \cos \pi (x + w_0)} \right] \\ &= -\frac{1 + \hat{\kappa}}{\pi} \frac{M}{\mathrm{G}} \ln \left| \frac{\sin[\pi (x - w_0)/2]}{\sin[\pi (x + w_0)/2]} \right|, \end{aligned} \tag{11.60}$$

and

$$\begin{aligned}
\bar{h}''_{01}(x) &= -\frac{1+\hat{\kappa}}{2}\frac{M}{\mathrm{G}}\Big\{\cot[\pi(x-w_0)/2] - \cot[\pi(x+w_0)/2\Big\},\\
\bar{h}''_{01}(0) &= \frac{M_\alpha(1+\hat{\kappa})}{\mathrm{G}}\cot(\pi w_0/2),\\
\bar{h}''_{01}(1) &= \frac{M_\beta(1+\hat{\kappa})}{2\mathrm{G}}\Big[\cot\pi(1+w_0)/2) - \cot\pi(1-w_0)/2\Big].
\end{aligned} \tag{11.61}$$

There are two unknown constants in the above outer solution, $\{y_{01}, d_{11,0}\}$, to be determined by the boundary conditions in the outer region. The smoothness conditions at the side-walls $x = 0, 1$,

$$\bar{h}'_{01,\alpha}(0) = \bar{h}'_{01,\beta}(1) = 0, \tag{11.62}$$

are automatically satisfied. The boundary condition at the tip of the $\alpha$-interface,

$$\bar{h}_{01,\alpha}(0) = 0, \tag{11.63}$$

yields

$$y_{01} = \frac{M_\alpha}{\mathrm{G}}P_{11}(0),$$

which gives a relationship between the constants $y_{01}$ and $d_{11,0}$. To fully determine these two constants, let us assume that the solution is valid up to $x = w_0$, so that the continuity condition can be applied to set $\bar{h}_{01,\alpha}(w_0) = \bar{h}_{01,\beta}(w_0)$ at the triple point $x = w_0$. This leads to the solution

$$\bar{h}_{01}(x) = \begin{cases} \bar{h}_{01,\alpha}(x) = \frac{M_\alpha}{\mathrm{G}}\big[P_{11}(0) - P_{11}(x)\big], & (0 < x < w_0);\\ \bar{h}_{01,\beta}(x) = \frac{M_\alpha}{\mathrm{G}}\big[P_{11}(0) - \frac{M_\beta}{M_\alpha}P_{11}(x)\big], & (w_0 < x < 1). \end{cases} \tag{11.64}$$

It is seen from (11.60) that the above solution has a singularity at $x = w_0$, yielding

$$\bar{h}'_{01}(w_0^-) = \infty; \quad \bar{h}'_{01}(w_0^+) = -\infty,$$

as shown in Fig. 11.6. This violates the slope conditions at the triple point. We therefore conclude that the solution (11.58) is applicable only in the region away from the triple point. Hence, it is called the *outer solution*.

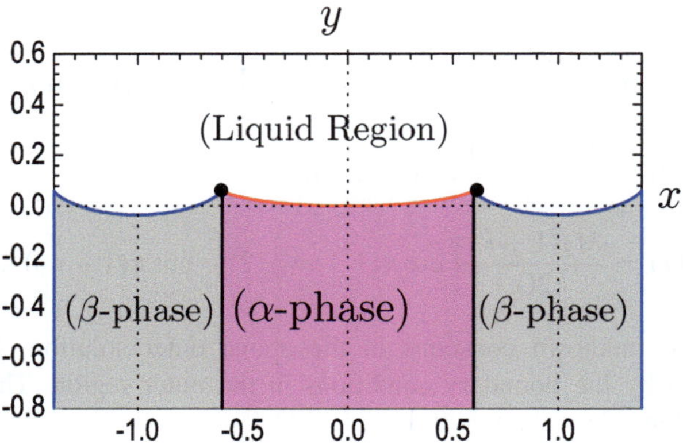


**Fig. 11.6** The interface shape of the outer solution for a typical case of steady eutectic growth with $C_\infty = -0.03$, $w_0 = 0.6$, $\varpi = 0.15$, $\hat{\kappa} = 1.0$, $\Gamma_\alpha = 3.0 \times 10^{-5}$, $\Gamma_\beta = 2.76 \times 10^{-5}$, $M_\alpha = 0.1$, $M_\beta = -0.2$, $\hat{\beta} = 3.0 \times 10^{-7}$, $\hat{v} = 3.0 \times 10^{-7}$, $\epsilon = 0.01$

## 11.5.2 The Inner Solution for the Interface Shape in the Inner Region near the Triple Point

In the inner region $|x - w_0| \ll 1$, we introduce the inner variables

$$\hat{x} = \frac{x - w_0}{\epsilon^{\hat{\alpha}}}, \quad \hat{y} = \frac{y}{\epsilon^{\hat{\alpha}}}. \tag{11.65}$$

In accordance with (11.65), we express the inner solution for the interface shape in the form $h_B = \epsilon^{\hat{\alpha}} \mathrm{h}(\hat{x}, \epsilon)$. In terms of the inner variables (11.65), the inner system can be derived from (11.1) to (11.5). The inner equation for the interface shape function can be obtained from the Gibbs–Thomson condition (11.3) as follows:

$$C_B((w_0 + \epsilon^{\frac{1}{2}}\hat{x}, \epsilon), 0, \epsilon) = \epsilon \frac{\mathrm{G}}{M} y_* - \epsilon^{1+\hat{\alpha}} \frac{\mathrm{G}}{M} \mathrm{h}$$
$$+\epsilon^{2-\hat{\alpha}} \frac{\bar{\Gamma}}{M} \frac{\frac{\mathrm{d}^2\mathrm{h}}{\mathrm{d}\hat{x}^2}}{\left[1 + \left(\frac{\mathrm{dh}}{\mathrm{d}\hat{x}}\right)^2\right]^{\frac{3}{2}}} + \mathcal{O}(\text{h.o.t.}). \tag{11.66}$$

To balance both sides of (11.66) in the leading-order approximation, it follows that $\hat{\alpha} = 1/2$, $\varpi = O(\epsilon^{1/2})$, and one may set $\varpi = \hat{\varpi}\epsilon^{1/2}$ with $\hat{\varpi} = O(1)$. Thus, one may write the inner variable $\hat{y} = \epsilon^{-1/2} y$, and $h_B(x) = \epsilon^{1/2} \mathrm{h}(\hat{x}, \epsilon)$.

Let us make the following inner expansion:

$$\mathrm{h}(\hat{x}, \epsilon) = \bar{\mathrm{h}}_{01}(\hat{x}) + \mathcal{O}(\text{h.o.t.}). \tag{11.67}$$

In the leading-order approximation, we obtain

$$\hat{\varpi}\bar{C}_{11}(w_0,0) = \hat{\varpi}P_{11}(w_0) = \hat{\varpi}\frac{G}{M}y_{01} - \frac{G}{M}\bar{h}_{01} + \frac{\bar{\Gamma}}{M}\frac{\frac{d^2\bar{h}_{01}}{d\hat{x}^2}}{\left[1+\left(\frac{d\bar{h}_{01}}{d\hat{x}}\right)^2\right]^{\frac{3}{2}}}. \tag{11.68}$$

The nonlinear equation (11.68) is the governing equation for the interface shape near the triple point. This equation can be solved analytically without much difficulty. Nevertheless, to better demonstrate the idea, here we simplify Eq. (11.68) as

$$\hat{\varpi}P_{11}(w_0) = \hat{\varpi}\frac{G}{M}y_{01} - \frac{G}{M}\bar{h}_{01} + \frac{\bar{\Gamma}}{M}\frac{d^2\bar{h}_{01}}{d\hat{x}^2}, \tag{11.69}$$

by neglecting the nonlinear term, $\left[1+\left(\frac{d\bar{h}_{01}}{d\hat{x}}\right)^2\right]^{3/2}$, on its right-hand side. This greatly simplifies the discussion without changing the qualitative nature of the solution. Noting that $\kappa, M, \bar{\Gamma}$ are piecewise constant, one needs to solve Eq. (11.69) in the subregion $(0, w_0)$ and $(w_0, 1)$ separately. By letting

$$k_s = \sqrt{\frac{G}{\bar{\Gamma}}} = \begin{cases} k_{s,\alpha} = \sqrt{\frac{G}{\bar{\Gamma}_\alpha}}, & (0 < x < w_0), \\ k_{s,\beta} = \sqrt{\frac{G}{\bar{\Gamma}_\beta}}, & (w_0 < x < 1), \end{cases} \tag{11.70}$$

the general solution of (11.69) has the form

$$\frac{1}{\hat{\varpi}}\bar{h}_{01}(\hat{x}) = \begin{cases} y_{01} - \frac{M_\alpha}{G}P_{11}(w_0) + \left[S_{01,\alpha}e^{k_{s\alpha}\hat{x}} + A_{01,\alpha}e^{-k_{s\alpha}\hat{x}}\right], \ (\hat{x} < 0); \\ y_{01} - \frac{M_\beta}{G}P_{11}(w_0) + \left[S_{01,\beta}e^{-k_{s\beta}\hat{x}} + A_{01,\beta}e^{k_{s\beta}\hat{x}}\right], \ (\hat{x} > 0). \end{cases} \tag{11.71}$$

The inner solution (11.71) contains the six arbitrary constants $S_{01,\alpha}$, $S_{01,\beta}$, $A_{01,\alpha}$, $A_{01,\beta}$, $d_{11,0}$ and $y_{01}$. These constants can be determined by the matching conditions with the outer solution and the connection conditions at the triple point $x = w_0, \hat{x} = 0$.

#### 11.5.2.1 Matching

The matching conditions of the inner solution (11.71) with the outer solution (11.58),

$$\lim_{x\to w_0} \hat{\varpi}\epsilon^{1/2}\bar{h}_{01}(x) \Longleftrightarrow \lim_{|\hat{x}|\to\infty} \epsilon^{1/2}\bar{h}_{01}(\hat{x}), \tag{11.72}$$

leads to

$$A_{01}(x) = 0.$$

As a consequence, the inner solution may be written as follows:

$$\frac{1}{\hat{\varpi}}\bar{\mathrm{h}}_{01}(\hat{x}) = \begin{cases} \frac{1}{\hat{\varpi}}\bar{\mathrm{h}}_{01,\alpha}(\hat{x}) = \bar{h}_{01,\alpha}(w_0)\big(1 - \mathrm{e}^{k_{s,\alpha}\hat{x}}\big) + S_{01,\alpha}\mathrm{e}^{k_{s,\alpha}\hat{x}}, \\ \quad (0 < x < w_0; \hat{x} < 0); \\ \frac{1}{\hat{\varpi}}\bar{\mathrm{h}}_{01,\beta}(\hat{x}) = \bar{h}_{01,\beta}(w_0)\big(1 - \mathrm{e}^{-k_{s,\beta}\hat{x}}\big) + S_{01,\beta}\mathrm{e}^{-k_{s,\beta}\hat{x}}, \\ \quad (w_0 < x < 1; \hat{x} > 0), \end{cases} \tag{11.73}$$

where

$$\bar{h}_{01}(w_0) = y_{01} - \tfrac{M}{G}\left[\tfrac{1}{2}d_{11,0} + \widehat{P}_{11}(w_0)\right],$$

$$\widehat{P}_{11}(w_0) = -\frac{w_0}{\pi}(1+\hat{\kappa})\ln 2 - \frac{1+\hat{\kappa}}{2\pi}\int_0^{2w_0} \ln(1-\cos\pi t)\mathrm{d}t,$$

as we previously derived. The constants $S_{01,\alpha}$, $S_{01,\beta}$, $y_{01}$, and $d_{11,0}$ are still to be determined.

#### 11.5.2.2 Connection Conditions at the Triple Point

The inner solution (11.73) must be subject to the following connection conditions at the triple point $x = w_0$:

1. the continuity condition:

$$\bar{\mathrm{h}}_{01,\alpha}(w_0^-, 0^-) = \bar{\mathrm{h}}_{01,\beta}(w_0^+, 0^+). \tag{11.74}$$

   It follows that

$$S_{01} = S_{01,\alpha} = S_{01,\beta}. \tag{11.75}$$

2. the slope condition:

$$\epsilon^{1/2}\frac{\mathrm{d}\bar{\mathrm{h}}_{01}}{\mathrm{d}\hat{x}}(w_0, 0) = h'_{\mathrm{B}}(w_0).$$

   Define $h'_{\mathrm{B}}(w_0^{\pm}) = \hat{\varpi}\hat{s}_1^{\pm} = \pm\tan\theta^{\pm}$, where

$$\theta^- = \theta_\alpha - \frac{\pi}{2} - \varphi, \quad \theta^+ = \theta_\beta - \frac{\pi}{2} + \varphi. \tag{11.76}$$

From (11.73), one derives that

$$
\begin{aligned}
k_{s,\alpha}\big[S_{01,\alpha} - \bar{h}_{01,\alpha}(w_0^-)\big] &= -\frac{1}{\hat{\varpi}} \tan\theta^- = \hat{s}_1^-, \\
-k_{s,\beta}\big[S_{01,\beta} - \bar{h}_{01,\beta}(w_0^+)\big] &= \frac{1}{\hat{\varpi}} \tan\theta^+ = \hat{s}_1^+.
\end{aligned}
\tag{11.77}
$$

In reality, the contact angle may be large; the slopes are $\tan\theta^\pm = \mathcal{O}(1)$. Therefore, the above connection conditions are balanced, and it follows that

$$
S_{01} - \bar{h}_{01,\alpha}(w_0^-) = \frac{\hat{s}_1^-}{k_{s,\alpha}}. \tag{11.78}
$$

From (11.75) and (11.78), we derive that

$$
\begin{aligned}
\tfrac{1}{2}d_{11,0} &= -\tfrac{\mathrm{G}}{M_\beta - M_\alpha}\left(\tfrac{\hat{s}_1^+}{k_{s,\beta}} + \tfrac{\hat{s}_1^-}{k_{s,\alpha}}\right) - \widehat{P}_{11}(w_0), \\
S_{01} &= \tfrac{\hat{s}_1^-}{k_{s,\alpha}} + \tfrac{M_\alpha}{\mathrm{G}}\left[\widehat{P}_{11}(0) - \widehat{P}_{11}(w_0)\right], \\
y_{01} &= \tfrac{M_\alpha}{\mathrm{G}}\left[\tfrac{1}{2}d_{11,0} + \widehat{P}_{11}(0)\right].
\end{aligned}
\tag{11.79}
$$

Thus, given the growth system and the parameters $(\epsilon, \varphi)$, the outer solution is fully determined as

$$
\bar{h}_{01}(x) = \begin{cases}
\bar{h}_{01,\alpha}(x) = \tfrac{M_\alpha}{\mathrm{G}}\big[\widehat{P}_{11}(0) - \widehat{P}_{11}(x)\big], \\
\quad (0 \le x < w_0); \\
\bar{h}_{01,\beta}(x) = \tfrac{M_\alpha}{\mathrm{G}}\big[\tfrac{1}{2}d_{11,0}\left(1 - \tfrac{M_\beta}{M_\alpha}\right) + \widehat{P}_{11}(0) - \tfrac{M_\beta}{M_\alpha}\widehat{P}_{11}(x)\big], \\
\quad (w_0 < x \le 1),
\end{cases}
\tag{11.80}
$$

$$
w_0 = (\hat{\kappa} - \hat{C}_\infty)/(1 + \hat{\kappa}). \tag{11.81}
$$

### 11.5.3 The Composite Solution for the Interface Shape

The uniformly valid composite solution for the interface shape reduces to the inner solution (11.71) in the inner region, while it reduces to the outer solution (11.64) in the outer region, as sketched in Fig. 11.7. A simple ad hoc form of the composite solution $\bar{\mathrm{h}}_{c,0}(x)$ valid with the condition $\sqrt{\mathrm{G}/(\epsilon\bar{\Gamma})} \gg 1$ may be given in the following form:

$$
\bar{\mathrm{h}}_{c,0}(x, \hat{x}) = \begin{cases}
\hat{\varpi}\bar{h}_{01,\alpha}(x)\big(1 - \mathrm{e}^{k_{s,\alpha}\hat{x}}\big) + S_{01}\mathrm{e}^{k_{s,\alpha}\hat{x}} & (0 < x < w_0), \\
\hat{\varpi}\bar{h}_{01,\beta}(x)\big(1 - \mathrm{e}^{-k_{s,\beta}\hat{x}}\big) + S_{01}\mathrm{e}^{-k_{s,\beta}\hat{x}} & (w_0 < x < 1).
\end{cases}
\tag{11.82}
$$

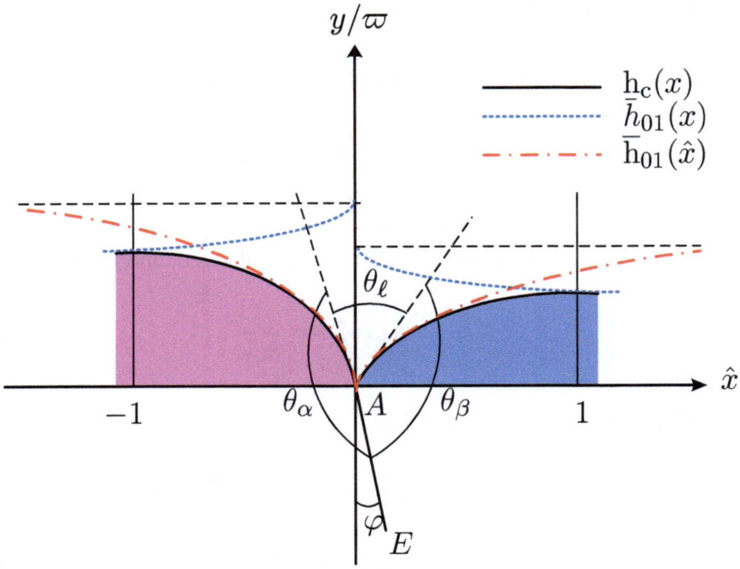


**Fig. 11.7** Sketch of the liquid open angle and the inner region near the triple point $A$. The *red long dash-dot lines* describe the interface of the inner solution; the *blue dashed lines* describe the interface of the outer solution; the *black lines* describe the interface of the composite solution. The *black solid line AE* represents the $\alpha$-$\beta$ interface

The function $\overline{\mathrm{h}}_{c,0}(x)$ approaches the inner solution as the outer variable $x$ approaches the inner region, namely $x \to w_0$. On the other hand, the function $\overline{\mathrm{h}}_{c,0}(x)$ approaches the outer solution $\{\bar{h}_{01,\alpha}(x), \bar{h}_{01,\beta}(x)\}$, as the inner variable $\hat{x}$ moves far away from the inner region from different sides, namely, $\hat{x} \to \pm\infty$.

Despite the fact that (11.82) satisfies the connect conditions at the triple point exactly, it does not fully satisfy the smoothness conditions at the two interface-tips. The errors that (11.82) contains in the tip conditions are exponentially small as $\epsilon \to 0$. Nevertheless, the numerical values of such errors may not be negligible when $\epsilon$ is not infinitesimally small. Therefore, here we call it an *ad hoc composite solution*.

For the problem under study, a more numerically accurate form of the composite solution is the *global interface solution* $\mathrm{h}_c(x, \epsilon)$, which is obtained by letting $y_{01} = y_{*c}$ and solving the equation

$$\hat{\varpi}\left[\frac{M}{\mathrm{G}}P_{11}(x) - y_{*c}\right] = -\mathrm{h}_c + \epsilon\frac{\bar{\Gamma}}{\mathrm{G}}\frac{\mathrm{d}^2\mathrm{h}_c}{\mathrm{d}x^2}. \tag{11.83}$$

Here $\mathrm{h}_c(x, \epsilon)$ is associated with the basic state as

$$h_{\mathrm{B}}(x) = \epsilon^{\frac{1}{2}}\mathrm{h}_c(x, \epsilon) + (\mathrm{h.o.t.}).$$

Indeed, it is seen that in the limit $\epsilon \to 0$, (11.83) reduces to the outer equation (11.57) in the outer region, while it reduces to the inner equation (11.69) in the inner region. The general solution of (11.83) may be obtained and written in the form

$$\frac{1}{\hat{\varpi}}\mathrm{h}_c(x,\epsilon) = H_c(x) + \Lambda_c + A_c \mathrm{e}^{\frac{k_s}{\sqrt{\epsilon}}x} + S_c \mathrm{e}^{-\frac{k_s}{\sqrt{\epsilon}}x}, \tag{11.84}$$

where $(A_c, S_c, \Lambda_c)$ are arbitrary piecewise constant functions, and

$$\begin{aligned} H_c(x) = \frac{M}{2\sqrt{\epsilon}\bar{\Gamma}k_s}\Big[ & \mathrm{e}^{\frac{k_s}{\sqrt{\epsilon}}x}\int_{w_0}^{x}\widehat{P}_{11}(x_1)\mathrm{e}^{-\frac{k_s}{\sqrt{\epsilon}}x_1}\mathrm{d}x_1 \\ & -\mathrm{e}^{-\frac{k_s}{\sqrt{\epsilon}}x}\int_{w_0}^{x}\widehat{P}_{11}(x_1)\mathrm{e}^{\frac{k_s}{\sqrt{\epsilon}}x_1}\mathrm{d}x_1\Big] \end{aligned} \tag{11.85}$$

and

$$\Lambda_c = -\left[\frac{1}{2}\frac{M}{\mathrm{G}}d_{11,0} - y_{*c}\right].$$

The three tip conditions at $x = 0, 1$ and the three connection conditions at the triple point $x = w_0$ together can uniquely determine the six constants $(A_{c,\alpha}, S_{c,\alpha})$; $(A_{c,\beta}, S_{c,\beta})$; $(y_{*c}, d_{11,0})$. The detailed derivations are given in Section 11.15, and the results are summarized below.

1.

$$\begin{aligned} S_{c,\alpha} &= -\left[\frac{H_c(0)}{2} - \frac{\sqrt{\epsilon}}{2}\frac{H_c'(0)}{k_{s\alpha}}\right] - \tfrac{1}{2}\Lambda_{c,\alpha}, \\ A_{c,\alpha} &= -\left[\frac{H_c(0)}{2} + \frac{\sqrt{\epsilon}}{2}\frac{H_c'(0)}{k_{s\alpha}}\right] - \tfrac{1}{2}\Lambda_{c,\alpha}. \end{aligned} \tag{11.86}$$

2.

$$\Lambda_{c,\alpha} = -\frac{\sinh\left[\frac{k_{s,\alpha}}{\sqrt{\epsilon}}w_0\right]H_c(0) + \cosh\left[\frac{k_{s,\alpha}}{\sqrt{\epsilon}}w_0\right]\sqrt{\epsilon}\frac{H_c'(0)}{k_{s,\alpha}} + \frac{\varpi}{\sqrt{\epsilon}k_{s,\alpha}}\hat{s}_1^-}{\sinh\left[\frac{k_{s,\alpha}}{\sqrt{\epsilon}}w_0\right]}. \tag{11.87}$$

3.

$$\begin{aligned} A_{c,\beta} = & \frac{1}{2}\mathrm{e}^{-\frac{k_{s,\beta}}{\sqrt{\epsilon}}w_0}\left[A_{c,\alpha}\mathrm{e}^{\frac{k_{s,\alpha}}{\sqrt{\epsilon}}w_0} + S_{c,\alpha}\mathrm{e}^{-\frac{k_{s,\alpha}}{\sqrt{\epsilon}}w_0}\right] \\ & + \frac{1}{2}\mathrm{e}^{-\frac{k_{s,\beta}}{\sqrt{\epsilon}}w_0}\left[\Lambda_{c,\alpha} - \Lambda_{c,\beta} + \frac{\varpi}{\sqrt{\epsilon}k_{s,\beta}}\hat{s}_1^+\right], \end{aligned}$$

$$S_{c,\beta} = \frac{1}{2}\mathrm{e}^{\frac{k_{s,\beta}}{\sqrt{\epsilon}}w_0}\left[A_{c,\alpha}\mathrm{e}^{\frac{k_{s,\alpha}}{\sqrt{\epsilon}}w_0} + S_{c,\alpha}\mathrm{e}^{-\frac{k_{s,\alpha}}{\sqrt{\epsilon}}w_0}\right]$$
$$-\frac{1}{2}\mathrm{e}^{\frac{k_{s,\beta}}{\sqrt{\epsilon}}w_0}\left[\Lambda_{c,\beta} - \Lambda_{c,\alpha} + \frac{\varpi}{\sqrt{\epsilon}k_{s,\beta}}\hat{s}_1^+\right]. \tag{11.88}$$

4.

$$\Lambda_{c,\beta} = \Lambda_{c,\alpha} + \left[A_{c,\alpha}\mathrm{e}^{\frac{k_{s,\alpha}}{\sqrt{\epsilon}}w_0} + S_{c,\alpha}\mathrm{e}^{-\frac{k_{s,\alpha}}{\sqrt{\epsilon}}w_0}\right] + \frac{\varpi\hat{s}_1^+}{\sqrt{\epsilon}k_{s,\beta}}\coth\left[\frac{k_{s,\beta}}{\sqrt{\epsilon}}(1-w_0)\right]$$
$$+\frac{\frac{\sqrt{\epsilon}}{k_{s,\beta}}H_c'(1)}{\sinh\left[\frac{k_{s,\beta}}{\sqrt{\epsilon}}(1-w_0)\right]}. \tag{11.89}$$

5.

$$d_{11,0} = \frac{2\mathrm{G}(\Lambda_{c,\alpha} - \Lambda_{c,\beta})}{M_\beta - M_\alpha}, \quad y_{*c} = \frac{\Lambda_{c,\alpha}M_\beta - \Lambda_{c,\beta}M_\alpha}{M_\beta - M_\alpha}. \tag{11.90}$$

Numerical calculations show that for the case $\sqrt{G/(\epsilon\bar{\Gamma})} > 6.0$, the differences between $\mathrm{h}_c(x)$ and $\bar{\mathrm{h}}_{c,0}(x)$ are invisible, as shown in Fig. 11.8. However, for more general cases, in which the value of $\sqrt{G/(\epsilon\bar{\Gamma})}$ is not so big, such as $\sqrt{G/(\epsilon\bar{\Gamma})} \approx 2.0$, the numerical discrepancy between the ad hoc composite solution $\bar{\mathrm{h}}_{c,0}(x)$ and

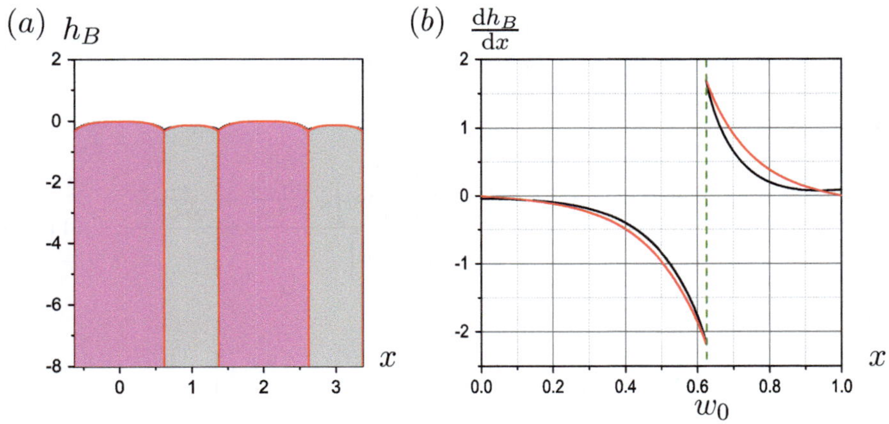


**Fig. 11.8** Comparisons of the steady global interface solution $\mathrm{h}_c(x)$ with the ad hoc composite solution $\bar{\mathrm{h}}_{c,0}(x)$ for the case $\epsilon = 0.01$, $\bar{\varphi} = 0$, $\varpi = 0.15$, $w_0 = 0.625$, $M_\alpha = 0.107$, $M_\beta = -0.218$, $G = 0.1073$, $\bar{\Gamma}_\alpha = 0.3$, $\bar{\theta}^- = 1.14$, $\bar{\theta}^+ = 1.04$, $R_c = 0.84$, and $C_\infty = -1.8644 \times 10^{-2}$, $\hat{v} = 3.0 \times 10^{-7}$, $\hat{\beta} = 3.0 \times 10^{-7}$, $\sqrt{\mathrm{G}/\epsilon\bar{\Gamma}_\alpha} = 5.98$. (**a**) The graph of the function $\mathrm{h}_c(x)$ (*red line*) in comparison with the graph of the function $\bar{\mathrm{h}}_{c,0}(x)$ (*black dotted line*). (**b**) The slope function of $\mathrm{h}_c(x)$ (*red line*) in comparison with the slope function of $\bar{\mathrm{h}}_{c,0}(x)$ (*black dotted line*)

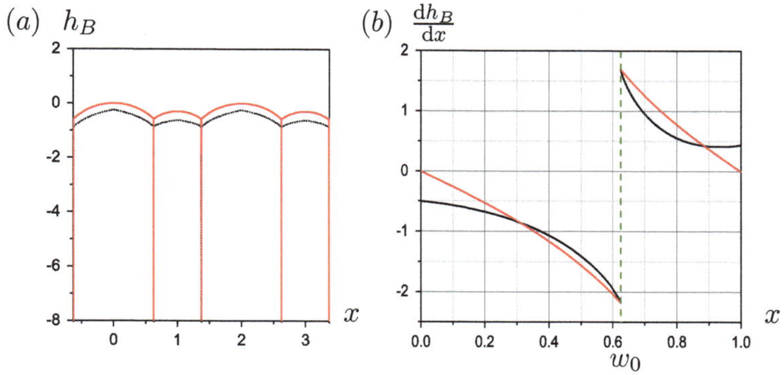


**Fig. 11.9** Comparisons of the steady global interface solution $h_c(x)$ with the ad hoc composite solution $\bar{h}_{c,0}(x)$ for the case $\epsilon = 0.01$, $\bar{\varphi} = 0$, $\varpi = 0.15$, $w_0 = 0.625$, $M_\alpha = 0.107, M_\beta = -0.218$, $G = 0.03577$, $\bar{\Gamma}_\alpha = 0.9$, $\bar{\theta}^- = 1.14$, $\bar{\theta}^+ = 1.04$, $R_c = 0.84$, and $C_\infty = -1.8644 \times 10^{-2}$, $\hat{v} = 9.0 \times 10^{-7}$, $\hat{\beta} = 3.0 \times 10^{-7}$, $\sqrt{G/\epsilon\bar{\Gamma}_\alpha} = 1.99$. (**a**) The graph of the function $h_c(x)$ (*red line*) in comparison with the graph of the function $\bar{h}_{c,0}(x)$ (*black dotted line*). (**b**) The slope function $h_c(x)$ (red line) in comparison with the slope function of $\bar{h}_{c,0}(x)$ (*black dotted line*)

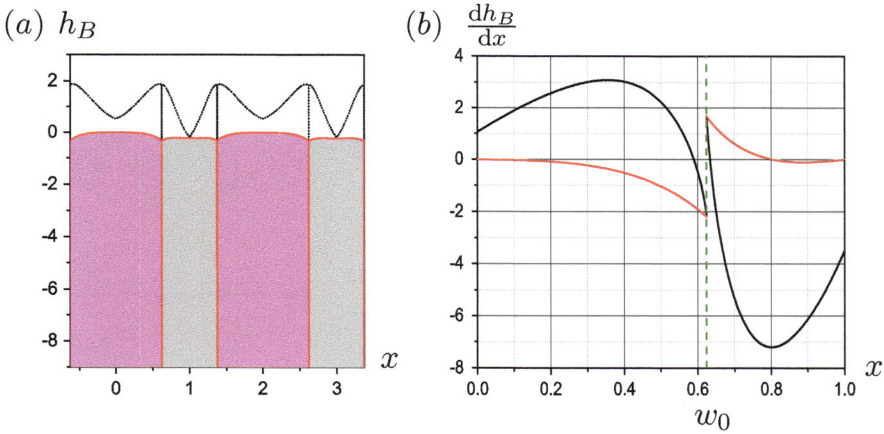


**Fig. 11.10** Comparisons of the steady global interface solution $h_c(x)$ with the ad hoc composite solution $\bar{h}_{c,0}(x)$ for the case $\epsilon = 0.00555$, $\bar{\varphi} = 0$, $\varpi = 0.15$, $w_0 = 0.625$, $M_\alpha = 0.107, M_\beta = -0.218$, $G = 0.0026176$, $\bar{\Gamma}_\alpha = 0.116964$, $\bar{\theta}^- = 1.14$, $\bar{\theta}^+ = 1.04$, $R_c = 0.84$, and $C_\infty = -1.8644 \times 10^{-2}$, $\hat{v} = 2.2 \times 10^{-8}$, $\hat{\beta} = 4.8774 \times 10^{-10}$, $\sqrt{G/\epsilon\bar{\Gamma}_\alpha} = 2.008$. (**a**) Graph of the function $h_c(x)$ (*red line*) in comparison with the graph of the function $\bar{h}_{c,0}(x)$ (*black dotted line*). (**b**) The slope function $h_c(x)$ (*red line*) in comparison with the slope function $\bar{h}_{c,0}(x)$ (*black dotted line*)

the global interface solution $h_c(x)$ may become significant, as shown in Figs. 11.9 and 11.10. Since the global interface solution $h_c(x)$ is considered numerically more accurate than $\bar{h}_{c,0}(x)$ for the description of the basic steady state, it will therefore be used for our further analysis.

The interfacial patterns of steady untilted and tilted eutectic growth calculated in terms of the global interface solution $h_c(x)$ for typical cases are shown in Fig. 11.11a and b, respectively. A variety of the interfacial patterns observed in experiments can be recovered by the global interface solutions obtained, which are affected by the parameters $\Gamma_\alpha$ and $\hat{\beta}$ significantly. Some additional samples of interfacial patterns for the cases of nontilted eutectic growth with different interlamellar spacings are shown in Figs. 11.12a, b and 11.13a, b.

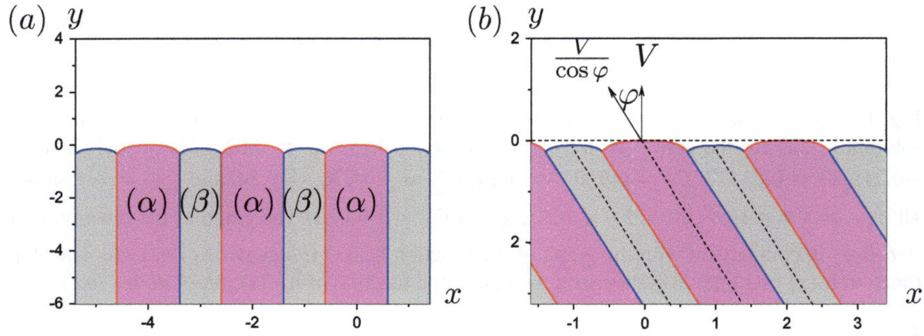


**Fig. 11.11** The interface shape of the eutectic growth described by the global solution $\overline{h}_c(x)$ for the case $C_\infty = -0.01$, $w_0 = 0.6$, $\varpi = 0.15$, $\hat{\kappa} = 1.33$, $G = 0.1$, $R_c = 0.84$, $M_\alpha = 0.1$, $M_\beta = -0.2$, $\hat{\beta} = 3.0 \times 10^{-7}$, $\hat{v} = 3.0 \times 10^{-7}$, $\epsilon = 0.01$, and **(a)** $\varphi = 0$, $\theta_\alpha = 155.3°$, $\theta_\beta = 149.5°$; **(b)** $\varphi = 24.5°$, $\theta_\alpha = 170.3°$, $\theta_\beta = 114.5°$

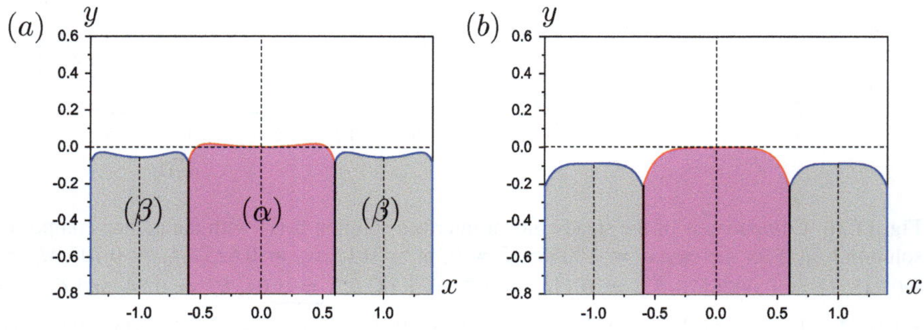


**Fig. 11.12** The interface shapes described by the global interface solutions for the cases $w_0 = 0.6$, $C_\infty = -0.03$, $\varpi = 0.15$, $\hat{\kappa} = 1.0$, $M_\alpha = 0.1$, $M_\beta = -0.2$, $\theta_\alpha = 155.3°$, $\theta_\beta = 149.5°$, $\varphi = 0$, $\theta^- = 65.3°$, $\theta^+ = 59.5°$, $\hat{\beta} = 3.0 \times 10^{-7}$, $\hat{v} = 3.0 \times 10^{-7}$, $\epsilon = 0.03$, $R_c = 0.84$, and **(a)** $\epsilon = 0.03$ ; **(b)** $\epsilon = 0.015$

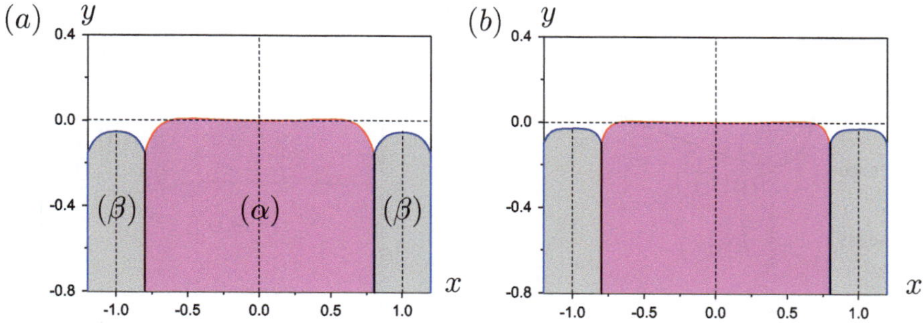


**Fig. 11.13** The interface shapes described by the global interface solutions for the cases $w_0 = 0.8$, $\epsilon = 0.02$, **(a)** $\hat{\beta} = 3.0 \times 10^{-7}$; and **(b)** $\hat{\beta} = 9.0 \times 10^{-7}$. Other parameters are the same as those in Fig. 11.12

## 11.6 Further Numerical Computations of Asymptotic Solutions

From the above analysis, as a good approximation, one may simply set $C_{\rm B}(x, y, \epsilon) \approx \varpi \hat{C}_\infty(1-{\rm e}^{-\tilde{y}})+\epsilon\varpi\bar{C}_{11}(x, y, \tilde{y})$, and $h_{\rm B}(x) \approx \varpi {\rm h}_c(x)$. It is seen that the steady state contains five parameters, $(\epsilon, {\rm G}, C_\infty, \Gamma_\alpha, \varphi)$, with two undetermined morphological parameters: the tilt angle $\varphi$ and interlamellar spacing $\ell_{\rm w}$. To show the dependence of the interfacial pattern of the steady state on the free parameters $\varphi$ and $\ell_{\rm w}$ under given growth conditions, we use $\ell_{\rm D}$ as the length scale and calculate the nondimensional curvatures of interface and other nondimensional physical lengths involved. In this way, we derive the formulas $\ell_{\rm D}/\ell_{{\rm tip},\alpha} = h''_{\rm B}(0)/\epsilon$, $\ell_{\rm D}/\ell_{{\rm tip},\beta} = h''_{\rm B}(1)/\epsilon$, for the local curvature at the interface-tip $x = 0$ and $x = 1$, respectively, where

$$\begin{cases} h''_{\rm B}(0) \approx \bar{h}''_{01}(0) = \dfrac{M_\alpha(1+\hat{\kappa})}{\rm G}\cot(\pi w_0/2), \\ h''_{\rm B}(1) \approx \bar{h}''_{01}(1) = \dfrac{M_\beta(1+\hat{\kappa})}{2{\rm G}}\Big[\cot\pi(1+w_0)/2) - \cot\pi(1-w_0)/2\Big]. \end{cases} \tag{11.91}$$

To better compare our solution with the flat interface approximation, we use $\widehat{y}_{*c}$ to denote the temperature of the $\alpha$-interface's tip and $\widehat{h}_{\rm B}(1)$ to denote the location of the $\beta$-interface's tip. We also use $\widehat{h}_{\rm B}(w_0)$ to denote the location of the triple point using the length scale $\ell_{\rm D}$. We derive that $\widehat{y}_{*c} = \epsilon y_{*c}$, $\widehat{h}_{\rm B}(1) = \epsilon h_{\rm B}(1)$, and $\widehat{h}_{\rm B}(w_0) = \epsilon h_{\rm B}(w_0)$. The numerical results of these quantities computed with given growth conditions and fixed tilt angle $\varphi = 0$ versus the Péclet number $\epsilon$ are shown in Fig. 11.14a, b. From Fig. 11.14a, we see that as the Péclet number $\epsilon$ approaches zero, the $y$-coordinates of the triple point and tips of the $\alpha$- and $\beta$-interfaces all converge to the same point $y = 0$. So, only for the case $\epsilon = 0$ might the flat interface approximation be locally applicable. However, for the realistic case $\epsilon > 0$, the $y$-coordinates of these three points on the interface diverge from each other, and a noticeable nonuniformity of the temperature distribution may appear on the

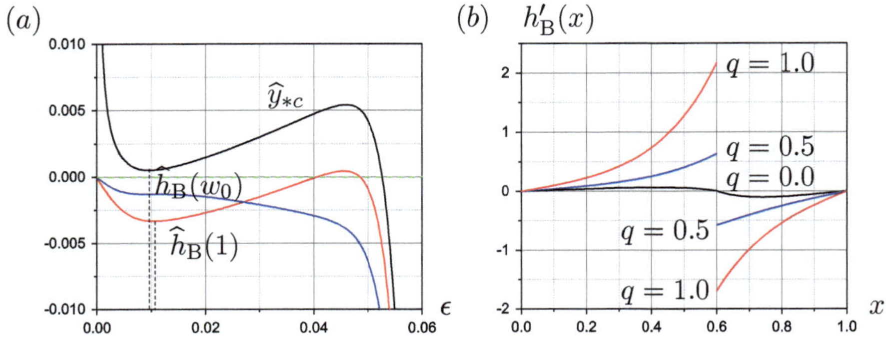


**Fig. 11.14** (**a**) The variations of $\widehat{y}_* = \epsilon y_{*c}$, $\widehat{h}_B(w_0) = \epsilon\, h_B(w_0)$, and $\widehat{h}_B(1) = \epsilon\, h_B(1)$ with $\epsilon$ for the case of $\varphi = 0$, $\varpi = \epsilon^{\frac{1}{2}}\hat{\varpi}$; $C_\infty = \epsilon^{\frac{1}{2}}\hat{\varpi}\hat{C}_\infty$; $\hat{\varpi} = 1.5$, $\hat{\kappa} = 1.0$, $\hat{C}_\infty = -0.2$, $w_0 = 0.6$. The *dashed green line* is the location of the $\alpha$-interface-tip. It describes the flat interface approximation imposed in the J–H theory. It is seen that $\min\{\widehat{y}_*\} = 0.5105 \times 10^{-3}$ at $\epsilon = 0.00970$, while $\min\{\hat{h}_B(w_0)\} = -0.3304 \times 10^{-2}$, at $\epsilon = 0.0107$. (**b**) The variation of the local slope distribution of the interface $h'_B(x)$ with contact angles for the case $\epsilon = 0.01$, $\theta^+ = q\times 65.3°$, $\theta^- = q\times 59.5°$, and $q = 0, 0.5, 1.0$

interface. To show the effect of the contact angles on the interface shape, we set $\theta^+ = q\times 59.5°$, $\theta^- = q\times 65.3°$. In Fig. 11.14b, we show the variation of $h'_B(x)$ with increasing value of the factor $q$. We see that for the case of the factor $0 \le q \ll 1$, our solutions display a flat or nearly flat interface. However, for the practical cases $q = 0.5, 1.0$, in which the contact angles are not small, our solutions show that in the region near the triple point, the magnitude of the slope of the interface shape will be of order unity, $|h'_B(x)| = \mathcal{O}(1)$. This implies that the nearly flat interface approximation will result in significant error.

In our analysis, the growth conditions $(V, (G)_D, C_\infty)$ have been described by the three independent nondimensional parameters $(\bar{\Gamma}_\alpha, G, C_\infty)$. The notation $\Lambda_{JH} = \ell_w V^{\frac{1}{2}}$, which was first introduced by Jackson–Hut and often used in the literature, can be related to the group of the parameter $\epsilon\bar{\Gamma}_\alpha$, as $\Lambda_{JH} = \ell_w V^{\frac{1}{2}} = \sqrt{\frac{\ell_c \kappa_D}{\epsilon \bar{\Gamma}_\alpha}}$. It should be pointed out that $\Lambda_{JH}$ is not a universal constant, since the selection of the value of $\epsilon\bar{\Gamma}_\alpha$ may depend on the growth parameters $(G, C_\infty)$ as well as the tilt angle $\varphi$.

Noting that the parameters $\epsilon$, $\bar{\Gamma}_\alpha$ and the parameter $G$ all depend on the pulling velocity $V$, it is sometimes better to explore the effect of each of the operation conditions $(V, (G)_D, C_\infty)$ on the behaviors of eutectic growth, for which we use the following new dimensionless parameters, which are similar to those used in the previous chapter:

- $\hat{\upsilon} = \frac{\ell_{c,\alpha}}{\ell_D} = \frac{\ell_{c,\alpha} V}{\kappa_D}$, which measures the pulling velocity $V$;
- $\hat{\beta} = \frac{\ell_{c,\alpha}}{\ell_{G,\alpha}} = -\frac{\ell_{c,\alpha}(G)_D}{m_\alpha C_e}$, which measures the temperature gradient $(G)_D$.

One may derive that

$$\hat{\upsilon} = \epsilon^3 \bar{\Gamma}_\alpha, \quad \hat{\beta} = \frac{\hat{\upsilon} G}{M_\alpha}. \tag{11.92}$$

Thus, the basic state can be considered to depend on three independent dimensionless parameters $\{\hat{v}, \hat{\beta}, C_\infty\}$ representing the growth conditions, and two free morphological parameters: the tilt angle $\varphi$ and Péclet number $\epsilon$. We have carried out the numerical computations of the solutions for given thermodynamic parameters: $\left(R_c = \Gamma_\alpha/\Gamma_\beta, M_\alpha, M_\beta\right)$, and various values of the growth parameters $(\hat{v}, \hat{\beta}, C_\infty)$ and the morphological parameters $(\epsilon, \varphi)$.

Recall that our solution is obtained under the restrictions $\bar{\Gamma} = \hat{v}/\epsilon^3 = \mathcal{O}(1)$ and $G = \mathcal{O}(1)$, so that for a given growth velocity $\hat{v}$, one may restrict the parameter $\epsilon$ in the range

$$\hat{v}^{1/3} \leq \epsilon \ll 1. \tag{11.93}$$

In Fig. 11.15a–c, we show the variations of location of the origin of the coordinate system $y_{*c}$ with the parameters ($\hat{\beta}, \hat{v}$ and $C_\infty$), respectively. The results

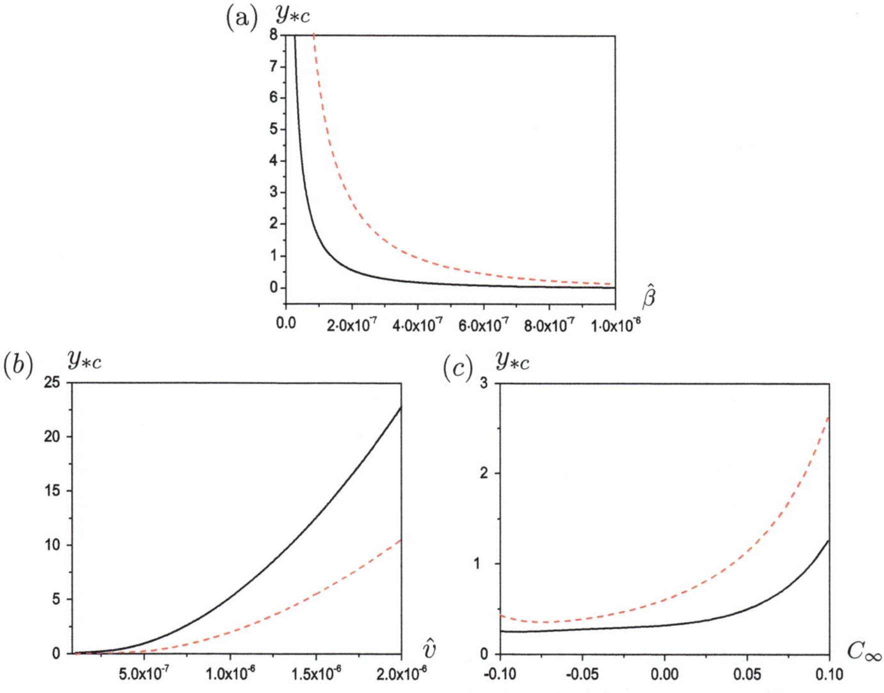


**Fig. 11.15** The location of the tip of the $\alpha$-interface evaluated for $y_{*c}$: (**a**) $y_{*c}$ versus $\hat{\beta}$ with $C_\infty = -0.03$, $\hat{v} = 3.0 \times 10^{-7}$ and $6.0 \times 10^{-7}$ from *bottom to top*; (**b**) $y_{*c}$ versus $\hat{v}$ with $C_\infty = -0.03$, $\hat{\beta} = 3.0 \times 10^{-7}$ and $6.0 \times 10^{-7}$ from *top to bottom*; (**c**) $y_{*c}$ versus $C_\infty$ with $\hat{v} = \hat{\beta} = 3.0 \times 10^{-7}$ and $\hat{v} = \hat{\beta} = 6.0 \times 10^{-7}$ from *bottom to top*. Other parameters are: $w_0 = 0.6$, $\Gamma_\alpha = 2.5 \times 10^{-5}$, $\Gamma_\beta = 2.1 \times 10^{-5}$, $R_c = 0.84$, $\kappa_\alpha = 0.85$, $\kappa_\beta = 1.15$, $\varpi = 0.15$, $M_\alpha = 0.1$, $M_\beta = -0.2$, G $= 0.3$, $\theta_\ell = 55.2°$, $\theta_\beta = 155.3°$, $\theta_\alpha = 149.5°$, $\varphi = 0$, $\epsilon = 0.012$, $\theta^+ = 65.3°$, $\theta^- = 59.5°$

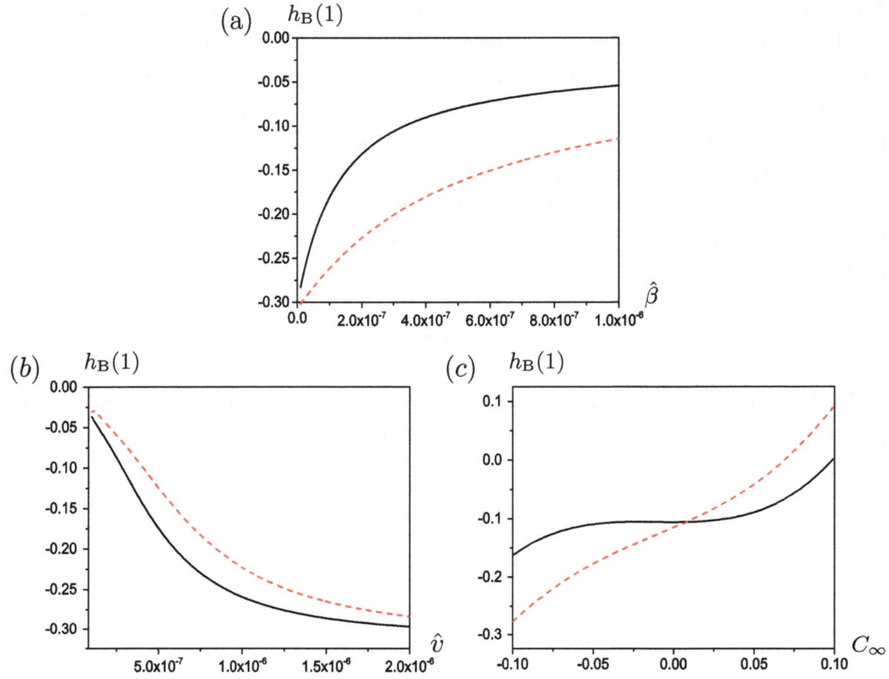


**Fig. 11.16** The location of the tip of the $\beta$-interface, $h_B(1)$ evaluated for the composite solution $h_c(x)$: (**a**) $h_B(1)$ versus $\hat{\beta}$ with $C_\infty = -0.03$, $\hat{v} = 3.0 \times 10^{-7}$ and $6.0 \times 10^{-7}$ from *top to bottom*; (**b**) $h_B(1)$ versus $\hat{v}$ with $C_\infty = -0.03$, $\hat{\beta} = 3.0 \times 10^{-7}$ and $6.0 \times 10^{-7}$ from *bottom to top*; (**c**) $h_B(1)$ versus $C_\infty$ with $\hat{v} = \hat{\beta} = 3.0 \times 10^{-7}$ and $\hat{v} = \hat{\beta} = 6.0 \times 10^{-7}$ from *bottom to top* on the *right side*. Other parameters are the same as given in Fig. 11.15

yield information on the temperature undercooling at the tip of the solid $\alpha$-interface. The variations of the location of the tip of the $\beta$-interface $h_B(1)$ with the parameters $(\hat{\beta}, \hat{v}, C_\infty)$ are respectively shown in Fig. 11.16a–c, while the variations of the triple point location $h_B(w_0)$ with the parameters $(\hat{\beta}, \hat{v}, C_\infty)$ are respectively shown in Fig. 11.17a–c.

## 11.7 Comparisons of Theoretical Solutions with Experimental Data

Typical experiments of steady lamellar eutectic growth with the system of $CBr_4$–$C_2Cl_6$ were performed by Seetharaman and Trivedi in 1988 [13], and by Mergy et al. in 1993 [10]. The data on the thermodynamic properties of the system provided by these two experimental groups show a large discrepancy. We adopt

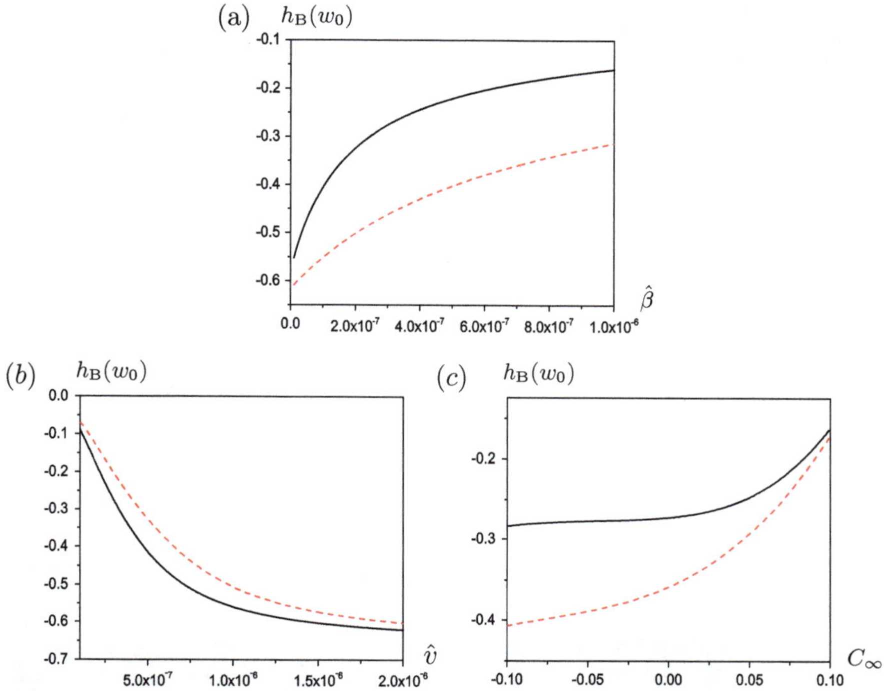


**Fig. 11.17** The location of the triple point $h_B(w_0)$ evaluated for the composite solution $h_c(x)$: (**a**) $h_B(w_0)$ versus $\hat{\beta}$ with $C_\infty = -0.03$, $\hat{v} = 3.0 \times 10^{-7}$ and $6.0 \times 10^{-7}$ from *top to bottom*; (**b**) $h_B(w_0)$ versus $\hat{v}$ with $C_\infty = -0.03$, $\hat{\beta} = 3.0 \times 10^{-7}$ and $6.0 \times 10^{-7}$ from *bottom to top*; (**c**) $h_B(w_0)$ versus $C_\infty$, with $\hat{v} = \hat{\beta} = 3.0 \times 10^{-7}$ and $\hat{v} = \hat{\beta} = 6.0 \times 10^{-7}$ from *top to bottom*. Other parameters are the same as given in Fig. 11.15

the more recent data given by Mergy in the comparisons with the theory (see Table 11.1).

For these cases, the parameter $\varpi$ is equal to 0.15. Hence, the global asymptotic solution is applicable. In Fig. 11.18, we show the experimental photos given by Mergy et al. [10] for two typical cases. For direct comparison, we have included the interface shapes predicted by the global solution in the same figures. It is seen that the experimental results are in excellent agreement with the theory.

In Fig. 11.19, we show the variation of the triple point's location with the parameter $1/(\epsilon\bar{\Gamma}_\alpha) = \frac{\ell_w^2 V}{\ell_{c\alpha}\kappa_D}$, and compare it with the experimental data. It is noted that the experimental data of the pulling velocity $V$ and temperature gradient $(G)_D$ have remarkable systematic errors up to 20%. The thermodynamic data of the system provided in the literature for $c_p$, $\ell_c$ and $\kappa_D$ also have a large uncertainty. Taking into account of all these errors and uncertainty, in calculating the experimental dimensionless parameter $\left[1/(\epsilon\bar{\Gamma}_\alpha)\right]_{exp}$ we adopted $\ell_{c\alpha}\kappa_D = 0.110\,\mu m^3/s$. Fig. 11.19 shows a reasonably good quantitative agreement between the theoretical predictions and the experimental data.

**Table 11.1** The thermodynamic characteristics of $CBr_4$–$C_2Cl_6$ (refer to [13])

| | |
|---|---|
| $T_e$ | 357.55 K |
| $T_{M0}$ | 365.65 K |
| $C_e$ | 0.118 mol |
| $C_\alpha$ | 0.088 mol |
| $C_\beta$ | 0.185 mol |
| $m_\alpha$ | −81.0 K/mol |
| $m_\beta$ | 165 K/mol |
| $\kappa_\alpha$ | 0.75 |
| $\kappa_\beta$ | 1.5 |
| $\kappa_D$ | 500 $\mu m^2/s$ |
| $\gamma_\alpha$ | $6.6 \times 10^{-15}$ $J/\mu m^2$ |
| $\gamma_\beta$ | $5.8 \times 10^{-15}$ $J/\mu m^2$ |
| $\theta_\alpha$ | 160° |
| $\theta_\beta$ | 157° |
| $T_{M0}\gamma_\alpha/\Delta H$ | 0.08 μm K |
| $T_{M0}\gamma_\beta/\Delta H$ | 0.114 μm Ks |

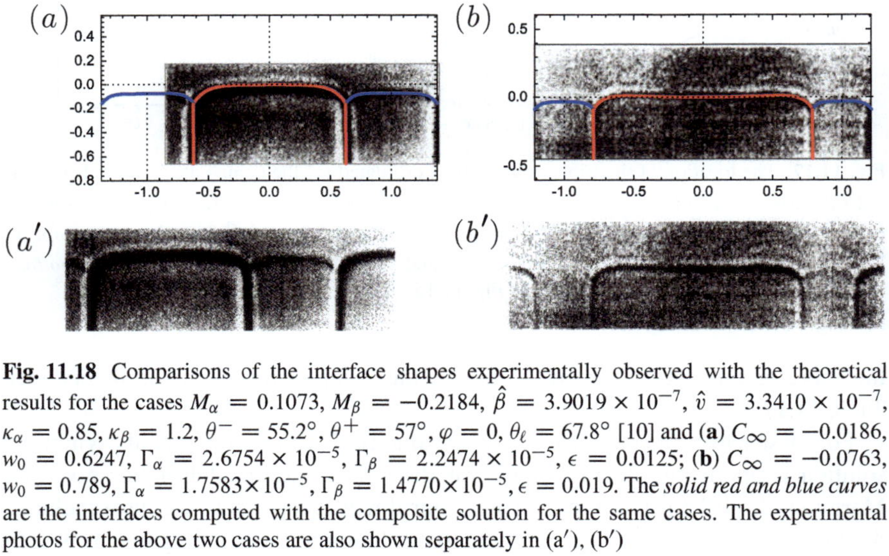


**Fig. 11.18** Comparisons of the interface shapes experimentally observed with the theoretical results for the cases $M_\alpha = 0.1073$, $M_\beta = -0.2184$, $\hat{\beta} = 3.9019 \times 10^{-7}$, $\hat{v} = 3.3410 \times 10^{-7}$, $\kappa_\alpha = 0.85$, $\kappa_\beta = 1.2$, $\theta^- = 55.2°$, $\theta^+ = 57°$, $\varphi = 0$, $\theta_\ell = 67.8°$ [10] and (**a**) $C_\infty = -0.0186$, $w_0 = 0.6247$, $\Gamma_\alpha = 2.6754 \times 10^{-5}$, $\Gamma_\beta = 2.2474 \times 10^{-5}$, $\epsilon = 0.0125$; (**b**) $C_\infty = -0.0763$, $w_0 = 0.789$, $\Gamma_\alpha = 1.7583\times10^{-5}$, $\Gamma_\beta = 1.4770\times10^{-5}$, $\epsilon = 0.019$. The *solid red and blue curves* are the interfaces computed with the composite solution for the same cases. The experimental photos for the above two cases are also shown separately in (a′), (b′)

Through the above studies, one may conclude that for given growth conditions and tilt angle $\varphi$, the system may yield a family of periodic steady-state solutions with different interlamellar spacing parameters $\epsilon$ in the range bounded below as given by (11.93). In other words, the "selected" pattern parameters $(\epsilon, \varphi)$ in experiments are not unique.

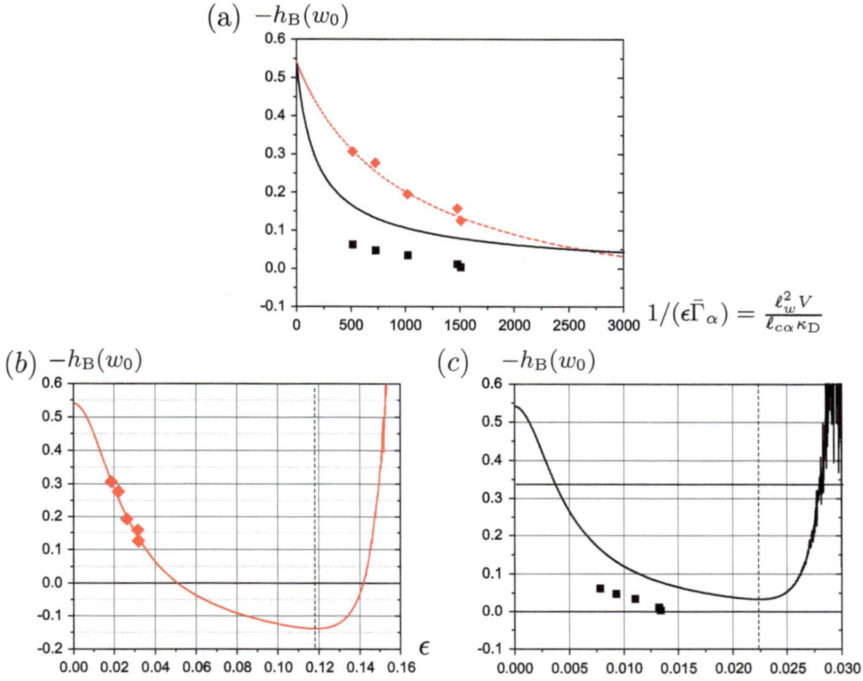


**Fig. 11.19** Comparisons of the experimental data (symbols) of the triple point location given by Mergy et al. [10] with the theoretical curves calculated with the composite solution for the case $C_\infty = -0.0186$, $w_0 = 0.6245$, $\varphi = 0$, $\bar{\Gamma}_\beta = 0.84\bar{\Gamma}_\alpha$, $M_\alpha = 0.1073$, $M_\beta = -0.2184$, $\hat{\beta} = 1.734 \times 10^{-7}$, $\kappa_\alpha = 0.85$, $\kappa_\beta = 1.2$, $\theta_\alpha = 60°$, $\theta_\beta = 57°$. (**a**) $-h_B(w_0)$ versus the parameter $1/(\epsilon\bar{\Gamma}_\alpha) = (\ell_w^2 V)/(\ell_{c\alpha}\kappa_D)$. The *red dashed line* and *symbols* correspond to the case $\hat{v} = 6.776 \times 10^{-7}$, while the *black line* and *symbols* correspond to the case $\hat{v} = 1.188 \times 10^{-7}$. (**b**) $-h_B(w_0)$ versus $\epsilon$ for the case $\hat{v} = 6.776 \times 10^{-7}$. (**c**) $-h_B(w_0)$ versus $\epsilon$ for the case $\hat{v} = 1.188 \times 10^{-7}$

## 11.8 Global Interfacial Instabilities of Eutectic Growth

As observed in experiments, under certain steady growth conditions, the eutectic system may display various spatially periodic eutectic patterns, which may be oscillating or not oscillating, tilted or not tilted, as shown in Fig. 11.20. Thus, just as before, questions are raised as to the origin and essence of eutectic oscillating pattern formation and how the various eutectic interfacial patterns relate to the properties of the system and growth conditions. The key to answering these questions is understanding the stability properties of the eutectic growth system.

The issues of linear instability mechanisms of eutectic growth have been investigated experimentally and numerically for decades by a number of researchers including Kassner et al. [9], Karma and Sarkissian [8], Ginibre et al. [6], Akamatsu et al. [1], and Plapp et al. [12]. There are some analytical works on stability analysis in the literature, including those by Chen-Davis [3] and by

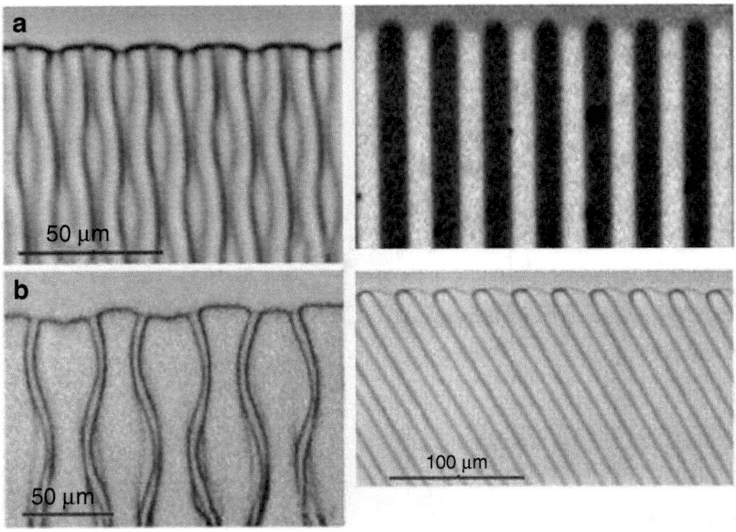


**Fig. 11.20** Some typical interfaces of neutral perturbed states observed in experiments [6]. (*Left side*): the interface shapes of neutrally stable oscillatory (AS) and (SA)-mode; (*right side*): the interface shapes of the steady nonoscillatory ST-mode

Datye-Mathur-Langer [4]. These analytical works used the steady solutions with planar interface as the basic state of the system, neglecting the significant effect of the triple point.

The global linear stability analysis around the basic steady state including the effect of the triple point can be conducted by following the approach developed in the previous chapters. In doing so, one may use the global steady solution derived in the previous section as the basic steady state and resolve the related linear eigenvalue problem for the perturbed states by applying the multiple variables expansion method.

The coordinate system $(x, y)$ moving along the tilted direction of $\mathbf{U} = V(-\tan\varphi \ \mathbf{i} + \mathbf{k})$ defined in the previous section is continuously adopted. The origin is located at the tip B of the solid $\alpha$-phase in the basic steady state, as shown in Figs. 11.3 and 11.4. But the tip of the solid $\alpha$-phase in the perturbed state may be oscillating around the origin.

We assume that the external temperature distribution along the device is fixed. Since the ratio of the mass diffusivity length and the thermal diffusivity length $\lambda_T$ is quite small, the temperature field in the unsteady perturbed state is unperturbed, remaining $T \approx T_e + (G)_D(y - y_*)$. We further assume that the interlamellar spacing $\ell_w$ is unperturbed. However, since the concentration field is perturbed, the location of the triple point $W_c$, the slope angles $\theta^{\pm}$, and tilt angle $\varphi$ may be subsequently perturbed. Moreover, the interface between the $\alpha$-phase and the $\beta$-phase for the perturbed state may oscillate along the flat $\alpha$–$\beta$ interface of the basic steady state, corresponding to the triple point oscillation.

The unknowns in the unsteady problem under investigation are the concentration field $C(x, y, t)$, the interface shape $h(x, t)$ between the liquid phase and the solid phase, the tilt angle $\varphi(t)$, and the location $W_c(t)$ of the triple point A.

### 11.8.1 Dimensionless Form of the System for Unsteady Eutectic Growth

The unsteady concentration field $C(x, y, t)$ is subject to the following dimensionless system:

$$\nabla^2 C + \epsilon\left(\frac{\partial C}{\partial y} - \frac{\partial C}{\partial t}\right) = 0. \tag{11.94}$$

The corresponding boundary conditions are as follows:

1. The far field condition: as $z \to \infty$,

$$C \to C_\infty = \frac{(C_\infty)_D}{C_e} - 1. \tag{11.95}$$

2. The interface-tip smoothness conditions at $x = 0, \pm 1$: for steady states,

$$\frac{\partial C}{\partial x} = \frac{\partial h}{\partial x} = 0, \tag{11.96}$$

while for the unsteady perturbed states, these conditions will be specified later.
3. The interface conditions between the liquid and solid phases, at $y = h(x, t)$:
   (a) the Gibbs–Thomson condition

$$C = \epsilon\frac{G}{M}y_* - \epsilon\frac{G}{M}h - \frac{\Gamma}{M}\mathcal{K}\{h\} + (\text{h.o.t.}), \tag{11.97}$$

where

$$\mathcal{K}\{h\} = -\frac{h_{xx}}{(1 + h_x^2)^{\frac{3}{2}}} \tag{11.98}$$

is the curvature operator and, as we designated before, the curvature satisfies $\mathcal{K} > 0$ when the interfacial finger points to the liquid phase side. Furthermore, $y_*$ is a function of $\epsilon$, and one can expect the asymptotic expansion

$$y_* = y_{*0} + \epsilon y_{*1} + \cdots, \quad \text{as} \quad \epsilon \to 0. \tag{11.99}$$

(b) the mass balance condition

$$\frac{\partial C}{\partial y} - h_x \frac{\partial C}{\partial x} + \epsilon(1-\kappa)(C+1)\left(1 + \frac{\partial h}{\partial t} + \tan\varphi \frac{\partial h}{\partial x}\right) = 0. \tag{11.100}$$

4. The connection conditions at the conjunction point, $x = \pm W_c$:

$$h'(W_c^+, t) = \tan\theta^+, \quad h'(W_c^-, t) = -\tan\theta^-. \tag{11.101}$$

5. The phase balance conditions at the triple point, $x = \pm W_c$, $y = h_B(x) = y_c$: Let $\gamma_\alpha$, $\gamma_\beta$, and $\gamma_{\alpha\beta}$ be the surface tensions on the interfaces of $\alpha$-liquid, $\beta$-liquid, and $\alpha$–$\beta$, respectively, while $\theta_\alpha$, $\theta_\beta$, $\theta_\ell$ denote the contact angles over the $\alpha$-phase, $\beta$-phase, and liquid phase, respectively. From the mechanical balance, it can be derived that

$$\begin{cases} \gamma_\alpha \cos\theta_\alpha + \gamma_\beta \cos\theta_\beta = -\gamma_{\alpha\beta}, \\ \gamma_\alpha \sin\theta_\alpha - \gamma_\beta \sin\theta_\beta = 0, \end{cases} \tag{11.102}$$

and $\theta_\ell = 2\pi - (\theta_\alpha + \theta_\beta)$. Hence, all the quantities $\gamma_\alpha$, $\gamma_\beta$, $\gamma_{\alpha\beta}$, $\theta_\alpha$, $\theta_\beta$, can be considered known constants determined by the thermodynamic properties of the system. However, as indicated before, due to the fact that the mass diffusivity in the solid state is negligibly small, the tangential line of the $\alpha$–$\beta$ interface may be tilted from the $y$-axis with an unknown tilt angle $\varphi(t)$, as shown in Fig. 11.5. As a consequence, in the coordinate system $(x, y)$, the slope angle $\theta^-$ of the interface between the $\alpha$-phase and liquid phase, and the slope angle $\theta^+$ of the interface between the $\beta$-phase and liquid phase, are not prescribed. These two angles are subject to the condition $(\theta^+ + \theta^-) = \pi - \theta_\ell$, and may be expressed in the form

$$\begin{cases} \theta^-(t) = \theta_\alpha - \pi/2 - \varphi(t), \\ \theta^+(t) = \theta_\beta - \pi/2 + \varphi(t). \end{cases} \tag{11.103}$$

6. The total mass conservation condition:

$$C_\alpha W_c + C_\beta(1 - W_c) + \int_0^{W_c} C\big[x, h(x,t), t\big]\kappa_\alpha \mathrm{d}x + \int_{W_c}^{1} C\big[x, h(x,t), t\big]\kappa_\beta \mathrm{d}x = C_\infty, \tag{11.104}$$

where

$$\kappa_\alpha = 1 + C_\alpha < 1; \quad \kappa_\beta = 1 + C_\beta > 1.$$

## 11.8.2 Basic States

We use the global steady solution found in the previous section as the basic state. The properties of the basic steady states are summarized as follows:

1. It is assumed that the Péclet number satisfies $\epsilon \ll 1$, and that $\varpi = 1 - \kappa_\alpha = \mathcal{O}(\epsilon^{\frac{1}{2}})$, $1-\kappa_\beta = \hat{\kappa}\varpi$, $\hat{\kappa} = \mathcal{O}(1)$; $\varpi = \hat{\varpi}\epsilon^{\frac{1}{2}}$, $\hat{\varpi} = \mathcal{O}(1)$, and also $(C_\infty)_{\mathrm{D}} \approx C_{\mathrm{e}}$, so that $C_\infty = \hat{C}_\infty\varpi$, $\hat{C}_\infty = \mathcal{O}(1)$.
2. The solution for the concentration field in the liquid phase,

$$\begin{aligned} C_{\mathrm{B}}(x, y, \epsilon) = {} & \varpi C_\infty(1 - \mathrm{e}^{-\epsilon z}) + \epsilon\varpi\left[\tfrac{1}{2}d_{11,0}\mathrm{e}^{-\epsilon z} + \bar{\Omega}_{11}(x, y)\right] \\ & + \cdots . \end{aligned} \tag{11.105}$$

3. The global solution for the interface shape is

$$\begin{aligned} h_{\mathrm{B}}(x, \epsilon) &= \epsilon^{\frac{1}{2}}\mathrm{h}_c(x) + \cdots \\ \mathrm{h}_c(x) &= H_c(x) + \Lambda_c + A_c\mathrm{e}^{\frac{k_S}{\sqrt{\epsilon}}x} + S_c\mathrm{e}^{-\frac{k_S}{\sqrt{\epsilon}}x}. \end{aligned} \tag{11.106}$$

The above basic state contains two free pattern parameters: Péclet number $\epsilon$ and the tilt angle $\bar{\varphi}$. The Péclet number $\epsilon$ is to be determined, whereas the tilt angle $\bar{\varphi}$ is believed to be related to the anisotropy of interfacial energy on the interface of two solid phases at the triple point. Since the anisotropy of the interface energy is not taken into account, here the tilt angle $\bar{\varphi}$ is considered a free parameter, whose value is to be determined experimentally. The important feature of the solution for the basic state is that its liquid–solid interface is intrinsically curved. Especially, as shown in Fig. 11.21, the slope of the liquid–solid interface near the triple point is noticeable (see Fig. 11.21a), even when the interface displacement itself is negligibly small (see Fig. 11.21b).

We find that due to the effect of the triple point,

$$\begin{aligned} h_{\mathrm{B}}(x, \epsilon) &\approx \epsilon^{\frac{1}{2}}\mathrm{h}_c(x, \epsilon) = \mathcal{O}(\epsilon^{\frac{1}{2}}), \\ h'_{\mathrm{B}}(x, \epsilon) &\approx \epsilon^{\frac{1}{2}}\mathrm{h}'_c(x, \epsilon) = \mathcal{O}(1), \quad (x \neq w_0), \end{aligned} \tag{11.107}$$

and

$$\begin{aligned} \tfrac{\partial C_{\mathrm{B}}}{\partial y}(x, 0) &\approx \epsilon\varpi\hat{\Theta}(x) = \mathcal{O}(\epsilon^{\frac{3}{2}}); \\ \tfrac{\partial^2 C_{\mathrm{B}}}{\partial y^2}(x, 0) &= -\tfrac{\partial^2 C_{\mathrm{B}}}{\partial x^2}(x, 0) = \mathcal{O}(\epsilon\varpi) = \mathcal{O}(\epsilon^{\frac{3}{2}}); \\ \tfrac{\partial^2 C_{\mathrm{B}}}{\partial x \partial y}(x, 0) &= \mathcal{O}(\epsilon\varpi) = \mathcal{O}(\epsilon^{\frac{3}{2}}). \end{aligned} \tag{11.108}$$

Recalling the definition of $\hat{\Theta}(x)$ in (11.33), one may write

$$\hat{\Theta}(x) = -\left(\tfrac{\partial\bar{C}_{10}}{\partial y} + \tfrac{\partial\bar{C}_{11}}{\partial y}\right), \quad 1 - \kappa = \epsilon^{\frac{1}{2}}\hat{\varpi}\hat{\Theta}(x).$$

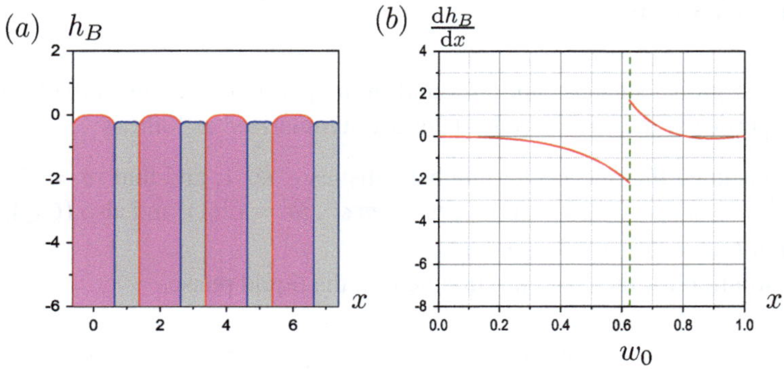


**Fig. 11.21** The features of the eutectic interface calculated with the steady global interface solution $h_c(x, \epsilon)$ for the typical case $\epsilon = 0.00555$, $\bar{\varphi} = 0$, $\varpi = 0.15$, $w_0 = 0.625$, $M_\alpha = 0.107$, $M_\beta = -0.218$, $\bar{\theta}^- = 1.14$, $\bar{\theta}^+ = 1.04$, $R_c = 0.84$, and $C_\infty = -1.8644 \times 10^{-2}$, $\hat{v} = 2.2 \times 10^{-8}$, $\hat{\beta} = 4.8774 \times 10^{-10}$. (**a**) The shape of the interface. (**b**) The slope of the steady interface shape function

## *11.8.3 Linear Perturbed States of Lamellar Eutectic Growth*

Let the unsteady state solutions be expressed in the following two parts:

$$\begin{aligned} C(x, y, t, \epsilon) &= C_B(x, y, \epsilon) + \tilde{C}(x, y, t, \epsilon), \\ h(x, t, \epsilon) &= h_B(x, \epsilon) + \tilde{h}(x, t, \epsilon). \end{aligned} \tag{11.109}$$

In the meanwhile, let the parameters $W_c$, $\theta^{\pm}$, and $\varphi$ be

$$\begin{aligned} W_c(t, \epsilon) &= \bar{W}_c + \tilde{W}_c(t, \epsilon), \\ \theta^-(t, \epsilon) &= \bar{\theta}^- + \tilde{\theta}^-(t, \epsilon), \\ \theta^+(t, \epsilon) &= \bar{\theta}^+ + \tilde{\theta}^+(t, \epsilon), \\ \varphi(t, \epsilon) &= \bar{\varphi} + \tilde{\varphi}(t, \epsilon). \end{aligned}$$

The first part above is the basic steady state of the system, while the second part comprises the perturbations around the basic state. The linear perturbed system is then derived as

$$\left(\frac{\partial^2}{\partial x^2} + \frac{\partial^2}{\partial y^2}\right)\tilde{C} + \epsilon\left(\frac{\partial \tilde{C}}{\partial y} - \frac{\partial \tilde{C}}{\partial t}\right) = 0, \tag{11.110}$$

with the following boundary conditions:

1. The far-field condition as $z \to \infty$,

$$\tilde{C} \sim 0, \tag{11.111}$$

2. The interface-tip conditions at $x = 0, \pm 1$,

   (a) for the symmetric S-modes:

$$\frac{\partial \tilde{C}}{\partial x} = \frac{\partial \tilde{h}}{\partial x} = 0; \tag{11.112}$$

   (b) for the antisymmetric A-modes:

$$\tilde{C} = \tilde{h} = 0. \tag{11.113}$$

3. The liquid–solid interface conditions: from (11.97) and (11.100), we have the following:

   (a) the Gibbs–Thomson condition:

$$\tilde{C} + \frac{\partial C_{\mathrm{B}}}{\partial y}\tilde{h} = -\epsilon \frac{\mathrm{G}}{M}\tilde{h} + \frac{\Gamma}{M}\frac{\mathrm{d}^2\tilde{h}}{\mathrm{d}x^2} + (\text{h.o.t.}). \tag{11.114}$$

   (b) the mass balance condition for the impurity:

$$\frac{\partial \tilde{C}}{\partial y} - h_{\mathrm{B}}'(x)\left(\frac{\partial \tilde{C}}{\partial x} + \frac{\partial^2 C_{\mathrm{B}}}{\partial x \partial y}\tilde{h}\right) + \frac{\partial^2 C_{\mathrm{B}}}{\partial y^2}\tilde{h}$$
$$+\epsilon(1-\kappa)\left(\tilde{C} + \frac{\partial \tilde{h}}{\partial t} + \tan\varphi\,\frac{\partial \tilde{h}}{\partial x}\right) = (\text{h.o.t.}). \tag{11.115}$$

4. The connection conditions at the triple point $x = W_c = w_0 + \tilde{w}_0$: From (11.101), we derive

$$\tilde{h}_\alpha(w_0) + \frac{\partial h_{\mathrm{B},\alpha}(w_0)}{\partial x}\tilde{w}_0 = \tilde{h}_\beta(w_0) + \frac{\partial h_{\mathrm{B},\beta}(w_0)}{\partial x}\tilde{w}_0; \tag{11.116}$$

$$\left[\frac{\partial \tilde{h}_\alpha}{\partial x}(w_0) + \frac{\partial^2 h_{\mathrm{B},\alpha}}{\partial x^2}(w_0)\tilde{w}_0\right]$$
$$= \left(\frac{\cos\bar{\theta}^+}{\cos\bar{\theta}^-}\right)^2\left[\frac{\partial \tilde{h}_\beta}{\partial x}(w_0) + \frac{\partial^2 h_{\mathrm{B},\beta}}{\partial x^2}(w_0)\tilde{w}_0\right]. \tag{11.117}$$

5. The total mass conservation condition: from (11.104), we derive that

$$\left[(\kappa_\alpha - \kappa_\beta) + \bar{C}_{\mathrm{B}}(\bar{W}_c, 0)\kappa_\alpha - \bar{C}_{\mathrm{B}}(\bar{W}_c, 0)\kappa_\beta\right]\tilde{W}_c$$
$$+ \int_0^{\bar{W}_c} \tilde{C}(x,0,t)\kappa_\alpha \mathrm{d}x + \int_{\bar{W}_c}^1 \tilde{C}(x,0,t)\kappa_\beta \mathrm{d}x = 0, \tag{11.118}$$

or

$$(\kappa_\alpha - \kappa_\beta)\Big[1 + \bar{C}_B(w_0, 0)\Big]\tilde{w}_0 + \int_0^1 \tilde{C}(x, 0, t)\kappa(x)\mathrm{d}x$$
$$= \mathcal{O}(w_0). \tag{11.119}$$

It will be shown later that $\tilde{C} = \mathcal{O}(\epsilon\tilde{h})$, whence $\tilde{w}_0 = \mathcal{O}(\epsilon\varpi^{-1}\tilde{h}) = \mathcal{O}(\epsilon^{\frac{1}{2}}\tilde{h})$.

One needs to solve the linear perturbed system (11.110)–(11.119) in the subregions $(0 < x < w_0)$ and $(w_0 < x < 1)$, separately. The asymptotic form of the solutions for the perturbed states will be derived in the limiting process $\epsilon \to 0$. Note that the basic steady state contained in the perturbed system depends on the small parameter $\epsilon$ in a very complicated manner. As a simplification, we assume that in the above limiting process $\epsilon \to 0$, *the small parameter* $\epsilon$ *in the basic state is "frozen"* as a constant.

### 11.8.4 Multiple Variables Expansion Form of the Perturbed System

It is presumed that the perturbed state involves multiple length scales. Hence as before, one may adopt the *multiple variables expansion (MVE) method* to find the solutions for the perturbed states. Let the variables $(x, y, t)$ be the slow variables and define the following fast variables:

$$\begin{aligned} t_+ &= \frac{t}{\epsilon^\alpha}, \\ x_+ &= \frac{\tilde{\Phi}(x,y)}{\epsilon^\alpha} = \frac{1}{\epsilon^\alpha}\int_{x_0}^{x} \tilde{k}(x_1, y, \epsilon)\mathrm{d}x_1, \\ y_+ &= \frac{\tilde{\Psi}(x,y)}{\epsilon^\alpha} = \frac{1}{\epsilon^\alpha}\int_0^{z} \tilde{k}(x, y_1, \epsilon)\mathrm{d}y_1. \end{aligned} \tag{11.120}$$

The exponent $\alpha > 0$ is to be determined later, and it will be shown that one must set $\alpha = 1/4$. The slowly varying wave number function $\tilde{k}(x, y, \epsilon)$ is to be determined; the lower limit of the integrals is set as $x_0 = w_0$, so that the continuity of the variable $x_+$ at the triple point is guaranteed. We deduce that

$$\tilde{\Phi}_x = \tilde{\Psi}_y = \tilde{k}(x, y, \epsilon), \quad \tilde{\Phi}(x_0, y, \epsilon) = 0, \quad \tilde{\Psi}(x, 0, \epsilon) = 0.$$

As before, let the pair of functions $(\tilde{\Phi}, \tilde{\Psi})$ be subject to the Cauchy–Riemann conditions and

$$\tilde{\Phi}_y = -\tilde{\Psi}_x = \tilde{g}(x, y), \quad \tilde{g}(x, 0) = \tilde{g}(x_0, y) = 0.$$

As a consequence, one has

$$\tilde{g}_x = \tilde{k}_y, \quad \tilde{g}_z = -\tilde{k}_x.$$

Let us formally treat the variables $(x, y, x_+, y_+, t_+)$ as independent variables and express the perturbed states in the following multiple variables form:

$$\begin{aligned} \tilde{C}(x, y, t, \epsilon) &= \tilde{\mathcal{C}}(x, y, x_+, y_+, t_+, \epsilon), \\ \tilde{h}(x, t, \epsilon) &= \tilde{\mathrm{h}}(x, x_+, t_+, \epsilon). \end{aligned} \tag{11.121}$$

By applying the following replacements of the derivatives in (11.110)–(11.118),

$$\begin{aligned} \frac{\partial}{\partial x} &\Longleftrightarrow \frac{\partial}{\partial x} + \frac{\tilde{k}}{\epsilon^\alpha}\frac{\partial}{\partial x_+}, \\ \frac{\partial}{\partial y} &\Longleftrightarrow \frac{\partial}{\partial y} + \frac{\tilde{k}}{\epsilon^\alpha}\frac{\partial}{\partial y_+}, \\ \frac{\partial}{\partial t} &\Longleftrightarrow \frac{\partial}{\partial t} + \frac{1}{\epsilon^\alpha}\frac{\partial}{\partial t_+}, \\ \frac{\partial^2}{\partial x^2} &\Longleftrightarrow \left[\frac{\partial}{\partial x} + \frac{\tilde{k}}{\epsilon^\alpha}\frac{\partial}{\partial x_+}\right]^2, \\ \frac{\partial^2}{\partial y^2} &\Longleftrightarrow \left[\frac{\partial}{\partial y} + \frac{\tilde{k}}{\epsilon^\alpha}\frac{\partial}{\partial y_+}\right]^2, \end{aligned}$$

the perturbed system (11.110)–(11.118) can be converted into the following system with the multiple independent variables $(x, y, x_+, y_+, t_+)$:

$$\begin{aligned} \tilde{k}^2\left(\frac{\partial^2}{\partial x_+^2} + \frac{\partial^2}{\partial y_+^2}\right)\tilde{\mathcal{C}} &= -2\epsilon^\alpha \tilde{k}\frac{\partial^2 \tilde{\mathcal{C}}}{\partial x \partial x_+} - 2\epsilon^\alpha \tilde{k}\frac{\partial^2 \tilde{\mathcal{C}}}{\partial y \partial y_+} \\ &\quad + \epsilon^{1+\alpha}\left(\frac{\partial \tilde{\mathcal{C}}}{\partial t_+} - \tilde{k}\frac{\partial \tilde{\mathcal{C}}}{\partial y_+}\right) + \text{(h.o.t.)}, \end{aligned} \tag{11.122}$$

with the following boundary conditions:

1 As $y_+ \to \infty$,

$$\tilde{\mathcal{C}} \sim 0. \tag{11.123}$$

2 At the interface-tips: $x = 0, 1$,

$$\left(\frac{\partial}{\partial x} + \frac{\tilde{k}}{\epsilon^\alpha}\frac{\partial}{\partial x_+}\right)\tilde{\mathcal{C}} = \left(\frac{\partial}{\partial x} + \frac{\tilde{k}}{\epsilon^\alpha}\frac{\partial}{\partial x_+}\right)\tilde{\mathrm{h}} = 0,$$

or

$$\tilde{\mathcal{C}} = \tilde{\mathrm{h}} = 0. \tag{11.124}$$

3 At the interface $z = y_+ = 0$, we define $\hat{k}(x,\epsilon) = \tilde{k}(x,0,\epsilon)$ and have

(a) the Gibbs–Thomson condition:

$$\tilde{\mathcal{C}} - \epsilon^{\frac{3}{2}}\hat{\varpi}\hat{\Theta}(x)\tilde{\mathrm{h}} = -\epsilon\frac{\mathrm{G}}{M}\tilde{\mathrm{h}} + \frac{\Gamma\hat{k}^2}{\epsilon^{2\alpha}M}\frac{\mathrm{d}^2\tilde{\mathrm{h}}}{\mathrm{d}x_+^2} + (\text{h.o.t.}), \tag{11.125}$$

(b) the mass balance condition for the impurity: noting that $(1-\kappa) = \epsilon^{1/2}\hat{\varpi}\hat{\Theta}(x)$, one may write

$$\hat{k}\frac{\partial\tilde{\mathcal{C}}}{\partial y_+} - \hat{k}h'_{\mathrm{B}}(x)\frac{\partial\tilde{\mathcal{C}}}{\partial x_+} + \epsilon^{\frac{3}{2}}\hat{\Theta}(x)\Big(\frac{\partial\tilde{\mathrm{h}}}{\partial t_+} + \hat{k}\tan\varphi\,\frac{\partial\tilde{\mathrm{h}}}{\partial x_+}\Big)$$
$$= (\text{h.o.t.}). \tag{11.126}$$

Let the perturbed states be expanded in the following asymptotic form in the limit $\epsilon \to 0$:

$$\begin{aligned}
\tilde{\mathcal{C}}(x,y,x_+,y_+,t_+,\epsilon) &= \epsilon^{\frac{3}{2}}\hat{\varpi}\left\{\tilde{\mathcal{C}}_0(x,y,x_+,y_+) + \epsilon^{\alpha}\tilde{\mathcal{C}}_1(\bullet) + \cdots\right\}\mathrm{e}^{\sigma t_+},\\
\tilde{\mathrm{h}}(x,x_+,t_+,\epsilon) &= \left\{\tilde{\mathrm{h}}_0(x,x_+) + \epsilon^{\alpha}\tilde{\mathrm{h}}_1(x,x_+) + \cdots\right\}\mathrm{e}^{\sigma t_+},\\
\tilde{W}_c(t_+,\epsilon) &= \left\{\tilde{w}_0 + \epsilon^{\alpha}\tilde{w}_1 + \cdots\right\}\mathrm{e}^{\sigma t_+},
\end{aligned} \tag{11.127}$$

and

$$\begin{aligned}
\tilde{k}(x,y,\epsilon) &= \tilde{k}_0(x,y) + \epsilon^{\alpha}\tilde{k}_1(x,y) + \cdots,\\
\hat{k}(x,\epsilon) &= \hat{k}_0(x) + \epsilon^{\alpha}\hat{k}_1(x) + \cdots,\\
\tilde{g}(x,y,\epsilon) &= \tilde{k}_0(x,y) + \epsilon^{\alpha}\tilde{g}_1(x,y) + \cdots,\\
\sigma(\epsilon) &= \sigma_0 + \epsilon^{\alpha}\sigma_1 + \cdots.
\end{aligned} \tag{11.128}$$

By substituting (11.128) into (11.122)–(11.126), one may obtain successively each higher-order approximate system.

### *11.8.5 Leading-Order Approximation*

In the leading-order approximation, to balance both sides of the interface condition (11.126), one derives that $\tilde{\mathcal{C}} = \mathcal{O}(\epsilon^{\frac{3}{2}}\tilde{\mathrm{h}})$. Furthermore, to balance both sides of the Gibbs–Thomson condition (11.125), one must set $\alpha = \frac{1}{4}$ and $\mathrm{G} = O(\epsilon^{1/2})$. We define

$$\Gamma = (\epsilon^2\hat{\varpi})\hat{\Gamma}, \quad \mathrm{G} = \epsilon^{\frac{1}{2}}\hat{\varpi}\hat{\mathrm{G}},$$

where $\hat{\Gamma} = \bar{\Gamma}/\hat{\varpi} = \mathcal{O}(1)$ and $\hat{G} = \mathcal{O}(1)$. The leading-order approximate system is written as

$$\left(\frac{\partial^2}{\partial x_+^2} + \frac{\partial^2}{\partial y_+^2}\right)\tilde{\mathcal{C}}_0 = 0, \tag{11.129}$$

with the following boundary conditions:

1 As $z \to \infty$,

$$\tilde{\mathcal{C}}_0 \sim 0. \tag{11.130}$$

2 The smoothness condition at the interface-tips of the solid phases: $x = 0, 1$:

(a) for the symmetric S-modes:

$$\frac{\partial \tilde{\mathcal{C}}_0}{\partial x_+} = \frac{\partial \tilde{h}_0}{\partial x_+} = 0; \tag{11.131}$$

(b) for the antisymmetric A-modes:

$$\tilde{\mathcal{C}}_0 = \tilde{h}_0 = 0. \tag{11.132}$$

3 The liquid–solid interface conditions at the interface $y = 0$: letting $\hat{k}_0(x) = \tilde{k}_0(x, 0)$, we derive

(a) the Gibbs–Thomson condition:

$$\tilde{\mathcal{C}}_0 - \hat{\Theta}(x)\tilde{h}_0 = -\frac{\hat{G}}{M}\tilde{h}_0 + \frac{\hat{\Gamma}\hat{k}_0^2}{M}\frac{d^2\tilde{h}_0}{dx_+^2}, \tag{11.133}$$

(b) the mass balance condition:

$$\hat{k}_0\frac{\partial \tilde{\mathcal{C}}_0}{\partial y_+} - \hat{k}_0 h_B'(x)\frac{\partial \tilde{\mathcal{C}}_0}{\partial x_+} + \hat{\Theta}(x)\left(\sigma_0\tilde{h}_0 + \hat{k}_0 \tan\bar{\varphi}\,\frac{\partial \tilde{h}_0}{\partial x_+}\right) = 0. \tag{11.134}$$

It is important to note that the second term, giving the function $h_B'(x)$ on the left-hand side of (11.134), describes the effect of the curved front of the basic state. Such a term will be missed under the assumption of a flat or nearly flat interface of the basic state.

4 The connection conditions at the triple point $x = w_0$:

$$\tilde{h}_{0,\alpha} = \tilde{h}_{0,\beta}; \quad \hat{k}_{0\alpha}\frac{\partial \tilde{h}_{0,\alpha}}{\partial x_+} = \hat{k}_{0\beta}\left(\frac{\cos\bar{\theta}^+}{\cos\bar{\theta}^-}\right)^2 \frac{\partial \tilde{h}_{0,\beta}}{\partial x_+}, \tag{11.135}$$

in which the constant $\left(\cos\bar{\theta}^+/\cos\bar{\theta}^-\right)^2$ describes the effect of the contact angles at the triple point on the perturbed states.

The above system can be solved analytically, since the slow variables $(x, y)$ in the system can are formally treated as constants. To proceed, we can consider the local normal modes solutions by following the approach developed in the previous chapters:

$$\begin{aligned}\tilde{\mathcal{C}}_0(x, y, x_+, y_+) &= \tilde{A}_0(x, y)\mathrm{e}^{\mathrm{i}x_+ - y_+}, \\ \tilde{h}_0(x) &= \hat{D}_0\mathrm{e}^{\mathrm{i}\bar{x}_+}.\end{aligned} \tag{11.136}$$

Let $\hat{A}(x) = \tilde{A}(x, 0)$. Here $\hat{A}$ is a piecewise continuous function, while $\hat{D}$ is piecewise constant. From the interface conditions (11.133) and (11.134), one may derive the local dispersion relationship. However, one may also take another way to obtain the same results, as described below. Define the complex variable $\zeta = (x_+ + \mathrm{i}y_+)$ and write the complex solutions for the concentration field in Fourier series form:

$$\tilde{\mathcal{C}}_0(\zeta) = \sum_{n=0}^{\infty} \tilde{d}_n \mathrm{e}^{\mathrm{i}n\pi\zeta}. \tag{11.137}$$

To satisfy the far-field condition $y_+ \to \infty, \tilde{\mathcal{C}}_0(\zeta) \to 0$, we require that the wave number function $\tilde{k}_0(x, y)$ contained in the fast variables $x_+, y_+$ be subject to the condition $\mathrm{Re}\{\tilde{k}_0(x, y)\} > 0$. From (11.137), we deduce that

$$\frac{\partial \tilde{\mathcal{C}}_0}{\partial x_+} = \tilde{\mathcal{C}}_0'(\zeta), \quad \frac{\partial \tilde{\mathcal{C}}_0}{\partial y_+} = \mathrm{i}\tilde{\mathcal{C}}_0'(\zeta) = \mathrm{i}\frac{\partial \tilde{\mathcal{C}}_0}{\partial x_+}.$$

Letting $y_+ = 0$ in the above formula, (11.133) and (11.134) can be rewritten as

$$\frac{\partial \tilde{\mathcal{C}}_0}{\partial x_+} = \hat{\Theta}(x)\frac{\mathrm{d}\tilde{h}_0}{\mathrm{d}x_+} - \frac{\hat{\mathrm{G}}}{M}\frac{\mathrm{d}\tilde{h}_0}{\mathrm{d}x_+} + \frac{\hat{\Gamma}\hat{k}_0^2}{M}\frac{\mathrm{d}^3\tilde{h}_0}{\mathrm{d}x_+^3} \quad (x \neq w_0), \tag{11.138}$$

and

$$\hat{k}_0\Big[\mathrm{i} - h_{\mathrm{B}}'(x)\Big]\frac{\partial \tilde{\mathcal{C}}_0}{\partial x_+} + \hat{\Theta}(x)\Big(\sigma_0\tilde{h}_0 + \hat{k}_0\tan\bar{\varphi}\,\frac{\partial \tilde{h}_0}{\partial x_+}\Big) = 0 \quad (x \neq w_0), \tag{11.139}$$

respectively. By combining (11.138) with (11.139), it follows that

$$\hat{k}_0\Big[\mathrm{i}-h_{\mathrm{B}}'(x)\Big]\left[\hat{\Theta}(x)\frac{\mathrm{d}\tilde{\mathrm{h}}_0}{\mathrm{d}x_+}-\frac{\hat{\mathrm{G}}}{M}\frac{\mathrm{d}\tilde{\mathrm{h}}_0}{\mathrm{d}x_+}+\frac{\hat{\Gamma}\hat{k}_0^2}{M}\frac{\mathrm{d}^3\tilde{\mathrm{h}}_0}{\mathrm{d}x_+^3}\right]$$
$$+\hat{\Theta}(x)\Big(\sigma_0\tilde{\mathrm{h}}_0+\hat{k}_0\tan\bar{\varphi}\,\frac{\partial\tilde{\mathrm{h}}_0}{\partial x_+}\Big)=0,\quad (x\neq w_0), \qquad (11.140)$$

or

$$\frac{\hat{\Gamma}\hat{k}_0^3}{M}\frac{\mathrm{d}^3\tilde{\mathrm{h}}_0}{\mathrm{d}x_+^3}-\hat{k}_0\left[\frac{\hat{\mathrm{G}}}{M}-\hat{\Theta}(x)\right]\frac{\mathrm{d}\tilde{\mathrm{h}}_0}{\mathrm{d}x_+}$$
$$+\frac{\hat{\Theta}(x)}{\mathrm{i}-h_{\mathrm{B}}'(x)}\Big(\sigma_0\tilde{\mathrm{h}}_0+\hat{k}_0\tan\bar{\varphi}\,\frac{\partial\tilde{\mathrm{h}}_0}{\partial x_+}\Big)=0\quad (x\neq w_0). \qquad (11.141)$$

This governing equation for the interface shape function allows the normal solution $\tilde{\mathrm{h}}_0(x)=\hat{D}_0\mathrm{e}^{\mathrm{i}x_+}$, which leads to the following local dispersion relationship:

$$\hat{\Gamma}\hat{k}_0^3+\hat{k}_0\left[\hat{\mathrm{G}}-M\hat{\Theta}(x)\right]+\frac{M\hat{\Theta}(x)}{\left[1+\mathrm{i}h_{\mathrm{B}}'(x)\right]}(\sigma_0+\mathrm{i}\hat{k}_0\tan\bar{\varphi})=0$$
$$(x\neq w_0). \qquad (11.142)$$

Let us now approximate the basic state solution $h_{\mathrm{B}}'(x)$ with the composite solution $\mathrm{h}_c'(x)$ in terms of the formula (11.107). Then (11.142) may be rewritten in the form

$$\sigma_0=\Sigma(\hat{k}_0,x)=\Lambda_0(x)\hat{k}_0\Big[M\hat{\Theta}(x)-\hat{\mathrm{G}}-\hat{\Gamma}\hat{k}_0^2\Big]-\mathrm{i}\hat{k}_0\tan\bar{\varphi}$$
$$=\Lambda_0(x)\hat{k}_0\left[M\hat{\Theta}(x)-\hat{\mathrm{G}}-\frac{\mathrm{i}\tan\bar{\varphi}}{\Lambda_0(x)}-\hat{\Gamma}\hat{k}_0^2\right], \qquad (11.143)$$

where we have defined the piecewise smooth function

$$\Lambda_0(x)=\frac{\left[1+\mathrm{i}\mathrm{h}_c'(x)\right]}{M\hat{\Theta}(x)}. \qquad (11.144)$$

The form of (11.143) is formally similar to the cubic dispersion formulas of the classic Mullins–Sekerka instability of the planar interface, as well as those encountered in the systems of dendritic cellular growth and viscous fingering formation. The term $-\hat{\Gamma}\hat{k}_0^2$ represents the effect of surface tension, which stabilizes the system in the short-wave region; the term $-\hat{\mathrm{G}}$ represents the effect of the temperature gradient at the interface, which stabilizes the system, since in the system of directional solidification, the parameter $\hat{\mathrm{G}}$ is always positive. Finally, the term $M\hat{\Theta}(x)$ describes the effect of the gradient of the concentration on the liquid side at the interface, which is positive and destabilizes the system. This term may be considered the constitutional undercooling effect.

Let us recall the *constitutional instability criterion* imposed by Rutter and Chalmer in 1955 for the system of cellular growth in the unidirectional solidification of a dilute binary alloy (2.191). It states that for $\hat{\Delta} = \frac{1-\kappa}{\kappa}M - \mathrm{G} < 0$, the flat liquid–solid interface is stable, while when the parameter $\hat{\Delta}$ is positive, the flat liquid–solid interface is unstable. Certainly, the constitutional instability criterion is not applicable to the system of eutectic growth under study, due to the fact that the current system is operating near the eutectic point and has two different curved liquid–solid interfaces: the $\alpha$-interface and the $\beta$-interface. However, to study the stability properties of the eutectic system, one may introduce a "modified" constitutional instability parameter $\Delta$ for the present system that is a piecewise smooth function defined as

$$\Delta = \begin{cases} \Delta_\alpha = M_\alpha \hat{\Theta}_\alpha - \hat{\mathrm{G}} & (0 < x < w_0), \\ \Delta_\beta = M_\beta \hat{\Theta}_\beta - \hat{\mathrm{G}}, & (w_0 < x < 1). \end{cases}$$

It can be seen that for the special cases of $w_0 \to 1$ (or 0) in which the $\beta$-interface (or the $\alpha$-interface) disappears, the parameter $\Delta_\alpha$ (or $\Delta_\beta$) reduces to $\hat{\Delta}$. By adopting the terminology of the constitutional criterion to the current eutectic system, we say that the interface with $\Delta = M\hat{\Theta}(x) - \hat{\mathrm{G}} < 0$ is $\mathcal{C}$*-stable*, and that the interface with $\Delta = M\hat{\Theta}(x) - \hat{\mathrm{G}} > 0$ is $\mathcal{C}$*-unstable*. As a consequence, one may distinguish the following four cases for the eutectic system:

(a) $\{\Delta_\alpha > 0, \Delta_\beta > 0\}$, in which both the $\alpha$- and $\beta$-interfaces are $\mathcal{C}$-unstable.
(b) $\{\Delta_\alpha < 0, \Delta_\beta < 0\}$, in which both the $\alpha$- and $\beta$-interfaces are $\mathcal{C}$-stable.
(c) $\{\Delta_\alpha > 0, \Delta_\beta < 0\}$, in which the $\alpha$-interface is $\mathcal{C}$-unstable, while the $\beta$-interface is $\mathcal{C}$-stable.
(d) $\{\Delta_\alpha < 0, \Delta_\beta > 0\}$, in which the $\alpha$-interface is $\mathcal{C}$-stable, while the $\beta$-interface is $\mathcal{C}$-unstable.

For a given complex number $\sigma_0$, from the local dispersion relationship (11.143), one can find the three roots $\{\hat{k}_0^{(1)}(x), \hat{k}_0^{(2)}(x), \hat{k}_0^{(3)}(x)\}$ as

$$\begin{cases} \hat{k}_0^{(1)}(x) = \widehat{M}\cos\Big[\frac{1}{3}\cos^{-1}\big(\frac{\sigma_0}{\widehat{N}\Lambda_0}\big)\Big] & \text{(short-wave (SW) branch)}, \\ \hat{k}_0^{(2)}(x) = \widehat{M}\cos\Big[\frac{1}{3}\cos^{-1}\big(\frac{\sigma_0}{\widehat{N}\Lambda_0}\big) + \frac{2\pi}{3}\Big], & \\ \hat{k}_0^{(3)}(x) = \widehat{M}\cos\Big[\frac{1}{3}\cos^{-1}\big(\frac{\sigma_0}{\widehat{N}\Lambda_0}\big) + \frac{4\pi}{3}\Big] & \text{(long-wave (LW) branch)}. \end{cases} \tag{11.145}$$

Here we have defined the following piecewise smooth functions,

$$\begin{aligned} \widehat{M} &= \sqrt{\frac{4}{3\hat{\Gamma}}}\Big[M\hat{\Theta}(x) - \hat{\mathrm{G}} - \frac{\mathrm{i}\tan\bar{\varphi}}{\Lambda_0(x)}\Big]^{\frac{1}{2}}, \\ \widehat{N} &= -\sqrt{\frac{4}{27\hat{\Gamma}}}\Big[M\hat{\Theta}(x) - \hat{\mathrm{G}} - \frac{\mathrm{i}\tan\bar{\varphi}}{\Lambda_0(x)}\Big]^{\frac{3}{2}}, \end{aligned} \tag{11.146}$$

and have designated that $\mathrm{Re}\{\hat{k}_0^{(1)}\} \geq \mathrm{Re}\{\hat{k}_0^{(3)}\} \geq 0.$ and $\mathrm{Re}\{\hat{k}_0^{(2)}\} < 0$.

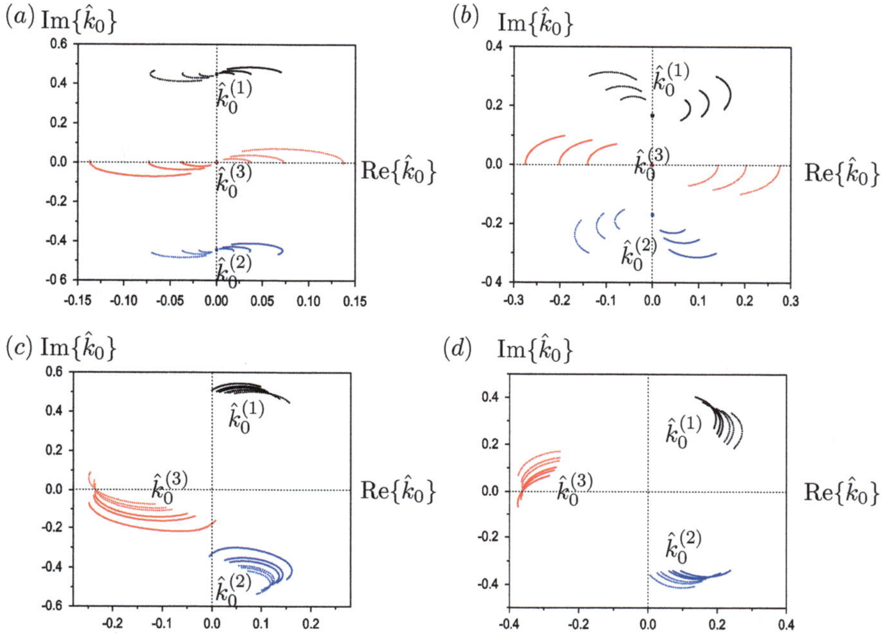


**Fig. 11.22** Figures showing the local dispersion relationship for typical cases of $\mathcal{C}$-stable $\alpha$- and $\beta$-interfaces, $\{\Delta_\alpha = M_\alpha\hat{\Theta}(x) - \hat{\mathrm{G}} < 0;\ \Delta_\beta = M_\beta\hat{\Theta}(x) - \hat{\mathrm{G}} < 0\}$ with $\varpi = 0.15$, $w_0 = 0.6$, $\epsilon = 0.01$, $\bar{\Gamma}_\alpha = 1.0$, $\bar{\Gamma}_\beta = 1.5$, $\hat{\kappa} = 1.0$, $M_\alpha = 2.0$, $\bar{\varphi} = 0$, $M_\beta = -3.0, C_\infty = -0.03, G = 0.5$, $\theta_\alpha = 155.3°$, $\theta_\beta = 149.5°$. (**a**) Typical dispersion curves for the $\alpha$-interface, with the real eigenvalues $\sigma_0 = -0.1, -0.05, -0.025, 0.0, 0.025, 0.05, 0.1$. (**b**) Typical dispersion curves for the $\beta$-interface, with the real eigenvalues $\sigma_0 = -0.1, -0.05, -0.025, 0.0, 0.025, 0.05, 0.1$. (**c**) Typical dispersion curves for the $\alpha$-interface, with the complex eigenvalues $\sigma_0 = 0.2 + \mathrm{i}\omega$, where $\omega = -0.1, -0.05, -0.025, 0.025, 0.05, 0.1$. (**d**) Typical dispersion curves for the $\beta$-interface, with the complex eigenvalues $\sigma_0 = 0.2 + \mathrm{i}\omega$, where $\omega = -0.1, -0.05, -0.025, 0.025, 0.05, 0.1$. In the figures, the *black line* describes $\hat{k}_0^{(1)}$, the *blue line* describes $\hat{k}_0^{(3)}$, the *red line* describes $\hat{k}_0^{(2)}$. The *dotted lines* describe the cases $\sigma_0 < 0$ in (**a**), (**b**) and describe the cases $\omega < 0$ in (**c**), (**d**)

For the case $\Delta = M\hat{\Theta}(x) - \hat{\mathrm{G}} < 0$ with the $\alpha$-interface and $\beta$-interface both $\mathcal{C}$-stable, the constants $\widehat{M}$ and $\widehat{N}$ in the formula (11.146) are imaginary numbers. The wave number functions $\hat{k}_0^{(i)}(x)$ $(i = 1, 2, 3)$ corresponding to a set of real numbers $\sigma_0$ are shown in the complex $\hat{k}_0$-plane in Fig. 11.22a, b, respectively, whereas the wave number functions $\hat{k}_0^{(i)}(x)$ $(i = 1, 2, 3)$ corresponding to a set of complex numbers $\sigma_0$ are respectively shown in Fig. 11.22c d.

Furthermore, for the case $\Delta = M\hat{\Theta}(x) - \hat{\mathrm{G}} > 0$ with the $\alpha$-interface and $\beta$-interface both $\mathcal{C}$-stable, the constants $\widehat{M}$ and $\widehat{N}$ in (11.146) are real numbers. The wave number functions $\hat{k}_0^{(i)}(x)$ $(i = 1, 2, 3)$ corresponding to a set of real numbers $\sigma_0$ are respectively shown in Fig. 11.23a, b, whereas the wave number functions $\hat{k}_0^{(i)}(x)$ $(i = 1, 2, 3)$ corresponding to a set of complex numbers $\sigma_0$ are respectively shown in Fig. 11.23c, d.

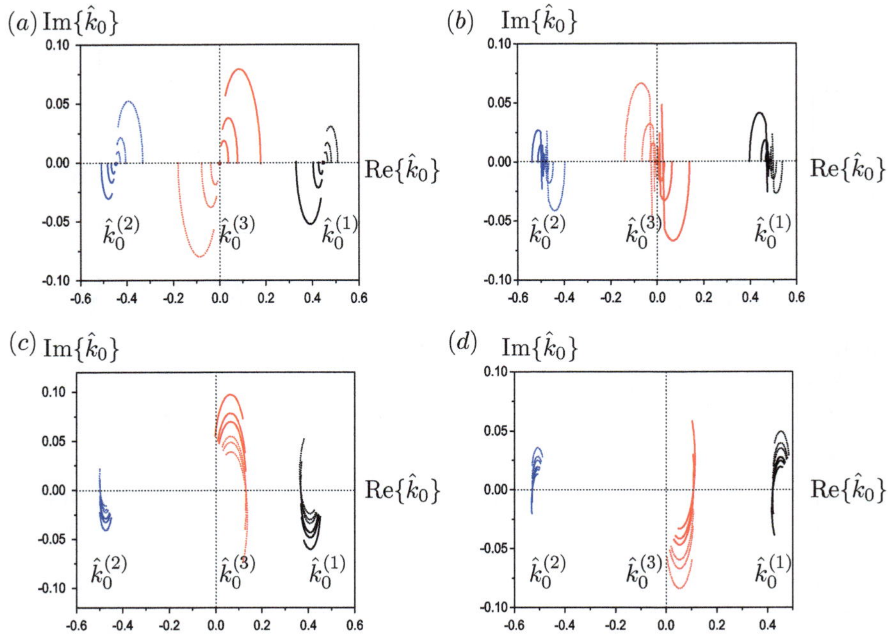


**Fig. 11.23** Figures showing local dispersion for typical cases of $\mathcal{C}$-unstable $\alpha$- and $\beta$-interfaces, $\{\Delta_\alpha = M_\alpha \hat{\Theta}(x) - \hat{\mathrm{G}} > 0;\ \Delta_\beta = M_\beta \hat{\Theta}(x) - \hat{\mathrm{G}} > 0\}$ with $\varpi = 0.15$, $w_0 = 0.6$, $\epsilon = 0.01$, $\bar{\Gamma}_\alpha = 1.0$, $\bar{\Gamma}_\beta = 1.5$, $\hat{\kappa} = 1.0$, $M_\alpha = 2.0$, $\bar{\varphi} = 0$, $\hat{\kappa} = 1.0$, $M_\beta = -3.0$, $C_\infty = -0.03$, $G = 0.1$, $\theta_\alpha = 155.3°$, $\theta_\beta = 149.5°$. (**a**) Typical dispersion curves for the $\alpha$-interface, with the real eigenvalues $\sigma_0 = -0.1, -0.05, -0.025, 0.0, 0.025, 0.05, 0.1$. (**b**) Typical dispersion curves for the $\beta$-interface, with the real eigenvalues $\sigma_0 = -0.1, -0.05, -0.025, 0.0, 0.025, 0.05, 0.1$. (**c**) Typical dispersion curves for the $\alpha$-interface, with the complex eigenvalues $\sigma_0 = 0.08 + \mathrm{i}\omega$, where $\omega = -0.04, -0.02, -0.01, 0.01, 0.02, 0.04$. (**d**) Typical dispersion curves for the $\beta$-interface, with the complex eigenvalues $\sigma_0 = 0.08 + \mathrm{i}\omega$, where $\omega = -0.04, -0.02, -0.01, 0.01, 0.02, 0.04$. In the figures, the *black line* describes $\hat{k}_0^{(1)}$, the *red* line describes $\hat{k}_0^{(3)}$, the *blue line* describes $\hat{k}_0^{(2)}$. The *dotted lines* describe the cases $\sigma_0 < 0$ in (**a**), (**b**) and describe the cases $\omega < 0$ in (**c**), (**d**)

The formula (11.145) cannot be applied in the case $\varpi M \hat{\Theta}(x) - \mathrm{G} = 0$, in which case $\widehat{M} = \widehat{N} = 0$ and $\hat{k}_0^{(1)}(x) = \hat{k}_0^{(2)}(x) = \hat{k}_0^{(3)}(x) = 0$.

From a stability analysis point of view, among the above four cases, case (a) will be of greatest interest to us.

The three wave number functions $\{\hat{k}_0^{(1)}(x), \hat{k}_0^{(2)}(x), \hat{k}_0^{(3)}(x)\}$ derived from the formula (11.145) respectively yield the three fundamental complex solutions

$$\tilde{\mathrm{h}}_0^{(1)}(x) = \tilde{D}_0^{(1)} \tilde{H}_1(x),$$
$$\tilde{\mathrm{h}}_0^{(2)}(x) = \tilde{D}_0^{(2)} \tilde{H}_2(x),$$
$$\tilde{\mathrm{h}}_0^{(3)}(x) = \tilde{D}_0^{(3)} \tilde{H}_3(x),$$

where

$$
\begin{aligned}
\tilde{H}_1(x) &= \exp\left\{\frac{\mathrm{i}}{\epsilon^{1/4}}\int_{w_0}^{x}\hat{k}_0^{(1)}(x_1)\mathrm{d}x_1\right\},\\
\tilde{H}_2(x) &= \exp\left\{\frac{\mathrm{i}}{\epsilon^{1/4}}\int_{w_0}^{x}\hat{k}_0^{(2)}(x_1)\mathrm{d}x_1\right\},\\
\tilde{H}_3(x) &= \exp\left\{\frac{\mathrm{i}}{\epsilon^{1/4}}\int_{w_0}^{x}\hat{k}_0^{(3)}(x_1)\mathrm{d}x_1\right\}.
\end{aligned}
\qquad (11.147)
$$

One sees from Figs. 11.22 and 11.23 that for all the unstable cases with $\mathrm{Re}\{\sigma_0\} > 0$, the two wave number functions $\{\hat{k}_0^{(i)}(x)\}$ $(i = 1, 3)$ have positive real parts, namely $\mathrm{Re}\{\hat{k}_0^{(i)}(x)\} > 0$ $(i = 1, 3)$, whereas the real part of the wave number function $\{\hat{k}_0^{(2)}(x)\}$ is negative, namely $\mathrm{Re}\{\hat{k}_0^{(2)}(x)\} < 0$. The fundamental solution $\tilde{\mathrm{h}}_0^{(2)}(x)$ corresponding to the wave number function $\{\hat{k}_0^{(2)}(x)\}$ must be ruled out, since the associated solution for the perturbed concentration field $\tilde{\mathcal{C}}_0(x_+ + \mathrm{i}y_+)$ will violate the far-field condition (11.130).

Thus, the local solutions for the perturbed interface shape $\tilde{\mathrm{h}}_0(x)$ can be expressed as

$$
\tilde{\mathrm{h}}_0(x) = \begin{cases}
\tilde{\mathrm{h}}_{0,\alpha}(x) = \tilde{D}_{0,\alpha}^{(1)}\exp\left\{\frac{\mathrm{i}}{\epsilon^{1/4}}\int_{w_0}^{x}\hat{k}_{0,\alpha}^{(1)}(x_1)\mathrm{d}x_1\right\}\\
\qquad +\tilde{D}_{0,\alpha}^{(3)}\exp\left\{\frac{\mathrm{i}}{\epsilon^{1/4}}\int_{w_0}^{x}\hat{k}_{0,\alpha}^{(3)}(x_1)\mathrm{d}x_1\right\}, \quad (0 < x < w_0);\\
\tilde{\mathrm{h}}_{0,\beta}(x) = \tilde{D}_{0,\beta}^{(1)}\exp\left\{\frac{\mathrm{i}}{\epsilon^{1/4}}\int_{w_0}^{x}\hat{k}_{0,\beta}^{(1)}(x_1)\mathrm{d}x_1\right\}\\
\qquad +\tilde{D}_{0,\beta}^{(3)}\exp\left\{\frac{\mathrm{i}}{\epsilon^{1/4}}\int_{w_0}^{x}\hat{k}_{0,\beta}^{(3)}(x_1)\mathrm{d}x_1\right\} \quad (w_0 < x < 1),
\end{cases}
\qquad (11.148)
$$

where $\tilde{D}_{0,\alpha}^{(1)}$, $\tilde{D}_{0,\alpha}^{(3)}$, $\tilde{D}_{0,\beta}^{(1)}$, and $\tilde{D}_{0,\beta}^{(3)}$ are arbitrary constants. The normal mode solutions (11.148) with $\sigma_0$ and $\hat{k}_0(x)$ satisfying the dispersion relationship satisfy the far-field condition and the interface conditions, but they may not satisfy the smoothness conditions of the interface-tip of the solid phase at $x = 0, 1$ or the connection conditions at the triple point. Associated with this situation, the coefficients and the eigenvalue $\sigma_0$ in (11.148) remain undetermined. Therefore, the solutions (11.148) are called the *local solutions*.

In the following sections, we shall first apply the interface-tip smoothness conditions and then apply the connection conditions at the triple point. The resultant solutions that satisfy **all** (not just some of) the boundary conditions are called the *global solutions*.

### *11.8.6 Perturbed State Solutions in the Subinterval $(0 \le x < w_0)$*

Let us apply the interface-tip conditions to the solutions $\tilde{h}_{0,\alpha}(x)$ of the perturbed state (11.148) in the subinterval $(0 \le x < w_0)$. The solutions $\tilde{h}_{0,\alpha}(x)$ may be either symmetric or antisymmetric about the tip $x = 0$.

(a) *Antisymmetric A-modes:*
The antisymmetric modes subject to the tip smoothness condition:

$$\tilde{h}_{0,\alpha}(0) = 0.$$

We obtain

$$\tilde{h}_{0,\alpha}(0) = \tilde{D}^{(1)}_{0,\alpha} \exp\left\{\frac{\mathrm{i}}{\epsilon^{1/4}} \int_{w_0}^{0} \hat{k}^{(1)}_{0,\alpha}(x_1)\mathrm{d}x_1\right\}$$
$$+\tilde{D}^{(3)}_{0,\alpha} \exp\left\{\frac{\mathrm{i}}{\epsilon^{1/4}} \int_{w_0}^{0} \hat{k}^{(3)}_{0,\alpha}(x_1)\mathrm{d}x_1\right\} = 0, \qquad (11.149)$$

$$\frac{\tilde{D}^{(1)}_{0,\alpha}}{\tilde{D}^{(3)}_{0,\alpha}} = -\exp\left\{\frac{\mathrm{i}}{\epsilon^{1/4}} \int_{0}^{w_0} \left[\hat{k}^{(1)}_{0,\alpha}(x_1) - \hat{k}^{(3)}_{0,\alpha}(x_1)\right]\mathrm{d}x_1\right\} = -\mathrm{e}^{\mathrm{i}\chi_\alpha},$$

where we have defined

$$\chi_\alpha = \frac{1}{\epsilon^{1/4}} \int_{0}^{w_0} \left[\hat{k}^{(1)}_{0,\alpha}(x_1) - \hat{k}^{(3)}_{0,\alpha}(x_1)\right]\mathrm{d}x_1.$$

Thus, one may write the local perturbed state solution in the subinterval $(0 \le x < w_0)$ as

$$\tilde{h}_{0,\alpha}(x) = \tilde{D}^{(1)}_{0,\alpha}\left[\tilde{H}_{1,\alpha}(x) - \mathrm{e}^{-\mathrm{i}\chi_\alpha}\tilde{H}_{3,\alpha}(x)\right]. \qquad (11.150)$$

(b) *Symmetric S-modes:*
The symmetric modes subject to the tip smoothness condition:

$$\tilde{h}'_{0,\alpha}(0) = 0.$$

We obtain

$$\hat{k}^{(1)}_{0,\alpha}(0)\tilde{D}^{(1)}_{0,\alpha} \exp\left\{\frac{\mathrm{i}}{\epsilon^{1/4}} \int_{w_0}^{0} \hat{k}^{(1)}_{0,\alpha}(x_1)\mathrm{d}x_1\right\}$$
$$+\hat{k}^{(3)}_{0,\alpha}(0)\tilde{D}^{(3)}_{0,\alpha} \exp\left\{\frac{\mathrm{i}}{\epsilon^{1/4}} \int_{w_0}^{0} \hat{k}^{(3)}_{0,\alpha}(x_1)\mathrm{d}x_1\right\} = 0, \qquad (11.151)$$

$$\frac{\tilde{D}^{(1)}_{0,\alpha}}{\tilde{D}^{(3)}_{0,\alpha}} = -\frac{\hat{k}^{(3)}_{0,\alpha}(0)}{\hat{k}^{(1)}_{0,\alpha}(0)} \exp\left\{\frac{\mathrm{i}}{\epsilon^{1/4}}\int_0^{w_0}\left[\hat{k}^{(1)}_{0,\alpha}(x_1) - \hat{k}^{(3)}_{0,\alpha}(x_1)\right]\mathrm{d}x_1\right\}$$

$$= -\mathrm{e}^{\mathrm{i}\chi_\alpha}\frac{\hat{k}^{(3)}_{0,\alpha}(0)}{\hat{k}^{(1)}_{0,\alpha}(0)}. \qquad (11.152)$$

Thus, one may write the local perturbed state solution in the subinterval $(0 < x < w_0)$ as

$$\tilde{\mathrm{h}}_{0,\alpha}(x) = \tilde{D}^{(1)}_{0,\alpha}\left[\tilde{H}_{1,\alpha}(x) - \mathrm{e}^{-\mathrm{i}\chi_\alpha}\frac{\hat{k}^{(1)}_{0,\alpha}(0)}{\hat{k}^{(3)}_{0,\alpha}(0)}\tilde{H}_{3,\alpha}(x)\right]. \qquad (11.153)$$

### *11.8.7 Perturbed State Solutions in the Subinterval ($w_0 < x \le 1$)*

Similarly, the local solution $\tilde{\mathrm{h}}_{0,\beta}(x)$ of the perturbed state (11.148) in the subregion $x \in (w_0, 1)$ is also either symmetric or antisymmetric about the tip $x = 1$.

(a) *Antisymmetric A-modes:*
The antisymmetric A-mode satisfies the tip smoothness condition

$$\tilde{\mathrm{h}}_{0,\beta}(1) = 0.$$

We obtain

$$\tilde{\mathrm{h}}_{0,\beta}(1) = \tilde{D}^{(1)}_{0,\beta}\exp\left\{\frac{\mathrm{i}}{\epsilon^{1/4}}\int_{w_0}^{1}\hat{k}^{(1)}_{0,\beta}(x_1)\mathrm{d}x_1\right\}$$
$$+\tilde{D}^{(3)}_{0,\beta}\exp\left\{\frac{\mathrm{i}}{\epsilon^{1/4}}\int_{w_0}^{1}\hat{k}^{(3)}_{0,\beta}(x_1)\mathrm{d}x_1\right\} = 0, \qquad (11.154)$$

so that

$$\frac{\tilde{D}^{(1)}_{0,\beta}}{\tilde{D}^{(3)}_{0,\beta}} = -\exp\left\{-\frac{\mathrm{i}}{\epsilon^{1/4}}\int_{w_0}^{1}\left[\hat{k}^{(1)}_{0,\beta}(x_1) - \hat{k}^{(3)}_{0,\beta}(x_1)\right]\mathrm{d}x_1\right\} = -\mathrm{e}^{\mathrm{i}\chi_\beta},$$

where we have defined

$$\chi_\beta = -\frac{1}{\epsilon^{1/4}}\int_{w_0}^{1}\left[\hat{k}^{(1)}_{0,\beta}(x_1) - \hat{k}^{(3)}_{0,\beta}(x_1)\right]\mathrm{d}x_1.$$

Thus, one may write the local perturbed state solution in the subinterval $(w_0 < x \leq 1)$ as

$$\tilde{\mathrm{h}}_{0,\beta}(x) = \tilde{D}^{(1)}_{0,\beta}\left[\tilde{H}_{1,\beta}(x) - \mathrm{e}^{-\mathrm{i}\chi_\beta}\tilde{H}_{3,\beta}(x)\right]. \tag{11.155}$$

(b) *Symmetric S-modes:*
The symmetric S-mode satisfies the tip smoothness condition

$$\tilde{\mathrm{h}}'_{0,\beta}(1) = 0.$$

We obtain

$$\hat{k}^{(1)}_{0,\beta}(1)\tilde{D}^{(1)}_{0,\beta}\exp\left\{\frac{\mathrm{i}}{\epsilon^{1/4}}\int_{w_0}^{1}\hat{k}^{(1)}_{0,\beta}(x_1)\mathrm{d}x_1\right\}$$
$$+\hat{k}^{(3)}_{0,\beta}(1)\tilde{D}^{(3)}_{0,\beta}\exp\left\{\frac{\mathrm{i}}{\epsilon^{1/4}}\int_{w_0}^{1}\hat{k}^{(3)}_{0,\beta}(x_1)\mathrm{d}x_1\right\} = 0, \tag{11.156}$$

so that

$$\begin{aligned}\frac{\tilde{D}^{(1)}_{0,\beta}}{\tilde{D}^{(3)}_{0,\beta}} &= -\frac{\hat{k}^{(3)}_{0,\beta}(1)}{\hat{k}^{(1)}_{0,\beta}(1)}\exp\left\{-\frac{\mathrm{i}}{\epsilon^{1/4}}\int_{w_0}^{1}\left[\hat{k}^{(1)}_{0,\beta}(x_1) - \hat{k}^{(3)}_{0,\beta}(x_1)\right]\mathrm{d}x_1\right\}\\ &= -\mathrm{e}^{\mathrm{i}\chi_\beta}\frac{\hat{k}^{(3)}_{0,\beta}(1)}{\hat{k}^{(1)}_{0,\beta}(1)}.\end{aligned} \tag{11.157}$$

Thus, one may write the local perturbed state solution in the subinterval $(w_0 < x \leq 1)$ as

$$\tilde{\mathrm{h}}_{0,\beta}(x) = \tilde{D}^{(1)}_{0,\beta}\left[\tilde{H}_{1,\beta}(x) - \mathrm{e}^{-\mathrm{i}\chi_\beta}\frac{\hat{k}^{(1)}_{0,\beta}(1)}{\hat{k}^{(3)}_{0,\beta}(1)}\tilde{H}_{3,\beta}(x)\right]. \tag{11.158}$$

There are two constants, $\tilde{D}^{(1)}_{0,\alpha}$ and $\tilde{D}^{(1)}_{0,\beta}$, that remain undetermined.

## 11.9 The Global Modes Solutions and Quantization Conditions

We now join the local solutions of the perturbed state in the intervals $(0 < x < w_0)$ and $(w_0 < x < 1)$ derived in the previous subsection at the triple point by applying the connection conditions (11.135). The resultant solutions are the eigenfunctions that satisfy all the boundary conditions. They are called the *global mode* solutions. The corresponding eigenvalues $\sigma_0$ are determined by the

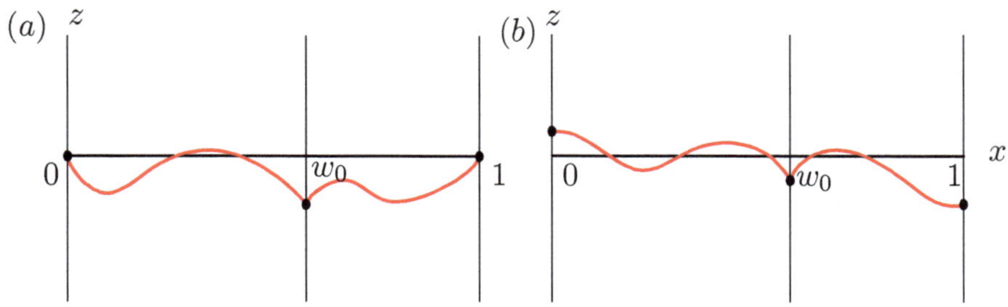


**Fig. 11.24** Analogy of the perturbed system to two vibrational systems connected to a mass: (**a**) antisymmetric mode; (**b**) symmetric mode

*quantization condition* obtained. All the coefficients retained in the global mode solutions are now determined except for one that remains free.

In some sense, the present perturbed system's behavior resembles a system of two springs (corresponding to the $\alpha$-interface and $\beta$-interface) connected with a mass at the junction point (corresponding to the triple point) as sketched in Fig. 11.24. In looking at these figures, keep in mind that the local perturbed state solution may be either the antisymmetric A-mode or the symmetric S-mode in the subintervals $(0, w_0)$ and $(w_0, 1)$. *Thus, via different combinations, the system will allow four types of global modes of perturbed states in a unit lamella* $(0 < x < 1)$. Namely, the (AA)-mode represents an A-mode of the $\alpha$-interface joined with an A-mode of the $\beta$-interface; the (SS)-mode represents an S-mode of the $\alpha$-interface joined with an S-mode of the $\beta$-interface; the (AS)-mode represents an A-mode of the $\alpha$-interface joined with an S-mode of the $\beta$-interface; and the (SA)-mode represents an S-mode of the $\alpha$-interface joined with an A-mode of the $\beta$-interface.

The whole eutectic pattern is obtained by extending the above mode solutions periodically from one unit lamella $(0 < x < 1)$ to $(-\infty < x < \infty)$. This yields the following types of patterns:

(1) $\{\cdots(AA)(AA)(AA)(AA)\cdots\}$, which has period $T = 1$;
(2) $\{\cdots(SS)(SS)(SS)(SS)\cdots\}$, which has period $T = 1$;
(3) $\{\cdots(AS)\overline{(AS)}(AS)\overline{(AS)}\cdots\}$, which has period $T = 2$;
(4) $\{\cdots(SA)\overline{(SA)}(SA)\overline{(SA)}\cdots\}$, which has period $T = 2$,

as sketched in Fig. 11.25. In the above, the notation $\overline{(AS)}$ represents the unit of the lamella created by the perturbed mode, $-(AS)$. Similarly, the notation $\overline{(SA)}$ represents the unit of the lamella created by the perturbed mode, $-(SA)$.

Associated with each of the above patterns, one may derive the global modes and corresponding quantization conditions:

**(1) Global (AA)-modes:**
By joining the A-mode of the perturbed state (11.150) on the subinterval $(0 < x < w_0)$ with the A-mode (11.155) on the subinterval $(w_0 < x < 1)$ in terms of

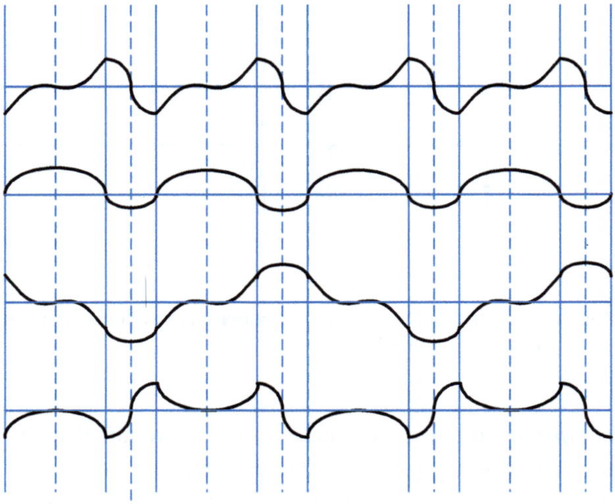

**Fig. 11.25** Sketch of the eutectic pattern types (1), (2), (3), (4) produced by the (AA)-, (SS)-, (AS)-, and (SA)-modes, respectively

the connection conditions (11.135), we obtain

$$\tilde{D}^{(1)}_{0,\alpha}\left[1-\mathrm{e}^{-\mathrm{i}\chi_\alpha}\right]=\tilde{D}^{(1)}_{0,\beta}\left[1-\mathrm{e}^{-\mathrm{i}\chi_\beta}\right], \tag{11.159}$$

and

$$\tilde{D}^{(1)}_{0,\alpha}\left[\hat{k}^{(1)}_{0,\alpha}(w_0)-\mathrm{e}^{-\mathrm{i}\chi_\alpha}\hat{k}^{(3)}_{0,\alpha}(w_0)\right]$$
$$=\left(\frac{\cos\bar{\theta}+}{\cos\bar{\theta}-}\right)^2\tilde{D}^{(1)}_{0,\beta}\left[\hat{k}^{(1)}_{0,\beta}(w_0)-\mathrm{e}^{-\mathrm{i}\chi_\beta}\hat{k}^{(3)}_{0,\beta}(w_0)\right]. \tag{11.160}$$

By combining (11.159) with (11.160), we obtain the quantization condition

$$\frac{\left[1-\mathrm{e}^{-\mathrm{i}\chi_\alpha}\right]}{\left[1-\mathrm{e}^{-\mathrm{i}\chi_\beta}\right]}=\left(\frac{\cos\bar{\theta}-}{\cos\bar{\theta}+}\right)^2\frac{\left[\hat{k}^{(1)}_{0,\alpha}(w_0)-\mathrm{e}^{-\mathrm{i}\chi_\alpha}\hat{k}^{(3)}_{0,\alpha}(w_0)\right]}{\left[\hat{k}^{(1)}_{0,\beta}(w_0)-\mathrm{e}^{-\mathrm{i}\chi_\beta}\hat{k}^{(3)}_{0,\beta}(w_0)\right]}. \tag{11.161}$$

**(2) Global (SS)-modes:**
By joining the S-mode (11.153) on the subinterval ($0 < x < w_0$) with the S-mode (11.158) on the subinterval ($w_0 < x < 1$) in terms of connection

conditions (11.135), we derive

$$\tilde{D}_{0,\alpha}^{(1)}\left[1-e^{-i\chi_\alpha}\frac{\hat{k}_{0,\alpha}^{(1)}(0)}{\hat{k}_{0,\alpha}^{(3)}(0)}\right]=\tilde{D}_{0,\beta}^{(1)}\left[1-e^{-i\chi_\beta}\frac{\hat{k}_{0,\beta}^{(1)}(1)}{\hat{k}_{0,\beta}^{(3)}(1)}\right] \tag{11.162}$$

and

$$\begin{aligned}&\tilde{D}_{0,\alpha}^{(1)}\left[\hat{k}_{0,\alpha}^{(1)}(w_0)-e^{-i\chi_\alpha}\hat{k}_{0,\alpha}^{(3)}(w_0)\frac{\hat{k}_{0,\alpha}^{(1)}(0)}{\hat{k}_{0,\alpha}^{(3)}(0)}\right]\\&=\left(\frac{\cos\bar{\theta}+}{\cos\bar{\theta}-}\right)^2\tilde{D}_{0,\beta}^{(1)}\left[\hat{k}_{0,\beta}^{(1)}(w_0)-e^{-i\chi_\beta}\hat{k}_{0,\beta}^{(3)}(w_0)\frac{\hat{k}_{0,\beta}^{(1)}(1)}{\hat{k}_{0,\beta}^{(3)}(1)}\right].\end{aligned} \tag{11.163}$$

By combining (11.162) with (11.163), we obtain the quantization condition

$$\begin{aligned}&\frac{\left[1-e^{-i\chi_\alpha}\frac{\hat{k}_{0,\alpha}^{(1)}(0)}{\hat{k}_{0,\alpha}^{(3)}(0)}\right]}{\left[\hat{k}_{0,\alpha}^{(1)}(w_0)-e^{-i\chi_\alpha}\hat{k}_{0,\alpha}^{(3)}(w_0)\frac{\hat{k}_{0,\alpha}^{(1)}(0)}{\hat{k}_{0,\alpha}^{(3)}(0)}\right]}\\&=\left(\frac{\cos\bar{\theta}-}{\cos\bar{\theta}+}\right)^2\frac{\left[1-e^{-i\chi_\beta}\frac{\hat{k}_{0,\beta}^{(1)}(1)}{\hat{k}_{0,\beta}^{(3)}(1)}\right]}{\left[\hat{k}_{0,\beta}^{(1)}(w_0)-e^{-i\chi_\beta}\hat{k}_{0,\beta}^{(3)}(w_0)\frac{\hat{k}_{0,\beta}^{(1)}(1)}{\hat{k}_{0,\beta}^{(3)}(1)}\right]}.\end{aligned} \tag{11.164}$$

**(3) Global (AS)-modes:**
By joining the A-mode (11.150) on the subinterval ($0 < x < w_0$) with the S-mode (11.158) on the subinterval ($w_0 < x < 1$) in terms of connection conditions (11.135), we obtain

$$\tilde{D}_{0,\alpha}^{(1)}\left[1-e^{-i\chi_\alpha}\right]=\tilde{D}_{0,\beta}^{(1)}\left[1-e^{-i\chi_\beta}\frac{\hat{k}_{0,\beta}^{(1)}(1)}{\hat{k}_{0,\beta}^{(3)}(1)}\right] \tag{11.165}$$

and

$$\begin{aligned}&\tilde{D}_{0,\alpha}^{(1)}\left[\hat{k}_{0,\alpha}^{(1)}(w_0)-e^{-i\chi_\alpha}\hat{k}_{0,\alpha}^{(3)}(w_0)\right]\\&=\left(\frac{\cos\bar{\theta}+}{\cos\bar{\theta}-}\right)^2\tilde{D}_{0,\beta}^{(1)}\left[\hat{k}_{0,\beta}^{(1)}(w_0)-e^{-i\chi_\beta}\hat{k}_{0,\beta}^{(3)}(w_0)\frac{\hat{k}_{0,\beta}^{(1)}(1)}{\hat{k}_{0,\beta}^{(3)}(1)}\right].\end{aligned} \tag{11.166}$$

By combining (11.165) with (11.166), we obtain the quantization condition

$$\frac{\left[1-\mathrm{e}^{-\mathrm{i}\chi_\alpha}\right]}{\left[\hat{k}^{(1)}_{0,\alpha}(w_0)-\mathrm{e}^{-\mathrm{i}\chi_\alpha}\hat{k}^{(3)}_{0,\alpha}(w_0)\right]} = \left(\frac{\cos\bar{\theta}-}{\cos\bar{\theta}+}\right)^2 \frac{\left[1-\mathrm{e}^{-\mathrm{i}\chi_\beta}\frac{\hat{k}^{(1)}_{0,\beta}(1)}{\hat{k}^{(3)}_{0,\beta}(1)}\right]}{\left[\hat{k}^{(1)}_{0,\beta}(w_0)-\mathrm{e}^{-\mathrm{i}\chi_\beta}\hat{k}^{(3)}_{0,\beta}(w_0)\frac{\hat{k}^{(1)}_{0,\beta}(1)}{\hat{k}^{(3)}_{0,\beta}(1)}\right]}. \tag{11.167}$$

**(4) Global (SA)-modes:**
By joining the S-mode (11.153) on the subinterval $(0 < x < w_0)$ with the A-mode (11.155) on the subinterval $(w_0 < x < 1)$ in terms of the connection conditions (11.135), we obtain

$$\tilde{D}^{(1)}_{0,\alpha}\left[1-\mathrm{e}^{-\mathrm{i}\chi_\alpha}\frac{\hat{k}^{(1)}_{0,\alpha}(0)}{\hat{k}^{(3)}_{0,\alpha}(0)}\right] = \tilde{D}^{(1)}_{0,\beta}\left[1-\mathrm{e}^{-\mathrm{i}\chi_\beta}\right] \tag{11.168}$$

and

$$\tilde{D}^{(1)}_{0,\alpha}\left[\hat{k}^{(1)}_{0,\alpha}(w_0)-\mathrm{e}^{-\mathrm{i}\chi_\alpha}\hat{k}^{(3)}_{0,\alpha}(w_0)\frac{\hat{k}^{(1)}_{0,\alpha}(0)}{\hat{k}^{(3)}_{0,\alpha}(0)}\right] = \left(\frac{\cos\bar{\theta}+}{\cos\bar{\theta}-}\right)^2 \tilde{D}^{(1)}_{0,\beta}\left[\hat{k}^{(1)}_{0,\beta}(w_0)-\mathrm{e}^{-\mathrm{i}\chi_\beta}\hat{k}^{(3)}_{0,\beta}(w_0)\right]. \tag{11.169}$$

By combining (11.168) with (11.169), we obtain the quantization condition

$$\frac{\left[1-\mathrm{e}^{-\mathrm{i}\chi_\alpha}\frac{\hat{k}^{(1)}_{0,\alpha}(0)}{\hat{k}^{(3)}_{0,\alpha}(0)}\right]}{\left[\hat{k}^{(1)}_{0,\alpha}(w_0)-\mathrm{e}^{-\mathrm{i}\chi_\alpha}\hat{k}^{(3)}_{0,\alpha}(w_0)\frac{\hat{k}^{(1)}_{0,\alpha}(0)}{\hat{k}^{(3)}_{0,\alpha}(0)}\right]} = \left(\frac{\cos\bar{\theta}-}{\cos\bar{\theta}+}\right)^2 \frac{\left[1-\mathrm{e}^{-\mathrm{i}\chi_\beta}\right]}{\left[\hat{k}^{(1)}_{0,\beta}(w_0)-\mathrm{e}^{-\mathrm{i}\chi_\beta}\hat{k}^{(3)}_{0,\beta}(w_0)\right]}. \tag{11.170}$$

Note that the above quantization conditions of the global modes are derived with given $C_\infty$, $\bar{\varphi}$, and fixed parameters $\hat{\Gamma}_\alpha = \mathcal{O}(1)$, $\hat{\Gamma}_\beta = \mathcal{O}(1)$, $\hat{\mathrm{G}} = \mathcal{O}(1)$. Hence, with given thermodynamic parameters of the system $\left(R_c = \hat{\Gamma}_\alpha/\hat{\Gamma}_\beta, M_\alpha, M_\beta\right)$, the parameters $(\hat{\Gamma}_\alpha, \hat{\mathrm{G}}, C_\infty)$ can be considered the control parameters. The eigenvalue $\sigma_0$ is calculated as a function of these control parameters and $\epsilon$. The dimensional growth parameters $V$ and $(G)_{\mathrm{D}}$ are determined after the parameter $\epsilon$ is selected.

As demonstrated in the previous section, to elucidate the effects of the operating parameters $((G)_{\mathrm{D}}, V)$ independently, it is better to adopt the dimensionless parameters $\{\hat{\upsilon}, \hat{\beta}\}$. With the formula (11.92), one may rewrite that

$$\hat{\upsilon} = \epsilon^3 \hat{\varpi} \hat{\Gamma}_\alpha; \quad \hat{\beta} = \epsilon^{\frac{1}{2}} \hat{\varpi} \frac{\hat{\upsilon}\hat{\mathrm{G}}}{M_\alpha}.$$

In terms of these relationships, the control parameters $(\bar{\Gamma}_\alpha, G, C_\infty)$ can be replaced by the growth parameters $(\hat{\upsilon}, \hat{\beta}, C_\infty)$.

To further explore the instability mechanisms, the most important thing is to seek the global growing unstable modes and neutrally stable modes. This is the objective of the next section. We focus on the case $\{\Delta_\alpha > 0, \Delta_\beta > 0\}$.

## 11.10 The Global Steady (ST) Mode of Perturbed States

We first prove that the system allows a set of steady perturbed state modes. To do so, let $\sigma_0 = 0$ in the quantization conditions. Thus, from the quantization conditions one can solve for some critical values of the real pattern parameters $\epsilon = \epsilon_{\mathrm{ST}}$ and $\bar{\varphi} = \bar{\varphi}_{\mathrm{ST}}$ with given growth conditions $V, G$, and $C_\infty$. We call these steady modes the ST-modes.

### *11.10.1 The Untilted ST-Modes*

Let us first seek the untilted steady modes with the pattern parameter $\bar{\varphi}_{\mathrm{ST}} = 0$. This case is possible only under some special growth conditions. It implies that given the pulling velocity $V$ and far-field concentration $C_\infty$, to obtain $\epsilon = \epsilon_{\mathrm{ST}}$ for the untilted ST-modes ($\bar{\varphi} = 0$), the temperature gradient parameter must take a special value, $\hat{\mathrm{G}} = \hat{\mathrm{G}}_*$, or equivalently, the parameter $\hat{\beta}$ must take some special value $\hat{\beta} = \hat{\beta}_*$.

In what follows, we restrict our discussion to the (AA) type of modes and the case $\{\Delta_\alpha > 0, \Delta_\beta > 0\}$. We find the values $\epsilon = \epsilon_{\mathrm{ST}}$ and $\hat{\beta}_*$ in analytical form.

We first solve the wave number functions of the ST-modes from the local dispersion formula (11.143), which is reduced to

$$0 = \Sigma(\hat{k}_0, x) = \Lambda_0(x)\hat{k}_0\Big[M\hat{\Theta}(x) - \hat{\mathrm{G}} - \hat{\Gamma}\hat{k}_0^2\Big]. \qquad (11.171)$$

We obtain

$$\hat{k}_0^{(1)}(x) = \mathrm{k}_0, \quad \hat{k}_0^{(3)}(x) = 0, \quad \mathrm{k}_0 = \sqrt{[M\hat{\Theta} - \hat{\mathrm{G}}]/\hat{\Gamma}}, \qquad (11.172)$$

where $\mathrm{k}_0(x)$ is a piecewise smooth function. We define

$$\chi_\alpha = \frac{1}{\epsilon^{1/4}}\int_0^{w_0}\left[\hat{k}^{(1)}_{0,\alpha}(x_1) - \hat{k}^{(3)}_{0,\alpha}(x_1)\right]\mathrm{d}x_1 = \frac{1}{\epsilon^{1/4}}\mathrm{k}_{0\alpha}w_0,$$

$$\chi_\beta = -\frac{1}{\epsilon^{1/4}}\int_{w_0}^{1}\left[\hat{k}^{(1)}_{0,\beta}(x_1) - \hat{k}^{(3)}_{0,\beta}(x_1)\right]\mathrm{d}x_1 = -\frac{1}{\epsilon^{1/4}}\mathrm{k}_{0\beta}(1-w_0)$$

$$= -\frac{\mathrm{k}_{0\beta}(1-w_0)}{\mathrm{k}_{0\alpha}w_0}\chi_\alpha;$$

and

$$X = \left(\frac{1}{\epsilon^{1/4}}\mathrm{k}_{0\alpha}w_0\right), \quad \mathcal{A} = \left(\frac{\mathrm{k}_{0\alpha}}{\mathrm{k}_{0\beta}}\right)\left(\frac{\cos\bar{\theta}-}{\cos\bar{\theta}+}\right)^2,$$

$$\mathcal{B} = \left(\frac{\mathrm{k}_{0\beta}}{\mathrm{k}_{0\alpha}}\right)\frac{(1-w_0)}{w_0}.$$

The variable $X$ is related to the parameter $\epsilon$, and $\mathcal{A}$ is related to the growth parameter $\hat{\mathrm{G}}$, which measures the effect of the triple point. Here $\mathcal{B}$ is a constant determined by the thermodynamic properties of the system and the growth parameter $\hat{\mathrm{G}}$. The quantization (11.161) for the (AA) type of global modes, as $\sigma_0 \to 0$, reduces to

$$\frac{\left[1-\mathrm{e}^{-\mathrm{i}\chi_\alpha}\right]}{\left[1-\mathrm{e}^{-\mathrm{i}\chi_\beta}\right]} = \left(\frac{\cos\bar{\theta}-}{\cos\bar{\theta}+}\right)^2\frac{\hat{k}^{(1)}_{0,\alpha}}{\hat{k}^{(1)}_{0,\beta}} = \mathcal{A}, \tag{11.173}$$

or

$$\left[1-\mathrm{e}^{-\mathrm{i}X}\right] = \mathcal{A}\left[1-\mathrm{e}^{\mathrm{i}\mathcal{B}X}\right]. \tag{11.174}$$

It follows that

$$\begin{cases} 1-\cos X = \mathcal{A}\left[1-\cos(\mathcal{B}X)\right], \\ \sin X = -\mathcal{A}\sin(\mathcal{B}X), \end{cases} \tag{11.175}$$

so that

$$\begin{cases} \dfrac{1-\cos X}{\sin X} = -\dfrac{1-\cos(\mathcal{B}X)}{\sin(\mathcal{B}X)}, \quad \dfrac{\sin X}{\sin(\mathcal{B}X)} = -\left(\dfrac{\sin X/2}{\sin(\mathcal{B}X)/2}\right)^2, \\ \mathcal{A} = -\dfrac{\sin X}{\sin(\mathcal{B}X)}. \end{cases} \tag{11.176}$$

Equation (11.176) has the solution

$$X = X_* = \frac{2(n+1)\pi}{\mathcal{B}+1}; \tag{11.177}$$

$$\mathcal{A} = \mathcal{A}_* = 1 \tag{11.178}$$

$(n = 0, 1, 2, \cdots)$. The solution (11.177) leads to the quantization condition

$$X = \frac{1}{\epsilon^{1/4}} \mathrm{k}_{0\alpha} w_0 = \frac{2(n+1)\pi}{\mathcal{B}+1} \quad (n = 0, 1, 2, \cdots), \tag{11.179}$$

which determines the critical number $\epsilon = \epsilon_{\mathrm{ST}}$ as a function of the growth conditions. The solution (11.178) yields

$$\mathcal{A} = \left(\frac{\mathrm{k}_{0\alpha}}{\mathrm{k}_{0\beta}}\right)\left(\frac{\cos\bar{\theta}^-}{\cos\bar{\theta}^+}\right)^2 = 1, \tag{11.180}$$

which leads to

$$\frac{\hat{\mathrm{G}} - M_\beta \hat{\Theta}_\beta}{\hat{\mathrm{G}} - M_\alpha \hat{\Theta}_\alpha} = R_c \left(\frac{\cos\bar{\theta}^-}{\cos\bar{\theta}^+}\right)^4. \tag{11.181}$$

The formula (11.181) shows that a steady ST-mode is allowable only under some special growth conditions.

As the matter of fact, given the thermodynamic parameters $(M, \varpi, \hat{\kappa}, \theta_\alpha, \theta_\beta, R_c)$, from (11.181) one obtains the critical number of $\hat{\beta}_*$ as a function of other two growth parameters $(C_\infty, \hat{v})$, namely,

$$\begin{aligned} \hat{\mathrm{G}} = \hat{\mathrm{G}}_* &= \frac{(M_\beta \hat{\Theta}_\beta \cos^4\bar{\theta}^+ - M_\alpha \hat{\Theta}_\alpha R_c \cos^4\bar{\theta}^-)}{\cos^4\bar{\theta}^+ - R_c \cos^4\bar{\theta}^-}, \\ \hat{\beta} = \hat{\beta}_* &= \epsilon^{\frac{1}{2}} \hat{\varpi} \frac{\hat{\mathrm{G}}_* \hat{v}}{M_\alpha}. \end{aligned} \tag{11.182}$$

Furthermore, recalling $\hat{\Gamma}_\alpha = \hat{v}/(\hat{\varpi}\epsilon^3)$ and taking the value of $\hat{\mathrm{G}}_*$ obtained above, from (11.172) one may write

$$\mathrm{k}_{0,\alpha} = \epsilon^{\frac{3}{2}} \sqrt{(M\hat{\Theta} - \hat{\mathrm{G}}_*)\hat{\varpi}/\hat{v}}.$$

Then, using the formula (11.179), we finally derive the following form of quantization condition of the ST-modes:

$$\epsilon^{\frac{5}{4}} = Q_0(n+1)\sqrt{\hat{v}/\hat{\varpi}}, \quad (n = 0, 1, 2, \cdots), \tag{11.183}$$

namely,

$$\epsilon = \epsilon_{\mathrm{ST}} = \hat{v}^{\frac{2}{5}} \hat{\varpi}^{-\frac{2}{5}} \Big[ Q_0(n+1) \Big]^{\frac{4}{5}}, \quad (n = 0, 1, 2, \cdots), \tag{11.184}$$

where

$$Q_0 = \frac{2\pi}{\left[ w_0 + (1 - w_0)(\cos\bar{\theta}^- / \cos\bar{\theta}^+)^2 \right] \sqrt{M\hat{\Theta} - \hat{\mathrm{G}}_*}}.$$

From (11.184), one derives a sequence of the critical numbers $\epsilon = \{\epsilon_{\mathrm{ST}}^{(n)}\}$: $\epsilon_{\mathrm{ST}}^{(0)} < \epsilon_{\mathrm{ST}}^{(1)} < \cdots$, as functions of the thermodynamic parameters and the two growth parameters $(C_\infty, \hat{v})$.

Figure 11.26a, b shows the variations of $\epsilon_{\mathrm{ST}}^{(n)}$ and $\hat{\beta}_*$ with $\hat{v}$ for the given material parameters and growth variable $C_\infty$, respectively. Figure 11.27 shows a typical interfacial pattern of the ST-mode.

In general, for a given set of the growth parameters $(\hat{v}, \hat{\beta}, C_\infty)$, and $\epsilon$, the eigenvalues $\sigma_0$ for AA-mode solutions can be solved only numerically from the quantization condition (11.161), in terms of the iteration method. As is known, in order for the iteration process to be convergent, it is crucial to have a good initial guess. The analytical solutions for the ST mode obtained above can be used for this important role. It means that starting with the analytical solution of steady untilted ST-modes, one can utilize the numerical continuation technique, successively changing the growth conditions step by step, until a convergent solution is obtained under the desired growth conditions.

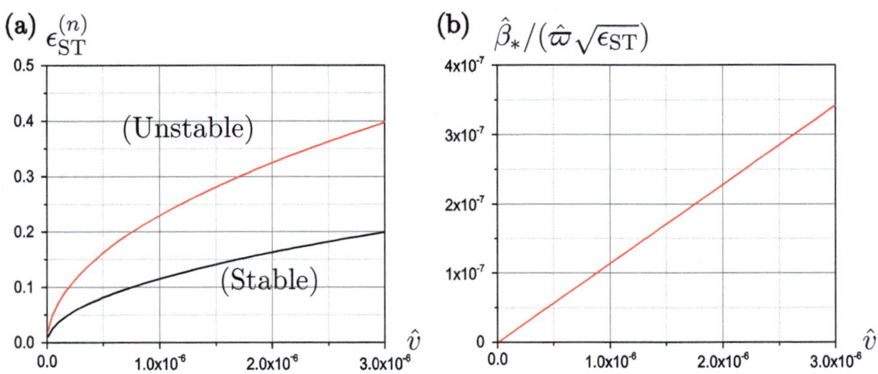


**Fig. 11.26** The critical numbers of $(\epsilon, \hat{\beta})$ of the ST-modes for the typical case $\{\Delta_\alpha > 0, \Delta_\beta > 0\}$ ($\mathcal{C}$-unstable $\alpha$- and $\beta$-interfaces), with $\varpi = 0.15$, $w_0 = 0.625$, $\bar{\theta}^- = 1.1397$, $\bar{\theta}^+ = 1.0385$, $\mathrm{M}_\alpha = 0.1073$, $\mathrm{M}_\beta = -0.06865$, $\hat{\kappa} = 1.333$, $R_c = 0.84$; and $C_\infty = -0.0186$. (**a**) The curves of the critical numbers of $\epsilon = \epsilon_{\mathrm{ST}}^{(n)}$ with $(n = 0, 1)$ versus $\hat{v}$ from *bottom to top*. (**b**) The curves of $\hat{\beta}_*$ versus $\hat{v}$

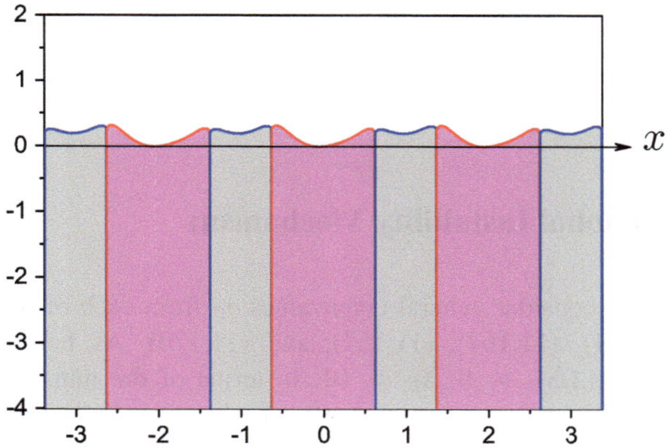


**Fig. 11.27** The eutectic pattern of the ST-mode calculated with $\bar{\varphi} = 0$ for the typical case $\hat{v} = 0.1200 \times 10^{-5}$; other parameters are the same as given in Fig. 11.26. The critical numbers for this case are obtained as $\epsilon = \epsilon = \epsilon_{\mathrm{ST}} = 0.1260$, $\hat{\beta} = \hat{\beta}_* = 2.280 \times 10^{-6}$

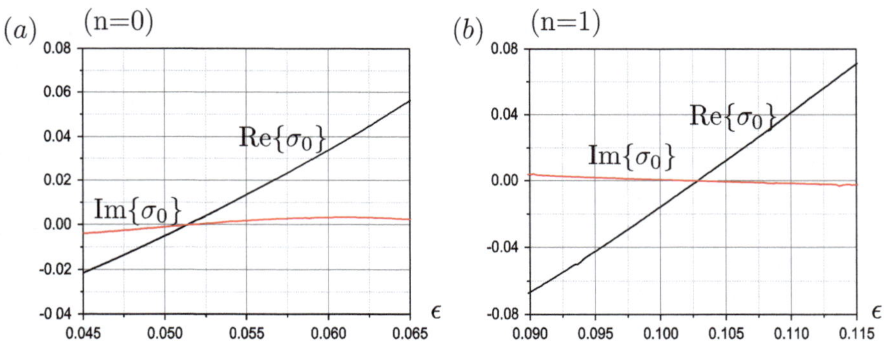


**Fig. 11.28** The eigenvalues $\sigma_0$ of the (AA)-modes for the case $\{\Delta_\alpha > 0, \Delta_\beta > 0\}$ ($\mathcal{C}$-unstable $\alpha$- and $\beta$-interfaces) with $\bar{\theta}^- = 1.1397$, $\bar{\theta}^+ = 1.0385$, $\mathrm{M}_\alpha = 0.1073$, $\mathrm{M}_\beta = -0.06865$, $\hat{\kappa} = 1.333$, $C_\infty = -1.8644 \times 10^{-2}$, $R_c = 0.84$, $G = 0.01224$, $\hat{v} = 2.0 \times 10^{-7}$, $\hat{\beta} = \hat{\beta}_* = 2.28 \times 10^{-8}$, and $\bar{\varphi} = 0$. (**a**) The variations of Re$\{\sigma_0\}$ and Im$\{\sigma_0\}$ of the branch of modes ($n = 0$) with $\epsilon$. (**b**) The variations of Re$\{\sigma_0\}$ and Im$\{\sigma_0\}$ of the branch of modes ($n = 1$) with $\epsilon$. It is seen that as $\epsilon$ increases, the system becomes more unstable

By adopting the above-described numerical procedure, we have calculated the solutions for $\sigma_0$ from the quantization condition (11.161) for some typical cases. In Fig. 11.28a, b, we respectively show the variations of Re$\{\sigma_0\}$ and Im$\{\sigma_0\}$ with $\epsilon$ for $n = 0$ and $n = 1$. It is seen that with fixed growth conditions and material parameters, the steady basic state is oscillatory and unstable in the region $\epsilon > \epsilon_{\mathrm{ST}}^{(0)}$. Hence, the steady ST-modes with ($n \geq 1$) are unstable and can be ruled out. A remark should be made here, that the effect on the stability property of the system caused by the parameter $\epsilon$ defined for the eutectic growth system is quite different from that of the parameter $\varepsilon$ defined for the systems of dendritic growth and cellular growth. *With a larger value of $\epsilon$, the system of eutectic growth is more unstable, while with a larger value of $\varepsilon$, the systems of dendritic growth and cellular growth are more stable.*

## 11.11 The Global Instability Mechanism

In this section we consider general eigenvalues $\sigma_0$ from each of the quantization conditions (11.161), (11.164), (11.167), and (11.170). As before, we restrict ourself to the case $\{\Delta_\alpha > 0, \Delta_\beta > 0\}$. In terms of the numerical techniques demonstrated in the previous sections, we are able to calculate the eigenvalues $\sigma_0$ as a function of the growth parameters $(\hat{v}, \hat{\beta}, C_\infty)$ from all these quantization conditions.

We find that the system allows both the real spectrum of eigenvalues and complex spectrum of eigenvalues. The ST-mode derived in the previous section is the neutral mode of the perturbed states with real eigenvalues. As a consequence, the ST-modes of perturbed states generate the so-called exchange stability mechanism. On the other hand, the complex spectrum of eigenvalues leads to four types of global oscillatory instability mechanisms, invoked by the unstable modes of the (AA), (SS), (AS), and (SA) types of perturbed states, respectively.

From each of the quantization conditions one can further determine the neutral modes corresponding to the critical number $\epsilon = \epsilon_*$, for which the real part of the eigenvalue $\mathrm{Re}\{\sigma_0\}$ is equal to zero. Certainly, the imaginary part of the eigenvalue corresponding to these neutral modes may not be zero, namely $\mathrm{Im}\{\sigma_0\} = \omega_* \neq 0$. Hence, these neutral modes are oscillatory modes.

We have calculated the critical number $\epsilon_*$ as a function of $(\hat{v}, \hat{\beta}, C_\infty)$, and $\bar{\varphi}$. In what follows, as examples, we shall show the critical number $\epsilon = \epsilon_*$ for each of the oscillatory neutral modes corresponding to given growth parameters $(\hat{v}, \hat{\beta}, C_\infty)$, and $\bar{\varphi} = 0$.

(a) Global (AA)-modes and the corresponding eigenvalues: In terms of the quantization condition (11.161), we calculate the critical number $\epsilon = \epsilon_*$ for the neutrally stable (AA)-modes with various values of $\hat{v}$. The resultant neutral curve of (AA)-modes in the $(\epsilon, \hat{v})$-plane is shown in Fig. 11.29.
(b) Global (SS)-modes and the corresponding eigenvalues: In terms of the quantization condition (11.164), we calculate the critical number $\epsilon = \epsilon_*$ for the neutrally stable (SS)-modes with various values of $\hat{v}$. The neutral curve of the (SS) type of modes with fixed $\bar{\varphi} = 0$ in the $(\epsilon, \hat{v})$-plane is shown in Fig. 11.30.
(c) Global (AS)-modes and the corresponding eigenvalues: In terms of the quantization condition (11.167), we calculate the critical number $\epsilon = \epsilon_*$ for the neutrally stable (AS)-modes with various values of $\hat{v}$. The neutral curve of the (AS)-modes with fixed $\bar{\varphi} = 0$ in the $(\epsilon, \hat{v})$-plane is shown in Fig. 11.31.

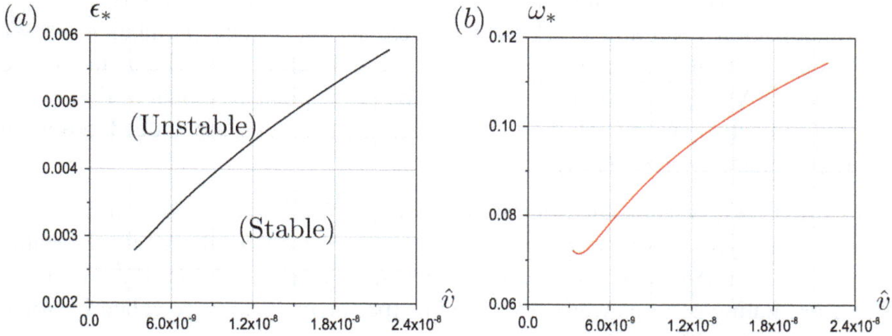


**Fig. 11.29** The neutral curves of (AA)-modes for the typical case $\{\Delta_\alpha > 0, \Delta_\beta > 0\}$ ($\mathcal{C}$-unstable $\alpha$- and $\beta$-interfaces), with $\bar{\varphi} = 0$, $\varpi = 0.15$, $w_0 = 0.625$, $\mathrm{M}_\alpha = 0.107, \mathrm{M}_\beta = -0.218$, $\bar{\theta}^- = 1.14$, $\bar{\theta}^+ = 1.04$, $R_c = 0.84$, and $C_\infty = -1.8644 \times 10^{-2}$, $\hat{\beta} = 4.8774 \times 10^{-10}$. (**a**) The variation of the critical number $\epsilon_*$ with $\hat{v}$. (**b**) The variation of the corresponding eigenfrequency $\omega_*$ with $\hat{v}$

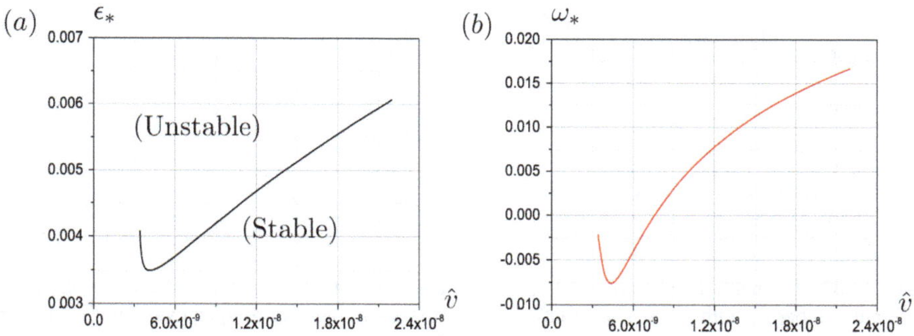


**Fig. 11.30** The neutral curves of (SS)-modes for the case $\bar{\varphi} = 0$: (**a**) the variation of the critical number $\epsilon_*$ with $\hat{v}$. (**b**) The variation of the corresponding eigenfrequency $\omega_*$ with $\hat{v}$. Other parameters are the same as those given in Fig. 11.29

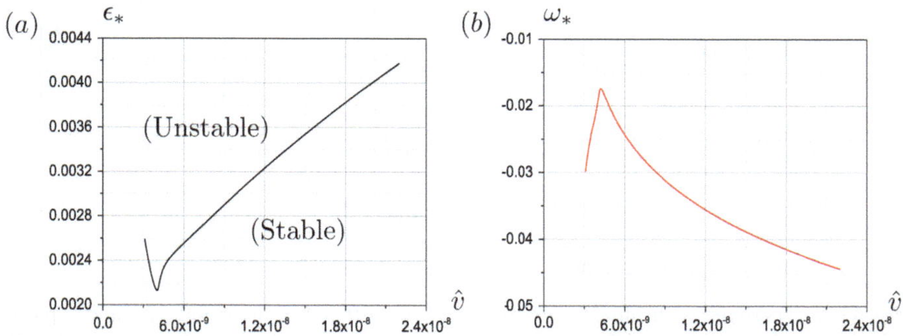


**Fig. 11.31** The neutral curves of (AS)-modes: (**a**) the variation of the critical number $\epsilon_*$ with $\hat{v}$. (**b**) The variation of the corresponding eigenfrequency $\omega_*$ with $\hat{v}$. Other parameters are the same as those given in Fig. 11.29

(d) Global (SA)-modes and the corresponding eigenvalues: In terms of the quantization condition (11.170), we calculate the critical number $\epsilon = \epsilon_*$ for the neutrally stable (SA)-modes with various values of $\hat{v}$. The neutral curve of the (SA)-modes with fixed $\bar{\varphi} = 0$ in the $(\epsilon, \hat{v})$-plane is shown in Fig. 11.32. The eutectic patterns of these neutrally stable modes are indeed observed experimentally (refer to [6]).

In Fig. 11.33, we plot the neutral curves of all oscillatory modes in the same plane. Some typical oscillating eutectic patterns created by the neutrally stable (AA)-, (SS)-, (AS)-, and (SA)-modes are shown respectively in Fig. 11.34a–d. These oscillating patterns were observed in experiments, and were called by some researchers 1λO, 2λO, and zigzag-like patterns, respectively.

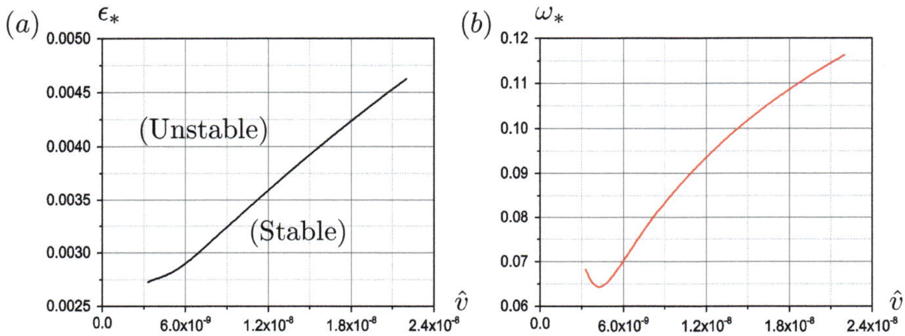


**Fig. 11.32** The neutral curves of (SA)-modes for the case of with $\bar{\varphi} = 0$: (**a**) The variation of the critical number $\epsilon_*$ with $\hat{v}$. (**b**) The variation of the corresponding eigenfrequency $\omega_*$ with $\hat{v}$. Other parameters are the same as those given in Fig. 11.29

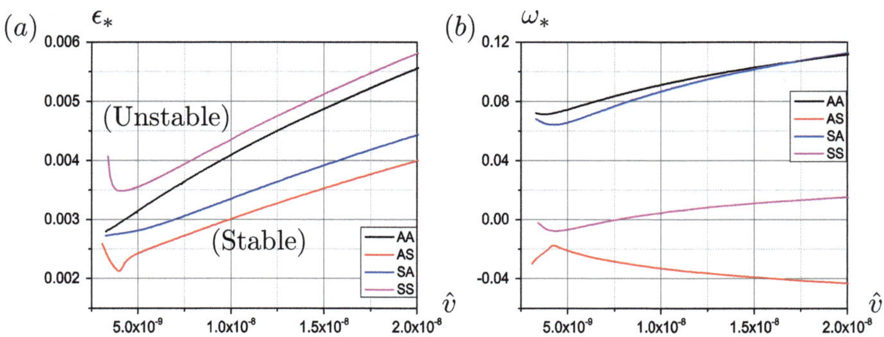


**Fig. 11.33** Four types of neutral curves for the case $\bar{\varphi} = 0$ plotted in the same parameter plane: (**a**) $\epsilon_*$ versus $\hat{v}$ for the (AS), (SA), (AA), and (SS) types of modes from bottom to top on the right-hand side, respectively. (**b**) $\omega_*$ versus $\hat{v}$ for the (AS), (SS), (AA), and (SA) types of modes from *bottom to top on the right-hand side*, respectively. Other parameters are the same as those given in Fig. 11.29

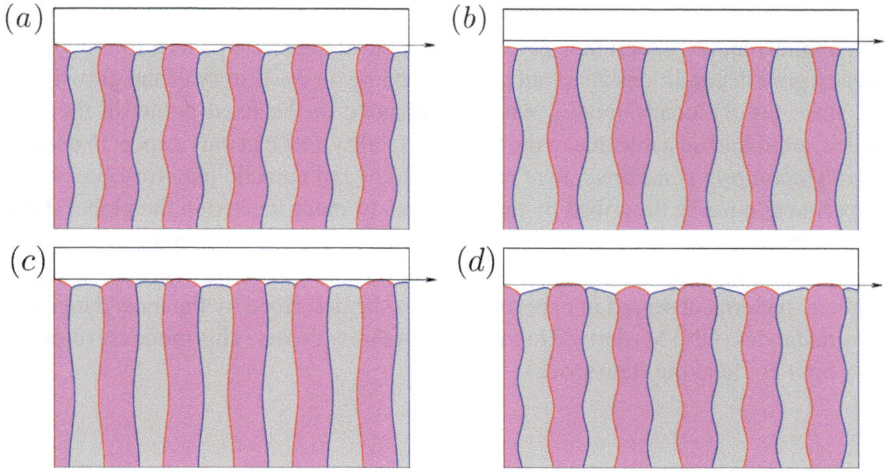


**Fig. 11.34** The eutectic patterns corresponding to different types of oscillatory neutrally stable modes with $\hat{\beta} = 0.48774 \times 10^{-9}$ and other parameters given in Fig. 11.29: (**a**) (AA)-mode with $\omega_* = 0.11446$, $\epsilon_* = 0.0057996$, $\hat{v} = 0.22 \times 10^{-7}$; (**b**) (SS)-mode with $\omega_* = 0.04111373$, $\epsilon_* = 0.0154808$, $\hat{v} = 0.156 \times 10^{-6}$; (**c**) (AS)-mode with $\omega_* = -0.0444385$, $\epsilon_* = 0.00417408$, $\hat{v} = 0.22 \times 10^{-7}$; (**d**) (SA)-mode with $\omega_* = 0.116273$, $\epsilon_* = 0.00462431$, $\hat{v} = 0.22 \times 10^{-7}$

Among the above four types of neutral modes, the most dangerous neutrally stable mode is the one with the smallest value of the critical number $\epsilon_*$. It is seen from Fig. 11.33 that for the case under study with given range of growth speed parameter, the most dangerous neutral mode is the oscillatory neutrally stable (AS)-mode. The critical number $\epsilon = \epsilon_{*,\min}$ for the most dangerous neutrally stable mode yields *the upper bound of the global stable region*. In other words, *all the basic steady states with* $\epsilon > \epsilon_{*,\min}$ *are globally unstable*.

## 11.12 Discussions on Selection of Eutectic Pattern Formation

The issue of selection of steady eutectic growth was initially raised by Jackson and Hunt [7] in 1966. With the assumptions of a flat interface and smallness of the Péclet number, they proposed a family of solutions with a free parameter related to the interlamellar spacing $\ell_w$ describing untilted steady eutectic growth. To determine this free parameter, they suggested that the spacing $\ell_w$ is *uniquely* selected via the "scaling law" $V\ell_w^2 = C$, where $C$ is the constant depending on the material properties. Jackson and Hunt also proposed as a hypothesis that the steady solution was selected at the maximum interfacial temperature, or minimum interfacial undercooling. However, such a hypothesis is not supported by later more detailed experimental evidence (see, for instance, [1, 6, 10]). Experimental results show that the idea of a *sharp selection principle* of interlamellar spacing for the the "nearly" steady and periodic eutectic pattern observed in experiments is not applicable; the experimentally "selected" steady eutectic pattern formed under steady growth conditions is not unique, depending on the history of the growth.

Since the ST-steady modes, which correspond to the fixed points in the state space, are found unstable due to the GTW instability, one can only expect that *under steady operating conditions*, the "nearly" steady and eutectic patterns observed in experiment must be described by the basic steady states located in the global stable region bounded above by the global neutral curve. One conclusion drawn from our analysis is that under steady operating conditions, the oscillating "nearly" periodic eutectic patterns observed in experiments must be described by the most dangerous neutral modes (GNSM) derived in the global stability theory. This mode corresponds to a limit circle in the state space.

### *11.12.1 Comparison with Experimental Data*

So far, most experimental investigations on stability of lamellar eutectic growth have been conducted using the system of $CBr_4$–$C_2Cl_6$. The thermodynamic property parameters for such a system have been shown in Table 11.1. To date, the experimental data obtained by Ginibre, Akamatsu, and Faivre (GAF) in 1997 appear to be the most accurate and complete. In their experiments, the primary component in the $\alpha$-phase is $CBr_4$, while the primary component in the $\beta$-phase is $C_2Cl_6$. The experimental conditions adopted by GAF satisfy the assumptions made in our analysis: $\mathrm{Pe} = \epsilon \ll 1$, $\mathrm{G} = \mathcal{O}(\varpi)$, $M = \mathcal{O}(1)$, and $\Gamma = \mathcal{O}(\epsilon^2)$. Therefore, GAF's results provided a good opportunity for us to examine the validity of the theoretical analysis. We calculate the theoretical solution for the experimental case corresponding to Fig. 11.20a. Under given experimental conditions, the most dangerous neutral mode (GNSM) is found to be the (AS)-mode. In Fig. 11.35a, we show the interface shapes of the (AS)-mode predicted by the theoretical solutions, while side by side in Fig. 11.35b, we show the experimental photo taken by GAF. Furthermore, for a better comparison, in Fig. 11.35c, we put the theoretical solutions and experimental results together in the same figure. It is seen from these figures that the theoretical predictions and experimental results are in excellent quantitative agreement. The mutual agreement in the shape of the $\alpha$–$\beta$ oscillatory interface shows that the frequency of oscillation of the eutectic system is in agreement with the eigenvalue of the most dangerous neutral mode (GNSM), whereas the mutual agreement in the shape of the liquid–solid interface shows that the interfacial pattern observed is described by the interface shape of the selected eigenmode: the most dangerous neutral mode (GNSM).

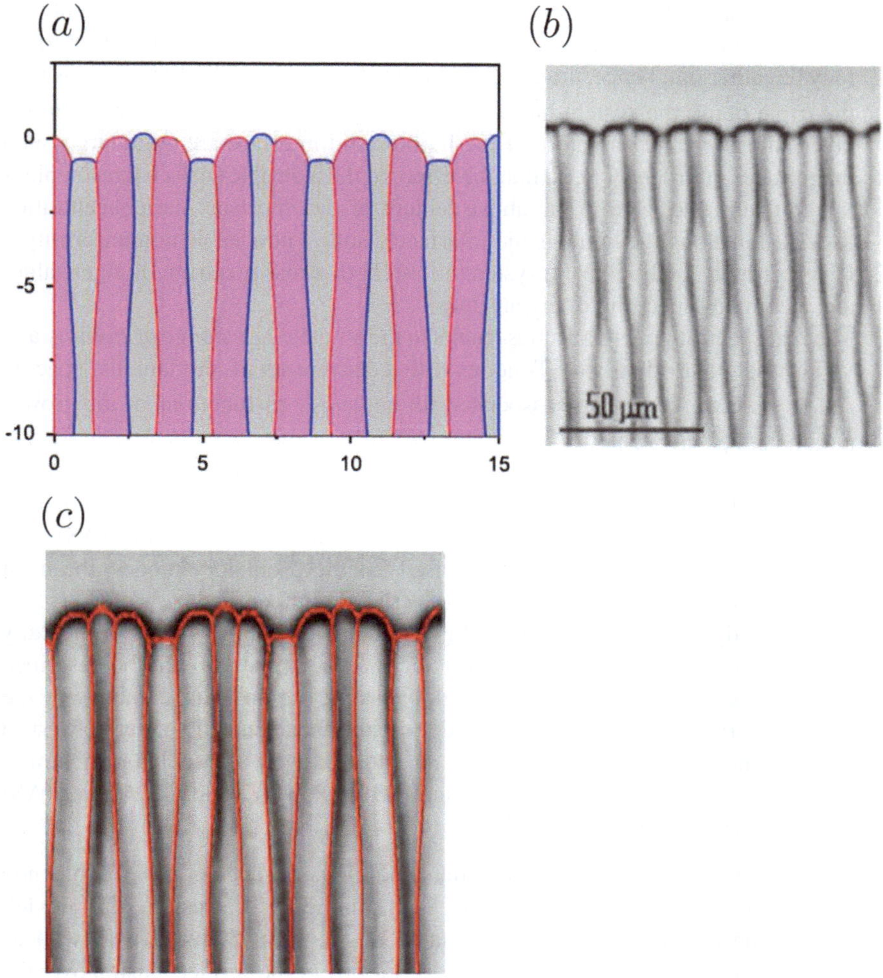


**Fig. 11.35** Comparison of the theoretical solution and experimental results of eutectic growth with the system $CBr_4$–$C_2Cl_6$ under the same growth conditions: $\epsilon = 0.03$, $\bar{\varphi} = 0$, $\hat{\kappa} = 2$, $\varpi = 0.25$, $w_0 = 0.475$, $M_\alpha = 0.138$, $M_\beta = -0.214$, $\bar{\theta}^- = 1.22$, $\bar{\theta}^+ = 1.17$, $R_c = 0.51$, and $C_\infty = 0.144$, $\hat{v} = 3.6433 \times 10^{-6}$, $\hat{\beta} = 1.003 \times 10^{-6}$. (**a**) The interface shape of the most dangerous neutral AS-mode calculated by the global stability theory. (**b**) The interface shapes observed in experiments. (**c**) Comparison of the theory with experimental observation

## 11.13 A Brief Summary

This chapter investigates steady spatially periodic pattern formation during eutectic growth for the case that the Péclet number $\epsilon$ is small and the segregation coefficient $\kappa$ is close to unity. The main results are summarized below.

1. The singularity of the system at the triple point plays a profound role in eutectic pattern formation. In the outer region away from the triple point, the interface may be rather flat; hence, the effect of surface tension is negligible, whereas near the triple point, the system has a boundary layer with thickness $\mathcal{O}(\epsilon^{\frac{1}{2}})$. In this layer, the slope and curvature of the liquid–solid interface are quite large, with magnitudes of order $\mathcal{O}(1)$, so that the effect of the isotropic surface tension plays a significant role. Due to the above feature of the interface, the concentration distribution along the liquid–solid interface displays noticeable nonuniformity.
2. Given growth conditions, the system allows both a real spectrum of eigenvalues and a complex spectrum of eigenvalues.
3. The real spectrum of eigenvalues results in the so-called *exchange of stability* and yields a set of steady tilted ST-modes with a discrete set of interlamellar spacing $\epsilon_{\text{ST}}^{(n)}$ $(n = 0, 1, \ldots)$ and the associated tilt angle $\bar{\varphi}_{\text{ST}}$ as functions of the growth parameters $(\hat{v}, \hat{\beta}, C_\infty)$.
4. The complex spectrum of eigenvalues results in oscillatory instability mechanisms and yields a set of neutrally stable oscillatory states. The oscillatory perturbed states have four types of modes, namely the (AA), (SS), (AS), and (SA). Among these four types of neutrally stable oscillatory modes, the most dangerous mode has for its minimum the critical number $\epsilon_* = \epsilon_{*,\min}$.
5. Under steady growth conditions and given the parameter $\bar{\varphi}$, the experimentally selected "nearly" steady spatially periodic interfacial pattern may be described by the basic steady state lying in the global stable region bounded above by the global neutral curve, whereas the oscillating interfacial eutectic pattern observed in experiments must be described by the most dangerous oscillatory neutrally stable mode, which may be of (AA) and (SS) type with period $T = 2$; or (AS) and (SA) type with period $T = 4$.

The current investigation focuses on understanding the essence of the phenomenon and underlying instability mechanisms. For simplicity, it uses the 2D model, neglecting the effect of three-dimensionality of the Hele–Shaw cell, as well as many other physical effects, including the anisotropy of the surface tension over the interfaces and the convective flow in the liquid phase. The results obtained suggest that the current 2D model is applicable to the typical cases of experiments whose results we have compared to those of our analysis. It is evident that the effects three-dimensionality, convective flow, and the anisotropy of surface tension may play an important role in eutectic pattern formation in some other cases. We leave these issues open for further studies in the future.

## 11.14 Appendix 1: The Derivation of the Basic Solution in the Far Field

One may use the method of separation of variables to solve (11.13). Letting

$$C_{\text{B}}(x, y) = A(x)B(y),$$

from Eq. (11.13), one may derive that

$$\frac{\partial^2 A}{\partial x^2} + \lambda^2 A = 0, \qquad \frac{\partial^2 B}{\partial y^2} + \epsilon \frac{\partial B}{\partial y} - \lambda^2 B = 0,$$

and obtain the solutions

$$A(x) = \hat{A}\cos(\lambda x), \quad B(y) = \hat{B}\exp(-ky),$$

and

$$C_{\mathrm{B}}(x, y) = \hat{a}\cos(\lambda x)\exp(-ky) + C_\infty,$$

where $k = \frac{\epsilon + \sqrt{\epsilon^2 + 4\lambda^2}}{2} > 0$. From the periodic boundary condition

$$\left(\frac{\partial C_{\mathrm{B}}}{\partial x}\right)_{x=\pm 1} = -\lambda \hat{a} \sin(\pm\lambda)\exp(-ky) = 0,$$

we obtain the eigenvalues

$$\lambda = \lambda_n = n\pi, \ (n = 0, \pm 1, \pm 2, \cdots).$$

The general solution $C_{\mathrm{B}}(x, y)$ is derived as

$$C_{\mathrm{B}}(x, y) = \sum_{n=0}^{\infty} \hat{A}_n \cos n\pi \exp(-k_n y) + C_\infty, \tag{11.185}$$

where

$$k_n = \frac{\epsilon + \sqrt{\epsilon^2 + 4n^2\pi^2}}{2},$$

and one has

$$k_0 = \epsilon, \quad k_n = n\pi + \mathcal{O}(\epsilon) \quad (n \geq 1).$$

We now apply the boundary condition at $y = 0$:

$$\frac{\partial C_{\mathrm{B}}}{\partial y}(x, 0) = -Q(x, \varpi) = -\sum_{n=0}^{\infty} \hat{A}_n(\epsilon, \varpi) k_n \cos(n\pi x),$$

which determines the Fourier coefficients

$$\hat{A}_0(\epsilon, \varpi) = \frac{1}{\epsilon}\int_0^1 Q(x)\mathrm{d}x,$$
$$\hat{A}_n(\epsilon, \varpi) = \frac{2}{k_n}\int_0^1 \cos(n\pi x)Q(x)\mathrm{d}x \quad (n = 1, 2, \cdots). \tag{11.186}$$

With the notation

$$\hat{Q}(\epsilon, \varpi) = \frac{1}{\epsilon}\int_0^1 Q(x)\mathrm{d}x,$$

we derive the far-field solution (11.14).

## 11.15 Appendix 2: The Fourier Series of Some Special Functions

$$\begin{aligned}
\sum_{n=1}^{\infty}\left[\frac{\cos(nx)}{n} + \mathrm{i}\frac{\sin(nx)}{n}\right] &= \sum_{n=1}^{\infty}\frac{\exp(\mathrm{i}nx)}{n} = \sum_{n=1}^{\infty}\frac{[\exp(\mathrm{i}x)]^n}{n} \\
&= -\ln(1 - \cos x - \mathrm{i}\sin x) \\
&= -\ln\sqrt{2 - 2\cos x} - \mathrm{i}\arctan\left[\frac{-\sin x}{1 - \cos x}\right] \\
&= -\frac{1}{2}\ln(2 - 2\cos x) - \mathrm{i}\arctan\Big[-\cot(x/2)\Big] \\
&= -\frac{1}{2}\ln(2 - 2\cos x) - \mathrm{i}\arctan\Big[-\tan(\pi/2 - x/2)\Big] \\
&= -\frac{1}{2}\ln(2 - 2\cos x) + \mathrm{i}(\pi/2 - x/2), \quad (0 < x < 2\pi).
\end{aligned} \tag{11.187}$$

1. One can derive from (11.187) that

$$S_1(z) = \sum_{n=1}^{\infty}\frac{\sin nz}{n} = \frac{\pi - z}{2}, \quad (0 < z < 2\pi), \tag{11.188}$$

which converges at $x = 0$, so that

$$S_1'(z) = \sum_{n=1}^{\infty} \cos nz = -\frac{1}{2} \quad (0 < z < 2\pi) \tag{11.189}$$

and

$$\frac{1}{2} + \sum_{n=1}^{\infty} \cos nz = \hat{D}(z) = 0 \quad (0 < z < 2\pi). \tag{11.190}$$

2. One may also derive from (11.187) that

$$D_1(z) = \sum_{n=1}^{\infty} \frac{\cos nz}{n} = -\frac{1}{2}\left[\ln 2 + \ln(1 - \cos z)\right], (0 < z < 2\pi), \tag{11.191}$$

$$D_1'(z) = -\sum_{n=1}^{\infty} \sin nz = -\frac{1}{2}\frac{\sin z}{(1 - \cos z)}, \ (0 < z < 2\pi), \tag{11.192}$$

so that

$$\hat{S}(z) = \sum_{n=1}^{\infty} \sin nz = \frac{1}{2}\frac{\sin z}{(1 - \cos z)} = \frac{1}{2}\cot(z/2) \quad (0 < z < 2\pi). \tag{11.193}$$

3. One can derive from (11.191) that

$$\begin{aligned} S_2(z) = \sum_{n=1}^{\infty} \frac{\sin nz}{n^2} &= \int_0^z D_1(x)\mathrm{d}x \\ &= -\frac{1}{2}\left[(\ln 2)z + \int_0^z \ln(1 - \cos x)\mathrm{d}x\right] \\ &= -\frac{z}{2}\left[\ln 2 + \ln(1 - \cos z)\right] + \frac{1}{2}\int_0^z \frac{z \sin z}{1 - \cos z}\mathrm{d}z, \\ &\quad (0 < z < 2\pi). \end{aligned} \tag{11.194}$$

4. One can derive from (11.188) that

$$\begin{aligned} D_2(z) = \sum_{n=1}^{\infty} \frac{\cos nz}{n^2} &= -\int_0^z S_1(x)\mathrm{d}x = \frac{\pi^2}{6} - \frac{\pi z}{2} + \frac{z^2}{4} \\ &(0 < z < 2\pi). \end{aligned} \tag{11.195}$$

5. One may derive from (11.195) that

$$S_3(z) = \sum_{n=1}^{\infty} \frac{\sin nz}{n^3} = \frac{\pi^2 z}{6} - \frac{\pi z^2}{4} + \frac{z^3}{12} \quad (0 < z < 2\pi). \tag{11.196}$$

6. One may derive from (11.196) that

$$D_4(z) = \sum_{n=1}^{\infty} \frac{\cos nz}{n^4} = \frac{\pi^4}{90} - \frac{\pi^2 z^2}{12} + \frac{\pi z^3}{12} - \frac{z^4}{48}$$
$$(0 < z < 2\pi). \tag{11.197}$$

## Appendix 3: The Derivation of the Global Interface Solution

In this section, we shall fully determine all unknown constants in the solution (11.84):

$$\frac{1}{\hat{\varpi}} \mathrm{h}_c(x, \epsilon) = H_c(x) + \Lambda_c + A_c \mathrm{e}^{\frac{k_s}{\sqrt{\epsilon}} x} + S_c \mathrm{e}^{-\frac{k_s}{\sqrt{\epsilon}} x}.$$

From the definition of the piecewise constant function $\Lambda$,

$$\Lambda_c = y_{*c} - \frac{1}{2} \frac{M}{\mathrm{G}} d_{11,0}$$
$$= \begin{cases} \Lambda_{c,\alpha} = y_{*c} - \frac{1}{2} \frac{M_\alpha}{\mathrm{G}} d_{11,0}, & x \in (0, w_0); \\ \Lambda_{c,\beta} = y_{*c} - \frac{1}{2} \frac{M_\beta}{\mathrm{G}} d_{11,0}, & x \in (w_0, 1), \end{cases} \tag{11.198}$$

one derives that

$$d_{11,0} = \frac{2\mathrm{G}(\Lambda_{c,\alpha} - \Lambda_{c,\beta})}{M_\beta - M_\alpha}, \quad y_{*c} = \frac{\Lambda_{c,\alpha} M_\beta - \Lambda_{c,\beta} M_\alpha}{M_\beta - M_\alpha}. \tag{11.199}$$

Furthermore, from the general solution of (11.84), we derive

$$\frac{1}{\hat{\varpi}} \mathrm{h}'_c(x) = H'_c(x) + \frac{k_s}{\sqrt{\epsilon}} \Big[ A_c \mathrm{e}^{\frac{k_s}{\sqrt{\epsilon}} x} - S_c \mathrm{e}^{-\frac{k_s}{\sqrt{\epsilon}} x} \Big], \tag{11.200}$$

where

$$H'_c(x) = \frac{M}{2\epsilon\Gamma} \left[ \mathrm{e}^{\frac{k_s}{\sqrt{\epsilon}} x} \int_{w_0}^{x} \widehat{P}_{11}(x_1) \mathrm{e}^{-\frac{k_s}{\sqrt{\epsilon}} x_1} \mathrm{d}x_1 + \mathrm{e}^{-\frac{k_s}{\sqrt{\epsilon}} x} \int_{w_0}^{x} \widehat{P}_{11}(x_1) \mathrm{e}^{\frac{k_s}{\sqrt{\epsilon}} x_1} \mathrm{d}x_1 \right].$$

In what follows, we are going to derive explicit expressions for the function $H_c(x)$ and its derivatives. Recalling that $\widehat{P}_{11}(x) = \sum_{n=1}^{\infty} d_{11,n} \cos(n\pi x)$, using the formula

$$\int_{w_0}^{x} \cos(n\pi x_1) e^{\frac{k_s}{\sqrt{\epsilon}} x_1} dx_1 = \frac{\epsilon}{k_s^2+\epsilon n^2\pi^2} \left\{ e^{\frac{k_s x}{\sqrt{\epsilon}}} \left[ \frac{k_s}{\sqrt{\epsilon}} \cos(n\pi x) + n\pi \sin(n\pi x) \right] \right.$$
$$\left. - e^{\frac{k_s w_0}{\sqrt{\epsilon}}} \left[ \frac{k_s}{\sqrt{\epsilon}} \cos(n\pi w_0) + n\pi \sin(n\pi w_0) \right] \right\},$$

we may derive that

$$\int_{w_0}^{x} \widehat{P}_{11}(x_1) e^{\frac{k_s}{\sqrt{\epsilon}} x_1} dx_1 = \sum_{n=1}^{\infty} d_{11,n} \int_{w_0}^{x} \cos(n\pi x) e^{\frac{k_s}{\sqrt{\epsilon}} x_1} dx_1$$
$$= \sum_{n=1}^{\infty} \frac{\epsilon\, d_{11,n}}{k_s^2 + \epsilon n^2 \pi^2} \left\{ e^{\frac{k_s x}{\sqrt{\epsilon}}} \left[ \frac{k_s}{\sqrt{\epsilon}} \cos(n\pi x) + n\pi \sin(n\pi x) \right] \right.$$
$$\left. - e^{\frac{k_s w_0}{\sqrt{\epsilon}}} \left[ \frac{k_s}{\sqrt{\epsilon}} \cos(n\pi w_0) + n\pi \sin(n\pi w_0) \right] \right\}, \tag{11.201}$$

and

$$\int_{w_0}^{x} \widehat{P}_{11}(x_1) e^{\frac{-k_s}{\sqrt{\epsilon}} x_1} dx_1 = \sum_{n=1}^{\infty} \frac{\epsilon\, d_{11,n}}{k_s^2 + \epsilon n^2 \pi^2} \int_{w_0}^{x} \cos(n\pi x) e^{\frac{-k_s}{\sqrt{\epsilon}} x_1} dx_1$$
$$= \sum_{n=1}^{\infty} \frac{\epsilon\, d_{11,n}}{k_s^2 + \epsilon n^2 \pi^2} \left\{ e^{\frac{-k_s x}{\sqrt{\epsilon}}} \left[ \frac{-k_s}{\sqrt{\epsilon}} \cos(n\pi x) + n\pi \sin(n\pi x) \right] \right.$$
$$\left. - e^{-\frac{k_s w_0}{\sqrt{\epsilon}}} \left[ \frac{-k_s}{\sqrt{\epsilon}} \cos(n\pi w_0) + n\pi \sin(n\pi w_0) \right] \right\}. \tag{11.202}$$

Hence, we obtain

$$H_c(x) = \frac{M}{2\sqrt{\epsilon}\bar{\Gamma} k_s} \left\{ e^{\frac{k_s}{\sqrt{\epsilon}} x} \sum_{n=1}^{\infty} \frac{\epsilon\, d_{11,n}}{k_s^2 + \epsilon n^2 \pi^2} \right.$$
$$\times \left[ e^{\frac{-k_s x}{\sqrt{\epsilon}}} \left( \frac{-k_s}{\sqrt{\epsilon}} \cos(n\pi x) + n\pi \sin(n\pi x) \right) \right.$$
$$\left. - e^{-\frac{k_s w_0}{\sqrt{\epsilon}}} \left( \frac{-k_s}{\sqrt{\epsilon}} \cos(n\pi w_0) + n\pi \sin(n\pi w_0) \right) \right]$$

$$-e^{\frac{-k_s}{\sqrt{\epsilon}}x}\sum_{n=1}^{\infty}\frac{\epsilon\, d_{11,n}}{k_s^2+\epsilon n^2\pi^2}\Bigg[e^{\frac{k_s x}{\sqrt{\epsilon}}}\Big(\frac{k_s}{\sqrt{\epsilon}}\cos(n\pi x)+n\pi\sin(n\pi x)\Big)$$
$$-e^{\frac{k_s w_0}{\sqrt{\epsilon}}}\Big(\frac{k_s}{\sqrt{\epsilon}}\cos(n\pi w_0)+n\pi\sin(n\pi w_0)\Big)\Bigg]\Bigg\},$$

or

$$H_c(x) = \frac{M}{2\sqrt{\epsilon}\bar{\Gamma}k_s}\Bigg\{\sum_{n=1}^{\infty}\frac{\epsilon\, d_{11,n}}{k_s^2+\epsilon n^2\pi^2}$$
$$\times\Bigg[-\frac{2k_s}{\sqrt{\epsilon}}\cos(n\pi x)-e^{\frac{k_s(x-w_0)}{\sqrt{\epsilon}}}\Big(\frac{-k_s}{\sqrt{\epsilon}}\cos(n\pi w_0)+n\pi\sin(n\pi w_0)\Big)\Bigg]$$
$$-\sum_{n=1}^{\infty}\frac{\epsilon\, d_{11,n}}{k_s^2+\epsilon n^2\pi^2}\Bigg[-e^{-\frac{k_s(x-w_0)}{\sqrt{\epsilon}}}\Big(\frac{k_s}{\sqrt{\epsilon}}\cos(n\pi w_0)+n\pi\sin(n\pi w_0)\Big)\Bigg]\Bigg\}.$$

It follows that

$$H_c(x) = -\frac{M}{\epsilon\bar{\Gamma}}\frac{\sqrt{\epsilon}}{k_s}\Bigg\{\sum_{n=1}^{\infty}\frac{\epsilon\, d_{11,n}}{k_s^2+\epsilon n^2\pi^2}\Bigg[\frac{k_s}{\sqrt{\epsilon}}\cos(n\pi x)$$
$$-\frac{k_s}{\sqrt{\epsilon}}\cosh\Big(\frac{k_s(x-w_0)}{\sqrt{\epsilon}}\Big)\cos(n\pi w_0)$$
$$+n\pi\sinh\Big(\frac{k_s(x-w_0)}{\sqrt{\epsilon}}\Big)\sin(n\pi w_0)\Bigg].\Bigg\}, \tag{11.203}$$

We finally derive that

$$H_c(x) = -\frac{M}{\epsilon\bar{\Gamma}}\Bigg\{\sum_{n=1}^{\infty}\frac{\epsilon\, d_{11,n}}{k_s^2+\epsilon n^2\pi^2}\Bigg[\cos(n\pi x)$$
$$-\cosh\Big(\frac{k_s(x-w_0)}{\sqrt{\epsilon}}\Big)\cos(n\pi w_0)$$
$$+\frac{n\pi\sqrt{\epsilon}}{k_s}\sinh\Big(\frac{k_s(x-w_0)}{\sqrt{\epsilon}}\Big)\sin(n\pi w_0)\Bigg]\Bigg\}, \tag{11.204}$$

and

$$H_c'(x) = \frac{M}{\epsilon\bar{\Gamma}}\left\{\sum_{n=1}^{\infty}\frac{\epsilon\, d_{11,n}}{k_s^2+\epsilon n^2\pi^2}\left[n\pi\sin(n\pi x)\right.\right.$$
$$+\frac{k_s}{\sqrt{\epsilon}}\sinh\left(\frac{k_s(x-w_0)}{\sqrt{\epsilon}}\right)\cos(n\pi w_0)$$
$$\left.\left.-n\pi\cosh\left(\frac{k_s(x-w_0)}{\sqrt{\epsilon}}\right)\sin(n\pi w_0)\right]\right\}. \tag{11.205}$$

From the above formulas, it follows that

$$H_c(0) = \frac{M}{\epsilon\bar{\Gamma}_\alpha}\sum_{n=1}^{\infty}\frac{\epsilon\, d_{11,n}}{k_{s,\alpha}^2+\epsilon n^2\pi^2}\left[-1+\cosh\left(\frac{k_{s,\alpha}w_0}{\sqrt{\epsilon}}\right)\cos(n\pi w_0)+\right.$$
$$\left.+\frac{n\pi\sqrt{\epsilon}}{k_{s,\alpha}}\sinh\left(\frac{k_{s,\alpha}w_0}{\sqrt{\epsilon}}\right)\sin(n\pi w_0)\Big)\right], \tag{11.206}$$

$$H_c'(0) = -\frac{M}{\epsilon\bar{\Gamma}_\alpha}\sum_{n=1}^{\infty}\frac{\epsilon\, d_{11,n}}{k_{s,\alpha}^2+\epsilon n^2\pi^2}\left[\frac{k_{s,\alpha}}{\sqrt{\epsilon}}\sinh\left(\frac{k_{s,\alpha}w_0}{\sqrt{\epsilon}}\right)\cos(n\pi w_0)\right.$$
$$\left.+n\pi\cosh\left(\frac{k_{s,\alpha}w_0}{\sqrt{\epsilon}}\right)\sin(n\pi w_0)\Big)\right], \tag{11.207}$$

$$H_c'(1) = \frac{M}{\epsilon\bar{\Gamma}_\beta}\sum_{n=1}^{\infty}\frac{\epsilon\, d_{11,n}}{k_{s,\beta}^2+\epsilon n^2\pi^2}\left[\frac{k_{s,\beta}}{\sqrt{\epsilon}}\sinh\left(\frac{k_{s,\beta}(1-w_0)}{\sqrt{\epsilon}}\right)\cos(n\pi w_0)\right.$$
$$\left.-n\pi\cosh\left(\frac{k_{s,\beta}(1-w_0)}{\sqrt{\epsilon}}\right)\sin(n\pi w_0)\right]. \tag{11.208}$$

We now apply all the boundary conditions to the solution (11.84):

1. The smooth tip conditions $h_c(0) = h_c'(0) = 0$ on the $\alpha$-interface, which result in

$$S_{c,\alpha} = -\left[\frac{H_c(0)}{2} - \frac{\sqrt{\epsilon}}{2}\frac{H_c'(0)}{k_{s,\alpha}}\right] - \frac{1}{2}\Lambda_{c,\alpha}, \tag{11.209}$$

$$A_{c,\alpha} = -\left[\frac{H_c(0)}{2} + \frac{\sqrt{\epsilon}}{2}\frac{H_c'(0)}{k_{s,\alpha}}\right] - \frac{1}{2}\Lambda_{c,\alpha}. \tag{11.210}$$

2. The smooth tip condition $h_c'(1) = 0$ on the $\beta$-interface, which leads to

$$\left[A_{c,\beta}e^{\frac{k_{s,\beta}}{\sqrt{\epsilon}}} - S_{c,\beta}e^{-\frac{k_{s,\beta}}{\sqrt{\epsilon}}}\right] = -\frac{\sqrt{\epsilon}}{k_{s,\beta}}H_c'(1). \tag{11.211}$$

3. The continuity condition at the triple point $\mathrm{h}_c(w_0^-) = \mathrm{h}_c(w_0^+)$, which leads to

$$\left[A_{c,\beta}\mathrm{e}^{\frac{k_{s,\beta}}{\sqrt{\epsilon}}w_0} + S_{c,\beta}\mathrm{e}^{-\frac{k_{s,\beta}}{\sqrt{\epsilon}}w_0}\right] + \Lambda_{c,\beta}$$
$$= \left[A_{c,\alpha}\mathrm{e}^{\frac{k_{s,\alpha}}{\sqrt{\epsilon}}w_0} + S_{c,\alpha}\mathrm{e}^{-\frac{k_{s,\alpha}}{\sqrt{\epsilon}}w_0}\right] + \Lambda_{c,\alpha}, \tag{11.212}$$

4. The slope conditions at the triple point: Noting that $h'_{\mathrm{B}}(w_0^{\pm}) = \pm\tan\theta^{\pm} = \mathcal{O}(1)$, $h_{\mathrm{B}}(x) = \epsilon^{1/2}\mathrm{h}_c(x)$, while $\mathrm{h}'_c(x) = \mathcal{O}(\epsilon^{-\frac{1}{2}})$, with the notation $\hat{s}_1^{\pm} = \pm\hat{\varpi}^{-1}\tan\theta^{\pm}$, we derive $\epsilon^{\frac{1}{2}}\hat{\varpi}^{-1}\mathrm{h}'_c(w_0^{\pm}) = \hat{s}_1^{\pm} = \mathcal{O}(1)$. It follows that

$$\epsilon^{\frac{1}{2}}\hat{\varpi}^{-1}\mathrm{h}'_c(w_0^-) = k_{s,\alpha}\left[A_{c,\alpha}\mathrm{e}^{\frac{k_{s,\alpha}}{\sqrt{\epsilon}}w_0} - S_{c,\alpha}\mathrm{e}^{-\frac{k_{s,\alpha}}{\sqrt{\epsilon}}w_0}\right] = \hat{s}_1^-, \tag{11.213}$$

$$\epsilon^{\frac{1}{2}}\hat{\varpi}^{-1}\mathrm{h}'_c(w_0^+) = k_{s,\beta}\left[A_{c,\beta}\mathrm{e}^{\frac{k_{s,\beta}}{\sqrt{\epsilon}}w_0} - S_{c,\beta}\mathrm{e}^{-\frac{k_{s,\beta}}{\sqrt{\epsilon}}w_0}\right] = \hat{s}_1^+. \tag{11.214}$$

It follows from (11.213) that

$$A_{c,\alpha}\mathrm{e}^{\frac{k_{s,\alpha}}{\sqrt{\epsilon}}w_0} - S_{c,\alpha}\mathrm{e}^{-\frac{k_{s,\alpha}}{\sqrt{\epsilon}}w_0} = \frac{\hat{s}_1^-}{k_{s,\alpha}}, \tag{11.215}$$

$$A_{c,\beta}\mathrm{e}^{\frac{k_{s,\beta}}{\sqrt{\epsilon}}w_0} - S_{c,\beta}\mathrm{e}^{-\frac{k_{s,\beta}}{\sqrt{\epsilon}}w_0} = \frac{\hat{s}_1^+}{k_{s,\beta}}. \tag{11.216}$$

By combining (11.215) with (11.210), one can determine $(A_{c,\alpha}, S_{c,\alpha})$ and $\Lambda_{c,\alpha}$:

$$\frac{\hat{s}_1^-}{k_{s,\alpha}} = A_{c,\alpha}\mathrm{e}^{\frac{k_{s,\alpha}}{\sqrt{\epsilon}}w_0} - S_{c,\alpha}\mathrm{e}^{-\frac{k_{s,\alpha}}{\sqrt{\epsilon}}w_0}, \tag{11.217}$$

$$S_{c,\alpha} = -\left[\frac{H_c(0)}{2} - \frac{\sqrt{\epsilon}}{2}\frac{H'_c(0)}{k_{s,\alpha}}\right] - \frac{1}{2}\Lambda_{c,\alpha}, \tag{11.218}$$

$$A_{c,\alpha} = -\left[\frac{H_c(0)}{2} + \frac{\sqrt{\epsilon}}{2}\frac{H'_c(0)}{k_{s,\alpha}}\right] - \frac{1}{2}\Lambda_{c,\alpha}. \tag{11.219}$$

Hence, we have

$$\frac{1}{k_{s,\alpha}}\hat{s}_1^- = -\mathrm{e}^{\frac{k_{s,\alpha}}{\sqrt{\epsilon}}w_0}\left[\left(\frac{H_c(0)}{2} + \frac{\sqrt{\epsilon}}{2}\frac{H'_c(0)}{k_{s,\alpha}}\right) + \frac{1}{2}\Lambda_{c,\alpha}\right]$$
$$+\mathrm{e}^{-\frac{k_{s,\alpha}}{\sqrt{\epsilon}}w_0}\left[\left(\frac{H_c(0)}{2} - \frac{\sqrt{\epsilon}}{2}\frac{H'_c(0)}{k_{s,\alpha}}\right) + \frac{1}{2}\Lambda_{c,\alpha}\right]. \tag{11.220}$$

We solve this and obtain

$$\Lambda_{c,\alpha} = -\frac{\sinh\left[\frac{k_{s,\alpha}}{\sqrt{\epsilon}} w_0\right] H_c(0) + \cosh\left[\frac{k_{s,\alpha}}{\sqrt{\epsilon}} w_0\right] \sqrt{\epsilon}\frac{H_c'(0)}{k_{s,\alpha}} + \frac{1}{k_{s,\alpha}}\hat{s}_1^-}{\sinh\left[\frac{k_{s,\alpha}}{\sqrt{\epsilon}} w_0\right]}. \tag{11.221}$$

Furthermore, one can determine $(A_{c,\beta}, S_{c,\beta})$ by combining (11.216) with (11.212):

$$\begin{aligned}
&\left[A_{c,\beta} e^{\frac{k_{s,\beta}}{\sqrt{\epsilon}} w_0} - S_{c,\beta} e^{-\frac{k_{s,\beta}}{\sqrt{\epsilon}} w_0}\right] = \frac{1}{k_{s,\beta}} \hat{s}_1^+;\\
&\left[A_{c,\beta} e^{\frac{k_{s,\beta}}{\sqrt{\epsilon}} w_0} + S_{c,\beta} e^{-\frac{k_{s,\beta}}{\sqrt{\epsilon}} w_0}\right] + \Lambda_{c,\beta}\\
&\quad = \left[A_{c,\alpha} e^{\frac{k_{s,\alpha}}{\sqrt{\epsilon}} w_0} + S_{c,\alpha} e^{-\frac{k_{s,\alpha}}{\sqrt{\epsilon}} w_0}\right] + \Lambda_{c,\alpha}.
\end{aligned}$$

It follows that

$$\begin{aligned}
A_{c,\beta} &= \frac{1}{2} e^{-\frac{k_{s,\beta}}{\sqrt{\epsilon}} w_0}\left[A_{c,\alpha} e^{\frac{k_{s,\alpha}}{\sqrt{\epsilon}} w_0} + S_{c,\alpha} e^{-\frac{k_{s,\alpha}}{\sqrt{\epsilon}} w_0}\right] +\\
&\quad + \frac{1}{2} e^{-\frac{k_{s,\beta}}{\sqrt{\epsilon}} w_0}\left[\Lambda_{c,\alpha} - \Lambda_{c,\beta} + \frac{1}{k_{s,\beta}} \hat{s}_1^+\right], \qquad (11.222)\\
S_{c,\beta} &= \frac{1}{2} e^{\frac{k_{s,\beta}}{\sqrt{\epsilon}} w_0}\left[A_{c,\alpha} e^{\frac{k_{s,\alpha}}{\sqrt{\epsilon}} w_0} + S_{c,\alpha} e^{-\frac{k_{s,\alpha}}{\sqrt{\epsilon}} w_0}\right]\\
&\quad - \frac{1}{2} e^{\frac{k_{s,\beta}}{\sqrt{\epsilon}} w_0}\left[\Lambda_{c,\beta} - \Lambda_{c,\alpha} + \frac{1}{k_{s,\beta}} \hat{s}_1^+\right]. \qquad (11.223)
\end{aligned}$$

One now can determine $\Lambda_{c,\beta}$ from the formula ((11.211)):

$$\begin{aligned}
A_{c,\beta} e^{\frac{k_{s,\beta}}{\sqrt{\epsilon}}} - S_{c,\beta} e^{-\frac{k_{s,\beta}}{\sqrt{\epsilon}}} &= -\frac{\sqrt{\epsilon}}{k_{s,\beta}} H_c'(1)\\
= \frac{1}{2} e^{\frac{k_{s,\beta}}{\sqrt{\epsilon}}(1-w_0)}&\left[A_{c,\alpha} e^{\frac{k_{s,\alpha}}{\sqrt{\epsilon}} w_0} + S_{c,\alpha} e^{-\frac{k_{s,\alpha}}{\sqrt{\epsilon}} w_0}\right]\\
- \frac{1}{2} e^{-\frac{k_{s,\beta}}{\sqrt{\epsilon}}(1-w_0)}&\left[A_{c,\alpha} e^{\frac{k_{s,\alpha}}{\sqrt{\epsilon}} w_0} + S_{c,\alpha} e^{-\frac{k_{s,\alpha}}{\sqrt{\epsilon}} w_0}\right]\\
+ \frac{1}{2} e^{\frac{k_{s,\beta}}{\sqrt{\epsilon}}(1-w_0)}&\left[\Lambda_{c,\alpha} - \Lambda_{c,\beta} + \frac{1}{k_{s,\beta}} \hat{s}_1^+\right]\\
+ \frac{1}{2} e^{-\frac{k_{s,\beta}}{\sqrt{\epsilon}}(1-w_0)}&\left[\Lambda_{c,\beta} - \Lambda_{c,\alpha} + \frac{1}{k_{s,\beta}} \hat{s}_1^+\right]. \qquad (11.224)
\end{aligned}$$

It follows that

$$(\Lambda_{c,\beta} - \Lambda_{c,\alpha})\left[\frac{1}{2}\mathrm{e}^{\frac{k_{s\beta}(1-w_0)}{\sqrt{\epsilon}}} - \frac{1}{2}\mathrm{e}^{-\frac{k_{s,\beta}}{\sqrt{\epsilon}}(1-w_0)}\right]$$

$$= \left[A_{c,\alpha}\mathrm{e}^{\frac{k_{s\alpha}}{\sqrt{\epsilon}}w_0} + S_{c,\alpha}\mathrm{e}^{-\frac{k_{s\alpha}}{\sqrt{\epsilon}}w_0}\right]\left[\frac{1}{2}\mathrm{e}^{\frac{k_{s\beta}}{\sqrt{\epsilon}}(1-w_0)} - \frac{1}{2}\mathrm{e}^{-\frac{k_{s\beta}}{\sqrt{\epsilon}}(1-w_0)}\right]$$

$$+\frac{1}{2k_{s,\beta}}\hat{s}_1^+\left[\mathrm{e}^{\frac{k_{s,\beta}}{\sqrt{\epsilon}}(1-w_0)} + \mathrm{e}^{-\frac{k_{s,\beta}}{\sqrt{\epsilon}}(1-w_0)}\right] + \frac{\sqrt{\epsilon}}{k_{s,\beta}}H_c'(1). \tag{11.225}$$

Hence, one obtains

$$\Lambda_{c,\beta} = \Lambda_{c,\alpha} + \frac{1}{\sinh\left[\frac{k_{s,\beta}}{\sqrt{\epsilon}}(1-w_0)\right]}$$

$$\times\left\{\left[A_{c,\alpha}\mathrm{e}^{\frac{k_{s,\alpha}}{\sqrt{\epsilon}}w_0} + S_{c,\alpha}\mathrm{e}^{-\frac{k_{s,\alpha}}{\sqrt{\epsilon}}w_0}\right]\sinh\left[\frac{k_{s,\beta}}{\sqrt{\epsilon}}(1-w_0)\right]\right.$$

$$\left.+\frac{\hat{s}_1^+}{k_{s,\beta}}\cosh\left[\frac{k_{s,\beta}}{\sqrt{\epsilon}}(1-w_0)\right] + \frac{\sqrt{\epsilon}}{k_{s,\beta}}H_c'(1)\right\}, \tag{11.226}$$

so that

$$\Lambda_{c,\beta} = \Lambda_{c,\alpha} + \left[A_{c,\alpha}\mathrm{e}^{\frac{k_{s,\alpha}}{\sqrt{\epsilon}}w_0} + S_{c,\alpha}\mathrm{e}^{-\frac{k_{s,\alpha}}{\sqrt{\epsilon}}w_0}\right] + \frac{\hat{s}_1^+}{k_{s,\beta}}\coth\left[\frac{k_{s,\beta}}{\sqrt{\epsilon}}(1-w_0)\right] +$$

$$+\frac{\frac{\sqrt{\epsilon}}{k_{s,\beta}}H_c'(1)}{\sinh\left[\frac{k_{s,\beta}}{\sqrt{\epsilon}}(1-w_0)\right]}. \tag{11.227}$$

## Appendix 4: A List of Principal Nomenclature

For the reader's convenience, we give here a list of the principal notation used in this chapter:

1. *Constants*

- $\mathrm{Pe} = \frac{\ell_w}{\ell_D} = \epsilon$: the Péclet number, where $\ell_w$ is half the interlamellar spacing.
- $\lambda_T = \frac{\ell_D}{\ell_T}$: the ratio of the two diffusion lengths $\ell_D$ and $\ell_T$.

- $\mathrm{G} = \frac{\ell_{\mathrm{D}}}{\Delta H/(c_{\mathrm{p}}\rho)}(G)_{\mathrm{D}} = \mathcal{O}(\epsilon^{1/2})$: the dimensionless gradient of the temperature, where $(G)_{\mathrm{D}} = \frac{T_{\mathrm{H}}-T_{\mathrm{C}}}{(L)_{\mathrm{D}}}$.
- $\varpi = 1 - \kappa_\alpha = \mathcal{O}(\epsilon^{1/2}), \quad \hat{\varpi} = \varpi/\epsilon^{1/2} = \mathcal{O}(1)$.
- $\hat{\mathrm{G}} = \mathrm{G}/(\epsilon^{1/2}\hat{\varpi}) = \mathcal{O}(1)$.
- $W_c = \frac{w_c}{\ell_{\mathrm{w}}}$: the dimensionless width parameter of the $\alpha$-phase, where $w_c$ is the location of the triple point.
- $C_\infty = \varpi \hat{C}_\infty = \mathcal{O}(\epsilon^{1/2})$: the dimensionless concentration in the far field.
- $\varphi$: The tilt angle.

2. *Piecewise constants and related quantities:*

- $M = -\frac{mC_{\mathrm{e}}}{\Delta H/(c_{\mathrm{p}}\rho)}$: the morphological parameter, where $m = (m_\alpha, m_\beta)$ is the slope of the liquidus.
- $\Gamma = \frac{\ell_{\mathrm{c}}}{\ell_{\mathrm{w}}} = \epsilon^2\bar{\Gamma}$: the interfacial stability parameter, where $\ell_{\mathrm{c}} = \frac{\gamma c_{\mathrm{p}}\rho T_{\mathrm{M0}}}{(\Delta H)^2}$.
- $R_c = \Gamma_\alpha/\Gamma_\beta$.
- $\hat{\Gamma} = \bar{\Gamma}/\hat{\varpi} = \mathcal{O}(1)$.
- $\lambda_{\mathrm{G}} = \frac{\ell_{\mathrm{D}}}{\ell_{\mathrm{G}}} = -\frac{\kappa_{\mathrm{D}}}{V}\frac{(G)_{\mathrm{D}}}{mC_{\mathrm{e}}} = \frac{\mathrm{G}}{M}$.
- $\hat{\Theta}(x) = (1-\kappa)/\varpi$.
- $\Delta = M\hat{\Theta}(x) - \hat{G}$.
- $\hat{v} = \frac{\ell_{\mathrm{c},\alpha}}{\ell_{\mathrm{D}}} = \epsilon^3\bar{\Gamma}_\alpha$: the parameter that independently measures the pulling velocity $V$.
- $\hat{\beta} = \frac{\ell_{\mathrm{c},\alpha}}{\ell_{\mathrm{G},\alpha}} = -\frac{\ell_{\mathrm{c},\alpha}(G)_{\mathrm{D}}}{m_\alpha C_{\mathrm{e}}} = \frac{\mathrm{G}\hat{v}}{M_\alpha}$: the parameter that independently measures the temperature gradient $(G)_{\mathrm{D}}$.

3. *Unknowns:*

   1. The general unsteady solution:

      - $C(x, y, t, \epsilon)$: the concentration field.
      - $h(x, t, \epsilon)$: the interface shape function.
      - $y_*(t, \epsilon)$: the temperature at the tip of the $\alpha$-interface.
      - $W_c(t, \epsilon)$: the location of the triple point.

   2. The basic steady solution in the outer region and its RPE (with slow variable $\bar{y} = \epsilon y$):

      - $C_{\mathrm{B}}(x, y, \epsilon) = \bar{C}(x, y, \bar{y}, \epsilon)$: $\{\bar{C}_{00}(x, y, \bar{y}), \bar{C}_{01}(x, y, \bar{y}), \bar{C}_{10}(\bullet), \cdots\}$.
      - $h_{\mathrm{B}}(x, \epsilon) = \bar{h}(x, \epsilon)$: $\{\bar{h}_{01}(x), \bar{h}_{10}(x), \cdots\}$.
      - $y_*(\epsilon)$: $\{y_{01}, y_{02}, \cdots\}$.
      - $W_c(\epsilon)$: $\{w_0, w_{01}, \cdots\}$.

   3. The inner solution of basic steady state near the triple point and its regular perturbation expansion (RPE) (with the inner variables $\{\hat{x}, \hat{y}\}$):

      - $C_{\mathrm{B}}(w_0 + \epsilon^{1/2}\hat{x}, \epsilon^{1/2}\hat{y}, \epsilon)$: the inner solution for the concentration field.
      - $\mathrm{h}(\hat{x}, \epsilon) : \{\bar{\mathrm{h}}_{01}(\hat{x}).\bar{\mathrm{h}}_{01}(\hat{x}), \cdots\}$: the inner solution for the liquid/solid interface shape function.

4. The composite solution for the liquid/solid interface shape function:
   - The global interface solution: $\{\mathrm{h}_c(x,\epsilon), y_{*c}\}$.
   - An ad hoc form of the composite solution: $\{\overline{\mathrm{h}}_{c,0}(x,\hat{x}), y_{01}\}$.
5. The solutions and MVEs for perturbed states:
   - $\{x_+, y_+, t_+\}$: the fast variables with factor $\epsilon^{1/4}$.
   - $\tilde{C}(x,y,t,\epsilon) = \tilde{\mathcal{C}}(x,y,x_+,y_+,t_+,\epsilon)$: $\{\tilde{\mathcal{C}}_0(x,y,x_+,y_+,t_+), \tilde{\mathcal{C}}_1(x,y,x_+,y_+,t_+),\cdots\}\mathrm{e}^{\sigma t_+}$: the perturbed concentration field.
   - $\tilde{h}(x,t,\epsilon) = \tilde{\mathrm{h}}(x,x_+,t_+)$; $\{\tilde{\mathrm{h}}_0(x,x_+), \tilde{\mathrm{h}}_1(x,x_+),\cdots\}\mathrm{e}^{\sigma t_+}$: the perturbed interface shape function.
   - $\tilde{W}_c(t_+,\epsilon); \{\tilde{w}_0, \tilde{w}_1,\cdots\}$: the perturbed location of the triple point.
   - $\tilde{k}(x,y,t); \{\tilde{k}_0(x,y), \tilde{k}_1(x,y),\cdots\}$: the wave number function.
   - $\hat{k}(x,\epsilon) = \tilde{k}(x,0,\epsilon)$; $\hat{k}_0(x) = \tilde{k}_0(x,0)$.
   - $\sigma$: $\{\sigma_0;\sigma_1;\cdots\}$: the eigenvalue.

# References


1. S. Akamatsu, M. Flapp, G. Faivre, A. Karma, Pattern stability and trijunction motion in eutectic solidification. Phys. Rev. E **66**, 030501(R) (2002)
2. S. Akamatsu, S. Bottin-Rousseau, G. Faivre, Determination of the Jackson–Hunt constants of the In–$In_2Bi$ eutectic alloy based on in situ observation of its solidification dynamics. Acta Mater. **59**, 7586–7591 (2011)
3. Y.J. Chen, S.H. Davis, Instability of triple junctions in lamellar eutectic growth. Acta Mater. **49**, 1363–1372 (2001)
4. V. Datye, R. Mathur, J.S. Langer, Stability of thin lamellar eutectic growth. Phys. Rev. B **24**(8), 4155–4169 (1981)
5. S.H. Davis, *Theory of Solidification*, Chap. 8 (Cambridge University Press, Cambridge, 2001)
6. M. Ginibre, S. Akamatsu, G. Faivre, Experimental determination of the stability diagram of a lamellar eutectic growth front. Phys. Rev. E **56**(1), 780–796 (1997)
7. K.A. Jackson, J.D. Hunt, Lamellar and rod eutectic growth. Trans. Metall. Soc. AIME **236**, 1129–1142 (1966)
8. A. Karma, A. Sarkissian, Morphological instabilities of lamellar eutectics. Metall. Mater. Trans. A **27**, 635 (1996)
9. K. Kassne, C. Misbah, Spontaneous parity-breaking transition in directional growth of lamellar eutectic structures. Phys. Rev. A **44**(10), 6533–6543 (1991)
10. J. Mergy, G. Faivre, C. Guthmann, R. Mellet, Quantitative determination of the physical parameters relevant to the thin-film directional solidification of the $CBr_4$–$C_2Cl_6$ eutectic alloy. J. Cryst. Growth **134**, 353–368 (1993)
11. G.E. Nash, A self-sistent theory of steady state lamellar solidification in binary eutectic systems. J. Cryst. Growth **38**, 155–180 (1977)
12. A. Parisi, M. Plapp, Stability of lamellar eutectic growth. Acta Mater. **56**, 1348–1357 (2008)
13. V. Seetharaman, R. Trivedi, Eutectic growth: selection of interlamellar spacings. Metall. Trans. A **19**, 2955–2964 (1988)
14. J.J. Xu, Y.Q. Chen, Steady spatially-periodic eutectic growth with the effect of triple point in directional solidification. Acta Mater. **80**, 220–238 (2014)
15. J.J. Xu, X.M. Li, Y.Q. Chen, Global steady state solutions for lamellar eutectic growth in directional solidification. J. Cryst. Growth **401**, 93–98 (2014)

# Bibliography


1. M. Abramovitz, I.A. Stegun (eds.), *Handbook of Mathematical Functions* (Dover, New York, 1964)
2. S.H. Davis, *Theory of Solidifcation* (Cambridge University Press, Cambridge, 2001)
3. M.E. Glicksman, *Principles of Solidification – An Introduction to Modern Casting and Crystal Growth Concepts* (Springer, New York, 2014)
4. C. Godreche (ed.), *Solids Far from Equilibrium* (Cambridge University Press, Cambridge, 1991)
5. P. Hartman (ed.), *Crystal Growth: An Introduction* (Elsevier Science/North–Holland, Amsterdam, 1973)
6. M.H. Holmes, *Introduction to Perturbation Methods*, 2nd edn. (Springer, New York, 2013)
7. D.T.J. Hurle (ed.), *Handbook of Crystal Growth, Volume 1: Fundamentals, Part A: Thermodynamics and Kinetics; Part B: Transport and Stability* (Elsevier Science/North–Holland, Amsterdam, 1993)
8. J. Kevorkian, J.D. Cole, *Multiple Scale and Singular Perturbation Methods*. Applied Mathematical Sciences, vol. 114 (Springer, New York, 1996)
9. C.C. Lin, *The Theory of Hydrodynamic Stability* (Cambridge University Press, Cambridge, 1955)
10. P. Manneville, *Dissipative Structures and Weak Turbulence*, ed. by H. Araki, A. Libchaber, G. Parisi. Perspectives in Physics (Academic, New York, 1990)
11. R.E. O'Malley, *Singular Perturbation Methods for Ordinary Differential Equations*. Applied Mathematical Sciences, vol. 89 (Springer, Berlin/Heidelberg, 1991)
12. P. Pelce, *Dynamics of Curved Front* (Academic, New York, 1988)
13. J.J. Xu, *Dynamical Theory of Dendritic Growth in Convective Flow* (Springer/Kluwer Academic, New York, 2004)





J.-J. Xu, *Interfacial Wave Theory of Pattern Formation in Solidification*, Springer Series in Synergetics, DOI 10.1007/978-3-319-52663-8

# Index





J.-J. Xu, *Interfacial Wave Theory of Pattern Formation in Solidification*,
Springer Series in Synergetics, DOI 10.1007/978-3-319-52663-8

MIX
Papier aus verantwortungsvollen Quellen
Paper from responsible sources
FSC® C105338

FSC
www.fsc.org

If you have any concerns about our products,
you can contact us on
**ProductSafety@springernature.com**

In case Publisher is established outside the EU,
the EU authorized representative is:
**Springer Nature Customer Service Center GmbH**
**Europaplatz 3, 69115 Heidelberg, Germany**

Printed by Libri Plureos GmbH
in Hamburg, Germany